4th International Conference on Research Methodology for Built Environment and Engineering (ICRMBEE 2019)

Synergising the Impacts of Globalization on the Built Environment and Engineering Research

IOP Conference Series: Earth and Environmental Science
Volume 385

Chulalongkorn, Malaysia
24 – 25 April 2019

Editors:

Siti Aekbal Salleh
Norazian Yusuwan
Norfashiha Hashim
Rostam Yaman

ISBN: 978-1-7138-0253-2
ISSN: 1755-1307

Printed from e-media with permission by:

Curran Associates, Inc.
57 Morehouse Lane
Red Hook, NY 12571

Some format issues inherent in the e-media version may also appear in this print version.

Institute of Physics
Dirac House, Temple Back
Bristol BS1 6BE UK

Phone: 44 1 17 929 7481
Fax: 44 1 17 920 0979

techtracking@iop.org

Additional copies of this publication are available from:

Curran Associates, Inc.
57 Morehouse Lane
Red Hook, NY 12571 USA
Phone: 845-758-0400
Fax: 845-758-2633
Email: curran@proceedings.com
Web: www.proceedings.com

TABLE OF CONTENTS

Acknowledgement

The organizing committee would like to acknowledge the following for their contribution and effort to this conference:

UNIVERSITI TEKNOLOGI MARA

จุฬาลงกรณ์มหาวิทยาลัย
CHULALONGKORN UNIVERSITY

F|S|P|U
Fakulti Senibina, Perancangan & Ukur
Faculty of Architecture, Planning & Surveying

Environment Development and Sustainability
Graduate School,
Interdisciplinary Program
Chulalongkorn University

IOP Conference Series
Earth and Environmental Science

(263468-V)
TETO engineering SDN BHD

EJ

esri Malaysia

IOP Conf. Series: Earth and Environmental Science **385** (2019) 011001 doi:10.1088/1755-1315/385/1/011001

Preface

ICRMBEE 2019

Chamchuri 10, Chulalongkorn University, Bangkok, Thailand

International Conference on Research Methodology for Built Environment and Engineering (ICRMBEE 2019) is to be held 24th and 25th April 2019 at Chamchuri 10, Chulalongkorn University, Thailand. ICRMBEE is the first international conference that will focus on the method adopted in the area of built environment and engineering. Faculty of Architecture, Planning and Surveying, Universiti Teknologi MARA (UiTM), Shah Alam, Malaysia will organize this conference with the theme, '*Synergising the Impacts of Globalization on the Built Environment and Engineering Research*' which co-host by Environmental Development and Sustainability (EDS), Chulalongkorn University, Thailand. ICRMBEE 2019 provides an opportunity for interaction and exchange of ideas on the most appropriate ways of building bridges across universities, research institutions, government, industry and experts from around the world.

Research development through invention and innovation have facilitated the methods and approaches in the field of built environment and engineering. Systematic research methodology is the most crucial aspects in yielding significant and reliable outcome and output of each study. Either quantitative or qualitative methodology or mixed methodology will influence research outcome and output. Being a conference that emphasizing particularly on the methodology of the research, 4th. ICRMBEE 2019 would pooled and assist prospectus research fraternity to source, understand, determine and recognize the available and updated research methodology varieties. The organizing and technical committees of 4th. ICRMBEE 2019 aims to put the conference expand to a renowned platform and annual avenue for trans-boundary intellectual discourse

It is a great honour and privilege to host the 4th ICRMBEE at the University of Chulalongkorn, Bangkok, Thailand. In past years, this growing city has played host to many international cultural, political and sporting events. Bangkok, Thailand's capital, is a large city known for ornate shrines and vibrant street life. The boat-filled Chao Phraya River feeds its network of canals, flowing past the Rattanakosin royal district, home to opulent Grand Palace and its sacred Wat Phra Kaew Temple. Nearby is Wat Pho Temple with an enormous reclining Buddha and, on the opposite shore, Wat Arun Temple with its steep steps and Khmer-style spire.

As this conference series is expected to enter its 5th edition next year, the organizing committee is committed and enthusiast to widen the researcher participation all over the globe in order to bring forward varies and interesting research methodology to academia fraternity and to the general. The Conference are looking forward to a more larger participations, higher quality of studies and another interesting conference venue in the region for the coming conference.

Guest Editors
Sr Dr Siti Aekbal Salleh
Sr Dr Norazian Yusuwan
Dr Norfashiha Hashim
Ts Dr Rostam Yaman

IOP Conf. Series: Earth and Environmental Science **385** (2019) 011002 doi:10.1088/1755-1315/385/1/011002

Peer review statement

All papers published in this volume of *IOP Conference Series: Earth and Environmental Science* have been peer reviewed through processes administered by the proceedings Editors. Reviews were conducted by expert referees to the professional and scientific standards expected of a proceedings journal published by IOP Publishing.

IOP Conf. Series: Earth and Environmental Science **385** (2019) 012001 doi:10.1088/1755-1315/385/1/012001

Prefabricated Prefinished Volumetric Construction: Key Constraints and Mitigation Strategies

W Q Liu[1], B G Hwang[2], M Shan[3] and K Y Looi[4]

[1]School of Civil Engineering, Central South University, 22 South Shaoshan Road, Changsha, China, 410075
[2]Department of Building, National University of Singapore. 4 Architecture Drive, Singapore, 117566
[3]School of Civil Engineering, Central South University, 22 South Shaoshan Road, Changsha, China, 410075
[4]Arcadis Singapore Pte. Ltd., 1 Magazine Road, #05-01 Central Mall, Singapore, 059567

Email: looi.kitying@arcadis.com

Abstract. Prefabricated prefinished buildings (PPVC) is a more innovative and cleaner way to reorganize production in the construction industry. Although the application of PPVC has been well received over the past few decades, the limitations of using PPVC have not been explored. Therefore, the purpose of this study is to investigate the main factors influencing the adoption of PPVC and propose a set of feasible mitigation strategies to address these factors. To achieve these goals, a comprehensive literature review was conducted first, followed by a structured questionnaire survey of 41 Singapore construction organizations. The analysis showed that the top five major limiting factors and found the top three most effective mitigation strategies. This study contributes to the knowledge system by studying the constraints and mitigation strategies of PPVC. At the same time, the results of this study also have some implications for the industry to understand and implement PPVC better.

1. Introduction

Prefabricated prefinished volumetric construction (PPVC) is a characteristic approach with high efficiency in off-site construction methods in which building elements, components, and modules will be manufactured and assembled in off-site factories and then they are transported to site for installation [1]. Even though PPVC has been widely recognized by construction industry worldwide, up to now, it's still in the start-up stage. A great deal of knowledge about PPVC is still unmined. Therefore, the purpose of the investigation is to search out significant constraints in adopting PPVC in construction projects as well as existing buildings, and to put forward a set of mitigation strategies which contribute to resolve these constraints. In addition, this study is also beneficial to the study of knowledge system of off-site construction study, particularly for the study of PPVC. At the same time, this study concluded the experience and lessons of PPVC project in real life, which is of significance to industry practitioners that want to adopt PPVC.

2. Background

2.1 Prefabricated Prefinished Volumetric Construction in Singapore

On the basis of the definition of BCA[2], PPVC is a kind of construction method in which independent volume modules (including finishes for walls, floors and ceiling) are manufactured and assembled in approved manufacturing facilities in the light of any approved manufacturing method and then transported to the site for installation to form a building. Unlike the traditional continuous construction methods of design, engineering and construction activities, PPVC can greatly accelerate the implementation of the project, thus improving the productivity [3]. Lately, PPVC has been popular around the world. Since 2010, Singapore's BCA has launched two rounds of the construction productivity roadmap (round 1 in 2010 and round 2 in 2015), for the aim of promoting prosperity in the local construction industry [4].

2.2 The Adoption of PPVC: Constraints and Mitigation Strategies

Literature review reveals that several studies have made outstanding achievements in concluding the limitations, challenges, obstacles and constraints of general off-site construction methods. For instance, Mao et al. [5] conducted a questionnaire survey to investigate the main obstacles hindering the adoption of off-site construction methods in Chinese construction industry. Eventually, 18 constraints that may impede the adoption of PPVC were extracted from the literature review, as shown in table 1.

Table 1. Constraints on Adopting PPVC

Code	Constraints
C1	Declined flexibility for future design changes
C2	Unsupportive decision made by designers
C3	Design restrictions because of traffic limitation (e.g. modules' size)
C4	Limited site distribution (e.g. lack of space for PPVC modules to be stored ,unloaded and moved)
C5	Less experiences design about PPVC
C6	Less installation experiences about PPVC
C7	Requirement for extra materials to protect PPVC modules
C8	Added logistics and transportation considerations
C9	Transportation restrictions due to rules and regulations
C10	Less awareness of PPVC's benefits among owners/developers
C11	Raised organizational demand (e.g. changes in the roles of project participants)
C12	Need for early commitment
C13	Requirement for extra project planning and design efforts
C14	More communication requested before and during construction
C15	Expensive initial costs of traditional construction method
C16	Expensive construction cost of the traditional construction method
C17	Requirement of more communication among all stakeholders
C18	Complicated process for inspection

In order to solve the above limitations, challenges, and obstacles, a variety of mitigation strategies have been put forward. For example, plenty of studies put forward to apply information technologies such as BIM, three-dimensional visualization, and product information model to solve these information barriers met with in implementing off-site construction projects. According to a literature review, there are totally 9 mitigation strategies determined to help deal with the constraints related to PPVC, and they are compiled in table 2.

Table 2. Mitigation Strategies for the Constraints on Using PPVC

Code	Mitigation Strategy	Targeted Constraint
MS1	Encouraging project stakeholders to collaborate closely in early period of project	C1, C2, and C3
MS2	Applying for temporary occupational permits (TOL) to build temporary sites near construction sites to overcome site layout limitations	C4
MS3	Apply of Just-in-Time delivery	C4
MS4	Providing project teams and workers with training sessions to enhance the skills and knowledge	C5, C6, and C7
MS5	Fabricating and assembling the module components closely to the construction site to decrease transportation workload	C8 and C9
MS6	Using Information Technologies (e.g. electronic file transfer) to overcome the additional demand of planning, coordination and communication	C10 and C11
MS7	Conducting feasibility study on PPVC method	C12, C13, C14, C15, C16, C17, and C18
MS8	Developing systems/computer programs for detailed economic analysis of PPVC and conventional construction methods	C12, C13, C14, C15, C16, C17, and C18
MS9	Applying BIM to promote coordination and increase communication among project stakeholders	C12, C13, C14, C15, C16, C17, and C18

3. Methods and Data Presentation

3.1 Data Collection and Presentation

This study adopted a variety of qualitative and quantitative research methods such as literature review, pilot interview and questionnaire survey.

In the study of construction engineering and management (CEM), interviewing experienced industry experts is a great chance for a research to absorb meaningful opinions from practice [7]. Therefore, after the literature review, the study conducted preliminary interviews with industry experts to verify the limitations and mitigation strategies determined in the literature review. We invited three experienced industry experts in local PPVC projects to conduct pilot interviews for this research. They are senior members of the architectural innovation group.

After that, this research adopts questionnaire survey to collect professionals' cognition of limitations and mitigation strategies. According to the results of the preliminary interview, a questionnaire was prepared. It was made up three parts, including the first part which recorded the respondents' background about their employer type and years of experience in construction and PPVC, the second part requiring respondents to use a five-point system (1 to 5 present different level of insignificance,1 is the lowest level and 5 is the significance level, and so on), and the third part requesting respondents to use another 5-point scale (1 to 5 present different level of efficiency,1 is the lowest level and 5 is the highest level, and so on).

The research determined 100 institutions with prefabricated engineering experience in Singapore as potential respondents and sent them questionnaires via email to collect data. Eventually, 41 respondents were received, showing a response rate of 41%.

3.2 Data Analysis Methods

This study carried out statistical test and analysis on the questionnaire survey data. Parametric statistical test and non-parametric statistical test are two popular statistical tests at present [8]. Firstly, the normality of data should be checked in data analysis. If the p-value of the test is less than the selected alpha level (e.g., 0.05), then rejecting the null hypothesis and concluding that the test data is not normally distributed

[9]. In this research, an ordinary alpha level of 0.05 was adopted. Meanwhile, SPSS statistical 17.0 was employed for the test.

Since the respondents come from all kinds of institutions, it's necessary to make inter-group comparisons to see if there are significant differences in the respondents' views on institutions. ANOVA and Kruskal-Wallis test were used for comparison between groups. Therefore, the normality of the data presented by the Shapiro-Wilk test will decide the method to use for inter-group comparison.

Finally, constraint/mitigation strategies were compared in pairs to determine the most important constraints as well as the best mitigation strategies in PPVC.

4. Results and Discussions

4.1 Significant Constraints on Adopting PPVC

Table 3 and table 4 show the respondents' evaluation of constraints and the results of statistical tests. The results of Shapiro-Wilk test in table 3 present that the constraint data are non-normal distribution. Thus, Kruskal-Wallis test was chosen for inter-group comparison, while Wilcoxon signed - rank test was employed for variable comparison. Results of Kruskal-Wallis test in table 3 indicate that, except for C7 " Requirement for extra materials to protect PPVC modules ", all constraints have no significant difference in the respondents' institutional backgrounds. Such results mean that the respondents' evaluations are basically consistent and can be analyzed as a whole. At the same time, the evaluation results indicated in table 4 that C14 " More communication requested before and during construction ", C13 " Requirement for extra project planning and design efforts ", C8" Added logistics and transportation considerations ",C12 " Need for early commitment ", and C15 "Expensive initial costs of traditional construction method". In addition, according to Wilcoxon signed - rank test results in table 4, C14 and C13 were statistically higher than most constraints, indicating that they were the most important constraints when adopting PPVC.

In addition, in table 3, the sign "*" present that the Shapiro-Wilk test was significant at the significance level of 0.05, indicating the data was non-normal distribution, while "**" the Kruskal-Wallist test was significant at the significance level of 0.05, indicating statistical difference among respondents from different institutions. In table 4, the sign "*" The Wilcoxon Signed-Ranks test was significant at the significance level of 0.05, indicating the two compared variables were statistically different.

Table 3. Evaluations and Statistical Test Resutls of Constraints

Code	Constraint	Mean	Rank	Shapiro-Wilk test (p-value)	Kruskal-Wallis test (p-value)
C14	More communication requested before and during construction	4.37	1	0.000*	0.344
C13	Requirement for extra project planning and design efforts	4.27	2	0.000*	0.810
C8	Added logistics and transportation considerations	4.10	3	0.000*	0.153
C12	Need for early commitment	4.00	4	0.000*	0.162
C15	Expensive initial costs of traditional construction method	3.95	5	0.000*	0.054
C3	Design restrictions because of traffic limitation (e.g. modules' size)	3.94	6	0.000*	0.677
C1	Declined flexibility for future design changes	3.93	7	0.000*	0.403
C9	Apply of BIM to improve coordination and facilitate communication among project stakeholders	3.80	8	0.000*	0.381
C4	Limited site distribution (e.g. lack of space for PPVC modules to be stored ,unloaded and moved)	3.78	9	0.000*	0.499

C16	Expensive construction cost of the traditional construction method	3.71	10	0.000*	0.424
C17	Requirement of more communication among all stakeholders	3.71	10	0.000*	0.242
C18	Complicated process for inspection	3.54	12	0.000*	0.405
C10	Less awareness of PPVC's benefits among owners/developers	3.51	13	0.001*	0.263
C5	Less experiences design about PPVC	3.49	14	0.000*	0.162
C6	Less installation experiences about PPVC	3.44	15	0.001*	0.829
C2	Unsupportive decision made by designers	3.44	15	0.002*	0.973
C7	Requirement for extra materials to protect PPVC modules	3.17	18	0.005*	0.026**
C11	Raised organizational demand (e.g. changes in the roles of project participants)	3.41	17	0.000*	0.203

Table 4. The Results of Constraints in Wilcoxon Signed-Ranks Test

	C14	C13	C8	C12	C15	C3	C1	C9	C4
C14	-	0.285	0.100	0.050	0.028*	0.056	0.014*	0.006*	0.016*
C13		-	0.213	0.114	0.065	0.133	0.031*	0.021*	0.014*
C8			-	0.715	0.399	0.516	0.382	0.058	0.154
C12				-	0.701	0.911	0.526	0.315	0.306
C15					-	0.987	0.809	0.325	0.466
C3						-	0.813	0.305	0.287
C1							-	0.498	0.474
C9								-	0.967
C4									-
C16									
C17									
C18									
C10									
C5									
C6									
C2									
C7									
C11									

	C16	C17	C18	C10	C5	C6	C2	C7	C11
C14	0.002*	0.001*	0.000*	0.001*	0.001*	0.000*	0.000*	0.000*	0.000*
C13	0.006*	0.001*	0.000*	0.001*	0.000*	0.000*	0.000*	0.000*	0.000*
C8	0.053	0.039*	0.002*	0.004*	0.005*	0.001*	0.006*	0.000*	0.000*
C12	0.157	0.073	0.027*	0.009*	0.007*	0.014*	0.032*	0.000*	0.002*
C15	0.040*	0.231	0.009*	0.027*	0.024*	0.014*	0.012*	0.000*	0.001*
C3	0.248	0.247	0.018*	0.035*	0.029*	0.013*	0.036*	0.001*	0.010*
C1	0.211	0.344	0.024*	0.066	0.019*	0.015*	0.029*	0.000*	0.005*
C9	0.699	0.802	0.147	0.145	0.167	0.083	0.615	0.081	0.643
C4	0.709	0.725	0.164	0.192	0.172	0.099	0.227	0.023*	0.129
C16	-	0.837	0.319	0.230	0.283	0.213	0.124	0.016*	0.070
C17		-	0.323	0.397	0.243	0.154	0.301	0.003*	0.081
C18			-	0.962	0.815	0.656	0.578	0.049*	0.535
C10				-	0.922	0.631	0.615	0.081	0.643
C5					-	0.790	0.868	0.025*	0.863
C6						-	0.973	0.179	0.867

C2	-	0.240	0.990
C7		-	0.140
C11			-

4.2 Effective Mitigation Strategies for Constraints

Table 5 and table 6 revealed the respondents' assessment of the proposed mitigation strategy and the related statistical test results. As shown in table 5, the results of Shapiro-Wilk test were non-normal data distribution, so inter-group comparison and variable comparison will choose Kruskal-Wallis test and Wilcoxon Signed-Ranks test. Kruskal-Wallis test presented that all mitigation strategies except MS1 were evaluated unanimously by respondents, indicating that respondents had basically the same views on mitigation strategies. The evaluation results in table 5 mean that MS1, MS9 and MS4 are the top three effective mitigation strategies to tackle PPVC constraints. Furthermore, the Wilcoxon signed - rank test results in table 6 show that evaluation of MS1, MS9, and MS4 are statistically higher than most mitigation strategies, meaning that they are the best mitigation strategies for PPVC constraints.

Meanwhile, the meanings of "*" and "**" in table 5 are same with those in table 3, and the implication of "*" in table 6 is same with that in table 4.

Table 5. Evaluations and Statistical Test Resutls for Mitigation Strategies

Code	Mitigation Strategy	Mean	Rank	Shapiro-Wilk test (p-value)	Kruskal-Wallis test (p-value)
MS1	Encouraging project stakeholders to collaborate closely in early period of project	4.17	1	0.000*	0.046**
MS9	Applying BIM to promote coordination and increase communication among project stakeholders	4.07	2	0.000*	0.437
MS4	Providing project teams and workers with training sessions to enhance the skills and knowledge	4.02	3	0.000*	0.544
MS3	Apply of Just-in-Time delivery	3.68	4	0.000*	0.684
MS5	Fabricating and assembling the module components closely to the construction site to decrease transportation workload	3.63	5	0.000*	0.310
MS6	Using Information Technologies (e.g. electronic file transfer) to overcome the additional demand of planning, coordination and communication	3.63	6	0.000*	0.839
MS8	Developing systems/computer programs for detailed economic analysis of PPVC and conventional construction methods	3.63	7	0.000*	0.673
MS7	Conducting feasibility study on PPVC method	3.56	8	0.000*	0.316
MS2	Applying for temporary occupational permits (TOL) to build temporary sites near construction sites to overcome site layout limitations	3.46	9	0.001*	0.425

Table 6. The Results of Mitigation Strategies in Wilcoxon rank test

	MS1	MS9	MS4	MS3	MS5	MS6	MS8	MS7	MS2
MS1	-	0.381	0.568	0.005*	0.014*	0.002*	0.003*	0.001*	0.001*
MS9		-	0.741	0.046*	0.039*	0.006*	0.007*	0.001*	0.004*
MS4			-	0.112	0.055	0.043*	0.030*	0.008*	0.012*
MS3				-	0.826	0.777	0.717	0.562	0.292

MS5	-	0.974	0.958	0.785	0.380
MS6		-	0.982	0.439	0.386
MS8			-	0.439	0.325
MS7				-	0.659
MS2					-

5. Conclusions and Recommendations

The literature review and empirical management investigation of industry experts was conducted. This study determined 18 important constraints (e. g, "previous extensive coordination requirements and construction period"), which hinder the adoption of PPVC. Besides, nine mitigation strategies to cope with constraints were proposed.

Even though the goal of this study has been achieved, there are still some limitations. First of all, the study collected subjective views of respondents according to their experience. Secondly, the sample quantity of questionnaire survey was not large enough. Finally, the results of this research can only be applied in Singapore, while it is not suitable for other countries. Nevertheless, the results of this research are still beneficial to the knowledge system, since they are the first to investigate the constraints as well as mitigation strategies of the PPVC methods. In addition, results of the research can provide the industry with a comprehensive understanding of the limitations and mitigation strategies of PPVC methods, thus benefiting the industry. In the future, it's necessary to conduct economic analysis on PPVC and to test the key risks and the influences on project aims when adopting PPVC methods.

6. References

[1] Goodier C and Gibb A 2007. Future opportunities for offsite in the UK. Construction *Management and Economics* **25** 585-595.
[2] BCA S 2017 Prefabricated Prefinished Volumetric Construction.
[3] Saad I 2016 BCA to expand buildability appraisal framework later this year.
[4] Navaratnarajah R 2016 BCA's new roadmap to raise construction productivity.
[5] Mao C, Xie F, Hou L, W P, Wang J and Wang X 2016. Cost analysis for sustainable off-site construction based on a multiple-case study in China. *Habitat International* **57** 215-222.
[6] Ramaji I J and Memari A M 2016 Product architecture model for multistory modular buildings. *Journal of Construction Engineering and Management* **142** 16-47.
[7] Li A S, Ling F Y Y, Low S P and Ofori G 2016 Strategies for foreign construction-related consultancy firms to improve performance in China *Journal of Management in Engineering* **32**
[8] Kim T K 2015 T test as a parametric statistic *Korean Journal of Anesthesiology* **68** 540-546
[9] Villasenor Alva J A and Estrada E G 2009 A generalization of Shapiro-Wilk's test for multivariate normality *Communications in Statistics - Theory and Methods* **38** 1870-1883

IOP Conf. Series: Earth and Environmental Science **385** (2019) 012002 doi:10.1088/1755-1315/385/1/012002

Developing a Knowledge-based Decision Support System for Prefabricated Prefinished Volumetric Construction

Y S Li[1], B G Hwang[2], M Shan[3] and K Y Looi[4]

[1]School of Civil Engineering, Central South University, 68 South Shaoshan Road, Changsha, China, 410004

[2]Department of Building, National University of Singapore. 4 Architecture Drive, Singapore, 117566

[3]School of Civil Engineering, Central South University, 68 South Shaoshan Road, Changsha, China, 410004

[4]Arcadis Singapore Pte. Ltd., 1 Magazine Road, #05-01 Central Mall, Singapore, 059567

Email: ming.shan@csu.edu.cn

Abstract. Prefabricated prefinished volumetric construction (PPVC) has a lot of advantages, including environmental protection, saving money, reducing on-site manpower and so on. Many countries advocate the development of PPVC, the decision making of PPVC construction becomes critical important. This study aims to developing a scoring approach based on Knowledge-Based Decision Support System (KBDSS) for PPVC implementation. To achieve this goal, 19 key decision-making factors (DMFs) were identified from literature reviews first, subsequently, a questionnaire survey was conducted on 41 construction organizations in Singapore and using a five-point rating scale. Then create a PPVC scoring approach by using the data from the questionnaire, this scoring approach will developing a program named KBDSS-PPVC, by using the Microsoft Excel and Microsoft Visual Basic for Applications (VBA). Finally, four industry experts will be invited to validate the feasibility of KBDSS-PPVC and scoring their current construction projects, this program can also give corresponding recommendations according the score. The results of validation showed the developed system really could provide some available suggestions for industry practitioners on the decision-making for PPVC implementation, this study complements the current knowledge system on PPVC decision-making and provides the basis for further research.

1. Introduction

Prefabricated prefinished volumetric construction (PPVC) is a kind of off-site construction method. It is a method in which the entire room is processed in a prefabricated factory to complete the structure and decoration including the ground, wall, ceiling and other parts and form a separate module, which is transported to the site for lifting. Due to its modular specificity, PPVC allows construction at the same time, and it can significantly shorten construction schedule to improve building productivity greatly. On the other hand, it will reduce the environmental pollution, the noise and dust pollution during the construction process will also be greatly reduced. As the production site transferred from on-site to the factory, the quality is guaranteed, furthermore, construction site for the worker is safer and pleasant.

Because of its benefits, more and more developers and contractors pay attention to PPVC and its implementation in projects. The focus of this study is developing a decision-making system of PPVC for implement in the construction project and provide an auxiliary tool of PPVC assessment for industry. In this regard, the first is to determine the key decision-making factors (DMF) for PPVC adoption in building and projects, and how will these factors affect the PPVC decision in the project.

Then based on these considerations, we hope to develop a convenient computer-aided tool for industry practitioners to provide some recommendations and promote their decisions. To achieve these goals, this study identified the 19 decision-making factors (DMFs) that affect the construction of PPVC from literature reviews, developing a scoring approach system named Knowledge-Based Decision Support System for Prefabricated Prefinished Volumetric Construction (KBDSS-PPVC), proposed an assessment methodology that could scoring the feasibility of using PPVC.

The background of this research is Singapore. In recent years, the Singapore government has vigorously promoted the construction of PPVC, the goal of local authorities is to use the PPVC method to build nearly 35% of public housing projects in Singapore by 2019 [1]. In addition, there are already some examples of successful use of the PPVC approach, including but not limited to the North Hill Residential Halls of Nanyang Technological University, and the Crown Plaza Changi Airport Hotel Extension. Therefore, the Singapore government pays great attention and has invested a lot of energy in the construction of PPVC, and this is why it is appropriate to choose Singapore as a research background.

2. Background

2.1. PPVC

PPVC has many advantages and is becoming more and more popular in the construction industry, although the use of repeated modular structures can cause the building to lose some aesthetics. PPVC is a new construction method in which prefabricated structures are manufactured in the factory and then transported to the construction site. It allows parallel work, thus speeding up the work schedule and increasing productivity. In addition, PPVC reduces cost of field manpower and equipment requirements compared to traditional methods. What's more, since the structural body is completed in the factory, the quality of the construction becomes easy to control and provides an environmentally friendly and safe working environment for workers.

Since 2010, BCA Singapore has launched Construction Productivity Roadmap with the aim of improving the productivity of the local construction industry [2]. BCA emphasized that prefabricated structures are one of the directions that the local construction industry should pay attention to. Apart from BCA, the Housing and Development Board (HDB) is another local authority interested in PPVC. HDB believes that PPVC is a useful tool for developing HDB apartments. Further, it is reported that the HDB has started research activities to fund the development of a comprehensive, automated PPVC production system [3]. In short, PPVC is actively promoting in Singapore.

2.2. Decision-Making Factors of PPVC

Due to the limited number of studies on PPVC, it is difficult to determine the decision-making factors (DMFs) for PPVC directly from the literature. However, there are quite a few documents about general off-site construction method, which have a high correlation with PPVC.

This study firstly conducted a comprehensive literature review on the research of general off-site construction methods to find out DMFs that may affect the adoption of PPVC, and we found many excellent research results about this. Murtaza et al. [4] divided the idea of modular construction into five categories: plant site selection, environmental and organizational factors, plant characteristics, labor considerations, and project risk. Chen et al. [5] divided DMF influencing the use of precast concrete building methods into four types: project features, site conditions, market attributes and local regulations. Zhai et al. [6] looked at factors affecting the off-site production of residential construction and found six main factors: implementation of constructability, social climate and attitude, building performance, costing, supply chain and preparation.

In addition, we conducted more research to investigate the drivers, constraints, challenges, barriers and limitations of the general off-site environment. Because these factors may affect the use of the construction method, they also include in the literature review, for the overall definition of the DMF. Gibb and Itsack [7] investigate the driving factors used in the construction project that affect the prefabricated construction method. They found that drivers include: reduced on-site congestion, saving time, cost, quality, productivity and on-site operations, improving building health and safety, and more predictable construction quality. Azar et al [8] identifies and evaluates constraints related to the adoption of modular structures. The constraints are: the key decision of construction methods made by designers without the participation of contractors, the owner's misunderstanding of the construction method, the unavailability of the prefabricated unit, site layout restrictions, the flexibility of changes during project implementation is reduced, and difficult to make on-site modifications. Kamali and Hewage [9] reviewed the challenges of using a modular construction approach. The main challenges identified were project planning, transportation constraints, negative impacts of public perceptions, site constraints, high initial cost, and the need for more coordination and communication. Mao et al. [10] reviewed the main obstacles preventing Chinese construction industry from using off-site construction methods. They found that the top three key barriers are lack of government incentives and supervision, high initial cost, and the reliance of practitioners on traditional architecture. Tam et al. [11] studies the barriers to application prefabrication. The study found that the top three hindrances were difficult design changes, lack of comprehensive information, and high initial cost compared to traditional buildings. Arashpour et al. [12] summarized the limitations of the use of off-site construction methods, including the lack of coordination between on-site requirements and off-site construction, resulting in extended construction periods. Expensive manufacturing, many types of customized products, low construction efficiency and high maintenance cost for users. This study conducted a comprehensive literature review of the general off-site construction methods to identify DMFs that may affect PPVC. On the basis of the literature review, we've identified 32 types of DMFs that affect the use of PPVC and after expert discussion, the 19 most relevant DMFs were finally obtained.

2.3. KBDSS

KBDSS is based on the development of expert systems and decision support systems. An expert system is a knowledge base containing expert with reasoning mechanisms specific problem areas, and a computer simulation program decisions. KBDSS is flexible and adaptable to adapt to changes in the environment and user decision-making methods. It has a friendly interactive user interface that can be easily used by anyone who is not good at computers. In view of these advantages, KBDSS has been widely used in various fields. KBDSS will be a good choice for developing PPVC decision tools.

3. Research Methodology

The entire research process consists of 5 steps. The first step is to determine the DMFs that are closely related to the PPVC decision. First, a comprehensive literature review is conducted to initially identify the 32 DMFs. Then interviews with industry experts are conducted, and 19 DMFs are retained, as shown in Table 1. The second step is to issue a questionnaire to industry practitioners to get their views on the 19 DMFs, using a five-point rating scale (i.e., 1 = very insignificant, 2 = insignificant, 3 = slightly significant, 4 = significant, and 5 = very significant). Finally, 41 responses from industry practitioners were received. In the third step, a PPVC scoring method was created, which can evaluate the feasibility of using PPVC in a project by scoring the DMFs obtained earlier. In step 4, using a scoring method in step 3 in conjunction with Microsoft Visual Basic for Application (VBA) programming language developed KBDSS-PPVC in Microsoft Excel platform. Finally, in step 5, four experts were asked to validate their current construction project with the developed KBDSS-PPVC, using the KBDSS-PPVC for the feasibility score of the PPVC project. In addition, they commented on the results of this system and provided suggestions for possible enhancements to KBDSS-PPVC.

Table 1. DMFs Affecting the Adoption of PPVC Approach.

Code	Decision-making Factors
DMF1	Applicability of PPVC design
DMF2	Repeat layout in the design
DMF3	Floor number
DMF4	Structural stability of independent modules and modular modules
DMF5	Inspection and supervision of the manufacturing process
DMF6	PPVC module's preparation time

DMF7	PPVC module's parameters
DMF8	Site layout (such as PPVC module storage space and sufficient unloading space)
DMF9	Site location
DMF10	Skilled workforce and experienced supervision
DMF11	Transport equipment
DMF12	Construction equipment
DMF13	Organizer's experience with the PPVC method
DMF14	The use of building information management technology
DMF15	The scope and nature of the project
DMF16	Shorten the construction period
DMF17	Provide a safe construction environment
DMF18	Reduce environmental pollution through fewer on-site activities
DMF19	Reduce interference and noise to the surroundings

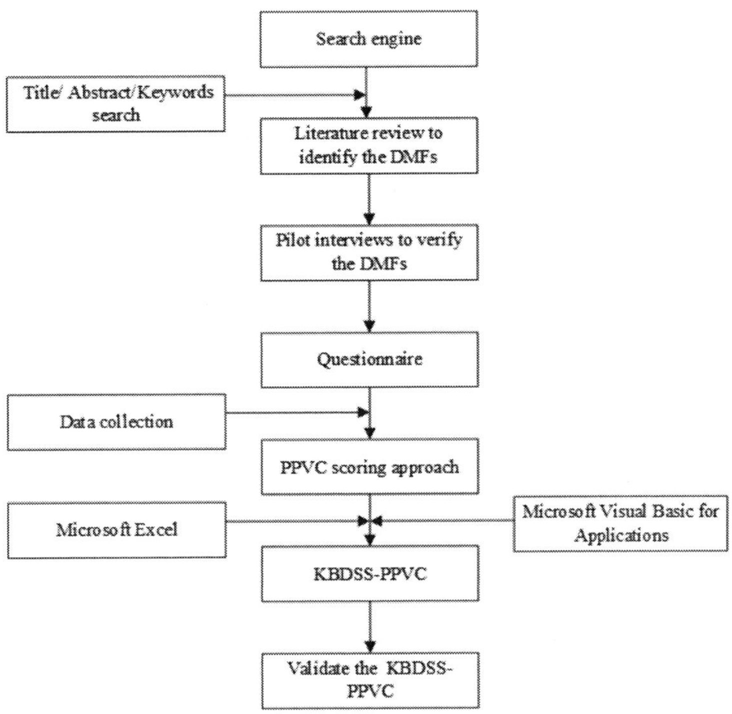

Figure 1. Research framework.

4. PPVC Scoring Approach

Based on the evaluation of 19 DMFs for a specific project, the PPVC scoring method can be derived. Set the evaluation level of all DMFs to 5 levels, and calculate the corresponding scores for each level. The highest achievable score of the project is set to 100, which means that the project is most suitable for PPVC method, and the lowest score is 19, indicates that the project is not suitable for this method and needs to be improved or reconsidered. By studying the method of Caldas, et al. [13], we calculated the maximum PPVC score that each DMF can obtain based on the average score given by 41 questionnaires.

In more detail, calculate the maximum PPVC score for the ith DMF using the following Formula (1) (Si5 is the maximum score of PPVC when the assessment level of DMF is 5, Mi is mean score). On this basis, score one point for the first level of all DMFs. Then, the 2, 3, and 4 levels are calculated using the following Equations (2), (3), (4). Finally, the scores of all 19 DMFs are added to get the PPVC score for a given item, Equations (5). Table 2 shows the PPVC scores for different DMFs at all evaluation levels.

$$S_{i5} = \frac{M_i}{\sum_{i=1}^{19} M_i} \times 100 \quad (1)$$

$$S_{i2} = \frac{S_{is}-1}{4} + 1 \quad (2)$$

$$S_{i3} = \frac{S_{is}-1}{4} + S_{i2} \quad (3)$$

$$S_{i4} = \frac{S_{is}-1}{4} + S_{i3} \quad (4)$$

$$PPVC\ Score = \sum_{i=1}^{19} S_i \quad (5)$$

Table 2. PPVC Scores of DMFs at Different Assessment Levels.

DMF	Mean	Maximum Score of PPVC	Assessment Levels of DMF				
			1	2	3	4	5
DMF1	4.34	6	1	2	3	5	6
DMF2	4.59	6	1	2	4	5	6
DMF3	3.80	5	1	2	3	4	5
DMF4	4.05	5	1	2	3	4	5
DMF5	3.73	5	1	2	3	4	5
DMF6	4.05	5	1	2	3	4	5
DMF7	3.95	5	1	2	3	4	5
DMF8	3.98	5	1	2	3	4	5
DMF9	3.56	5	1	2	3	4	5
DMF10	3.90	5	1	2	3	4	5
DMF11	3.73	5	1	2	3	4	5
DMF12	3.80	5	1	2	3	4	5
DMF13	3.90	5	1	2	3	4	5
DMF14	3.76	5	1	2	3	4	5
DMF15	3.76	5	1	2	3	4	5
DMF16	3.95	5	1	2	3	4	5
DMF17	3.80	5	1	2	3	4	5
DMF18	3.80	5	1	2	3	4	5
DMF19	3.85	5	1	2	3	4	5

5. KBDSS-PPVC

5.1. Purpose of the KBDSS-PPVC

The specific goal of KBDSS-PPVC is to provide a reference tool for practitioners in the construction industry to assess the feasibility of implementing a PPVC construction method in a particular project and to provide the project team with relevant information. Then give a comprehensive recommendation based on the evaluation results.

5.2. Structure of the KBDSS-PPVC system

The KDSS-PPVC system consists of three parts as shown in Figure 1, including a knowledge base, decision support system, and user interface. The picture below shows more details.

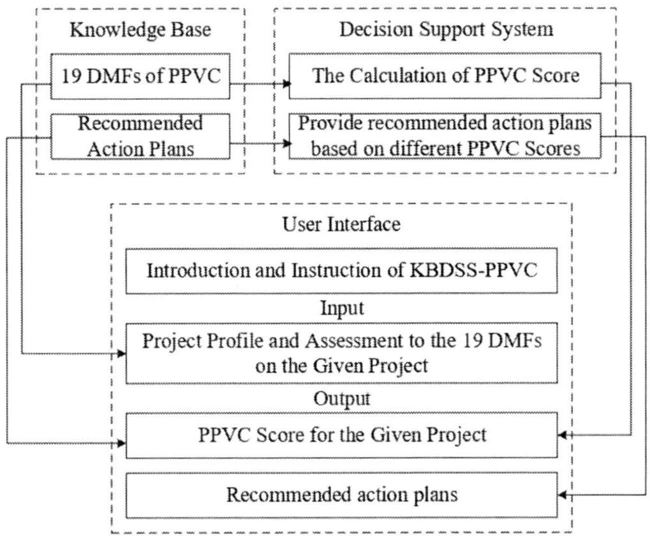

Figure 2. Structure of the KBDSS-PPVC system.

KBDSS-PPVC is made in Excel and its user interface is very interactive. It consists of two parts. The first part introduces the KBDSS-PPVC overview and user guide. Users can also enter brief information about the project to be evaluated. The second part is a score sheet for 19 DMFs, and the user evaluates the project using the five-point Likert rating scale. The calculated PPVC score is divided into three levels through the decision support system, and different suggestions recommended to the user are given according to the calculation result of the PPVC, as shown in Table 3. Moreover, the second part also contains a dialog message box, which is shown in Figure 2 for different operational items recommended by the user based on the calculated PPVC score.

Table 3. Recommended Action Items for Different score segments.

Computed PPVC Score	Action Items
< 60 (Basic)	Recommended for traditional construction approach
60 – 80 (Intermediate)	Revisit building design and modify accordingly if needed Appoint PPVC specialist builders and experts with relevant experiences Use past projects (local and/or overseas) as references
> 80 (Advanced)	Active participation and commitment of project stakeholders in the early stages of the project Understand the environment before construction and production, arrange on-site and factory visits Appoint industry experts with extensive PPVC experience to establish a quality management system for the standards and dimensions of overseas prefabricated modules.

IOP Conf. Series: Earth and Environmental Science **385** (2019) 012002	doi:10.1088/1755-1315/385/1/012002

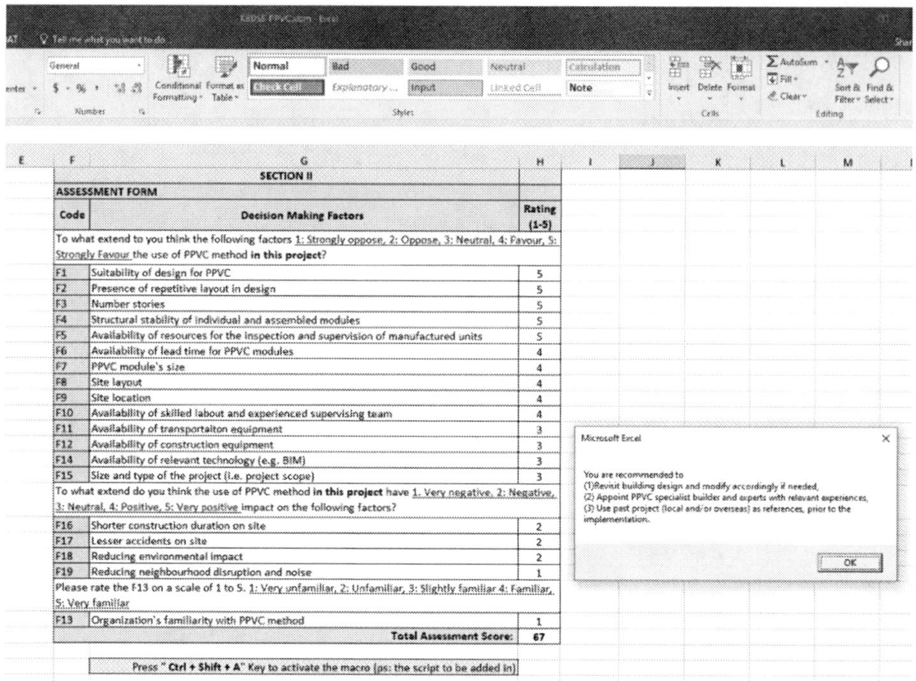

Figure 3. Output and Recommended Action Plans Interface

5.3. Validation of Developed System

We invited four experts from the Singapore Building Innovation Panel to validate the developed KBDSS-PPVC. Experts were asked to rate their existing projects using KBDSS-PPVC. Experts A and B use the PPVC method to engage in a residential construction project at the Nanyang Technological University, Singapore, which had adopted the PPVC method. Expert C did not use the PPVC method to engage in a multi-purpose stadium construction project at the National University of Singapore. Expert D also did not use PPVC in his research on condominium construction projects. Table 4 lists the PPVC scores calculated by KBDSS-PPVC. It is pointed out that the results of the evaluation are consistent with the decision of the project team at the beginning of the construction project, indicating that the system has certain reference, it shows that the developed KBDSS-PPVC can provide some reliable suggestions for users in the industry.

All experts agree that this KBDSS-PPVC system will help them better understand what is lacking in the construction of the PPVC project and what preparations can be made before implementation. The experts also put forward some opinions and suggestions. Expert A is more concerned about the importance of prefabricated module manufacturer capabilities, while Expert B recommends that the authorities need to establish a consistent quality assurance and quality control system, in order to ensure that the overseas prefabricated modules meet the standards and requirements of a predetermined size.

Table 4. Validation Results of Developed KBDSS-PPVC.

Expert	Project Name	PPVC Project	PPVC Score
A	Nanyang Technological University Residential Construction Project	Yes	92
B	Nanyang Technological University Residential Construction Project	Yes	85
C	National University of Singapore Multi-purpose Stadium Construction Project	No	57
D	Condominium Construction Project	No	51

6. Conclusions and Recommendations

This study explores the decision-making problem of a typical off-site construction method called PPVC. First, the study identified the 19 kinds of DMFs may affect PPVC used in construction projects, according to a comprehensive literature review, and pilot interviews of three experienced construction experts.Then, a PPVC scoring method was established to evaluate the feasibility of using PPVC in a specific project, according the data from 41 questionnaires for DMFs evaluation. Subsequently, using the research of PPVC scoring method, a knowledge-based decision support system called KBDSS-PPVC was developed, which can help industry practitioners make PPVC decisions. Finally, four experts verified the KBDSS-PPVC. Experts believe that it is a convenient and useful tool for making PPVC decisions in construction projects. Therefore, this study can provide some advice to the industry.

Despite the goals achieved, the conclusions drawn from the results still have some limitations. For example, the sample size of the questionnaire is small. Also, the recommendations given in the KBDSS-PPVC system are not specific and detailed, and in practice should be selected in conjunction with the characteristics of the project itself.

This research is to supplement the current knowledge system and contribute to further research. It is necessary to carry out relevant practical work on the best decision-making methods for PPVC in the future, and it is also expected that this study will be used to evaluate more projects.

References

[1] Channel NewsAsia Singapore 2017 HDB to expand use of prefabrication building methods in BTO projects (accessed March 23, 2018)

[2] Navaratnarajah R 2016 BCA's new roadmap to raise construction productivity (accessed May 13, 2017)

[3] Lim P J 2016 Industry practices need to change to benefit from advanced pre-fab techniques: Experts. (accessed May 12, 2017)

[4] Murtaza M B, Fisher D J and Skibniewski M J 1993 Knowledge-based approach to modular construction decision support *Journal of Construction Engineering and Management* **119** (1) (1993) 115-30

[5] Chen Y, Okudan G E and Riley D R 2010 Decision support for construction method selection in concrete buildings: prefabrication adoption and optimization *Automation in Construction* **19** (6) (2010) 665-75

[6] Zhai X, Reed R and Mill A 2014 Factors impeding the offsite production of housing construction in China: an investigation of current practice *Construction Management and Economics* **32** (1-2) (2014) 40-52

[7] Gibb A and Isack F 2003 Re-engineering through pre-assembly: client expectations and drivers *Building Research & Information* **31** (2) (2003) 146-60

[8] Azhar S, Lukkad M Y and Ahmad I 2013 An investigation of critical factors and constraints for selecting modular construction over conventional stick-built technique *International Journal of Construction Education and Research* **9** (3) (2013) 203-25

[9] Kamali M and Hewage K Life cycle performance of modular buildings: a critical review *Renewable and Sustainable Energy Reviews* **62** (2016) 1171-83

[10] Mao C, Shen Q, Pan W and Ye K 2015 Major barriers to off-site construction: the developer's perspective in China *Journal of Management in Engineering* **31** (3) (2015) 04014043

[11] Tam V W Y, Tam C M, Zeng S X and Ng W C Y 2007 Towards adoption of prefabrication in construction *Building and Environment* **42** (10) (2007) 3642-54

[12] Arashpour M, Wakefield R, Abbasi B, Lee E W M and Minas J 2016 Off-site construction optimization: Sequencing multiple job classes with time constraints *Automation in Construction* **71** Part 2 (2016) 262-70

[13] Caldas C H, Kim J Y, Haas C T, Goodrum P M and Zhang D 2014 Method to assess the level of implementation of productivity practices on industrial projects *Journal of Construction Engineering and Management* **141** (1) (2014) 04014061

Microclimate Environmental Model for New Built Environment and Design Complex UiTM Puncak Alam

A Misni[1], S N A Buyadi[1], F Ahmad[1], M F Jani[1], I A Zakaria[1] and N I Nordin[1]
[1] Faculty of Architecture, Planning and Surveying, Universiti Teknologi MARA, Puncak Alam, Selangor, Malaysia

Email: alamahmisni@gmail.com

Abstract. Microclimate is a smaller area within a general climate zone that has its own unique climate. In a particular area, there can be a myriad of smaller climates within the larger climate zone. ENVI-met is a computer program that enables to stimulate the microclimate of an environment. It is also can be used to estimate the effect of changing climate conditions on the thermal comfort within the area. The aim of this study was to measure the microclimate of an area been received within, by generating the microclimate environmental model using ENVI-met software. This simulation module is used to understand and quantify the interactions between the man-made and natural environment, and the outdoor microclimate in new development. The microclimate area that has been selected is at Built Environment Complex, UiTM Puncak Alam, a new campus area located in a hot and humid tropical climate. The physical elements involved were vegetation, multi-storey buildings, roads, paving areas and open/green areas. The simulation model shows that the highest temperature recorded was 33°C, which is at 17.00 hours. The relative humidity is below 58% during this peak time with wind speed 1.3 m/s. This high temperature occurs due to the pre-mature tree, which only can provide a minimum shade and make the area become hot and uncomfortable to users.

Keywords: Microclimate, environmental model, ENVI-met, tropical climate, development.

1. Introduction

The microclimatic environmental quantification is conducted to identify the range of neutral thermal environment of the development areas in a hot and humid tropical climate. The study area is located in the new campus development includes the open space and green area. There are two main microclimate evaluations in order to gain the microclimate character of the urban areas. Namely as microclimate ground measurement and microclimate simulation model [1]. This study focused on numerical simulation model of ENVI-Met, supporting with the ground data obtained from the Meteorology Department. Basically, the evaluation methods cover the real-time or continuous monitoring at hourly interval in tropical climatic condition. This is to enables to monitor the surrounding new green space microclimate modification surround the campus layout. The overall microclimate condition around the new campus will influence the comfort level of users especially in outdoor areas. The basic characters of outdoor thermal environment space usually should have covered by natural shades. The shades can be provided by mature trees, small trees, shrubs, and groundcovers [2]. Water elements and natural earth surfaces can provide evaporation process together with transpiration from plants. All these elements usually will be supported by wind flow to comfort the outdoor spaces.

2. Literature review

Urban microclimatic models can be used to evaluate the interactive effects between green space and development areas such as in new campus development. Data related to microclimate components are air temperature, relative humidity, wind, solar radiation, and evapotranspiration. There is a significant microclimate model that can be used for estimating the campus microclimates pattern, such as the ENVI-met model which was developed by M. Bruse from the University of Mainz, Germany in 2004. ENVI-Met is a microclimate model for measuring those microclimate parameters. ENVI-Met is a three-dimensional numerical model that can simulate the surface-plant-air interactions of urban environments with a typical resolution of 0.5-10 m in space from a single building up to a neighbourhood providing 250 grids as a maximum [3]. ENVI-Met is one of the only models of its kind that is actually considered for vegetation in a detailed manner, using single trees to compose, or grasses, paved areas, roads, buildings and so on. Study by [3] also explained the basic data structure of ENVI-met as a simulation program includes; i) simulate the complete coupled climate system; ii) provides a high resolution model resolving single buildings; iii) adequate simulation of surface-vegetation-atmosphere processes; iv) use state of the art computational techniques; and v) easy to use interface and input/output data handling.

The strength of this model is that ENVI-Met has implemented a soil-plant-air sub-model and its plant database, which depends on the plant numerical physiological representation using height, albedo, leaf area density, stomata resistance, leaf surface temperature water content, etc. Simulation result in the previous studies indicated that the location of buildings near or surrounded by different greenery levels has showed different ambient temperatures than those located with less density of the green area [4] [5]. It is scientifically believed that adding more vertical green cover within the built-up area is one effective way to lower the ambient temperature [5]. A similar model was used to predict different scenarios and the results showed that by adding more green cover, the campus environment becomes markedly cooler than it is at present [3].

Some studies on microclimate were also conducted in order to identify the temperature impact of vegetation in the urban environment with respect to the reduction of air temperature. Hence, the microclimatic effect of vegetation underwent ground measurement for verification and computer simulation. These two approaches were conducted on the institutional campus of Saga University, Japan, including an on-site measurement and a numerical simulation model that used ENVI-Met [6]. The on-site measurement program was designed and the real-time microclimate conditions were measured in the campus environment at the pedestrian level. Then, the level of greening was increased via a simulation model of ENVI-Met. Results from this study presented that these green areas had a pronounced low temperature and reduced the ambient outdoor summer temperature. It has been shown the importance of greening as a potential method for passive cooling and for use in reducing ambient air temperatures, especially at a pedestrian level in university campus.

The microclimatic simulation model by ENVI-Met can provide more detailed features for evaluating trees and vegetation effects to the surrounding urban areas. The study aims to model an environmental microclimate for new Built Environment and Design Complex UiTM Puncak Alam. ENVI-Met is used because this software capable of simulating built environments from a microclimate scale in a local area at a campus location.

3. Methodology

3.1 Site study

The selected site for this study is located in the new campus of Built Environment and Design Complex, which are consist of Faculty of Architecture, Planning and Surveying (FSPU), and Faculty of Art and Design (FSSR), Universiti Teknologi MARA, Puncak Alam, Selangor. The visual image is shown in Figure 1. The age of these new developments was around two years. The area is located in the hot and humid tropical climate, at 3°N Latitude and 101°E Longitude with average elevation 159 meters. The total site study areas are around 40,000 sq m. The microclimate environmental model was built on 24th September 2017. The campus building area that has been selected to be stimulated was supported by other buildings such as hostels, library, vehicle depot, cafeteria and *musallah*. The physical elements involved in this selected area are vegetation, multi-storey buildings, roads, paving areas, courts, and green/open areas. Natural or soft-scape element such as water elements (lake) and small secondary forest located on the east side. Most of trees planted in this campus include around the lake and park, roadside and parking were in small size and in pre-mature condition.

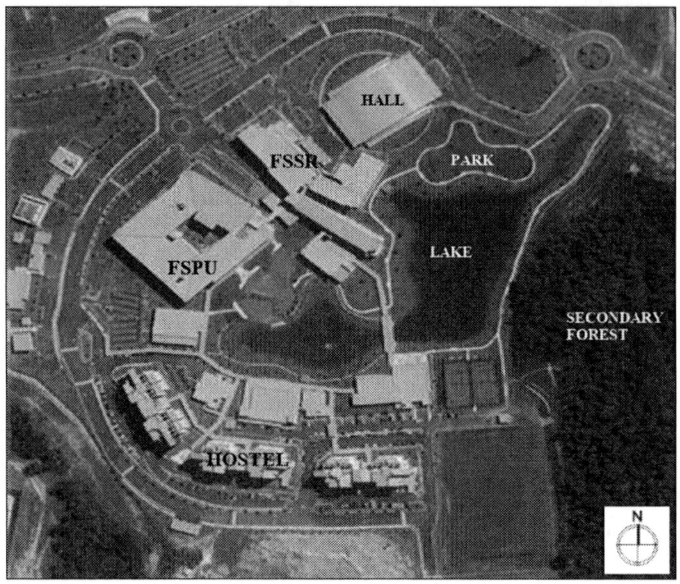

Figure 1. The site study of new Built Environment and Design complex, UiTM Puncak Alam

3.2 ENVI-met microclimate simulation model

ENVI-met is a multidisciplinary simulation consultancy with a core services portfolio, including architecture, building physics and microclimate, landscape architecture and garden design. The ENVI-met Tree Pass is a comprehensive overview of plants' requirements and the site conditions match to obtain the best possible thermal benefits for landscape designs and to ensure optimal growing conditions. The climatic factors include temperature, humidity and wind speed simulation were set up using the ENVI-Met software as well as to evaluate the accuracy of the simulation model. The simulated climatic maps were used to analyse the pattern. For this study, hourly climatic simulation for a period of 24 hours was carried out. The hourly data obtained from the simulated process were later compared to the weather data gained from the Malaysian Meteorological Department station in Subang Jaya, Malaysia. The real-time comparison was carried out between the permanent weather station and simulated weather data, which was of the same date and same time observation. The models generated from the ENVI-Met numerical model were essential to identify the heat pattern and its effects to the surrounding campus environment at any time.

The ENVI-Met modelling program has four user interfaces. Firstly, editing the input and preparing digital maps of the layout. Secondly, only the environmental domain is chosen to work and high-resolution aspects of the program enable the users to go into finer details in smaller scales (0.5) or to be coarser (and less detailed) in lower scales. For example, a 10m square area in the campus study area is represented by 1km square in a model. Third, databases for soil types, humidity, temperature, temporal input, etc. are entered as the modelling area and where additional parameters are present, the modelling process takes place. Fourth, the output data can be interpreted and visualized in LEONARDO. ENVI-Met generates hourly fields of climatology simulation (i.e. temperature, relative humidity (RH), radiation) variables [7]. Bruse also defined that the integration of ENVI-Met 4.0 and LEONARDO 2017 has the capacity to project small to large-scale climatic impacts. The model can evaluate future parkways in the areas of optimal outdoor comfort, optimal citizen use, and minimal environmental damage. As a non-hydrostatic model that simulates surface-plant-air interactions inside urban environments on a three-dimensional rectangular grid with variable spacing in x-, y-, and z-directions, the model functions over a range of spatial scales

4. Results and Discussion

The open space within the build-up areas can resulted low temperature. The role of open spaces in alleviating an urban microclimate has been examined all over the globe. The green space within this Built and Environment campus complex consists of large coverage area of small tree's age around two years old, supported by a small amount of shrubs, water bodies, and turf to cover earth surfaces. This coverage of pre-mature trees will possible in providing a minimum shade and evaporative process. The minimum cooling effects of the open space will affect the thermal surrounding environment of PSPU and FSSR areas. Evaluating the open space cooling consequences around this new building complex were validated via simulation models.

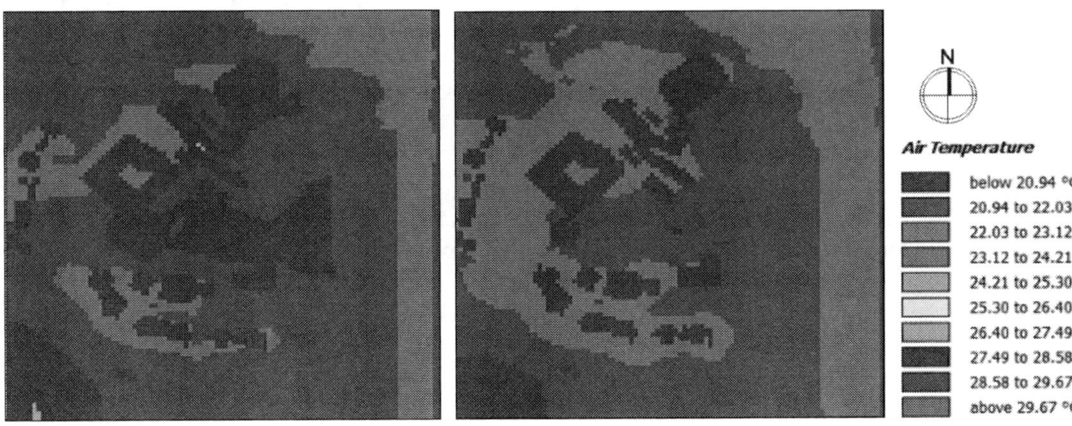

Figure 2. Temperature during peak time of the day, 15.00 hours (left) and 17.00 hours (right)

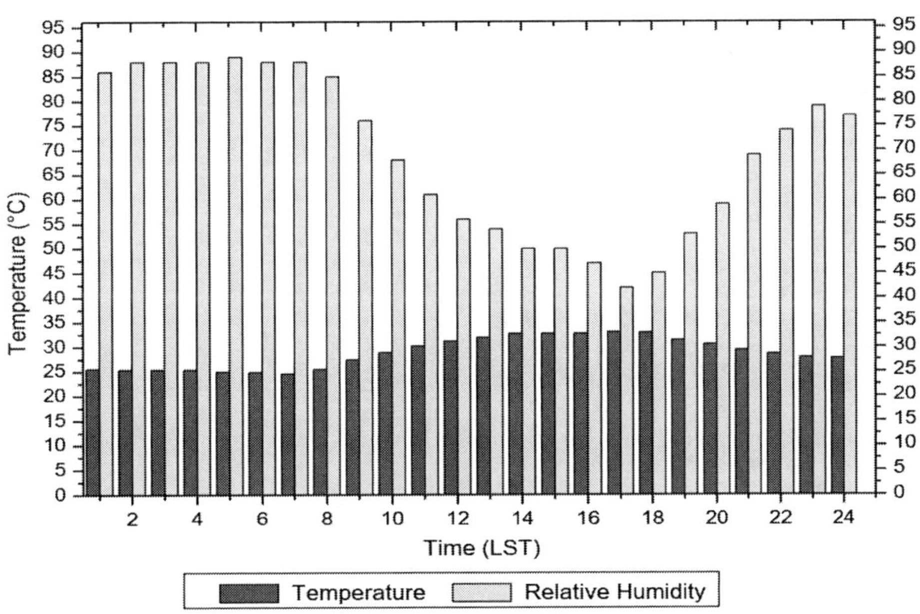

Figure 3. Temperature and RH in 24 hours. Peak time of the day at 17.00 hours.

The simulation models were generated 6 times in 24 hours. The simulation shows that the lowest temperature was 25.5°C which is 8.00 hours in the morning. The temperature began to rise around the building blocks, which is the data recorded around 26.94°C to 29.98°C, represented as paved areas which are parking lots, roads, building entrances and plaza. This is due to the trees that are small size and premature within the site. The physical of trees have a very minimum canopy size and those trees were replanted and not yet reach to the age of mature trees. Therefore, it's unable to give the shades as well as to lower the temperature. Conversely, made the areas become hot and uncomfortable for people. While Figures 2 and 3 also shows that at 13.00 and 17.00 hours, the temperature of surrounding building recorded in 31.90°C to 33.00°C. The high temperature range is due to the paved areas absorbing the heat during day time and emit the heat in the evening when the surrounding starting being cool. The other area within the park, which most of the areas were covered by turf and small trees, surrounded by water element was lower it ambient temperature as much as 1°C compared to built-up areas. The surrounded building with small and premature trees exposed the direct sunlight towards buildings as well as unable provide adequate shade over peak hours of the day. They received and absorbs heat from direct sunlight and reached its heat storage capacity as much as 33.00°C at 17.00 hours in the evening.

When the sun rises early in the morning, the RH is usually high due to low temperature overnight near the dew point. Similarly, the highest RH was during the early morning at 8.00 hours as much as 87.12%. The RH was gradually decreasing until reach the lowest level of around 45% between 15.00 and 17.00 hours located around building blocks as shown in Figures 3 and 4. The highest RH recorded in green areas which is 75.74%. This is because of the areas were surrounded by water/lake and plantation. Water element can generate the high RH 60% toward the surrounding. The combination of water, soil and premature trees with turf cover can generate evapotranspiration process during the day and contributed to the reading of RH. The reading of RH is lower as much as 10%. Meanwhile, premature trees and small shrubs that located surround parking lots, entrance and buildings generate very low RH reading during its photosynthesis process. The dark blue colours located at the parking space and road and surround buildings recorded at lower RH because of the built-up and paved area made the areas hot and dry.

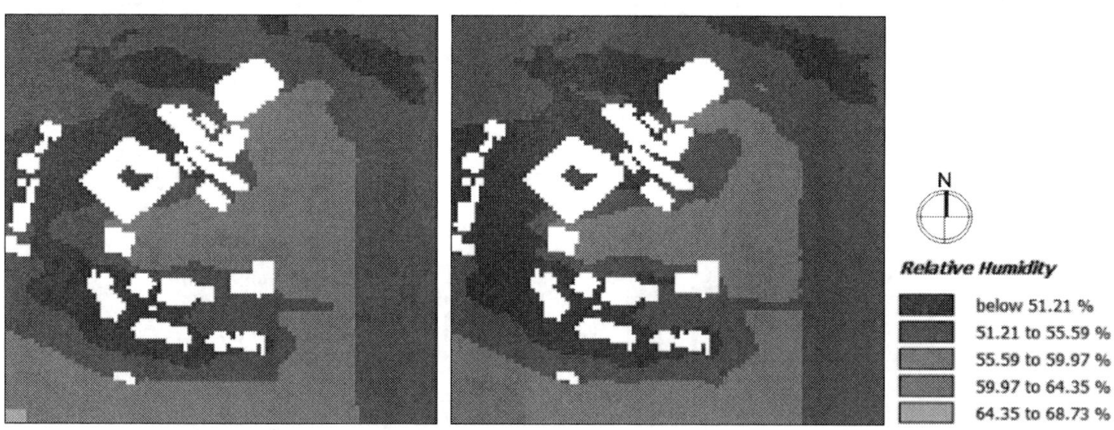

Figure 4. RH during peak time of the day, 15.00 hours (left) and 17.00 hours (right)

According to meteorology report, this is the most suitable time wind flow along the day in this area between 1.1 and 3.5m/s. Wind flow/speed in the study area is stable below 3m/s from the south, southwest, and west directions. Wind speed in surrounding the buildings during peak time of the day is the lowest reading around 0.33–1m/s as shown in Figures 5 and 6. While the surrounding green area the speed is more stable, around 1–3m/s. This is because of all of tree sizes are small and unable to provide the proper movement of the wind in this environment. If the evapotranspiration and shade from the tree are well produced, with this stable wind flow, the cool and comfortable environment will

IOP Conf. Series: Earth and Environmental Science **385** (2019) 012003 doi:10.1088/1755-1315/385/1/012003

be provided, even during peak hours. The findings of wind flow further revealed that most of the high wind flow is located at surrounding the site study.

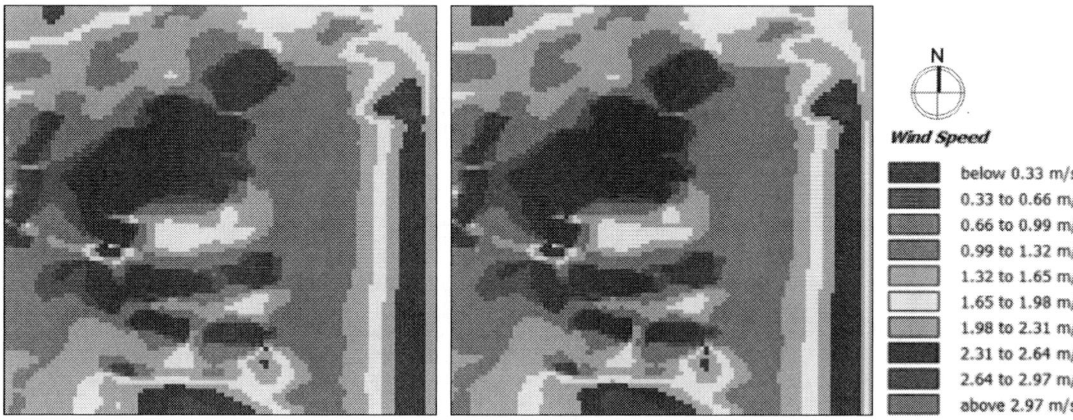

Figure 5. Wind flow during peak time of the day, 15.00 hours (left) and 17.00 hours (right)

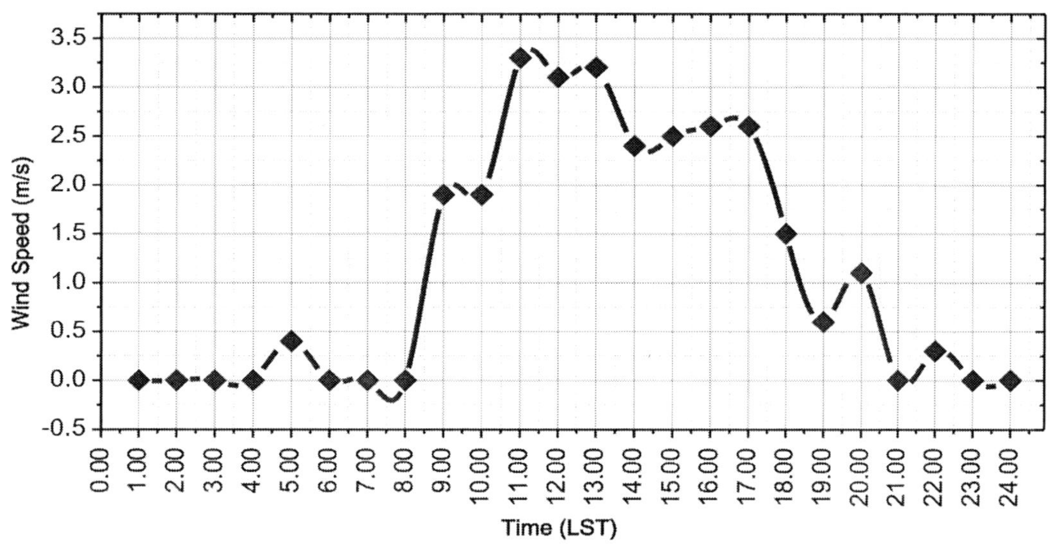

Figure 6. Wind speed in 24 hours; higher at 11.00 to 17.00 hours

The findings show that the increasing of temperature moving surrounds the green area. The built-up areas were recorded with higher temperature readings as compared to green spaces. It reveals a dramatic effects of greens park (tree, shrubs and turf) in lessening the ambience temperature of the campus areas. Furthermore, it also indicates that mature tree with sufficient canopy size and have dense of the leaves are more effective in reducing the temperature as compared to small vegetation such as shrubs and grass. Thus, in order to reduce the temperature in this campus complex, greener area should be preserved and planting more shade tree to reduce the surrounding temperature. Trees planted with other vegetation, located in a strategic location around buildings are the most effective mitigation strategy to reduce high temperature.

The microclimate model for this campus complex shows the highest temperature during peak time of the day is 33˚C at 17.00 hours in the evening, with RH 45% and wind speed around 2.5–2.8m/s. This temperature reading is high because of surrounding shade trees are in premature conditions is unable to contribute shade and evapotranspiration cooling. The matured park in tropical climate can provide an ambient temperature as low as 31˚C during peak time of the day [8][9].

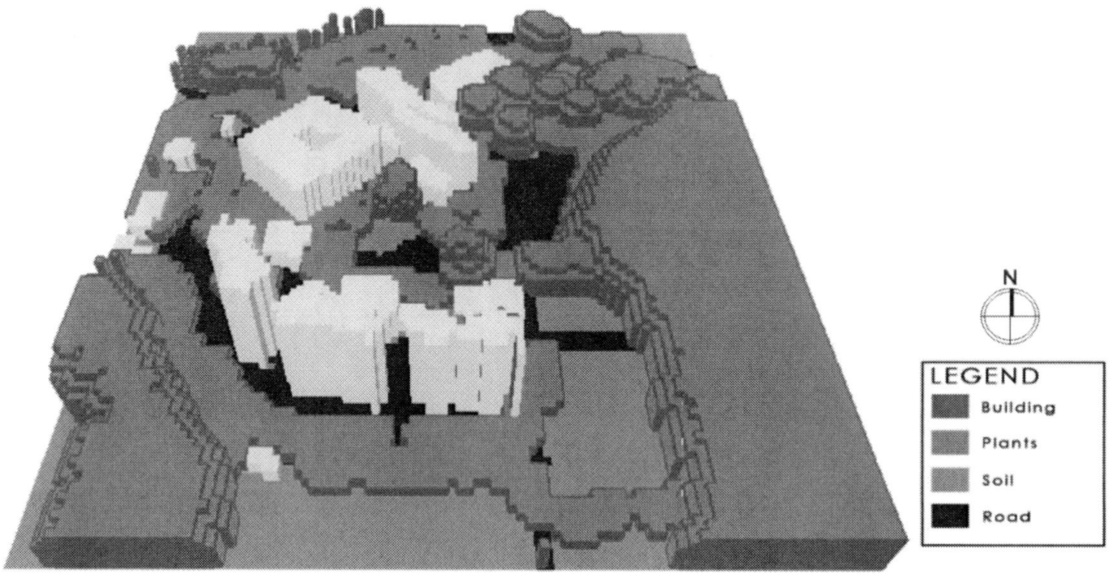

Figure 7. 3D Model of the FSPU and FSSR campus complex with surrounded by a park with mature trees combined with water elements will be generated after 10 to 15 years after development

Figure 7 shows a 3D microclimate model of the FSPU and FSSR campus complex surrounded by a park with mature trees and water elements will be generated after 10 to 15 years after development. These mature green surrounding can reduce as much as 3°C ambience temperature. The low temperature will be supported by maximum shaded by mature trees, the evapotranspiration process of vegetation, water element and earth surfaces as well as encourage the wind flow surround the buildings and the park. While the secondary forest that located at the east and southwest of the campus complex, should be maintained. The network of these forests/green areas will be continued into the campus building complex forming a harmony and sustainable greenbelt, and furthermore friendly to wildlife. Cool and comfort campus environment is suitable for outdoor learning activities and for the students work out in the evening.

5. Conclusion

ENVI-met is a software to stimulate the microclimate of an environment. The microclimate simulation by ENVI-met can be used to estimate the effect of changing climate conditions in the outdoor comfort within the area. As this study shows, green spaces and are normally considered to help improving the outdoor comfort by reducing air temperature. The microclimate model for this campus complex shows the highest temperature during peak time of the day is 33°C at 17.00 hours in the evening, with RH 56% and wind speed around 1-3 m/s. This temperature reading is high because of surrounding shade trees are in premature conditions can not contribute to reduce the hot and humid tropical climate in surrounding them. The predicted/future 3D microclimate model illustrates the huge green spaces contribute a significant impact in the microclimate. The cooling effect of park can dominate the immediate surroundings, especially when the evapotranspiration process is in the maximum. It mainly consequences by natural shading, lower ambience temperature and humidity level in the environments. Finally, the cooling performance in the park depends on the density of the foliage in the green space.

References

[1] Shashua-Bar L and Hoffman M E 2004 Quantitative Evaluation of Passive Cooling of the UCL Microclimate in hot Regions in Summer, Case Study: Urban Streets and Courtyards with Trees *Build. Environ.* **39** 1087–1099

[2] Misni A 2013 Modifying the Outdoor Temperature around Single-Family Residences: The

Influence of Landscaping *Procedia - Soc. Behav. Sci.* **105** 664–673
[3] Ozkeresteci I, Crewe A J, Brazel K and Bruse M 2003 Use and Evaluation of the Envi-Met Model for Environmental Design and Planning: An Experiment on Linear Parks in *21st International Cartographic Conference (ICC) in "Cartographic Renaissance* 10–16
[4] Boukhabl M and Alkam D 2012 Impact of Vegetation on Thermal Conditions Outside, Thermal Modeling of Urban Microclimate, Case Study: The Street of the Republic, Biskra *Energy Procedia* **18** 73–84
[5] Jesionek K and Bruse M 2004 Impacts of Vegetation on the Microclimate: Modeling Standardized Building Structures with Different Greening Levels Europe
[6] Srivanit M and Hokao K 2013 Evaluating the Cooling Effects of Greening for Improving the Outdoor Thermal Environment at an Institutional Campus in the Summer *Build. Environ.* **66** 158–172
[7] Bruse M 2004 ENVI-Met 3.0: Updated Model Overview 2004.
[8] Buyadi S N A, Wan Mohd W M N and Misni A 2014 Impact of vegetation growth on urban surface temperature distribution *IOP Conf. Ser. Earth Environ. Sci.* **18** 12104
[9] Misni A, Baird G and Allan P 2013 The Effect of Landscaping on the Thermal Performance of Housing *Int. Rev. Spat. Plan. Sutainable Dev.* **1** 29–48

IOP Conf. Series: Earth and Environmental Science **385** (2019) 012004 doi:10.1088/1755-1315/385/1/012004

Revitalizing Rural Areas in Malaysia: A Framework for Economic Revitalization

M F Rashid[1], I Ngah[1] and S H Misnan[1]
Department of Urban and Regional Planning, Faculty of Built Environment and Surveying, Universiti Teknologi Malaysia (UTM), 81310 Johor, Malaysia

Email:pku.fadhli@gmail.com

Abstract. Rural, poor and marginalized community often face many challenges in achieving economic equality and attaining socio-economic sustainability, especially among rural communities who have limited sources of income. In countries with emerging economies such as Malaysia, rural populations constitute the majority of citizens. They are generally poorer and more disadvantaged than their urban counterparts. In Malaysia, rural communities and their development are important aspects of a country's political and economic stability in which revitalization of the rural economic is part and parcel of a nation's development. Rural revitalization is an outcome of the interplay between many sources of influence and it is also a response to the opportunities and challenges which are presented by rural decline. This paper provides an overview of revitalization of rural areas. Besides that, it also explains the factors differences in economic performance and socioeconomic differentiation in rural areas in order to achieve a better outcome for rural revitalization. Finally, this paper discusses the framework to revitalize rural areas and the methods to use this framework that measure rural economic performance.

1. Introduction

Today, more than 46 percent of the world's live in rural areas [21]. Rural communities and their development is an important aspect of political and economic stability of a country whereby it has always been considered as important in the agenda of national development [3]. Many programs and strategies were introduced to promote the well-being of rural community ranging from the development of the agriculture sector, resettlement schemes, provision of public facilities and infrastructure to community development [10]. However, scholars and international organizations were highlighted issues on the cycle of rural decline which happening in most the rural areas around the world today. Majority of rural areas faced the issue on the rural decline which directly affects to the rural communities and towards a complicated issue of breaking the cycle of declining [3] [11] [15] [16]. A new challenge has been faced by rural areas and its communities due to an issue on globalization and rapid change of economic. This poor and isolated community who lived in rural areas often has to deal with many issues in achieving economic equality and accomplishing sustainability of socio-economic [7]. They have seen little income growth because they are largely involved informal sector and distress from stagnant agricultural productivity [16]. This economically disadvantaged have the most happened to the rural communities [6] [20]. In order to overcome these various problems faced by rural areas, revitalizing economic of rural areas is one of the best and suitable strategies to be implemented. Therefore, rural revitalization is a process and strategy involving all aspect of rural life-economic, social, human, cultural and environmental which seeks to reverse rural decline, to develop a more resilient, sustainable and diversified local economy, and to enhance the quality of life of rural communities [9] [13]. Thus, the aim of this paper is to identify factors of rural economic performance contributing to the framework for economic revitalization involving spatial scale of rural and household.

2. Differences in Economic Performance in Rural Areas

The picture of rural areas seems to emerge should be seen rather in terms of 'a new mosaic of rural areas' with winners, in-betweens, and losers [14]. This mosaic of rural areas directly raises questions about driving forces behind this pattern. Of course, such questions are merely a variant of the often-posed question in economic literature: why do economic growth rates differ among countries or regions? Insight into the driving factors behind the economic performance of rural areas is not only scientifically of interest, but it is also highly politically relevant. There is a suggestion that the determinants behind the differences of rural economic performance and socioeconomic differentiation are related to the interplay of local and global factors, in which population dynamics, territorial dynamics, and the current globalization process are thought to be main determinants. It is a significant exercise to elaborate on these determinants, as they give a bird's-eye view of economic of rural areas and socio-economic dynamics of rural populations (Table 1).

Table 1. Overview of factor/indicator selection method applied by researchers.

Source of Benchmark	Spatial Level	Factor/Indicator				
		Economic	**Social**	**Human**	**Cultural**	**Environmental**
Terluin (2003)	Region	Employment growth, Economic activities	Capacity of policymakers, Internal and external networks, Communities capacity	Local knowledge, Population growth, Innovation	Local character	Natural resources, Rural amenities, Transport infrastructure, Soft infrastructure
Courtney and Moseley (2008)	Region	Businesses existence, Investment, Economic linkages	Networking, Trust, Norms, Quality of governance	Health, Skill, Education, Rest taking	Attitudes, Local history, Customs and heritage	Natural asset, Peripherality, Accessibility to facilities
Agarwal *et al* (2009)	Region & Village	Employment, Enterprise business	Participation rates, Engagement	Skills, Education	Resilience	Transportation, Infrastructure, Location, Natural beauty
Klok (2011)	Village	Investment, Labour resources, Properties, Economic sector	Social infrastructure, Involvement of authorities	Knowledge, Experience	Historical environment, Hospitality	Attractiveness of environment, Location
Sánchez-Zamora *et al* (2014)	Village & Housheold	Income, Employment, Economic structure, Infrastructure	Public-Private sector network, Community co-operative	Demography, Skill, Education, Access to service	Identity, Heritage, Civic engagement	Peripherality, Natural resource, Environmental quality
Straka and Tuzova (2016)	Village & Household	Employment, Income, Number of businesses, Property ownership	Political engagement	Demography, Education, Skill, Health	Resilience, Attitudes	Infrastructure, Telecommunication

Deeper discussions on this issue of rural economic had established to highlight their research on this matter. As Terluin elaborated that population dynamics, territorial dynamics and globalization process are the major factors of the economic performance differences between rural areas using method of pattern-matching on factors of rural economic performance [19]. The continuity of Terluin's work has been shifted to a new paradigm whereby the introduction of five types of capital - economic, human, social, cultural and environmental by Courtney and Moseley [2] and Agarwal *et al.* [1] as both of them used district level analyses to evaluate economic performance of region level. Their focus has changed from investigations of single issues to an analysis of multiple issues particularly based on these five capitals. Klok [8] and Sánchez-Zamora *et al.* [17] suggested that this approach proved to be useful for rural areas to measure its economic performance in the context of village and households levels. Therefore, it's important in order to explain the differences of economic performance factors using five type of capital in each level of rural areas especially in the developing countries as Straka and Tuzova [18] explains that the importance of village and household level in assessing the economic performance of rural areas using correlation matrix analyses. These five type of capital that started by Agarwal *et al.* is significance to the Terluin's work on the territorial innovation model (theory of mixed endogenous and exogenous) which would help future research on working new framework in measuring rural economic performance. In addition, it does consider the interplay between internal and external determinants in measuring differences in economic performance (village level) and socioeconomic differentiation (household level) of rural areas [1] [2] [8] [17] [18] [19].

3. Rural Areas in Malaysia

An issue of population decline has influenced most of the rural areas in Malaysia. It is shown that the population has been gradually declined from years to years [22]. Agriculture productivity loss, lacking job opportunities in rural areas, rural-urban migration and lacking government incentives on rural products are the main related issues affecting population decline. Main determinants have been driving forces behind this pattern is rural-urban migration. Migration from rural areas to urban areas is the only way to find suitable jobs and better salary whereby it contributes to the population decline in rural areas. Besides that, rural households faced challenges which are household incomes disparities between urban areas and rural areas. It is shown that the total average monthly household income for urban and rural areas, although total revenue for the two regions is seen to increase from year to year, the household income gap is very significant [5]. Even though the gap of urban and rural income has been reduced from year to year due to several current policies implemented by the government to improve the wellbeing of peoples, but the effects of the development of urban areas has spread out unintentionally to the rural areas physically and economically. Deeper insight on differences in economic performance of rural areas should be intended to elaborate involving variations of determinants incorporates between and within the rural areas in the relevance of Malaysia context.

Table 2. Distribution of population and gross monthly income of household by strata.

Year	Population (Percentage)		Household Income (RM)	
	Rural	Urban	Rural	Urban
2010	29.1	70.9	2545	4705
2011	28.3	71.7	-	-
2012	27.6	72.4	5742	3080
2013	26.9	73.1	-	-
2014	26.3	73.7	3831	6833
2015	25.7	74.3	-	-
2016	25.0	75.0	4359	7671
2017	24.5	75.5	-	-
2018	23.9	76.1	Unpublished	Unpublished

Source: Household Income Survey Report, Department of Statistics Malaysia (DOSM)

4. Framework for Rural Revitalization

The existence of a various well-established body of literature on revitalizing rural economic performance had come to a specific research into this issue in rural areas began in the late 1990s. The research on the determinants of economic performance in rural areas around the world has evolved from investigations of single issues, to analyses of multiple issues particularly of the five types of capital - economic, human, social, cultural and environmental [1] [2] [8] [17] [18] [19]. Agarwal *et al.*

recommended that this framework proved to be useful for rural areas to measure the differences in economic performance and socioeconomic differentiation in rural areas especially in the developing countries [1].
This framework is used to measure the differentiation of rural economic performance in rural areas and socioeconomic differentiation of rural household. Thus, this framework was designed specifically to revitalize the economy of rural areas. In this research, this framework was established to emphasize relationships between all contributory factors within, and between, the five types of capital. For the village level, it involves Economic (5 factor); Social (3 factor); Human (4 factor); Cultural (4 factor); and Environmental (4 factor). Meanwhile, for the household level, it involves Economic (9 indicator); Social (8 indicator); Human (8 indicator); Cultural (7 indicator); and Environmental (9 indicator). Based on the discussion of rural economic performance above, the framework set out in Table 3.

Table 3. Evaluation of factors of economic performance in rural areas and socioeconomic differentiation of rural households.

Capital	Factors (Village Level)	Indicator (Household Level)
Economic	Occupations and Income	Increase income every year
		Stable in occupation
		Able to support a family well
		Able to get an additional source of income
	People Employed in Households	Good job and a balanced salary by family members
	Remittance from Families	Money transfer by family members really helps family life
	Assistance from Government and Private Agencies	Financial and welfare assistance by the government
		Business/agricultural/fishery assistance by the government or private sector help to increase income
	Asset and Property Ownership	The cultivated land can generate considerable yields
Social	Trust and Norms	Trust in neighbors
	Membership and Participation in Community	Involved in social organizations such as the JKKK/mosque committee/youth/women's organization
		Engaging in community activities and gotong royong
	Collective Action and Neighbourhood Connection	Good contacts to facilitate a process in obtaining financial capital
		A good relationship with community leaders (Village Head)
		A good relationship with the Wakil Rakyat
		A close relationship with successful entrepreneurs
		No discrimination between the people inside the village
Human	Health	Good level of health
		Able to do hard work
	Education	Have a perfect formal education
	Skill	Follow training and skills
		Skills inherited by previous generations
		Skills are shared with the younger generation
		New skills by younger generation shared to an older generation
	Confidence and Leadership	Like to be a leader in an organization
Cultural	Attitudes	Good attitude and personality by surrounding community
	Religious	Regularly to the synagogue to fulfill my duty as a believer
		Praying and put fate and help from God
	Cultural and Way of Life	Exercise regularly (walking, jogging, sports)
		Local cultural activities involvement

		Religious activities involvement
	Resilience	Financial assistance to relatives/neighbors/villagers
Environmental	Natural Environment	Attractive natural resource in the village
		Well maintained of natural resources in the village
		No natural disasters such as floods and storms occur
		Frequently visited by visitors/tourists
	Soil Fertility and Environmental Quality	No pollution problems such as water, air and others
		Good soil fertility level in the village
	Accessibility to Facilities	Good basic infrastructure like electricity/water/road
		Public transport services and village are linked
	Location	Easy to get the goods and services in town/city center

5. Methodology

This case study follows the framework for revitalizing rural areas based on factors of rural economic performance, in which data is collected in the field, the questionnaire form, observation, and in-depth interview are the primary data collection instruments. Meanwhile, the sampling method is purposive where the survey involved the head of households, and both quantitative and qualitative research methods are used to capture views on differences in economic performance of rural areas and socioeconomic differentiation of rural households. This paper explained the research on differences in economic performance of rural areas which is not associated with the macro scale of study but it concerned with large-scale population data whereby it involves every family who lived in traditional villages in rural areas. On top of that, the socioeconomic differentiation of rural household studies was also involved selected successful and less successful head of households who lived in that village.
Three type of research instrument was used which are the questionnaire, observation, and in-depth interview. The questionnaire form is the main research tool to collect data and information of rural households and perception of five capitals involved in this research that covered the endogenous and exogenous factor which influence the economic performance of rural areas and socioeconomic differentiation of rural households through field survey. Meanwhile, observation form through field survey is used to gather data of surrounding the selected village in terms of the physical aspect of development that relates to the five capitals which influenced the economic performance of the villages and its households. Other than that, in-depth interview session will be used to support the structured questionnaire and obtain important information of life story analysis of selected successful and less successful rural households.
There are two level of spatial scale involved in explaining performance (1) Village level - Differences in economic performance of rural areas; (2) Household level - Socioeconomic differentiation of rural households. These two spatial levels will be analyzed separately which each of them have their own analysis to measure differences in economic performance and socioeconomic differentiation. Each analysis must follow exactly to the aspect of the study. Most importantly, to use this framework for revitalizing rural areas based on factors of rural economic performance, it must be translated to the instrument of research like questionnaire, observation, and in-depth interview. Likert-scale for the questionnaire is the most appropriate techniques to measure each of the elements or measurement in these five capitals.

Table 4. Aspects of study for analysis in the economic performance of rural areas.

Level of Analysis	Aspects of Study
Factors for differences in economic performance within the rural areas (village)	• Village level studies. • Data collection from every head of household in the village involving selected villages in selected rural region or in Malaysia known as "mukim" based on rural density of 0-150 people/km^2. • Collection of data is from household's survey of a village and village observation of physical aspects through field survey.

	• Quantitative data analysis (Mean and Spearman Correlation).
Factors for socioeconomic differentiation of the households in rural areas (household)	• Household level studies. • Data collection from the selected households in the village involving selected successful and less successful households lived in the villages according to their household income level. • Collection of data is from household's survey of selected household using an in-depth interview to reveal the process of transformation and how the process of change (dynamic process). • Quantitative and qualitative data analysis (Mean, Spearman Correlation and life story analysis).

This analysis provides relevance information about the differences in economic performance of rural areas and socioeconomic differentiation of its households where it measures the framework for revitalizing rural areas based on factors of rural economic performance using descriptive data analysis (Mean) and inferential data analysis (Spearman Correlation) between factors. For the village and household level, it will measure all five capitals involving Economic (5 factor, 9 sub-factor); Social (3 factor, 8 sub-factor); Human (4 factor, 8 sub-factor); Cultural (4 factor, 7 sub-factor); and Environmental (4 factor, 9 sub-factor). Thus, in order to revitalize rural areas, each factor in all five capitals must be counted and analysis. Once the descriptive data analysis shows the low rate of performance for selective factors influencing village and it's household, then appropriate suggestion or action plan must be carried out to solve that issues relating to the factor involved. Instead of that, Spearman correlation were used to evaluate the relationship of factors in five capitals with economic performance of village and socioeconomic of housheholds. This framework is straightforward and relevance government/private agencies or committee of villagers can conduct this framework and implemented the action plan once all the factors in the framework are being analyzed and the findings have revealed.

6. Discussion and Conclusion

Differences of economic performance in rural areas caused by different factors were drive in the relevant agencies to develop solutions to overcome the decline of rural areas. One of these solutions is rural revitalization. Successful rural revitalization requires a holistic approach, which includes components such as the local economy, physical environment, political environment and the social environment [9]. In respect to that, rural revitalization is linked to the economic planning and strategic spatial planning whereby it could play a key role in the revitalization process [12]. Significantly, this framework for economic revitalization of rural areas was used as a comprehensive approach to address rural issues and challenges involving two level of spatial scale which are village level and household level which consists of five types of capital, 20 factors and 41 indicators. This framework will play a part in enhancing the quality of the community's life in rural areas. Besides that, this framework provides understanding of the determinants of economic performance which largely underpinned by a complex interplay of internal and external forces based on five capitals. Thus, this framework is very relevance in revitalizing the economy of rural areas and can be applied as important tools to assess the factors for differences in economic performance and socioeconomic differentiation of rural areas.

Acknowledgments
The authors would like to acknowledge assistance and appreciation for the support of the sponsors from Government of Malaysia, State Economic Planning Unit of Johor, and Universiti Teknologi Malaysia for this project - Factor for Differences in Economic Performance of Rural Areas in Johor, Malaysia (Project No. PY/2017/01514).

IOP Conf. Series: Earth and Environmental Science **385** (2019) 012004 doi:10.1088/1755-1315/385/1/012004

References

[1] Agarwal S, Rahman S and Errington A 2009 Measuring the determinants of relative economic performance of rural areas. *Journal of Rural Studies*. **25**(3):309-21.

[2] Courtney P and Moseley M 2008 Determinants of local economic performance: experience from rural England. *Local Economy*. **23**(4):305-18.

[3] Heimann C 2010 *Comprehensive rural development programme* InPresentation by Chief Director of Dept of Rural Development and Land Reform. ISAD multi-stakeholder forum 2010 (Vol. 25).

[4] Howe J, McMahon E T and Propst L 1997 *Balancing nature and commerce in gateway communities* (Washington DC: Island Press)

[5] Hussain N E, Abdullah N and Abdullah H 2014 The relationship between rural-urban migration, household income and unemployment: Malaysia case study. *Economic Planning*. **3**(3.5):2-9.

[6] Isa H 2010 *An effective institutional arrangement in rural development: Malaysia* (Doctoral dissertation, University of Glasgow).

[7] Iwala O S 2004 Achieving sustainable poverty reduction and rural development in Nigeria through local economic development strategies. *American Journal of Rural Development*. 2(1):13-9.

[8] Klok N 2011 The detection of main factors that influence on the development of rural tourism. *Socio-economic Research Bulletin*. **41**:37-40.

[9] Meyer D F 2014 Exploration of solutions for revitalisation of rural areas in South Africa. *Mediterranean Journal of Social Sciences*. **5**(4):613-25.

[10] Ngah I 2012 Rural transformations development: A review on Malaysia's transformation program. *Int. Conf. on Social Sciences & Humanities (ICOSH-UKM 2012)*.

[11] OECD 2006 *Investment in priorities for rural development* (Edinburgh: Scottish Executive).

[12] OECD 2007 *Competitive cities: A new entrepreneurial paradigm in spatial development* (Paris: OECD).

[13] Page-Adams D and Sherraden M 1997 Asset building as a community revitalization strategy. *Social work*. 42(5): 423-34.

[14] Persson L O and Westholm E 1994 Towards the new mosaic of rural regions. *European Review of Agricultural Economics*. **21**(3-4):409-27.

[15] Phillipson J, Shucksmith M, Turner R, Garrod G, Lowe P, Harvey D, Talbot H, Scott K, Carroll T, Gkartzios M and Hubbard C 2011 *Rural economies: incubators and catalysts for sustainable growth*. (Centre for Rural Economy, Newcastle University, in collaboration with the UK Research Councils' Rural Economy and Land Use Programme).

[16] Salborn E 2015 *Revitalization of rural economies for inclusive development* (Taipeh: Asian Productivity Organization).

[17] Sánchez-Zamora P, Gallardo-Cobos R and Ceña-Delgado F 2014 Rural areas face the economic crisis: Analyzing the determinants of successful territorial dynamics. *Journal of Rural Studies*. **35**:11-25.

[18] Straka J and Tuzová M 2016 Factors affecting development of rural areas in the Czech Republic: A literature review. *Procedia-Social and Behavioral Sciences*. **220**:496-505.

[19] Terluin I J 2003 Differences in economic development in rural regions of advanced countries: an overview and critical analysis of theories. *Journal of Rural Studies*. **19**(3):327-44.

[20] Vavrek B 1995 *Rural and Small Libraries: Providers for Lifelong Learning* (Washington DC: National Institition on Postgrade Education)

[21] World Bank 2015 *Rural areas facing problems of declining* (Web Accessed-http://www.data.worldbank.org).

[22] Yaakob U H and Nor N N 2013 The process and effects of demographic transition in Penang, Malaysia. *Kajian Malaysia*. **31**(2):37-64.

IOP Conf. Series: Earth and Environmental Science **385** (2019) 012005 doi:10.1088/1755-1315/385/1/012005

A study on the constraints of implementing Information and Communication Technology (ICT) in Malaysian Construction Industry

Y L Lew[1], T C Toh[1], K L Lim[1], F Y Y Yan[1] and L P Yow[1]
[1]Department of Surveying, Lee Kong Chian Faculty of Engineering & Science, University Tunku Abdul Rahman, Sungai Long Campus, Jalan Sungai Long, Bandar Sungai Long, Cheras, 43000 Kajang, Selangor

Email: lewyl@utar.edu.my

Abstract. Information and Communication Technology (ICT) refers to a combination of hardware, software and communication facilities. Often, ICT is introduced to equip construction stakeholders with information and analytical tools for a better control throughout the entire construction delivery process. The aim of this study is to investigate the issues relating to implementation of ICT in Malaysian construction industry with the objectives of: (i) to identify the benefits of adopting ICT in construction industry; (ii) to determine the constraints of implementing ICT in construction industry; and (iii) to identify the methods to encourage construction industry in adopting ICT. A questionnaire survey was conducted in Johor and Klang Valley of Malaysia, a total of 70 completed questionnaires were collected from contractors and quantity surveying consultation firm. Based on the analysis conducted, the top three benefits of adopting ICT in Malaysian construction industry ranked by the respondents are improvement on accuracy of measurement works, minimization on usage of paper by digitalization and allowed rapid sharing of information among project team. The barriers in implementing ICT in Malaysian construction industry identified included unaffordable initial capital and long term investment required, difficulties encountered to recruit capable and trustworthy in-house IT specialist, and security issues on confidential issues. Training and vendor's efforts to improve their software have been agreed as the most achievable and practical methods to encourage ICT uptake. This study had explore the implementation and acceptance level of ICT in Malaysian construction industry. The views from both contractor and quantity surveying consultation firm were revealed. The findings from this study could be used as the baseline for future development of ICT in the local construction industry.

1. Introduction

Information and Communication Technology (ICT) plays an important role in nation's well-being and prosperity due to its enormous and profound socioeconomic benefits. The practice of people live, work, learn and communicate with each other has been improved by the introducing of ICT [1]. Since ICT assists in transmitting data and messages among individual, new knowledge can be developed and shared in order to achieve innovation [2]. Adoption of ICT as an integral part of construction process can improve the current work practices in order to be more client orientated, more competitive as well as productive [3]. Advanced technologies tools can improve project performance in terms of budget and schedule, as well as the coordination within team member being improved, transferring and handling of document being facilitated, bottlenecks in communication reduced and updated records of organization enhanced [4, 5].

In Malaysia, Construction Industry Development Board (CIDB) has developed a Construction Industry Master Plan (CIMP) to overcome some of the inherent weaknesses existing in the construction industry [6]. One of the strategic thrusts identified included leverage on ICT in construction industry. Nevertheless, there was limited evidence of adoption by the company due to various barriers presence in the industry. Therefore, this study is initiated to expose the issues relating to implementation of ICT in Malaysian construction industry.

2. Literature Review

ICT refers to a combination of hardware, software and communication facilities [7]. The adoption of ICT improves information management especially on delivery of project information from offices to construction sites [8]. Construction industry has been defined as an industry which is vulnerable to various technical, sociopolitical and business risks [9]. Due to the high risk prone of typical construction industry, the introduction and implementation of ICT within this industry can micrify the severe impacts beget from its uncertainty environment.

Benefits generated from ICT adoption in construction industry were innumerable and involved many aspects. Exchange of information supported by ICT application eliminates the mistakes occur during data input or transfer [10]. The automation in generating list of materials eliminate the human error in taking off, especially careless mistakes or any arithmetical errors which might have severe impacts on the construction cost. ICT is able to predigest an organization with high hierarchical structure into a flatter structure and coordination among staff can be improved [11].

Yet, the speed of internet connection in Malaysia still needs improvement as according to Malaysiandigest.com [12], Malaysia has been mentioned as one of the Top Ten Slowest Internet Speed Country in the world. Incompleteness of infrastructure works in suburbs areas discourage people from adopting ICT unlike more stable and higher speed of internet connections available in capital area. Government and industry bodies support have an immediate impact and responsible to popularize the application of ICT within industry [13].

Previous surveys have shown that the widespread "technophobia" is still happening even in the year 2010 [14]. In order to allow users to change accordingly, a realistic objectives and benefits must be informed. When individual understands how the ICT tools can improve company as well as individual work load and daily activities, this will make individuals less reluctance to change. The users' input and comments on a new introduced product are useful and practical. Based on Kuncinas [15], Malaysia's budget in year 2015 has been targeted to reinforce the ICT industry by improving local infrastructure. The prime minister has allocated budget for the expansion of high-speed broadband in urban areas. As the local network connection improved, the frequency of ICT uptake in organizations will be increased as well. CIDB [16] has launched ICT programme and targets to enhance the information exchange between each industry player via the use of ICT in respective industry. The collaboration of CIDB with Persatuan Industri Komputer Malaysia (PIKOM) and Multimedia Development Corporation (MDeC) aimed to increase the opportunity of ICT companies to involve in construction industry. In order to prove the effort being put in by CIDB to pursue a high usage of ICT in construction industry, CIDB has improved their administration works through ICT. The actively adoption of ICT by CIDB allowed it functions to assemble all the information and data in the form of digital information to allow CIDB's external and internal usage.

3. Research Methodology

A questionnaire survey using five-point Likert scales was adopted for data collection. In order to overcome the low feedback from respondents, the combination of distribution methods were adopted, the questionnaires were distributed in the form of hardcopy and through an online survey tool – Survey Face. The questionnaire forms were distributed among the QS consultant firms and contractors within Kuala Lumpur, Selangor and Johor Bahru. Johor Bahru was included as the targeted area due to

the rapid growth of construction industry over these few years. Based on CIDB Construction Statistics Bulletin, the total number of registered contractors is 6,383 which is the third highest number while Kuala Lumpur and Selangor have the top two highest numbers of registered contractors as described in the quarterly report [16]. The targeted respondents for the questionnaires of this study are engineer, quantity surveyor, project manager, contract manager from both quantity surveyor (QS) consultant firms and contractors. These targeted respondents were selected as they were the group of professionals that mostly deal with the QS related software that studied in this study.
A total of 100 sets of questionnaires were distributed to the contractor and QS consultant firm selected by convenient sampling method with the estimated rate of more than 60%. The use of convenient sampling is ensure the high response rate for this exploration study. The questionnaire was divided into four sections, namely Section A, Section B, Section C and Section D. Section A included the profile and demographic details of the respondents. Section B is used to investigate the benefits of ICT adoption in organization, Section C is used to identify the possible constraints and limitations faced by individual while adopting ICT in construction industry whilst Section D is used to identify the methods available to encourage the ICT uptake in construction industry. Likert scale were used in the questions in Section B, C and D as it allowed the respondents to convey the degree of their agreement or disagreement about the research questions or statements [17].
Statistical Package for Social Science (SPSS) was adopted in this study to perform data analysis. Cronbach's Alpha Coefficient is used to ensure the internal consistency among each individual variable in the research. Cronbach's Alpha Coefficient equal or more than 0.700 is considered acceptable [18].
Spearman's Rank Correlation Coefficient (rho) is used to measure the difference in ranking between two groups of respondent's scoring a number of issues, attitudes or factors [19]. Use of non-parametric test is suitable for any data with no restriction of the normality of the data distribution. This test is used to find out whether there is any significant difference in the ranking between two groups of respondents. The respondents were grouped according to the type of organization, QS consultant firm or contractor. Correlation coefficient (rho) range from -1.000 to +1.000 indicating the significance relationship (degree of agreement) between two sets of results from two respondents groups. If the rho is nearer to 0, this interpret that there is only limited of degree of agreement towards the variables from two groups of respondents. While, when the rho is closer to -1.000, it shows that there is perfect disagreement towards the variables and if the rho is closer to +1.000, it shows perfect agreement towards the variables between two groups of respondents.

4. Results and Discussions

A total of 70 questionnaires were collected out of the 100 sets distributed. The respondents comprised of 55 personnel with work position of quantity surveyor (78.6%), 7 project manager (10.0%), 6 contract manager (8.6%), and 2 engineer (2.9%) with majority position of respondents are senior position. The personal details and company information of the respondents were not collected to ensure the guarantee of anonymity and confidentiality. There are 31 respondents (44.3%) who have working experience of 2 to 5 years, almost half of the total respondents are belong to this category. There are 20 respondents (28.6%) who have less than 2 years working experience and followed by the group of having 10 or above 10 years experience, which there are total of 12 respondents (17.1%) belong to this category. Lastly, there are only 7 respondents (10.0%) who have 6 to 9 years of working experience.
The respondents were asked to rank the benefits on ICT adoption, constraints of ICT adoption and methods to encourage ICT adoption using a 5-point Likert scale (5-Strongly Agree, 4-Agree, 3-Neutral, 2-Disagree and 1:Strongly Disagree). Cronbach's Alpha analysis was carried out in order to measure the internal consistency or average correlation of items in a survey instrument in order to calculate the reliability of the Likert scale used [20]. It was found that all the scale used indicated good consistency with Cronbach's alpha coefficient of 0.793, 0.710 and 0,816 respectively.

IOP Conf. Series: Earth and Environmental Science **385** (2019) 012005 doi:10.1088/1755-1315/385/1/012005

It was found that there were common agreements between two groups of respondent towards the top three benefits of ICT adoption (summarized in Table 1). Both respondents from contractor and consultant rated that the improved accuracy and preciseness of work done brought by different kind of software was the most beneficial results by adopting ICT. According to Autodesk [21], Quantity Take Off (QTO) software allowed users to expedite on taking off for similar components or materials on the drawings by using "Search Takeoff" function, hence a higher accurateness on compiling the same components can be attained within short time frame.

Table 1. Ranking of Benefits on ICT Adoption

Benefits	**Consultant Rank**	**Contractor Rank**	**Overall Rank**
Sharing of information within project team	3	3	3
Improve accuracy of measurement works	1	1	1
Better communication	3	5	5
Globalization	6	6	6
Project management tools	5	4	4
Digitalization	2	2	2

The second benefit which ranked by both consultant and contractor respondents is digitalization. According to Matsumura & Lewis [22], digitalization and dematerialization are the indispensable functions of ICT as different technologies able to convert the physical into digital or virtual. At the simplest level, electronic transmission of business documents created large savings in paper usage and postage [23].

The sharing of information among team members and minimized the errors during transferring of data has been ranked as the third benefit could be obtained by respondents by adopting ICT in their workplace. ICT tools facilitated efficient handling of large amounts of documents that originate from difference sources and in different formats [24], hence mistakes occurred during the data transmission process being eliminated.

In contrast, the lowest ranking of the benefit by both groups of respondents was the adoption on ICT able to improve globalization by materials and specialist procurement in order to enhance the quality of the end products. Globalization is closely linked with the international trade of goods and services and is closely related to the procurement on technical specialist and materials in construction industry [25]. This lowest ranking may due to the low awareness on the importance of globalization in Malaysia construction industry.

Table 2 summarised the value of correlation coefficient value obtained based on the ranking between respondents from contractor and consultant firm is 0.899 at a significance level of 0.05 (95% confidence level) for a two tailed test. This value indicated that the ranking between respondents from consultant and contractor has high and positive correlation. Thus there is a high degree of agreement between them and hence there is no significance difference in their opinions.

Table 2. Spearman Rank Correlation Coefficient between Respondents from Contractor and Consultant toward the Benefits on ICT Adoption

		Consultant	**Contractor**
Consultant	Correlation Coefficient	1.000	.899*
Contractor	Correlation Coefficient	.899*	1.000

*. Correlation is significant at $p < 0.05$ (2-tailed).

This agreement is due to the fact that benefits brought by ICT implementation are the same by almost all industry professionals regardless of their organization nature. The top three benefits ranked by two groups of respondents were mainly concentrated on the Triple Constraint (time, cost and quality) in

construction industry, which represented a model of the constraint of project management. By adding benefits to each element in the triple constraint, professionals either from contractor or consultant able to master the project in better ways and bring the project to success by different parties.
In Table 3, it was shown that there were contrary in the ranking of constraints on ICT implementation. From the view of consultants, the barrier that they encountered mostly was due to large capital and huge investment cost were vital to implement ICT tools at initial stage as well as a high budget need to be allocated for the employee training. Besides than the procurement and installation cost, the annual licensing cost and maintenance cost have contributed on the total cost of ownership become an extra burden of cost for consultant firms.

Table 3. Ranking of Constraints on ICT Implementation

Constraints	**Consultant Rank**	**Contractor Rank**	**Overall Rank**
Create changes in the lines of communication and authority	4	7	6
Large capital and investment cost required	1	3	1
Inability to recruit in-house IT specialist	2	4	2
Issues arise from low speed of internet connection	5	4	5
Security issues	3	1	3
Unstandardised usage of ICT by different parties	7	4	7
Encounter difficulties when deal with new ICT tools	5	2	4

The second ranked barrier by respondents from consultant was the inability to recruit an in-house IT specialist group to support the users from time to time. Most of the firms need to depend on the outsourcing services for ICT implementation which would increase the overall cost for ICT adoption [26]. Besides, the user friendliness of ICT has been defined as one of the important barrier followed by high capital cost required. User friendliness often depends on the vendor's ability to offer software support and training to the new users in order to maximize the number of staff who predominating particular software.
Whilst the respondents from contractors have different opinions towards the barriers on ICT implementation as they have ranked the security issues as the top constraint. The second highest ranked constraint by contractor is encounter difficulties and insufficient of grace period to implement and wholly accustomed to the application of ICT in working life. As mentioned by Collins [27], many organizations neglected the importance of providing adequate training to employees, even though companies offered training to employees when new software was installed, but failed to train the new employees when turnover occurred.
Similar to respondents from consultants, contractors also emphasized on the minimum administrative cost in order to achieve the largest overall profit generated from each project. However, contractors focused on the financial impact of ICT only came after the security issues and insufficient training provided. The number of software licenses required in a contractor company was relatively lesser than a consultancy firm, therefore, contractors might discovered other issues such as the insecurity and unreliability of the ICT applications before the necessary cost to implement ICT.
Based on Table 4, the correlation coefficient value obtained based on the ranking from two groups of respondents is 0.072 at a significance level of 0.05 (95% confidence level) for a two tailed test. This value at almost equal to zero indicated that the ranking between respondents from consultant and contractor category has no linear relationship and implying no relationship between them.

Table 4. Spearman Rank Correlation Coefficient between Respondents from Contractor and Consultant toward the Constraints of ICT Implementation

		Consultant	Contractor
Consultant	Correlation Coefficient	1.000	0.072
Contractor	Correlation Coefficient	0.072	1.000

*. Correlation is significant at $p < 0.05$ (2-tailed).

The different opinions given by both groups of respondents proved that different types of hindrance have been encountered by professionals from different nature of organizations. Consultants have rated the financial considerations as the first and most important barrier during the uptake of ICT but contractors have prioritized the security issues of their confidential data as the first barrier which hindered the uptake of ICT within their organization. The other related cost might be cost of supporting infrastructure and computer ancillaries cost which these cost must be related to expected income and return on investment, QS consultant firms often incurred a relatively higher license cost for QS related software compared to contractor firms. Consequently, QS consultant firms placed the financial considerations as the first encumbrance whereas this barrier had less impacted on contractor and only being ranked at third place by contractor.

Based on Table 5, consultants believed that the most practical way to improve the uptake of ICT within industry shall depends on the vendors' effort to improve the user's negatively perceptions towards the advanced software. Various training prepared for professionals would have significant contribution to the implementation of ICT within industry. The software vendors should owned the initiative to arrange various kind of training in order to allow user to fully comprehend and achieve the potential advantages of the software. In contrast, contractors have recognized that the importance of training available within industry was the main and prior concern in promoting ICT usage in the industry. Training on the latest ICT tools and software in the industry was equally important and effective regardless the nature of the professionals working with and hence vendors shall execute their responsibility in promoting their products in order to achieve win-win situation. Lastly, all respondents from contractors and consultancy attained common agreement that government's effort to improve the local network connection was the least effective method in advocating the application of ICT within industry.

Table 5. Ranking of Methods to Encourage ICT Adoption

Methods to Encourage	Consultant Rank	Contractor Rank	Overall Rank
Training provided by software vendors	2	1	1
Vendors should improve user's perceptions	1	2	1
Government to improve the local network connection	4	4	3
Government provides special funding	3	2	4

Based on Table 6, the correlation coefficient value obtained based on the ranking from two groups of respondents is 0.632 for a two tailed test. This value indicated that ranking between respondents from consultant and contractor category has positive and moderate relationship between them. The moderate relationship between two rankings also represented that there is only limited level of disagreement between two groups of respondents and hence there are certain similarities in their opinions on the methods to encourage and promote ICT.

Table 6. Spearman Rank Correlation Coefficient between Respondents from Contractor and Consultant toward the Methods to Encourage ICT Adoption

		Consultant	Contractor
Consultant	Correlation Coefficient	1.000	0.632
Contractor	Correlation Coefficient	0.632	1.000

*. Correlation is significant at $p < 0.05$ (2-tailed).

The contractors and consultants have almost the same opinions towards the methods available to promote ICT usage as they agreed that training was inevitable and give priority to promote ICT awareness as well as education amongst senior staff. Respondents from either contractors or consultants organization concluded that training and efforts done by software vendors able to make largest improvements and more practical to achieve the high adoption rate of ICT in the industry within shorter time frame. Efforts done by software companies were functional and able to realize the objectives in a shorter time frame compared to those implementation programme introduced and launched by government which relatively longer duration and resources were needed. Contractor and consultancy firms nowadays required the most efficient, highest return and method with certain assurance to implement a drastic change or evolution in their organizations.

5. Conclusions

The top three most significant benefits being ranked by respondents are focused on the accuracy of workdone, minimize usage on the paper and ICT allowed sharing of latest information among team members within shortest duration. This strongly supported that the importance of work quality, cost allocation and time management in every single project, regardless the nature of organization within industry. The ideal benefits pursued by professionals were almost the same and owned the ultimate objective of achieving project success, within the allocated budget, time frame given and also realized the clients' requirements.

The constraints and barriers encountered by professionals were relatively different due to the nature of organization, size of organizations, individuals' experiences, individuals' perceptions and so on. The challenging constraints agreed by the overall respondents in this study were the unaffordable initial capital and long term investment cost required, difficulty to recruit capable and trustworthy in-house IT specialist to support software and unwilling to expose their confidential data at a high risk environment have hindered organizations to adopt and fully utilized the software in the industry. Whereas, the inability to recruit capable in-house IT support team may be due to the current trends in market as most of the firms relied on outsourcing IT consultant services. While, for the security issues on the confidential data can be improved by enhance the ICT knowledge among employees to ensure a better handling process on company data.

Based on the respondents' suggestions, the importance of workshop and seminar cannot be neglected as well. Next, the incorporation of ICT content in tertiary education syllabus especially for construction professionals helped to enhance the quality and ICT knowledge of future fresh graduates and consequently assist in disseminating ICT knowledge within the industry. Besides that, clients' role in ICT implementation cannot be overlook as well, clients' interest on consultants and contractors' ICT capacity became a driving factor to urge the ICT adoption within organizations. The rapid evolution of ICT in construction industry required more research to be carried out continuously in order to identify the latest issues in this challenging environment.

References

[1] Ganah and Kamara 2013 A Qualitative Method For Assessing The Impact Of ICT On THe Architectural Design Process.

[2] Gann, D., 2000 Building innovation: complex constructs in a changing world. Thomas Telford.

[3] Weippert, A, Kajewski, S.L. and Tilley, P.A., 2003 The implementation of online information and communication technology (ICT) on remote construction projects. Logistics information management, 16(5), pp.327-340.

[4] O'Connor, J.T. and Yang, L.R., 2004. Project performance versus use of technologies at project and phase levels. Journal of Construction Engineering and Management, 130(3), pp.322-329.

[5] Nitithamyong, P. and Skibniewski, M.J., 2006. Success/failure factors and performance measures of web-based construction project management systems: professionals' viewpoint. Journal of construction engineering and management, 132(1), pp.80-87.

[6] Sundaraj, G., 2007. The way forward: construction industry master plan 2006-2015. Master Builders, pp.48-51.

[7] Berce, J., Lanfranco, S. and Vehovar, V., 2008. Egovernance: Information and communication technology, knowledge management and learning organisation culture. Informatica, 32(2).

[8] Chen, Y. and Kamara, J., 2008, June. The mechanisms of information communication on construction sites. In Forum Ejournal (Vol. 8, No. 1, pp. 1-32).

[9] Ehsan, N., Mirza, E., Alam, M. and Ishaque, A., 2010, July. Notice of Retraction Risk management in construction industry. In Computer Science and Information Technology (ICCSIT), 2010 3rd IEEE International Conference on (Vol. 9, pp. 16-21). IEEE.

[10] Barba-Sánchez, V., Martínez-Ruiz, M.D.P. and Jiménez-Zarco, A.I., 2007. Drivers, benefits and challenges of ICT adoption by small and medium sized enterprises (SMEs): a literature review. Problems and Perspectives in Management, 5(1), pp.103-114.

[11] Tatari, O. and Skibniewski, M.J., 2009. Business benefits framework of enterprise systems in construction: An exploratory factor analysis. Managing IT in Construction/Managing Construction for Tomorrow, p.455.

[12] Malaysiandigest.com 2014 Malaysia is in The World's Top 10 Slowest Internet Connection. [Online]. Available: http://malaysiandigest.com/features/484394-malaysia-is-in-the-world-s-top-10-slowest-internet-connection.html. [Accessed 11 8 2015].

[13] Alam, S.S. and Noor, M.K.M., 2009. ICT adoption in small and medium enterprises: An empirical evidence of service sectors in Malaysia. International Journal of Business and management, 4(2), p.112.

[14] Ode, F. J. 2010 Software Implementation- Why users are the key to success CFMA Building Profits.

[15] Kuncinas, P. 2014 Budget boost for Malaysia's ICT sector [Online]. Available: http://www.theborneopost.com/2014/10/25/budget-boost-for-malaysias-ict-sector/. [Accessed 17 3 2015].

[16] CIDB 2014 Construction Quarterly Statistical Bulletin [Online]. Available: http://www.cidb.gov.my/index.php/en/component/content/article/72-kos-pembinaan/performance-appraisal-forecasting-market-watch-perform/buletin-statistik-pembinaan-suku-tahunan-tahun/234-construction-statistics-quarterly-bulletin-2014?Itemid=140 [Accessed 17 3 2015].

[17] Creswell, J.W. and Creswell, J.D., 2017. Research design: Qualitative, quantitative, and mixed methods approaches. Sage publications.

[18] Gliem, J.A. and Gliem, R.R., 2003. Calculating, interpreting, and reporting Cronbach's alpha reliability coefficient for Likert-type scales. Midwest Research-to-Practice Conference in Adult, Continuing, and Community Education.

[19] Naoum, S., 2012. Dissertation research and writing for construction students. Routledge.

[20] Santos, J.R.A., 1999. Cronbach's alpha: A tool for assessing the reliability of scales. Journal of extension, 37(2), pp.1-5.

[21] Autodesk 2010 Autodesk Quantity Take Off Brochure [Online]. Available: http://images.autodesk.com/adsk/files/Autodesk_Quantity_Takeoff_Brochure.pdf. [Accessed 26 7 2015].

[22] Matsumura, T. and Lewis, A. 2015 The Power of ICT to Solve Global Sustainability Issues [Online]. Available: http://www.sustainability-perspectives.com/article/the-power-of-ict-to-solve-global-sustainability-issues/. [Accessed 26 7 2015].

[23] Hore, A., 2006. Use of IT in managing information and data on construction projects-A prespective for the Irish construction industry. Information Technology in construction Project Management Engineers Ireland Project management society..

[24] Ugwu, O.O., 2006, May. Information and communication technology for collaborative working in project management. In Proceedings of the One day Seminar on “Recent Developments in (Vol. 7).

[25] Mrak, M., 2000. Globalization: trends, challenges and opportunitess for countries in transition. UNIDO.

[26] Parida, V., Johansson, J., Ylinenpää, H. and Braunerhjelm, P., 2010. Barriers to information and communication technology adoption in small firms.

[27] R. Collins 2006 Needed: Training - Lack of Training Biggest Hindrance to Software Success [Online]. Available: http://tilt-up.org/tilt-uptoday/2006/04/01/needed-training-lack-of-training-biggest-hindrance-to-software-success/. [Accessed 8 8 2015].

Investigation on the Performance of Construction Companies in Malaysia with Entropy-TOPSIS Model

W S Lam[1,2,3], K F Liew[1,2] and W H Lam[1,2,3]
[1]Department of Physical and Mathematical Science, Faculty of Science, Universiti Tunku Abdul Rahman, Kampar Campus, Jalan Universiti, Bandar Barat, 31900 Kampar, Perak, Malaysia
[2]Centre for Mathematical Sciences, Universiti Tunku Abdul Rahman, Kampar Campus, Jalan Universiti, Bandar Barat, 31900 Kampar, Perak, Malaysia
[3]Centre for Business and Management, Universiti Tunku Abdul Rahman, Kampar Campus, Jalan Universiti, Bandar Barat, 31900 Kampar, Perak, Malaysia

Email: lamws@utar.edu.my

Abstract. Construction industry plays an important role to a nation due to its contribution to the economic growth. The objective of this paper is to propose an Entropy-TOPSIS model in evaluating and comparing the financial performance of the construction companies in Malaysia. In this study, the listed construction companies in Malaysia are investigated using Entropy-TOPSIS model. The results of this study show that ZECON, DKLS, GADANG, ECONBHD and MELATI are ranked as top five construction companies based on the proposed model. This study is significant because it helps to determine the financial performance of the construction company as well as identify the importance of the decision criteria in the evaluation on the financial performance of the construction companies in Malaysia with the proposed conceptual framework based of entropy-TOPSIS model.

1. Introduction

In Malaysia, construction is one of the vital sectors that contribute to the economic growth of a nation. Moreover, the government shows high interest to the construction industry as the construction industry is an investment-led sector. Nowadays, the government strives to contract with the construction industry in order to develop and provide the modern and advanced infrastructure that related to education, transport, health and housing [1]. Therefore, the financial performance of the construction company in Malaysia is investigated in this study based on the important financial ratios. TOPSIS is a useful and practical model to rank the decision alternatives as well as select the best decision alternative [2]. TOPSIS model is seeking the shortest distance to the positive ideal solution (PIS), and the longest distance from the negative ideal solution (NIS) for the best decision alternative [3, 4]. The entropy weight method is one of the prevalent methods that used for weighing the criteria, whereas TOPSIS is a popular decision tool that utilized to rank the alternatives.

Entropy-TOPSIS model has been studied by the past researchers in the field of safety evaluation [5], cell phone evaluation [6], environment risk assessment [7], automobile [8], supplier selection [9], high-tech industry [10], electric vehicle charging stations [11], manufacturing enterprises [12] and urban rail transit system [13]. The objective of this paper is to propose a conceptual framework in evaluating and comparing the financial performance of the construction companies in Malaysia with Entropy-TOPSIS model. The significance of this study is to evaluate the financial performance of the construction companies in Malaysia by considering the important financial ratios. The structure of this paper is

IOP Conf. Series: Earth and Environmental Science **385** (2019) 012006 doi:10.1088/1755-1315/385/1/012006

organized as follows. In section 2, data and methodology of this study are discussed. Section 3 presents the empirical results of this study. Finally, section 4 concludes the paper.

2. Data and Methodology

In this study, the financial performance of the construction companies listed in Malaysia stock market is analysed [14]. The data of this study are obtained from the financial annual report of the construction companies in year 2016. Table 1 displays the proposed conceptual framework to evaluate the construction sector companies' financial performance in Malaysia with Entropy-TOPSIS model.

Table 1. Proposed Conceptual Framework.

Level		
Objective	Evaluation on the construction companies' financial performance	
Decision criteria	Current ratio (CR)	
	Debt to assets ratio (DAR)	
	Debt to equity ratio (DER)	
	Earnings per share (EPS)	
	Return on asset (ROA)	
	Return on equity (ROE)	
Decision alternatives	AZRB	MELATI
	BENALEC	MERGE
	BPURI	MITRA
	BREM	MLGLOBAL
	CRESBLD	MUHIBAH
	DKLS	PESONA
	ECONBHD	PRTASCO
	EKOVEST	PSIPTEK
	FAJAR	PTARAS
	GADANG	SUNCON
	GAMUDA	SYCAL
	GBGAQRS	TRC
	HOHUP	TRIPLC
	HSL	TSRCAP
	IJM	VIZIONE
	IKHMAS	WCHEHB
	KERJAYA	WCT
	KIMLUN	ZECON
	LEBTECH	

Based on the past studies, there are total of six important financial ratios considered in this study. These financial ratios consist of current ratio (CR), debt to assets ratio (DAR), debt to equity ratio (DER), earnings per share (EPS), return on asset (ROA) as well as return on equity (ROE) [15-22]. Among these financial ratios, DAR and DER are required to be minimized while CR, EPS, ROA and ROE are needed to be maximized.

The Entropy-TOPSIS model consists of the following steps [2]. According to Shannon and Weaver [23], the entropy weight method is utilized to determine the weights of the decision criteria.

Step 1: Determination of weight of the decision criteria via entropy weight method. Compute the proportion "p_{ij}" of index value of alternative m under criteria n.

$$p_{ij} = \frac{x_{ij}}{\sum_{i=1}^{m} x_{ij}}, i = 1, 2, ..., m, j = 1, 2, ..., n \quad (1)$$

Step 2: Computation of the entropy "e_j" of alternative m.

$$e_j = -k \sum_{i=1}^{m} p_{ij} . \ln(p_{ij}), j = 1, 2, ..., n \quad (2)$$

Where

$$k = \frac{1}{\ln(m)}$$

Step 3: Computation of the entropy weight "w_j" of alternative m.

$$w_j = \frac{1 - e_j}{\sum_{j=1}^{n} (1 - e_j)}, j = 1, 2, ..., n \quad (3)$$

Step 4: Construct the decision matrix ($(x_{ij})_{m \times n}$):
A $m \times n$ decision matrix is formed. m refers to the alternatives and n refers to the criteria.

Step 5: Normalize the decision matrix:
Normalization method is used to form a normalized decision matrix $R = (r_{ij})_{m \times n}$.

$$r_{ij} = \frac{x_{ij}}{\sqrt{\sum_{i=1}^{m} x_{ij}^2}}, i = 1,2,..., m, j = 1,2,..., n \quad (4)$$

Step 6: Calculate the weighted decision matrix (T):
The weighted normalized decision matrix is formed.

$$\mathrm{T} = (t_{ij})_{m \times n} = (w_j r_{ij})_{m \times n}, i = 1,2,..., m \quad (5)$$

Step 7: Determine the positive ideal solution (PIS) and negative ideal solution (NIS):

$$\text{PIS} = \{< \min(t_{ij} \mid i = 1,2,..., m) \mid j \in J_{-} >, < \max(t_{ij} \mid i = 1,2,..., m) \mid j \in J_{+} >\} \equiv \{t_{bj} \mid j = 1,2,..., n\}, \quad (6)$$

$$\text{NIS} = \{< \max(t_{ij} \mid i = 1,2,..., m) \mid j \in J_{-} >, < \min(t_{ij} \mid i = 1,2,..., m) \mid j \in J_{+} >\} \equiv \{t_{wj} \mid j = 1,2,..., n\}, \quad (7)$$

Where J_{+} is associated with the positive impact criteria and J_{-} is associated with the negative impact criteria.

Step 8: Calculate the distance of each alternative from PIS and NIS:

$$d_{ib} = \sqrt{\sum_{j=1}^{n}(t_{ij} - t_{bj})^2}, i = 1,2,...,m \quad (8)$$

$$d_{iw} = \sqrt{\sum_{j=1}^{n}(t_{ij} - t_{wj})^2}, i = 1,2,...,m \quad (9)$$

Step 9: Obtain the coefficient of the relative closeness to the ideal solution (s_{iw}):

$$s_{iw} = \frac{d_{iw}}{d_{ib} + d_{iw}}, 0 \le s_{iw} \le 1, i = 1,2,...,m \quad (10)$$

Step 10:
Select the decision alternatives with the highest value of s_{iw}. The larger the value of s_{iw}, the better the decision alternatives.

3. Empirical Results

Figure 1 presents the weights of the financial ratios on the construction companies' financial performance evaluation in Malaysia.

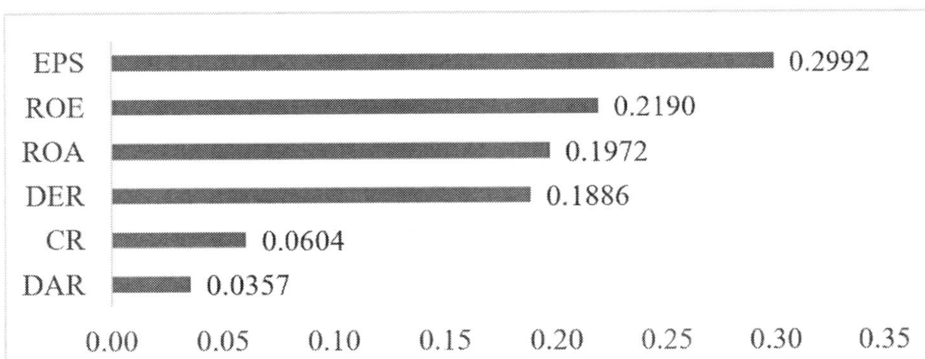

Figure 1. Weights of financial ratios on the construction companies' financial performance evaluation in Malaysia.

As presented in Figure 1, the weights of financial ratios on the construction companies' financial performance evaluation in Malaysia have been determined by the entropy weight method. The EPS gives the highest weights of 0.2992 followed by ROE (0.2190), ROA (0.1972), DER (0.1886), CR (0.0604) and lastly DAR (0.0357). This indicates that EPS, ROE, ROA and DER are the most important financial ratios on the construction companies' financial performance evaluation in this study.
The PIS and NIS for each decision criterion are determined by using the equation (6) and (7) respectively. Figure 2 shows the PIS and NIS for each decision criterion.

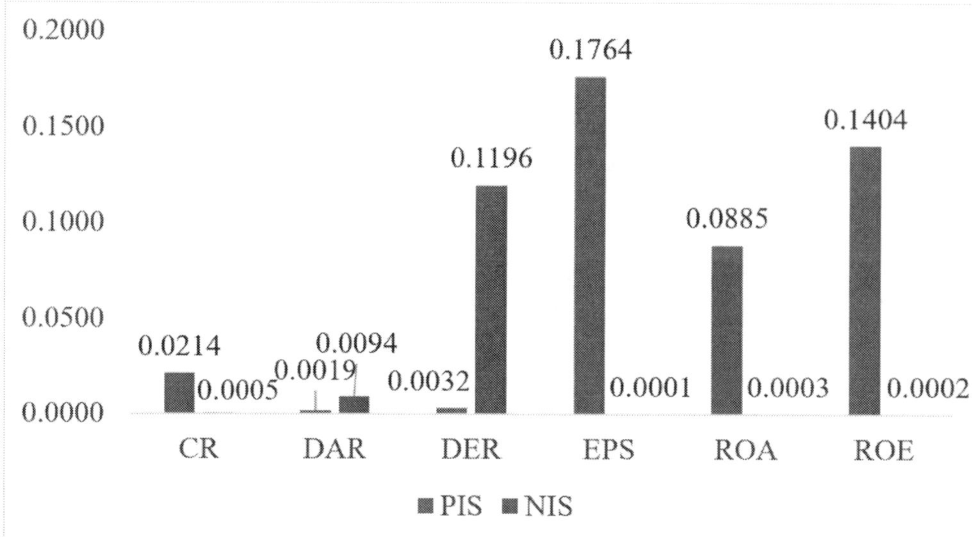

Figure 2. The PIS and NIS for each decision criterion.

As shown in Figure 2, the PIS that determined by the Entropy-TOPSIS model for CR, DAR, DER, EPS, ROA and ROE are 0.0214, 0.0019, 0.0032, 0.1764, 0.0885 and 0.1404 respectively. On the other hand, the NIS for CR, DAR, DER, EPS, ROA and ROE are 0.0005, 0.0094, 0.1196, 0.0001, 0.0003 and 0.0002 respectively.

Table 2. presents the empirical results of the overall performance and ranking of the construction companies in Malaysia.

Table 2. Overall Performance and Ranking of the Construction Companies.

Company	s_{iw}	Rank
ZECON	0.794435	1
DKLS	0.561749	2
GADANG	0.495555	3
ECONBHD	0.487243	4
MELATI	0.467980	5
KIMLUN	0.453050	6
MITRA	0.439362	7
KERJAYA	0.435783	8
TRIPLC	0.435227	9
HOHUP	0.434927	10
MUHIBAH	0.419538	11
GAMUDA	0.417372	12
IJM	0.412251	13
SUNCON	0.385459	14
HSL	0.381459	15
PTARAS	0.372171	16
FAJAR	0.362631	17
BREM	0.355798	18
EKOVEST	0.354441	19
PESONA	0.342139	20
TRC	0.329608	21
PRTASCO	0.324532	22
LEBTECH	0.324400	23

VIZIONE	0.319137	24
MERGE	0.316246	25
GBGAQRS	0.313126	26
PSIPTEK	0.311180	27
IKHMAS	0.309185	28
TSRCAP	0.307799	29
BENALEC	0.307397	30
SYCAL	0.306031	31
MLGLOBAL	0.305399	32
WCT	0.296347	33
CRESBLD	0.286531	34
WCHEHB	0.272415	35
BPURI	0.190942	36
AZRB	0.075163	37

As shown in Table 2, the s_{iw} for each construction company and the overall ranking of companies are determined by the proposed conceptual framework with Entropy-TOPSIS model. ZECON achieves the first ranking among the construction companies with s_{iw} of 0.794435 which is the highest among the companies. The s_{iw} for DKLS, GADANG, ECONBHD, MELATI, KIMLUN, MITRA and KERJAYA are 0.561749, 0.495555, 0.487243, 0.467980, 0.453050, 0.439362 and 0.435783 respectively. As a result, DKLS, GADANG, ECONBHD and MELATI obtained the second, third, fourth and fifth ranking respectively. On the other hand, the s_{iw} for WCT, CRESBLD, WCHEHB, BPURI and AZRB are 0.296347, 0.286531, 0.272415, 0.190942 and 0.075163 respectively. As a result, these companies obtain the lowest ranking in this study. This study can serve as reference to those investors who interested to invest in the construction companies. In conclusion, ZECON is the best construction company in term of financial performance and can be considered by the investors since ZECON obtains the first ranking in the evaluation.

4. Conclusion

In this study, the conceptual framework has been proposed in evaluating the financial performance of the construction companies in Malaysia with Entropy-TOPSIS model. The results of this study show that ZECON, DKLS, GADANG, ECONBHD and MELATI are ranked as top five construction companies based on the proposed model. Furthermore, EPS is the most important decision criterion that considered in this study, followed by ROE, ROA, DER, CR and finally DAR. This study is significant because it helps to determine the financial performance of the construction company as well as determine the weights of the decision criteria in the evaluation on the financial performance of the construction companies in Malaysia.

Acknowledgments
The authors express gratitude to Universiti Tunku Abdul Rahman (UTAR) for the sponsorship.

References

[1] Khan, R.A., 2008. Role of construction sector in economic growth: Empirical evidence from Pakistan economy. *First International Conference on Construction In Developing Countries. Pakistan*, pp. 279-290.
[2] Hwang, C.L. and Yoon, K., 1981. *Multiple Attribute Decision Making: Methods and Applications*. Springer-Verlag: New York.
[3] Jahanshahloo, G.R., Hosseinzadeh, L.F. and Izadikhah, M., 2006. Extension of the TOPSIS method for decision-making. *Applied Mathematics and Computation*, vol. 181, no. 2, pp. 1544-1551.

[4] Wang, T.C. and Lee, H.D., 2009. Developing a fuzzy TOPSIS approach based on subjective weights and objective weights. *Expert Systems with Applications*, vol. 36, no.5, pp. 8980-8985.
[5] Li, X., Wang, K., Liu, L. and Xin, J., 2011. Application of the Entropy Weight and TOPSIS method in safety evaluation of coal mines. *Procedia Engineering 26*, pp. 2085-2091.
[6] Akyene, T., 2012. Cell phone evaluation base on Entropy and TOPSIS. *Interdisciplinary Journal of Research in Business*, vol. 1 no. 12, pp. 9-15.
[7] Jozi, S.A., Shafiee, M., MoradiMajd, N. and Saffarian, S., 2012. An integrated Shannon's Entropy-TOPSIS methodology for environmental risk assessment of Helleh protected area in Iran. *Environmental Monitoring and Assessment*, vol. 184, no. 11, pp. 6913-6922.
[8] Tewari, P.C., Prakash, U., Khanduja, D. and Sandeep, 2015. Ranking of sintered material for high loaded automobile application by applying Entropy-TOPSIS method. *Materials Today: Proceedings*, vol. 2, no. 4-5, pp. 2375-2379.
[9] Mavi, R.K., Goh, M. and Mavi, N.K., 2016. Supplier selection with Shannon entropy and fuzzy TOPSIS in the context of supply chain risk management. *Procedia-Social and Behavioral Sciences 235*, pp. 216-225.
[10] Chen, G., 2017. An Entropy-TOPSIS method for evaluation innovation performance of the high-tech industry. *Boletin Tecnico/Technical Bulletin*, vol. 55, no. 3, pp. 155-163.
[11] Xu, W. and He, F., 2017. Entropy-TOPSIS method for selecting locations for electric vehicle charging stations. *Advances in Transportation Studies*, vol. 3, pp. 83-96.
[12] Zhu, J., Zhu, J. and Zhu, Y., 2017. Assessment on the innovation environment of manufacturing enterprises in China based on Entropy-TOPSIS model. *Journal of Harbin Engineering University*, vol. 38, no. 5, pp. 808-814.
[13] Huang, W., Shuai, B., Sun, Y., Wang, Y. and Antwi, E., 2018. Using Entropy-TOPSIS method to evaluate urban rail transit system operation performance: The China case. *Transportation Research Part A: Policy and Practice*, vol. 111, pp. 292-303.
[14] Bursa Malaysia (n.d.). Company Announcements | Bursa Malaysia Market. Retrieved from http://www.bursamalaysia.com/market/listed-companies/company-announcements/#/?category=all
[15] Price, J.E., Haddock, M.D. and Brock, H.R., 1993. *College Accounting*. 10th ed. New York: Macmillan/McGraw-Hill.
[16] Akguc, O., 2010. *Financial Statement Analysis*. 13th ed. Istanbul: Arayis Publication.
[17] Khrawish, H.A., 2011. Determinants of commercial banks performance: Evidence from Jordan. *International Research Journal of Finance and Economics*, vol. 81, pp. 148-159.
[18] Liew, K.F., Lam, W.S. and Lam, W.H., 2017. An empirical evaluation on the efficiency of the companies in Malaysia with Data Envelopment Analysis model. *Advanced Science Letters*, vol. 23, no. 9, pp. 8264-8267.
[19] Lam, W.H., Din, M.A., Lam, W.S., Chen, J.W., 2018. Evaluation on the performance of suppliers in Malaysia with TOPSIS model. *Journal of Fundamental and Applied Sciences*, vol. 10, no. 6, pp. 406-415.
[20] Lam, W.H., Lam, W.S. and Liew, K.F., 2017. Improvement on the efficiency of technology companies in Malaysia with Data Envelopment Analysis model. *Lecture Notes in Computer Science*, vol. 10645, pp. 19-30.
[21] Lam, W.S., Liew, K.F. and Lam, W.H., 2018. An optimal control on the efficiency of technology companies in Malaysia with Data Envelopment Analysis model. *Journal of Telecommunication, Electronic and Computer Engineering*, vol. 10, no. 1, pp. 107-111.
[22] Lam, W.S., Liew, K.F. and Lam, W.H., 2018. Investigation on the efficiency of financial companies in Malaysia with Data Envelopment Analysis model. *Journal of Physics: Conference Series*, vol. 995, no. 1, pp. 1-10.
[23] Shannon, C.E. and Weaver, W., 1947. *The Mathematical Theory of Communication*. Urbana: The University of Illinios Press.

IOP Conf. Series: Earth and Environmental Science **385** (2019) 012007 doi:10.1088/1755-1315/385/1/012007

Evaluating community participation in planning decision-making in Iskandar Malaysia: Thematic analysis of public planner's perception

K Zanudin[1], I Ngah[1], S H Misnan[1]
[1]Department of Urban and Regional Planning, Faculty of Built Environment and Surveying, Universiti Teknologi Malaysia, Skudai, Malaysia.

Email: khalidzanudin@yahoo.com

Abstract. The purpose of this study is to evaluate the effectiveness of community participation in planning decision-making in Iskandar Malaysia. The metropolitan region consists of four local planning authorities that are responsible for governing the spatial planning in the region. Rhetorically, effective community participation helps the local planning authority in making better planning decision, hence provided a sustainable and quality living environment to the community. Thus, it leads to the questions; what constitutes to the effectiveness of community participation in the planning process? To what extent does the current participatory avenue offer a genuine opportunity to the community to incorporate their interest in the planning decision? This phenomenological study used thematic analysis to evaluate the insight postulated by selected public planners which gained through a series of in-depth interview. Following the 12 interviews performed; a total of 37 themes and 68 sub-themes are transpired. A table is used to illuminate the factors that influence the effectiveness of the participatory process in Iskandar Malaysia. The findings suggested that public planners in Iskandar Malaysia believed the effectiveness of community participation in the region is significantly associated with the operational than the structural and cultural dimensions of the process.

1. Introduction

A local authority is responsible for providing quality living to its community; as well as accountable for their decision. However, the impact of urbanisation on major cities in Peninsular Malaysia has seen local authorities face with a daunting challenge to accommodate the increasing demand for basic amenities and affordable housing [1]. Besides, the local authority cum local planning authority needs to provide more opportunity for the community to participate in planning decision-making [2].
Community participation in urban planning in Peninsular Malaysia (except Kuala Lumpur) is guided by the Town and Country Planning Act 1976 (Act 172) [3]. Rhetorically, community participation is essential in achieving a sustainable and quality living environment thru incorporating the present and future interests of the community [3]. Thus, the questions raised in this study are; "what constitutes to the effectiveness of community participation in the planning process?" and "to what extent does the current participatory avenue offer a genuine opportunity to the community to incorporate their interest in the planning decision?"
This phenomenological study has employed in-depth interview for data collection and thematic analysis to analyse the qualitative data. The sample size is determined using the concept of data saturation with several works of literature that offered the guideline to estimate the sample size for

phenomenological research is reviewed. The purpose of this paper is to evaluate the effectiveness of community participation in planning decision-making in Iskandar Malaysia. The evaluation is elucidated by thematically analyse the public planner's perception. Findings suggested that the effectiveness of community participation in planning decision-making is significantly associated with the operational than to structural and cultural dimensions of the process. The paper concluded with the current community participation in the planning process in Iskandar Malaysia has pointed to the ineffectiveness of the process; thus hinder the ability of the community to incorporate their interest in planning decision genuinely.

2. Methods

2.1. Participants

Public planners of local planning authorities in Iskandar Malaysia are initially approached to gain their consent to be interviewed. The selection of public planners is through the purposive sampling, a non-probability sampling method based on three criteria; local authority, position and scope of work. However, in this study, several lower-ranked officers are also selected based on the recommendation by higher-ranked officers. Following the concept of the key informant by [4], these criteria ensure that public planners with vast knowledge of the planning procedure were selected to provide comprehensive insight of the current state of community participatory in the planning process in the metropolitan region. The sample size is determined using the concept of data saturation. [5] defined data saturation as "the point in data collection and analysis when new information produces little or no change to the codebook". Saturation is explained further in other subsection.

2.2. Data Collection

A series of in-depth interviews with key informants are performed to gain qualitative data from the public planners in Iskandar Malaysia. The questions asked are circled their perception on the community participatory in the planning process. The selection of in-depth interviews for data collection is based on the ability of this technique to allow the researcher to gather a greater picture of a phenomenon through a small number of respondents that are based on their experiences [6]. Furthermore, [7] in his definition of phenomenological research highlight the employment of in-depth interview as one of the techniques in understanding people's motivation and action, henceforth their experience. Subsequently, in-depth interviews with key informants are fundamental in gathering valuable information on the current setting of urban planning in Iskandar Malaysia.

Interviews are performed in the form of semi-structured to offer to both researcher and interviewee flexibility and fluidity of ways a question been put out and answered. It gives interviewee more freedom in responding to each question hence lead to interesting findings [4]. In other words, the semi-structured interview provides the opportunity to both researcher and interviewee to discuss and expand the knowledge of both parties.

At the start of each interview, the researcher briefly explains the purpose of the study. To put the interviewee to ease, the researcher has agreed to retain his confidentiality. Permission to record the interview is also sought in every session although one of the respondents did not approve. The following variables were recorded from the interviews - intergovernmental relationship in planning; access to information and process; coordination and collaboration between stakeholders; legislative framework; capacity of local authority; skill and attitude of public officer; influencing capability; transparency and accountability; participatory method; public awareness and knowledge; residents committee; and community representation.

2.3. Saturation of Data

Scholars have used the concept of saturation as the yardstick in determining the sample size for qualitative research [5]; [8]. Scholars had defined data saturation as the point when the data collection and analysis did not produce new information [5]; [8]. Although the idea of achieving saturation in data collection and analysis is useful for qualitative researchers, it provides a little guideline to estimate the sample size prior to the data collection [5]; [9]. It is because; most of the researches require an estimation of sample size during the proposal stage.

Subsequently, some literature that has provided a guide to estimate the sample size for phenomenological research has been reviewed (**Table 1**). From the literature review, it is suggested the data saturation in a phenomenological study is probably achieve between 10 to 12 interviews. Therefore, in performing this study researcher has estimate a total of 12 in-depth interviews as the yardstick to reach saturation. Similar to [9], the researcher believes that it is better to overestimate rather than to underestimate the sample size.

Table 1. Summary of Data Saturation in Phenomenological Researcher from Empirical Studies

Authors	***Definition of Saturation***	**Findings**
Creswell (1998)	n/a	Ten in-depth interviews to reach saturation
Morse (2000)	n/a	It requires 6 to 10 interviews to reach data saturation
Guest et al., (2006)	"..the point in data collection and analysis when new information produces little or no change to the codebook".	12 interviews to reach 92% data saturation
Coenen et al., (2012)	"Saturation refers to the point at which an investigator has obtained sufficient information from the field".	Nine interviews to reach saturation in a deductive approach; and 12 interviews in an inductive approach

2.4. Data Analysis

Thematic analysis is used by researchers whose tend to employ a low-level interpretation. This method is suitable to analyse the information from the in-depth interview of public planners in Iskandar Malaysia. It is justified through the argument by [10] that thematic analysis able to produce an insightful analysis of research questions and facilitates in answering the research questions by capturing the key themes of the data. This study has followed the [10] framework to explain the analysis process.

Step 1 – Transcribe and familiarise the data: The audio files from the interviews are listened several times and transcribed using the Microsoft Office program. Nearly all interviews are performed in Bahasa Malaysia with several English terminologies are used. The researcher has employed edited transcription approach in transcribing the audio files. **Step 2 – Generate initial codes:** At this stage, the researcher only coded data that are relevant to the research questions. The code in this study is in the form of statement deliberated by respondents. Similarities in responses between interviews are highlighted and organised using the Microsoft Office. Transcripts are reviewed several times with new or modified codes is updated. **Step 3 – Search for themes:** All relevant coded data are organised into different themes. A table is used to classify and display the relationship between themes and codes that based on three dimensions – operational, structural and cultural as adapted from [11] in evaluating the effectiveness of community participation in planning decision-making. Each dimension comprises several variables (as stated in 2.2). The table helps the researcher in reviewing the level of themes, hence led to the emergence of sub-themes. Also, codes are linked to one theme or more.

Step 4 – Reviewing themes: In this stage, all themes, sub-themes and codes are revised and modified to established coherent connections between them. At this point, the fundamental question is, does each connection seem relevance? The researcher used the 'cut and paste' function in Microsoft Office to reorganise the data in the transcript. **Step 5 – Define themes:** The main questions asked in this stage are; what are the relations built between themes and sub-themes? How do these relations attain the research questions? Contrary to [10] thematic map, the researcher has used the table to illuminate the variables (factors) that possibly influence the effectiveness of the participatory process in Iskandar Malaysia (**Table 2**). **Step 6 – Reporting:** Researcher reporting the findings.

3. Result and Discussion

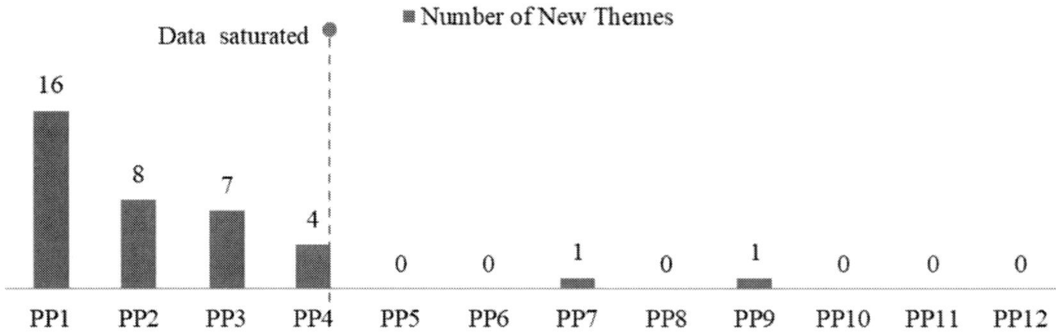

Figure 1. Themes emerge during analysis of 12 in-depth interviews with public planner (PP)

Figure 1 shows the whole themes emerged throughout the analysis of 12 in-depth interviews with public planners in Iskandar Malaysia. **A total of 37 themes** are transpired from **124 codes (statements)**. **94.5% (35)** of the themes emerged by the fourth interviews. In other words, saturation of data in this study achieved at the fourth interviews. Furthermore, **68 sub-themes** are derived from the themes and codes.

Following the analysis, themes and sub-themes that pertain to the research questions are corroborated through the table below (**Table 2**). The regularities of both themes and sub-themes emerged in every interview are recorded to determine the scale of impact causes by corresponding themes and sub-themes to the participatory process. The scale of impact is divided into three levels – **low impact (1-4 incidents); medium impact (5-8 incidents); and high impact (9-12 incidents)**.

From the thematic analysis, **83% of the respondents** believed the current urban planning system in the Iskandar Malaysia, as well as Peninsular Malaysia, is more centralised (**Table 2**). The upper-tier planning authorities have a more significant influence on the planning decision than the local planning authority [PP2; PP3; PP4; PP5; PP6; PP8; PP9; PP10; PP11; PP12]. As a result, a decision made might not represent the interest of the local. Moreover, **75% of the respondents** agreed that local planning authorities in the region still depend on the conventional medium in sharing information and engaging with the community in strategic planning [PP1; PP2; PP3; PP6; PP7; PP10] and operational planning [PP3; PP4; PP9; PP10; PP12]. **11 from the 12 respondents** stated that the current platform for collaboration between stakeholders is restricted within the framework of Act 172 [PP1; PP2; PP3; PP7; PP8; PP10; PP12]. Additionally, the community are legislatively excluded from planning permission process whereas only principle submitting person (PSP), applicant and authorities are engaging in the process [PP4; PP5; PP6; PP9; PP10; PP12]. Consequently, the limitation of access to planning information and platform for collaboration among the community may lead to the gap in knowledge and participation among stakeholders, hence benefited to certain stakeholders and community segments.

The analysis has highlighted the importance of the planning statutory and guideline as the main references to the state and local authorities in making strategic and operational planning decisions [PP1; PP2; PP3; PP4; PP5; PP6; PP7; PP8; PP9; PP10; PP11; PP12]. **All respondents** have concurred that technical factor is the main reason behind any decision taken. However, some of them believed the community unable to participate effectively in preparing the strategic plan due to the technicality of the process and document [PP3; PP11].

Table 2. Factors that Influence the Effectiveness of Community Participation in Planning Decision-Making

Dimension	Variable	Theme	Sub-theme	PP1	PP2	PP3	PP4	PP5	PP6	PP7	PP8	PP9	PP10	PP11	PP12
Operational	Intergovernmental relationship in planning	Translation of development strategy	Top-down approach		●	●	●	●	●		●	●	●	●	●
			Top-down & bottom up approach	●						●	●				●
		Relationship between authorities	In decision-making		●	●	●		●	●	●	●	●	●	●
			Intervention in decision-making			●									
	Coordination & collaboration between stakeholders	Factors considered in making decision	Technical factor	●	●	●	●	●	●	●	●	●	●	●	●
			Political factor				●		●	●		●	●		●
			Market factor	●				●		●	●	●			
		Incorporate community interest	Statutory instrument	●	●	●	●	●	●	●	●	●	●	●	●
			Through engagement	●	●	●	●	●	●	●		●	●	●	●
			LPA's initiative				●	●					●		●
		Collaboration between stakeholders	Planning platform	●	●	●	●	●	●	●	●	●	●		●
			Non-planning platform	●		●	●	●		●	●	●		●	
	Access to information & process	Access to information	Limited medium	●	●	●	●		●	●		●	●		●
			Role of community representative		●		●				●		●		
		Access to planning process	Knowledge related access		●	●		●	●			●			
			Selected location		●	●					●		●		
		Technicality of document	Gap in understanding			●					●		●	●	●
Structural	Capacity of local authority	Administration capacity	Lack of manpower		●						●				
		Financial capacity	Lack of capital		●					●				●	
	Skill & attitude of public officer	Treatment of community participation	In development planning	●	●	●		●	●	●	●		●	●	●
			In planning permission				●		●			●	●	●	●
		Public officer professionalism & skill	Public planner's professionalism		●	●	●	●	●	●	●	●	●		●
			Public planner's initiative			●									●
		LPA's perception	Participation is time-consumed			●	●		●						
			Participation is associated with knowledge			●								●	●
			Participation is associated with psychological factor	●	●	●			●	●		●	●	●	●
	Influencing capability	Private developer's domination	Influence in planning decision		●	●					●				
			Attitude & perception				●	●			●	●	●		●
	Influencing capability	Empower the community influence	Through knowledge			●							●		
			Through access to process			●									
			Through association			●							●	●	●
		Local councillor's influence	-				●	●	●	●		●	●		●
		LPA's influence	-									●			
Structural	Legislative framework	Guideline for public participation	Lack of guideline			●				●		●		●	
		Content of document	Lack of localizing the local development plan				●			●				●	
	Transparency & accountability	Incorporate transparency	Notify planning decision		●	●	●			●			●		●
		Incorporate accountability	Evaluate the local development plan							●	●			●	●
	Participatory method	Current method	Within Act 172 framework	●	●	●			●	●		●	●	●	●
			In strategic planning	●	●	●				●	●		●		●
			In operational planning				●	●	●			●	●	●	
Cultural	Public awareness & knowledge	Awareness & knowledge level	Related to interest	●	●	●		●	●	●		●	●	●	●
			Unrelated to income	●							●		●		
		Willingness to participate	Associate with capacity	●	●					●				●	
			Associate with psychological factor	●	●			●		●	●		●	●	
			Associate with ethnicity		●										
	Community representation	Efficiency of local councillor	Instrument to perform			●									
			Attitude of local councillor			●			●	●			●		●
			Proficient of planning procedure			●				●	●	●			●
			Equitable in representation			●	●								
		Role of local councillor	In strategic planning	●	●					●	●		●		
			In operational planning				●	●	●			●	●		●
			As mediator	●	●	●	●		●	●	●		●		●
		Local councillor background	Appointment of local councillor	●						●	●				●
		Initiative by local councillor	Facilitate with information & process		●	●					●				
	Residents committee	Role of residents committee	-				●	●							

Relationship between theme and sub-theme that has high impact to the effectiveness of the community participation

IOP Conf. Series: Earth and Environmental Science **385** (2019) 012007 doi:10.1088/1755-1315/385/1/012007

All respondents concurred that technical factor is the main reason behind any decision taken. However, some of them believed the community unable to participate effectively in preparing the strategic plan due to the technicality of the process and document [PP3; PP11]. Thus, the legitimacy of both statutory and guideline to represent the community interest is questionable. Furthermore, **91.6% of respondents** highlighted the current approaches undertaken by local planning authorities to include the community interest in planning decision. Initially, during the development planning, a series of focus group discussion is conducted to gather public opinions [PP1; PP2; PP3; PP7; PP10]. As for planning permission, it is down to the professionalism of a public planner to advise the developer to consult with the community if their proposal potentially affects them [PP4; PP5; PP9]. In the event of an objection, after permission is granted, a dialogue between community, developer and local planning authority is held to resolve the dispute [PP4; PP5; PP6; PP9; PP10; PP11; PP12].

Ways of community participation are treated by the local planning authority has also greatly influenced the effectiveness of the process. Following the interviews, it is comprehended that any objection and suggestion received throughout the publicity period of the development plan is considered based on its relevance to the notion of urban planning [PP1; PP2; PP3; PP7; PP8; PP10; PP11; PP12]. Unfortunately, the majority of the respondents **(75%)** believed the community did not have the technicality sound in urban planning [PP1; PP2; PP3; PP5; PP6; PP7; PP9; PP11; PP12]. This will dampen the capacity of the community to participate, hence influence the planning decision effectively. **9 of 12 respondents** concurred the community participation is associated with psychological factor [PP1; PP2; PP3; PP6; PP7; PP9; PP10; PP11; PP12]. They believed the public planner's perception towards the community had influenced the way they handle the participatory process [PP2; PP3; PP7; PP11]. They too believed that our society is still immature in making objection and suggestion concerning urban planning [PP1; PP2; PP3; PP7; PP9; PP11; PP12].

The capability of the community to participate effectively is also substantially rest on the professionalism of a public planner in handling the process. **58% of the respondents** believed that public planner should be more thoughtful and understand their area well, mainly associated with the community interest. The public planner should put extra effort to go to the ground and flexible in delivering information and implementing publicity process to ensure all community segments relish access to the information and process [PP3; PP4; PP5; PP6; PP7; PP8; PP12]. Operationally, the public planner will not grant permission until all the technical requirements are fulfilled by the applicant [PP2; PP4; PP8; PP9; PP10]. The professionalism will ensure the community interest documented in the development plan and guidelines are protected and realised.

Structurally, the current platform for community participatory which is limited to development planning and within the framework of Act 172 has led to the community mainly the low-income segment to have little access, hence capacity to engage in the planning process due to the scarce of opportunities [PP1; PP2; PP3; PP6; PP7; PP9; PP10; PP11; PP12].

From the cultural aspect of the process, more than half of the respondents **(58%)** stated that the community only willing to participate if it involves their interest. This has become a challenge for the authority to attract their involvement especially the low-income segment [PP2; PP3; PP6; PP7; PP9; PP10; PP11]. Substantially, public planners do believe there is a connection between awareness of planning procedure and the ownership of property [PP1; PP2; PP3; PP5; PP6; PP7; PP10; PP11]. Therefore, there is a gap in awareness between the private and large property owner and ordinary people. **75% of the respondents** believed that the community generally did not know anything about urban planning [PP1; PP2; PP3; PP5; PP6; PP7; PP9; PP11; PP12]. Allegedly, there is a gap in participation between urban stakeholders as current procedure only serves certain segments [PP11].

The final factor that significantly influences the effectiveness of the community participation is the role of local councillor as the representative of the community. As the mediator between the local authority, community and private developer, it is essential for the local councillor to continually share information on planning with their community [PP1; PP2; PP3; PP4; PP6; PP7; PP8; PP10; PP12]. Also, the community can indirectly exert their interest in planning decision through the local councillor. The ability of local councillor to represent their community depends on their competency especially in relative to their attitude [PP3; PP6; PP7; PP10; PP12] and proficiency of planning procedure [PP3; PP7; PP9; PP12].

The findings from thematic analysis suggested that the current community participation in planning decision-making in Iskandar Malaysia required improvement mainly to its operational dimension. It is corroborated with the majority of respondents concurred that operational variables such as intergovernmental relationship in planning, collaboration and coordination between stakeholders and access to information and process are fundamental to the successful community participation. From the elaboration of the relationship between themes and sub-themes, it is plausible to conclude the improvement of operational aspect will prospectively improve both structural and cultural aspects of the process in long-term. By mitigating the operational-related shortcomings, it will initiate further enhancement to other aspects and create a planning milieu. Subsequently, it will spur for greater collaboration between stakeholders; thus empower the capacity of the community to participate in the planning decision-making effectively.

4. Conclusion

This phenomenological study anticipates in the continuous debate on the effectiveness of community participatory in urban planning as the vehicle to incorporate the community interest in the planning decision, hence produce greater living to the community. Thematically, researcher has derived the factors that influenced the effectiveness of the current practices in Iskandar Malaysia which is based on the perception of public planners of local planning authorities. The current state of community participatory in Iskandar Malaysia highlighted the inability of the present avenue to provide genuine participation to all community segments. The prominent limitation highlighted is associated with the operational aspect of the process. Prior to the finding, it helps decision-maker to address the shortcomings of current practice, henceforth promote genuine participation to the community.

Acknowledge

This research is funded by Academic Training Scheme for Bumiputera (SLAB) from Ministry of Higher Education Malaysia and Young Lecturer Scheme from Universiti Teknologi MARA (UiTM).

References

[1] Razak M R Johar F & Abd Khalil R A 2016 The Impact of Iskandar Malaysia Development on Urban Amenities *Journal of the Malaysian Institute of Planners* **Special Issue IV** p 121-134

[2] Phang S 2011 *Public Administration in Southeast Asia: Thailand, Philippines, Malaysia, Hong Kong and Macao* E M Berman Ed CRC Press Boca Raton FL

[3] Omar D Leh O 2009 Malaysian Development Planning System: Kuala Lumpur Structure Plan and Public Participation *Asian Social Science* **5(3)** p 30–36

[4] Edwards R and Holland J 2013 What is Qualitative Interviewing? *'What is?' Research Methods Series* Bloomsbury Academic London p 2-31

[5] Guest G Bunce A and Johnson L 2006 How many interviews are enough?: An experiment with data saturation and variability *Field Methods* **18(1)** p 59-82

[6] Guion L Diehl D and McDonald D 2006 Conducting an in-depth interview University of Florida Institute of Food and Agriculture Sciences

[7] Creswell John W 1998 Qualitative inquiry and research design: Choosing among five traditions Sage Publications Inc. Thousand Oaks p 51-65

[8] Mason M 2010 Sample size and saturation in PhD studies using qualitative interviews *Forum: Qualitative Social Research* ***11(3)*** p 1-13

[9] Morse J M 2000 Determining sample size *Qualitative Health Research* ***10(1)*** p 3-5

[10] Braun V and Clarke V 2006 Using thematic analysis in psychology *Qualitative Research in Psychology* ***3*(2)** p 77–101

[11] Tosun C 2000 Limits to community participation in the tourism development process in developing countries *Tourism Management* ***21*(6)** p 613–633

IOP Conf. Series: Earth and Environmental Science 385 (2019) 012008 doi:10.1088/1755-1315/385/1/012008

Application of Principal Component Analysis: Deriving Influencing Factors of Teenagers' Recycling Practice

S M Kamaruddin[1], A Misni[1], P Ahmad[1] and R Yaman[1]
[1]Faculty of Architecture, Planning and Surveying, Universiti Teknologi MARA, Puncak Alam, Selangor, Malaysia

E-mail: Sitim065@salam.uitm.edu.my

Abstract. This study aims to derive the significant factors which influence adolescents' participation in recycling in schools through Principal Component Analysis. This method transforms statistically a set of observations of correlated variables into a set of values of linearly uncorrelated variables for further analysis. It is a classical feature extraction and data representation technique. Using a questionnaire, data were collected from 328 willing participants and Principal Component Analysis (PCA) was applied to group the initial variables into smaller, interpretable underlying factors through the use of statistical software. The factors derived were then analysed using regression analysis. The results indicated that social influence in schools involvement with teachers and other schools in recycling programs and having environmenta knowledge have the greatest influencing factors towards teenagers' involvement in recycling programs in schools.

1. Introduction

The future of the environment lies in the hands of every human being. The concern for the environment is relevant to all sections of the community from preschoolers, adults to senior citizens. It is also a fact that teenagers will eventually become leaders or policy makers responsible for the environment and therefore, must be encouraged to be environmentally aware [1]. Understanding what makes them more informed or what influences their participation in a specific waste management behaviour, such as recycling, can contribute to more-relevant programme design and implementation [2]. In the context of sustainable waste management (SWM) and teenagers, the types of activities in school have a role in the good practice of SWM [3]. Recycling involves processing used materials (waste) into new, useful products to reduce the use of raw materials that would have been used. Recycling also uses less energy and a good way of controlling air, water and land pollution. Effective recycling starts with household or the place where the waste was created. In addition, environmental awareness and knowledge about environmental conservation were found to affect recycling attitude positively.

In this study, the main research question is concerned with explaining *what* factors significantly influence adolescents' participation in SWM mainly recycling in schools. The data was analysed using the method of Principal Component Analysis (PCA)[4]. PCA is best known for possible reducing the number of variables while retaining information of the original data set and is widely known and used technique for dimension reducing [5]. The principal components are then used in a regression model to help answer the research question.

2. Methodology and Findings

Quantitative data were obtained from 328 willing urban secondary students aged 13-18 from Subang Jaya, District of Petaling Selangor. Through a stratified sampling method, the researchers obtained an 80% response rate. The data was subsequently screened and keyed into a computer worksheet. Data screening is necessary to minimize or eliminate inaccurate data [6]. Students' basic demography, the perception of waste and relevant social aspects of solid waste management was recorded in questionnaires.

The first section was related to the respondents' age and gender using categorical or nominal scales. The second consisted of statements designed to evaluate the students' self-reported attitudes toward recycling. An attitude toward a specific act can give an indication of an individual's behavior [7]. It has been found that when teenagers are concerned about the environmental issues they feel they can make a difference in the local environment [8].A scale to measure attitudes towards recycling is necessary since attitude and awareness about recycling cannot be directly observed [9]. Taking into consideration the situational aspects involved (i.e. that the school's recycling activities are ongoing, infrastructure for recyclables had been put in questions related to different aspects of recycling were formed. Programmes and infrastructure i.e. "intervention strategies" [10] are believed to increase recycling behaviour. Other psychological intervention strategies that promote household recycling include information, rewards, commitment, behavior modeling and environmental changes [11]. A study confirmed that a lack of effective recycling facilities does impose an obstacle to recycling participation [12]. In the third section, the questionnaire included statements of the respondents' general perceptions of current waste collection and management situation. In the fourth section, questions were constructed to obtain respondents' views about engagement with other organisations involved with the recycling program in school. It has been found that teenagers social interactions with other groups concerned with the environment further strengthens their concern for the environment [13]. The questions were also constructed based on past interviews with local municipal officers in Selangor [14] and review from the relevant literature [15]. A Likert-type scale was used as a measurement indicator for all items: (1 = Highly Disagree, 2 = Disagree, 3 = Undecided, 4 = Agree, 5 = Highly Agree).

Based on relevant literature, questions were formed related to attitudes, the perception of waste and incentives in recycling. In the first step, principal component analysis (PCA) grouped the initial variables into fewer factors. Kaiser-Meyer-Olkin (KMO) sampling adequacy test and Bartlett's test of sphericity can be applied to assess the suitability of the data for analysis [16]. The KMO sampling adequacy test predicts if data will factor well, and a KMO measure value of more than 0.6 is considered acceptable [9]. The data analysed was suitable for analysis for all three groups i.e. the attitudes toward and perception of waste and perceived involvement in recycling. Secondly, the researchers used the Kaiser eigenvalue criterion and the scree test as recommended [17] to decide which factors to retain. If the eigenvalue is greater than 1, the factor is retained. The next step is a scree plot, where the eigenvalue magnitudes on the vertical axis are plotted against the number of components on the horizontal axis. A graphical interpretation will show how many factors should be retained. The same procedure was conducted for items of "perception of waste". Only one factor is to be retained based on its larger variance. The third component analysis of items of "involvement with others" resulted in two factors. In addition, Varimax rotation test [19] to rotate the retained factors in each application of analysis was conducted. The loadings are indicated in all Tables below. Rotating factors increases their interpretability and proportion of variance [18]. The variables with greater than 0.5 loadings and highest loadings were the basis for assigning the new variable's description. Cronbach's coefficient alpha was used to test the reliability of the items' consistency. For an exploratory study, constructs that indicate higher than 0.5 are considered acceptable [17].

In the first PCA application of this study, three new factors were derived. They are labeled *Attitude, Social influence in school and familiarity with logistics (SISFL)* and *Indifference to the programme* (Indifference) (Table 1.1). The first factor describes the students' attitudes toward recycling with perceived constraints on the behaviour. The second factor gives an indication of the students' familiarity with recycling infrastructure (the three coloured recycling bins) its location, family environmental practices and expectation of monetary reward. The third factor is called Indifference to the programme (Indifference) and relates to the students' indifference toward the recycling programme. The variance of the three components account for 58%; P-value = 0.000 (Bartletts test); KMO = 0.7; Range of Cronbach's alpha 0.5 to 0.7.

Table 1.1 Factor analysis -*Attitudes*

Component 1: Factor Attitude	Loadings
(Variance : 30%; Cronbach's alpha: 0.7)	
• I will send recyclable waste to the centre if it is near.	0.81
• I will separate recyclable waste using the bins if instructed by law	0.80
• I only send recyclables to school when I have time	0.72
Component 2: Factor Social Influence in School – SISFL	
(Variance: 15%; Cronbach's alpha: 0.7)	
• Many teachers and friends store recyclables in the recycling bins at school.	0.76
• The bins are visible in my school compound.	0.70
• I know where the recycling bins are in my neighbourhood.	0.62
• My family and I send recyclables to recycling centres.	0.58
• I look forward to getting some pocket money from selling recyclables.	0.52
Component 3:Factor Indifference	
(Variance : 13.4 %; Cronbach's alpha: 0.5)	
• Although my school has some kind of recycling programme, I have not started to recycle.	0.82
• Although my school has some kind of recycling programme, I have not encouraged my family to recycle.	0.74

- Loadings (post Varimax rotation)

The students who participated in recycling activities in school were influenced by both their teachers and friends at school and their family members' environmental behavior. As significant change agents towards students' recycling participation, it is suggested that parents can start or continue to practice separation of waste and sending recyclables to recycling bins for their children to emulate. Parents or adult family members could further instill the notion of 'trash to treasure' by starting crafting projects such as fashion jewelry made from unused plastics or initiating composting. The recyclable product could be sold online on easily accessible online social media platform such as Facebook or Instagram. Children should also be applauded whenever they make any effort to curb waste and reuse or recycle. Parents too can further extend their pro-environmental ideas and actions with the school through their Parents and Teachers Association annual meetings and create some incentives or rewards to the family with the most recyclables collected and sold to recycling vendors. The PTA could make 'green actions' as one of the compulsory agenda in their annual plan to involve more students, teachers, and parents involved with recycling and environmental behavior.

In the second PCA application of this study, student' perception of waste was evaluated through similar statistical procedures (Table 1.2). Applying Cronbach's alpha test of reliability, one factor *Perception* was accepted and met the requirement.

Table 1.2 Factor analysis - *Perception of waste*

Component: Factor Perception of waste (Perception) variance 25%; P value (Bartlett's test) = 0.000; KMO 0.6; Alpha Cronbach's 0.77.	Loadings
• Waste in landfills cause environmental pollution.	0.84
• Waste has value	0.83
• Waste can be turned into resource	0.82

- Loadings (post Varimax rotation)

It is proposed that students perception of waste as a resource can be heightened through activities such as recycling local resources optimally to achieve improvements, for example, recycling or harvesting of rainwater for the school plants or gardens. This could raise interest in the environment which brings a culture of Community participation which increases the sense of responsibility for solving waste problems [3]. Thus, strategic planning for waste management program is required. The school management can create an environmental program that sets a clear aim. The elements in the program must outline the focus and achievement of the program, the benefits, why it is important, when is the timeline, which is the best method and implementation. For example through a composting program the school children and school community can learn composting skills and knowledge on how much food waste is required to turn waste into resource materials (compost) while getting involved with recycling enterprises to obtain reasonable benefits or monetary incentives from this initiative.

Table 1.3 Factor analysis - *Incentives*

Components:	Loadings
Component 1 : Factor Involved with Others -InWOB (Variance : 55 %; Cronbach"s alpha: 0.9)	
• My school's recycling programme participate with other schools' recycling programmes	0.77
• My school's recycling programme help generate my environmental awareness	0.77
• It's fun to participate in my school's recycling programme	0.75
• I gain more knowledge on how to protect the environment by participating in myschool's recycling/environmental programme	0.63
• My school gets revenue by selling recyclables to outside vendors	0.63
• I think the government's recycling rate is realistic.	0.63
• I think more people should participate in environmental community programmes	0.63
• Neighbourhood and school recycling programmes are good environmental discussion platforms.	0.63
Component 2: Factor **Social concerns (SC)** (Variance 8%; Cronbach's alpha: 0.9)	
• Community recycling centres are beneficial to society	0.74
• If more individuals participate in environmental pro-community programmes, more will care for the environment.	0.72

- Loadings (post Varimax rotation)

The third PCA application highlighted two factors. The first component *Involvement with others* gave an indication that respondents' had fun and gain environmental awareness through participation. The second component *Social Concerns* highlighted students' perception of the public's involvement toward recycling participation. The variance of the two components are 63% with KMO = 0.94; Bartlett test of P-value = 0.000; Cronbach's alpha ranges 0.63 to 0.74. Involvement with others involved can be seen as a process in which community members are engaged at different stages and degrees of intensity in the project cycle with the objective to build the capacity of the community [8]. This element is important to maintain services created during the project after the facilitating organisations with the school eg. Environmental NGO has left. Other organisations such as the local authority could also directly involve schools in their organic waste composting projects. Students can be taught ways or techniques to run their organic waste disposal (domestic waste and garden waste) with the assistance of the local authority. The Subang Jaya City Council, for example, could extend their provision of the 70-liter compost bin provided free of charge to residents to implement organic waste composting activities. As an example from Japan, after the Food Waste Recycling Law was enacted in 2001, many innovative projects have been put in operation such as The Sapporo Kitchen Garbage Recycle Center [23]. This centre transforms 50 tons of food waste collected daily from 188 local schools, hospitals and companies into 10 tonnes of dehydrated cattle feed using a special process. The resulting compost fertilizer can be used or sold by the school to gain some income.

Recycling in school should be expanded to involve water and possibly food waste recycling (composting activities). Food waste, garden waste, market waste, used cooking oil to new forms that have a value and benefits can support the implementation of the program currently ongoing such as Integrated Biomass Center of MPSJ in Bukit Puchong Town and MPSJ Biomass Center in Seri Serdang.

A subsequent evaluation of logistic regression was conducted with the six factors derived from the PCA. Two other variables obtained from the data (i.e. Information received and Perceived financial incentive) were included. The regression model was analysed by statistical software (SPSS17) using the Pearson and deviance goodness of fit test [19]. It was statistically significant ($p < .05$). The factor scores for each individual were averaged by summing raw scores corresponding to the items that loaded on the factor. This method is considered acceptable for most exploratory research [20].

The regression analysis confirmed that three variables (factors) *Social Influence in School (SISFL), Involvement with Others (InWOB,)* and *Perception* compared to the other five variables had significant impacts on those who responded 'Yes' to recycling. This finding indicates that those who participate in recycling programs in schools are those that are influenced by their teachers and friends' environmental actions, have fun while participating and have environmental knowledge. It is interesting to note that financial incentive, although a common strategy to encourage pro-environmental behavior, has small to medium effect on pro-environmental behavior as other scholars promote moral norms as important predictors to recycling behavior [21].

In summary, schools as an organisation can attempt to integrate change techniques [22] or intervention strategies. From the findings in this research, the school could first list out the key influences towards recycling participation by establishing a 'fact sheet' which is visible to all students. This information or 'waste audit' provides the information of the volume of recyclables that each group of students either by their individual class or collective classes have collected weekly or monthly, sold and gained any income according to a certain time period. The Principle of the school can acknowledge their efforts and highlight that the student's consistent small acts can bring about change and benefit the school financially and environmentally. Currently, schools within Subang Jaya Municipality are annually involved with Green Challenges sponsored by corporate or financial institutions concerned with innovative environmental products and green living and gain some financial incentive for their school and could be channeled to help others for example, to charity homes. A recycling reward system of issuing food tokens could also encourage and motivate some students to behave pro-environmentally. Schools should continue expanding further collaborating with other corporate sponsors to improve existing programs and incorporate the "fun" facto. The corporate sector in return can gain on its environmental corporate social responsibility through its sponsorships to promote pro-environmental behavior. The school community for example with the involvement of an Environmental NGO and the Coordinator of the Environmental Club and possibly the Parents Association can engage with Government Agencies that support an Integrated Solid Waste Management (SWM) to establish an effective and efficient environmental program. The principal task of the NGO would be to coordinate activities that can accelerate efforts toward efficiency and effectiveness of integrated SWM, monitoring, and advice to the stakeholders that are directly involved. This is due to the evidence from this study that students are found to be highly influenced by the presence and involvement with others in recycling participation compared to gaining financial incentives although studies have found it may be a realistic option to produce the desired result of increasing recycling participation.

The schools can move forward by improving and adding more effective and "fun" infrastructure and making them more visible and accessible to everyone in the school. Clear signage of the types of waste considered recyclables are equally important on the recyclable bins provided and the design of these signages can be part of the Green challenges organized. Recycling bins and devices can be made to be more attractive and creative to encourage anti recyclers. Examples include recycling devices or bins that could 'crush' recyclables such as plastic bottles or cans so more could be stored. Other interesting efforts that could involve school students include making the bins more colourful with art deco inspirations having different sizes of bins with interesting designs to attract staff and students to use the bins.

Recent waste recycling innovation from students' efforts in more developed countries in the U.S for example indicate that teenage students of the 21st Century are capable of inventing reusable products such as insulation material and fertilizers from recyclables wastes such as plastic. This is parallel with science becoming more important to solve environmental issues and technological advancement especially the internet allowing heightened access to information. Students' innovative efforts in scientific competitions towards producing environmentally sustainable products from recyclables lead to new technology patents Recycled plastic products are now used in the construction industry [23]. Currently, in Malaysia, several schools in Selangor compete to collect the most recyclables while being awarded for the environmental effort by gaining monetary rewards from sponsorships of consumer product companies. Other examples of innovative recycling bin designs incorporate artistic efforts that add colour to form interesting urban hardscapes, as implemented by the City of Chicago's Green Initiatives 2013 [24] can be emulated by students towards heightened environmental awareness.

Finally, it is highly recommended that the process to sustain the recycling program or any environmental program in school needs an evaluation of its successes and the constraints faced by the students and the school. Current practices of reporting students' activities and contribution in recycling programs to the community, encouraging innovations and creative products by students using recyclables and their participation in environmental competitions would further strengthen their self-efficacy [19] and heighten their concern for the environment. Thus, there is a need for continuous improvement evaluation with involvement from both the teachers as the current guardians, other organisations for support of the programs and students as future leaders and active participants.

3. Conclusion

This study aims to derive the significant factors that encourage teenagers to participate in the schools recycling program. This could be used towards the formulation of more effective pro-environmental programs related to recycling participation. Data was collected through questionnaires from students in secondary schools in the District of Petaling, Selangor. The student's age ranges from 13 – 18 years old. The data collected were cleaned and manipulated through several statistical tests relevant to Principal Component Analysis. This method derived several factors that met the validation criteria. The basic protocol including the tests relevant to PCA was conducted to derive the significant factors and subsequently analysed using regression analysis. The results indicated that social influence in schools and visibility of infrastructure/recycling bins and involvement with teachers and other schools and other organisations in recycling programs, having environmental knowledge are the significant influencing factors towards teenagers' behavior to participate in recycling programs in schools. Parents and families' 'green actions or pro-environmental behavior are a significant influence on students compared to the monetary reward. Thus parents should continue to applaud any recycling efforts made by their children. Parents and teachers association could also foster networking with other stakeholders that support the drive towards recycling and conservation of the environment.

Consistent with the current literature, the social context within the school environment pose a great influence on students recycling and general pro-environmental behavior and participation. It is recommended thus that pro-environmental programs be retained as a consistent feature in the school and more importantly must engage students, parents and the school community to ensure participation. Based on current trends and environmental practices at both local and international levels, students of the 21st.

Century is more aware of sustainable waste management but require continuous support from all levels of the community to sustain their interest, creativity, and participation in SWM generally and recycling programs specifically.

Acknowledgement

We thank Universiti Teknologi MARA (UiTM) for funding this study through the LESTARI UiTM Research Grant 600-IRMI 5/3/LESTARI(053/2018).

References

[1] Uitto J and Puri J, Van den Berg 2015 Evaluating Climate Change Action for Sustainable Development: Introduction. Springer Verlag
[2] Mtutu and Thondhlana 2015 Encouraging pro-environmental behaviour: Energy use and recycling at Rhodes University, South Africa *Habitat Internationa*l **53** 142-150
[3] Rada E C, Bresciani C, Girelli E, Ragazzi M, Schiavon M and Torretta V 2016 Analysis and Measures to Improve Waste Management in Schools. *Sustainability* **8** 840.
[4] Kambhatla, Nandakishore L, Todd K 2017 Dimension Reduction by Local Principal Component Analysis *Journal of Environmental Psychology* **54** 57-6
[5] Jolliffe I 2011 Principal Component Analysis. In: Lovric M. (eds) International Encyclopedia of Statistical Science. Springer, Berlin, Heidelberg
[6] Hair Jr J F, Sarstedt M, Hopkins L, Kuppelwieser V G 2014 European Partial least squares structural equation modeling (PLS-SEM): An emerging tool in business research *Business Review* **26** 106-121
[7] Valle P O D, Rebelo E, Reis E and Menezes J 2005 Combining behavioural theories to predict recucling involvement *Environment and Behavior* **37** 364-396
[8] Anna U, Boeve-de Pauw J and Saloranta S 2015 Participatory school experiences as facilitators for adolescents' ecological behavior *Journal of Environmental Psychology* **43** 55-65
[9] Siddique S F, Lupi, F and Joshi S V 2009 The effects of behavior and attitudes on drop-off recycling activities *Resources, Conservation and Recycling*
[10] Schultz P W, Oskamp S and Mainieri T 1995 Who recycles and when? A review of persona and situational factors *Journal of Environmental Psychology* **15** 105-121
[11] Varotto A and Spagnolli A 2017 Psychological strategies to promote household recycling. A systematic review with meta-analysis of validated field interventions *Journal of Environmental Psychology* **51** 168-188
[12] Zhang S, Zhang M, Yu X, Ren H 2016 What keeps Chinese from recycling: Accessibility of recycling facilities and the behavior *Resources, Conservation and Recycling* **1** 176-186.
[13] Duarte R, Escario J J and Sanagustín M V 2017 The influence of the family, the school, and the group on the environmental attitudes of European students *Environmental Education Research* **23** 23-42
[14] Kamaruddin S M and Omar D 2010 Waste Management And The Role Of Waste Administrators In Selangor, Malaysia WIT Transactions on Ecology and the environment._ https://www.witpress.com/elibrary/wit-transactions-on-ecology-and-the-nvironment/148/22968
[15] Matthies E, Selge S and Christian A 2012 Klöckner,The role of parental behaviour for the development of behaviour specific environmental norms – The example of recycling and re-use behaviour *Journal of Environmental Psychology* **32** 277-284
[16] Hair Jr J F et al. 1998 Multivariate Data Analysis with Readings. Englewood Cliffs, NJ: Prentice-Hall.

[17] Nunnally J C and Bernstein I H 1997 *Psychometric Theory* New York, McGraw-Hill.
[18] Wang F, Rob K and Nigel T 2009 Factor Analysis and Principal-Components Analysis. *International Encyclopedia of Human Geography.* Oxford: Elsevier
[19] Meinhold J L and Malkus A J 2005 Adolescent environmental Behaviours: Can knowledge, attitudes and self-efficacy make a difference *Environment and Behaviour* **37** 511-532.
[20] DiStefano 2009 *Practical assessment, research & evaluation* **14** Issue 20 (2009) ISSN: 1531-7714
[21] Johansson K 2012 Understanding Recycling Behavior: A Study Of Motivational Factors Behind Waste Recycling *WIT Transactions on Ecology and the Environment* 401-414
[22] Loughland T, Reid A K, Tim K and Walker P 2010 Factors Influencing Young People's Conceptions of *Environment Environmental Education Research* 3-19
[23] Singh N, Hui D, Singh R, Ahuja I P S, Feo L, Fernando 2017 Fraternal Recycling of plastic solid waste: A state of art review and future applications *Composites Part B:Engineering* **115** 409-422
[24] Zhang X 2013 Going green: Initiatives and technologies in Shanghai World Expo, *Renewable and Sustainable Energy Reviews* **25** 78-88s

Impact of Construction Project Planning on Contractor's Profit

O Akinradewo[1,2] and C Aigbavboa[1]
[1]SARChl in Sustainable Construction Management and Leadership in the Built Environment, University of Johannesburg, South Africa.
[2]Department of Quantity Surveying, Federal University of Technology, Akure, Nigeria

Abstract. Planning is very crucial and important for any activity to be executed to know beforehand the nature of the task, risk and the resources that will be required in order to attain the objectives of the organization. Planning in the construction industry involves establishing the most effective sequence of events necessary to accomplish a project as well as the allocation of resources in a way that could guarantee the effectiveness of the construction firm. The aim of this research was to evaluate the impact of construction project planning on Contractor's profit in the Nigerian construction industry. The research reviewed what planning in the construction industry entails as well as the effects of planning in the construction industry. Qualitative and Quantitative research method was adopted in carrying out the research work. Interview was carried out to gather necessary information from respondents after which well-structured questionnaire was used to validate the information gathered from the interview conducted. The professionals for this research work were mainly Quantity Surveyors, Builders, Engineers and Architects. From the survey conducted, the result revealed that there is a relationship between construction planning and the profit a contractor makes on a particular construction project. It was also observed in the result of the study that well-planned contract reduces waste on the construction site, ensure efficient use of labour workforce, aids completion to time, cost and quality which will eventually improve the contractor's profit.

1. Introduction

The importance and essence of management in any construction firm is to describe how effectively and efficiently men and other resources - plant, finance, and materials can be planned, organized, controlled, motivated and managed to achieve the organizational goals and objectives easily and readily. Project failures, bankruptcy of contractors and financial loss recorded in the Nigerian Construction industry are mostly as a result of inadequate use of formal contract planning and management principles and techniques by contracting organization. This becomes more evident frequent in renovation and refurbishment projects in which unexpected and unforeseen additional works, scope management issues and excessive requirements are inevitable problems [1]. The objective of any organization is very crucial to its existence which is the commonest and most important reason without which no one will exist. The objective(s) to be achieved in any organization should, therefore, make planning a critical process in the organisation. Planning is very crucial and important for any activity to be executed to know beforehand the nature of the task, risk and the resources that will be required in order to attain the objectives of the organization. Planning in the construction industry involves establishing the most effective sequence of events necessary to accomplish a project as well as the allocation of resources in a way that could guaranty effectiveness of the construction firm. There is need to put proper checks and balances in place in the construction

firm to make sure that projects are completed on time, within the schedule of cost and to the client's required, desired and satisfactory quality. To achieve this, planning techniques come in which can be applied in varying degrees depending on the state at which it has been carried out [2].
Project planning technique serves as a basis of determining the success of a project and also its reality. Most building or construction project experiences delay in various ways leading to either an extension in the time of project delivery due to the negligence of the management of the resource-oriented factors, i.e. the components such as material, labour and plant as they even have a negative effect on the cost of the project if not properly looked into [3]. The impression is that planning techniques are very complex, time-consuming and involve difficult procedures. It is against this background that this research seeks to evaluate the impact of project planning on actualization and the delivery of projects in the Nigerian Construction Industry considering contractors' profit. The major limitation to this research study is the unavailability of published articles on improvement of contractors profit on construction project.

2. Literature Review

The construction industry is essentially a service-oriented industry, whose responsibilities and obligations are to convert plans and specifications into a finished product. Al-Jibouri [4] from his own perspective described the construction industry to include residential construction, industrial and public buildings construction, civil engineering jobs and the building materials industry all functioning together essentially to provide services responsible to convert plans and specifications into a finished product. Harris and McCaffer [5] defined and described it from the functional perspective that construction industry is made up of the client, consultants, contractor, materials, manufacturers, distributor, financial institution and labour force. The Nigerian form of construction steadily remains in the form of block work construction for the majority of her building construction steel or concrete framed buildings cladded or infilled in whatever type of constructions [2]. He postulated further that though the construction industry in Nigeria and her educational system are fashioned alongside that of the United Kingdom, her performance has not been very impressive when compared to the international standard obtained in the United Kingdom. It was argued that most construction firms are very small, ill-equipped and have no sophistication adequately enough to rise up to face the challenges of rapid industrialization and urban development as compared to the international growth obtained in the industrialized world because of the ugly experiences of political and social instability in the country since the civil war ended in 1970.
Despite the effectiveness of the industry to the economy of the country, it has been faced with difficulty in winning the fight against delay that arises in construction projects undertaken within the country. Jagboro [6] has defined that, "a delay is a time during which some part of the construction project has been extended or not performed due to an unforeseen circumstance." So, it is essential to define who and what is responsible for this delay and at which stage is this delay mostly felt in construction projects. It is therefore suggested that the lack and improper utilization of various planning techniques could be the cause of this delay experienced by the construction industry. Construction Industry is faced with a lot of risks which results in financial loss. These risks are events that generally influence any or all of the project's objective which a contracting organisation globally has for a particular construction project and as such the Nigeria Construction Industry contractors are not exempted. Risks events could result in positives (opportunities) and/or negatives (threats) to contractor's set project objectives [7].
However, the construction industry had witnessed a tremendous rise in the number of firms since the end of the civil war, but statistics show that the rate of failure of construction firms in Nigeria when compared with other industries records one of the highest bankruptcy rates due to incompetency (43%) and lack of managerial experience (50.4%) [2]. The indication of this is that success in construction business operations will depend entirely upon the type of management organization operated by the firm which in turn will depend on the effective and efficient application of management and planning techniques and tools such as in this work study.

2.1. Planning in the Construction Industry

Every project is unique on her own with predetermined commencement and delivery dates including prepared Bill of Quantities i.e. strict budget. The physical construction of a Four-bedroom apartment does not qualify it as a project except the added constraints of quality, time and cost with the concerted efforts by the client/project manager re-allocating adequate and timely resources to keep within those constraints that truly distinguished it as a project [8]. The constraints imposed on the project make it necessary to plan its execution so as to keep it within those constraints. The checks and balances put in place in the construction industry to facilitate the completion of projects within contractor's schedule and to the client's desired time, quality and cost as planning techniques are applied in varying degrees depending on the stage at which it is carried out [9]. As planning is one of the most important five managerial functions that need to be exercised to achieve the set objectives of any company, other four managerial functions include organizing, co-coordinating, motivating and controlling. Planning is not only the most basic of all the managerial functions above, but it determines how the four other functions will be implemented. As a result, planning plays a pivotal role in which managerial decision-making process seats [10].

According to Hendrickson and Au [11], Contract planning is a "fundamental and challenging activity in the management and execution of construction projects which involves the choice of technology, the definition of work tasks, the estimation of required resources and the duration of individual tasks, and the identification of any interaction among the different work tasks". Planning as defined by [4] is "determining what the organization's position and the situation should be at sometimes in future and deciding how best to bring that situation to fulfilment". Project Planning is more than estimating time and cost, it is a question of preparing schedules and charts, it is primarily a management function that creates a framework of agreement on who supplies what, to whom, where and when [12]. Hensey [12] further argued that planning is not a secondary function performed by computers or specialized planners. It is a vital activity to be carried out at all levels of the project organization. Planning proceeds all other managerial functions and it is acclaimed as the most important of them all in any organization enterprise, or establishment. Wonder that the saying goes that: "No man plans to fail but man fails to plan". Construction planning involves proper management of the available construction planning tools as they have an effect on the time and delivery. Planning, therefore, permits, harmony within an establishment. Furthermore, it aids control which is necessary to ensure that resources are allocated and utilized prudently to make individual and groups perform to achieve maximally the stated goals and objectives. It states what are to be accomplished and controlled systems are created and developed to evaluate how plans are progressing [3].

2.2. Effect of Planning on Project Delivery

Working out a construction plan is a critical task when it comes to management of construction project as it determines the layout of how the construction project will be executed within the scheduled period of time to ensure cost and expenses is not exceeded while maintaining the specified quality. The inability of the contractor to complete the aforementioned works at any stage prior to practical delivery can be classified or categorized as an abandoned project. Sundaraj [13] analysed and described the abandoned project as the one that was once initiated, but on which construction work for one reason or the other stopped mostly as a result of improper planning or absence of planning. Resumption of work on such a project may almost be nil. In addition, [13] further categorized abandoned projects into two, viz: partial and complete abandonment. Suspended project or partial abandonment are those projects on which work was once in progress, but because of certain considerations, work had to stop for some time and after a while, it will pick up again. Projects of this nature in this category are classified as partially abandoned. They are not completely abandoned, but rather are undergoing some financial metamorphosis or adjustment programmes to meet up with the realities of time before they commence again. Secondly, the complete or totally abandoned projects in the 2nd category had progress in the site formerly but had to stop completely, not to resume again because of certain factor(s) and consideration(s). Both categories of project abandonment have a common economic trait of wastage both to the clients and the contractor executing the project. Such wastage comes in form of cost and time overrun. With project planning, utilisation of administrations, hierarchical strategies, inside operations and new items can be devised [14]

3. Research Methodology

The main purpose of this study is to evaluate the impact of construction project planning on Contractor's profit in the Nigerian construction industry. Therefore, Qualitative and Quantitative research method was adopted in carrying out the research work. This is because interview will be carried out to gather necessary information from respondents and then a well-structured questionnaire was used to validate the information gathered from the interview conducted. The professionals used for this research work were mainly Quantity Surveyors, Builders, Engineers and Architects who are working with contracting firms and have more knowledge about construction project planning in Nigeria and could provide adequate information needed for the research. A purposive sampling technique was used in carrying out this research, this is because this sampling technique satisfies the principle of statistical regularity [15]. Descriptive statistics (percentile and frequencies) were used to analyse the demographic information of the respondents. Interview conducted was analysed using Content analysis and the questionnaire for validating the information from the interviews was analysed using Mean Item Score (MIS). In order to ensure that the content analysis was done correctly, the four stages of content analysis according to [16] was followed which are: "*the decontextualisation, the recontextualisation, the categorisation,* and *the compilation*". A total of 70 professionals were interviewed and a total of another 158 professionals responded to the questionnaire survey.

4. Findings and Discussions

Analysis of the background information of the respondents shows that the years of establishment of the respondent's firm with 6-10years accounted for 33.8% while 11-15years, 1-5years, 16-20years and above 20years for 29.60%, 15.5%, 12.7% and 8.5% respectively. 33.8% has handled 11-15 projects, 25.4% has handled 6-10 projects followed by 16.90% who has handled above 20 projects then 15.5% for 1-5 projects, the least was 8.5% for 16-20 projects. Most of the respondents are Quantity Surveyors with 28.2%; Builders 21.1% while Architects, Engineers and Project Managers are 16.9% each. Highest academic qualification of the respondents indicates that B.Tech/B.Sc had the highest value with 29.6% followed by M.Sc/M.Tech holders 25.4%; HND holders 21.1% while PGD holders are 21.1%. Professional registration status of the respondents with NSE membership is 16.9%; NIOB membership is 25.4%; NIQS membership is 40.8% and NIA membership is 16.9%. Years of working experience, 23.9% have 1 – 5years of experience; 33.8% have 6 – 10years of experience; 25.4% have 11 – 15years of experience; 12.7% have 16 – 20years of experience and 4.2% have above 20years of experience in the construction industry. 32.4% of the respondents work with a small-scale firm while 33.8% each works with both medium and large-scale firm. From all these data, it is evident that the respondents possess adequate educational qualification, professional qualification and also with the wealth of experience, there is confidence in the response of the professionals as regards the evaluation of the impact of contract planning techniques on Contractor's profit in the Nigerian construction industry.

Table 1. Does Contract planning affect contractor's profit?

Variables	Frequency	Percent
Yes	48	68.6
No	22	31.4
Total	70	100

From table 1 above, it can be concluded that most of the respondents (68.6%) are of the opinion that contract planning has effects on contractor's profit while the remaining 31.4% claimed that contract planning techniques do not affect contractor's profit on a particular construction project. The 68.6% were asked further to suggest ways in which contract planning can improve contractor's profit and their opinions are reported as follows – with minor editing – not minding the orderliness as this is to preserve the genuineness of the response. From the analysis of their responses, all the 48 respondents were of the opinion that

i. Well-planned contract reduces waste on the construction site which will eventually improve the contractor's profit.
ii. It reduces excessive spending on labour, material, and plant because these things have been carefully planned for;
iii. It helps to avoid double handling which is another aspect of construction contractor lose money;
iv. It reduces expenditures and cost;
v. It reduces the time spent on construction projects;
vi. It brings efficiency in labours on site;
vii. It makes the project come out better thereby attracting another job which will bring more profit;
viii. It saves material consumption and so results in saving money;
ix. It aids completion to time, cost and quality which helps the contractor to get his retention on time.
x. It allows on-time delivery of materials to site because the purchasing would have been done based on the plan that has been scheduled.

Having suggested the following ways in which construction planning improve the profit made by contractors on construction projects, a further survey was carried out to validate the information gathered as shown in table 2.

Table 2. How construction planning improves contractor's profit

Variables	MIS	Rank
It reduces waste on construction site	4.43	1
It reduces expenditures and cost	4.21	2
It reduces time spent on construction projects	4.20	3
It saves material consumption and so results in saving money	4.17	4
It allows on-time delivery of materials to site	4.00	5
It brings efficiency in labours on site	3.89	6
It aids completion to time, cost and quality which helps the contractor to get his retention on time	3.84	7
It helps to avoid double handling	3.81	8
It reduces excessive spending on labour, material, and plant	3.80	9
It makes the project come out better thereby attracting another job which will bring more profit	3.72	10

From the table above, it is evident that professionals are of the opinion that construction planning affects the profits made by contractors on a construction project with all the identified factors ranking above the average Mean Item Score of 3.00. Reduction in waste ranks highest with an MIS of 4.43 followed by reduction in expenditures & costs and Reduction in time spent on construction project with MIS of 4.21 and 4.20 respectively. This shows that construction project delivery to time and cost coupled with lean construction practice on site will assist the contractor make the targeted profit on the construction project. The lowest ranked factor talked about construction planning making a project come out better thereby attracting another job which will bring more profit with an MIS of 3.72. Even though it was ranked lowest among the suggested factors, a Mean Item Score of 3.72 is still a strong statistical value to show that it thus contributes to the means of achieving profit on a construction project while should be taken seriously by contractors.

With proper planning put in place before the commencement of construction project, wastes are reduced on construction site which makes lean construction practicable such that expenditures and costs are reduced. This is in tandem with the research study of [8, 17] who found out that project planning benefits the organisation in reducing waste, achieving strategic goals, reduction in time spent on site and making efficiency profit. Having made planning for construction projects, it helps avoid double handling and enhance labourers' efficiency which supports the findings of [18] that poor project planning results in rework due to poor labour handling.

5. Conclusion and Recommendation

This study has been able to assess the positive impacts of construction project planning on contractor and it is evident that most contracting organisations are not aware of these impacts. Proper construction planning can ensure contractors get more profit on construction projects executed and this can be achieved in ways suggested by the respondents some of which are material wastage reduction, labourers' efficiency and on-time material acquisition which will speed up the quick completion of the construction work. It is therefore recommended that proper awareness is carried out to sensitized professionals in the construction industry about the different contract planning techniques available and their suitability for each construction project. Further research can be carried out on the impact of Construction project planning on the efficiency of labourers on site.

References

[1] Chong N B, Uden L and Naaranoja M 2007 Knowledge management system for construction projects in Finland. *Int. Jou. of Kno. Manag. Stu.* **1**(3-4) p 240-60.
[2] Ibrahim I I, Daniel S and Ahmad A 2014 Investigating Nigerian indigenous contractors project planning in construction procurement: An Explanatory Approach. *Int. Jou. of Civ. & Env. Eng. IJCEE-IJENS.* **14**(4) p 16-25.
[3] Andawei M M, Enenimiet A K A and Openebo O B 2007 Project planning in the construction industry: A study of some selected project in river state of Nigeria. *The Qty. Sur.* **55**(3) p 9-16
[4] Al-jibouri S 2002 Effects of resource management regimes on project schedule. *Int. Jou. of Pro. Manag.* **20**(4) p 271-7.
[5] Harris F and McCaffer R 2013 *Modern Construction Management.* John Wiley & Sons.
[6] Jagboro G O 1998 An Evaluation of the factors that cause delays on Construction Project in Nigeria. *Afric. Jou. of Dev. Stu.* **1**(1-2) p 105-14.
[7] Aibinu A A and Jagboro G O 2002 The effects of construction delays on project delivery in Nigerian construction industry. *Int. Jou. of Pro. Manag.* **20**(8) p 593-9.
[8] Kazie M R 1987 The development of indigenous contracting in Nigeria. *The Nig. Eng.***4**(2) p 41-6.
[9] Ogunsemi D R and Jagboro G O 2006 Time-cost model for building projects in Nigeria. *Constr. Manag. and Eco.* **24**(3) p 253-8.
[10] Fayol H 2016 *General and industrial management.* Ravenio Books.
[11] Hendrickson C and Au T 1989 *Project management for construction.* Pittsburgh: Prentice Hall.
[12] Hensey M 1993 Essential tools of total quality management. *Jou. of manag. in eng.* **9**(4) p 329-39.
[13] Sundaraj G 2007 The way forward: construction industry master plan 2006-2015. *Master Builders.* p 48-51.
[14] Naeem S, Khanzada B, Mubashir T and Sohail H, 2018 Impact of Project Planning on Project Success with Mediating Role of Risk Management and Moderating Role of Organizational Culture. *Int. Jou. of Bus. and Soc. Sci.*, **9**(1).
[15] Kothari C R 2004 *Research methodology: Methods and techniques.* New Age International.
[16] Bengtsson M 2016 How to plan and perform a qualitative study using content analysis. *NursingPlus Open.* **2** p 8-14.

[17] Atkinson R 1999 Project management: cost, time and quality, two best guesses and a phenomenon, its time to accept other success criteria. *Int. Jou. of Pro. Manag.* **17**(6) p 337-42.
[18] Laird, D.J., 2016. The impact of planning and other organizational factors on the success of small information technology projects (Unpublished Doctoral dissertation, University of Pittsburgh).

Introducing ARC (ask, record and confirm) as the new validating technique in real-time

Z Mohd[1]
[1]Department of Architecture, Faculty of Design and Architecture, Universiti Putra Malaysia, Serdang, Selangor, Malaysia

Email: m_zairul@upm.edu.my

Abstract. The ARC technique (ask, record and confirm) is introduced to perform validation and member verification procedures for focus-group data in real-time. During the focus-group sessions, response keywords were initially recorded on Post-it notes and were then pasted as a category onto an A1-sized sheet of paper. To illustrate the nature of the ARC technique, several steps were taken during the focus group and post-focus groups using this method to gauge feedback on the proposed business model (BM) of affordable housing. Next, the retrieved information was transferred into ATLAS.ti 7 for coding analysis and to generate meaning from the transcription. This paper is concluded with a discussion concerning the advantages and disadvantages of using the ARC technique for future studies.

1. Introduction

Typically, focus groups are linked with interactions or discussions that involve a group of respondents and mediators [1], [2]. Focus groups have become a popular means of exploring novel ideas, gathering opinions from a collaborative group, restructuring one's view towards issues discussed, and interpreting culture (Gibbs, 1996; Kidd & Parshall, 2000; [3]; [4]. Nevertheless, what is missing from this method is validation of focus-group session data in real-time. As such, this article addresses this issue by discussing the weaknesses of validation strategies in the conventional focus-group strategies and by providing empirical evidence on conducting validation of focus groups using the ARC technique, in order to gain feedback for the proposed business model (BM) meant for affordable housing in Malaysia.

In increasing the validity of focus groups, the ARC technique is recommended, especially to clarify member verification procedures. Employing ARC in member verification allows the confirmation of discussion topics in real-time. Upon reaching end of the session, issues at hand, new ideas, and discussion summary are tentatively re-visited based on those written on Post-it notes for further clarification and validation. Additionally, the findings are summarised, and debriefing session is performed via snowballing techniques to assess strong opinions and ideas, as well as to determine areas for enhancement for the following sessions. The ARC technique assists inexperienced facilitators clarify important points that they may have missed during the session at the confirmation stage.

The next section elaborates the methodology and the sampling employed for the general research. The triangulation method ensures that the data sources do not cause conflict and that the moderator's interpretation is validated in real-time during the confirmation stage. The techniques applied in one of the focus groups are elaborated in this section. Finally, this article is summarised by presenting the advantages and the drawbacks of using this technique.

2. Research Methodology

This study employed the triangulation approach by incorporating case studies, a design workshop for architects, and data gathered from the focus group sessions [5]. Besides, the triangulation approach was embedded in this study to enhance data reliability [6], [7]. The focus in this article lies on the validation of the focus group sessions using the ARC technique. In the ARC technique, data gathered in the preliminary round were presented to the respondents in order to evaluate their responses and to make changes if required. Most importantly, the snowballing technique fed the outcomes from a session into other sessions to gain more information. As such, more questions were posed, and outputs opined from prior session were further discussed in upcoming sessions.

Initially, the Malaysians students in Netherlands became the pilot respondents. The following focus group sessions were carried out in Malaysia by incorporating three groups of respondents from varied backgrounds. The emphasis of three focus group sessions with young starters had been on the selection of financial and design flexibility. Meanwhile, the topics discussed in focus groups 2 (entrepreneurs and private developers) and 3 (government agencies) were related to finance and supply chain. In particular, the focus group topics matched the expertise and the interests amongst the respondents [8]. The idea of having varies sampling for the study is to get in-depth responses on the new concept and to investigate deeper on the feedback from the different sampling.

Group 1: young starters (Malaysians aged between 25 and 34 years old)
(1 session of pilot study and 3 focus-group sessions held in Malaysia)
Group 2: private developers, entrepreneurs, and NGOs addressing housing
(1 session)
Group 3: government agencies addressing housing, i.e., PR1MA and KPKT (1 session)

2.1 Group 1 (young starters)

A pilot session and three focus group sessions were conducted amongst the young starters to gain their opinion regarding the proposed BM, particularly about the services and goods offered. A feedback analysis was performed to determine if the proposed BM i) addressed the aspect of flexibility in the present housing choice, ii) addressed the financial issues experienced by young starts, and iii) served as an alternative to the present housing model within the industry.

Additionally, controls had been applied to gather the demographic profiles of the respondents (age, occupation, income, and residence) to comprehend how the respondents addressed the questions posed, mainly because comprehension of the BM was dictated by their demographic factors. A set of criteria that was developed to select the focus group respondents is as follows: 1) young professionals aged between 23 and 30; 2) household income (joint income) below RM 6000; 3) presently renting and residing at the city or staying with parent(s); and 4) have considered housing ownership or have been rejected by bank or other financial institution.

In the second session, the young starters refined the questions for the following focus group sessions. This allowed the researcher to examine the responses obtained from the prior session to ascertain their comprehension and acceptance towards the proposed BM. In fact, the focus group sessions were repeated by embedding the snowballing technique so that the retrieved findings could enhance the proposed BM. Both oral and written feedbacks gathered from the initial session were used to add, delete, and improve on the topics for the following sessions.

2.2 Group 2 (private developers, entrepreneurs, and NGOs addressing housing)

The sampling technique applied for focus group 2 differed from that employed for focus group 1. The respondents for focus group 2 consisted of private developers and active entrepreneurs of the Malaysian housing domain and were selected via purposive sampling. This cluster of respondents was chosen as they represent the housing industry. The topic introduced to this group was finance and supply chain.

2.3 Group 3 (government agencies)

Focus group 3 involved government agencies that looked into matters related to housing in Malaysia. The topic discussed in this group was legal and financial matters. The improved BM, after incorporating feedbacks retrieved from focus groups 1 and 2, was revised with the respondents from focus group 3. These government agencies drew a bridge between the entrepreneurs and the young starters.

3. Methods Procedure

The housing affordability rates for young individuals appear to be a concern, hence the need for customisation to meet present and future requirements concerning quality-related issues that affect the current conventional housing stock. Focus group interview was performed to explore the effectiveness of the proposed BM. The focus groups involved three sessions held with young individuals, as well as sessions with private housing developers, housing entrepreneurs, NGO representatives, academicians, and representatives of government-associated business arrangements with housing. This study heavily relied on focus groups to develop understanding towards the BM structure, including service proposition offers by companies (design), affordability rates for the target group (financial), and means of improving the quality of housing through industrialised production (supply chain). The research began with three priori codes (design, financial, and supply chain). The sessions revealed several inductive codes, which sparked a debate in the discussion chapter in the thesis.

The focus group approach was performed to refine a draft BM into a final BM that can accommodate Malaysia's mass housing industry. Based on the findings, the final BM considered implementation gaps discovered in the Malaysian context, as well as interpretations of the choices and reasons behind the feedbacks provided. In exploring new ideas for a new BM, several methodological challenges were faced due to difficulty in expressing knowledge and the non-standard practice amongst the respondents. The protocol for the focus group implemented a 20-minute presentation regarding the BM, as illustrated in Figure 1. Each respondent was given marker pen and shared an A1 paper with description on BM based on three deductive codes, which were design, financial, and supply chain, that functioned as guideline. The proposed BM was printed for the respondents to write their opinions and feedbacks. Different coloured Post-it notes were given to the respondents so that each and every member contributed to the discussion [2].

Figure 1. Presentation of the conceptual business model

The discussion was initiated by adhering to the following steps: 1) asking the respondents regarding the associated issues using open-ended questions; 2) recording answers given by the respondents, as well as requesting them to jot down comments and reflections on the Post-it notes to be placed later at the right side of the A1 paper; 3) confirming twice the ideas, discussions, codes, and responses. A second moderator was included to determine the verbal and non-verbal aspects for further analysis and to transcribe the gathered data in the next stage. It was predicted that the respondents might offer extra information regarding the BM as they were asked to narrate their experiences associated to housing ownership and the impact of this new BM on the housing industry.

The proposed BM was introduced during the presentation stage to elaborate how housing manufacturers increasingly retain ownership over housing components when there was a chance to do so. These manufacturers served as service providers, instead of selling consumption products, assisting in progression of better 'return policies', and generating durable products to facilitate disassembly, re-use, re-distribution, refurbishment, and remanufacturing. All focus group sessions were video-

IOP Conf. Series: Earth and Environmental Science **385** (2019) 012010 doi:10.1088/1755-1315/385/1/012010

recorded and audio-recorded. The ARC method was applied to ascertain that accurate information was gathered from the focus group respondents via Post-it note intervention. A group debate was allowed to verify the feedback. Additionally, verification of members was performed in real-time to ensure data validity and to hinder misunderstanding.

Upon generating transcripts, several questions served as guideline: 1) Should all topics or only key session issues be transcribed? 2) Should sessions be transcribed verbatim or summarised? and 3) Should observations (silence/audible behaviour) be recorded [9]? Despite the usefulness of complete transcription, it may not solely meet the research goals [10]. As in this study, the Post-it notes facilitated the whole process of transcription by summarising the findings and addressing questions linked to the study. Besides, this step had considered some emerging codes obtained from the focus group approach. As displayed in Figure 2, photos were captured to serve as proof of the study conducted. The outcomes from the transcriptions were transferred into a survey table via Microsoft Excel.

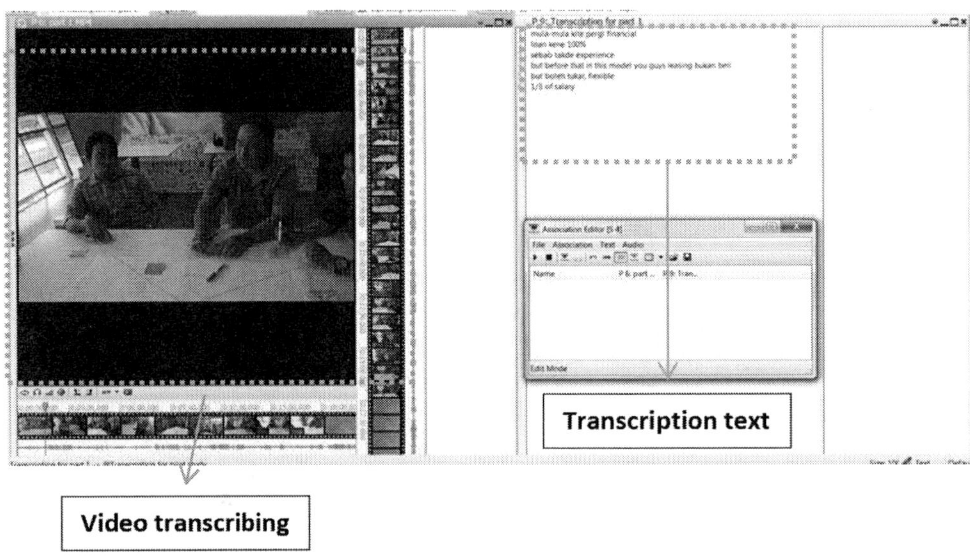

Figure 2. Transcription process using ATLAS.ti 7

Microsoft Excel was applied to develop the survey form, which embedded the background information to function as the study guideline. As portrayed in Figure 3, the questions asked during the focus group were a blend of single and multiple choices. The respondents were requested to complete the information sheet in the attempt of gathering their background information.

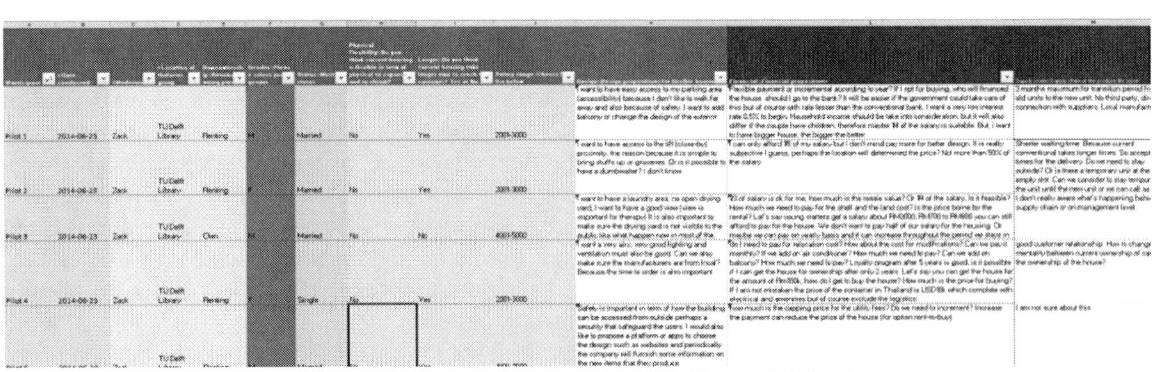

Figure 3. Survey form created in Microsoft Excel

The survey tables were interpreted differently by using ATLAS.ti. Prefixes were applied to ascertain that the information gathered from the survey fit the document in ATLAS.ti. For instance, prefixes ending with "!" reflected primary documents, as illustrated in Figure 4 that shows the primary document of each respondent.

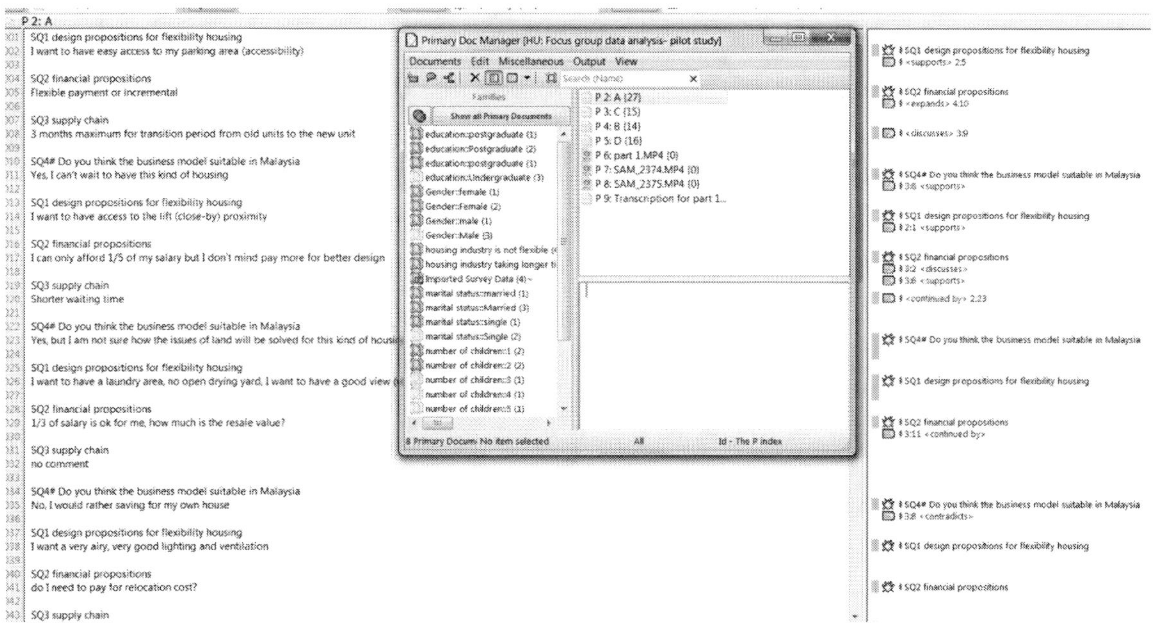

Figure 4. Primary documents in ATLAS.ti 7

Next, the information was used to generate codes, as presented in Figure 5. The information was keyed-in after two colons (::) were attached to the object comment area, which were part of the coding process or resulted in PD families. The network view shows that the correlations of the quotations generated active networks between the quotations and similar codes. This network view enhances understanding regarding the variety of needs linked with major issues, as well as how the respondents discussed and debated about a particular topic.

Table 1 presents a summary of the focus group sessions in condensed manner to ease readability, especially in comprehending how the posed questions were addressed. The focus group sessions were divided into three broad categories, which were: 1) proposition of design and value, 2) production and supply chain, as well as 3) financial and terms of payment.

The co-occurrence function found in ATLAS.ti was applied to analyse the coding responses frequency, which reflected deductive codes identified earlier. The following figures show that young individuals responded mostly to matters related to finance in conjunction to design. The co-occurrence table illustrated in Figure 6 displays the outcomes of density, as well as a discussion pertaining to issues. The sessions with the government agencies, however, excluded topics associated to design due to lack of relevance and were covered in prior sessions. Discussions on response interpretations, nonetheless, were discussed at length in the discussion chapter of the thesis, as they are excluded from this paper.

IOP Conf. Series: Earth and Environmental Science **385** (2019) 012010 doi:10.1088/1755-1315/385/1/012010

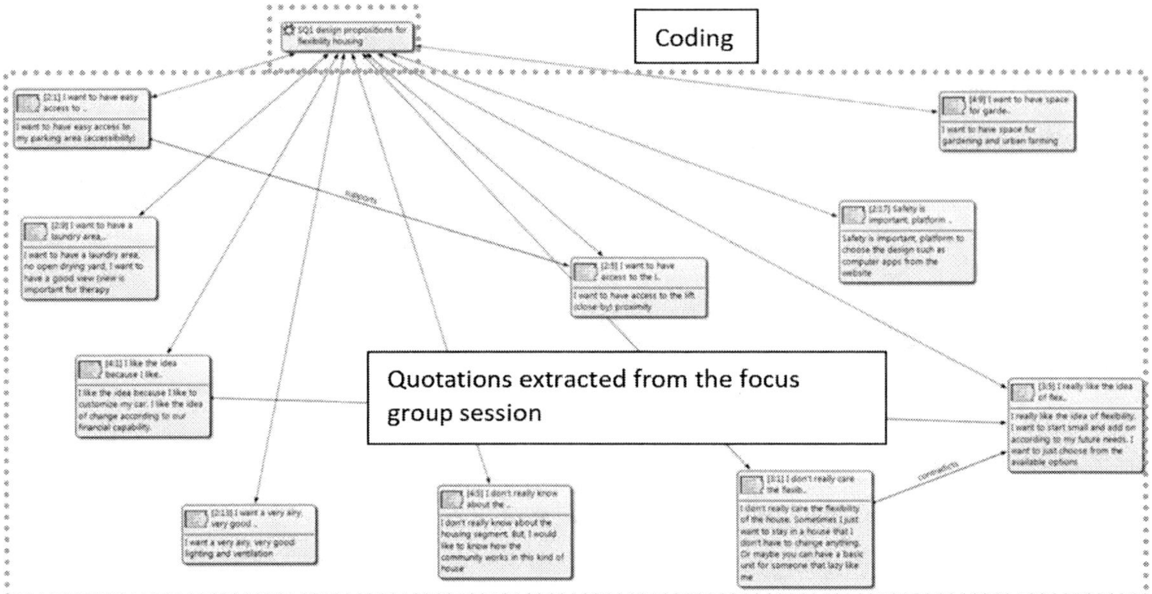

Figure 5. Codes drawn from the survey table

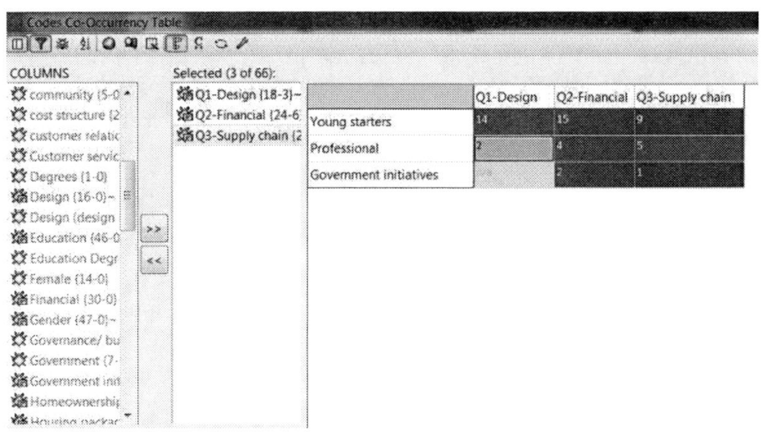

Figure 6. ATLAS.ti co-occurrence tool

IOP Conf. Series: Earth and Environmental Science **385** (2019) 012010 doi:10.1088/1755-1315/385/1/012010

Table 1. Transcription summary

PILOT GROUP : Young starters			!Coding categories
			A = Postgraduate students
Design / Flexibility / Products / Services			**B = Undergraduate students**
What type of house, services and product do you need?			**C = Academicians**
			D = Others
!Category code	^Participant ID	:Gender	::Responses
A	1.1		I want to have easy access to parking
A	1.2		I want to have access to the lift
A	1.3		I want my own laundry area
A	1.4		I want to have good view
A	1.4		I want my interior to have good lighting and ventilation
A	1.2		Can I have more accessibility than others (I pay more)?
A	1.4		My priority is safety
A	1.3		I want to choose design options from an easy platform (online shopping)
Supply chain / Time leads / Waiting time / Sustainability / Quality products / Services			
How long do you think is the appropriate time for waiting? Is it necessary to have sustainable housing?			
!Category code	^Participant ID	:Gender	::Responses
A	1.1		3 months maximum for transition period
A	1.2		I want to have direct connection with the suppliers
A	1.3		Easy to order and customize
Financial / terms of payment / credit? / Loan / Mortgage / Rent to buy options / Incentives			
How much can you afford to pay / How do you like to pay?			
!Category code	^Participant ID	:Gender	::Responses
A	1.4		I can afford only 1/5 of my salary for beginning
A	1.2		Provide more incentives for loyal customers
A	1.3		Refurbishments cost?
A	1.2		Do I need to pay for moving out?
A	1.4		Resale value? Is it increasing?
A	1.3		How much are the utility fees?
A	1.1		Flexi credit with more options to pay

4. Drawbacks and directions for future research

An important lesson learned from this study is that a researcher's degree of competence can affect focus-group moderation outcomes. As explained above, the ARC technique facilitated the validation processes and member verification procedures in real-time. However, moderators must develop their skills in facilitating groups and in helping passive members to actively engage in the discussions. Based on the author's experience, when a focus group is initially conducted, more lines of transcripts are spoken, and groups tend to become more spontaneous and conversational. One solution is to be completely prepared prior to the real sessions, e.g., by conducting several pilot studies in order to gain experience.

Another issue of concern is that the respondents might not be honest in their responses regarding the topic due to the holistic nature of the outputs and agreed upon by all members. The respondents, hence, become cautious with their opinions and thoughts, especially when their thoughts contradict to those uttered by the other group members. Therefore, the aim of conducting focus group is to gain insight and detailed information pertaining to particular ideas and products holistically and not by individual experience, as in the case of face-to-face interview.

Upon examining the methods associated to translating the discussion into keywords, some respondents appeared to be succinct in their expressions. When attempting to summarise the discussed ideas, the moderators seemed to interrupt the discussion and suggest keywords, hence creating biasness and influencing the final codes. Thus, some degree of distance should be maintained between the respondents and the moderators. Despite the importance of moderator roles during focus group discussions, the respondents have the final responses.

Another issue that emerged refers to homogeneity and heterogeneity of the background of the respondents. Upon sharing similar backgrounds, group dynamics seemed to function better. Nonetheless, those with varying backgrounds generated broader and fascinating views. The pilot study involved focus group amongst Malaysian PhD students at TU Delft. The homogeneity of the respondents contributed to a session that was relaxed and sociable as they appeared to support opinions given by one another, while some sessions went to extended time due to the causal behaviour of the respondents. Respondents from a variety of backgrounds (graduate students, young lecturers, and young professionals) were included in the following sessions and resulted in dominance by a particular group and astounding outcomes. A wider range of issues was discussed, and new codes emerged with the multi-background group. This offered richer data, hence worthwhile exploring.

In one of the sessions, those older and professional respondents requested the moderator to jot down the keywords from them. This required classic goals of rapport, especially when the respondents have established connections prior to the focus group discussion. In this case, it was easier to validate the outcomes and required less time because the moderator was in control of the whole situation.

Keywords provided from the previous session, which functioned as indicators on how the opinions were collected, had been discussed in the following sessions, making the sessions more precise and shorter. This is attributed to discussions that disregarded extended elaboration, as the initial focus group. Data saturation was also contributed by pooling of ideas [11]. The primary topic in further exploration of the ARC method is related to the success of sessions with not more than two moderators. While one moderator controls the session, the other looks into keywords that emerge during debates, apart from re-addressing them during member verification session. In fact, the second moderator may further seek clarification about the keywords to prevent misinterpretation. Such clarification may be performed by incorporating feedback via ATLAS.ti 7 video transcriptions. Despite the recording techniques employed, it is crucial for the second moderator to be alert, especially when the discussion turns lively and when several respondents speak at once.

5. Conclusion

When is the ARC technique most likely to be valuable? Based on the author's experience, this technique is useful when the researcher has insufficient time to send a final transcript to each member for verification. In this study, member verification was required to ascertain that the input was indeed usable within the context of Malaysia, i.e., material costs, standard operating cost for similar activities, appropriate sizing of the prefabrication area, as well as the technical inputs to determine resources and activities linked with the housing industry.

Since this technique appears to be new, new cases of usage would demand more exploration and further improvement. However, at this stage, this technique proved to save time during validation, hence increasing the internal validity of the research findings. As demonstrated by conventional processes of member verification after completing transcription, some respondents are unwilling to validate transcriptions, and logistical issues can arise when respondents are unavailable. Finally, this technique offers data collection triangulation, which involves voice and video recording, direct observations, and ARC techniques. Furthermore, sessions can be completed after all respondents agree on the summaries obtained from the sessions. The snowballing method in the ARC technique may differ from the Delphi method because the information is snowballed to a different group, and not with the same participants, as performed in the Delphi method. Additionally, the ARC technique will help inexperienced facilitators clarify important points that they may have missed during the session at the confirmation stage and avoid disagreement on the data collected after the session. Overall, it is believed that this new technique will benefit focus groups and qualitative researchers.

Acknowledgments

This paper is part of my thesis on flexZhouse: a new business model for affordable housing in Malaysia. I acknowledge the Ministry of Education, Malaysia (MOE) and Universiti Putra Malaysia; my promotor Prof. Hans Wamelink; co-promotor Prof. Vincent Gruis; and Dr. John Heintz for their continuous support of my PhD study and related research and for their patience, motivation, and immense knowledge.

References

[1] Gibbs A 1997 Focus groups *Pediatr. Nurs.*
[2] Kidd P S and Parshall M B 2000 Getting the focus and the group: Enhancing analytical rigor in focus group research *Qual. Health Res.*
[3] Kitzinger J 1995 Qualitative Research: Introducing focus groups *BMJ*
[4] Liu N F and Carless D 2006 Peer feedback: the learning element of peer assessment *Teach. High. Educ.* **11** 279–290
[5] Yin R 1981 The case study crisis: some answers *Adm. Sci. Q.* **26** 58–65
[6] Crabtree J L, Scott P J, and Kuo F 2016 Peer Observation and Evaluation Tool (POET): A Formative Peer Review Supporting Scholarly Teaching *Open J. Occup. Ther.* **4**
[7] Stange K and Miller W 1994 Multimethod research *J. Gen. Intern. Med.* **9** 278–282
[8] Zairul M 2017 flexZhouse New business model for affordable housing in Malaysia
[9] Jirwe M 2011 Analysing qualitative data *Nurse Res.* **18** 4–5
[10] Zhang Y and Wildemuth B 2009 Qualitative analysis of content
[11] Bench S D, Day T, and Griffiths P 2011 Involving users in the development of effective critical care discharge information: a focus group study *Am. J. Crit. Care* **20** 443–52

Comparison of Development Control Process for Serviced Apartment (Commercial) and Residential Apartments. Case Study: Shah Alam City Council

M A Marzukhi[1], O L H Leh[1], Y A Abdullah[1], N S Khalid[1], H Zainol[2] and R A Abdul Apet[1]
[1]Faculty of Architecture, Planning and Surveying, Universiti Teknologi MARA, Selangor Branch, Puncak Alam, Malaysia
[2]Faculty of Architecture, Planning and Surveying, Universiti Teknologi MARA, Sri Iskandar, Perak, Malaysia

Email: marlyana@salam.uitm.edu.my

Abstract. The development of serviced apartments has increased throughout the years in Malaysia. However, the lack of knowledge on the development control process of serviced apartments and residential apartments has causes developers took advantage by placing unreasonable house prices in respect to profits generation. In order to overcome this issue, the difference of the development control process between serviced apartments and residential apartment must be identified. Hence, the aim of this study is to compare the serviced apartment and residential apartment in Shah Alam from the development control aspect, and the impact of the development post certification. Analysis is done based on the data obtained from relevant Acts, Manuals, Checklists and interviews with personnel that is involved in the process. The results of the study indicates differences between these two development in terms of land matters, legal matters, development requirements, development charges, development application and approval process, assessment rates and utility rates.

1. Introduction

Serviced apartments are often regarded as 'trendy' as it suites current modern contemporary lifestyle. Equipped with facilities such as swimming pool, gymnasium and retail outlets, serviced apartments become more of popular choice among young professionals. Sometimes, serviced apartment development is incorporated with office tower or shopping complex as a mixed development which further adds to its value. It also provides extra security and cleanliness as it is maintained by a designated management body. Now, in the era where the internet is a necessity, the service apartment will incorporate the most sophisticated communication system for the users.

Serviced apartments popularity rises around the 1970s in South East Asia and around late 1980s in Malaysia. But at that point of time serviced apartments are only used for temporary stay mainly as tourism attractions. The concept is similar to hotel, where investors or owners will buy the property and fully furnish the rooms to rent it to expatriates or tourists. Many prefer serviced apartments because it is considered cheaper than staying in a hotel. But since the property crisis in 2008 and the rise of housing selling price continues, majority of the low to medium income group opted to purchase a smaller and cheaper property, in this case one of the choice being serviced apartments. Therefore, serviced apartments were purchased as residential units - for them to live in. Due to this factor, the popularity of the service apartments increases the demand for this property. Hence, the developers begin to develop more serviced apartments over time. Simultaneously, the local authority plays a huge role in the development control process, most importantly to set requirements for submission for Planning Permission and Building Plan (BP) approval for service apartments.

This research will cover the approval process, procedures, and the statutory requirements of the Serviced apartments and the Residential Apartment in the study area, which is Majlis Bandaraya Shah Alam (MBSA). Due to its 'commercial' title, developers would instruct planners or architects to propose a mixed development, where the serviced apartment has to be built on a podium with shopping complex below, or merge the service apartment with office blocks and retails. Residential Apartment, on the other hand is exclusively built on 'Residential' titled land for residential use only. Hence, it is important to study and differentiate between Serviced apartments and Residential Apartment in terms of design guidelines, planning and building standard from the development control aspect.

2. Literature Review

2.1 Development Control in Malaysia

The understanding of the key aspects of development control which includes the development requirements, the submission and approval process and procedures, land matters, legal matters, and development charges are important for the development of the Serviced Apartment and Residential Apartment. According to Section 2 (1) of the Town and Country Planning Act 1976 (Act 172), development control is define as "the carrying out of any building, engineering, mining, industrial, or other similar operation in, on, over, or under land, the making of any material change in the use of any land or building or any part thereof, or the subdivision or amalgamation of lands; and "develop" shall be construed accordingly" [1]. Meanwhile, control means the process of exercising power over ones' area of jurisdiction. The laws, procedures and guidelines pertaining to the land and property development process in Malaysia are quite extensive. Some of the main written land policies and laws to guide the land development activities as derived from the Malaysian Constitution are the National Land Code - Act 56 (1965), an Act to amend and consolidate, ensuring uniformity of law and policy relating to land and land tenure, registration of titles and collection of revenue and matters connected therewith, Land Acquisition Act which describes compulsory purchase and compensation, and Town and Country Planning Act – Act 172 which contains planning and related development regulations.

2.2 Definitions of Residential apartments and Service Apartments

Nowadays, houses not only perceived as physical shelter but also become symbol of urbanization, social acceptance, and achievement [2]. The types of houses are ranging from terraced houses to semi-detached houses, bungalows, and apartments to serviced apartments. According to Marcus, apartment is defined as;

> "[A] uniquely urban form of housing that combine a relatively private spaces of an individual apartment units with a common spaces of shared entrances, staircases, and part walls, the apartment house embodied the continuity between domestic and urban, private and public spaces" [3].

Hirsh claimed that;

> "Over the past 30 years, serviced apartment complexes have become one of the defining architectural typologies of Asian urbanization. Relatively unknown before the 1980s, these complexes are designed to accommodate foreign business people, knowledge workers and students for a fixed period, typically three to six months. But since the turn of the 21st century, the fastest-growing segment of the serviced apartment industry has been facilities serving what might be called an expanding class of migrant workers: mid-level employees whose jobs require frequent relocation, creating a new professional class of nomads" [4].

This statement explains the evolvement of serviced apartments in other countries across Asia, where serviced apartments is meant for stay for a fixed period of time before the 1980s but became more a temporary stay after the millenium. In Malaysia, the serviced apartment is one of the terms for 'apartment' but located in the 'commercial' land use zoning. In the *Manual Garis Panduan dan Piawaian Perancangan Negeri Selangor*, Serviced Apartment is defined as 'a concept of commercial development that embodies features of residential furnished with designated facilities packages for each unit' [5]. To date, there is only definition for Serviced apartments in the official documents. But in Malaysia, there are more terms with similar definition, but only used as developers' marketing strategies which is SOHO - Small Office Home Office, SOFO - Small Office Flexible Office or Flexi-Office, and SOVO - Small Office Versatile Office. The demand of this types of houses is influenced by purchasers preferences after high level of decision making process [6]. In fact, home buyers are more concerned on the quality characteristics of a house such as modern and trendy facilities [7,8].

3. Research Methodology

This study used a case study method and in-depth interview to investigate the approval process, procedures, and the statutory requirements of the serviced apartments and the residential apartments in Shah Alam City Council.

3.1 Case Study

In order to conduct a thorough analysis, one residential apartments and two serviced apartments located in Shah Alam are chosen as case study, as follows:

1. Menara U1 - located in Section 13, Shah Alam on commercial land use zone and is classified as Service Apartment.
2. Plaza Azalea - located in Section 14, Shah Alam. It is also located on commercial land use zone. It is a flexi-office or SOFO.
3. Pangsapuri Danaumas - Residential Apartment located in Section 7 Shah Alam. The apartment has obtained its Strata Title.

All apartments in this study is located in Blok Perancangan Kecil (BPK) 1 in the MBSA Local Plan. Figure 1 shows the location of these apartments.

3.2 Semi Structured Interview

In order to obtain the related information to this study, the semi-structured interview were conducted with 7 representative from MBSA from respective departments (Table 1).

Table 1. List of Interviewees from MBSA

No	Department	Designation	Experience
1	OSC	Officer	7 Years
2	OSC	Technical Assistant	9 Years
3	Planning	Officer	14 Years
4	Planning	Technical Assistant	8 Years
5	Building	Architect	7 Years
6	Building	Technical Assistant	16 Years
7	Valuation	Evaluation Officer	11 Years

Interviews are also conducted with representatives from the management bodies of the serviced apartments to obtain information on the management of the serviced apartments (Table 2).

Table 2. List of Interviewees from the Apartment Management

No	Management	Apartment
1.	Joint Management Body (JMB)	Menara U1
2.	Joint Management Body (JMB)	Plaza Azalea
3.	Management Corporation (MC)	PangsapuriDanaumas

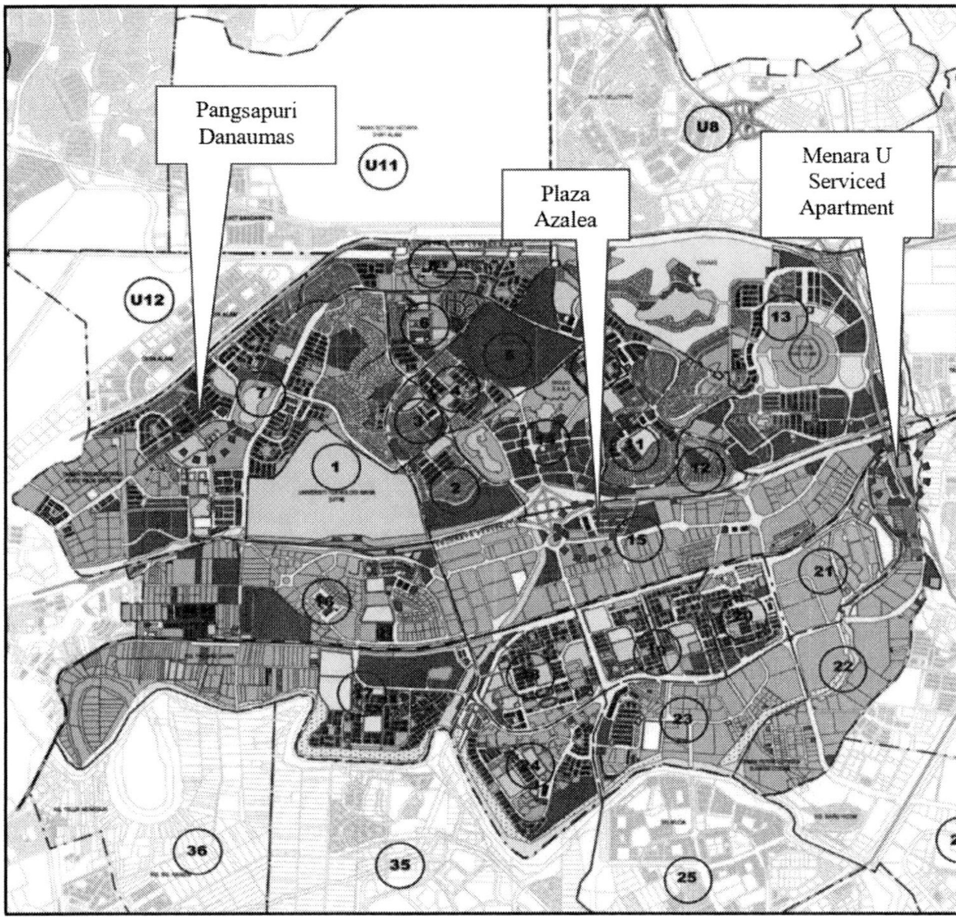

Figure 1. Location of Menara U1, Plaza Azalea, and Pangsapuri Danaumas
Source: Shah Alam Local Plan, 2012 [5]

In addition, interviews are also conducted with experienced consultants and project managers to obtain additional information on the difficulties to comply with the requirements and the developers strategies for the development of Serviced Apartments (Table 3)

Table 3. List of Interviewees from consultant and developer firms

No	Designation	Firm
1.	Land Surveyor	PerundingUkur Iman
2.	Project Manager	I&P Group SdnBhd
3.	Project Manager	PermodalanNegeri Selangor Berhad
4.	Architect	PertubuhanArkitek Malaysia and Rehda Committee Member

3.3 Content Analysis

Content analysis was used as the primary analysis technique of the collected data. This technique extracted and categorized information from documents that provides new insights, increases a researcher's understanding of a particular phenomena, or informs practical actions. In this study, analysis was conducted by extracting the secondary data obtained from data collection, i.e. mainly from the laws, acts, and submission checklists (Table 4).

Table 4. List of contents to be analyzed

No	Matter	Content
1.	Land Matters	NLC 1965, Land Title
2.	Legal Matters	HDA 1966, STA 1986, SMA 2013
3.	Development Requirements	Manual Garis Panduan Dan PiawaianNegeri Selangor Edisi 3 2016, Manual OSC MBSA, Senarai Semak Cadangan Baru Jabatan Bangunan MBSA
4.	Utility & Assesment Charges	MBSA Valuation Department Website, Assesment Receipts, TNB Website, SPAN Website, IWK Website

After obtaining the data relevant to the development of Serviced Apartment and Residential Apartment, comparative study is done to compare the two types of apartment.

4. Discussion and Findings

As mentioned earlier, the four key elements of development control aspect that is analyzed are the legal matters, the development requirements, the submission and approval process, the legal matters, and the charges involved in the development of serviced apartments and residential apartments. In the legal aspect, the category of the two types of apartments are the same, but the expressed condition or implied condition stated the different building use, whether to be used as commercial or serviced apartment or for residential use only. The main difference between this two types of apartments are the development requirements. In terms of design guideline and planning standard, the main difference can be distinguished by the land use zone, where the serviced apartments i.e. the serviced apartment, SOHO, SOVO, and SOFO are in the commercial land use zone and the residential apartments are in the residential land use zone.

Besides that, the difference between the apartments is the development form. Serviced apartments can be developed as a part of mixed development, as a free standing building, or within a shopping complex, whereas residential apartment should contain residential units only except the required facilities for the residents. The retails or facilities for residential apartment are for the residents only and not allowed for public use. By looking at the development form, the developers can benefit more from building the serviced apartments because they are able to develop more properties within one site. The developers can profit from the sales and operations of shopping complex, retails, or other facilities that they can provide at the podium level at the same time profit from the sales of serviced apartment units.

The height control for commercial-type apartments is determined by the plot ratio calculation, whereas the height for residential apartments are determined by the density. The allowable plinth area are the same for both types of apartments which is 60%. Accordingly, the plot ratio requirements will result in more number of units compared to the density calculation. Therefore, the developers will gain more profit by developing the serviced apartments or SOHO compared to residential apartments. As for the classification, in the *Manual Garis Panduan dan Piawaian Perancangan Negeri Selangor* [5], serviced apartments and SOHO is located on different type of development but still belongs to commercial development zone. Therefore, the submitting person will submit the planning permission and building plan application following the requirements for 'office' instead of for serviced apartments or SOHO which is more specific.

The residential apartment, on the other hand, is divided into four types, which is the Low Cost, Low Medium Cost, Medium Cost, and High End Apartment. These four types of residential apartments' requirements are same in terms of development form, service road widths, and building setbacks. The only difference is the minimum floor area and the parking requirements. The minimum floor area for low cost units is 700 ft^2, 750 ft^2 for low-medium and medium cost apartments, and 850 ft^2 for high end apartments. Only one car parking lot required for low cost apartment whereas minimum 2 car parking lots are required for low-medium, medium, and high end apartments.

Requirements for serviced road width for residential apartments is 66', which are more than serviced apartments which only requires 55' width. Minimum ingress and egress road width are 50' for residential apartments but not specified for serviced apartment and SOHO. Requirements for setbacks are different between commercial apartment and residential apartment. Commercial apartments are usually consists of 1 building per site or connected by a podium at ground floor level whereas residential apartments are consists of separate buildings or blocks within one site. The residential apartment buildings may be at the same height or differs to one another. Therefore, there are additional requirements for setbacks in between buildings within the same site. This will result in lesser permissible plinth for development of residential apartment due to more setbacks and minimum road width requirements.

Some of the supporting facilities i.e. praying area, mini library, restaurants or food court, management office, and refuse chamber are required for serviced apartments. Exclusive requirements for SOHO and serviced apartment also differ to the exclusive requirements for SOFO and SOVO. This is due to the commercial-type apartments has dual usage because it is usually within a mixed development or above shopping complexes, therefore there must be a differentiation of usage between the public or commercial customers with the residents of the apartment above. Exclusive facilities are intended to provide specifically for the residents' use. Provisions for open space and recreation and perimeter planting are the same between Commercial Apartment and Residential Apartment.

Building standards which are the set of requirements for Building Planning Approval follows closely the requirements stated in the design guideline and planning standard. But, in the Shah Alam City Council Building Department Checklist for Building Submission, the Principal Submitting Person/Submitting Person (PSP/PSP) are only required to follow the requirements for 'apartment' in general. There are no specific requirements for serviced apartments, SOHO or residential Apartment. This situation may cause inconsistency in requirements set by the Department of Town and Country Planning and the Building Department. Some PSP/SP may refer to the JPPD guideline and some may refer to the Shah Alam City Council Building Department's checklist. Usually the officers in the Building Department will only refer to the Building Requirements Checklist alone. This checklist is more lenient compared to the guidelines by Department of Town and Country Planning, therefore developers may be able to avoid certain requirements. Procedures of submission to obtain the planning permission and building plan approval are the same for both serviced apartments and residential apartment. The submission procedures is not only the same for both type of apartments, but for any type of development in the Shah Alam City Council designated area. This is due to Shah Alam City Council as local authority only follows to the procedures as per stated in the Street, Drainage and Building Act (Act 133), TCPA (Act 172) and Uniform Building By-Law Selangor 1986. Moreover, as the implementation of new standardized submission procedure OSC 3.0 introduced, the same submission procedure is implemented all over Malaysia.

As for the legal matters, residential apartments and serviced apartments belongs to the stratified development because both types are 'shared' properties. The development of both properties are vastly affected by the implementation of Strata Management Act and amendment of Housing Development (Control and Licensing) Act in July 2015 where the developers must issue Strata Title together with the handover of Vacant Possession together with the issuance of Certificate of Completion and Compliance (CCC) and signing of Sale and Purchase. The residential apartment, serviced apartment and SOHO are bound to the Housing Development (Control and Licensing) Act therefore the developers for these three types of developments have to obey the procedures for application of the Advertising Permit and Developers License. Assessment rates and utility bills are higher for serviced apartments because they follow the commercial rates. Owners of commercial properties will use the property to allocate and operate business premises or firms. The owners will obtain profits from the businesses therefore the rates are considered reasonable. But in this case of commercial apartment, the owners use the apartments as residents, therefore they are burdened by the high utility bills and assessment rates.

5. Conclusions and Recommendations

Comparative study has been done to compare the development control process for commercial apartments and residential apartments. Shah Alam City Council being the local authority of Shah Alam is the administrator the planning permission and building plan application. In this study, it is recognized that the planning permission and building plan application process and building standard are similar for both types of development, it cannot be avoided due to regulations stipulated in Act 133 and Uniform Building By-Law Selangor 1986 under the OSC 3.0.

From the analysis and comparative study, it is discovered that the development of serviced apartment brings more benefit to the developers, especially through the development of flexi-offices such as SOFO and SOVO. This is proven to be beneficial to the developers because they are able to avoid stringent development requirements for housing development, i.e. the minimum floor area, usage of density calculation. The developers are able to construct more units and at the same time benefit more from the construction of mixed development. Simultaneously, due to the required location for commercial apartments which happens to be near the central business district area or public transportation enable developers to sell the properties for higher price. In contrast to the advantages that the developers benefit, from the analysis and comparative study it is discovered that the commercial apartment burdened the residents through higher assessment rates and utility charges. Accordingly, it is timely for the government to review and revise the Acts, rules, and requirements. This can be seen through the review of the third edition of *Manual Garis Panduan dan Piawaian Negeri Selangor* whereby the provisions for SOHO is included in the guidelines. The government's effort towards improvement of the development control can also be seen from the revision of Strata Title Act and Strata Management Act.

Acknowledgements

The authors would like to thank Universiti Teknologi MARA (UiTM) for the support and partly funding the study and publication through the LESTARI research grant (600-IRMI/MYRA 5/3/LESTARI (K) (237/2017)). The authors are also thankful to all the departments, organisations, and individual who had contributed to this study.

References

[1] Town and Country Planning Act 1976 2016 Laws of Malaysia. Kuala Lumpur: *Percetakan Nasional Malaysia Berhad.*

[2] Buang S 2010 Malaysian Torrens System (2nd ed.). Kuala Lumpur: *Dewan Bahasa dan Pustaka.*

[3] Marcus S 1999 Apartment Stories: City and Home in Nineteenth Century, Los Angeles: *University of California Press.*

[4] Hirsh M 2015 The shifting demographics of the serviced apartment industry in South East Asia. *South East Asia Research* **23** 187–204.

[5] Jabatan Perancangan Bandar dan Desa Semenanjung Malaysia 2016 Manual Garis Panduan dan Piawaian Perancangan Negeri Selangor. Kuala Lumpur: *Jabatan Perancang Bandar dan Desa Semenanjung Malaysia.*

[6] Shah Alam City Council 2012 Shah Alam City Council Local Plan 2020.

[7] Tan T H 2011 Neighborhood preferences of house buyers: the case of Klang Valley, Malaysia, *International Journal of Housing Markets and Analysis* **4** 58 – 69.

[8] Tan T H 2012 Meeting first-time buyers' housing needs and preferences in greater Kuala Lumpur. *Journal of cities: The international journal of urban policy and planning* **29** 389-96.

IOP Conf. Series: Earth and Environmental Science **385** (2019) 012012 doi:10.1088/1755-1315/385/1/012012

Motives for solar photovoltaic (PV) adoption in urban Nigeria

A I Ugulu [1, 2] and C Aigbayboa [1]

[1] Department of Construction Management and Quantity Surveying, Faculty of Engineering and the Built Environment, University of Johannesburg, South Africa

E-mail: anthonyugulu@yahoo.com; aiu30@hw.ac.uk

Abstract. Microgeneration technologies like residential solar photovoltaic (PV) systems have been shown to have immense potential for energy security and climate change mitigation. At over 50% private power systems ownership, the majority of urban Nigerian households rely on conventional self-generation utilising expensive petrol and diesel-powered generators. Despite the high upfront costs of a sizeable PV module, some households are choosing such sustainable alternative power. This paper investigates the motives for PV adoption in urban Nigeria following innovation diffusion and adoption theories. Using interviews, PV adopter data were collected on key drivers for uptake and results analysed. Field survey analysis established the key motives for uptake as power outages, energy cost-savings, including generator use fuel fraud, awareness and access to finance while key barriers include high capital costs and lack of finance. The results point to the need for regulatory and political intervention. Effective PV awareness creation campaigns and promotional strategies would also be necessary in the changing face of electricity supply in Nigeria.

1. Introduction

One of the most urgent societal challenges today is how to increase electricity and energy access while reducing atmospheric emissions. Over-reliance on unsustainable combustible fuels in energy supply especially for electricity generation is a major contributor to climate change and global warming [1]. A UN report estimates that globally 66% of aggregate electricity production comes from fossil fuels contributing to approximately 60% of GHG emissions which is responsible for climate change [2]. The bulk of this electricity and total energy demand takes place in buildings particularly during its operational life. Since buildings account for up to 40% of energy use as well as 30% of the share of CO_2 worldwide [2], most green energy policies and initiatives focus on improving energy use in buildings especially dwellings.

In Nigeria, the bulk of total energy demand comes from the domestic sector representing about 57% in 2014. A large proportion of this is used to meet household demand for electricity. Clearly, the sector is an important one in relation to electricity supply, demand and for the promotion of micro-generation technologies (MGTs). MGTs refer to the off-grid generation of electricity of <50kWp and/or heat of <45kWp using low-carbon sources such as solar hot water (SHW), solar photovoltaic systems and micro-wind turbines [3].Such renewable off-grid applications have huge prospects particularly in developing countries like Nigeria where the operational grid network capacity at 6GW [4], is underdeveloped and unable to meet consumer demand. As the Nigerian government desperately seeks to improve power supply, PV present opportunities that could be taken advantage of. Although Nigeria has a favourable solar radiation resource that ranges between 3.5-7kWh/day [5], there are presently no government monetary or fiscal incentives for PV promotion. But, there has been clear interest in this technology for over 3 decades [5]. However, there is a renewed interest in diversifying the energy portfolio in Nigeria and the government have now set target of 1GW of power to be generated from solar by the year 2020 [4].

This paper investigates the motives for PV uptake by households and private consumers in urban Nigeria. It does this to reveal the key determinants and identify suitable avenues to promote increased adoption and thereby improve supply. This paper adopts the following structure. The next section discusses innovation diffusion theory. This is followed by the data and methods used. Next is a discussion of the results. Finally, the conclusions, implications and recommendations are presented.

2. Innovation diffusion theory

Different theories have been used to explain the determinants of PV adoption and MGTs in general. The most common theories that have found application in literature include the innovation diffusion theories, consumer attitudes or response theories and socio-technical systems [6]. The diffusion of innovations model is a social psychological perspective developed by Rogers [7]. Under this model, diffusion is defined as the process through which an innovation is communicated via certain channels over time through a social system [8]. This innovation may be an idea, product, process (or a combination of these) but does not necessarily have to be new, as long as it is new to the adopting community [9]. The innovation diffusion concept can be examined from various perspectives including product attributes, time and speed of uptake and characteristics of adopting population. Diffusion is the relative speed with which an individual adopts a product or service compared to other members of the same social system or society. Because of this time dimension, the speed of adoption is taken to be a function of certain attributes that the innovation possesses. These include: relative advantage, perceived risk, complexity, compatibility, trialability and observability [6]. Because every individual does not adopt the innovation at the same time, Rogers' [7] divided the stages of adoption into five categorises according to consumer personalities, values and behaviour [10]. Within this five-stage category and in ranking order are the innovators, early adopters, early majority, late majority and laggards. The first two stages i.e. innovators and early adopters are considered crucial as they indicate consumers who show the most initiative in the adoption process. See Figure 1. However, they are often smaller in number. Their innovativeness in the adoption process is attributed to their level of awareness and higher education [10]. As the cost of purchasing a sizeable PV is still relatively high, this paper investigates the motives for PV uptake by consumers, mainly households in urban Nigeria.

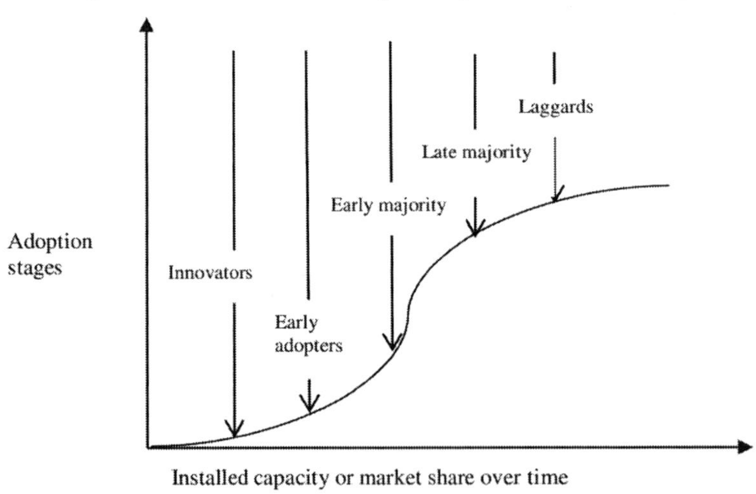

Figure 1. Adapted Rogers' Innovation Diffusion S-Curve

IOP Conf. Series: Earth and Environmental Science **385** (2019) 012012 doi:10.1088/1755-1315/385/1/012012

3. Data and methods

Unlike the traditional method largely deployed in engineering and the built environment studies, this paper adopts an approach that takes into cognisance the impact of human factor in energy production and consumption in buildings. This multidisciplinary approach treats electric power technology as socio-cultural and therefore includes the effects building occupant behaviours and decisions can have on energy demand and supply. This paper therefore applies a qualitative method of enquiry encompassing all the socio-cultural and technical aspects of a MGT.

Based largely on grounded theory, a qualitative research method was used in the form of interviews. As a deductive method, grounded theory is useful where there is lack of theory and concepts and when data collection, analysis and theory development and testing are intermingled [11]. Dearth of empirical research and literatures on urban PV adoption (both residential and commercial) in Africa as a whole supported grounded theory [12].

3.1. Purposive Sampling

Sample size decisions should be based on the nature of the topic under investigation, scope of the data and research methods [13]. Purposive sampling a generally acceptable technique in qualitative studies [11] was used. Here, the researcher's judgment is relied upon based on study interest and rationale. Three of the PV adopters were identified from the questionnaire survey from a different but related study. Others were selected via referral, which is a situation whereby an interviewed adopter refers the researcher to other adopters. In most cases, adopters knew someone else using solar PV and may have had their panels installed by the same technician. Because PV is one of the most popular MGT, its appeal and acceptance has meant that it is a topic that many people are somewhat familiar with due to its high visibility. 14 adopters were interviewed. Experiential data requires fewer participants [11]. Bahaj and James did a similar study in Southampton UK using 9 households [14].

3.2. Interviews, questions and data

Open-ended semi-structured interviews were carried out and the interviews recorded using an MP3 recorder. The interview questions format took the order of the opening questions, the intermediate and the final questions. The opening questions enquired of the adopting households' PV capacity and duration of use. It was very important to establish duration of PV use, as a very recent installation would not have been of great benefit to the study. The intermediate questions largely examined the role of feedback mechanisms such as PV monitor. The final questions covered maintenance issues, challenges and likelihood of recommending PV for other households.

Interviews can be useful when the investigation as to the rationale for a decision is sought. Research has shown that interviews are good strategies in consumer attitudes to green power [15] and adoption promotion studies [16]. For added clarity the recorded interviews were transcribed using a data analysis software. The human factor questions asked during the data analysis included:

- What *led* to the adopters' decision?
- What were the PV households' *experiences* from using the device?
- Based on their experience, would they *recommend* PV to other households?

Respondents were primarily drawn from Lagos State, Nigeria. Due to the small number of PV usage in Lagos, respondents were taken from other states i.e. cities in Nigeria such as Abuja (Federal Capital City) Delta, and Edo.

4. Results and analysis

The motives for PV adoption in Nigeria were broadly found to be socio-economic and environmental. The most significant motive was households' need for reliable electricity supply. In other words, frequent power outages were driving the demand. Others are energy cost savings and familiarity with PV generated power. They are explained accordingly.

4.1. Reliable supply

Power outages and the need for a more reliable supply source represented 80% of motives for PV adoption as Figure 2 indicates. All the interviewed found PV to be very reliable and better than grid electricity and petrol and diesel-generated electricity. The innovative adopters described PV as rugged, regular, uninterruptible, efficient and the most rational source of power supply.

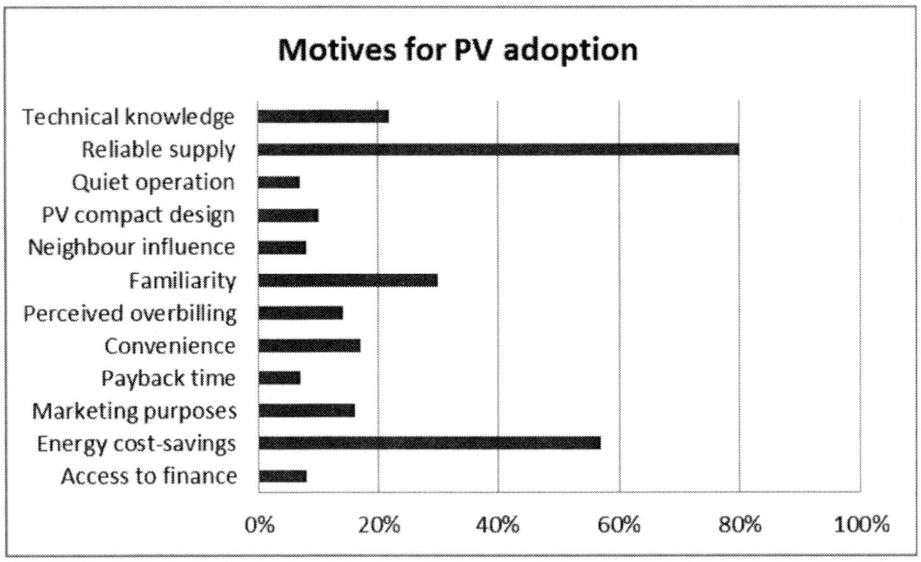

Figure 2. Motives for PV adoption in Nigeria

What was even more remarkable was that adopters whose systems eventually failed after a short time held this same optimistic view. Two of the interviewed adopters had systems that lasted for 3 and 6 months respectively as can be seen here in Figure 3. Overall, the opinions were positive.

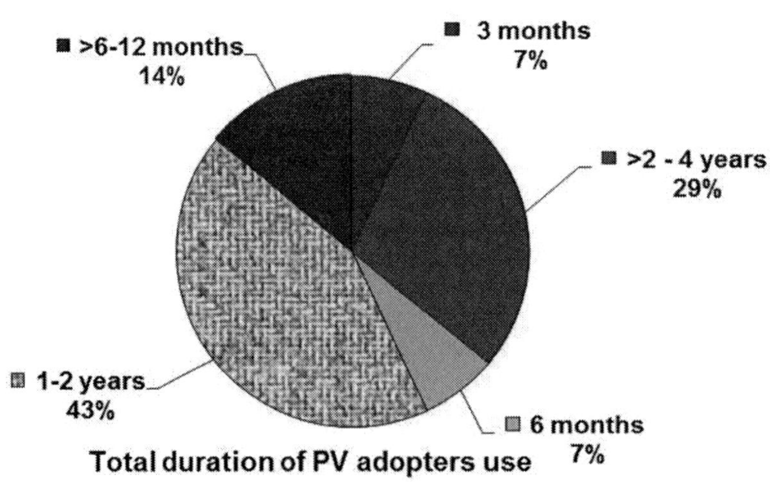

Figure 3. Total duration of adopters' PV ownership prior to survey

IOP Conf. Series: Earth and Environmental Science **385** (2019) 012012 doi:10.1088/1755-1315/385/1/012012

The short lifespan did not create a dislike for or disapproval of PV as would be expected. The adopters instead believed that the issues arose from not having sufficient funds to pay for a fully functional PV system. They also attributed the early PV breakdown to having installed an undersized inverter commensurate to the panels' capacity.

4.2. Energy cost savings

Representing about 60%, energy cost savings was the second most important motive for PV uptake. This finding was surprising because of the widely acknowledged high initial cost of PV modules. Most of the adopters clearly thought that in the long-run, PV generated electricity was more economical than private generators and grid distributed electricity, with some specifically given a breakdown *of the cost-savings*. A notable point made in the interview regarding PV was that its installation turned out to be a mechanical solution to the problem of generator-use fuel fraud. An adopter narrated how his PV installation helped curb being defrauded by his employees. They said:

> "I tell you something. In my shops when I am away, I have been having this issue where my staff members do not buy fuel for generator use but tell me that they did buy fuel. There might have been central electricity supply and they will say oh there was no *light*[1]; that they bought fuel when they didn't buy fuel. So, I have used my PV installation to cut them off. It is actually saving me a lot of money. They can tell you that they bought fuel and they used generator from morning to night when they didn't use it." *-F. I.*

The above described scenario was made possible because the PV adopter made use of a switchover connection allowing the system to switch automatically to grid electricity when central power was restored and vice versa. Therefore, depending on how the connections are made, PV can help safeguard adopters and users from being defrauded by employees and family members

4.3. Familiarity and technical knowledge

As identified in studies, awareness precedes uptake in the adoption decision. Not only the awareness of the technology but also some basic technical knowledge about electricity. Sovacool [17] pointed out how basic electricity literacy was very low even in the United States. Familiarity represented about 30% of views. Many adopters said that they deliberately made certain electrical connections in their dwellings to accommodate such modern energy systems like PV. Awareness and familiarity also meant that some adopters were able to question the quality and number of modules, batteries and general work before deciding whether to go ahead with the installation. While the role of marketers influenced PV uptake as well as perceived overbilling from grid supplied electricity distributors was a driver of the change, they were less significant. Understandably, payback time was the least of the motives. This is expected for consumers who suffer frequent power outages and its associated inconvenience.

Lastly, studies reveal that tenancy can negatively affect MGT adoption decisions [18]. In this study, views about PV reliability, energy cost-savings and familiarity were relatively the same across tenant groups. This means that there was no significant difference between home owners and those that were renting. Figure 4 below illustrates. It is reasonable to expect that home owners would not object to their tenants installing PV on their property. This signals that the tenant or renter is willing to stay in the property for a long period of time hence guranteeing rental income for the property owner.

[1] Note that in Nigeria and most of Africa, electricity is generally referred to as "light."

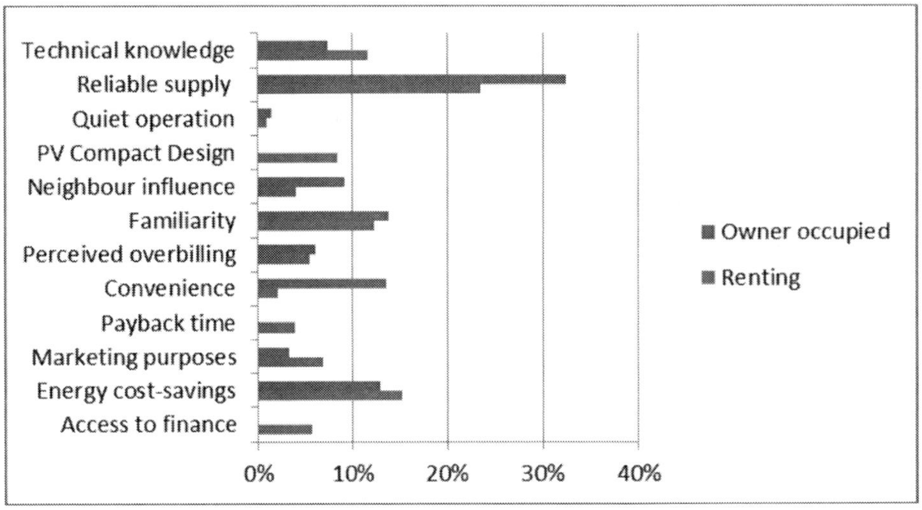

Figure 4. Adoption motives by tenant groups across categories

5. Conclusions and some recommendation

This paper has shed light on the motivations for urban household-level PV uptake in Nigeria. The key factors identified as driving adoption are national power outages and the need for regular power and households' intention to reduce energy costs. The rise in fuel and electricity prices in Nigeria following the recent removal of subsidies has led the self-generating households to seek alternative power. In terms of reliability, the adopters found PV to be the best power supply technology and would recommend PV. There is also evidence that the cost of PV has been steadily declining in the last 10 years.

The identified motives for PV uptake in this paper provide evidence of the vital role of the private sector towards transforming the power sector in Nigeria and similar countries. Like Adhekpukoli, [4] rightly pointed out in his paper on the democratisation of power, meaningful change can only come in the Nigerian power sector only when it is decentralised and handed back to the people. Until, the end-users are seen as a key part of the power supply and demand process, the power access, affordability, sufficiency and efficiency challenge will remain unsolved. Developed nations have realised the power of prosumers[2]. So, should Nigeria, as it considers grid expansion and renewables with the intent to increase grid access to 75% of the population by 2020. Some of the hindrances to PV adoption include lack of finance, low product quality and poor technical know-how.

There is a need to regulate the nascent PV industry so that only quality modules are imported and used in Nigeria. The presence of monetary and non-fiscal incentives would encourage potential adopters who by their actions are contributing to CO_2 reduction. This would also help support businesses as well as grow the economy as new markets are created. As a public good with net environmental benefits, PV will be instrumental for transforming the Nigerian power sector while contributing towards resolving the twin challenge of ensuring energy efficiency and reduced emissions.

Acknowledgement

We acknowledge the financial support of the Global Excellence and Stature (GES) fellowship in this research.

[2] These are self-generators or investors who produce own power for own use or for export from a combination of micro-generation technologies.

References

[1] Stern N 2006 *The Economics of Climate Change: The Stern Review* (Cambridge: Cambridge University Press)

[2] UNEP 2018 Renewable Energy: Renewables on the rise *United Nations* [Online]. Available: https://www.unenvironment.org/explore-topics/energy/what-we-do/renewable- energy. [Accessed: 29-Jul-2018]

[3] Balcombe P, Rigby D and Azapagic A 2013 Motivations and barriers associated with adopting microgeneration energy technologies in the UK *Renew. Sustain. Energy Rev.* **22** pp 655-66

[4] Adhekpukoli E 2018 The democratization of electricity in Nigeria *Electr. J.* **31** (2) pp 1-6

[5] Ohunakin O S 2010 Energy utilization and renewable energy sources in Nigeria *J. Eng. Appl. Sci.* **2** (5) pp 171-77

[6] Claudy M C, Michelsen C and O'Driscoll A 2011 The diffusion of microgeneration technologies: assessing the influence of perceived product characteristics on home owners' willingness to pay *Energy Pol.* **39** (3) pp 1459-69

[7] Rogers, E M 1995 *Diffusion of innovations* (New York: The Free Press)

[8] Faiers A, Cook M and Neame C 2007 Towards a contemporary approach for understanding consumer behaviour in the context of domestic energy use *Energy Pol.* **35** (8) pp 4381-90

[9] Friebe C A, Von Flotow P and Täube F A 2014 Exploring technology diffusion in emerging markets - the role of public policy for wind energy *Energy Pol.* **70** pp 217-26

[10] Faiers A and Neame C 2006 Consumer attitudes towards domestic solar power systems," *Energy Pol.* **34** (14) pp 1797-06

[11] Robson C 2002 *Real World Research: A Resource for Social Scientists and Practitioner-Researchers* (Oxford: Blackwell Publishers Inc.)

[12] Soon J J and Ahmad S A 2015 Willingly or grudgingly? A meta-analysis on the willingness-to-pay for renewable energy use *Renew and Sustain. Energy Rev.* **44** pp 877-87

[13] Creswell J W 2007 *Research Design: Qualitative, Quantitative and Mixed Method Aproaches* (California: SAGE Publishers)

[14] Bahaj A S and James P A B 2007 Urban energy generation: The added value of photovoltaics in social housing *Renew and Sustain Energy Rev.* **11** (9) pp 154-65

[15] Scarpa R and Willis K 2010 Willingness-to-pay for renewable energy: Primary and discretionary choice of British households' for micro-generation technologies *Energy Econ.* **32** (1) pp 129-36

[16] Sovacool B K, D'Agostino A L and Jain Bambawale M 2011 The socio-technical barriers to Solar Home Systems (SHS) in Papua New Guinea: 'Choosing pigs, prostitutes, and poker chips over panels' *Energy Pol.* **39** (3) pp 1532-42

[17] Sovacool B K 2009 Rejecting renewables: The socio-technical impediments to renewable electricity in the United States *Energy Pol.* **37** (11) pp 4500-13

[18] Abdullah S and Jeanty P W 2011 Willingness to pay for renewable energy: Evidence from a contingent valuation survey in Kenya *Renew. Sustain. Energy Rev.* **15** (6) pp 2974-83

IOP Conf. Series: Earth and Environmental Science **385** (2019) 012013 doi:10.1088/1755-1315/385/1/012013

Analyzing and Improving Natural Daylight in Educational Buildings Using Skylights, Victoria International School of Sharjah as a Case Study

E Mushtaha[1], H Eiz[1], L Janbih[1], and R E Rifai[1]
[1]Department of Architectural Engineering, College of Engineering, University of Sharjah, United Arab Emirates

Email: emushtaha@sharjah.ac.ae

Abstract. Due to the great importance of natural daylight in interior spaces, architects keep experimenting different techniques to achieve a good luminous environment. One of the techniques used in building design is the skylight. This study focuses on the impact of the skylight on internal spaces in schools with regards to daylight factor, glare, and cooling loads. Thirteen simulation cases, including the existing case, were simulated and analyzed using several software packages such as Revit, Ecotect and Radiance. The thirteen cases are categorized into three main groups: A, B, and C. Each category has set some fixed parameters in order to reach an optimum design compared with the existing case. After conducting the simulation and analysis, the sub-cases of Group-A failed to achieve a good luminous environment and therefore the results were rejected due to the large increment in cooling load. On the other hand, the sub-cases of Group-B showed considerable improvement in the cooling loads, but some of these cases had caused glare in the space. The sub-cases of Group-C were better than of those in Group-B. Herein, some of the results of this study were satisfactory compared with the existing case study. However, one of the cases in Group-B had the optimum case among the examined cases where all the results are within the accepted range.

1. Introduction

1.1. Research Background

The three largest uses of energy in buildings are heating, cooling and lighting. 84% of the heating and cooling loads, and most of the electricity used of artificial lighting, are generated using fossil fuels which has its large environmental impact. Thus, the importance of introducing daylight in building, while carefully considering its impact on heating and cooling loads, can have substantial effect on the building, not only in terms of environmental and economic factors, but also in its contribution to the wellbeing of the occupants [1]. The importance of natural light is seen in human visual function test, as higher visual comfort and effect is found to be obtained in natural light compared with artificial light [2]. Despite the need of daylighting in all buildings in general, daylighting in educational building particularly is of high importance. It has been demonstrated that the presence of daylight has a direct link to the performance of the students, considering the fact that human health and mental function are influenced by the light exposure's intensity and duration throughout the day [3]. Different techniques have been used by architects to offer sufficient daylight for internal spaces; however, each climate has its own constraints and challenges. One of the greatest challenging climates, when daylight is considered, is desert climate. Challenge with this climate regarding daylight is mainly the heat gain due to the solar radiation. Natural light usually reaches the internal spaces in three forms: direct sunlight, direct skylight, and reflected light, either sunlight or skylight, from the ground and nearby surfaces. The entering should be controlled and investigated to ensure the quality of light in the interior spaces [4]. One method of introducing natural light to interior spaces is the use of skylights. Roof skylight system is a construction technique used to allow natural light to pass through the roof and reach internal spaces. Skylights are being frequently used by modern architects as a solution to for internal spaces where sufficient daylight could not be achieved by the use of windows only. Although skylights are very favorable solution in cold climate where solar radiations through skylight help in reducing heating loads, it is challenging to use it in hot climates where this solar radiation can increase the cooling load significantly [5].

1.2. Case Study Background

Victoria International School of Sharjah (VISS) is an Australian school located in Ta'awon, Sharjah. It is designed by Taylor Oppenheim architects, a firm specialized mainly in educational facilities. The architecture design of school uses different techniques to minimize the power consumption of the school, including skylights, shading devices, and roof shading. Most of the skylights are used in corridors or interior rooms where no exterior windows are available, while few only used in classes with exterior windows. This study aims to investigate the effect of skylight on the luminous environment of internal classrooms as well as the extra cooling load caused by the presence of the skylight. A classroom located in the North-West without skylight is considered as the Case-0 (Figure.1 and 2). The field measurement of the illuminance of the class using a light-meter has indicated the reading as an average of 150 lux, which is considerably lower than other classes where some reached up to 600 lux. In this study the authors are trying to create a classroom lit naturally within a thought of no artificial light is needed during the daytime. However, visual comfort and energy consumption of cooling load would be taken into consideration.

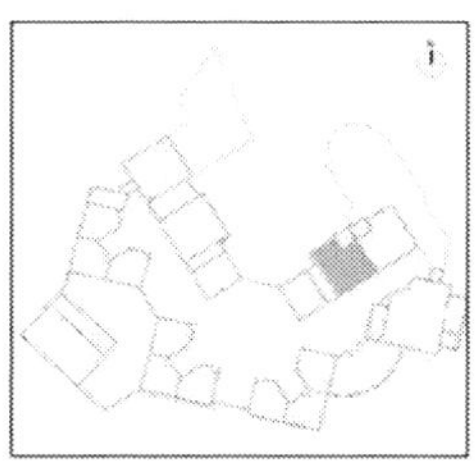

Figure 1. Case study first floor plan

Figure 2. Case study Interior view of the class

2. Objectives of The Study:

The objectives of this study include the following:

- Analyze the natural daylight and cooling load in the existing classroom.
- Analyze the effects of different skylights with different parameters on daylight factor, glare, and cooling loads study.
- Find the best adjustment for the existing case.

3. Research Methodology

This study was conducted using Revit for 3d modelling, Ecotect for lighting and thermal analysis, and Radiance for glare study. Field study has been conducted using light meter and the results were compared with the simulated model. Thirteen cases were studied and analyzed using the mentioned software. These cases are divided into three groups; A, B and C. Each group has one or more common parameters between its cases. The groups differ from one another in terms of shape of skylight and/or type of glass used. Group A cases have one 30° skylight (Figure 4). Group B cases have one or more 90° skylight (Figure 5) with clear glass, while Group C cases have one or more 90° skylight with translucent glass. Cases within the same group differs from one another in terms of number of skylights, orientation of skylight(s) (Figure 6), and how many faces of skylight are glazed. Table 1 indicates the simulation parameters of the 13 cases.

Table 1. Cases' parameters

Case	Orientation of Skylight	Slope degree	Types of glass	Glazing sides	No. of skylights
Base Case	No skylight	-	-	-	-
Case A-1	North-South	30°	Clear	All sides	1
Case A-2	East-West	30°	Clear	All sides	1
Case A-3	East-West	30°	Translucent	All sides	1
Case A-4	East-West	30°	Clear	Only north-facing side	1
Case B-1	East-West	90°	Clear	All sides	1
Case B-2	East-West	90°	Clear	Only north-facing side	1
Case B-3	East-West	90°	Clear	All sides	2
Case B-4	East-West	90°	Clear	All sides [a]	2
Case B-5	North-South	90°	Clear	All sides	2
Case C-1	East-West	90°	Translucent	All sides	1
Case C-2	East-West	90°	Translucent	All sides	2
Case C-3	North-South	90°	Translucent	All sides	3

[a] South side with horizontal shading

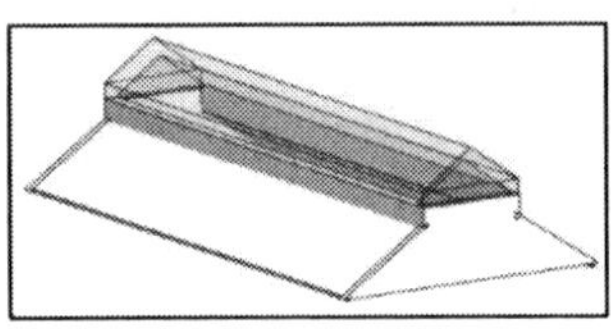

Figure 3. 30° skylight

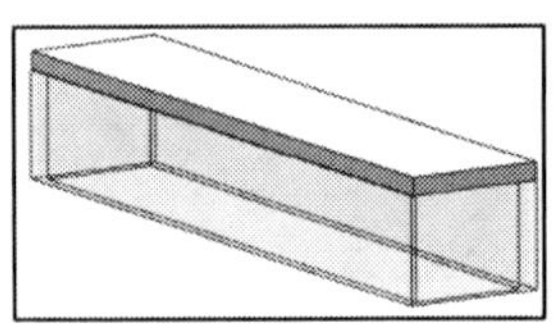

Figure 4. 90° skylight

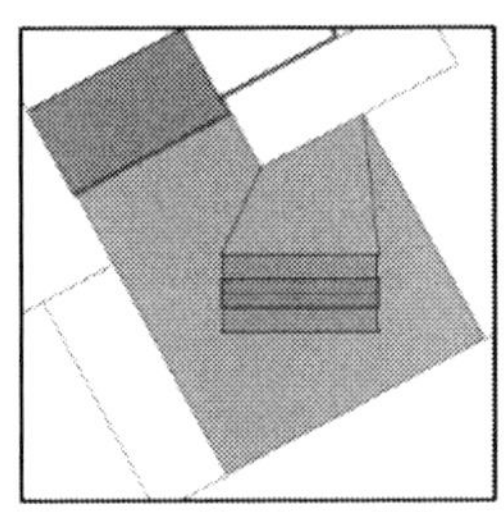

Figure 5. East-West oriented

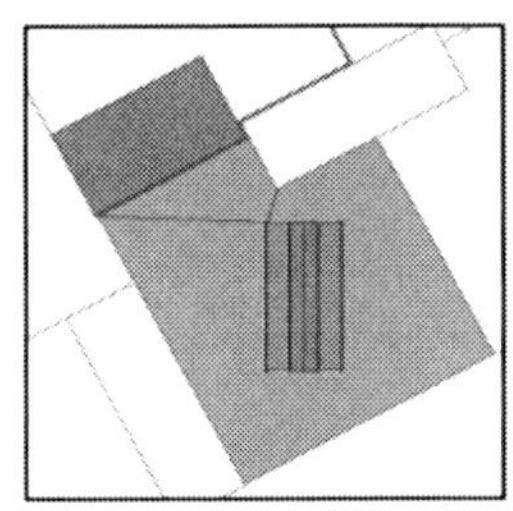

Figure 6. North-South oriented

All the simulated data used the following fixed parameters:

- Weather data file: Abu Dhabi, UAE
- Design sky illuminance: 11,000 lux
- Visible transmittance (0-1): Clear glass: 0.611 and Translucent glass: 0.658
- U-Value (W/m^2k): Walls: 1.72, Floor: 0.88, Clear glass: 2.41, Translucent glass: 1.40
- Room area: 93.6 m^2

4. Simulation Analysis

4.1 Base case – Existing

It can be observed, by both field measurement and virtual analytical model, that the natural daylight in the existing case is not satisfactory. Average daylight factor is 1.67% which is considerably low. Illuminance in the center is around 120 lux. The field measurement showed an average of 150 lux at the center which is very similar to the calculated *value. This* similarity validates the accuracy of the model. The cooling load was also simulated for the comparison with the later cases.

4.2 Group A - Skylights with 30° angle

All cases in Group-A have significant increment in terms of daylight factor compared to the existing case. Glare was not avoided in case A-1 and A-2. All cases are rejected due to the considerably higher cooling loads comparing to the existing case study (Charts 1 *and* 2). Changing the orientation of the skylights from North-East, case A-1 (Figure *8*), to East-West, case A-2 (Figure *9*), has increased the cooling loads insignificantly and thus it is neglected. Replacing the clear glazed with translucent, case A-3 (Figure 1*0*), has reduced the cooling loads by 5%. The same results of case A-4 (Figure 1*1*) in terms of cooling load were achieved. Although cases 3 and 4 were similar in cooling loads and in reducing glare in the space, case A-4 with only north-facing clear glazed showed better results in terms of daylight factor for using translucent panel. Since all Group-A was failed, the original shape of the 30°-slopped-skylight in Group-A has been switched to Group-B, 90° angle skylight with clear glazing panels. This proposal is expected to reduce the cooling loads which were the common problem in all Group-A cases as well as to reduce glare in the space.

4.3 Case B - Skylights with 90° angle skylight with clear glass

In Group-B the cooling loads area considerably lower than those of Group-A due to the skylight's 90° inclination rather than the 30°. Daylight factor is less than Group-A but it is still accepted in most of the cases and it is greater than existing. Case B-1 is significantly better than those of Group A in terms of cooling loads (only 30% greater than existing case). However, small amount of glare is caused, and visual comfort is not achieved. (Figure 12). Light was more evenly distributed in Case B-2 (Figure 13) where south side is blocked; however, daylight factor was not sufficient. Glare was caused in Case B-3 (Figure 14) and Case B-5 (Figure 16). Blocking the skylight from the south resulted in obvious decrement in glare. Changing the orientation of the skylight from east-west to south-north increased the amount of glare in the space. By adding more skylights with all sides clear glazed and adding shading horizontal device instead of blocking the south-faced side of skylight, better results are obtained, and the resulted case is considered satisfactory. Cooling load is also reduced, compare to Group A, and the new cooling load is acceptable in all cases except Case B-5, which suggested adding three skylights to increase the amount of daylight in the space since; although daylight factor reached very good levels, glare was caused, and cooling load was increased rapidly. Better results can be achieved by multiple trial and error experiments but Case B-4 (Figure 15) is a good start for future development (Chart 1 – 2).

4.4 Case C - Skylights with 90° with translucent glass:

Clear glass was replaced with translucent panel in Group C. This was an attempt to create space with diffused light that will prevent the glare. Although it has small effect in minimizing cooling load, the problem of glare was not resolved (Figure 17, Figure 18 and Figure 19) the results were not as expected (Chart 1-2).

5. Analysis and Discussion

All cases with skylight increased the cooling loads of the existing case but with different amounts. Initially, the authors added 30° slopped skylight (Group A) similar to the skylights used in other parts of the school. The results of adding slopped skylight with 30° was not satisfactory in all cases. Cooling loads were increased significantly; it reached almost to triple the existing case in most scenarios. Visual comfort was not obtained either; most scenarios caused glare and failed to give uniform light distributions in the space.

Due to the results of Group A, the authors offered another design of skylight with 90° angle (Group B), which is a monitor design. This design has given better results regarding the cooling loads which were gradually developed to reach the optimum scenario.

IOP Conf. Series: Earth and Environmental Science **385** (2019) 012013 doi:10.1088/1755-1315/385/1/012013

Although some of the previous scenarios were satisfactory regarding the amount of heat gain and the increment in daylight factor, most of them failed to obtain visual comfort. Glare was the main problem in the previous cases; thus, the clear glass used in Group B was replaced with translucent panels in Group C. However, the glare was not solved, and the results were rejected.

For a better understanding, Chart 1 compares the thirteen cases in terms of energy consumption due cooling loads where Group A is the highest. Chart 2 compares the daylight levels of all cases. Among all the thirteen cases, Case B-2 has the lowest increment in daylight factor which is the main purpose of this research and thus rejected.

Figure 7. Base case – perspective view

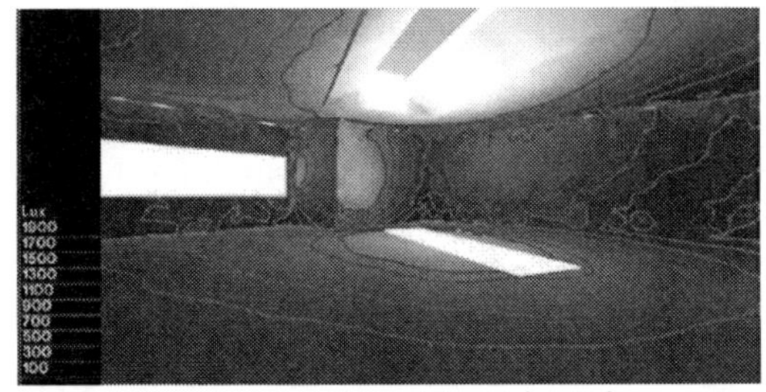

Figure 8. Case A-1 – perspective view

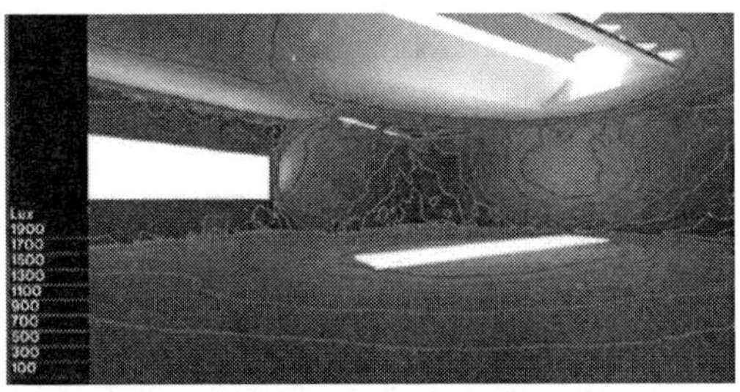

Figure 9. Case A-2 – perspective view

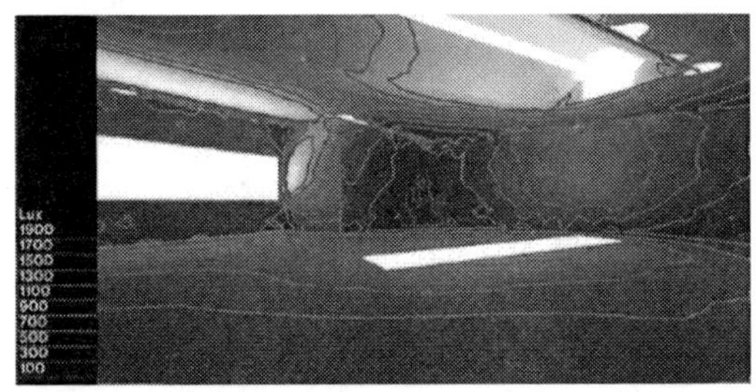

Figure 10. Case A-3 – perspective view

Figure 11. Case A-4 – perspective view

Figure 12. Case B-1 – perspective view

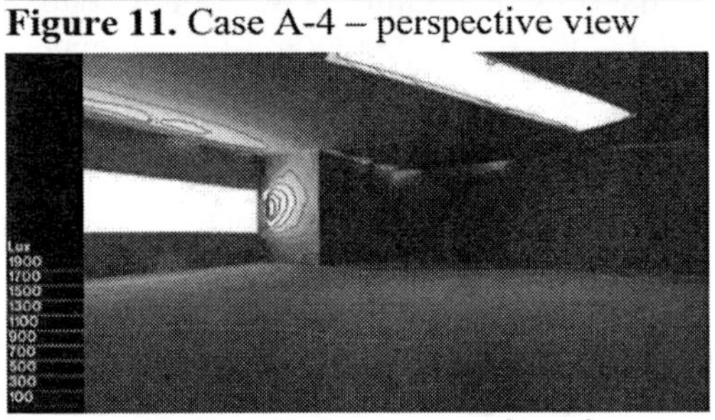

Figure 13. Case B-2 – perspective view

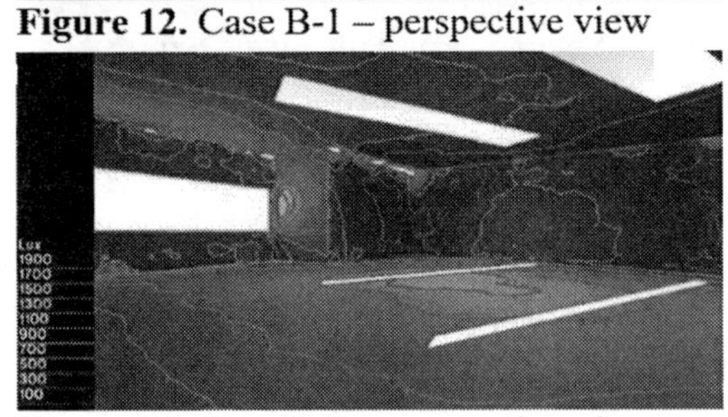

Figure 14. Case B-3 – perspective view

Figure 15. Case B-4 – perspective view

Figure 16. Case B-5 – perspective view

Figure 17. Case C-1 – perspective view

Figure 18. Case C-2 – perspective view

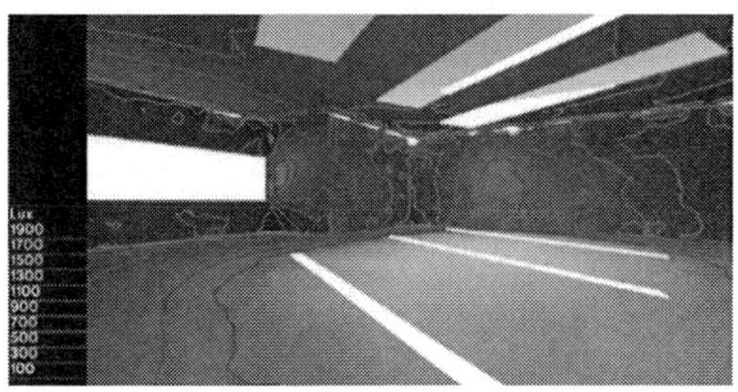

Figure 19. Case C-3 – perspective view

Below are charts showing all the studied cases in terms of energy consumption due to cooling loads as well as in terms of daylight levels.

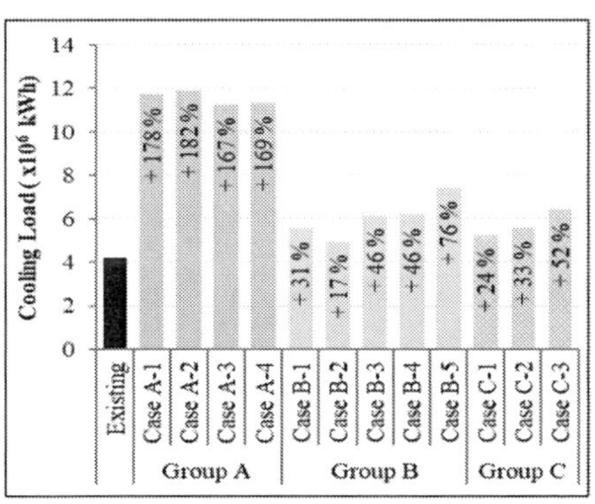

Chart 1: Comparisons between all the studied cases in terms of energy consumption due to cooling loads. The percentage of increment compared to the existing case is indicated on each bar.

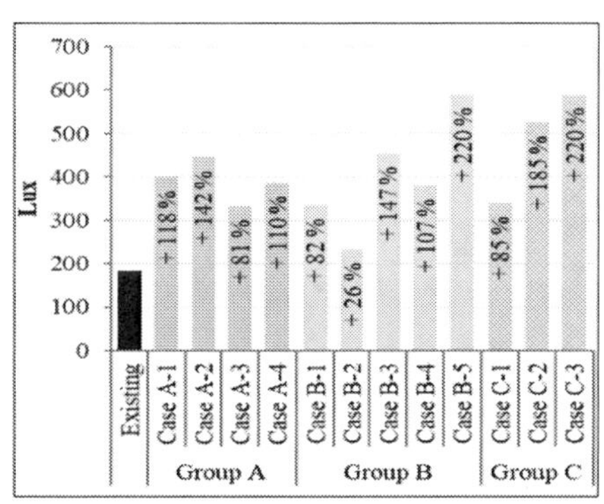

Chart 2: Comparisons between all the studied cases in terms of energy consumption due to cooling loads. The percentage of increment compared to the existing case is indicated on each bar.

6. Conclusion

The following conclusion can be drawn:

- Group A failed to offer good alternative for the existing case as the cooling loads were not accepted.
- Changing the geometry of skylight had great impact on cooling loads without compromising the amount of daylight entering the space.
- Replacing clear glass with translucent panels minimized the cooling load, decreased the glare; however, daylight factor was decreased as well. Tested scenarios were not satisfactory but further study could lead to better results.
- Case B-4 is the optimum case among the examined cases where all the results are within the accepted range.
- The possibility of integrating a skylight into classes could be an option if the coaling loads were studied carefully.

References:

[1] Gutiérrez RU, Du J, Ferreira N, Ferrero A, Sharples S 2019 Daylight control and performance in office buildings using a novel ceramic louvre system *Building and Environment* **151** 54-74

[2] Zhu T, Li R and Li C 2017 The Analysis of Natural Lighting Simulation and Study on Energy Saving in Cigarette Factory *Procedia Engineering* **205** 895-901

[3] Pellegrino A, Cammarano S and Savio V 2015 Daylighting for Green schools: a resource for indoor quality and energy efficiency in educational environments *Energy Procedia 78 3162-3167*

[4] Alshaibani K A 2015 Planning for Daylight in Sunny Regions *International Conference on Environment and Civil Engineering (ICEACE'2015)* April 24-25, Pattaya (Thailand), Retrieved from http://ia-e.org/siteadmin/upload/1741IAE0415410.pdf

[5] Acosta I, Navarro J, & Sendra, J J 2013 Towards an analysis of the performance of lightwell skylights under overcast sky conditions *Energy and Buildings* **64** 10-16.

The hidden benefits of stand-alone PV module efficiency limits

A I Ugulu[1] and C Aigbayboa[1]

[1]Department of Construction Management and Quantity Surveying, Faculty of Engineering and the Built Environment, University of Johannesburg, South Africa

E-mail: anthonyugulu@yahoo.com; aiu30@hw.ac.uk

Abstract. Energy and cost savings from a PV installation in a dwelling is influenced by a number of variables including occupant behaviour and the feedback mechanism accompanying the PV system. This monitoring device can reveal the energy efficiency opportunities present in real-time. The same way an in-home smart metre shows how much energy is being consumed by a household, a PV metre highlights both the generated power and the amount being used at any point in time. Results from a field survey revealed that post-PV installation, there was increased electricity awareness and energy conservation amongst the households. It was further found that in addition to the presence of a feedback metre, conscious energy management efforts arising from the part of the PV adopters and users led to the energy and cost savings. Most importantly, it was revealed that the energy savings were primarily driven by PV efficiency limits. These findings were used to develop a PV efficiency cycle demonstrating how the energy conservation and use efficiency occurred. The study showed the implications and proposed ways by which PV can be promoted in emerging economies.

1. Introduction

The deployment of green energy technologies and energy efficiency are the two core tenets of the transition to a low-carbon future. The goal is to advance renewables and encourage resource-use efficiency especially in the transport, industry and residential sectors which constitutes the bulk of energy use globally [1]. It is generally accepted that following this path would ensure the security of vital energy systems via the use of more sustainable energy sources. This is because conventional fossil-based sources are environmentally harmful.

The decision to limit atmospheric greenhouse gases (GHGs) and promote cleaner energy systems is linked to the first UN Framework Convention on Climate Change (UNFCCC). Subsequently, the Paris agreement held in December 2015 led to a renewed commitment towards this agenda. Today, over 170 nations including Nigeria have ratified this agreement [2]. Notwithstanding, many economies worldwide still rely on heavily polluting conventional fuels predominantly coal, crude oil and natural gas for their electricity production. The UN estimates that around 80% of global energy and 66% of electricity production comes from fossil fuels and contributes an estimated 60% of greenhouse gas (GHG) emissions responsible for climate change [3].

Since buildings account for up to 40% of energy use as well as 30% of the share of CO_2 worldwide [4], most green energy policies and initiatives focus on improving energy use in buildings especially dwellings. One of the most deployed green power technologies in dwellings is solar photovoltaic (PV)

systems. Evidence from studies reveal that the uptake of decentralised low carbon technologies such as PV, can result in energy savings for the household [5], [6]. Keirstead [7], reported a 6% savings from the use of such modern energy technologies in a UK study. The presence of a feedback monitor led to the savings in Keirstead's study. Through monitoring of occupants' energy use in a PV adopted social housing, Bahaj and James [8] reported a correlation between occupant behaviour and energy consumption.
In addition to examining the experiences, challenges and energy-saving potential of installing PV in urban dwellings in Nigeria, this paper goes further by investigating the processes by which PV adopters saved energy and costs, if at all. This paper is structured as follows. The next section discusses the patterns of electricity use and energy consumption in Nigeria including PV generated electricity use. Next, the research methods are discussed. Subsequently, a discussion of the results is detailed. Finally, the conclusions, recommendation and implication of research are presented.

2. Patterns of electricity use and energy consumption in Nigeria

Energy demand in Nigeria is a function of rising population and economic growth. The outcome of rise in population and growth in gross domestic product (GDP) of 6% between 2006 to 2015 is increased demand for electricity [9]. Electricity demand in Nigeria is broadly categorised into industrial, residential and commercial sectors. However, demand from the residential sector has in the past decades taken up the bulk of final electricity consumption [10]. Figure 1 illustrates.

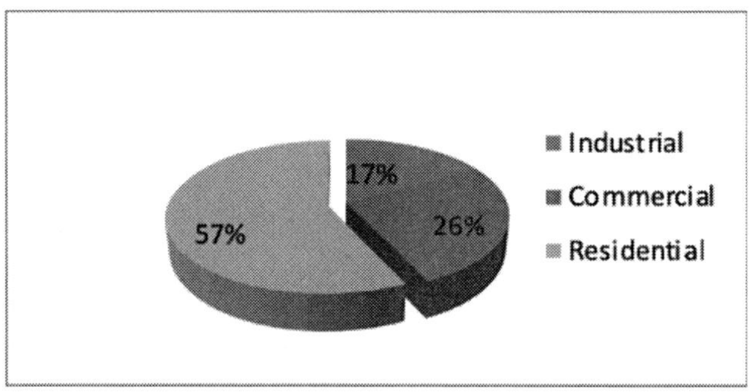

Figure 1. Total electricity demand in Nigeria by sector 2011-2013 Source: Ugulu, 2016.

Some explanations for the residential sector constituting over 50% of total energy demand stems from the fact that most urban households substitute central electricity with auto-generation and kerosene while rural households rely mainly on kerosene due to the low grid connectivity rate to this group. The unreliable grid infrastructure operating under 6GW [11] in Nigeria for a population of over 185million people [12] has meant that electricity is rationed, load shedding is commonplace and power outages is frequent. Secondly, in order to minimise damage to equipment from grid power outages, the commercial and industrial sectors depend heavily on private power plants and less on central electricity. This situation has made most Nigerian households' and general end-users "experienced" conventional private power generators.

2.1. Solar PV, monitoring device and feedback

In a typical domestic rooftop PV system, a monitoring device is attached to the inverter as part of the installation. This in-home metre helps to monitor performance in terms of power being generated and used at any particular time and helps users ascertain peak periods for scheduled and other high demand usage. With grid-connected PV systems that support feed-in tariffs (FITs), this monitor or metre aids the export and import arrangement [7]. The device helps to coordinate the feeding-in and the buying back process from the grid (in the case of net metering) hence cutting energy costs for consumers.

2.2. Energy efficiency and conservation

Energy efficiency and conservation from PV use has been previously reported in studies but for different reasons. Hast et al, [6] revealed savings from a consumer response study in China citing the motivation as environmental concerns and energy savings. In the UK, Keirstead [7] found that the presence of monitors or metres resulted to a 6% savings in energy cost for the PV households. Using results obtained from nine social housing residents, Bahaj and James [8] presented proof of a reduction in energy use in PV households but the source of savings was unclear as Stedmon et al, [13] also noted. This study aims to fill this gap from the perspective of microgeneration technology users in developing countries where such studies are lacking.

2.3. End-user behaviour and the rebound effect

While there is a link between domestic PV use and energy savings, in some cases, an improvement in supply using renewables can result in a rebound effect for households. Under a rebound, improvement in technical efficiency and falling prices lead to increased consumption; thus, offsetting the reduction in energy consumption earlier experienced [14]. In the case of a PV installation, a rebound would mean that post-PV adoption, energy consumption increases rather than dropping compared to pre-retrofit work. This is often due to changes in user behaviour as some dwelling occupants tend to add more appliances to the existing ones defeating the purpose of efficiency of use and savings.

3. Research methods

Following similar consumer response studies, a qualitative approach was taken. Interviews were conducted and aside questions relating to socio-demographics, questions generally covered relevant aspects of a PV such as system size, drivers, role of PV monitor feedback, energy use efficiency, payback time and challenges and suggestions for increased uptake. Questions were essentially designed to discern if a feedback mechanism such as PV metre can lead to changes in energy demand by examining the pre and post-purchase behaviour of the adopting households. This is to gain insight into the effects of PV uptake on the adopters' overall energy consumption. Understanding the behavioural modifications and demand management techniques employed would be the basis for any energy efficiency bid.

3.1. Sampling and sample size

Purposive sampling was used. This is a permissible technique in qualitative studies (Robson, 2002). In this type of sampling, the researcher's judgment is relied upon based on study interest and rationale. Some PV adopters interviewed were identified from a questionnaire survey for a different but related study. Others were selected via referral. This is a situation whereby an interviewed adopter refers the researcher to other users. A total of 14 adopters were interviewed as experiential data require fewer participants [15]. Bahaj and James did a similar PV study in Southampton UK using 9 households [9]. Respondents were primarily drawn from Lagos State, Nigeria.

Due to the small number of PV users here, respondents were taken from other states i.e. cities in Nigeria such as Abuja (Federal Capital City) Delta, and Edo. Open-ended semi-structured interviews were carried out and the interviews recorded using an MP3 recorder. The interview questions largely examined the role of feedback mechanisms such as PV monitor. This was done to understand the relationship between PV use and energy savings as identified in literatures [6]–[8]. In addition, it was designed to ascertain if the variables at play are interconnected and if the reports were consistent when applied to a developing country with unreliable grid electricity supply.

4. Results and analysis

In order to ascertain whether PV could help bring about household energy use efficiency and management, its contribution towards helping to modify household energy consumption behaviour was examined. Through the inclusion of questions bordering around whether the adopter's installation

had a monitoring metre attached, the location of the device and how frequently the monitor was viewed, this important aspect was analysed. The findings were positive and revealed that PV use can contribute to increased energy use awareness and in consequence reduce overall demand, improve supply and encourage de-consumption.
One reason behind the energy use management uncovered was the presence of a feedback metre on the inverter. This supports Keirstead's [7] finding. All but one of the adopters stated having a display monitor on their PV inverter. With most of the adopters' inverter located inside the dwelling, there were more regular checks by the energy conserving and efficient households.
The presence of an inverter monitor can influence the number of checks on a PV system which can in turn lead to more energy use conservation and efficiency of use. There was an almost five-fold increase in energy use monitoring and conservation following the introduction of PV. See Figure 2. Therefore, the use of PV directly impacts on households and consumers energy demand and fosters interaction between the users and the system. This interaction is the direct result of PV output limitations.

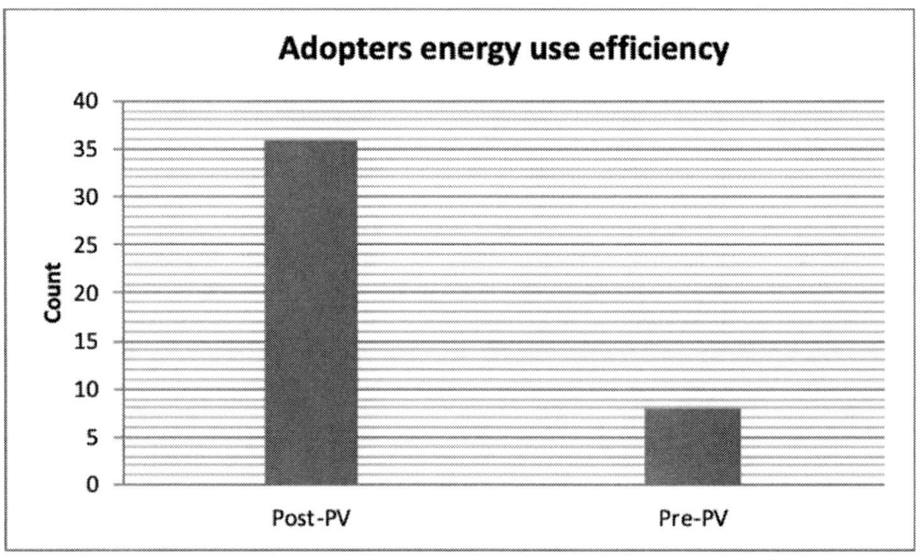

Figure 2. Energy use and savings pre and post-PV

4.1. The hidden benefits of PV module power limitations
One of the most interesting discoveries of this study was that household PV adoption leads to increased energy use awareness and subsequently energy conservation and efficiency. An almost fivefold increase in energy management was reported by the adopting households' post-PV. They said the presence and location of the feedback metre helped. Like smart metres, PV metre does not only provide useful information for adopters, it served to place control in the hands of the users in a way centrally-supplied electricity does not. This directly impacted household's energy demand and fostered interaction between the users and the system. The energy management opportunity was therefore a function of the high costs of a sizable PV, the presence of an in-house metre and most essentially, the output limits of PV. It is this latter feature of PV that makes for its distinctiveness. Figure 3 details the energy-saving processes taken by the PV adopters and users.

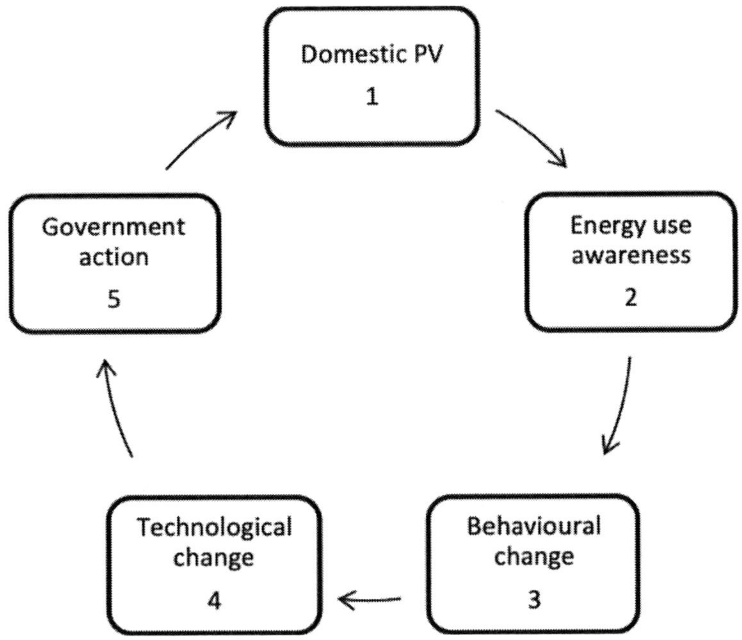

Figure 3. The PV efficiency cycle

The continuous processes illustrated in Figure 2 above serves to indicate the crucial steps that PV-adopting households in urban Nigeria take to achieve energy use efficiency. Starting from Step 1 when the PV is installed, the PV electricity generating households start to notice the improvements to their energy demand which PV has enabled. This excitement stems from the fact that it may be the first time the users have had uninterrupted supply for a while, without the need to use alternative sources like fossil-based generators that require constant fuel purchases. This excitement, plus metre/monitor checks, draws the users closer to the workings of a PV unit.

Clockwise from point 1, at Step 2, PV design limitations start to bear on the users as they realise that they cannot use their PV for every appliance but mainly for essentials. Due to its non-dispatch design, the users then realise that they have to conserve power to allow the efficient running of the device. At this stage, energy use management becomes the sole prerogative of the PV adopters as they make sure that unnecessary plugging in of every household appliance is avoided and discouraged. Increased electricity and energy use awareness arises from the restrictive use of PV and prevents overconsumption habit formation by households. This new energy use awareness creates the "conditioning effect" resulting in behavioural changes (Step 3) which is the most important role of a PV module as it pertains to energy cost savings, efficiency and environmental sustainability.

It thus falls on the households to begin to search out other ways they can reduce consumption in order to make the device function more effectively and efficiently. This leads to greater awareness and the consideration and uptake of modern power saving household appliances and building technologies (Step 4). Also, other possible avenues to minimise energy consumption is sought. Here, energy efficiency ideas like purchasing more energy saving appliances and retrofitting the building for passive cooling and heating, to allow for natural airflow and solar gains to the dwelling, become more realistic. This is a form of climate change building adaptation that PV utilisation can stimulate.

Some households would voluntarily make these building changes having had the opportunity to observe first-hand the energy dynamics of a PV unit. For other non-adopting households, the government (Step 5) can assist with support instruments (with particular emphasis on efficiency) toencourage PV adoption e.g. by mandating the building construction industry and private developers to incorporate low-carbon technologies in dwellings such as the German pioneered Passivhaus standard.

It should be noted that not in all cases is the PV cycle sequential as it appears on the diagram. For example, those individuals with high technical knowledge might bypass the energy awareness and conditioning phase (behavioural change) by deliberately installing their PV to be used with certain selected appliances in the home. They could include a PV controller or exclude power guzzling devices like multiple A/Cs from the unit to better manage the system. Nevertheless, their use of PV will still demand that they routinely check their energy consumption as solar radiation and daylight intensity varies daily and hourly. Notwithstanding, the point of the cycle is that most adopters will pass through the 4 initial phases and then decide if there is further need for additional adjustments such as more building retrofit works to accommodate the proper function of the solar panels.

5. Conclusions, recommendations and research implications of findings

This paper is evidence that a simple decision to install solar PV in order to improve power supply to a home can bring about a whole range of unforeseen positive behavioural changes and long-term benefits for the adopting households and society at large. Also, energy and power shortages can lead to adjustment, sometimes unconsciously. This is because, with scarcity, serious reflection begins. It is highly unlikely that a rebound situation would arise as the surveyed households were very conscious of their energy use made possible by the PV feedback metre and decades of energy management.

The PV efficiency limits served as a reminder to households that electricity needs to be conserved for regular power to be guaranteed and for the system to function optimally. The metre readings and PV capacity limitations were what compelled the users to make the necessary changes. While it may appear as an inconvenience, it was the reason for the savings and the adopters did not find it inconveniencing. Not yet identified in discourses is this "conditioning effect" of PV. In fact, this aspect of PV is often described using negative connotation. For example, the well-known issue of PV efficiency receives this description. Meanwhile, silicon module and cell efficiencies have exceeded 20% [16], [17], while in a recent study researchers Burlingame et al [17] and others [18] reported an efficiency level of >15% for Organic PV (OPV). However, PV output limits is frequently cited as a hindrance. While one cannot deny the existence of output limitations and capacity factor issues, there seem to be an understated usefulness of solar PV in this regard. Dismal as PV efficiency limits may appear, published research gives credence to its immense prospects.

Since the aim of promoting renewables is to limit emissions and for energy security, PV is one of the best micro-generation technology (MGT) for this purpose because it acts as a "conditioner" for households to adjust. This notable feature can be best visualised in the energy demands of households in developing nations, where citizens experience long hours of power outages. Having experienced load-shedding, power rationing for a long time, adjusting to PV-supplied power becomes much easier. The long periods of national power shortages have prompted energy frugality which PV deployment helps to preserve. Unlike in OECD countries where citizens are used to decades of constant power to their homes, widespread PV use becomes difficult because it is certainly not going to meet their very high energy needs. This is probably the reason why the hidden benefits of PV output limitations are yet to be established in more developed countries. However, it is only such limitations that can help reintroduce de-consumption and reinforce demand reduction and eventually help accomplish energy efficiency goals. If national and global emissions targets are to be met, there is an urgent need to support widespread uptake and diffusion of residential PV in emerging economies using fiscal and non-monetary incentives.

Acknowledgement

The financial support of the Global Excellence and Stature (GES) fellowship in this research is gratefully acknowledged. Opinions expressed and conclusions reached are those of the authors.

IOP Conf. Series: Earth and Environmental Science **385** (2019) 012014 doi:10.1088/1755-1315/385/1/012014

References

[1] IEA, 2018 World Energy Investments (WEI) 2018 *World Energy Investments (WEI)*, 2018. [Online]. Available: https://webstore.iea.org/download/direct/1242?filename=wei2018.pdf.

[2] UN Treaty Collection (UNTC), 2018 Environment: Ratifications status, Available: https://treaties.un.org/Pages/ViewDetails.aspx?src=TREATY&mtdsg_no=XXVII-7-d&chapter=27&clang=_en. [Accessed: 26-Jun-2018]

[3] UNEP, 2018 Renewable Energy: Renewables on the rise *United Nations*, 2018. [Online]. Available: https://www.unenvironment.org/explore-topics/energy/what-we-do/renewable-energy. [Accessed: 29-Jul-2018]

[4] Asif M and Muneer T 2007 Energy supply, its demand and security issues for developed and emerging economies *Renew. Sustain. Energy Rev.* **11** (7) pp 1388–413.

[5] Balcombe P, Rigby D and Azapagic A 2013 Motivations and barriers associated with adopting microgeneration energy technologies in the UK *Renew. Sustain. Energy Rev.* **22** pp 655-66

[6] Hast A, Alimohammadisagvand B and Syri S 2015 Consumer attitudes towards renewable energy in China - The case of Shanghai *Sustain. Cities Soc.* **17** pp 69-79

[7] Keirstead J 2007 Behavioural responses to photovoltaic systems in the UK domestic sector *Energy Pol.* **35** (8) pp 4128–41

[8] Bahaj A S and James P A B 2007 Urban energy generation: The added value of photovoltaics in social housing *Renew and Sustain Energy Rev.* **11** (9) pp 154-65

[9] World Bank, 2015 Nigeria: Country Overview *World Bank*, 2015. [Online]. Available: http://www.worldbank.org/en/country/nigeria/overview. [Accessed: 01-May-2015]

[10] Ugulu A I 2016 *The determinants of decentralised photovoltaic (PV) adoption in urban Nigeria and a verified model for rapid diffusion* Doctoral Thesis, Heriot-Watt University, UK

[11] Adhekpukoli E 2018 The democratization of electricity in Nigeria *Electr. J.* **31** (2) pp 1-6

[12] World Bank, Nigeria: Population as at 2016 *World Bank*, 2016. [Online]. Available: https://data.worldbank.org/country/nigeria?view=chart. [Accessed: 17-Jun-2018]

[13] Stedmon A W, Winslow R and Langley A 2013 Micro-generation schemes: user behaviours and attitudes towards energy consumption *Ergonomics* **56** (3) pp 440–50

[14] Buchanan K, Russo R and Anderson B 2015 The question of energy reduction: The problem(s) with feedback *Energy Pol.* **77** pp 89-96

[15] Robson C 2002 *Real World Research: A Resource for Social Scientists and Practitioner-Researchers* (Oxford: Blackwell Publishers Inc.)

[16] Mints P 2012 The history and future of incentives and the photovoltaic industry and how demand is driven in *Progress in Photovoltaics: Research and Applications* **20** (6)

[17] Burlingame Q, Coburn C, Che X, Pamda A, Qu Y and Forestn S R (2018) Centimetre-scale electron diffusion in phtoactive organic heterostructures *Nature* **554** 77-80

[18] BBC News, 2018 Organic solar cells set 'remarkable' energy record [Online]. Available: https://www.bbc.co.uk/news/science-environment-45132427. [Accessed: 10-Aug-2018]

IOP Conf. Series: Earth and Environmental Science 385 (2019) 012015 doi:10.1088/1755-1315/385/1/012015

Capabilities of UAV-Based Watershed Segmentation Method for Estimating Tree Crown: A Case Study of Oil Palm Tree

S Abdullah S[1], K N Tahar[2], M F A Rashid[1], M A Osoman[3]
[1]Faculty of Architecture, Planning and Surveying, Universiti Teknologi MARA, Seri Iskandar Campus, Perak Branch, 32610 Perak, Malaysia.
[2]Faculty of Architecture, Planning and Surveying, Universiti Teknologi MARA, Shah Alam, 40450 Selangor Malaysia,
[3]Geoinfo Services Sdn. Bhd, No.31, Jalan Bandar 2, Taman Melawati, Kuala Lumpur

Email: suzan156@perak.uitm.edu.my

Abstract. Currently, the tree crown delineation has become an important part of the tree parameter measurement in particular for the forest management plan. Moreover, the requirement for the advanced information on forest resources has headed researchers to built-up more effective approaches. With the advanced in the technology of imagery, an Unmanned Aerial Vehicle (UAV) has become one of the appearance tools that offers an affordable, cheaper and faster technology in acquiring data for numerous applications. More importantly, this technology has its own efficient approaches for real-time acquisition of high resolution to produce three-dimensional (3D) information. The digital surface model was extracted and used to estimate tree crown. The influences of different filtering method such as Gaussian and Laplacian filter were investigated on tree crown estimation. The result shows the differences in tree crown delineation that were confirmed by processing methods. This study has potential for future use by tree parameters that related researcher, industries and farmers to estimate tree height and tree counting.

1. Introduction

The inventory of the tree information such as height and crown parameters has always been the important works regarding forest management plan. That kind of the parameter is very important not only for the future planning and managing of the forest resources but also for database development purposes. However, the conventional method is quite expensive due to the involvement of the data processing and data acquisition [5]. In fact, the popular conventional technology like LiDAR has also faced the challenges in the data processing and filtering especially for a high-level area and mixture of geographic features. In this case for example, the classification of point cloud is one of the major problems in producing the high quality of digital terrain data [11]. In addition to that, the conventional technology involves a well-trained person especially at the stage of data collection and the requirement of an expensive camera [16]. Also, the technology is not suitable to perform in a small area as it will produce a very high-resolution data [12]. Moreover, there are several techniques of estimating tree height and determining tree crown efficiently and accurately. Traditionally, tree height is evaluated through a comparative and costly field surveillance techniques and labor-intensive or the combination of ground measurement and modelling [2]. This technique, however, requires high cost due to the involvement of user expertise in handling the instrument to measure stand height [8].

Therefore, UAV has become a new advanced technology with more cost-effective as compared to airborne LiDAR and RADAR technologies [8], in terms of controllable and achievable to obtain the high-resolution results [12] and produce 3D image in data processing [15]. Undoubtedly, the images derived from the UAV is widely used in determining tree parameters such as crown diameter, canopy volume, tree height, species types and many more. The image has emerged as a feasible for 3D structure of forests and monitoring [14]. Indeed, this technology has broadened for 3D data acquisition that becomes more affordable, perceived as a reasonable cost [3] and can remotely be operated from ground surface [7]. It is noteworthy, recently, the UAV has been put forth in multiple applications such as agriculture and forest monitoring and management plan. Inspired by this, the current paper attempts to demonstrate the capabilities of UAV-based watershed segmentation method for estimating tree crown of oil palm tree. That is an important ecological trait in the forest observation variable [13]. Therefore, this research can be the extension of the previous studies which focused on estimation of forest [8], forest assessment of planning and design [10], 3D model of forest structure [4] and model the forest canopy surface [6].

2. Methodology

2.1 Study Area

In this study, the UAV image of oil palm trees located along the street in the District of Parit, Perak (4.407997 latitude, 100.928463 longitude) was captured (Figure 1). The tested area is a private oil palm plantation project which has been started since late of the 2000s. In order to optimize the image processing, only a small site that is approximately 75m x 85m has been selected in the crown tree estimation process.

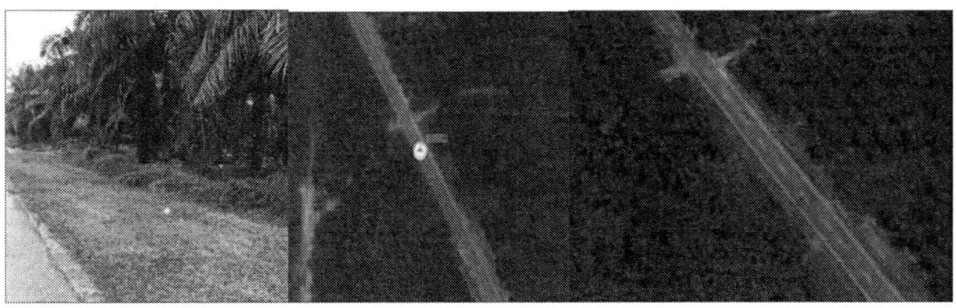

Figure 1. Location of The Test Area

2.2 Process and Methods

The overall process of estimating the crown tree of oil palm is illustrated in Figure 2. It starts with UAV flight planning that involves the determination of a case study and site location, data captured, image processing and generating DSM and up to tree crown identification. This section explains the procedures and methods involved in the data capture, image processing and generating the DSM as well as the application of the UAV-based watershed segmentation method in the crown tree estimation. Furthermore, the result will be presented and interpreted in the results and discussion section afterwards.

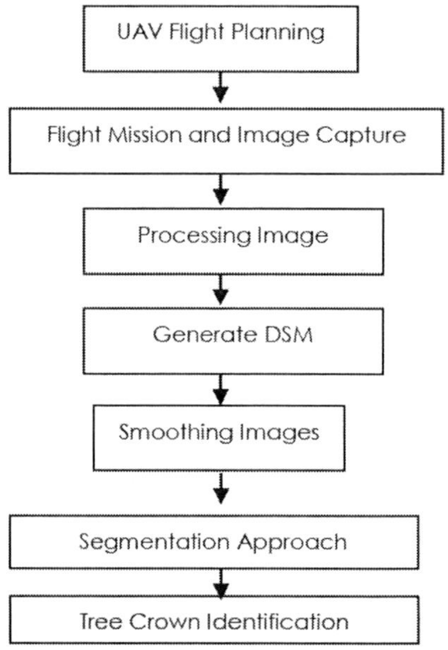

Figure 2. Methodology Process

2.2.1 UAV Flight and Data Capture. The flights were conducted using a multirotor DJI Phantom 4 pro UAV fitted with a 20mp resolution camera and capable of produced 4k at 60 frames per second video. The UAV camera is mounted with a pair of sensors that will automatically respond and avoid the front barrier by sideways, fly higher, lower altitude, and even stop in the air. These sensors also function as an antenna which receive commands from the smartphone and also equipped with the ability to spy on objects.

The flight mission was conducted to obtain imagery of approximately 0.89cm Ground Sampling Distance (GSD). The UAV was flown at an altitude of 30m with an image overlapping 80% side and 70% front for high quality oblique images. The DJI Phantom 4 pro UAV works on the principle of on-board direct geo-referencing where the UAV component, equipped with a GPS that geotagged coordinate for each acquired image. This indicates that the acquired images have been directly geo-referenced by GPS capability during flight mission.

In this experiment, the flight path was designed by using DJI GS pro software. This software is interconnected with UAV through wireless of remote controller that operate at 2.4GHz. The flying altitude is configured based on user's requirement as the flying time depends on the flight altitude and the percentage of image overlapping required. In this case, the images were captured every two second during data acquisition. Figure 3 shows the flight mission during data acquisition process that emerged on the iPad attached to the UAV's remote controller.

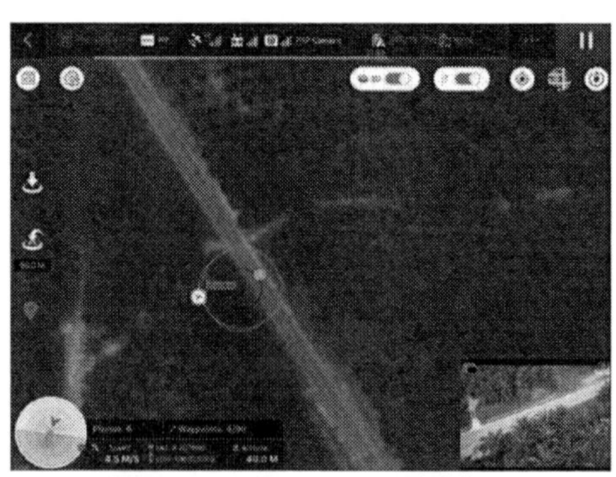

Figure 3. Data Acquisition

2.2.2 Image Processing and Generating DSM. In this experiment, the flying height, image overlap and range of camera to the object were specified. All the UAV components have been tested to ensure all are functioning properly before flight mission. The UAV flies autonomously with the aid of GPS fitted on it. The DJI GS pro navigates the UAV autonomously during flight mission and the attitudes of UAV also can be seen in this apps.
The images from the flight conducted over the tested area were used to build point cloud and produce digital surface model (DSM) using Agisoft photoscan software. This software computes the 3D position based on automatic aerial triangulation methods. The stitching process becomes more efficient since the images had an overlapping of 80%. All the processes were performed under the requirement settings and in a fully automated approach. A total of 149 images from the flight conducted were used to generate point cloud. All the images obtained will be imported into the software to be processed. Align photos is the process to identify the same point of each image and match the same point from two or more images. In this process, the selection of accuracy is based on the highest, high, medium and low processes which involved the time of processing align photos. The image matching and aerial triangulation were performed to create dense point cloud as in Figure 4. The process of build dense cloud was performed based on the estimated camera position which is to be edited and classified before proceed to 3D generation.

(a) (b)

Figure 4. (a) Align Photo (b) Build Dense Cloud

2.2.3 The Application of UAV-Based Watershed Segmentation Method. The tree crown delineation is provided by the segmentation approach namely watershed segmentation algorithm. The segmentation technique is required to determine the diameter of tree crown and tree height derived from digital image [1]. In this paper, the SAGA (System for Automated Geoscientific Analyses) software is used to determine tree crown by implementing the watershed segmentation using different techniques of filtering. The Gaussian and Laplacian filter were implemented by specifying standard deviation and radius of the image as illustrated in Figure 5.

(a) (b)

Figure 5. Filtering Images (a) Gaussian Filter (b) Laplacian Filter

Watershed algorithm is the well-known method for segmenting in order to determine tree crown and tree height [9]. This segmentation is applied to all images that have been captured through the two filter processes namely Gaussian and Laplacian filter. The image segmentation for Gaussian filter process shows that, the image was segmented by color partitioning into multiple segment, while for

IOP Conf. Series: Earth and Environmental Science **385** (2019) 012015 doi:10.1088/1755-1315/385/1/012015

Laplacian filter detected the edges of image to reduce the amount of pixels in order to maintain the shape of the image. Figure 6 shows the watershed segmentation of the image for both of filtering processes.

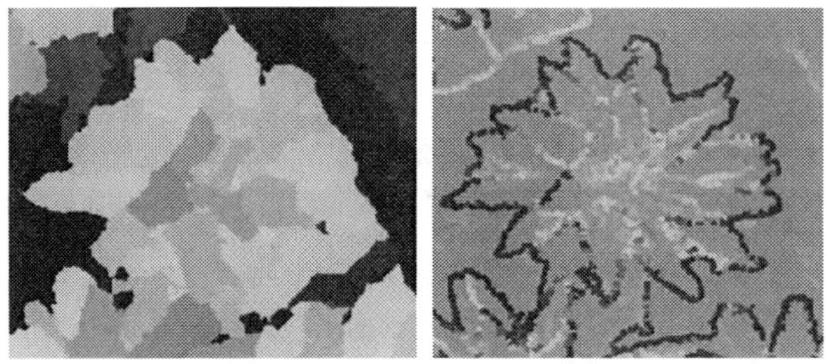

(a) (b)

Figure 6. Watershed Segmentation (a) From Gaussian Filter (b) From Laplacian Filter

3. Results and Discussions

Figure 7 shows the single tree crown delineation obtained with the watershed segmentation method to extract the single tree crowns, the segmentation method was carried out for both Gaussian and Laplacian filter processes. The Gaussian filter is used to blur and remove noise from the images while the Laplacian filter is used to capture the area around images. After performing the watershed segmentation algorithm, the resultant images were converted into shape files which to be used in ArcGIS software. The polygon parameter and the size area of the single tree crown were derived.

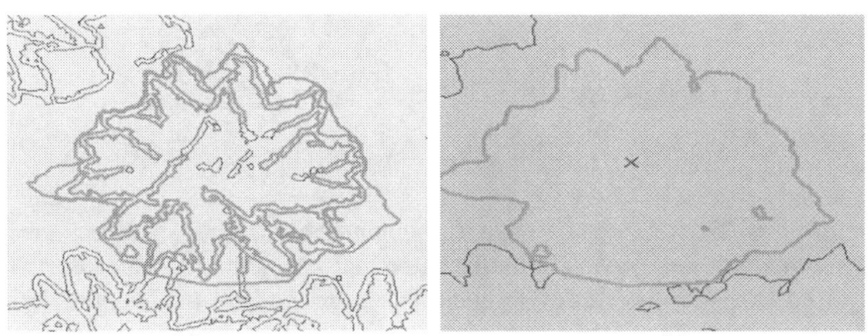

(a) (b)

Figure 7. Single Tree Crown (a) Derived from Laplacian (b) Derived from Gaussian

The tree crown delineation was recognized since the watershed segmentation is applied for both filter processes. However, the tree crown delineation which derived from Laplacian filter shows the shape around the edge of the tree crown. Regarding to that appearance, the value of shape area is based on the calculation for the edge of tree crown, which is not for the whole tree crown. In contrast to the usage of Gaussian filter process, it provides the shape area value of tree crown as a whole. In this paper, the single tree crown was extracted at the same location derived from segmentation method, by using different filtering method.

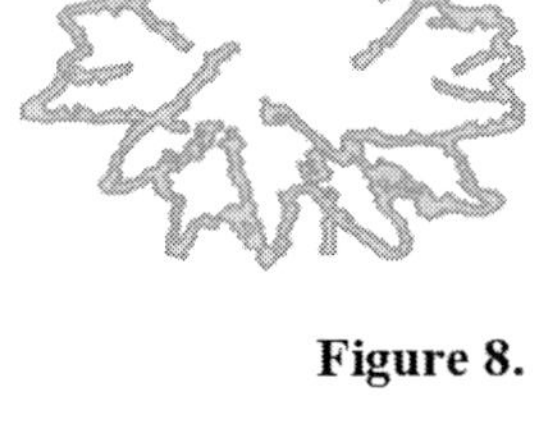

OBJECTID *	Shape *	Id	gridcode	Shape_Length	Shape_Area
3209	Polygon	3209	0	0.671402	0.026743
3210	Polygon	3210	0	0.66319	0.020084
3211	Polygon	3211	0	0.794385	0.031042
3212	Polygon	3212	0	0.294081	0.0046
3213	Polygon	3213	0	1.011869	0.03134
3214	Polygon	3214	0	0.377711	0.008532
3215	Polygon	3215	1	1.105149	0.064626
3216	Polygon	3216	0	0.415469	0.004373
3217	Polygon	3217	0	1.717037	0.116823
3218	Polygon	3218	0	29.886288	9.113703
3219	Polygon	3219	1	1092.323828	4753.620491
3220	Polygon	2326	1	4.571405	0.340504
3221	Polygon	2326	1	180.4489	17.759547
3222	Polygon	2326	1	343.70297	35.709162

Figure 8. Shape Area from Laplacian Filter Process

Figure 8 shows the graphical representation of segmentation result for Laplacian filter with shape value of single tree crown is 17.759 square meter. Meanwhile, the Gaussian filter with reference to Figure 9, shows 76.524 square meter with better value attained in the shape area as compared to Laplacian filter. This probably due to the fact that the watershed segmentation treats raster image to determine tree crown by using the appropriate filtering process. It becomes more challenging to delineate the tree crown that is close to each other.

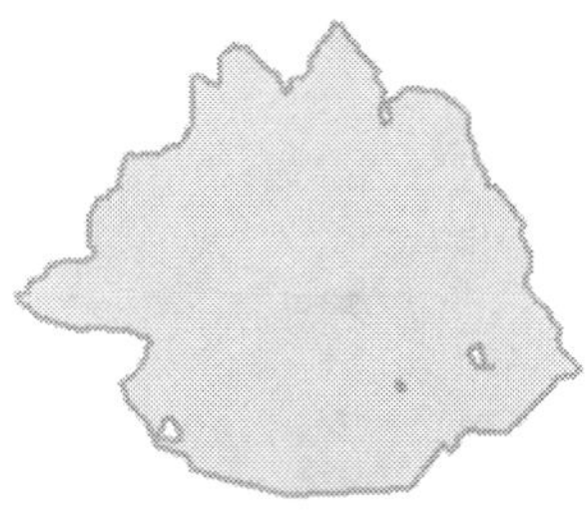

OBJECTID *	Shape *	Id	gridcode	Shape_Length	Shape_Area
282	Polygon	282	0	3.175057	0.293779
283	Polygon	283	0	1.104967	0.067937
284	Polygon	284	0	0.822489	0.037304
285	Polygon	285	1	1458.985096	10049.874522
287	Polygon	285	1	1.913089	0.133385
288	Polygon	285	1	2.717158	0.187967
289	Polygon	285	1	46.2442	76.524301
290	Polygon	285	1	1.672458	0.011186

Figure 9. Area Value Derived from Gaussian Filter Process

4. Conclusion and Recommendation

In conclusion, it is important to highlight that an UAV has the potential to provide huge amounts of information in forest management plan and might be used for surveying and mapping. At present, the use of UAV has been widespread in many industries. Sometimes, the usage of UAV will be limited as there is no method of processing or analysis on the data obtained. Therefore, this study shows that in oil palm trees, tree crown is possible point cloud to be generated, thus improving management decision. The present study also applied the method that produces accurate result with low cost technology and does not require UAV trained specialists. In addition, this study inadvertently able to assist the agricultural sector especially in plantation for further planning and farm managing. The further study will investigate the extraction of tree height based on tree crown in different method. The future work would focus on different types of segmentation algorithm in order to determine the tree crown and tree height for closure ratio among the trees.

4. References

[1] Alberto B, Miranda M, Ribeiro S P and Megill P W 2018 Drone-Based Remote Sensing Application in the Field of Forestry, (July).

[2] Gatziolis D, Fried J S and Monleon V S 2010 Challenges to estimating tree height via LiDAR in closed-canopy forests: A parable from Western Oregon. *Forest Science* **56** 139–155.

[3] Hung C, Bryson M and Sukkarieh S 2012 Multi-class predictive template for tree crown detection *ISPRS Journal of Photogrammetry and Remote Sensing* **68** 170–183

[4] Iizuka K, Yonehara T, Itoh M and Kosugi Y 2018 Estimating Tree Height and Diameter at Breast Height (DBH) from Digital surface models and orthophotos obtained with an unmanned aerial system for a Japanese Cypress (Chamaecyparis obtusa) Forest. *Remote Sensing* **10**

[5] Kim N V, Bodunkov N E, Cherkezov R I 2015 Ensuring the accuracy of traffic monitoring using unmanned aerial vehicles vision systems. *Modern Applied Science*, **9** 323–336

[6] Lisein J, Pierrot-Deseilligny M, Bonnet S and Lejeune P 2013 A photogrammetric workflow for the creation of a forest canopy height model from small unmanned aerial system imagery. *Forests* **4** 922–944

[7] Mohan M, Silva C A, Klauberg C, Jat P, Cardil A, Hudak A T and Dia M 2017 Individual Tree Detection from Unmanned Aerial Vehicle (UAV) Derived Canopy Height Model in an Open Canopy Mixed Conifer Forest **8**

[8] Mweresa I A, Odera P A, Kuria D N and Kenduiywo B K 2017 Estimation of Tree

Distribution and Canopy Heights in Ifakara , Tanzania Using Unmanned Aerial System (UAS) *Stereo Imagery* **6** 187–200

[9] Panagiotidis D, Abdollahnejad A, Surový P, Chiteculo V 2017 Determining tree height and crown diameter from high-resolution UAV imagery. *International Journal of Remote Sensing*, **38** 2392–2410

[10] Ritter B 2014 Use of Unmanned Aerial Vehicles (UAV) for Urban Tree Inventories.

[11] Salleh M R M, Ismail Z and Rahman M Z A 2015 Accuracy Assessment of Lidar-Derived Digital Terrain Model (Dtm) With Different Slope and Canopy Cover in Tropical Forest Region. *ISPRS Annals of Photogrammetry, Remote Sensing and Spatial Information Sciences*, **II-2/W2**(October), 183–189

[12] Seul L Y, Hien L P, Soo P J, Hee L M, Wook P M and In K J 2015 Calculation of Tree Height and Canopy Crown from Drone Images Using Segmentation Calculation of Tree Height and Canopy Crown from Drone Images Using Segmentation. *Journal of the Korean Society of Surveying Geodesy Photogrammetry and Cartography* **33** 605–613

[13] Singh M, Evans D, Tan B S and Nin C S 2015 Mapping and characterizing selected canopy tree species at the Angkor world heritage site in Cambodia using aerial data. *PLoS ONE*, **10** 1–26.

[14] Strigul N S, Gatziolis D and Vogs A 2015 Complementing Forest Inventory Data With Information From Unmanned Aerial Vehicle Imagery and Photogrammetry, 346–351.

[15] Tuominen S, Balazs A, Saari H, Pölönen I, Sarkeala J and Viitala R 2015Unmanned aerial system imagery and photogrammetric canopy height data in area-based estimation of forest variables **49** 1–19.

[16] Zarco-Tejada P J, Diaz-Varela R, Angileri V and Loudjani P 2014 Tree height quantification using very high resolution imagery acquired from an unmanned aerial vehicle (UAV) and automatic 3D photo-reconstruction methods. *European Journal of Agronomy* **55** 89–99

Compatibility Issues of Industrialized Building System (IBS) in Interior Construction: A Case Study in Kuala Lumpur

R Yaman[1], K Ali[1], H Abdullah[1], F Muhamad Halil and N Ahmad[2]
[1] Faculty Architecture, Planning & Surveying, Universiti Teknologi MARA, 40450 Shah Alam, Selangor, Malaysia.
[2] Kulliyah of Architecture & Environmental Design, International Islamic University of Malaysia, Gombak, Selangor, Malaysia.

Email: rostamyaman1@yahoo.co.uk

Abstract. The Construction Industry Development Board Malaysia (CIDB) defines Industrialized Building System (IBS) as a construction technique in which components are manufactured in a controlled environment (on or off-site), positioned, transported and assembled into a structure with minimal additional site works. Such as, component that are being pre-fabricated in a controlled environment on-site are also considered as IBS. Main problem is when the contractors have bad workmanship while using pre-fabricated panels at site and IBS components are deliberated to be inflexible with respect to be changes or renovations. The aim of this research to identify the issues of compatibility of Interior construction in IBS components. This study will concentrate on interior construction works using of IBS components. The importance of this research is to establish design for solution the issues in using IBS components in interior constructions. This research also to develop design criteria of IBS components that have advantage and disadvantage in interior construction industry in Malaysia. According to this study, the availability of knowledge and information on IBS components for interior construction will be contribute to others. This research also develops new design criteria benefits to designer that using IBS components at site. In long term this research will implement as set of guidelines and additional knowledge that may encourage usage of user and CIDB.

1. Introduction

Construction Industry Development Board Malaysia (2001) imply IBS as a technique in building construction whereby the material components are pre-fabricated off-site or in a regulated setting, organized, transported and constructed into a completed construct with minimum fixing and installation work at the construction site [1]. CIDB further explained that component of building construction that being pre-fabricated in the working site under a controlled setting is also recognized as an IBS system. Eventhough there are various research being conducted in defining the Industrialized Building System (IBS) concept, the concluded findings suggested that IBS construction method concerned with prefabricated components or it being mass-produced off-site [1, 2].

The major problem with IBS system is that the contractors are poor in workmanship when installing pre-fabricated components at construction site. This can cause problems in the termination joints, hence resulted water leaks and eyesores. In the tropical country such as

Malaysia whereby precipitations virtually falls almost daily, the issues is very critical. To address all these problems, the designers need to diligently examine thoroughly the finished units upon handover in order to verify that the developer addressed all defect works.
Another issue related with pre-fabricated components is that IBS system are considered as inflexible with respect to alterations or reconstructions. Most people likely want to make installation holes for wall fixtures or may add other feature on the wall. But certainly, in renovation works there might be hacking in order to provide additional conduits and electrical wirings that might be compromised the systems or components. Or it might arise a problems in trying to demolish down certain walls in order to combine two units to become larger one unit if units is contiguous to each other. The deficiency of methods and tools in the constructions defining a value. The main point is to develop a criteria or guidelines for designers decisions making when using the system. Each sustainable criteria performance must be address to solve the hidden prospects of implementing IBS construction system [2].
According to [3, 4] the limited knowledge about IBS construction system, plenty of cheap inexpert labors and huge initial investment outlay some of the barriers in implementing IBS construction systems [3, 4] These rendered causes and difficulties will affect designers and constructors to opt for conventional approach instead of implementing IBS construction systems. According to Hussein, 2009, he added that key stakeholders in construction industry have undesirable views towards implementing IBS construction systems [5]. To homogeneously and broadly define the sustainability concept and the vital aspects in built environment is to consider sustainability evaluation of the built facilities [6].
The research problem is lack of knowledge by the designers in implementing IBS construction system. This study findings will assist the designers in identifying and managing the economic, social and environmental risks in a holistic manners. The conclusions will validate the guidelines and help the designers in determining the advantages and limitations of utilizing IBS material components and maximized the use of the system.

2. Research Framework

The research type used in this study is explanatory method to determine the current situation, issues, and problems in IBS system components and obtained a new understanding from another perspectives. In this study, the correlated variables are recognized and hypotheses in order to generate future study [7].
From the concept of this study, the conceptual framework was synthesized as shown in Figure 1.

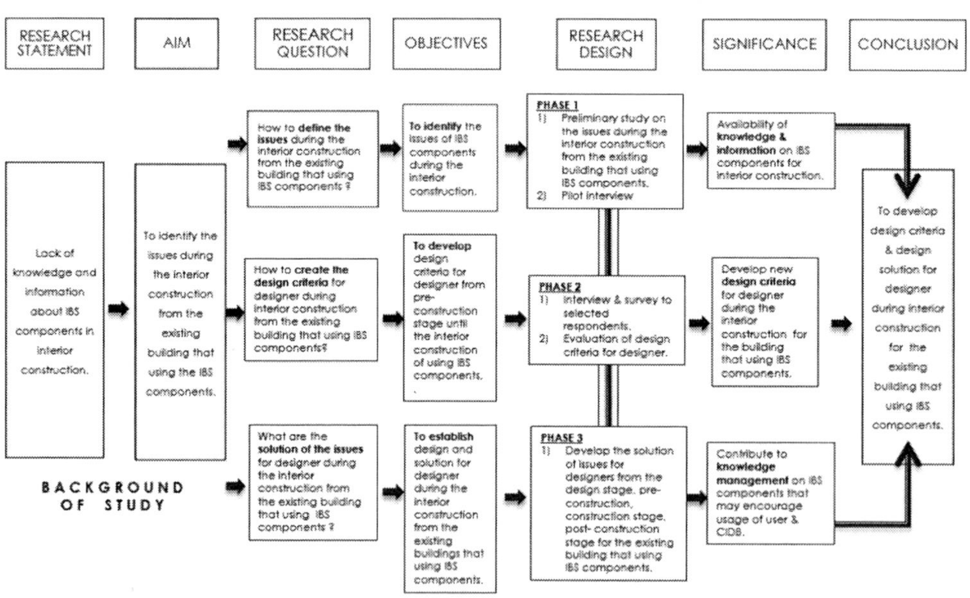

Figure 1. Framework of Study.

2.1 Objectives

The objectives of this research are to distinguish the significant understanding of stakeholders in IBS system components throughout the period of construction. The information and application are expended to develop preliminary study framework. There are three key point questionnaires are being conducted:

1) **To identify** the problem statement concerning compatibility of interior construction in Industrialized Building System (IBS).
2) **To develop** design criteria in interior construction of using Industrialized Building System (IBS) material components.
3) **To establish** design for solution the issues of Industrialized Building System (IBS) in interior construction.

3. Methodology

The methodology chosen must matched the addressed research questions. In this study, a combined data will deliver the utmost comprehensive problem statement analysis. The main problem of this study will lead to answer the addressed study question. For appropriate and thorough analysis, this study will develop a decision making guideline for designer understanding in implementing IBS system components in building construction particularly in interior works. Figure 2 below indicated the study method framework.

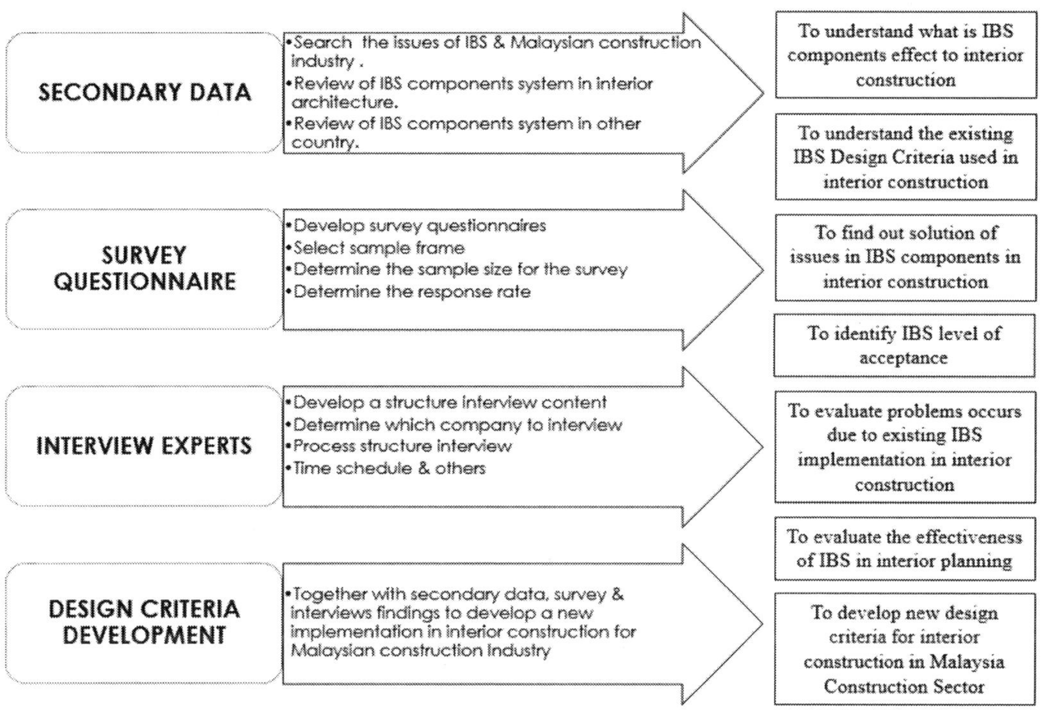

Figure 2. The Study Method Framework

This research involves a number of sampling in answering the questionnaires. Besides the survey questionnaires, this study also used structured interview in order to obtain data and input from varies experienced and expert respondents in enhancing the comprehension towards solving the main study problems.

The sequential of research design, combined both quantitative and qualitative data [8]. The entire investigation is presented as a combination of both types of gathered data. The sequential strategies of the study is as indicated in the Figure 3.

IOP Conf. Series: Earth and Environmental Science **385** (2019) 012016 doi:10.1088/1755-1315/385/1/012016

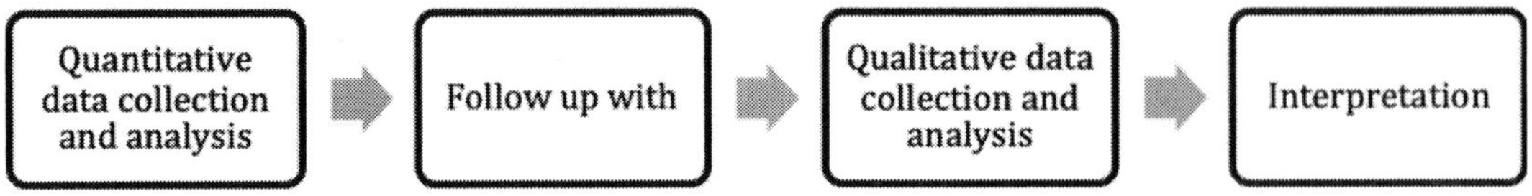

Figure 3. Explanatory Progressive Research Design
Adapted from: Creswell & Plano Clark, 2011 [9].

In this study, the developed guidelines are based on every issue in-depth investigation via semi-structure study interviews. Thus, it will improvise understanding on the positive and negative aspect and contexts within proper investigation. Table 1 below indicated the summary of the selected method.

Table 1. Summary of the Selected Method

Study Method	Key Aspects	Objectives
Survey Questionnaires	• All targeted respondents • A variables quantity to be quantified and analyzed • No physical context	• Examine the probable IBS stakeholders • obtaining consensus from various types of stakeholders • statistically analyzed the respondents feedback.
Semi- Structured Interviews	• investigate directly the recognized probable factors • provide discussion on devising the efficiency of IBS system components • allow researcher to pursue and ratify the addressed issues	• validating the study framework grounded based statistical examination • develop the decision-making guidelines for improving the understanding of IBS components • obtain opinions by responding to every vital factors
Case Studies	• allow researcher to investigate the comprehended framework and guidelines developed • Indicates the procedures, needed actions and required activities concerning the issues in IBS system components in interior construction	• investigate on how the formulated guidelines were used in addressing problems of IBS system components • explore the procedures, process and activities affected the decisions-making • measure the effectiveness and practicality of the formulated guidelines

To improve the IBS system components understanding within IBS stakeholders, the decision-making guidelines are develop as the final findings of this study. This sections is to explain the study process implicated in developing the guidelines. The study process was based from the preceding study by Riduan Yunus (2012) [10], whereby the presented step is embraced in developing decision-making guidelines for IBS stakeholders. To further improve the understanding for best decision among IBS stakeholders, SWOT analysis method was adopted.

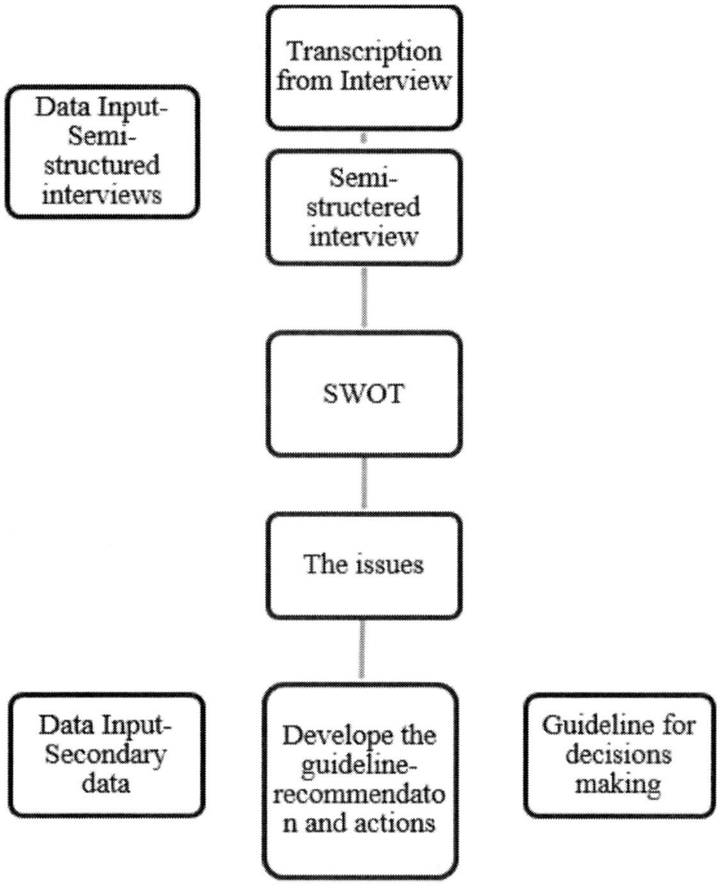

Figure 4. Development Guideline Process
Adopted from: Riduan Yunus, 2012 [10].

4. Result

Survey Questionnaires. The surveys are conducted to examine the stakeholders' understanding on Industrialized Building System (IBS) in the field of interior construction. Founded on the survey conducted, the following findings can be concluded:

1. The stakeholders understood that IBS system components are not easily installed or use for interior construction.
2. The stakeholders understood that IBS systems are more advantageous in term of speed, quality and safety during the period of construction.
3. The stakeholders are still comparing the conventional system and IBS system based the problems that they encounter throughout the construction phase.
4. The IBS systems need to be more detail during the planning and design phases.
5. The stakeholders are still unwilling to recommend the IBS systems in their future coming projects due to higher cost during construction, knowledge lacking among the workers, failure design during the implementation and not clear guidelines in Malaysian construction industry.
6. The authority or government should reward more incentive via the related construction bodies.
7. The related construction bodies should come up with more campaign or conferences in encouraging the designers to specified design based on IBS systems.

Interview. This method present the analyze result based on semi-structured interviews. It provide in-depth understanding concerning IBS systems from respondents who answer the questions. The critical circumstances during phases of construction were deliberated and validated by the respondents. This method is important in order to ensure the vital factors and

problems in IBS systems during construction phases can be identified and improvised in the near future. Based on given view, the respondents provide the possibility to stakeholders in gaining more knowledge on IBS systems. The SWOT investigation and commendations is formulated based on the feedback response in order to develop criteria guidelines in assisting designers to appraise the issues and principles on IBS systems in design process.
Case Studies. The outcome from this method in this study is to develop an answer or guidelines to help designers improve their understanding on IBS systems during interior design and construction phases. The on-site observations of real projects is to help researcher in improvising the understanding on IBS systems. Thus, to ensure the results which translated as guidelines can be used as deciding tools for designer. The process involved when using guidelines can assist the designers in attaining the appropriate decision at the early planning and design phases. It will act as a reference guide for the designer to understand the systems before implementation or construction.

5. Conclusion

The support from the Construction Industry Development Board Malaysia really facilitate the researcher in identifying the stakeholders that implementing IBS systems in their projects [11]. The conceptual framework focused on vital issues which are highlighted from beginning of the research stage. The identified critical issues is to help elevating the awareness between industry key players, understanding the systems, the concern issues and how to manage it on-site. In this study, the SWOT analysis is to redefine the identified issues whether internal or external aspects concerning IBS systems. The SWOT analysis also will assist the designers or related users in making appropriate system components selection for interior construction phase. In this study, majority of respondents prefer to have a reference guide on IBS systems for interior construction. This study is also to provide a result and guides for space designers who are the decision maker during design phase.
The limitation of this study is that throughout the research, the focus is towards types of IBS systems and the demanding issues of current IBS implementation. From study planning and recommendation from chosen respondents, are based on their experience and knowledge in building industry. During construction phase of projects, many persons involved in decision making even though the same objectives is to complete the project. Every respondents need to call for a decision making which can be improvised and optimized. Hence, there might be non-uniformity on the feedback as the respondents in this study are varies in background and implementation process but yet to deliver decision making.
For future research, this study has embark in more opportunities to investigate on the development of IBS systems and for interior design construction. It's recommended that the findings in this study can used to review and redefine the guidelines for IBS implementation. Focus should be emphasize on the materials, standardization of procedure and work skills. Consideration need to be given for refurbished spaces or buildings that implementing IBS systems. Future study should also investigate the way to minimize the problems pertaining IBS system during interior fit-out or refurbishment in building. Other aspects is to relate electronic tools for guideline information, so that it will be more systematic in analyzing and processing the IBS system input. The auto-generated report ease the procedure and process via the electronic guideline tool. Knowledge and data from the IBS stakeholders also can be easily shared for wider awareness of the systems and to promote sustainable construction.

References

[1] CIDB Malaysia 2001 Manual for Assessment of Industrialised Building Systems", *CIDB, Kuala Lumpur*
[2] Construction Industry Development Board 2003 *Industrialised Building System (IBS) Roadmap 2003-2010*. Kuala Lumpur: CIDB.
[3] Blismas N G, Pendlebury M, Gibb A and Pasquire C 2005 Constraints to the use of off-site production on construction projects. *Architectural Engineering and Design Management* **1** 153-162

Blismas N and Wakefield R 2009 19 Engineering sustainable solutions through off-

[4] site manufacture. *Technology, design and process innovation in the built environment* **10** 355.

[5] Nawi M N M, Lee A and Nor K M 2011 Barriers to implementation of the industrialised building system (IBS) in Malaysia. *The Built & Human Environment Review* **4** 34-37

[6] Pearce A R and Vanegas J A 2002 Defining sustainability for built environment systems: an operational framework *International Journal of Environmental Technology and Management* **2** 94-113

[7] Fellows R and Liu A 2008 Impact of participants' values on construction sustainability In *Proceedings of the Institution of Civil Engineers-Engineering Sustainability* **161** Thomas Telford Ltd.

[8] Creswell J W 2009 *Research Design: Qualitative, Quantative and Mixed Methods Approaches* (3rd ed.). Los Angeles: SAGE Publications, Inc.

[9] Creswell J W and Plano Clark V L 2011 *Designing and conducting mixed methods research* (2 ed.). California: SAGE Publications, Inc.

[10] Yunus R 2012 *Decision making guidelines for sustainable construction of industrialised building systems* (Doctoral dissertation, Queensland University of Technology).

[11] CIDB Malaysia & APEC Technology Foresight Centre, Bangkok 2000 Malaysian Construction Industry: Technology Foresight Report *Proceedings of the CIDB Workshop on Technology Foresight, CIDB, Kuala Lumpur, 21- 22 March 2000.*

IOP Conf. Series: Earth and Environmental Science **385** (2019) 012017 doi:10.1088/1755-1315/385/1/012017

A S.W.O.T. Analysis of Green Procurement Implementation in Construction Projects

Z A Bidin[1], A A M Bohari[1], S L A Rais[1] and M M Saferi[1]
[1]Faculty of Architecture, Planning and Surveying, Universiti Teknologi MARA, Cawangan Sarawak, Kampus Samarahan, Kota Samarahan, Sarawak, Malaysia

Email: zafikha936@sarawak.uitm.edu.my

Abstract. Green-related practices are widely implemented in various countries to ensure that problems related to the environment can be reduced. The concept of green-related practices and strategies were introduced and identified in providing the green procurement can be fully implemented in the Malaysian construction project. Green procurement concept is presented as one of the initiatives to support in reducing environmental issues throughout construction execution. Hence, this paper focuses on analysing the strengths and weaknesses of green procurement as well presented the opportunities and threats towards construction. Focus group discussion was conducted among green expert's participant who shares their experiences and point of views related to green procurement. The outcomes of this research findings offer plans and a clearer picture for practitioners in implementing green procurement practices in construction execution.

1. Introduction

Construction activities give significant impacts on the environment, social and economics. There are some significant issues related to construction activities such as pollution, degradation of natural resources and carbon footprint [1]. Hence, there are some initiatives to ensure these environmental problems can be conserved and minimised [2]. The introduction of green-related practices has been widely endorsed to alleviate these issues.

The Malaysian government has encouraged construction stakeholders to implement green procurement approach to help and ensure that the project execution is being greened [1]. Green procurement was introduced under the National Green Technology Policy 2009 as one of the strategies to conserve the environment [3]. Malaysia is one of the developing countries that are rapidly building physical development to cater to facilities such as building and infrastructure for communities, and these activities will contribute toward indoor and outdoor environmental pollution [1]. As the green procurement concept is still very new in the Malaysian construction industry, it is crucial to address this towards the construction stakeholders to ensure they implement it to procure green construction projects [1].

Procurement is one of the essential elements in managing the construction project. Sustainability in the project life cycle is crucial in ensuring environment impact can be minimised. However, there are issues in the current procurement practices that disregard project lifecycle sustainability. [4] To help and manage environmental issues, green procurement is introduced as useful tools. Implementing green procurement means construction stakeholders need to ensure that environmental problems can be minimised from the initial towards the completion of construction activities [1]. The objective of this paper to identify strength, weaknesses, opportunities and threats of green procurement practices using SWOT analysis and assessed green's perceptions on these matters in the local context of the Malaysian construction industry.

2. Green Procurement

There is several authors' defined green procurement as an effective way of purchasing products or services that can minimise and reduce environmental impacts. Moreover, [5] mentioned that green procurement is a process of purchasing products, works and services which the process is interrelated with environmental factors.

According to [6], green procurement is a process of assessing the purchasing products at lifecycle stages and its environmental effect from purchasing raw materials, transporting, handling, manufacturing, storage, consumptions and disposals. This is also supported by [7], green procurement is the process at the lifecycle stages. Apart from that [8] mentioned that it is essential to select and set the environmental requirements in the contract. Several green procurement initiatives have been introduced such as LEEDS certification that focuses on the standards of energy use, toxic material specifications, recycle of material, works safety procedures and social benefits [4]. [14] mentioned that green procurement capable in giving benefits towards environment. The implementation of green procurement can increase more demand for terms of green products, materials and services. Moreover, in Malaysia, MyHijau were introduced under the Ministry of Energy, Green Technology and Water (KeTTHA) to reassure green technology and green purchasing were implemented in construction projects in Malaysia [1]. According to [9], Eco-labels is essential tools to promote sustainable procurement, and it is crucial to strengthen green project management knowledge among construction project stakeholders [10].

3. Methodology

3.1 Research design

A comprehensive literature review was commenced to gather data on green procurement, the challenges and success factors of green procurement in construction projects. Data and information from the literature review were used to support the primary findings. A focus group discussion (FGD) was conducted with green experts to validate the objectives of this paper and to obtain a better knowledge of green procurement in Malaysia. The discussions were structured according to the themes using research questions prepared in advance before the discussions started and questions for the session were given to the moderator as guidelines.

3.2 Expert discussion

Expert group discussion was conducted to validate and obtain useful insights and knowledge of green procurement practices in the construction industry. Selection of green experts' participants was based on their rich experience and background in green procurement practices in the construction industry. A total of 18 participants in the focus group and they were among experts and academicians. Focus group discussion (FGD) approach was used to collect qualitative data for this research.

3.2.1 Arrangements of FGD

- **Invitations** – Construction companies and academicians who involved in green construction and experts on green procurement was contacted in advance, and they were explained about the discussion for the FGD. Invitation letters were sent to each participant that agreed to join the FGD.
- **Venue** - The FGD was conducted at Dewan Perpaduan III 1st Floor, Merdeka Palace Kuching on 6th February 2018. Duration of the event took place at 2:00 pm – 5:00 pm.
- **Sharing session** – A brief sharing session by Dr Khairul Naim bin Adhar on Green Construction Projects through Construction Procurement before the breakout sessions started.

- **Group Discussion composition** – The FGD were conducted using face to face discussion consisted of a group of green experts and academician. Each participant was assigned according to the themes of the discussion. The themes include the challenges and success factors of green procurement in construction projects. A brief introduction, purpose and scope of the FGD are explained to the participants, and they were asked to give a short introduction and background about themselves. All participants are given a chance to participate during the discussion and discussed relevant topics according to the themes. The discussion was led by one main moderator to ensure the discussion run smoothly and steer the in-depth topic of discussion with the participants.
- **Seating arrangements** – Circular seating arrangement was set to ensure clear interaction among participants and each participant provided with table name tags to enable moderator and rapporteur recognise the participants.
- **Recording and note-taking** - A rapporteur was assigned to take extensive notes during the discussion, and the discussion was audio recorded. All the input gathered will be further analysed. The session of the FDG lasted for 50 -60 minutes to complete. All participants were informed that all responses and discussion were kept strictly confidential which only be used for research purposes only. All transcribed audio record will be kept anonymously, and the final discussion results were then emailed to the participants.
- **Light refreshments, gifts, and photo session** – Light refreshments and gifts for the participant were given as a token of appreciation for willingness to spend their time to participate in the FGD and group photo session was taken after the discussion.

3.3 SWOT Approach
SWOT (Strength, Weakness, Opportunities, Threats) analysis is a technique that widely used to conduct a strategic study. A detailed SWOT approach was conducted to attain the understanding of the strength of green procurement, the characteristic of green procurement and the benefits of green procurement in construction project execution. Moreover, the drawbacks of green procurement were further identified and what can be improved. The changes in government policy, trends in construction-related with green procurement and the challenges of green procurement in construction project were further analysed using the SWOT approach.

4. Analysis and Findings
4.1 SWOT analysis
Different authors highlighted that based on countries that implement green procurement, green procurement could minimise environmental damages, less waste, less energy consumption, financial and environmental of construction can be improved, reduce pollution, raised awareness among staff on green practices and policies as well as improve company's positive image. Supply chain networks also can be enhanced.

Table 1. SWOT analysis by different authors.

Ref.	Authors	Strength
1.	ICLEI European Secretariat, 2007, Green Council 2010; Varnas et al., 2009), Hutchison, 1998; Salam, 2008; Cruz and Wakolbinger, 2008; European Commission, 2011; Hutchison, 1998	• Less energy consumption products, • Less waste • Lower carbon footprints • Minimise environmental damages • Environment and financial performance of construction and building process can be improved. • Reduce pollution, reduce environmental handling costs. • Raised and increased staff awareness and knowledge of green practices and green policies. • Improve the company's positive public image (corporate market competitiveness) • Supply chain networks can be enhanced, transaction costs, risks and emissions can be reduced.

Ref.	Authors	Weaknesses
1.	Essa & Fortune, 2008	• Lack of knowledge (social sustainability factors)
2.	Carter and Fortune, 2007	• Lack of structured frameworks (assists project teams)
3.	Faith-Ell et al., 2006	• Lack of information and the inability to supervise after contract award. (intimidating tasks of practical implementation of environmental criteria)
4.	Sourani and Sohail, 2011; Walker and Hampson, 2008; Ruparathna and Hewage, 2015	• Lack of funding, lack of awareness, lack of research and development, lack of understanding, information, commitment, and demand, insufficient policies,

		regulations, incentives and lack of leadership

Ref.	Authors	Opportunities
1.	Green Council, 2010	• Increased and create business opportunities with international investors
2.	Office of the Auditor General Canada, 2005	• Build green procurement expertise, expands tools and set requirements.

Ref.	Authors	Threats
1.	Pitt, M., Tucker, M., Riley, M., & Longden, J., 2009; Shiers et al., 2006	• Lack of client demand • Reluctance in contractor

As gathered through extensive literature review, these statements were further validated through focus group discussion that has been conducted. The focus group discussion data were analysed by transcribing focus group comments. Analysing the transcripts and then identify responses that relevant to the questions asked. The answers of the participants were themes and categories according to the research questions. SWOT analysis and findings based on focus group discussion were summarised as follows;

Table 2. SWOT analysis based on FGD comments.

Themes	Data extract	Participants	S, W, O, T
Policies & guidelines	"No available alternative in green practices."	P1	T
	"existing policies not enough yet to drive green procurement."	P6	W
Green Specification	"must be integrated into tendering and contract."	P4	W
Sponsorship	"Lack of sponsorship and stakeholder not ready to change."	P4	T
Knowledge/ Awareness	"Local construction industry still not mature in term of green procurement."	P5	W
	"Everybody wants to go green, but not many understand what the going green means is"	P7	W
	"Not serious and not clear in		W

	implementing green procurement."		
	The government need to create awareness, info, input or training that provide knowledge to industry practitioners	P8	W
	"people did not see the impacts of using certain materials and relate it with environmental aspect rather than giving them profit"	P6	T
Training	"Providing training to construction stakeholders."	P1	O
Government support	"Government play important roles, in policy enforcement such as green certification."	P8	O
	"Impose the use of green procurement as mandatory, if government say do it, everyone will follow."	P9	O
Cost	"we concern about the budget. when we talk about green procurement, we will think about the higher cost."	P10	W
Local supply	"Main concern, supplier for green material and products, are still lacking, mostly we need to resource from Semenanjung most of the time we have a problem such as transport and local supply."	P8	W
Expertise	"In term of resources, yes, we have it but in term of human resources and expertise is still lacking."	P11	T
Incentives	"Incentives from the Government as initiatives for the next step so that the stakeholders make use and take advantage from it."	P12	O

4.2 Summary way forward/improvement

From the analysis, it can be concluded that from the perspective of local contexts, there is no strength of green procurement been highlighted by the participants and this shows minimal knowledge and awareness of green procurement strengths and benefits. Different authors highlighted that based on countries that implement green procurement, green procurement could minimise environmental damages, less waste, less energy consumption, financial and environmental of construction can be improved, reduce pollution, raised awareness among staff on green practices and policies as well as improve company's positive image. Supply chain networks also can be enhanced.

From a focal point of local contexts, the weaknesses of green procurement are the policies and guidelines not enough yet to drive local construction players to implement green procurement. Previous researchers also mentioned that there are insufficient policies available. Moreover, green specification not integrated into tendering and contract. It is crucial for the government to enforce the policy and make it as mandatory as a driver towards implementation of green procurement in construction projects. [16] also highlighted that policy pressure such as changes in building codes will make the developers refused to implement green procurement practice.

Moreover, this incurred a higher cost and lack of resources and supply of green products. According to [11], [12]; [13]; [4], lack of awareness, lack of knowledge, lack of funding are the weaknesses of green procurement implementation. Without awareness and knowledge, the construction players will not implement this in construction projects because giving knowledge regarding future benefits of environmental, economic and social benefits are essential to ensure that they aware of the benefits of using the green procurement. Lacking human resources and expertise will create problems in the implementation of green procurement. According to [15], it is important to provide technological basis such as green techniques to ensure the implementation of green procurement. Moreover, support and collaboration from both government and industry is important to increase the awareness in construction. It is also highlighted by the participants on the lack of awareness and knowledge of green procurement among local construction player. Giving training will provide opportunities regarding building green procurement expertise, expands its tools and set requirements. Implementation of green procurement will increase and creates business opportunities with international investors. From the group discussion, one of the participants highlighted that there are a lot of activities were introduced to educate contractor and stakeholder on the importance of taking care of the environment such as seminar to expose them on the environmental aspect and safety at the construction site. The green campus has also been introduced as an early exposure to students and staff to raise awareness of the green element in construction.

From the participant's perspectives, green procurement incurred a higher cost to implement in the construction project. They need government support regarding giving incentives, sponsorship or funding as a driver to perform green procurement in the construction project. Lacking regarding funding will make contractors reluctant and not ready to change. Top management supports are also essential to ensure green procurement can be implemented.

5. Conclusion

As a conclusion, the awareness and knowledge of green procurement need to be exposed towards construction players to ensure that they know about the benefits and function of green procurement in construction. The participants have not highlighted the strengths of green procurement because of minimal awareness and knowledge about it. The drivers to implement green procurement in construction are essential to ensure that it can be slowly being implemented in the construction projects. Policies and guidelines need to be clear, specifics and enforced by the government as mandatory to ensure that construction players implement it in the projects. More training required to form more expertise and give them knowledge on green-related practices and procurement. Furthermore, the need for financial support from the government is essential to show they support green procurement implementation in construction projects. These are the important factors in ensuring the implementation of green procurement in construction projects among Malaysian.

References

[1] Bohari A A, Skitmore M, Xia B, Teo M, Zhang X and Adham K N 2015 The path towards greening the Malaysian construction industry *Renewable and Sustainable Energy Reviews* **1** 1742-8

[2] Dodds F, Strauss M 2012 Only one Earth: The long road via Rio to sustainable development. *Routledge*

[3] Kahlenborn W, Mansor N and Adham K N 2013 Government green procurement (GGP) Short-term action plan 2013-2014 *Malaysia: Economic Planning Unit*
[4] Ruparathna R and Hewage K 2015 Sustainable procurement in the Canadian construction industry: current practices, drivers and opportunities *Journal of Cleaner Production* **16** 305-14.
[5] Günther E and Scheibe L 2006 The hurdle analysis. A self-evaluation tool for municipalities to identify, analyse and overcome hurdles to green procurement. *Corporate Social Responsibility and Environmental Management* **13** 61-77
[6] Salam M A 2008 An empirical investigation of the determinants of adoption of green procurement for successful green supply chain management *4th IEEE International Conference* 21 September 2018 1038-1043
[7] Ofori G 1999 Satisfying the customer by changing production patterns to realise sustainable construction *Joint Triennial Symposium of CIB Commissions W65* **55** 41-56
[8] Srinivas H 2018 Sustainability Concepts: Green Procurement [Internet]. Gdrc.org. 2018 [cited 9 December 2018]. Available from: http://www.gdrc.org/sustdev/concepts/14-gproc.html
[9] Mosgaard M, Riisgaard H and Huulgaard R D 2013 Greening non-product-related procurement–when policy meets reality *Journal of cleaner production* **1** 137-45
[10] Hwang B G and Ng W J 2013 Project management knowledge and skills for green construction: Overcoming challenges *International Journal of Project Management* **1** 272-84
[11] Essa R and Fortune C 2008 Pre-construction evaluation practices of sustainable housing projects in the UK *Engineering, Construction and Architectural Management* **7** 514-26
[12] Sourani A Barriers to addressing sustainable construction in public procurement strategies.
[13] Walker D and Hampson K 2008 Procurement strategies: A relationship-based approach *John Wiley & Sons*
[14] Bohari A A, Skitmore M, Xia B, Teo M 2017 Green oriented procurement for building projects: Preliminary findings from Malaysia *Journal of cleaner production* **1** 690-700.
[15] Wong J K, San Chan J K and Wadu M J 2016 Facilitating effective green procurement in construction projects: An empirical study of the enablers *Journal of Cleaner Production* **1** 859-71
[16] Shen L, Zhang Z and Zhang X 2017 Key factors affecting green procurement in real estate development: a China study *Journal of Cleaner Production* **1** 372-83

IOP Conf. Series: Earth and Environmental Science **385** (2019) 012018 doi:10.1088/1755-1315/385/1/012018

Post Occupancy Evaluation Measures of Green Interiors Tools Energy Efficiency (EE) Core Criteria

R Yaman[1], J Abdullah[1], H Adnan[1], F Z Ismail[1] and N Ahmad[2]

[1] Faculty Architecture, Planning & Surveying, Universiti Teknologi MARA, 40450 Shah Alam, Selangor, Malaysia

[2] Kulliyah of Architecture & Environmental Design, International Islamic University of Malaysia, Gombak, Malaysia

Email: rostamyaman1@yahoo.co.uk

Abstract. Interior spaces sustainable rating system in Malaysian is rather new. Even though Green Building Index Interior Tools (GBI-IT) has been developed and implemented but there is lack of post-occupancy evaluation being conducted in assessing the sustainability level of the certified project, particularly on the Space Energy Intensity (SEI). Thus, embarking on the research problem whether the certified interior space sustainable level is according to the pre-occupancy scoring benchmark as vision by the designers upon occupancy, particularly on Energy Efficiency (EE). The research objective is to examine the SEI, to identify similarities and discrepancies of SEI upon pre-occupancy evaluation and post occupancy evaluation and to assess the key EE sub-criteria that affect Space Energy Intensity (SEI) and to improve continuing sustainable EE upon post occupancy. The Space Energy Intensity equation approach and implementation method is used in this research in order to measure total space energy consumption per year/interior space area. The research design will be formulated into three key stages, which are; 1. Content analysis of GBI Interior Tools EE core criteria; 2. Comparative data analysis between pre-occupancy and post-occupancy EE sub-criteria scoring and; 3. Assessment of key EE sub-criteria that affect Space Energy Intensity (SEI) and to improve continuing sustainable EE upon post occupancy. The findings for research problem based on Space Energy Intensity (SEI) on-site measure from case study will improve sustainability practice and continuing sustainable agenda particularly on EE. The study outcome will facilitate and enhance the current and future sustainable approach in interior SEI.

1. Introduction

Malaysia is a Party to the United Nations Framework on Climate Change and has ratified the Kyoto Protocol on 4 September 2004 [1]. Hence, Malaysia has moved towards a developed nations by the year 2020 and it needs to fulfill the requirements of being a developed country not only in economic and technology aspect but also in varieables aspects of sustainable development. Beside the legislation of interior design professions that leads to development of GBI interiors rating beside the existing available tools. Malaysia Government support the campaign towards green building and technology as announced in 2010 Budget which comprehended [2]: the fund of not less than 1.5 billion Malaysian Ringgit (MYR) to be disburse as soft loans, tax exemption to building proprietors who acquire the new GBI certification, purchasers of buildings with GBI certification will also exempted from stamp duty which valid between Oct. 2009 until Dec 31 . 2014 [3]. The government interventions and supports in promoting sustainable development are very important. It creates synergies and drives in the efforts of rapid modernization without compromising the environment.

Green Building Index (GBI) was introduced in Malaysia in 2009 as Malaysia first green building rating tool [4]. GBI was jointly developed by two professional bodies, the Malaysian Institute of Architect (PAM) and the Association of Consulting Engineers Malaysia (ACEM) [4]. The recognized indexing rating tool for green buildings is Malaysia's industry is Green Building Index (GBI). Different from other rating system such as LEED, GBI Malaysia assessment criteria is designed specifically for the tropical climate while integrate Malaysia's current social, infrastructure and economic development [5]. Hence, a new framework for developing GBI assessments criteria and methods is needed to ensure that green and sustainable interiors will reduce environmental impacts while being design-effective and socially responsive.

Green or sustainable design ideations in interior design project are gaining momentum in the Malaysia because of the high awareness on societies and the extra marketable values that it carry along the extensive and diverse project development in capital Kuala Lumpur. This is evidenced by joint effort development of Malaysian Institute of Interior Designers (MIID) and Malaysia Green Building Confederation (MGBC) in establishing the framework of Green Building Index (GBI) Tools for interior design project. Such joint effort to address the gap required in supporting the ever complex. Sustainable lifestyle and development in Malaysia especially in public interior buildings and amenities operation is imperative. The way of building designed, constructed and operated is significant to human health and environment impact. The fact that buildings required maintenance procedures, air ventilations, state-of-art equipment and 24hrs operation elevated the energy consumptions [6]. The resources used in building stressed on the importance of assessment criteria's in Malaysia green ratings and served as an initial point for the optimum and systematic green and sustainable interior practice in Malaysia [7].

2. Research Framework

GBI for Interiors initiative aims to assist the building industry in its march towards sustainable development in interior design, architecture and urban planning as a holistic approach. The aims is to set common accepted standards; promoting integrated interior and based building from inception of the project; recognised and reward green initiatives; and relevancy of interior projects in the future. However, this pre-occupancy evaluation measures and guides for certified Green Interior Project is evaluated during design, construction and completion stage. Hence the evaluation was not done and carried out after the space is occupied by the end-users. Thus, embark to the research problem whether the certified interior space sustainable level is according to the pre-occupancy scoring benchmark as vision by the designers upon occupancy, particularly on Energy Efficiency (EE).

From the problem of this study, the framework of the study was formulated as shown in Figure 1.

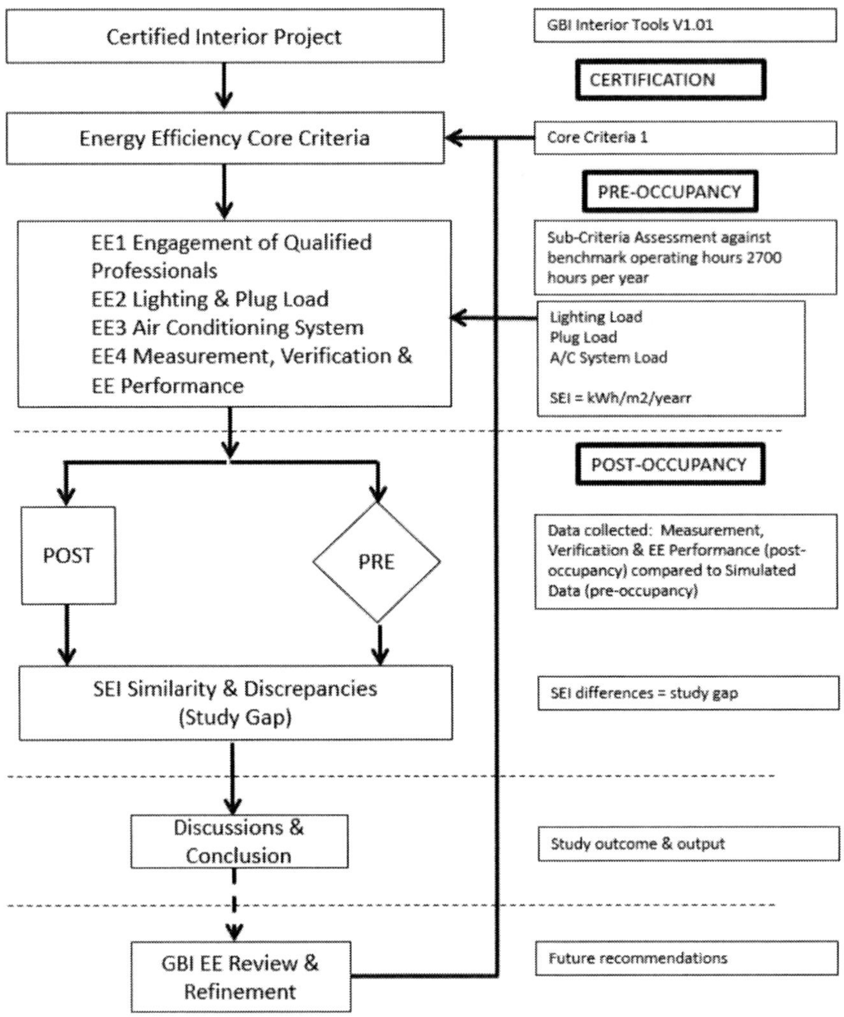

Figure 1. Framework of Study.

2.1 Research Objectives

The study objectives are to establish the gap between pre-occupancy and post-occupancy SEI of certified interior project. There are three key objectives of the study:

1. To examine the SEI upon post occupancy in GBI certified Interior Project.
2. To identify similarities and discrepancies of SEI upon pre-occupancy evaluation and post occupancy evaluation.
3. To assess the key EE sub-criteria that affect Space Energy Intensity (SEI) and to improve continuing sustainable EE upon post occupancy

2.2 Research Questions

The formulated research questions are to address the main objectives of the study. There are three main study questions:

1. What is the SEI upon post occupancy in GBI certified Interior Project.
2. Why similarities and discrepancies of SEI upon pre-occupancy evaluation and post occupancy evaluation need to be identified
3. How to measure key EE sub-criteria that affect Space Energy Intensity (SEI) in order to improve continuing sustainable EE upon post occupancy

IOP Conf. Series: Earth and Environmental Science **385** (2019) 012018 doi:10.1088/1755-1315/385/1/012018

3. Methodology

The scope of the study focus on GBI-IT Energy Efficiency (EE) core criteria and certified GBI Interior project in Malaysia which were occupied for more than one year. The research methodology adopt Space Energy Intensity equation approach in order to measure total space energy consumption per year/interior space area.

$$\text{SEI} = \sum (l^1 + l^2 + l^3)/F(m^2) \tag{1}$$

Where;
Space Energy Intensity is defined follows:
SEI = Total Space Energy Consumption per year ($\sum$) / Interior Space Area (F)
Where,
Total Space Energy Consumption per year
Total amount of energy used per year from all services required to operate the space [Air Conditioning (l^1), Electrical Lighting (l^2), and Plugged Load (l^3)]
SEI is to be calculated within the official operating hours.
The SEI may be rationalized for benchmarking against the industry through the following;
Benchmark operating hours 2700 hours per year as indicated in Table 1.

Table 1. Factors of Operation Hours

OFFICIAL OPERATING HOURS	OPERATION HOUR FACTOR
2100	1.29
2400	1.13
2700	1.00
3000	0.90
3300	0.82

The Space Energy Intensity (SEI) to gain benchmarking scores are proportionate against Building Energy Intensity (BEI) consumed for the total building measurement of energy use. In pre-occupancy evaluation, whereby the internal spaces are using centralized air conditioning systems, the energy consumption from air conditioning system used is to be acquired through the base building operation, and to be included in the SEI calculation. If this is not possible to attain, the simulated calculation may be applied in order to assess the consumed energy using centralized air conditioning system.

SEI benchmark values are derive as of formulation indicated in Table 2.

Table 2. SEI Benchmarking based on 2700 Hours of Office Operation

Building Energy Intensity	Space Energy Intensity	A/C Energy Load	Lighting Energy Load	Plug Energy Load
100%	55%	50%	20%	30%
150	83	41	17	25
140	77	39	15	23
135	74	37	15	22
120	66	33	13	20
110	61	30	12	16
100	55	28	11	17
90	50	25	10	15

The research design will be formulated into three key stages, which are;

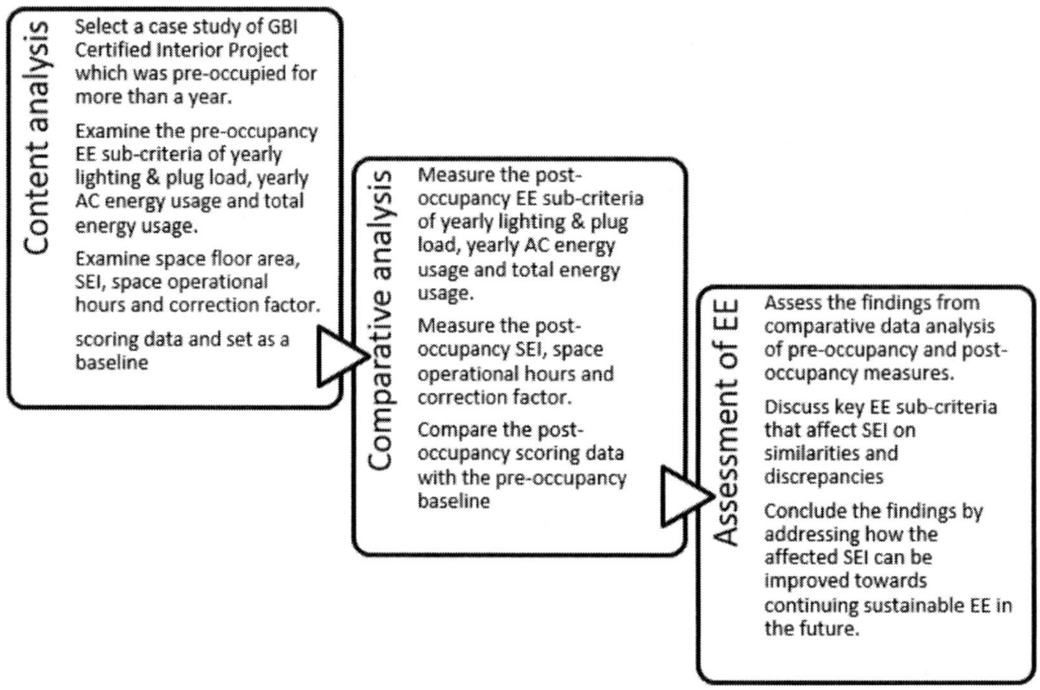

Figure 2. Sequence process of the study.

4. Discussions

In this study, the sequence of research process is designed in a way to achieve the objectives which derived from the main research problem statement. Based on sequence research process as indicated in Figure 2, there are three main process are conducted in concluding the findings in order to pursue continuing sustainable agendas particularly bay addressing the energy efficiency core criteria. This is vital and highly significant in sustainable measure whereby the projected EE evaluation and certification during pre-occupancy assessment is compared against the real post-occupancy assessment on-site. The yielded outcome will validate the similarity and the discrepancies between the ideal designer's vision and the actual habitual consumption of energy in the interior spaces be it for working, living or any other routines.

The first research process, content analysis of EE core criteria in GBI Interior Tools t is to identify and choose a case study of GBI Certified Interior Project. The data is obtained by examining the pre-occupancy EE sub-criteria of yearly lighting & plug load, yearly AC energy usage and total energy usage. Next is to examine space floor area, SEI, space operational hours and correction factor. This data is projected data of EE measurement and verification which was projected for Completion and Verification Assessment (CVA) submission to GBI for certification application. The SEI equations is employed in calculating this data. This scoring data is set as a baseline for the study.

The second process is to comparative data analysis of pre-occupancy EE sub-criteria scoring benchmark against post occupancy EE sub-criteria scoring benchmark. The selected case study of this research is GBI Certified Interior project which was occupied for more than 1 year. The justification for a minimum 1 year post-occupied GBI Certified Interior Project is to get the data of energy used based on actual data collection and calculations. The same method of data collection is used in measuring and calculating data of the same case study after it being occupied for more than 1 year. The gathered actual data is calculated using the same SEI equations in order the get consistency and validated results. This is very important as the research gap and research objectives on similarity and discrepancies of the variables can be identified and verified. Comparison of variables can be distinguish clearly on the EE sub-criteria energy used between pre-occupancy and post-occupancy. This is done by comparing the post-occupancy scoring data with the pre-occupancy baseline data.

Assessment of key EE sub-criteria that affect Space Energy Intensity (SEI) and to improve continuing sustainable EE upon post occupancy.
The final stage is to assess the findings from comparative data analysis of pre-occupancy and post-occupancy measures. From the findings the key EE sub-criteria that affect SEI on similarities and discrepancies based on pre-occupancy baseline and upon post-occupancy measures will be used to validate the study. The findings is then addressed how the affected SEI can be improved towards continuing sustainable EE in the future. The continuing sustainable measure on-site and at real-time, particularly on energy efficiency is vital in promoting sustainable agenda holistically and thoroughly.

5. Conclusion

Green Building Index (GBI) Tools for interior is gaining acceptance as indexing work structure in implementing green interior benchmarking issues, disagreement and management requirements. The main goal of mainstreams green management for Interior and related assessment tools is to achieve sustainable use of design resources and user experience [8]. Green Building Index (GBI) for interior project is a sustainable rating system developed by MGBC (Malaysia Green Building Confederation) and MIID (Malaysian Institute of Interior Designers). The developed GBI for Interiors will be first Malaysia's comprehensive assessment system for assessing the sustainable design and performance of interior design project upon six (6) main factors which is energy efficiency (EE), indoor environment quality (EQ), sustainable planning management (SM), materials resources (MR), water efficiency (WE), and innovation (IN). The heading criterias set is to maintain all of standards scoring criteria in all previosly developed tools in GBI rating framework system [9] The GBI for Interiors is developed specifically for the Malaysian tropical weather, environmental and developmental context, cultural and social needs [10]
Based on the discussion above, it can be concluded that projected measure of EE during pre-occupancy is important in promoting and parallel with the establisment of GBI Green Interior Tools as an effort towards sustainable development. However, it is equally important that this effort is extended by reassessing the certified green interior project particularly in EE. This vital study is to addressed on any similarities or discrepancies of EE variables between pre-occupancy and post-occupancy. The designers vision of pre-occupncy assessment are based on projected energy used of ideal scenario of building occupancy. Logically, buildings or any built environment is a 'living organism', it may expand to grew or declined, or even died. Hence, the actual measurement of certified green space is to get the real on-site data is important to assess the green 'health' of certified green space. The comparative analysis of post-occupancy data against pre-occupancy data which set as a baseline will offer similarity or discrepancy EE variables for future diagnosed.

Acknowledgement

I would like to acknowledge and extend heartfelt gratitude to Universiti Teknologi MARA (UiTM) for funding this study under the Lestari Research Grant: 600-IRMI 5/3 LESTARI (052/2018) and International Islamic University Malaysia (IIUM).

References

[1] Ari I and Sari R 2017 Differentiation of developed and developing countries for the Paris Agreement *Energy Strategy Reviews* **18** 175-182
[2] MGBC 2010 Green Building Index Logo & Malaysia Green World Launched 21 May 2009 in Green Transformation. IGBC Penang. Source: http://www.greenbuildingindex.org/Resources/20100513%20%20IGBC%20Penang/20100513%20-%20IGBC%20Penang%20-%20Green%20Transformation%20-%20Ar%20Boon%20Che%20Wee.pdf

[3] Hamid A A and Embi M R 2012 Factors for Implementing Green Building Rating Tool to Government Building in Malaysia in *Asian Conference on Sustainability, Energy and the Environment Proceeding* Osaka, Japan.

[4] MGBC 2010 Malaysian Prime Minister supports to Green Building Index in Green Transformation. IGBC Penang. Source: http://www.greenbuildingindex.org/Resources/20100513%20%20IGBC%20Penang/20100513%20-%20IGBC%20Penang%20-%20Green%20Transformation%20-%20Ar%20Boon%20Che%20Wee.pdf

[5] Shaikh P H, Nor N B M, Sahito A A, Nallagownden P, Elamvazuthi I and Shaikh, M S 2017 Building energy for sustainable development in Malaysia: A review *Renewable and Sustainable Energy Reviews* **75** 1392-1403

[6] Yu W, Li B, Yang X and Wang Q 2014 A development of a rating method and weighting system for green store buildings in China. Renewable Energy **73** 123-129

[7] Sahamir S R and Zakaria R 2013 Green Assessment Criteria for Public Hospital Building Development in Malaysia *Procedia Environmental Sciences* **20** 106 – 115

[8] Rashida Y R, Sulaiman M S, Aziz A, Selamat H, Yani A H M and Kandar M Z 2011 Greening government's office buildings: PWD Malaysia Experiences *Procedia Engineering* **11**

[9] Lowe C and Ponce A 2012 UNEP-FI/SBCI's Financial & Sustainability Metrics Report: An international review of sustainable building performance indicators & benchmarks. Retrieved February 28th, 2012, from UNEP-FI website: http://www.unepfi.org/fileadmin/documents/metrics_report_01.pdf

[10] Kohler N 1999 The relevance of Green Building Challenge: an observer's perspective *Building Research & Information* **27** 4-5

IOP Conf. Series: Earth and Environmental Science **385** (2019) 012019 doi:10.1088/1755-1315/385/1/012019

The Regaining of Traditional Street: Combining Street Network Configuration with Temporary Urban Interventions

N S Khalid[1], R N Raja Othman[1], I Che Abdullah[1], M A Marzukhi[1] and N Nasrudin[1]

[1]Centre of Ekistics and Human Settlement, Faculty of Architecture, Planning and Surveying Universiti Teknologi MARA, 42300 Puncak Alam, Selangor, Malaysia

Email: shakilakhalid@salam.uitm.edu.my

Abstract. Observations and Axial analysis models are combined into Space Syntax to study the corresponding relationship between a street network and how temporary urban interventions in a traditional shopping street in Little India, Kuala Lumpur, may affect public life. Little India was chosen as the case study because of its historical and cultural significance. The study aims to explore the most important temporary urban interventions - activity-based interventions in traditional shopping street through the spatial logic of street network. Static snapshots and people-following were conducted in Little India, to observe activity and movement during a whole day on a weekday and a weekend day. Two (2) parameters in Space Syntax were chosen including global integration and local integration to measure its urban interventions, activities, and movement among visitors in traditional shopping streets. The result indicates that there is a certain correlation among these parameters with the following spatial configurative analyses; all lines analyses and segment based angular analyses. It finds significant relationships between global integration and local integration, and the activity pattern. It is proposed in this paper that streets with a high degree of integration and adjacent to main streets; Jalan Tun Sambanthan is a key factor for carrying forward temporary urban interventions in Little India. More attention should be paid to the street network to preserve diversity, make the streets active, and importantly to improve public life.

1. Introduction

Concerning what lies behind a city, [1] reveals that it is a street. If a city's streets look interesting, the city will ideally looks interesting. But if they look dull, the city looks dull. Furthermore, [2] provide the definition of streets, quoting [3] as linear three-dimensional spaces enclosed on contrasting sides of the building, as part of public realm. An extensive description of the roles of streets in most of the cities is provided by [4] who view streets as not only the corridors of movement for public and goods, but streets showcase the community and connect people. Here, streets play a vital role as a social space and as connectors to emphasize the quality of public life [5]. In details, they represent how people use different spaces in urban areas, diverse facilities, and how urban design improve the strong relationship between physical space and way of use. Having given those broad perspectives of the roles of streets, it is fair enough to say that today street increasingly recognised as an effective urban public space to health, happiness, democracy, and improved urban economy.

Evidently, small-scale, quicker, improvements are progressively achieving the short-term commitment of making a place and local economy, the shop owners themselves achieve these short-term improvements through temporary urban interventions – the extensions to build urban fabric. Such

IOP Conf. Series: Earth and Environmental Science **385** (2019) 012019 doi:10.1088/1755-1315/385/1/012019

elements in temporary urban interventions are likely to support the economic environment, infrastructure upgrade and cultural enrichment, while supporting a strong sense of place to maximize quality of place as confirmed by [4].

Great changes of urban appearance have been resulted from urbanization in recent 30 years in Kuala Lumpur. Many historical buildings with iconic characteristics disappeared dramatically. To say this, there is the issues concerning on the weakened for the role of the street network in efficiently organizing urban space and public activities. Many large-scale development projects and tallest buildings can be found everywhere. It is still quite serious for enclosed management of city blocks as well as for the phenomenon of encroachment of public space. Street network alike capillary of human-being is being erased for massive urban regeneration. The flattening trend of urban spatial hierarchy also weakened richness and diversity of urban life. Besides, increasing automobiles aggravated energy issues and crucial climate change. Resulting pedestrians are pushed to the sides of road - streets become wider and less parts are left available for people to walk around. With Kuala Lumpur as an example, therefore, technique of space syntax is used to explore the relationship between the street network and temporary urban interventions, and what impact of the interventions on public life. More insightfully, a traditional shopping street in Kuala Lumpur in Little India, Brickfields, will serve as a case study to observe and measure the streets integration and connectivity. The aim of this research therefore is to examine how the relationship between the activity-based interventions and the spatial layout of the traditional shopping streets in urban area. This thus answer the research question of what and how temporary urban interventions contribute to the process of place making in traditional shopping street in Kuala Lumpur.

2. Literature Review

2.1. Streets as a Stage for Public Practice

In most urbanized cities, streets are the origin of dynamic commercial activities and have become the culmination of city infrastructure because of its flexibility to replicate facilities, social interactions and economy of people [6]. That is because most of the informal shopping activities, bazaar, temporary and mobile activities are established around the multiple needs of people. This proved that street as the physical carrier for the occurrence of urban activities. Beyond acting as thoroughfares for motor vehicles, urban streets often double as public space, based upon the cultural perspective of a particular place. In cultural perspective, urban streets are active public space and provide a place for variety social, cultural and political activities that create vibrant urban spaces and add to the experience of public life [2], [3]. People's public life occurs in those public spaces. These spaces capable to take in diverse human behaviour, uses and activities such as shopping, walking, gathering, and even using the amenities provided to socialize, as a daily routine and during the seasonal events.99r

The streets is no longer just perceived as a tool which facilities transportation between buildings and networks as mentioned by [1]. This thus supported by [7] that no longer is the street a tool that can only facilitate transportation between buildings, but streets more than that, it provides link to adjoining networks. The street is now seen as a public realm. The effective making of commercial streets is the way to improve the quality of place with concentration of activities and many amenities over a period of time. These quality shopping streets have a strong sense of place. In fact, street culture and bazaar naturally get unsettled and discontented. People who continue with the everyday city routine have enhanced the unpredictable kinetic atmosphere.

2.2 Street Networks and Temporary Urban Interventions

Comprehensive series of activity patterns based shopping streets are pioneer for a powerful change of traditional shopping streets, which provide opportunities to strengthen the sense of place [2], [3]. Different layers of street such as permanent, temporary and hybrid have great potentials to invite people and create welcoming spaces for all ages. Therefore, to uplift the quality of places that contribute to the vibrant public life – necessary to bring social, cultural, and economic potential in a single space.

IOP Conf. Series: Earth and Environmental Science **385** (2019) 012019 doi:10.1088/1755-1315/385/1/012019

Places are for people and improve public life, thus should be always flexible for various activities, users, and functions in relation to its components [4]. To added temporary urban interventions have a potential of adding more activity and at the same time, enhancing existing activities to these shopping streets, which will provide new opportunity for regenerating the place and its functions. Examples of temporary urban interventions are activity-based interventions, structural interventions and sensory interventions in shopping streets as placemaking elements. Because of the time limitation, this paper only gives focus on activity-based interventions which include seasonal activities, promotional activities, bazaar type activities, and pop-up activities in shopping streets as shown below.

Figure 1. Pop-up activities in Seoul, Korea, can lend richness to social, civic and economic fabric of local communities.

Source: To and Chong (2017)

Figure 2. Pop-up activities and bazaar type activities during the festivals in Malaysia.

Source: Tourism Malaysia, 2014

Figure 3. Example of the benefits of human-scaled streets in market streets in Wuhan, China.

Source: Kpeinhardt, PPS.org.

3. Research Methodology

To analyse the spatial quantitative, the following space syntax measures were taken into account - to describe global and local network configuration of streets design in Kuala Lumpur within 3 km from Little India. By according configurational measurements to the several streets segment, relationships between street morphology and individual activity patterns can be measured. The focuses will be given as [8] found that the presence of activity on street especially when there have temporary activities, thus high integration value of streets will shows a higher number of pedestrian and car movements [8].

IOP Conf. Series: Earth and Environmental Science **385** (2019) 012019 doi:10.1088/1755-1315/385/1/012019

At this point, the study has drawn an attention of how integrated and connected the traditional shopping streets within 3 km radius are in the Kuala Lumpur as a broad scale using axial map (see Figure 4). According to [9] and [10], an axial map is a network of intersecting lines that comprises the longest sets of lines of sight that pass through all the spaces in a study area. Integration is an indicator of how easily one can reach a specific line of the axial map [9]. This state that the higher the integration value of a line, the lower the number of axial lines needed to reach that line. In order to perform this, the researcher employed the parameters in space syntax that uses to measure the local integration (the local network) and global integration (for entire system). For the global network of analysis, space syntax take into account every possible relationship in the system – from anywhere to anywhere, whereas for the local integration analysis, only examine a certain local catchment.

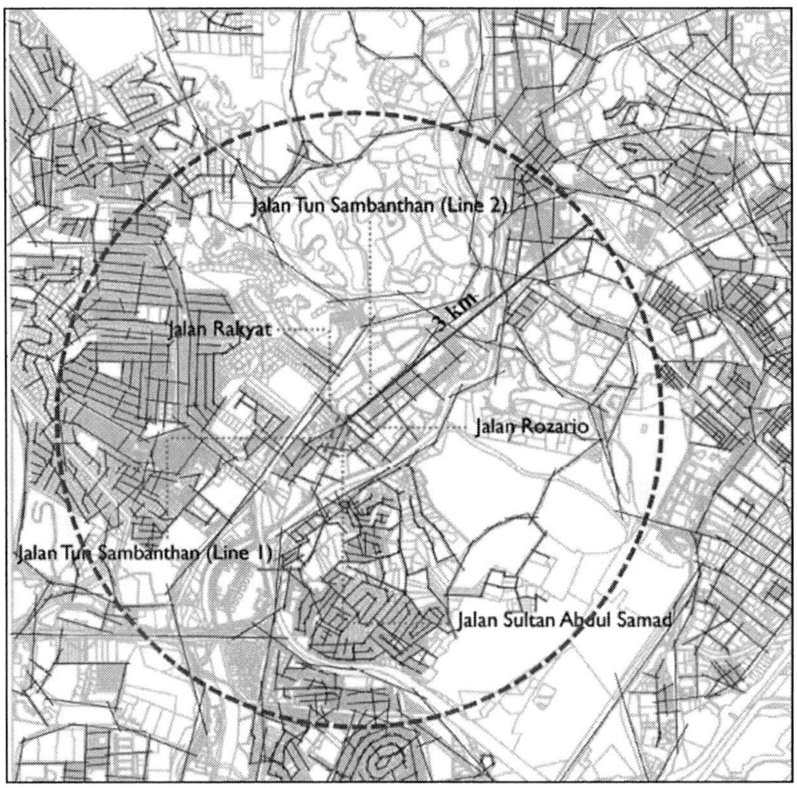

Figure 4. Axial map of the pedestrian and vehicle network for the 3 km radius from Little India, Brickfields. The axial map reveals the significant balance for both global and local integration value.
*Note: Black lines represent the global integration and local segregated lines represents in light grey.

Because of city is very large, dense and well-populated, the purposes of individual moving about the urban landscape do not constitute the sum-total of activity contained them [11]. Their collective activity gives rise to a pattern of use and movement that is independent of the individual behavioural. In this kind of situation, static snapshots and people-following method were conducted in Little India during a whole day on a weekday and a weekend day. The observations were taken during seasonal activities such as Deepavali festival when they have bazaar type activities. Additionally, along Little India, the visitors will see several informal activities and pop-up booths that offers various local goods such as telco promotion booth and drinks stall. Specifically chosen sites were observed and photographed through static snapshot and people-following methods in order to understand the environment setting, crowding, types of visitors, their activities and movement pattern, and temporary events (temporary urban interventions) as shown in Table 1.

Table 1. The framework of research design for gathering data of how people use space in study area.

Temporary Urban Interventions	Observation Methods	Outcomes
Activity-based i. Seasonal activities ii. Informal activities iii. Pop-up activities iv. Bazaar types activities	Static snapshot	Mapping activity patterns in shopping streets
	People-following	Mapping pedestrian movement pattern of the shopping streets

When comparing the results from the static snapshots in Figure 4, the results comply with the dispersal of the integration values. This explains the debates by [8] that the more spatially integrated the street is, the more people on street. On the contrary, the more spatially segregated the streets are, resulted in the fewer people on street. It is described that space syntax is a set of tool linking space and people, and a set of techniques for analyzing spatial configuration [8].

4. Analysis and Results

For the purpose of this research, two (2) analyses were conducted in Little India to read the pattern of activities. One of the analysis was a global measurement (Rn), which helped examine the relationship between each axis and all other axes, and to show the degree of integration. The second analysis was a local measurement (R3), which helped to identify the relationship of an axis with its connected axes (up to three steps away). The global and local analyses are very useful methods of looking at different scales of a spatial system of Kuala Lumpur.

4.1 Analysis on activity pattern in shopping streets

From the observation, there are many people on the street in Little India both during a weekday and a weekend. The activities took place in Little India are such as walking, shopping, taking the photo, eating, standing, and sitting. However, there is no large group of young people except those who come with their family. Most of the visitors are female. This explains that women more frequently be on street because they want to buy the groceries. For the observed activity patterns, few people stand or sit in front of the shops – just having a conversation and lingering. Most people are walking and travel through Little India from KL Sentral in order to reach the school, church, and offices. Also, there are people who take a photo around Little India especially after the upgrading of the urban image of Little India. Most of them are foreigners. Regardless of when comparing the results between during a weekday and a weekend, and the static snapshots analyses to evaluate the activity pattern, no significant differences can be seen. Since the degradation of business activities in Little India and the crowdedness of traffic, then these results are not surprising.

When revealing the flow of people to conclude the activity pattern on a weekday during the Deepavali festival, the largest concentration is on the western part of Little India (refer Figure 6). This is caused by a location of Indian restaurant at the edge and a temporary bazaar that sell the festive goods. There are most people concentrate on the temporary bazaar because the goods are cheaper and variety compares to the formal shop lots. In this scenario, the most silent streets/parts are in the formal shop lots. In the weekend, most activities take place particularly shopping and eating along Little India, even though the formal shop lots are closed. People visit Little India to buy the goods from the temporary bazaar. Based on the observation on a normal weekday, there are fewer people on the streets than in a weekend, except for the purpose of most of the people visit Little India is to find the Indian foods. Since most of the shops are closed on Sundays, it was assumed to find more people in Little India on weekday compare to on a weekend day.

IOP Conf. Series: Earth and Environmental Science **385** (2019) 012019 doi:10.1088/1755-1315/385/1/012019

When informal activities such as pop-up stalls take place in front of the formal shop lots, people will more attract. The pop-up stalls normally open only during the weekend. This type of temporary urban intervention offers activity mostly young people and children such as entertainment and eating local foods. This explains why space and society are importantly linked. Human activities, from the least to most complex, create spaces that reflect closely what they do and how they live in the space.

Legend:

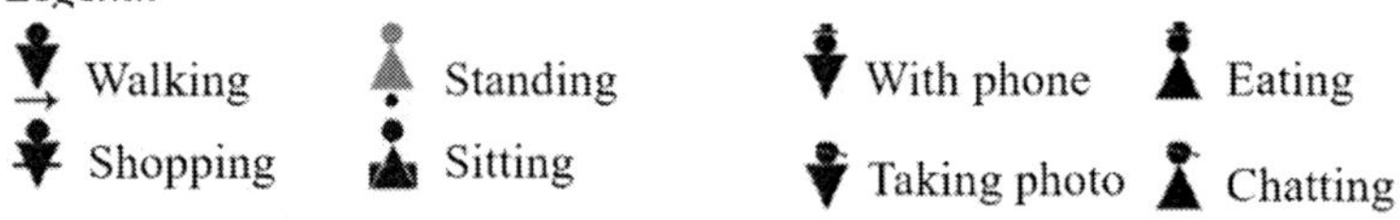

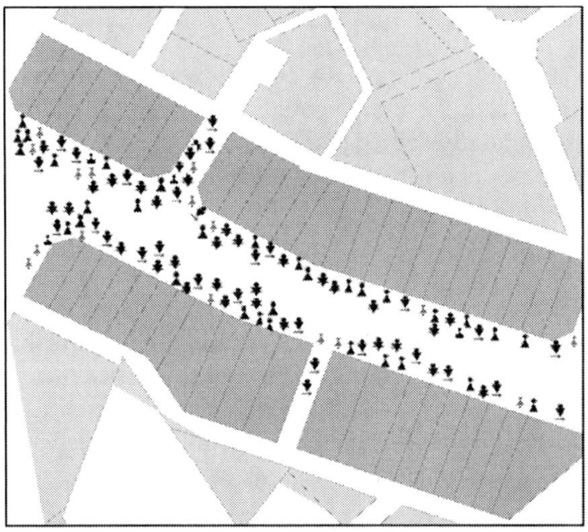

Figure 6. Concentration of activities during Deepavali Festival is quite scattered along Little India.

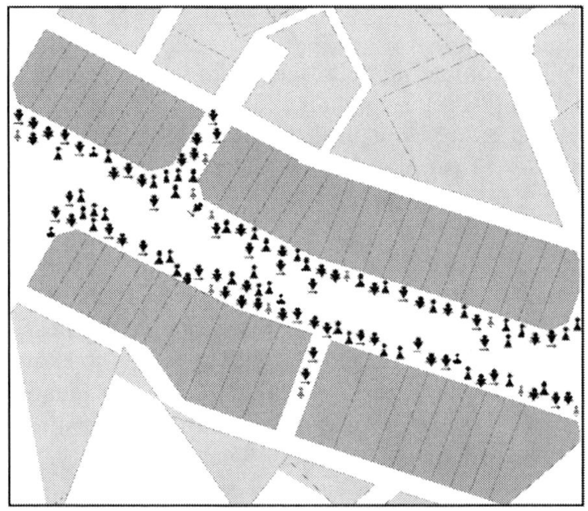

Figure 7. Concentration of activities when there have the informal activities in Little India.

*Note: The diagrams show a single round of observations.

4.2 Analysis on pedestrian flow of the shopping streets

To examine the differences between pedestrian and vehicular networks in explaining people movement behaviour, an axial map for the pedestrian and vehicular street network were developed, to represent the spatial configuration of the traditional shopping street network. The pedestrian network includes all alleys, paths and pedestrian-only lane in addition to the vehicle network. The global integration values result for the pedestrian axial maps is as shown in Figure 8.

Based on Figure 8, it is obvious that the more globally integrated streets, shown with the red lines, are more accessible streets within the network, whereas the more globally segregated streets, shown with the blue lines, are the fewest accessible streets. As described by [10], there is a strong relationship between spatial configuration and how people move within the city and the patterns of vehicular movements.

Figure 8. The global integration map shows that the highest integrated is the red line which is along Jalan Tun Sambathan.

Figure 9. The local integration map shows that the highest integrated are Jalan Hang Tuah and Jalan Imbi.

The axial line with the highest value of global integration, Jalan Tun Sambathan would be the one that could be accessed with the fewest number of turns from all other axial lines in 3 km radius of Kuala Lumpur. In contrast, an axial line that requires many turns to get to it from all other lines in the system is considered to have low global integration value, which is Jalan Sultan Abdul Samad and Jalan Tun Sambanthan 3. An axial line with the highest local integration value such as Jalan Hang Tuah and Jalan Imbi is a line that accessible with the fewest number of connections from all other lines in its surrounding. This result reflects the debate that the higher is the integration value, the nearer is the space to all spaces within the network.

Another way of using space syntax differently is by combining space syntax and observation method to track different pedestrian movement patterns in the street that could argue the integration value. To prove that, people-following observation were conducted during a weekday from 8 am to 9 am

IOP Conf. Series: Earth and Environmental Science **385** (2019) 012019 doi:10.1088/1755-1315/385/1/012019

(morning rush hour) and a weekend from 10 am to 12 pm (mid-morning period). The people following method has the ability to investigate the pattern of movement from a specific location, and to measure the average distance people will walk. Three (3) different points was chosen (A, B, and C) as a starting point to follow the people. The total number of people followed is 35 people during a weekday and 42 people during the weekend. They are various types of people – male, female, school children and elderly.

Table 3. Syntactical values of Little India for pedestrian and vehicular networks.

		Pedestrian street network	**Vehicular street network**
Total number of axial lines		251	395
Global Integration (R*n*)	Mean	0.791	0.724
	Std. Dev	0.344	0.313
	Maximum	3.211	3.637
	Minimum	0.464	0.296
Local Integration (R3)	Mean	1.087	1.590
	Std. Dev	0.535	0.571
	Maximum	3.017	3.637
	Minimum	0.251	0.333

The results reveal that pedestrian pattern of movement during the weekday is rather static. Most of the people exit from KL Sentral (from point A and B) to go to the services (on the western part of Brickfields) such as schools, and offices, as a short distance to access the services. Some of them went to the restaurant to have their breakfast (see Figure 10) because only restaurants operate in the early morning. The rest of the shops only open at 10 am every morning. This limits the movement of people in the streets. People movement pattern from point C is linear and they only go to the post office in Jalan Tun Sambanthan (in front of KL Sentral) which is the distance is 700 m.

While during the weekend (Saturday), the people movement seems scattered. The people-following observation was undertaken from 10 am whereas the shops are opened, but most of them are restaurants and grocery stores. A few of people use the back lane after Zen Hotel to go to the church. From the Figure 11, the results can indicate that the average distance of people can walk is 300 metre and the farthest place that they can reach from point A is Taj Garden Restaurant and church from point B. It also reveals that the longest line in the syntactical analysis is seems a semi-linear of street morphology.

Pedestrian enjoys walking along Jalan Tun Sambanthan because of the elements of built environment such as the small blocks, the street furniture, façade, pop-up booth and stalls, and the connected pathways. Additionally, pedestrian feel safe to walk because of the wide pathways (segregated from the vehicle lane), and there are many activities occur in front of the formal shop lots. – active public life.

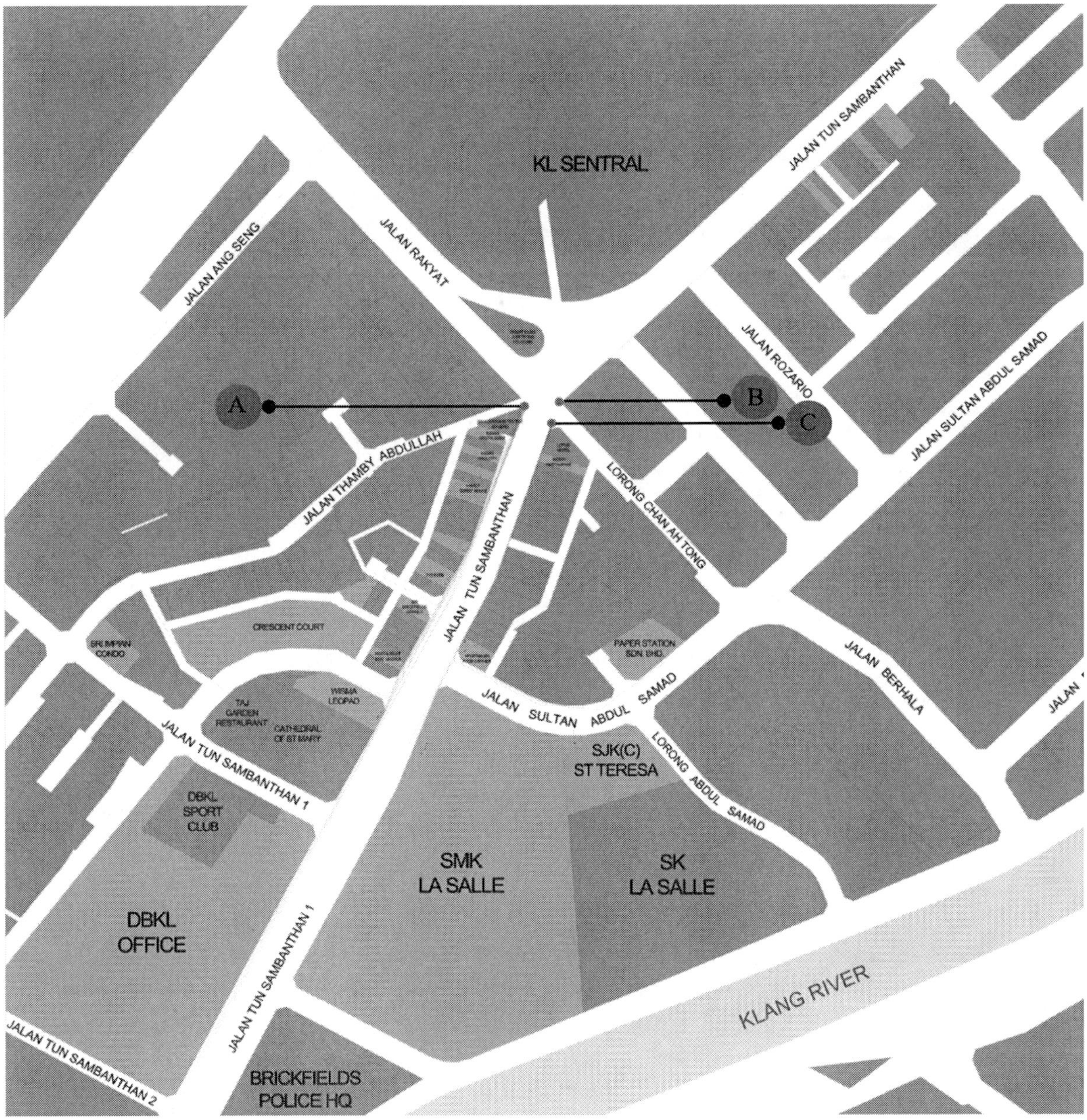

Legend:

The people followed destination

Figure 10. The pedestrian movement patterns from three (3) different points during the weekday.

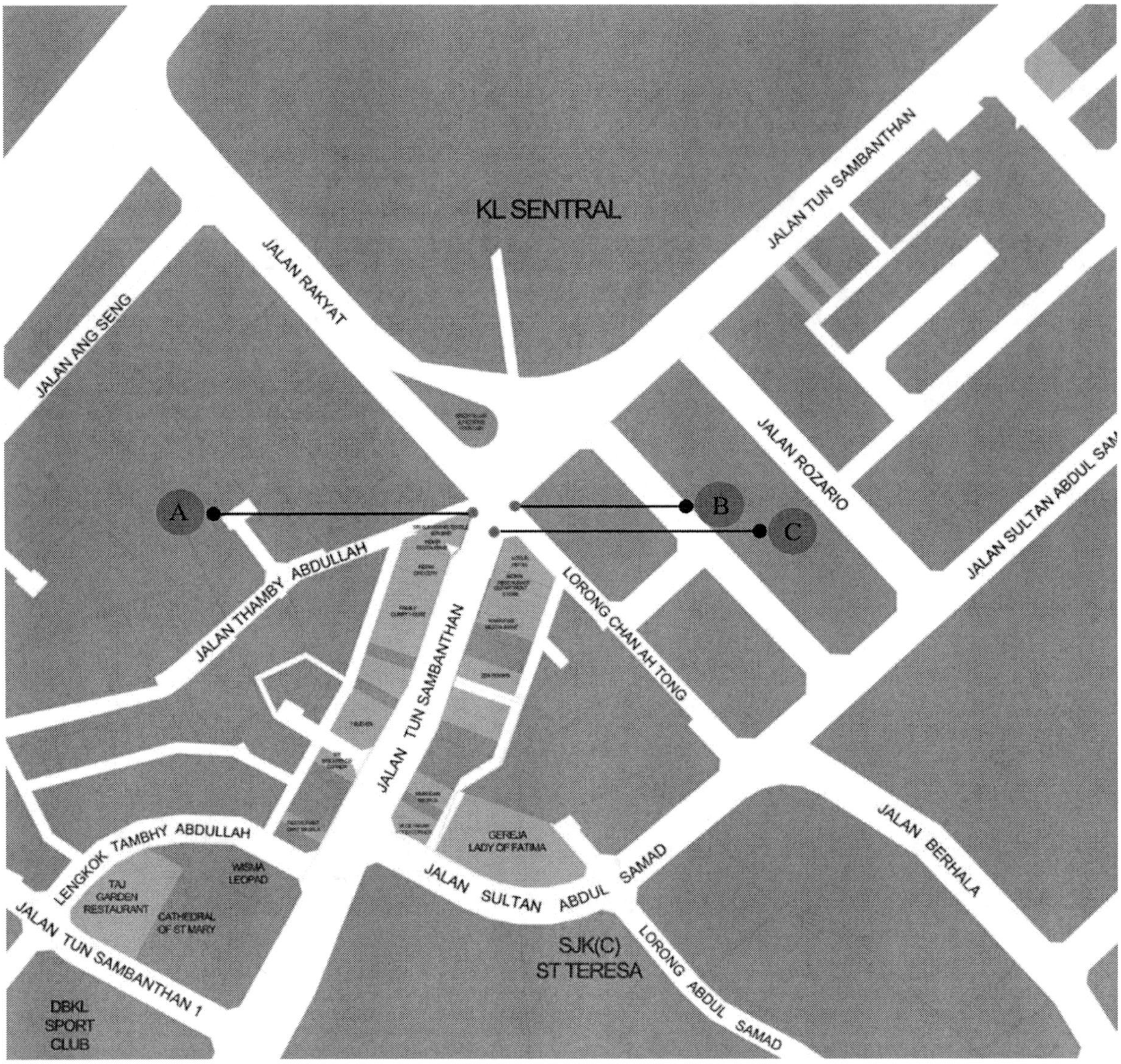

Legend:

The people followed destination

Figure 11. The pedestrian movement patterns from three (3) different points during the weekend.

5. Discussion and Conclusions

This research employed space syntax as an advanced tool to analyze the traditional shopping street network in Little India, Kuala Lumpur. The results analyze that the axes with the lowest global integration values were concentrated 0.464 for pedestrian network while the vehicular network is 0.296. This may significantly improve the overall experience of the street for the pedestrian, as illustrated by improvements in the R*n* value from 0.464 to 3.211 as the R3 value would increase from 0.251 to 3.017. The level of space network could increase by 3.552. Here, it is emphasized that, while this is not a significant numerical change, the space network positively can change the application of the suggested urban design process. The results are also vital because of the space syntax method demonstrated a great potential to extend our knowledge of how people move through space, encounter each other and form social groups.

This research has shown that the quality of local and global relational properties of street morphologies play an important role in measuring human behavior (how people response to space, how people move), and activity patterns as socio-spatially. The results are not surprising since the relational descriptions developed by space syntax theory have a solid behavioral basis. Research findings suggest that urban planners, architects and traffic engineers should give more consideration of how design in built environment influences human behavior and activity patterns beyond the local concern. The findings also contribute to the social contents in space as references for actions what and where we could perform. A broader conception of shop lot blocks and streets, one that emphasizes fit within the traditional shopping street and its temporary urban interventions, likely supported for making a traditional area more meaningful. This paper supports the city vision suggesting an integrated modelling approach in space syntax analysis with the hope to encourage space syntax community for further researches towards putting the people on safer streets and making an impact to the "place".

Acknowledgements
The authors gratefully acknowledge the support of Universiti Teknologi MARA and Institute of Research Centre under the Lestari grant 600-IRMI 5/3/Lestari (040-2018).

References

[1] Mehta V 2014 The Street. Routledge Taylor & Francis Group, London
[2] Sivam A and Karuppannan S 2013 The Role of Streets within Placemaking in Cross-Cultural Contexts: Case Studies from Adelaide, Australia and Georgetown, Malaysia
[3] Ujang, N 2014 Place Meaning and Significance of the Traditional Shopping District in the City Centre of Kuala Lumpur, Malaysia *International Journal of Architectural Research: ArchNet-IJAR* **8** 66-77
[4] Madanipour A and Hull A 2017 The Governance of Place: Space and Planning Processes. Routledge
[5] Zukin S 1998 Urban Lifestyles: Diversity and Standardizations in Spaces of Consumption. *Urban Studies* **35** 825-839
[6] To K and Chong K H 2017 The Traditional Shopping Street in Tokyo as a Culturally Sustainable and Ageing-Friendly Community *Journal of Urban Design* **22** 637-657
[7] Goodsell C T 2003 The Concept of Public Space and Its Democratic Manifestations *American Review of Public Administration* **33** 361–83
[8] Hillier B, Penn A, Banister D and Xu J 1998 Configurational Modelling of Urban Movement Networks *Environment and Planning B: Planning and Design* **25** 25-84.
[9] Bafna S 2003 Space Syntax: A Brief Introduction to its Logic and Analytical Techniques. *Environment and Behavior* **35** 17-29
[10] Hillier B 2007 Space is the Machine: A Configurational Theory of Architecture. Space Syntax.
[11] Chiaradia A 2007 Emergent Route Choice Behaviour, Motorway and Trunk Road Network: the Nantes Conurbation *6th International Space Syntax Symposium*

IOP Conf. Series: Earth and Environmental Science **385** (2019) 012020 doi:10.1088/1755-1315/385/1/012020

The effects of lateral boundary conditions resolution for heat island studies in tropical urban of Kuala Lumpur

N A Isa[1], S A Salleh[1,2*], W M N Wan Mohd[1], M C G Ooi[3], A Chan[4] and M A Islam[4]

[1]Centre of Studies of Surveying Science and Geomatics, Faculty of Architecture, Planning and Surveying, Universiti Teknologi MARA, 40450 Shah Alam, Selangor, Malaysia
[2]Applied Remote Sensing and Geospatial Research Group, Faculty of Architecture, Planning and Surveying, Universiti Teknologi MARA, Shah Alam, Selangor, Malaysia
[3]Department of Atmospheric Sciences, National Central University, 32001 Chung-Li, Taiwan
[4]Department of Civil Engineering, Faculty of Engineering, University of Nottingham, Malaysia Campus, Jalan Broga, 43500 Semenyih, Selangor, Malaysia

Email:aekbal@salam.uitm.edu.my (S A Salleh)

Abstract. Lateral boundary condition (LBC) is one of the key features included in the Weather Research and Forecasting (WRF) simulation model. Nowadays, numerous LBCs were developed with various spatial (grid) and temporal resolutions for a wide range of applications. Choosing the most suitable LBCs to ensure appropriate representation of climates should be properly conducted. Thus, this paper analysed the effects of the LBC resolutions on the regional climate downscaling for heat island studies. The comparisons were made on the performance of respective LBCs to regenerate the near-surface temperature distributions within the Kuala Lumpur city. NCEP GDAS/FNL 0.25 Degree Global Tropospheric Analyses dataset with higher spatial and temporal resolution was found to perform better than the other LBC during intermonsoon season. However, both datasets were determined to give reliable representations of urban climate condition within the city as both datasets depicted close results in determining the impact of urbanization on the thermal environment.

1. Background

Studying climate has many challenges and one of them is obtaining the climate datasets. As for today, general circulation models (GCM) are the most common tool used for climate simulation and assessments [1]. Abundant GCMs are developed in order to cater the needs in climate studies around the world to ensemble the main characteristics of the general circulation pattern for high simulation performances. However, the applications of the GCMs for regional climate studies are very disappointing due to typically poor spatial resolutions (1° to 3°). The GCMs only sufficient to resolve large-scale forcings but fail to resemble the local impacts [2].Therefore, the regional climate models (RCM) are employed to downscale the GCMs into finer resolutions which are able to account the local impacts by nesting a higher resolution into limited area model (child domain) over the area of interest [3]. To achieve this, lateral boundary conditions (LBC) are employed for boundary condition initialization in the parent domain.

Lateral boundary condition (LBC) represents the boundary condition of the Earth surfaces which is required for dynamical downscaling from mesoscale model into regional model. In nested domains, the LBC is provided by parent domains for its nest(s). Previous studies have disputed the errors resulted by the LBCs, however, several recommendations were made to minimize these errors [2], [4]–[6] including spatial and temporal resolution selections. Recent study by Davies [7] has determined that, the errors resulted by the LBCs can be controlled and only consists of a tiny part of the overall errors. Even though the previous studies have enlightened the performance of LBCs, these studies were conducted in the middle to high latitude regions with very little information regarding this matter within the tropical regions [2], [4]–[7]. Due to this, site-specific study is essential in choosing the best LBCs for the particular region especially Malaysia.

The present study analysed the impact of two different global LBC datasets and their performance in determining the urbanization impacts towards thermal environment. These LBCs are freely available in the National Centre of Atmospheric Research (NCAR) archive, developed with different spatial and temporal resolutions. The comparisons made can be used for future research with a similar climate background. New research within this region are urged to bring the uncertainty of LBCs issues to light especially when very high resolution simulations are needed.

Historically, the climate was naturally changing over a long period of time, ranging from decades to centuries. Back then, natural phenomena such as solar output variations and volcanic eruptions were the sole cause of the climate changes [8]. However, after the Industrial Revolution in 1700s, urbanization was one of the significant causes that contributes to the changes in global climate and the change rate was rapidly increasing in the 21st century due to the explosive of human population [8], [9]. Therefore, it can be concluded that the urbanization is significantly affecting the climate as documented previously. Thus, this study also investigated the effects of urbanization towards the thermal variation in determining any supporting evidence on the issues within the tropical context.

2. Methods

The methods employed during this study are explained as follows:

2.1. WRF-ARW Simulation

Table 1. LBCs' characteristics.

	NCEP FNL Operational Model Global Tropospheric Analyses, continuing from July 1999	NCEP GDAS/FNL Global Tropospgeric Analyses and Forecast Grids
Data Type	Grid	Grid
Data Format	WMO_GRIB1 WMO_GRIB2	WMO_GRIB2
Temporal Resolution	6 hours	3 hours
Spatial (Grid) Resolution	$1^{\circ} \times 1^{\circ}$	$0.25^{\circ} \times 0.25^{\circ}$
Data Coverage	From 0E to 359E and 90N to 90S	From 0E to 359E and 90N to 90S

Two widely used global LBCs were chosen for the comparisons, namely NCEP FNL Operational Model for Global Tropospheric Analyses which provided from 1999 onwards (refers as Simulation 1) and NCEP GDAS/FNL for Global Tropospheric Analyses and Forecast Grids which provided from 2015 onwards (refers as Simulation 2). Table 1 shows the comparisons between the LBCs' characteristics. The datasets were employed due to their common applications in representing the urban climate condition within regional context especially in tropical simulation modelling [10]–[14]. The dataset employed were dated 17th April 2017 to coincide the Landsat satellite images utilized in this study. This date was also chosen due to intermonsoon season phase which has less synoptic forcing as compared to monsoon seasons and another intermonsoon season in October [15].

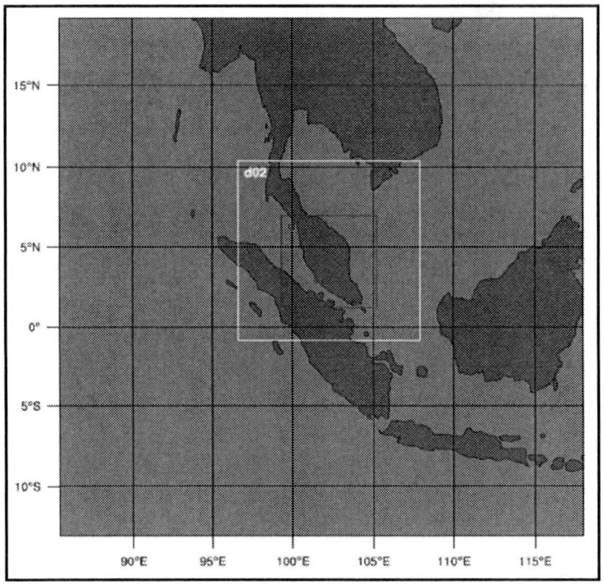

Figure 1. The domains design and setup

Table 2. Physics schemes used during the simulation process.

Physic Option	Physic Scheme
Microphysics	WRF Single-Moment 3-class scheme
Longwave Radiation	RRTM scheme
Shortwave Radiation	Dudhia scheme
Surface Layer	MM5 similarity
Land Surface	Noah Land Surface Model
Planetary Boundary Layer	Yonsei University
Cumulus Parameterization	Kain-Fritsch

The regional dynamic downscaling were performed through Weather Research and Forecasting Model with Advanced Research WRF (WRF-ARW) core solver version 3.8. Using four nested domains with the grid size of 37.5km (40 × 40), 12.5km (88 × 88), 2.5km (201 × 201) and 0.5km (151 × 151), the

IOP Conf. Series: Earth and Environmental Science **385** (2019) 012020 doi:10.1088/1755-1315/385/1/012020

near surface temperature were produced. The dimensions in the parentheses are the grid dimensions designed in the easting and northing direction. Based on Figure 1, the centre of all four domains were focused on the location of the Kuala Lumpur city to provide relaxation zone to the edge of each domain, intended to reduce the errors by employing the LBC[16]. The largest domain (D01) covered a part of Southeast Asia region and the smallest domain (D04) was designed to focus on the Klang Valley region where the Kuala Lumpur city is located. Since the WRF-ARW offers a variety of multiple physics options ranging from simple to very sophisticated schemes, a well-tried scheme employed by Isa et al [17] was used to configure the physics options available in the model as shown in Table 2. Similar domain design and physics schemes were employed in both simulations for the two different LBCs.

2.2. Remotely-Sensed Data Extraction

To extract the urbanized area, the present study has utilized the Landsat 8 Operational Land Imager (OLI) satellite image. The image was dated 17th April 2017 and chosen due to the clear sky view which reduce the chance of atmospheric errors to occur [17]. Radiometric calibration were employed to correct the image as suggested by the data provider [18] by converting the pixel values into surface reflectance as shown in Equation 1. Then, sun angle correction was performed using Equation 2.

$$\rho\lambda' = M_\rho \times Q_{CAL} + A_L \tag{1}$$

where $\rho\lambda'$ = surface reflectance without sun angle correction
$M\rho$ = band-specific multiplicative rescaling factor from metadata
Q_{CAL} = Quantized and calibrated standard product pixel value
A_ρ = band-specific additive rescaling factor from metadata

$$\rho\lambda = \frac{\rho\lambda'}{\cos(\Theta_{SZ})} = \frac{\rho\lambda'}{\sin(\Theta_{SE})} \tag{2}$$

where Θ_{SZ} = Local sun elevation angle
Θ_{SE} = Local solar zenith angle

This study employed a combined algorithm conducted by Bhatti and Tripati [19] to extract the built-up areas (urbanized areas) since both study areas share the same tropical climate characteristics. [19] has suggested that the original NDBI algorithm by [20] should be calibrated. This is supported by [21]–[24] which suggested that the confusion to separate the built-up areas from green cover and water features can be eliminated or at very least reduced using Normalized Difference Vegetation Index (NDVI) and Modified Normalized Difference Water Index (MNDWI). [19] also suggested that, the NDBI layer should be enhanced using Principal Component Analysis (PCA) technique to assign the best pixel values for the MIR wavelength. Equation 4, Equation 5 and Equation 6 show the algorithm of NDBI, NDVI and MNDWI respectively.

$$Built-up\ Area = NDBI - NDVI - NDWI \tag{3}$$

$$NDBI = \frac{(PCA\ Band\ 6{,}7 + PCA\ Band\ 10{,}11) - Band\ 5}{(PCA\ Band\ 6{,}7 + PCA\ Band\ 10{,}11) + Band\ 5} \tag{4}$$

$$NDVI = \frac{Band\ 5 - Band\ 4}{Band\ 5 + Band\ 4} \tag{5}$$

$$MNDWI = \frac{Band\ 3 - Band\ 7}{Band\ 3 + Band\ 7} \tag{6}$$

2.3. Model Validation

The performance of both simulations which utilized the two different LBCs was evaluated and presented using statistical error indices, including mean absolute error (MAE), root mean square error (RMSE) and linear agreement (R^2). The performance of both simulations in reproducing the near-surface temperature were tested against the ground observations provided by the Malaysian Meteorological Department (MMD). Three ground stations were involved and located around the Kuala Lumpur city (in red line), namely MARDI Serdang, FRIM Kepong and Subang as shown in Figure 2.

Figure 2. Location of MMD ground stations

2.4. Data Sampling and Inferential Statistics

The original raw data has been filtered to ensure it is free from outliers. The exclusion of data outliers was performed using quartiles and boxplots as suggested by [25], [26]. Then, descriptive data examination were performed prior to analyses for determining the data distribution nature. In this study, the whole population was used as data samples. Since the data nature was found to be approximately normally distributed; indicated by the ratio of kurtosis and skewness value against the standard error, a series of parametric statistical analyses was chosen to analyze the data samples. Therefore, the interactions of the urbanized area towards the thermal environment were examined using inferential statistics including the Pearson's correlation test to identify their relationships and one-way Analysis of Variance (ANOVA) with the control of Welch's and Brown-Forsythe's tests to determine whether or not the impact is significant.

3. Result and analysis

The findings and discoveries of the present study are discussed and presented as follows:

3.1. Simulations Performance

Table 3 shows the MAE, RMSE and R^2 values for the two simulations. Based on the results, it can be concluded that Simulation 2 excels in regenerating the actual near-surface temperature profile with the MAE of ±1°C, RMSE of ±1.4°C and R^2 of 0.881 as compared to Simulation 1 which has the MAE of ±1.5°C, RMSE of ±1.8°C and R^2 of 0.763. The diurnal pattern of air surface temperature in the three ground stations are shown in Figure 3. In both simulations, similar patterns were identified. In both simulations, the models tend to underestimate the near-surface temperature in two stations which are MARDI and Subang despite some overestimations during night time, whereas, in FRIM Kepong, the

IOP Conf. Series: Earth and Environmental Science **385** (2019) 012020 doi:10.1088/1755-1315/385/1/012020

near-surface temperature was overestimated except during the highest peak. Comparing both simulations, large residuals were identified in Simulation 1 during day time when the temperature was increasing rapidly.

Table 3. MAE, RMSE and R^2 of Simulation 1 and Simulation 2

	NCEP FNL Operational Model Global Tropospheric Analyses, continuing from July 1999 (Simulation 1)	NCEP GDAS/FNL Global Tropospgeric Analyses and Forecast Grids (Simulation 2)
MAE	±1.5°C	±1.0°C
RMSE	±1.8°C	±1.4°C
R^2	0.763	0.881

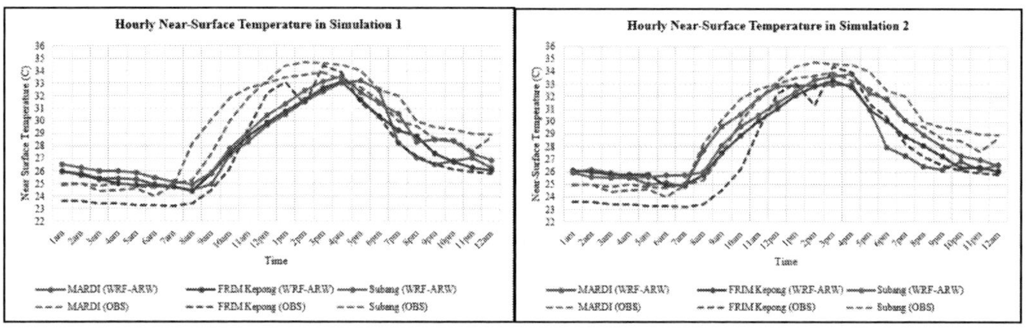

Figure 3. Hourly profiles of near-surface temperature for both simulations in three different stations

3.2. Urbanization Effects

In this study, the urbanization effects towards the thermal environment were also investigated and the results are discussed as follows:

3.2.1. Spatial distribution of urbanized areas and near-surface temperature profile. Based on Figure 4, the Kuala Lumpur city was dominated more than half by built-up areas which covers approximately 85% of the city. The highest built-up area percentage is 99.5%, which indicates that the surface area is almost impervious whereas the lowest built-up percentage is 0%, which indicates that there are no man-made features constructed in this area. The highest built-up area percentage was found in Sentul which compacted with residential areas. The lowest built-up percentage was found in forest reserves in the city such as Bukit Tabur, Bukit Gasing and Bukit Besi.

By examining the spatial distribution of the built-up areas, the study has found that the higher built-up area percentages were found in the northern, southern and central region of the city. The higher built-up area percentages often found within developed areas such as Sentul (residential), Kepong (industries), Central of Kuala Lumpur (business) and Bukit Bintang (business). These areas are expected to have higher thermal variation as compared to other areas. Lower built-up percentages were found in western and eastern regions of the city such as Bukit Tunku and Mont Kiara and are expected to have lower thermal variation.

IOP Conf. Series: Earth and Environmental Science **385** (2019) 012020 doi:10.1088/1755-1315/385/1/012020

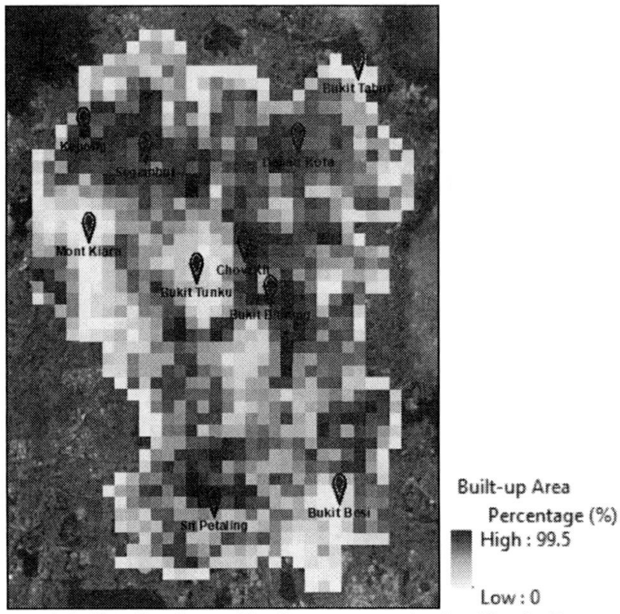

Figure 4. Spatial distribution of built-up areas in the Kuala Lumpur city

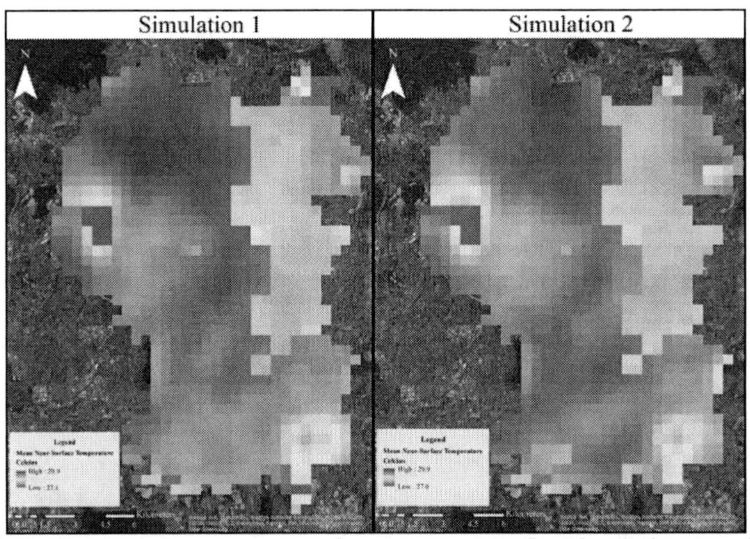

Figure 5. Spatial distribution of mean near-surface temperature for Simulation 1 and Simulation 2

Figure 5 shows the spatial distribution of the daily mean of near-surface temperature within the Kuala Lumpur city produced by Simulation 1 and Simulation 2. In Simulation 1, the lowest mean temperature was 27.1˚C and the highest mean temperature was 29.9˚C. A close resemblance was also determined in Simulation 2 with the lowest mean temperature was 27.0˚C and the highest mean temperature was 29.9˚C. Based on the figure, the spatial distribution and pattern of the mean near-surface temperature in both simulations were similar to each other where the eastern regions posses lower mean temperature as compared to other regions across the city.

By delving deeper into Figure 5, it was found that the lowest near-surface temperature was identified in the mountainous region of Bukit Tabur which located in the north-eastern region of the Kuala Lumpur city. Other than Bukit Tabur, several areas were also identified to have lower near-surface temperature distributions which are Bukit Kiara, Bukit Tunku and Bukit Besi. The highest near-surface temperature was identified in Sri Petaling which located in the southwestern region. Higher near-surface temperatures were also found in other areas as well, namely Kepong, Danau Kota, Bukit Bintang, Chow Kit, Kampung Baru and Segambut. Based on this, it can be said that the areas which are more urbanized contribute to the increase in the near-surface temperature.

3.2.2. Pearson's correlation test result. By comparing both simulations, it was identified that Simulation 1 obtained a close result as Simulation 2. The correlation between the near-surface temperature and the urbanized area were 0.662 and 0.648 for Simulation 1 and Simulation 2 respectively. Portrayed in Figure 6, the relationships were indicated in increasing trendline. The trendline in Simulation 1 slanted more as compared in Simulation 2 which explained a slightly higher correlation result obtained. To conclude, in presenting the correlation, both simulations gave reliable performances. In other words, both LBCs can be employed for similar studies in tropical regions for regional downscaling.

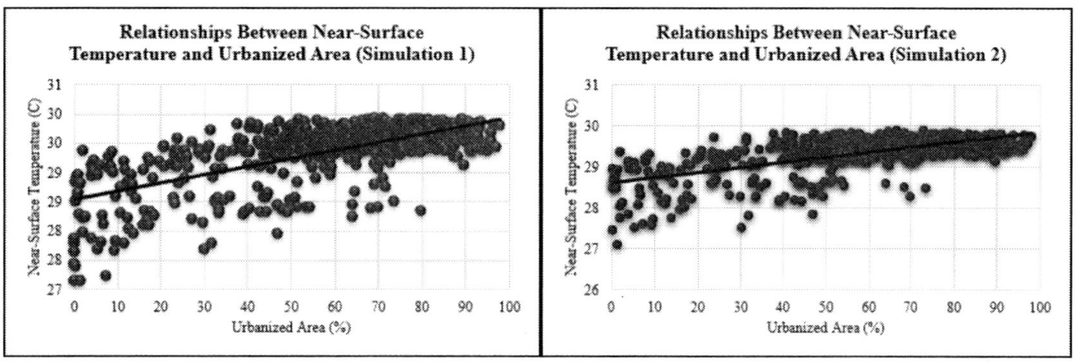

Figure 6. Relationships between near-surface temperature and urbanized area for Simulation 1 and Simulation 2

Supporting the previous studies, it was identified that the urbanization was affecting the thermal environment with a positive correlation. This advocates that urbanization is indeed responsible in the increase of the ambient temperature within the city based on the fact that the increase of built-up area coverage will increase the near-surface temperature.

3.2.3. One-way ANOVA test result. In investigating the performance of both LBCs employed, both simulation results undergo the same one-way ANOVA procedures. By utilizing the datasets produced by both simulations, the investigation of the effects of urbanization towards the thermal environment returned similar results for the one-way ANOVA test performed where every test was rejecting the null hypothesis, suggesting the same conclusion.

The variations of the means within the samples studied revealed the effects of urbanization towards the thermal environment was significant as the p-value rejected the null hypothesis of the one-way ANOVA. The test has confirmed that there was at least one significant difference in the comparison between the group means suggesting the changes in urbanized area coverage will significantly affect the thermal variation within the Kuala Lumpur city.

4. Conclusion

This study analysed the performance of two different LBCs in initializing the boundary conditions during WRF-ARW simulations to regenerate the near-surface temperature profile. It has been determined that the LBC with higher spatial and temporal resolution offer better performance of regional climate downscaling. Even though this is the case, both LBCs were capable to offer reasonable results pertaining to interactions of the urbanized area on the thermal environment within the Kuala Lumpur city. Thus, it is suggested that both of the LBCs can be employed in simulating the regional climate. Future studies can be conducted to examine the spatial and temporal resolution impacts on regional climate downscaling separately for detailed investigation.

Acknowledgements
The authors would like to express their gratitude to the Universiti Teknologi MARA (UiTM) for funding this project under Geran Bestari Perdana 600-IRMI/DANA 5/3 BESTARI (P) (074/2018).

References

[1] IPCC 2001 Climate change 2001: synthesis report
[2] Amengual A, Romero R, Homar V, Ramis C and Alonso S 2007 Impact of the lateral boundary conditions resolution on dynamical downscaling of precipitation in Mediterranean Spain *Clim. Dyn.* **29** 487-499
[3] Giorgi F and Mearns L 1999 Introduction to special section: regional climate modelling revisited *J. Geophys. Res.* **104** 6335–52
[4] Warner T, Peterson R and Treadon R 1997 A tutorial on lateral boundary conditions as a basic and potentially serious limitation to regional numerical weather prediction *Bulletin of the American Meteorological Society* 2599–2617
[5] Denis B, Laprise R and Caya D 2002 Sensitivity of a regional climate model to the resolution of the lateral boundary conditions *Clim. Dyn.* **20** 107–126
[6] Dimitrijevic M and Laprise R Validation of the nesting technique in a regional climate model and sensitivity tests to the resolution of the lateral boundary conditions during summer *Clim. Dyn.* **25** 555–580
[7] Davies T 2014 Lateral boundary conditions for limited area models *Q. J. R. Meteorol. Soc.* **140** 185–196
[8] EPA 2016 Climate change indicators: weather and climate *United States Environmental Protection Agency* [Online]. Available: https://www.epa.gov/climate-indicators/weather-climate. [Accessed: 05-Nov-2018]
[9] Asimakopoulos D 2011 Climate and climate change *Energy and Climate in the Urban Built Environment*, 2nd ed., M. Santamouris, Ed. New York, USA: Routledge Taylor and Francis Group 19–32
[10] Morris K 2016 Effect of vegetation and waterbody on the garden city concept : An evaluation study using a newly developed city , Putrajaya , Malaysia *Comput. , Environ. Urban Syst.* **58** 39–51
[11] Trihamdani A, Lee H, Kubota T and Phuong T 2015 Configuration of green spaces for urban heat island mitigation and future building energy conservation in Hanoi Master Plan 2030 *Buildings* **5** 933–947
[12] Isa N, Wan Mohd W and Salleh S 2017 The effects of built-up and green areas on the land surface temperature of the Kuala Lumpur City *ISPRS Archives* **42** 107–112
[13] Lima F, Martins F, Pereira E, Lorenz E and Heinemann D 2016 Forecast for surface solar irradiance at the Brazilian Northeastern region using NWP model and artificial neural networks *Renew. Energy* **87** 807–818
[14] Van Doan Q, Kusaka H, and Ho Q 2016 Impact of future urbanization on temperature and thermal comfort index in a developing tropical city: Ho Chi Minh City *Urban Clim.* **17** 20–31
[15] Ooi *et al.* 2018 Comparison of WRF local and nonlocal boundary layer physics in Greater Kuala Lumpur, Malaysia in *IOP Conference Series: Earth and Environmental Science* 1–7
[16] NCAR, *Weather Research and Forecasting: ARW version 3 modelling system user's guide January 2016*. 2016.
[17] Isa N, Wan Mohd W, Salleh S and Ooi M 2018 The effects of green areas on air surface temperature of the Kuala Lumpur city using WRF- ARW modelling and Remote Sensing technique *IOP Conf. Ser. Earth Environ. Sci.* **117** 1–8
[18] USGS 2018 Using the USGS Landsat Level-1 Data Product [Online]. Available:

https://landsat.usgs.gov/using-usgs-landsat-8-product. [Accessed: 07-Dec-2018].
[19] Bhatti S and Tripathi N 2014 Built-up area extraction using Landsat 8 OLI imagery," *GIScience Remote Sens.* **51** 445–467
[20] Zha Y, Gao J, and Ni S 2003 Use of normalized difference built-up index in automatically mapping urban areas from TM imagery *Int. J. Remote Sens.* **24** 583–594
[21] He C, Shi P, Xie D and Zhao Y Improving the normalized difference built-up index to map urban built-up areas using a semiautomatic segmentation approach *Remote Sens. Lett.* **1** 213–221
[22] Lee J, Lee S and Chi K 2010 Development of an urban classification method using a built-up index *Sel. Top. power Syst. Remote Sens.* 39–43
[23] Varshney A 2013 Improved NDBI differencing algorithm for built-up regions change detection from remote-sensing data: An automated approach *Remote Sens. Lett.* **4** 504–512
[24] Xu H 2008 A new index for delineating built-up land features in satellite imagery *Int. J. Remote Sens.* **29** 4269–76
[25] Tukey J 1977 Exploratory data analysis *Biometrical J.* **23** 413–414
[26] Hoaglin D, Iglewicz B and Tukey J 2012 Performance of some resistant rules for outliers labeling *J. Am. Stat. Assoc.* **81** 991–999

IOP Conf. Series: Earth and Environmental Science **385** (2019) 012021 doi:10.1088/1755-1315/385/1/012021

Modelling landscape aesthetic of planting composition influencing visual quality and well-being: PLS-SEM approach

N Mt Akhir[1] S R Md Sakip[1,2] M Y Abbas[3,4] N Othman[4]
[1]Faculty of Architecture, Planning and Surveying, University Teknologi MARA, Perak Branch, Seri Iskandar Campus, 32610 Perak, Malaysia
[2]Green Safe Cities Research Group, Universiti Teknologi MARA, Shah Alam Campus, 40450 Selangor, Malaysia
[3]Head, Centre for Environment-Behaviour Studies (cE-Bs), Faculty of Architecture, Planning & Surveying, Universiti Teknologi MARA, Shah Alam Campus, 40450 Selangor, Malaysia
[4]Faculty of Architecture, Planning and Surveying, University Teknologi MARA, Selangor Branch, Puncak Alam Campus, 42300 Selangor, Malaysia

Email: noriz102@perak.uitm.edu.my

Abstract. Planting composition (PC) is designed to be aesthetically pleasing, functional utilizing and ecologically sustaining. In landscape research, a plant is a powerful element in assessing visual quality. Visual quality for each landscape depends on the visual composition of plants. The study on landscape assessment in a campus is commonly correlated with green space. However, the composition and design of planting in the green spaces on campus grounds have not been considered yet. Therefore, this study aims to explore the influence of PC towards visual quality and its relationships with the well-being of students. The conceptual model is composed of endogenous latent variables (LVs) according to the pattern of visual composition namely coherence, complexity, legibility and mystery. The exogenous LVs is PC and well-being. Positive emotion, engagement, relationships, meaning and accomplishment (PERMA) are the items used to test the relationship with the pattern of PC. This study applied confirmatory factor analysis (CFA) with pilot data to measure the relationship between LVs. The result validates that PC should consider 5 pillars of well-being which is PERMA to serve students with better visual access to landscape on campus.

1. Introduction

A beautiful and tranquil landscape is important. When people are surrounded by lush greenery it can stimulate more positive emotions **Error! Reference source not found.**. The psychological study of environmental aesthetic hypothesized that people tend to judge a scenically beautiful landscape with positive preference **Error! Reference source not found.**. The quality of landscape should be considered because low quality landscape can induce negative reactions, while, high quality landscape evoke positive responses **Error! Reference source not found.**. In the planting design, landscape properties like planting composition (PC) is important to secure the whole landscape as being unified in design. Indeed, the composition could enhance aesthetic value resemble of colour, texture, form, rhythm, order, emphasis and other related formal aesthetic principles. PC may consist of groundcovers, shrubs and trees which are truly important in reviving the condition of the surrounding environment **Error! Reference source not found.**. Among these group of plants, trees significantly act as a visual accent to the scenery while supported by other plant categories. Therefore, these group of plants should be managed carefully with a good composition to create better visual interest.

The primary objectives in this study are; (1) to explore the characteristics of PC related to visual quality; and (2) to examine the relationship between PC preferences and the well-being of students. The theoretical explanations for PC characteristics builds upon a data base from earlier research by Kaplan and expands the subject pool used in that theory to encompass a range of principles on PC.

1.1. Theory in landscape aesthetic

The perception of individuals towards a landscape is associated with certain variables of visual quality preferences. The different arrangement of plants, forms different spatial setting. Hence, the different spatial variables affect aesthetic and landscape preferences **Error! Reference source not found.**. The preference matrix **Error! Reference source not found.** is an evolutionary theory that adopts the aesthetic appraisal to encourage the selection of habitat **Error! Reference source not found.**. The preference matrix in the Information Processing Theory have determined four visual response variables; coherence, complexity, legibility and mystery (see Table 1). According to the theory, a good understanding of the physical environment was crucial to human survival, but an exploration of a new habitat conveyed adaptive benefits as well **Error! Reference source not found.**. This support the hypothesis by Kaplans that the perceptual process involves extracting information from an individual's environment **Error! Reference source not found.**.

Table 1. The preference matrix of Kaplans' Information Processing theory

	Understanding Making sense	**Exploration** Being involved
Immediate The visual array	**COHERENCE** Related with PC principles like order, balance, repetition and simplicity	**COMPLEXITY** Related with PC principles like complexity, harmony and contrast
Inferred Future, promised Three-dimensional space	**LEGIBILITY** Related with PC principles like emphasis, focalisation, scale and proportion	**MYSTERY** Related with PC principles like rhythm, sequential and chaotic

Based on the theory, this study addresses the following hypotheses:

H1: Perceived characteristics of planting composition is positively associated with coherence, complexity, mystery and legibility.
H2: There are positive relationship between well-being and the preferences of students

1.2. Visual quality in planting composition

Plants properties are able to attract the attention of the viewers through colour, form, texture and line. In planting design practices, a real difference in landscape aesthetic is when visual properties and basic design principles are well accomplished **Error! Reference source not found.**. With assistance from planting design principles, designers can influence the perception of viewers and their aesthetic satisfaction **Error! Reference source not found.**. Findings from previous scholars show, the compositions of plants and their colours and plant species diversity are the main elements of the visual quality of landscape areas **Error! Reference source not found.Error! Reference source not found.**. In concurrence with the preference matrix stated above, it parallels with the complexity preference predictor which can be hypothesized as a significant factor in judging landscape visual quality.

1.3. Influencing pattern in well-being

Based on the pattern in landscape visual assessment, the natural setting has become a major concern of preference predictor **Error! Reference source not found.Error! Reference source not found. Error! Reference source not found.**. There are some studies which confirm that contact with nature is fundamental to health and well-being **Error! Reference source not found.Error! Reference source not found.Error! Reference source not found.**. However, in the context of study students are continuously connected with technology devices **Error! Reference source not found.** and crucially have less contact with nature **Error! Reference source not found.**. Spending more time indoors has predicted to lead to a negative impact on students **Error! Reference source not found.Error! Reference source not found.**. Therefore, this study is evocative to encourage outdoor connectedness. Before that, better planting design must be reviewed to ensure the effectiveness in visual preferences. The advantages of plant view have correlated with enhance positive emotions **Error! Reference source not found.Error! Reference source not found.**,stress reduction **Error! Reference source not found.Error! Reference source not found.**, mental health **Error! Reference source not found.**, well-being **Error! Reference source not found.Error! Reference source not found.** as well as students' performance **Error! Reference source not found. Error! Reference source not found.**.

2. Relationship between planting composition, visual quality and well-being

In parallel with the theoretical study and landscape properties, a similar relationship can be found between variables for PC pattern and well-being. According to Kaplans' theory, coherence, complexity and mystery are among the significant variables in the context of nature, while legibility has less significance in landscape aesthetics. Previous studies have found that plant views are able to evoke positive emotions among people when engaging with the green environment **Error! Reference source not found.Error! Reference source not found.Error! Reference source not found.**. Furthermore, nature contact also affects as adaptive behaviour **Error! Reference source not found.** which are able to motivate the students to nature connectedness. The well-being model by Seligman is known as positive emotion, engagement, relationship, meaning and accomplish hence the PERMA acronym used in this study. Although this PERMA model is derived from the psychology field, the meaning behind the model is allied with landscape preferences. The question in PERMA has been adjusted according to a landscape scene to relate with the preferences pattern and psychology outcome.

3. Methodology

Data collection was conducted through a pilot survey. In the first phase, the respondents were selected at random from different faculties with design based and non-design based students. This selection is necessary to generalise the questionnaire as well as to obtain diverse perception views. The total number of respondents involves 50 samples. In the pilot study, 10 to 30 samples are sufficient **Error! Reference source not found.**. Convenient sampling technique was used to meet the users of the campus with green spaces as respondents. The survey was divided into three sections. Part A includes personal information question in nominal scale, while part B and C were in the form of attitude rating scale. Many researchers used this scale to measure subjective phenomena such as satisfaction, perception, expectation and awareness **Error! Reference source not found.**. Part B consists of 38 numbers of PC images that represent faculties with green spaces. Part C comprises of well-being items which relate to the most preferred PC images in part B. Five point of the Likert scale was used for respondents to answer the question in part B and C. The five point scale was typically used by many researchers in various field such as management, business and social sciences. The alternative scale were assigned from 1 (strongly dislike) to 5 (strongly like) for part B and 5 (strongly agree) to 1 (strongly disagree) for part C. Then, the data was analysed using PLS-SEM 3.0 to test the relationship between preference pattern with PC and well-being. The path coefficients (β) which beta value represent a strength of correlation between LVs **Error! Reference source not found.** were measured. The higher β value indicates the most influential relationship of exogenous LVs towards endogenous LVs which is close to 1, whereas if the β value that close to 0 indicate weak relationship **Error! Reference source not found.**.

4. Result, analysis and discussion

In order to investigate the relationship between PC and well-being in landscape visual quality, the reflective model of that relationship have been recorded. To validate the reflective measurement model, there are two types of validity that should be assessed which are convergent validity and discriminant validity **Error! Reference source not found.**. In the convergent validity, there are three components to be reported accordingly: 1) composite reliability (CR), 2) average variance extracted (AVE) and 3) factor loadings. Table 2 demonstrates the result of CR and AVE. Both results have achieved the minimum guidelines with CR > 0.70-0.90 and AVE > 0.50. The result for factor loadings are shown in Figure 1. The recommended result is 0.70, but for the loadings >0.70 to 0.40 is adequate if other items have high scores and complement to AVE and CR **Error! Reference source not found.**.

Table 2. Convergent validity result for AVE and CR

Construct	AVE	CR
PC	0.459	0.922
WB	0.502	0.950
Coherence	0.519	0.842
Complexity	0.562	0.834
Legibility	0.566	0.935
Mystery	0.585	0.848

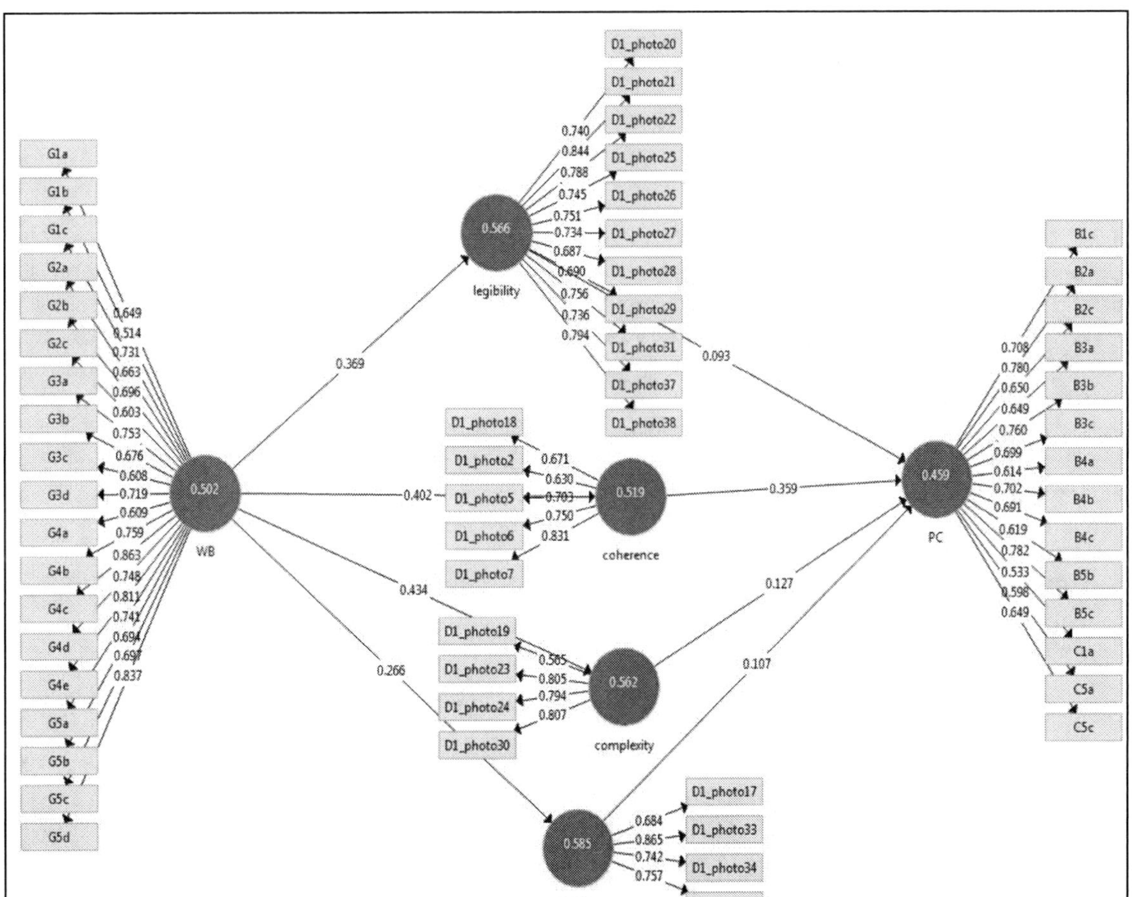

Figure 1. Model relationship between planting composition, visual quality and well-being

Table 3. Fornell-Lacker Criterion Result

Construct	PC	WB	Coherence	Complexity	Legibility	Mystery
PC	**0.677**					
WB	0.481	**0.709**				
Coherence	0.446	0.402	**0.721**			
Complexity	0.230	0.434	0.115	**0.750**		
Legibility	0.361	0.369	0.487	0.431	**0.753**	
Mystery	0.256	0.266	0.253	0.195	0.358	**0.765**

The following Table 3 shows a result of discriminant validity. There are three methods to assess, Fornell-Lacker criterion, cross loading and Heterotrait-Monotrait Ratio (HTMT). Based on the criterion the indicators should load more strongly on their constructs rather than on other constructs in the model **Error! Reference source not found.** as well as cross loading. HTMT values close to 1 indicate a lack of discriminant validity **Error! Reference source not found.**. The result show there is no discriminant issues between constructs. Table 3 indicate that all constructs exhibit sufficient discriminant validity. Table 4 depicts a method result of HTMT.

Table 4. HTMT Result

Construct	PC	WB	Coherence	Complexity	Legibility	Mystery
PC						
WB	**0.509**					
Coherence	**0.519**	**0.449**				
Complexity	**0.277**	**0.448**	**0.337**			
Legibility	**0.364**	**0.362**	**0.558**	**0.491**		
Mystery	**0.297**	**0.345**	**0.378**	**0.349**	**0.358**	

In determining the relationship between PC and well-being the path coefficient was assessed. Path coefficient value should exceed 0.10 to measure the impact of a certain model **Error! Reference source not found.**. However, if the value less than 0.10 also can be accepted for the different result of a weak relationship or negative relationship depending on the theory **Error! Reference source not found.**. The result shown in Figure 1 illustrates the relationship between LVs. The model illustrates the strong relationship between well-being, complexity and coherence pattern. With regard to Kaplans' theory, the most preferable indicator in landscape aesthetic always highly correlated with complexity and coherence matrix. Table 5 shows the relationship between LVs and hypothesis remarks. For the relationship between PC and preference pattern, coherence shows the most strength than other constructs. This result detected that coherence is the significant and strongly accepted variable because people can easily cognitively understand a scene with an orderly arrangement. Legibility has a weak relationship with 0.093. As supported by Kaplans' theory, legibility variable is difficult to assess and does not have a significant role in landscape preference. However, legibility can clearly be measured for urban setting.

Table 5. Path Coefficient between Well-Being, PC and Preference Pattern

Hypothesis	Path coefficient	Remarks
H1: Perceived characteristics of planting composition is positively associated with:		
a) Coherence	**0.359**	Strongly accepted
b) Complexity	0.127	Accepted
c) Legibility	0.093	Slightly accepted

d) Mystery	0.107	Accepted
H2: There are positive relationship between well-being and students' preferences on:		
a) Coherence	0.402	Accepted
b) Complexity	**0.434**	Strongly accepted
c) Legibility	0.369	Accepted
d) Mystery	0.266	Accepted

Overall, the research model in Figure 1 accounted that landscape aesthetic variables are able to influence the PC pattern and well-being. All of the hypothesis are accepted according to the path coefficient relationships. The relationship proves that coherence and complexity are the most significant indicators in evaluating landscape properties like the composition of plants in a campus setting.

5. Conclusion

Based on the findings discussed above, the evaluation of PC towards well-being are the areas that should receive careful attention from campus landscape management and campus community. The research is significant to enhance the quality of life of the campus community by viewing or exploring the good composition of plants. As supported by previous scholars research, viewing high quality landscape is able to evoke a positive response which increases the happiness index and reduce the level of stress **Error! Reference source not found.Error! Reference source not found.Error! Reference source not found.**. Therefore, the setting of PC should integrate the preference indicators to enhance the quality and impact of a green surrounding in a campus area. In conclusion, by identifying the relationship of PC for the well-being of students towards a high quality landscape, this study is beneficial to all universities in assisting the identification of green spaces that need improvement. Since the respondents of this study are only students, therefore it is recommended to extend the findings with various types of users in the future.

Acknowledgement

We would like to thank UiTM Cawangan Perak and Research Management Unit (RMU) for the support and the incentive in *Geran Khas Insentif Penyeliaan Perak (GKIPP).*This research is a part of a PhD study of the corresponding author.

References

[1] Speake J, Edmondson S and Nawaz H 2013 Everyday encounters with nature: students' perceptions and use of university campus green spaces. *Human Geographies – Journal of Studies and Research in Human Geography* 7 21-31

[2] Daniel T C and Vining J 1983 Methodological issues in the assessment of landscape quality In: Altman I, Wohlwill J (Eds.) *Human Behavior and Environment* 7 39-84

[3] Han K T 2010 An exploration of relationships among the responses to natural scenes *Environment and Behavior* **42** 243-270

[4] Noriah O, Masbiha M I, Noralizawati M and Ramly H 2015 Street Planting Compositions: The Public and Expert Perspectives. *Procedia - Social and Behavioral Sciences*, **170** 350-358

[5] Polat A T and Akay A 2015 Relationships between the visual preferences of urban recreation area users and various landscape design elements. *Urban Forestry and Urban Greening* **14** 573-582

[6] Kaplan R, Kaplan S and Brown T 1989 Environmental Preference: A Comparison of

Four Domains of Predictors *Environment and Behavior* **21** 509-530
[7] Van der Jagt A P N, Craig T, Anable J, Brewer M J and Pearson D G 2014 Unearthing the picturesque: The validity of the preference matrix as a measure of landscape aesthetics *Landscape and Urban Planning* **124** 1-13
[8] Lothian A 2000 Landscape quality assessment of South Australia. University of Adelaide
[9] Yılmaz S, Özgüner H and Mumcu S 2018 An aesthetic approach to planting design in urban parks and greenspaces *Landscape Research* 1-19
[10] Hartig T, Mitchell R, Vries S de and Frumkin H 2014 Nature and health. Annual Review Public Health.
[11] Kaplan S 1979 Perception and Landscape: Conceptions and Misconceptions. In *National Conference on Applied Techniques for Analysis and Management of the Visual Resource* 241-248
[12] Kaplan R 1984 Impact of urban nature: A theoretical analysis *Urban Ecology* **8** 189-197
[13] Ulrich R S 1979 Visual landscapes and psychological well-being *Landscape Research* **4** 17-23
[14] Ulrich R S 1983 Aesthetic and affective response to natural environment In: Altman I, Wohlwill J (Eds.) *Human Behavior and Environment* **6** 85-125
[15] Ulrich R S 2002 Health Benefits of Gardens in Hospitals. In Plants for People: *International Exhibition Floriade* 1-10
[16] Bellur S, Nowak K L and Hull K S 2015 Make it our time: In class multitaskers have lower academic performance *Computers in Human Behavior* **53** 63-70
[17] Crump N 2015 Digital Natives: The Nature of Technology on College Student Mental Health James Madison University
[18] Flora Wei F Y, Ken Wang Y and Klausner M 2012 Rethinking College Students' Self-Regulation and Sustained Attention:Does Text Messaging During Class Influence Cognitive Learning? *Communication Education* 1-20
[19] Wentworth D K and Middleton J H 2014 Technology use and academic performance *Computers and Education* **78** 306-311
[20] Ballew M T and Omoto A M 2018 Absorption: How nature experiences promote awe and other positive emotions *Ecopsychology* **10** 26-35
[21] Kaiserman J 2017 Urban Nature for Well-being: Design Recommendation for Psychological Benefits in Urban Public Spaces. University of Washington.
[22] Li D and Sullivan W C 2016 Impact of views to school landscapes on recovery from stress and mental fatigue *Landscape and Urban Planning* **148** 149-158
[23] Wee C 2017 Assessing the perception of campus green space and stress levels among students at Michigan State University. Michigan State University
[24] Coles R and Millman Z K 2013 Landscape, Well-Being and Environment New York: Routledge Taylor and Francis Group
[25] Matsuoka R H 2008 High school landscapes and student performance. University of Michigan
[26] Isaac S and Micheal W B 1995 Handbook in Research and Evaluation 3rd edition San Diego CA: EdITS
[27] Zamalia M 2009 Handbook of Reseach Methodology: A simplified version UiTM Shah Alam: UPENA
[28] Ramayah T, Cheah J, Chuah F, Ting H and Mumtaz Ali M 2018 Partial Least Squares Structural Equation Modelling (PLS-SEM) using SmartPLS 3.0 (Second Edition) Kuala

Lumpur: Pearson Malaysia Sdn Bhd.
[29] Mohd Hanafi A O 2017 Structural Equation Modelling by using Partial Least Squares with SmartPLS Dewan Al-Ghazali FSKM UiTM Shah Alam
[30] Fornell C and Larcker D F 1981 Evaluating Structural Equation Models with Unobservable Variables and Measurement Error *Journal of Marketing Research* **18** 39-50
[31] Hair J F, Ringle C M and Sarstedt M 2011 PLS-SEM: Indeed a Silver Bullet *The Journal of Marketing Theory and Practice* **19** 139-151

Decorative Elements of Traditional Malay House: Comparative Study of Rumah Limas Bumbung Perak (RLBP) and Rumah Limas Johor (RLJ)

M S Ab Rashid[1], M N Baharuddin [1], K Alauddin [1] and I A H Choo [1]
[1]Knowledge & Understanding of Tropical Architecture & Interior (KUTAI) Research Group, Faculty of Architecture, Planning & Surveying, Universiti Teknologi MARA, Perak Branch, Seri Iskandar, 32610 Perak, Malaysia

mohds637@uitm.edu.my

Abstract. In traditional Malay architecture, decorative elements are defined as non-structural elements and as "additional member" of the building that contributes to aesthetic and functional aspects of the architecture. Its ornamentation, form and pattern particularly in woodcarving and bamboo weaving reflect the uniqueness of local craftsmanship and the master carvers. There are at least 16 (sixteen) decorative elements of the traditional Malay house that have been identified by previous researcher. Architecturally in Malaysia traditional houses were categorised based on roof design, spatial organization and geographical setting which finally this categorization was named after the states. Each traditional house is in the state of Malaysia signifying particular style and architectural characteristics of that particular state. This article discusses the study of those decorative elements on the traditional Malay houses in the state of Perak known as Rumah Limas Bumbung Perak (RLBP) and traditional house from the state of Johor known as Rumah Limas Johor (RLJ). With the objective is to investigate the type, terms and characteristics of decorative elements site observations were done on selected twenty-five (25) of RLBP houses and another twenty-five (25) of RLJ houses. The analysis on visual data and verbal data were carried out to established architectural terminology and typology based on form and style of every decorative element. The study found that nine (9) decorative elements come to be the major characteristics of RLBP whilst seven (7) for RLJ; some with different name, form and style from other states. These similarity and dissimilarity contribute to the extension of the aesthetical theory and philosophy of the Malay traditional house in Malaysia.

1. Introduction

Decorative elements in Malay traditional architecture are a manifestation of the understanding of integration between functionalism and aesthetics. It represents the uniqueness and beauty of Malay symbols, arts, philosophy as well as great design work by local craftsmanship or carpenter. In fact, according to Ezrin [1], despite of traditional architecture being treated as primitive or outdated, it has always been the main source of references in the social and cultural studies by architects, historian and anthropologists. Thus, the term of 'architecture is the manifestations of the society and their cultural practise' will reflects the society's way of life, their belief and philosophy; 'all encompassing; forms important components in the uniqueness of a culture'. The existence of decorative elements in traditional Malay houses represents the integration between functionalism and aesthetics of architecture. Its ornamentation, form and pattern particularly in woodcarving and bamboo weaving reflect the uniqueness of local craftsmanship and the master carvers. According to Nursuriani and Ismail [2], a decorative elements or architectural elements in Malay Traditional House were designed intentionally to facilitate and express the respect for Malay culture through the design structure of ornamentation.

An Overview of Traditional Malay House (Rumah Limas Bumbung Perak)

The development and expansion of the traditional Malay villages and towns at Perak during the colonial era outlined an interesting collection of architectural study. Architectural study along Sungai Perak established two types of architecture which are:

a) Traditional Malay architecture
The traditional Malay architecture found along the riverbank of Sungai Perak is mostly in the form of old Malay palaces, traditional houses including Rumah Limas Bumbung Perak (RLBP) and Rumah Kutai (RK) or Rumah Bumbung Melayu and mosques. Based on study by Anuar [3], Norhasandi [4] and Mohd Sabrizaa [5] found that RK were the earlier architectural style of the Perak Malay houses in comparison to RLBP. Apart from the timeline, the roof design and construction detail also mark the different between the two houses. The earliest house of RLBP found was built circa 1900s and continue until 1970s with certain architectural design development.

b) Colonial architecture
Colonial architecture are mostly located in several towns along the river such as Kuala Kangsar, Parit, Bota, Pasir Salak, Kampung Gajah and Teluk Intan where many houses, schools, offices, rest houses, mosques and hospital still stands and in use to this day.

An Overview of Traditional Malay House (Rumah Limas Johor)

Majority of 'Rumah Limas Johor' were architecturally influence from the Riau-Lingga in Indonesia, Dutch and English style Mastor Surat et-al [6]. The people of Bugis were migrated after the arrival of the Dutch East India Company in 1669 with the goal of monopolising the spice trade. Apart from that, more Bugis migrated to the other parts of Indonesia and other countries including Malaysia. Hence, most of the architectural styles in Johor were affected based on their history.

Several of traditional and hybrid Malays house can be classified mainly by their basic roof shapes Mastor Surat et-al [6]. They further explain that there are two (2) categories of Malays Traditional House which are:

a) Rumah Bumbung Panjang
The most common houseform is the bumbung panjang, characterised by a long gable roof. The bumbung panjang houses are the oldest identified in Peninsular Malaysia, many of them being over a hundred years old and still in good condition. It has a simple gable roof, supported by kingposts. The most common roofing material used for the bumbung panjang is the attap (a thatch made from nipah and other palm trees found in the local natural vegetation) Mastor Surat et-al [6].

b) Rumah Bumbung Lima and Limas
It was transformed from Rumah Bumbung Panjang over the years. The roof shapes of Rumah Berbumbung Limas was architecturally derived from pyramidal concept and fitted with 'V' shape 'tebar layar' at the top of the roof as to allow natural ventilation to coolest the house.

IOP Conf. Series: Earth and Environmental Science **385** (2019) 012022 doi:10.1088/1755-1315/385/1/012022

2. Literature Review

A few studies have been done on decorative elements of traditional Malay houses. With reference to Diagram 1, they were listed sixteen (16) decorative elements and further explain in detail in Diagram 2 within a particular traditional Malay house which is:

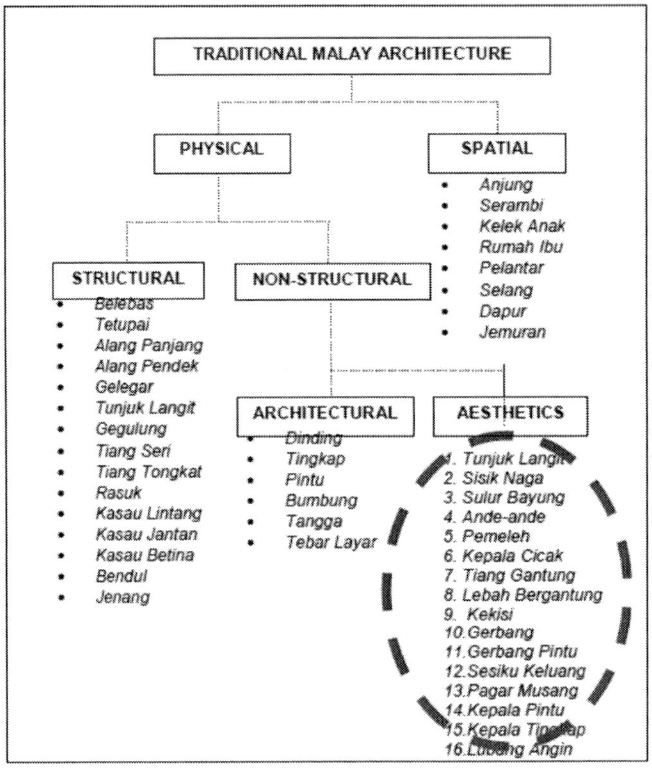

Figure 1. Decorative Elements of Traditional Malay House
Source: Mohd Sabrizaa and Sufian [7]

Decorative Elements at Malay Traditional House

a) Tunjuk Langit (literally means pointed to the sky)

It is a piece of decorative element in an upright position at the tip of the gable end; commonly used in houses and palaces in Peninsula Malaysia. In general, this type can be further categorised into few basic types; i. batang (rod), ii. bulat (rounded) iii. silang (cross) and iv. bunga (floral). The batang type is a piece of decorative timber or rod and is commonly used for Bumbung Limas roof. It has been observed that the rounded and floral is commonly used on bumbung perabung lima or pyramidal type of roofing and is positioned in the middle top of the roof. Most are made of carved timber solids but of late there has been practices found to have used the concrete mix similar to the one used for the Malay gravestones. Commonly called *buah buton* in Kelantan and *buah gutung* in Terengganu

b) Sisik Naga (literally means dragon fin)

Sisik naga refers to the decorative carvings along the ridge of a roof *(perabung and rabung atap)*. Sisik naga are decorative architectural elements can be found in palace architecture, residential and religious buildings in Malaysia, Indonesia and Indochina. Other names are *kemuncak, naga-naga, puncak rabung*.

c) Sulur Bayung

Sulur bayung refer to a decoration on the corner of a roof edge. Other names are *sulo bayung, sayap layang-layang, sayap layangan, ekor Itik, anjong balla', som*. In the traditional architecture of Bugis Makasar it is also called *anjung balla*, representing the creeping plants, Mohd Sabrizaa and Sufian [7].

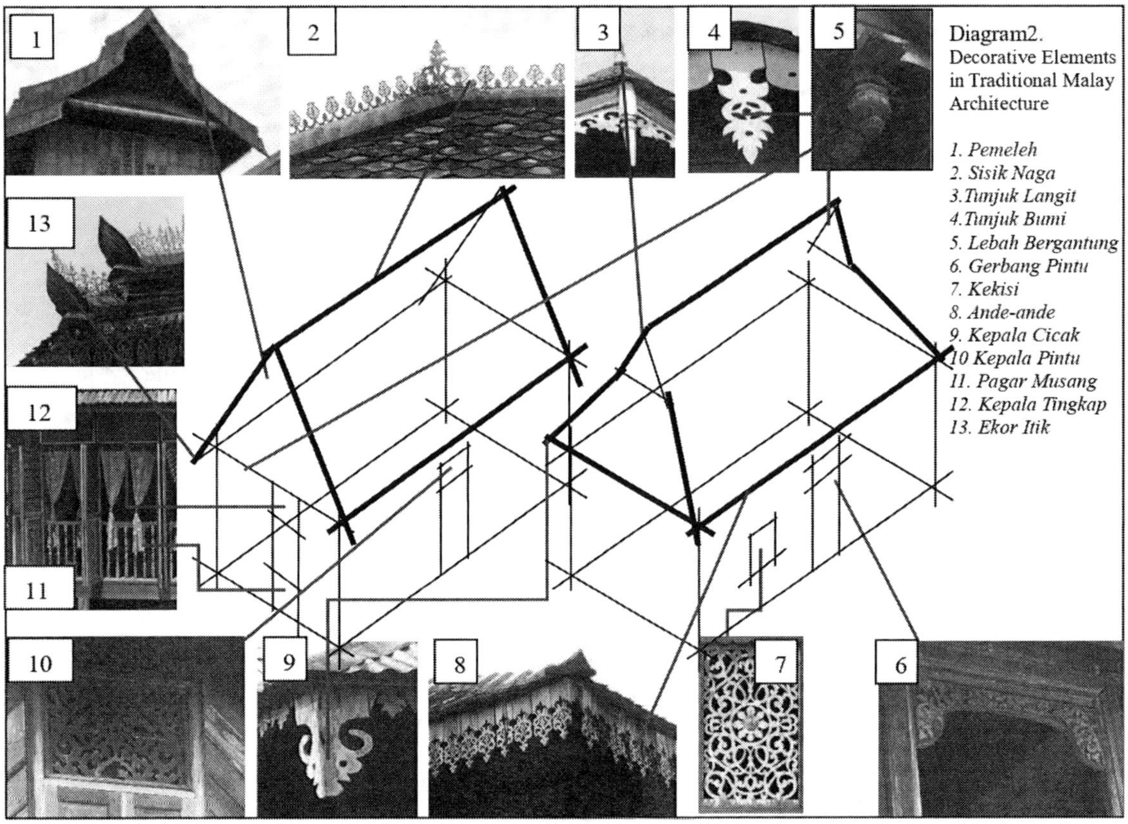

Figure 2. Decorative Elements of Traditional Malay House

d) Ande-Ande

Ande-ande is the term used in Kelantan and Terengganu for fascia board also known as *papan manis* is the horizontal decorative timber roof eaves covering the timber rafters (kayu kasau). Other names ande-ande are *papan meleh, papan pator, papan cucuran atap, papan kaki atap, kening, papan tumpu kasau, papan cantik.* There are at least two types of ande-ande which are the single piece type with carvings on a single long timber piece and multiple pieces) with carvings on multiple timber pieces arranged side by side, Mohd Sabrizaa and Sufian [7].

e) Pemeleh

Other names for *kayu pemeleh* are *peles, pemeles, papan layan*. Kayu pemeleh is actually a pair of timber pieces located on two fascia ends of a gable roof (tebar layar). The ones covering kasau lintang or kasau jantan is called peles or pemeleh in Kelantan and Terengganu but called papan layang elsewhere. Kayu pemeleh or peles is shaped slightly curving and sharpens towards the top as a sign of strength in the design language of houses in Kelantan, Terengganu on the East coast and also in Cambodia and Pattani in Thailand, Yusoff and Rashid [8].

IOP Conf. Series: Earth and Environmental Science **385** (2019) 012022 doi:10.1088/1755-1315/385/1/012022

f) Kepala Cicak (literally means lizard head)

Hanafi [9] identified kepala cicak as a type of end decoration to tumpu kasau or papan meleh. There are two types of kepala cicak which are the "L" type with the end timber pieces positioned abutting to one another at corners of the roof eaves *(bucu atap)* and the "I" type with a single end timber piece positioned on chamfered corners of roof eaves.

g) Tiang Gantung (literally means hanging column)

Other names for tiang gantung are *saka bentung, tiang guntung, tiang sotoh*. In traditional houses of Negeri Sembilan it looks like a small carved hanging column positioned at the tip of the alang that is supporting the main kasau jantan serambi and is called tiang guntung. Also called saka bentung in the traditional Javanese architecture and on the house of Negeri Sembilan it is known as tiang sutoh.

h) Lebah Bergantung (literally means hanging bees)

Halim [10] , Norhasandi et. al [11] and Jahn et. al [12] uses lebah bergantung to describe it in his narration of Rumah Kutai(RK) houses in Perak; as a style of hanging column at the end of *alang panjang*, rumah ibu, that is constructed as a set (left and right) on the bottom of the gable end of the roof (tebar layar). It is round in shape the size of the local water vessel (labu air) looking very similar to the bees nest.

i) Kekisi (other names Kisi-kisi, kekipas, jeriji, jeruji, terali)

Decorative piece of timber or rod used at the opening of a window function as window grill.

j) Gerbang (Archway)

Decorative piece of timber as an arch at entrance of a house or used for dividing space usually between Anjung and Rumah Ibu of the main house.

k) Gerbang Pintu (Arch door)

Decorative piece of timber as an arch at entrance of a house or used for dividing space usually between Anjung and Rumah Ibu of the main house.

l) Sesiku Keluang

Decorative piece of timber normally in shape of triangle fix to the column and sometime function to support the beam. Otherwise it is for decorative purpose.

m) Pagar Musang

Balustrade In the form of timber rod or timber plank often beautifully carved and arranged to decorate the window opening. It is function as safety features apart from its aesthetical role. The other names also known as *Jeriji, Jerejak, Pagar Tinggalung.*

n) Kepala Pintu (literally means door head)

This is a decorative piece of timber on top of the door. Normally carved with floral pattern, Quranic verses calligraphy or simple motif of jejala (lattice work). Apart from aesthetical reason its function is to allow sufficient cross air ventilation to flow into the house.

o) Kepala Tingkap (literally means Window head)
A decorative piece of timber located on top of the window. Normally carved with floral pattern, Quranic verses calligraphy or simple motif of jejala (lattice).

p) Lubang Angin (literally means air hole)
A ventilation hole or panel located mostly at two locations such as wall and roof. Lubang Angin at roof area is actually located at gable's end whilst Lubang Angin of the wall is commonly found at the upper level of wall in the form of carving or latticework known as *jejala*.

3. Research Methodology

This study aims to investigate the existence of decorative elements at RLBP and RLJ. As discussed, RLBP are located in the state of Perak whilst RLJ in the state of Johor at the southern of the Peninsula. The selection of RLBP and RLJ were done based from previous study by the author and the research team member. An exploratory research was carried out through site observation on traditional villages at both states during the previous study by Mohd Sabrizaa et al. [13] and current study. Thirty (30) RLBP and twenty seven (27) RLJ were observed and documented from that particular. For the purpose of this study twenty-five (25) houses were selected from each state for RLBP and RLJ. The selection was done based on their best level of authenticity, architectural type, current physical condition, accessibility and geographical factor. All the houses than were coded accordingly. Many houses were found have done major renovation by the owner and has lost its ingenuity in terms of traditional architectural style. Few houses also found to be abandoned and left over by the owner and make it difficult to access for the study. Many houses were found heavily damaged and ruined. These are some of the factors that contribute to the limitation of the scope of the study. Documentations of visual data were collected using EOS DSLR Canon 6D camera whilst semi-structured interview were done to collect verbal data from the owner of the house or the occupancies. Both data were used to investigate and identify the existence of the type of decorative elements at selected RLBP and RLJ. Results were discussed based on the visual tabulation and inventory analysis produced from the study.

4. Discussion and Analysis

This research is intended to execute a Comparative Study of Decorative Elements of Rumah Limas Bumbung Perak (RLBP) and Rumah Limas Johor (RLJ) as a main subject matter. Hence, before proceed to the analysis of the decorative elements; identifying a location of the houses is critical. According to Mohd Sabrizaa et al. [13], the best of twenty-five (25) of RLBP were selected for this study and were coded as (Table 1) while this study observation identify the best of twenty-five (25) of RLJ house and were coded as (Table 2).

Table 1. RLBP location and GPS Positioning

CODE	House Address	Area	GPS
PLH1	195A, Jalan Besar Utara Selatan, Kampung Bendang Susur, 33000, Kuala Kangsar, Perak.	Kuala Kangsar	(4.797586, 100.947479)
PLH2	5460, Jalan Besar Utara Selatan, Kampung Kuala Chempias, 33000, Kuala Kangsar, Perak.	Kuala Kangsar	(4.787324, 100.942841)
PLH3	5460, Jalan Besar Utara Selatan, Kampung Kuala Chempias, 33000, Kuala Kangsar, Perak.	Kuala Kangsar	(4.791505, 100.944878)
PLH4	Jalan Besar Utara Selatan, Kampung Kuala Chempias, 33000, Kuala Kangsar, Perak.	Kuala Kangsar	(4.791723, 100.944916)
PLH5	Jalan Besar Utara Selatan, Kampung Kuala Chempias, 33000, Kuala Kangsar, Perak.	Kuala Kangsar	(4.792849, 100.944893)
PLH6	17, Jalan Besar Kampung Bendang Kering, Kampung Kuala Sayong Hulu, 33000, K.K.	Kuala Kangsar	(4.769046, 100.964989)
PLH7	Jalan Besar Utara Selatan, Kampung Padang Ampang, 33000, Kuala Kangsar, Perak	Kuala Kangsar	(4.809263, 100.953140)

PLH8	Jalan Besar Utara Selatan, Kampung Padang Ampang, 33000, Kuala Kangsar, Perak	Kuala Kangsar	(4.810440, 100.954338)
PLH9	Jalan Besar Utara Selatan, Kampung Padang Ampang, 33000, Kuala Kangsar, Perak	Kuala Kangsar	(4.810744, 100.954491)
PLH10	Kampung Sayong Lembah, Kuala Kangsar, Perak	Kuala Kangsar	(4.771876, 100.948547)
PLH11	2010, A3, Kampung Suak Petai, 33800, Manong, Perak.	Manong	(4.595965, 100.881981)
PLH12	2010, A3, Kampung Suak Petai, 33800, Manong, Perak.	Manong	(4.601509, 100.880867)
PLH13	A3, Kampung Ulu Piul, 33800, Manong, Perak.	Manong	(4.572912, 100.874680)
PLH14	A17, 32800, Parit, Perak. (Jalan Parit)	Parit	(4.443276, 100.905777)
PLH16	A17, 32800, Parit, Perak	Parit	(4.454128, 100.907036)
PLH17	Jalan Parit, Kampung Senin, 32800, Parit, Perak.	Parit	(4.386045, 100.901413)
PLH18	Jalan33A, Siputeh, Parit, Tanjung Ara,32800, Perak	Parit	(4.474646,100.913315)
PLH19	Jalan Parit, 32600, Bota, Perak	Bota	(4.336138, 100.895859)
PLH20	Jalan Parit, 32600, Bota, Perak.	Bota	(4.363739, 100.892860)
PLH21	Jalan Parit, 32600, Bota, Perak.	Bota	(4.364304, 100.893463)
PLH22	Jalan Parit, 326600, Bota, Perak.	Bota	(4.363926, 100.893311)
PLH23	Jalan Teluk Intan, Kampung Padang Changkat Bota, 32600, Bota, Perak.	Bota	(4.324942, 100.907959)
PLH24	5A, Jalan Teluk Intan, Bota Kanan, 32600, Bota, Perak.	Bota	(4.346352,100.884758)
PLH25	8089, Jalan Bota Kiri-Teluk Sena, Bota Kiri, 32900, Lambor Kanan, Perak	Lambor Kanan	(4.348048, 100.874062)

Table 2. RLJ location and GPS Positioning

Code	Owner's of Traditional House	Location	GPS
RLJ 1	En Mokhtar Masnot	Muar 1	1.9839767217636108,102.62134552001953
RLJ 2	Hj Hassan Bin Ali	Muar 2	1.9896643161773682,102.6170883178711
RLJ 3	Hj Adnan Hj Kasan	Muar 3	2.010014295578003,102.63230895996094
RLJ 4	Hjh Mastijah Hj Hashim	Muar 4	2.0694141387939453,102.57777404785156
RLJ 5	Rumah Perjumpaan Dato' Onn	Muar 5	2.0673136711120605,102.57295227050781
RLJ 6	Hj Sirat Bin Wongso	Muar 6	2.074946403503418,102.56961822509766
RLJ 7	Mohd Noor Bin Mumba	Muar 7	2.047588, 102.597622
RLJ 8	Tn Hj Ismail Bin Sharif	Muar 8	2.0516326427459717,102.57233428955078
RLJ 9	Hjh	Tangkak 1	2.1494321823120117,102.5612564086914
RLJ 10	Hj Ali Haji Husin	Tangkak2	2.1592586040496826,102.5587387084961
RLJ 11	Hj Hamzah Hj Hassan	Tangkak3	2.1592586040496826,102.5587387084961
RLJ 12	En Johari Abdul Rahman (OKU)	Tangkak4	2.1644864082336426,102.55673217773438
RLJ 13	Hj Yusof Abdullah	Tangkak5	2.0862557888031006,102.53794860839844
RLJ 14	Tn Hj Bakar Hj Harun	Tangkak6	2.0826587677001953,102.54164123535156
RLJ 15	Kakak Jaafar Onn	Tangkak7	2.0796127319335938,102.5452651977539
RLJ 16	Hj Rostam Hj Abd Kadir	Benut 1	1.6430389881134033,103.29680633544922
RLJ 17	Hj Abd Rahman Kasan (Kini En Zaki Hj Yusof) – Yg bagi Kayu Ukir	Benut 2	1.6880640983581543,103.28093719482422
RLJ 18	Daeng Marpunah (Kini En Mohamad Noor Bin Daeng Marpunah)	Benut 3	1.6742193698883057,103.22491455078125
RLJ 19	Parit Karimun	Benut 4	1.6798202991485596,103.23563385009766
RLJ 20	Parit Makueseng	Benut 5	1.623819351196289,103.2294692993164
RLJ 21	Hj Omar Abu Hasan (Kini Hj Md Noh Bin Hj Omar)	BatuPahat 1	1.8820103406906128,102.78423309326172
RLJ 22	Maan Bin Ahmad	BatuPahat 2	1.8837379217147827,102.78175354003906
RLJ 23	Isnin Bin Ujan (Kini Khairul Rizal Bin Ibrahim Bin Isnin)	BatuPahat 3	1.8789409399032593,102.78873443603516
RLJ 24	En Rosli Bin Latif	BatuPahat 4	1.8784593343734741,102.78955841064453
RLJ 25	Mat Som Hj Ahmad (Kini Fatimah Bt Mat Som, KL) – sebelah Z&D Homestay	BatuPahat 5	1.870325, 102.801342

A Comparative Study of RLBP and RLJ

Table 3. RLJ location and GPS Positioning

Nos	Decorative Elements / Coding Code	House Code Rumah Limas Bumbung Perak (RLBP)									House Code Rumah Limas Johor (RLJ)								
A	Tunjuk Langit *(literally means pointed to the sky)* / **TL**	1	2	3	4	5	6	7	8	9	1	2	3	4	5	6	7	8	9
		10	11	12	13	14	15	16	17	18	10	11	12	13	14	15	16	17	18
		19	20	21	22	23	24	25			19	20	21	22	23	24	25		
B	Sisik Naga *(literally means dragon fin)* / **SN**	Not Available									Not Available								
C	Sulur Bayung / **SB**	Not Available									Not Available								
D	Ande-Ande / **AA**	1	2	3	4	5	6	7	8	9	1	2	3	4	5	6	7	8	9
		10	11	12	13	14	15	16	17	18	10	11	12	13	14	15	16	17	18
		19	20	21	22	23	24	25			19	20	21	22	23	24	25		
E	Pemeleh / **P**	1	2	3	4	5	6	7	8	9	1	2	3	4	5	6	7	8	9
		10	11	12	13	14	15	16	17	18	10	11	12	13	14	15	16	17	18
		19	20	21	22	23	24	25			19	20	21	22	23	24	25		
F	Kepala Cicak *(literally means lizard head)* / **KC**	1	2	3	4	5	6	7	8	9	Not Available								
		10	11	12	13	14	15	16	17	18									
		19	20	21	22	23	24	25											
G	Tiang Gantung *(literally means hanging column)* / **TG**	Not Available									Not Available								
H	Lebah Bergantung *(literally means hanging bees)* / **LB**	Not Available									Not Available								
I	Kekisi *(other names Kisi-kisi, kekipas, jeriji, jeruji, terali)* / **K**	1	2	3	4	5	6	7	8	9	1	2	3	4	5	6	7	8	9
		10	11	12	13	14	15	16	17	18	10	11	12	13	14	15	16	17	18
		19	20	21	22	23	24	25			19	20	21	22	23	24	25		
J	Gerbang *(Archway)* / **G**	1	2	3	4	5	6	7	8	9	1	2	3	4	5	6	7	8	9
		10	11	12	13	14	15	16	17	18	10	11	12	13	14	15	16	17	18
		19	20	21	22	23	24	25			19	20	21	22	23	24	25		
K	Gerbang Pintu *(Arch door)* / **GP**	Not Available									Not Available								
L	Sesiku Keluang *(other names Tupai-tupai)*/ **SK**	Not Available									Not Available								
M	Pagar Musang / **PM**	1	2	3	4	5	6	7	8	9	1	2	3	4	5	6	7	8	9
		10	11	12	13	14	15	16	17	18	10	11	12	13	14	15	16	17	18
		19	20	21	22	23	24	25			19	20	21	22	23	24	25		
N	Kepala Pintu *(literally means door head)* / **KP**	1	2	3	4	5	6	7	8	9	1	2	3	4	5	6	7	8	9
		10	11	12	13	14	15	16	17	18	10	11	12	13	14	15	16	17	18
		19	20	21	22	23	24	25			19	20	21	22	23	24	25		
O	Kepala Tingkap *(literally means Window head)* / **KT**	1	2	3	4	5	6	7	8	9	1	2	3	4	5	6	7	8	9
		10	11	12	13	14	15	16	17	18	10	11	12	13	14	15	16	17	18
		19	20	21	22	23	24	25			19	20	21	22	23	24	25		

5. Findings

This study established and highlighted the existence of decorative elements on both RLBP and RLJ. The aim of this study is to look into the architectural characteristics by investigating the similarity and differences of decorative elements of both types of houses (refer Table 3). Based on inventory analysis and comparative analysis done on both type of houses it can be concluded that:

A) Most of the decorative elements studied were found on both types of houses. There were 9 decorative elements that found on RLBP and 7 decorative elements on RLJ. 2 decorative elements which do not exist at RLJ are Kepala Cicak (KC) and Tunjuk Bumi (TB). This indicates that KC and TB are two major decorative elements that shape the characteristic of RLBP.

B) Kepala Tingkap (KT) was found at both RLBP and RLJ. As discussed earlier the existence of KT was not only for aesthetic reasons but also function for air flow. This indicates that both types of houses produce good cross ventilation by maximizing wall openings. This study found that the designs of KT at RLBP are slightly different from RLJ. Both have the semi-circular or fan like shape with different composition.

C) Both RLBP and RLJ were built on stilt which elevates the house from the ground. This is common for a vernacular house in the tropical climatic zone. This study found that most of RLBP are much higher from the ground (around 2 – 2.5m) compared to RLJ (around 1.5m). The lower part of the house at the ground is called *Kolong*. For RLBP this space is used as a place to relax during daytime. This study found that decorative fan was used to cover the bottom part of the house called Pagar Musang (PM) or Kekisi (KK) for RLJ. It is also known as *Banji*. This decorative element enhances the architectural characteristics of RLJ.

D) The window design for both RLBP and RLJ is called Tingkap Labuh (literally means long window) simply because it is a full height opening from floor level to the above. Both RLBP and RLJ used Pagar Musang (PM) as decorative elements and function as balustrade to protect people from falling down. This study found that 4 inches to 5 inches width of carved timber piece were used at RLJ compared to carving array used at RLBP.

E) Openable louvred timber window is popularly used as *Daun Tingkap* or the window piece used at RLJ. This compared to RLBP that used solid timber plank as their window piece. These elements contribute to the characteristics of RLJ.

F) The used of Gerbang only can be found at RLBP. This element become one of the major characteristics of RLBP

As discussed earlier decorative elements were considered or defined as additional or extra elements of the more functional structure elements. As a traditional architecture decorative element play important role in shaping the characteristics and identity of a particular house. This study contributes to the development of the aesthetical theory and concept of the Malay architecture especially the RLBP and RLJ.

Acknowledgement
I would like to acknowledge and extend heartfelt gratitude to The Ministry of Higher Education and Universiti Teknologi MARA (UiTM) for funding this study under the Fundamental Research Grant Scheme (FRGS). Research code FRGS/1/2017/SS/11/UITM/02/6. My sincere thanks to all KUTAI members for their support and provided valuable comments of this article.

References

[1] Ezrin A. Ke Arah Jatidiri Senibina Dalam Konteks Nusantara, Prosiding Simposium Nusantara . UiTM Cawangan Perak, Kampus Seri Iskandar. 21 Jun 1997.

[2] Norsuriani S. and Ismail S. Types of Floral Motifs and Patterns of Malay Woodcarving in Kelantan and Terengganu. Social and Behavioral Sciences Procedia, Elsevier, 2013

[3] Anuar T., The Perak Kutai Asli Malay House: Form, Function and Meaning, unpublished PhD thesis, Institut Pengajian Siswazah, Universiti Teknologi MARA, October 2005.

[4] Norhasandi M. Ciri-Ciri Fizikal Ruang Dalaman Rumah Melayu: Kajian Kes Rumah Bumbung Melayu Perak, unpublished PhD thesis, Institute of Graduate Studies, USM, 2010.

[5] Mohd Sabrizaa A.R. Rumah Kutai : Documentation of Memories, Institut Darul Ridzuan : Ipoh, 2017

[6] Mastor S. and Nik H.S. Rumah Limas Johor, Yayasan Warisan Johor : Johor Baharu, 2006

[7] Mohd Sabrizaa A.R and Sufian C.A. The Traditional Malay Architecture: Between Aesthetics and Symbolism in Mark Taylor (ed.), Interior Design and Architecture: Critical and Primary Sources, Vol 1.Bloomsbury Publishing: London, 2013

[8] Yusoff, R and Rashid, M.S. The Analysis of the "Pemeleh": Decorative Roof Element of Malay Traditional Architecture, Proceedings of The 3rd International Conference on Heritage & Sustainable Development, HERITAGE 2012 , Porto, Portugal, June 2012

[9] Hanafi Z. Pola-pola hiasan di dalam bangunan tradisional Melayu, Dewan Bahasa dan Pustaka : Kuala Lumpur, 2000

[10] Halim N. Traditional Malay Wood Carving, Dewan Bahasa dan Pustaka : Kuala Lumpur, 1987

[11] Norhasandi Mat et al., The Architecture of Perak Traditional Malay Roof Houses (PTMRH) in the Year of 1820-1940: A Chronological Analysis of the Arabesque Types on Wall of Rumah Ibu, Prosiding Simposium Nusantara 9 (SIMPORA 9:2012), Dewan Al- Khawarizmi, UiTM Perak. 11 & 12 Disember 2012.

[12] Jahn S. et. al. The Resilience of Tradition: Malay Allusions in Contemporary Architecture. (S. J. Kashim, Ed.). Penang: Areca Books, 2017

[13] Sabrizaa R, Iryani AHC, Rohaslinda R, Mohd Nurfaisal B, Kartina A. Decorative Elements Of Traditional Malay Houses: Case Study Of Rumah Limas Bumbung Perak (RLBP). JSSR. 2018; special issue(6):105-115

IOP Conf. Series: Earth and Environmental Science **385** (2019) 012023 doi:10.1088/1755-1315/385/1/012023

What are the factors behind the distinct character of Mosques in Turkey and Iran?

Tamara Kelly[1]

[1] School of Engineering Abu Dhabi University, United Arab Emirates.

Email: Tamara.kelly@adu.ac.ae

Abstract. Cities or its urban fabric and the characteristics architecture in many regions during various timescale is shaped by numerous factors. Among those factors are intangibles as political motivation, inspirations from other civilizations, religious influence. Whereas physical sways such as climate or geographical consideration or availability of materials and skilled labor are behind other great architecture creation. Therefore this article examines the impact of these factors in shaping the religious architecture in two vast neighboring countries Iran and Turkey. Iran is an immense country and its weather is variable and key factors in shaping a specific type of architecture while in Turkey, other influences are more dominant. The methodology of this paper is based on a literature review of various references and a theoretical analysis of several case studies in Iran and Turkey. This paper is an attempt to help students and researchers to understand how architects creates distinct architecture by acknowledgement of intangible forces.

1. Introduction

In today's modern age with the selection of technological means being at arm's reach and the development of new construction materials, buildings are extremely costly to build and maintain, let alone being uncomfortable to live in at times especially when human comfort is more favored. As a result, constructional methods are based on making the best use of consumptions of resources as well as a production of waste. However, in the past builders had minimal knowledge as well as resources and had to make the best use of factors and elements around them to overcome certain obstacles. For instance the environment; how they had to maintain a comfortable living setting during extreme hot and cold weathers. Looking back at urban studies the city was known as an 'organism' it was an outcome of the relationship between a system of rules, land division, building aggregation and road network that all simultaneously worked together, strictly connecting with climatic and solar conditions. The design of the city was never a coincidental effect and the original structure was made of parts as scalar sub-systems of strongly inter-connected organisms which thrived in an unlimited growth without the need of interrelations contrary to belief with the environment. The architecture in Iran formed according to climate constraints, Iran is diverse in its climate and typology, the northern part is divided into two areas; coastal section situated beneath the Caspian Sea and featured by dense forests. The climate is hot and cold according to the seasons with high humidity level throughout the year with farming being the main activity in this region Fig. 1. Second northern plateau is mountainy stretched along the borders with Turkey and Azerbaijan, series of Zagros Mountains extend westwards causing cold winter and mild dry weather in those parts. Central Iran is cold in winter with over 70 % humidity and hot dry in summer. Moreover, two deserts are located in this section (Dashet – e Lut, Dashet e Kavir). Last part is the southern coastal above the Arabic Gulf and Oman Sea, the climate here is similar to neighboring countries hot and humid in summer mild and cold in winter. The various weather stipulations in Iran is the main reason behind opting Iran to study its sacred architecture. Iran is Islamic country but Zoroastrian religion was before Islam and fire temple existed in form of layered Ziggurat nevertheless Islam replaced this atheist practice. Zoroastrian religion was the state religion of the pre-Islamic Iranian empires for over a millennium, from around 600 BCE to 650 CE. Zoroastrianism was overturned from the 7th century onwards following the Muslim conquest of Persia of 633–654 and many mosques replaced demolished temples.

IOP Conf. Series: Earth and Environmental Science **385** (2019) 012023 doi:10.1088/1755-1315/385/1/012023

The mosque generally in Iran accommodates the typical components of prey hall courtyard and colonnade around it with Minarets but the opening in an external envelope and the court may vary according to a location of the mosque. In the following section, the impact of the climate on the mosque's layout should be examined and case studies will be used to reflect the various treatments exerted to overcome the harsh weather.

2. Mosques in various parts of Iran

The contrast and variety in weather conditions in Iran overshadowed the identity of sacred buildings and as early mentioned the coastal northern regions and mountainy lands are cold in winter hot humid in summer. Hence logically most of the winter pray halls are bigger covered with pitched low ceiling, openings in the external envelope are small and limited to maintain the warm air inside.

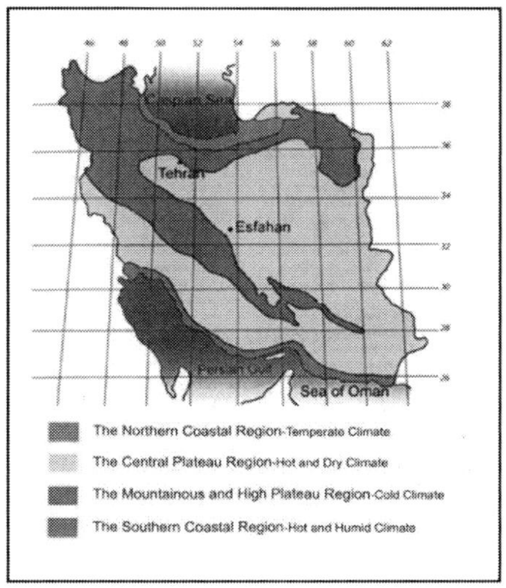

Figure 1. Plateaus of Iran

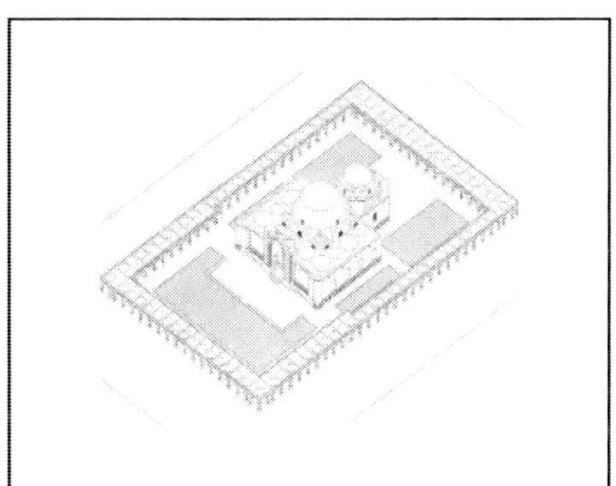

Figure 2. The Blue Mosque in Tabriz, Iran

Furthermore, mosques are inwards looking and typical courtyard with Evans found in almost every mosque is preventable in this part. In Astara the coastal city and Tabriz county mosques are without the courtyard and Blue Mosque in Tabriz is a great sample of mosques in mountain cities Fig. 2. The external envelope of this mosque is concealed or windows are minimized in size and likewise the middle court with its Evans are omitted. However occasionally in different coastal cities the court employed to provide ventilation. In middle platue winter is cold and in summer the heat is excessive with dust or sand storms moreover the difference in temperature between summer, winter and even the day or night

A

is too high. Hence the mosques most provide comfort through the seasons. Consequently, the courtyard is essential to provide ventilation and openings in an external envelope is diminished because most of the openings are around the internal court.

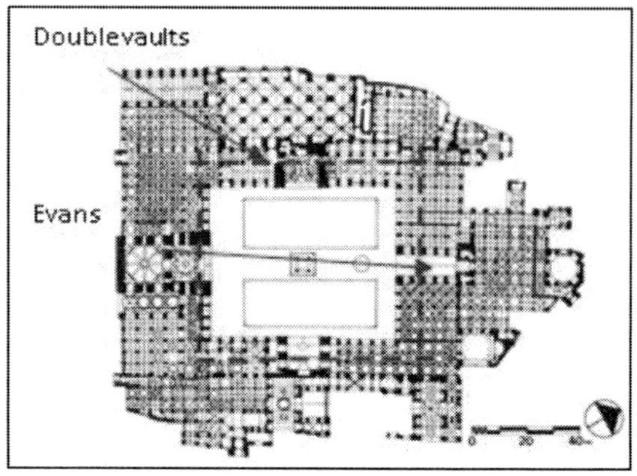

Figure 3. Great Mosque, Esfahan

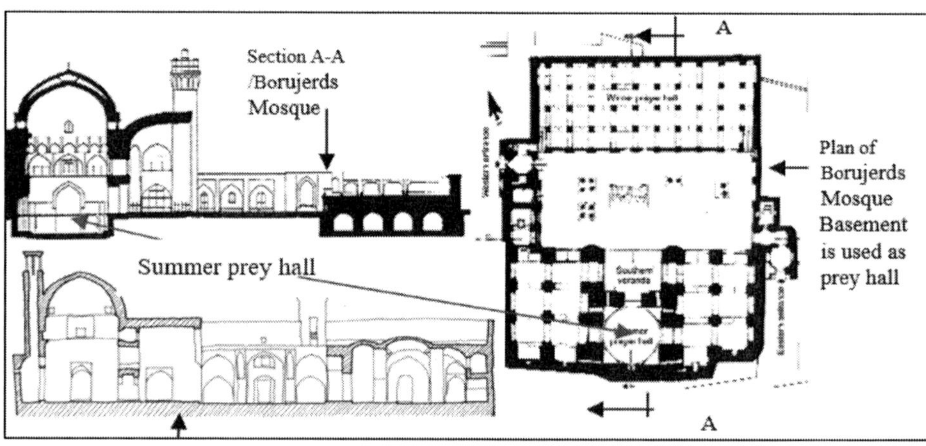

Figure 4. Firooz – Abad mosque, Iran, usage of wind catchers.

A water pond is used as a cooling component and commonly situated inside the courtyard providing a breeze and enhance the quality of air. The Evans is crucial feature of mosques in Iran and commonly big in Middle plateau's mosques and regarded as part of comfortable shaded breezy spaces that can be used to give lessons.

Additionally, double or cross-vaulted arcade are commonly located around the courtyard providing shelter from the sun during hot summer and Great Mosque in Esfahan is one of the sample of central Iran Architecture Fig.3. Other distinct components known as wind catchers added into the mosques in Yazd city in a central region, the wind towers enhance air circulation or ventilation and provide breezed air into summer pry hall and Firooz –Abad mosques reflects this approach Fig 4. A basement is another feature of the mosques in central Iran and can be used as summer prey hall since it is a cold shaded area. In southern coastal cities the mosques were smaller since the population is less than other parts, this affected the size of the mosques and hence courtyard is unnecessary. Likewise the economy was weaker and fund insufficient to build elaborate architecture. Consequently, the mosques were basic and even less ornamented. Vahid Ghobadian in his book mentioned that mosques with central courtyard were less common here in compared to other regions of Iran and there were very few mosques of the central courtyard type [1]. Most of the mosques in this part are constructed over a 2 m platform to overcome the floor dampness Figure 5, and other measures employed to reduce the effect of the heat such as:

i. light colures coating internally and externally
ii. Shaded arcade around the building
iii. Bigger opening oriented towards breeze and wind direction and a row of a small opening is common near the ceiling to enhance ventilation inside prey hall Figure 6.
iv. Prey hall ceiling is high allowing additional air circulation and fans commonly pending from the roof. Mosques are detached from surrounding urban fabric allowing air movement Figure.7

Figure 5. Omar mosque, usage of higher opening.

Figure 6. Omar Mosque, Ventilation system through court and small openings

Figure 7. Mosque in Bender city shows the premises built over a platform

In fact the weather is one of numerous components which characterized the architecture of Iran, but other political decisions were behind some unusual form of Mosques in northern Iran. Vahid Ghobandian in his book the sustainable traditional building of Iran stated that the three highest sacred buildings can be found in the north where the Mongol occupied the region between 128 till 1334.Those proposals are; Mausoleum of Ul-Jaitu in Soltaniye near Zanjan built in 1305, the Friday mosque of Urmia built in; 1277 and the Mosque of Arg- e Ali Shah in Tabrizbuilt in 1312. The Mongolians reflect their triumph and strength by the construction of high leveled mosques contradicting prior approaches.

The first part of the paper discussed the impact of a tangible factor the weather in defining the identity of mosques in Iran nevertheless only two imperceptible aspects were behind some monumental premises in North Iran. And likewise, two reasons including low population and weaker economy resulted in rudimentary layout mosques in southern regions of Iran. The following part addresses religious architecture in Turkey, the history of Turkey examined and other aspects will be tested to identify the main influences behind shaping the Ottomans Mosques.

3. Mosques in Turkey, Historical Evaluation

Location of Turkey is a key factor in shaping the Ottomans architecture, Turkey is a middle basin between various cultures. Christianity from west contradicting with Islam from southern territories and Turkey represents the discourse between these religions. Christianity spread eastwards after the collapse of Rome and Istanbul (Constantinople city) was the capital of new evolving empire which lasted over one millennium till 1455B.C the East part beyond the Constantinople city was divided into many weak nations then and there the Seljuk arrived and established their Empire between 1037- 1194. Therefore From 11th till 15th the Turkish sacred premises forge its identity by taking various component from surrounding cultures and in particular Byzantine Empire, Seljuk state and the Islamic

Nations. Mohamed Zein Al Abidine in his publication the Architecture of Ottomans Mosques underline to study the history of Turkey to develop an accurate understanding of Turkish architecture. According to Mohmoud Zein Aabidin, the history of Ottomans backdated to a small tribe Kayi escaped the horror of Mongol and migrated in long search of land to inhabit in the 13th century [2]. Finally the Seljuk leader granted a northeast border district with Byzantine Empire to Kayi tribe as gift since they supported him against the Turkestan sultan. In fact, they were exploited as barrier against the Christians in the west, but the Kayins were strong and brave led by founder of Ottoman Empire Osman who inherited the power after the death of his father Ertogrul and was the initiator of the Ottoman Empire. Orhan the son of Osman took the lead in 1326 and enlarged his province by conquering Bursa but Orhan's son Murad I subjugated Edirne and announced it as a capital of Ottomans Empire after passing away of Orhan. Murad I was assassinated and his son Beyazet expanded the empire southeast whereas Sultan Mehmed named conqueror defeated Christian and transferred the capital of Ottoman Empire from Edirne to Constantinople (Istanbul), and established the foundation of the Empire. The capture of Istanbul marked the end of Byzantine Empire, which had lasted for nearly 1,500 years. The conquest of Constantinople resulted in massive blow to Christendom, as the Muslim Ottoman armies subsequently were left unhindered to advance into Europe without an adversary to their rear. Mohamed II in 1453 when he defeated Christians and entered the city he converted church of Aya Sofia into a mosque and the church was a great inspiration for Ottomans architects. The multi-domed roofing framework of Aya Sofia dominated by central spherical dome adopted by Ottomans and became a great feature of Ottoman mosques. Another characteristic of early Ottoman architecture was the use of Byzantine technique of alternate stone and brick courses which replaced the ashlar faced walls of Seljuk period in Central and western Anatolia.

In early era of Ottomans Empire, the churches were converted to Mosques because resources of Ottomans were employed for defense purpose or to conquer further cities. Therefore lack of funds could be the leading reason behind using of pre-existing church and recycling its components. Degan Turkish authors to avow that the Sultan take fifty present of the revenues or may demand to take all of it for his personal uses and 10 % dedicated to city councils. This can be a second reason behind the usage of churches as mosques. In his book- p356 Degan mentioned that sultan and his courtiers employed the money allocated to public works and urban development to create a comfortable environment for them [2].

However before the Ottomans, Seljuk Empire was dominating east Turkey and in Turkistan with Islamic states in Anatolian east of Turkey, locals were engaged in construction premises heavily influenced by Seljuk architecture. The result was many mosques imitating the Seljukian major features of; elaborate decoration around doorways, ornamental stone buildings of elegantly simple design and harmonious proportion. Yet, the dominant element of design in the Seljuk period wasn't the dome but the portal hence the gates of Ottomans mosques emphasized manifestly Fig.8. But after the collapse of Seljuk and Ilkhani, eastern and western Antonia were united by Mehmed the Conqueror and for the first time a new monumental approach employed in Yakutiya madrasa by adding enumerate twin Minaret erected in frontal façade of Yakutiya madrasa 1310 Fig.9. Dogan described the twin Minaret as a synthesis non- existent in the 13th century and the combination of twin- minaret façade with a roofed madrasa created a step toward monumental architecture. He believes that the 14th century witnessed the erection in Eastern Antonia of imposing monument reminiscent of the tradition of the proceeding century [3].

Figure 8. Portals of the school reflects influenced by Seljuk architecture

Figure 9. Monumental design attempt by adding the Columns

Though, the stonework of Yakutiya madrasa reflects the deep influence of Christian architecture on local traditions. Dogan wrote that from the reign of Murad I onwards, the Ottomans, having gained wealth and power as the result of in Balkans, and turned to a new monumental style on the basis of their 14th-century experience and potentials [3]. The monumental buildings in Europe cities were subsidized by religious power presented by the church whereas in Turkey the willpower of sultans is the crucial factor behind assemble of spectaculars monumental buildings.

After the 14th century, the Ottomans gained more power hence there was an urge to construct monumental buildings reflecting the triumph of Ottomans Empire. Most of the key buildings were constructed in main political capitals, the political structure existed mainly in Manisa, Bursa, Amasya, Edirne, and Istanbul. Thus, the core of imperial administration aroused in these provinces and all other high ranked houses, social, cultural and commercial activities were gathered.

In the following section several case studies considered to assess the development of Ottoman style in sacred Architecture. Religious architecture in Turkey as well as history of Turkey with other aspects, will be tested to identify the main influences behind shaping the Ottomans architecture.

4. Case studies

4.1. Ulu Mosque, 1396th in Bursa City, Sultan Beyazet

Bursa is an essential state in the political Ottoman fabric and the foundation of the Empire was laid in this early capital including the organizations and institutions of the nation. The city was transformed from small Constantin city into formal Ottoman capital for hundred year and many monumental buildings constructed in the city. Furthermore new building types existed such as baths, covered markets, courts, learning school, and *Zaviya* which is a center combining between learning activity and distribution of food into poor citizens. Bursa was a critical state and commercial capital during the Byzantine Empire and intensely connected with Western culture however it stayed the same after the collapse of the Byzantine Empire.

John Freely mentioned in his book that the political power surpassed religious and economic factors in Turkey nonetheless it acknowledged the cultural inspiration [4]. Consequently during the Ottoman era, the growth of the city wasn't randomly but based on the desire of the sultan who instructed the construction of many Zaviye, Public baths and great mosque of Bursa *Ulu Mosque*. This mosque is first symbol of the Islamic Ottoman capital, and regarded a shift in the Ottomans architecture, the size of central dome reduced and the single dome became multiples stretching over the pry hall Fig 10. Aptullah Kuran described *Ulu Mosqe* as one of the very early types of Ottomans mosque constructed after the order of Beyazet in 1396, the plan is rectangle shape and the central dome was replaced by twenty homogenous hemispheric domed units [5]. The central row of domes are higher emphasizing the fountain underneath Fig 11. The Seljuk influence is obvious in particular through; simple design of the mosque, elaborate main gates, bricked Minaret, and stone structure. This layout was commonly used in six century of Ottomans Empire.

Figure 10. Dome became multiples stretching over the pray hall.

Figure11. The fountain in the middle of prey hall

4.2. Beyazet Mosque 1501th in Istanbul City, Sultan Beyazet II

During the 14th centuries, the design of mosques advanced and the central spherical domed supported by half-domed replaced the multiple columns in prey hull. The numerous columns in pry hall exited in previous case study divided internal spaces, separating the rows of prayers and disguised the vision but the new notion of domed mosques provided one specious prey zone without divisions. Beyazet II ordered to build a new mosque to commemorate him and khare Al deen was the architect who adopted this new approach of undivided prey space allowing a better connection between prayers and the person provides Friday speech. The mosque layout was significantly changed from rectangle shape into cross and the two asymmetrical wings in west and east considered new distinct features of Ottomans mosques. The roof over east wings comprised four small domes with single bigger central dome whereas the west accommodates two extra domes thus the name of this mosque is non – homogeneous multi-unit mosque Fig. 12. This mosque represents the ottoman approach and was a great inspiration for many ottoman architects. The courtyard was a new component but the entire roofing envelope over the prey hull was inspiration from Byzantine architecture however the layout of this mosque and in particular the roof was advanced by Sinan the famous Ottoman architect Fig. 13.

The roofing damaged by the earthquake and rebuilt under the supervision of Sinan and his style is observable thru the elegant domed towers, half-domed around central and the windows in the rotunda.

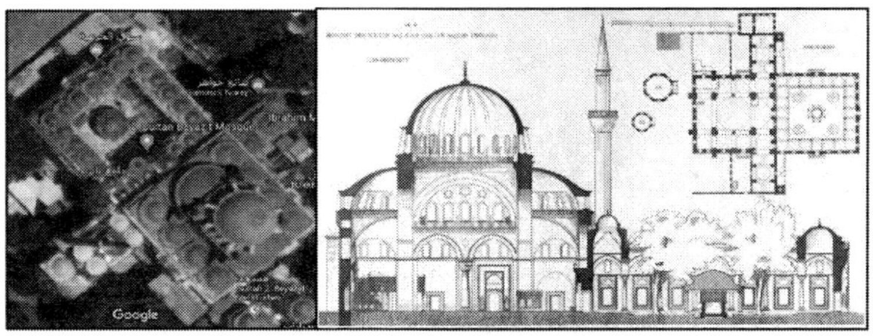

Figure 12. Usage of Semi domes

Figure13. Roofing layout.

4.3. Uc Serefeli Mosque 1447th in Edirne City, Sultan Murad II.

In Edirne the closer state to Europe, the Üç Şerefeli Mosque erected by Murad II 1447th as a symbol aiming to announce Edirne the capital of Ottomans state. The design of the mosque a step towards monumental architecture by augmenting the height of the mosque and this was achieved thru: 1.increasing the size of central single dome which is supported with multiple four lower domes emphasizing central one, 2. height of the mosque gradient towards central dome underlining prey hall, 3. Minarets situated next to prey hall are taller and consist of three layers. But the influence of Byzantine architecture is less vivid because the single dome is dominated over the various small cupolas and without semi - circle ones. Uc Serefeli Mosque consists of open courtyard Fig. *14* surrounded by arcades and two various minarets located at the corners of the mosque whereas the higher pair of Minarets stressed prey hall Fig. 15. Edirne was one of the first cities witnessed the conversion of churches into Mosques and the influence of Byzantine is apparent in next case study Selimiye Mosque designed by Sinan the famous Ottoman architect.

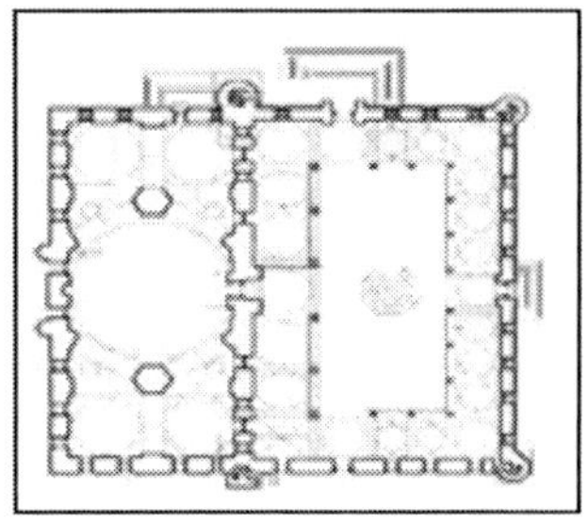

Figure 14. Open courtyard

Figure 15. Vertical emphasize by various components.

4.4. Şehzade Mosque 1544th and Süleymaniye Mosque 1557th in Istanbul City by Sinan.

Gülru Necipoglu conveys that Sinan was a vital Factor in evolving the Architecture of Ottomans, he joined the army in 151 and was a soldier in the Sultan Salem Military. Sinan participated in many Ottoman battles in Europe, Asia, and Africa and during the battles he provided advises to construct bridges in shorter period allowing the army to advance or concurred further cities [6]. Throughout his military career with Sultan Salem and his sun Suleymen he had had the opportunity to study the architectural monuments in various conquered cities and was in charge of converting churches into Mosques. In later stages, he was the Architect of much crucial architecture such as Şehzade mosque built to commemorate the death of a beloved son of Sultan Suleymen, the urge for monumental architecture led Sinan to arise different method in the creation of a historical building. He was inspired by Aya Sofia but came up with bigger central dome size thru; 1.supporting pendentives of the central dome by four tower located at the corners of the central dome, 2. Four semi Domes are situated north, east, west and south of central dome taking the load into massive buttressed walls Fig. 16. The four towers crowned by elegant small domes and the massive buttressed walls hidden wittily by small domes Fig. 17. The domes sizes are variable to break monotony and descent to emphasize central dome creating the Turkish triangle Fig. 18.

This was a crucial feature and progressive in Süleymaniye Mosque thru double story corridors topped by domes and pitched roof to break dullness of same size domes.

- 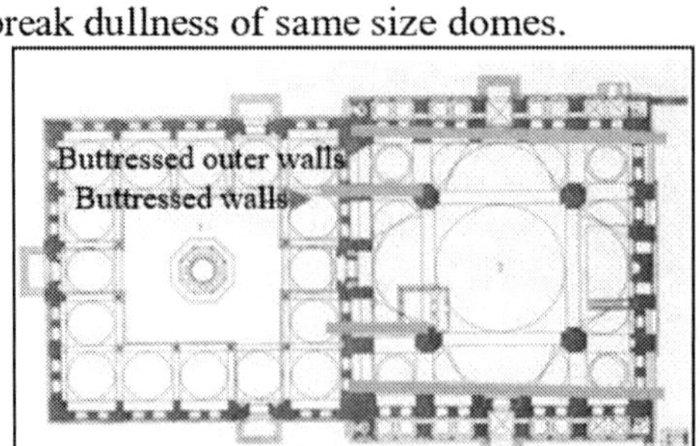

Figure 16. Four Semi circle domes

Figure 17. Domed towers

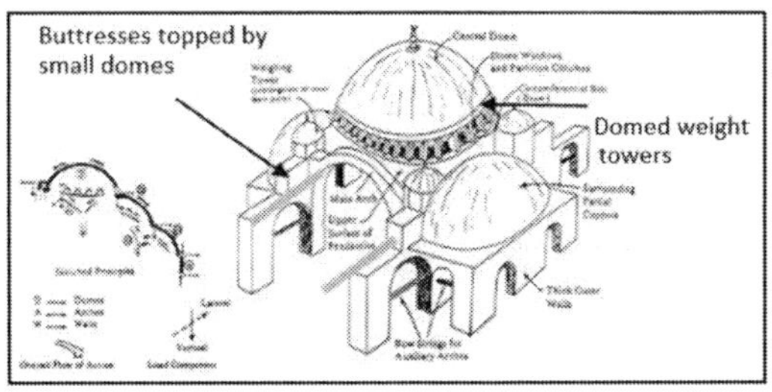

Figure 18. Organic Roofing system

Süleymaniye Mosque is a great icon design by Sinan, It was built on the command of Sultan Süleyman atop one of İstanbul's seven hills and providing a landmark for the entire city asserted Suleyman's historical importance. The roofing envelope over the prey hall innovative, Sinan replaced the two semi-dome north and south-central cupola by five domes. Three of them are bigger located at the center and other two over the corners alternating with two smaller domes to break the monotony. The dome supported by two half dome and four pillars similar to shezadah mosque, the mosque is the second largest mosque in Istanbul blended Islamic and Byzantine architecture Fig. 19. Symbolism stressed by using four Minarets representing Suleymen as fourth Ottoman Sultan since the collapse of Byzantine empire and the ten layers in the four Minarets articulate that Suleymen is the tenth sultan since the Ottoman empire establishment Fig. 20 .

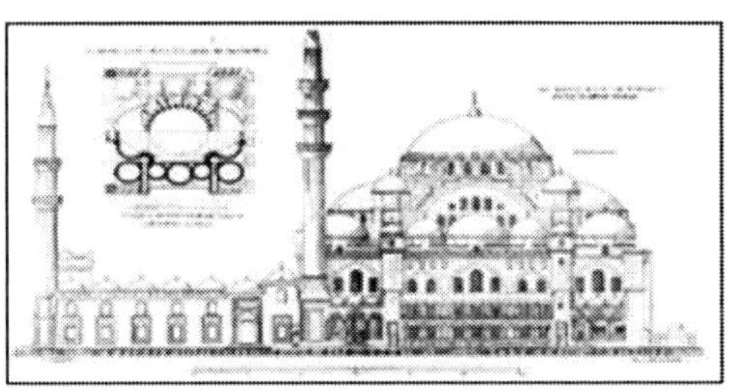

Figure 19. Semi-domes around central

Figure 20. Roofing structure

4.5. Selimiye Mosque1569th in Edirne City, Sultan Salem II

The mosque designed by Sinan the great Ottoman architecture, the mosque was at the center of many amenities including a hospital, Islamic school, library, baths, and a row of shops. Sultan Salem ordered the construction of the mosque hence Sinan chosen highest site in Edirne permitted the mosque as focal point seen from different locations in Edirne. Sinan in his design attempt to defeat the rumors spread by Christian engineers who claimed Muslims are incapable to construct a domed mosque bigger than Aya Sofia. Sinan employed an octagonal supporting system through eight pillars incised in a square shell of walls allowing the dome to be bigger than Aya Sofia. The buildings represent political power and the growth of economic and cultural functions, in fact, Sinan succeeded in creating a bigger dome but his proposal was an inspiration version of Aya Sofia and many components were taken from Aya Sofia including; the octagon piers spherical central dome, small sized half-domed around the central dome Fig. 21, multiple domed, a gradient in height towards the central dome and finally the preference of high land. Sinan was 84 years during the construction of Seilimiye mosque and managed to overcome the need for four immense semi-domed around middle dome by replacing the square walls below the dome with octagon piers Fig 22. The result was massive spacious internal dome and visibility enhanced in prey haul, whereas monumental and symbolism stressed by the usage of six layers:

i. A base layer is an arcade around the open courtyard.
ii. 2-Second level is a transmitted corridor between prey hull and rear court
iii. Third and fourth strata connect the roofing envelope with vertical supports and accommodate domed weight towers or buttresses. The roofing of this layer is an alteration between pitched, towers and smaller semi-domes roofing.
iv. Fifth height is the central Dominant dome surrounded by eight pinnacles.
v. Last echelons represented by four minarets Fig 23
vi. The sixth layers peaks towards the prey hull creating the vertical emphasize or the Turkish triangle Fig.24.

Figure 21. Alya Sofa Mosque

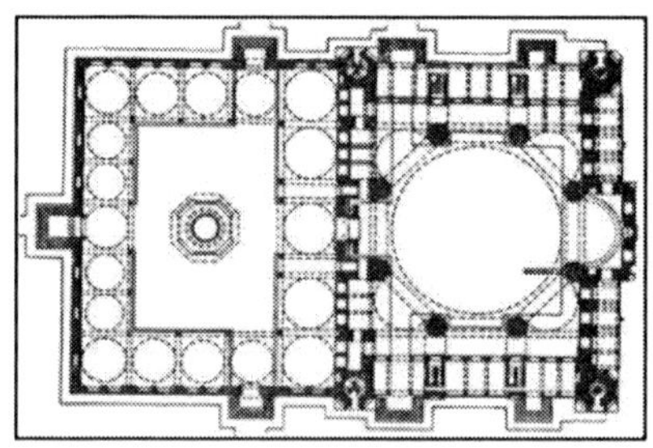

Figure 22. Eight octagon supporting piers

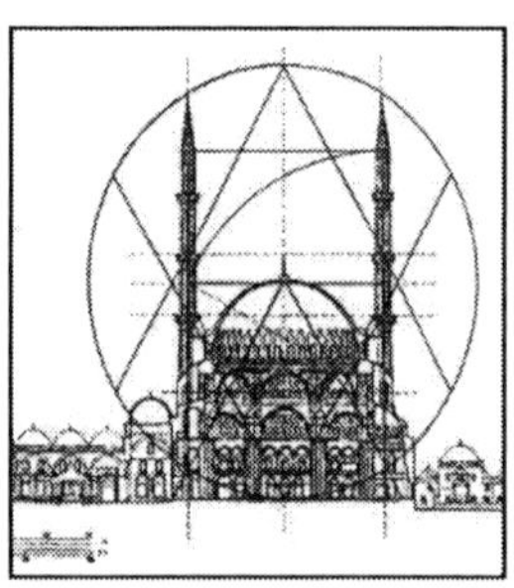

Figure 23. Four Minarets

Figure 24. Turkish triangle

5. Conclusion

The architecture before the establishment of Ottoman Empire profoundly influenced by Seljuk, Islamic and Byzantine approaches. But after the Ottomans domain instituting in around 13th-century numerous cities within and beyond turkey borders concurred by Sultan Osman and his sons or grandsons therefore the Ottoman state named after Sultan Osman. The new Ottoman leaders occupied parts of Asia, Europe, and Africa, consequently these achievements urged to forge a monumental architecture. The mosques were at the focus of Ottomans Sultans who claimed themselves leaders of Islamic nations hence the mosques symbolized political power rather than a worshiping place. The political power or will is the main engine behind most of the monumental architecture in Ottoman Empire. In fact during the research other intangible factors identified behind the creation of magnificent mosques such as the great knowledge of the architect Sinan and his will to challenge Christian engineering. At the early eras of Ottoman nation lack of fund resulted in converting churches into Mosques but excessive attainment or triumph of Ottomans army urged Sinan to come up with monumental mosques surpassing byzantine structure. Dome used before in East and West in sacred premises but commonly was single and crowning worshiping area beneath and particularly in Islamic or even in ancient architecture. The Byzantines were the first who employed half dome around the central cupola and used variously sized dome beside the main one and this treatment adopted by Sinan. Most of Sinan's proposals severely stimulated by Byzantine and Greek architecture nevertheless he significantly advanced the roofing framework in Selimiye and Süleymaniye mosques by changing the drum of the dome. The connection between the dome and supportive vertical wall underneath is a crucial point and Sinan smartly provides organic soft transition zone thru various sizes of domes. The bolsters side walls and massive square under the central dome of Aya Sofya is rigid despite the usage of row windows to soften solid plane but Sinan covered the massive wall by stretching the domes along the side walls creating a series of curved outline around the center of the Mosque and this solution came in a later stage of his life. In point of fact, the research concluded that Sinan and the Ottoman sultans or their triumphs in expanding the empire territories are key factors in shaping the features of Turkish architecture and created crucial iconic mosques combining Islamic and Christian architecture. In Iran, the climate is a key factor in shaping the external envelope and components of religious premises however generally mosque layout is similar to a typical mosque in most of the Islamic regions apart from Wind catcher and Evan. Evan is a distinct feature of Iranian mosque but the wind catcher usually an element employed in houses and the necessitate for air movement resulted in utilizing them in mosques. In Iran, except the weather, and through the literature review none of other intangible factors influenced the layout of the mosques and at variance, with Ottoman sacred architecture the impact of political power, expansion of Empire, the setting and the architect Sinan were key elements in shaping Ottoman mosques.

References

[1] Vahid Ghobadian, *Sustainable Traditional Buildings of Iran: A Climatic Analysis - Vahid Ghobadian - Google Books*. Iran: Islamic Azad University, 2009.

[2] Mohmoud Zein Aabidin 2005 *The architecture of Ottoman Mosques* (Dar Qabess House publisher) 15-70.

[3] Dogan Kuban 2010 *Ottoman Architecture* (Antique collectors Clun Ltd Publisher),60-350.

[4] John Freely 2011 *A history of Ottoman Architecture* (WIT Press publisher), 70-150.

[5] Aptullah Kuran 1968 *The Mosque in Early Ottoman Architecture* (University of Chicago Press), 50-100.

[6] Gülru Necipoglu 2007 *The age of Sinan: Architectural culture in the Ottoman empire* (Reaction book Ltd, 30-120.

IOP Conf. Series: Earth and Environmental Science **385** (2019) 012024 doi:10.1088/1755-1315/385/1/012024

Isovist and Visibility Graph Analysis (VGA): Strategies to evaluate visibility along movement pattern for safe space

F Othman[1], Z Mohd Yusoff[1] and A R Abdul Rasam[1]
[1]Faculty of Architecture Planning and Surveying, Universiti Teknologi MARA Shah Alam, Selangor, Malaysia

Email: fazzamiothman@gmail.com

Abstract. Quality of life is the main interest for people nowadays. Live in safe place with strong spatial economical will bring city liveable and among valuable assets in a city. Many innovations in town planning and design have been launched with great optimism over the years to addressing the people's concern including crime reduction. High development in city nowadays will minimize the visual capability that can seriously cause fear of crime. Many studies proved that man-made and natural environment affect the visibility degree thus affect the crime. With the emphasizing of physical environmental such as contour, trees, and building height into the space visibility analysis could inherent the actual and real-time phenomenon. Isovist and Visibility Graph Analysis (VGA) could be methods to evaluate the visibility in environmental space. Thus, proposed approach which consider the evaluation of visibility in 2D and 3D with incorporate the movement pattern will allow the evaluation based on the human visual perspective.

1. Introduction

The sustainable of cities is the main interest for a planning process and designing the physical environment. Good environmental design and space will encourage high surveillance, social interaction as well as increase safety level. Human perception always be a major influence to be measured or quantified while visuality is often thought to be one of the significant human's perceptions in environmental study. According to Jacob's (1961) idea on "eye on the street", explained the safest urban place is one that is continuously watched by human being. This idea then has been used as one of the principles in Crime Prevention Through Environmental Design (CPTED) called natural surveillance. Natural surveillance deals with location of physical features and people activity as well as to maximise the visibility naturally. Basically, resident will feel more secure in area consist of high surveillance for easier monitoring. [1] in the study on accessibility against crime believed the high integration of street will encourage number of user and create good visibility thus reduce crime.

Good visibility consists of better visual environment without any obstacle which inherent to unsafe activity. It is a crucial design process with rapid development nowadays especially in urban area where it can deteriorate the human visual. Physical feature and environment such as contour, vegetation, and building height and shape are among the contributor to affect. Study made by [2], recommended that contours of visual areas, useful and needed in the upcoming study of the relationship between humans and environments. [3] who conducted study on visibility and fear of crime using space syntax method urged to capture and consider the curvature of the surface. This recommendation supported by [4], who feel that environmental height or as the change in the topography of the surface, would allow researchers to more accurately in study of urban space in the future.

There is a long tradition of addressing crime in problem areas by removing vegetation to ensure the original line of sight designed do not get obscured over time. As a rule, the belief is that vegetation facilitates crime because it hides perpetrators and criminal activity from view. In visibility study, vegetation affects significantly in terms of height and density. Studied from [5] which focused on environment and crime towards the inner city of Chicago shows that the greener a building's surrounding was, the fewer the crime reported. This finding supported by [6] where they admitted that tree canopy and crime index are correlated in waterloo, USA. According to [7] which conducted study to estimate the relationship between trees and crime in Portland, Oregon have found that several types, heights and sizes of tree can affected the human's visual thus it can reduce or increase crime opportunity. Visibility should always be considered when designing or planning spaces. Some of man-made design would affect the sight line to be clear. For example, sudden changes of grade on walkways can also create the blind spot thus encourage the entrapment area. Other than that, street design and configuration also give an impact towards visibility [4],[8]-[11]. The street with high integration and connectivity can reduces the number of crimes. It shows that the higher integrated and connectivity of street will increase the number of street users, thus, increase the surveillance for that area and reduces the crime incident. Man-made and natural environment thought to give an impact to sight line. Thus, quantitative visual analysis should take part to give a good visual management and real-time process prior to real-life situation and environment. For this reason, this paper will explore on the use of visual analysis techniques of Isovist to promote safe space.

2. Isovist and Visibility Graph Analysis (VGA)

Isovist is an analysis to evaluate or measure the visual experience and analyse on physical body bound by closed polygon. It is an intuitively attractive way of thinking about a spatial environment, because they provide a description of the space `from inside', from the point of view of individuals, as they perceive it, interact with it, and move through it [12]. This method is practically used for the analysis of architectural space and linked to pedestrian movement distribution. A set of visibility analysis can be constructed from a path to show experiences in human perspective as walking through the space from point of origin to a destination. Isovist could provide an alternative technique to evaluate space for representation of urban systems with GIS. In many studies, isovist analysis is limited to two dimensions in a horizontal plane [2],[13]. The integration of Isovist into GIS would stimulate study oriented towards visual analysis in different level. Spatial data analysis and 3D visualization capabilities in GIS will help to support urban studies thus provide different experimental analysis. Isovist is practically used for the analysis of architectural space and linked to pedestrian movement distribution for urban morphology research.

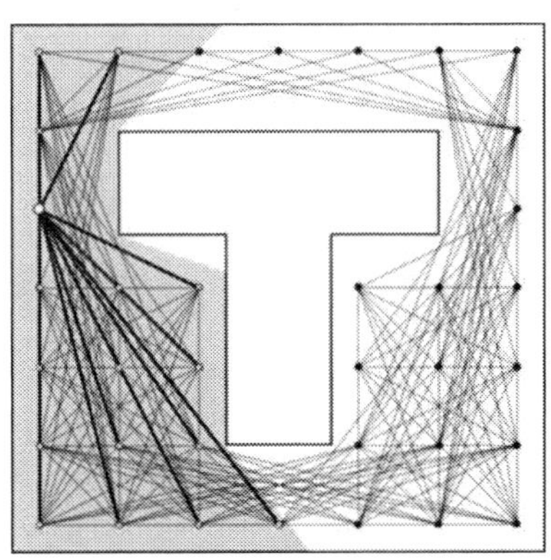

Figure 1. An example of visibility graph that showing the connection of nodes (Turner et al., 2001).

The integrating of analysis evaluation technique in space will provide important advantages from both systems and user point. Isovist which based on space syntax theory of spatial configuration will enhances

IOP Conf. Series: Earth and Environmental Science **385** (2019) 012024 doi:10.1088/1755-1315/385/1/012024

the functionality of GIS into urban morphology research. Rich of geo-referenced data and visualization analysis technique in GIS will promotes additional information in visibility analysis especially into human perspective. [14], then have constructed an undirected graph (figure 1) to connect all visible point in human scale grid called Visibility Graph Analysis (VGA). This graph is notated as a node and inter-visibility to linking all point together. The visual relationship can be calculated using different measure provided in the system.

The main topological measure for VGA that explain the inter-visibility of space based on urban morphology are clustering coefficient, global measure of integration and local measure of control. All the measure used depend on the neighbourhood size. Clustering coefficient is derived from the local configuration and calculate the degree of visible nodes from the observer. This measurement indicates the visual information, and any loses visual when moving. Next is integration which computes the mean shortest path from all nodes in the graph. The shortest path in graph is the number of linking nodes to be traversed to reach every node. The mean value shows the average value of all shortest path from the entire nodes. Lastly is a control that calculate the area of the neighbourhood with respect to the adjoining neighbourhood. This measure is suitable for large analysis of visual. Since, the VGA is suitable for evaluate visibility on path or movement, it can be used to promotes the safety since several literature emphasize that natural environment and man-made can affect the sight line.

Isovist and VGA is a suitable tool to evaluate the visibility on a movement pattern. Some of the researches have done the combination techniques to the indoor environment. [15] have done the experiment using that combination tool and compared to the behavioural data. From the result, it was found that several different descriptors can be used as measure describing the environmental complexity such as jaggedness, clustering coefficient, revelation and openness ratio. Thus, the combination of techniques was so benefit for the critical issues regarding behavioural. According to [16], using the combination of this techniques, the average number of turns and movement in space can allow to evaluate how the variation of volume influences the visibility. As the result from the study, it found that the movement and turn in space effect the visibility degree.

3. Visibility Analysis for Three Dimension Layout

Visibility analysis is a crucial process for designing better visual environment thus many techniques have come out. Study on visibility must consider on human perspective and real-time situation as to accurately measure. In the environmental crime studies, visibility analysis was used to identify the capability of human vision as a pedestrian, residents, driver, etc. to minimize the fear of crime as well as opportunity from being a victim. Study made by [17] which to identify the relationship between visibility and fear of crime has shown that low visibility degree reported higher levels of fear of crime. In addition, the visible distance was a significant element that affects fear of crime. However, the study was analysed with two visibility dimensions of visual connectivity of school layout. Fast development progress on digital software and technique can help to capture and analyse on height, density, and visibility degree including the features of building, terrain, trees and the social interaction.

Visibility analysis for human perspective should be made in 3D view. The visual significant of terrain is heavily influenced by the vertical dimension and distance. The capability of GIS to analyses line of sight in 3D such as viewsphere which proposed to measure the visible urban space quantitatively is a well-known enough even though the analysis has been conducted mainly in 2D before. A test case of Singapore's urban space was conducted by evaluating the visibility of three alternative urban design scenarios and their potential impacts on the visual quality of open space [18]. As the result, the proposition that 3D visibility is more significant and effective as compared to 2D because the changes of height (z-dimension). This study has applied both directional and non-directional approaches to the mapping based on 2D and 3D indices.

According to [19], the easiest method to analyse the visibility in 3D by using the Line of Sight (LOS) technique in GIS. This tool can identify the volume of visible object from the observer to a target. With the integrate data of DEM, the evaluation of visibility could be more realistic base on the human perspective vision.

4. Elevation data for 3D mapping

By using appropriate data of Digital Elevation Model (DEM) such as Light Detection and Ranging (LiDAR), accurate model of layout can be created as well as analyst on visibility degree from a specific location. Many researches use the GIS and LIDAR technology for visibility analysis which based on the point clouds. According to [20], point cloud from LiDAR can improve the visualization in quantitative analysis on vegetation blockage. It has been proving that using the point cloud either by mobile or airborne LiDAR, the actual interpretation of vegetation physical can be drawn accurately for visibility analysis [21]-[23]. Thus, the analysis from a point cloud is the best way to find any obstacles between the driver, pedestrian, cyclist including physical environment of vegetation and man-made features.

5. Proposed approach

Based on the literature review before, about the visibility analysis technique and relation to the crime or safety purpose, a method on how to evaluate the visibility on the movement pattern for safety with relation to safety will be propose. Good visibility believes to reduce the fear of crime. To increase the visual capacity, the location or space need to analyse specifically to find out any feature that will minimize the vision capability. In an earlier theory of environmental planning, [24] noted human vision based on the following, (1) from 3 to 10 feet is in a "close relationship to us," (2) up to 40 feet "we can distinguish facial expression," (3) up to 80 feet "we can recognize a friend's face," (4) up to 450 feet "we can discern body gesture," and (5) 4,000 feet is the "maximum distance for seeing people." The idea of [25] "eye on the street" will be using to create the method on how to measure the visibility based on real-time and human-based vision.

To get the real-time situation for visibility analysis, 3D layout needs to develop so that all vertical height including man-made can be observed using the DEM and point cloud from LiDAR. Movement pattern also need to trace so that the visibility can be analyst on the path to as for real situation sight view. The visibility analysis using the GIS and Isovist with maximum vision boundary buffer of 450 feet (discern body gesture) have to analyst and integrated with an overlay 3D layout. Movement pattern will be used as the starting point of isovist to find out any features that affect to visual degree. From this analysis, the output of visibility degree needs to correlate with crime incident using statistical method.

5.1. Observation method for movement pattern

Visibility analysis for environmental crime must observed anything that related to a real-time situation. Thus, incorporate with movement pattern in this study is a good technique to know what is seeing by pedestrian. Based on Jacob's idea "eye-on street", where more people (eye) on street, the more surveillance will be in that location thus, will increase the safety level. The movement pattern should be observed on-site at every street and junction. Observations allow retrieving something that might be considered as an objective view of human behaviour in the built environment.

5.1.1 Gate Counts. This count usually helps to observe the pedestrian's density and vehicle movement flow in urban layout. This technique allows researchers to collect data and represented in graphic or statistically. The method as show in figure 2, must be applied consistency at many locations that cover area under study.

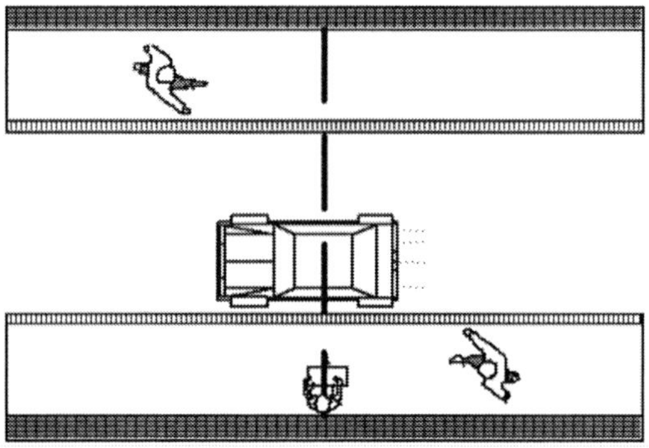

Figure 2. Observer position in relation to observed gate [26]

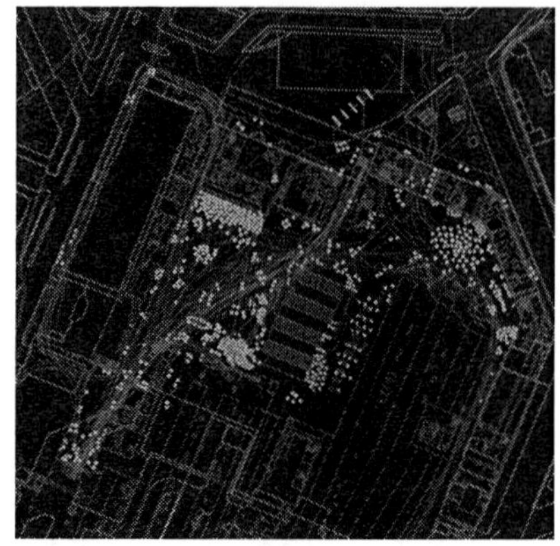

Figure 3. Movement trace been mapped to see the pattern [27]

5.1.2 Movement traces. This method is useful to tracking and mapping flow dynamics in specific area (figure 3). Green color in the map shows the movement pattern of pedestrian captured within several minutes while yellow point shows the location of people. Movement trace will help the researcher to understand the movement pattern from one location to another. The observer can collect and trace the pattern manually or using the other equipment (recorder) based on the area. Normally, observer will stand over in one location which maximize their vision and trace every 5 to 10 minutes (based on the people crowd) at several time interval. Observer also can capture the pattern by using drone to record the video on specific time and area based on needed. Every movement then will be mapping to see the pattern.

5.2. Measuring visibility for safe environment
GIS offer many analyses for visibility and most common tool is view dome tool. This tool which can found in CityEngine scene allow user to specified visible and non-visible feature in some extent. User can visualize visible and hidden area from an omnidirectional observer. Using the 3D layout, visibility for safety purpose can be analyse using this tool by set the limit of human vision to 450 feet in globe to identify how many percentages that observer can see from the starting point as shown in (figure 4). [28], have used dome visibility tools to identify 360 degree of the visibility from a single spot. Any obstacle within the globe that can blocking the view will be shown in red colour and anything that can be seen from the observer point will turn to green. The observation point can be move from one location to another and recommend testing on the traced movement path (pedestrian or vehicle). This technique must implement to see the relationship between visibility and crime in human perspective.
In environmental crime studies, identifying the ways for crime prevention always related to the spatial configuration. That is the reason of using the isovist-based visibility analysis to measure the visibility index. Isovist was use for visibility analysis in 2D and that is no doubt that, that is the limitation of the technology. However, with the needed of 3D to measure visibility for human-perception, the integrated analysis of visibility for Isovist into GIS is needed. Using the isovist, path and scan was the most suitable tools for studying the visibility and safety. Path analysis need to test on the movement pattern either pedestrian or vehicle. The values of visibility will dynamically be updating into the line-chart that was drawn across the viewport. However, scan analysis will allow to produce high resolution of isovist measure as shown in (figure 5). This tool will analyse the layout in 2D to identify in every angle which location is visible. [29] with the experimental test conducted to measure the visibility at Hans Hollein's Monchengladbach museum. From the tested, several volumes of area are shown in several colour's scheme. From red to blue, the map was notified the area with high accessible (red) to low accessible (blue). It has shown that the high accessible is much easier for the observer to see and visible.

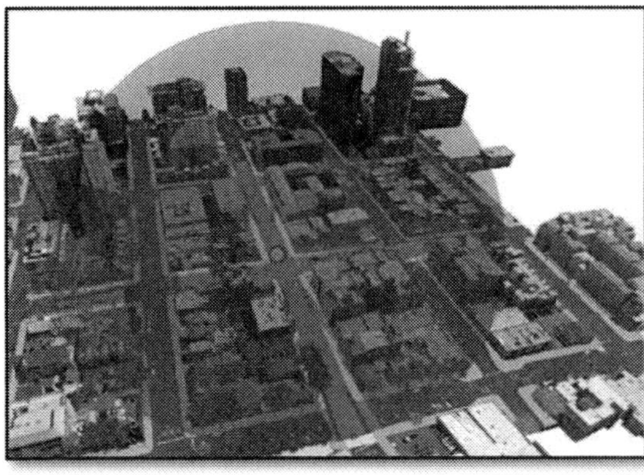

Figure 4. View dome tool one of the visibility analysis tool in GIS to identify 360-degree visibility from single spot [26].

Figure 5. Reviewing measure fields within Hans Hollein's Monchengladbach Museum [27]

6. Conclusion

Several strategies to analyse and evaluate the visibility on movement patter are evaluated. Each method allows to extract the important spatial and topological information with 2D and 3D environment. The obstacle will promote some unsafe phenomenon thus increase fear of crime. That is why the evaluation of visibility is important to identify poor sight line. Studying the visible space for safety purpose must go wide with the interference of physical environment such as terrain, vegetation and buildings height. Building on related work of point cloud-based, show that it is still lack of integrate visibility analysis for urban spaces. The proposed approach which consider the evaluation of visibility in 2D and 3D with incorporate the movement pattern will allow the evaluation is based on the human visual perspective. Setting up the limit of human sight and height will make it the observation and measurement as real. Future work will investigate, the parameter of visibility that can influence the crime activity in urban area with the same visibility evaluation method and relate to the spatial location of crime based on the idea of [25], 'eye on the street'.

Acknowledgments

This research was funded by the Universiti Teknologi MARA (UiTM) under GIP grant with file no: 600-IRMI 5/3/GIP (060/2018). Special thanks to the Royal Police Malaysia, Town and Regional Planning Department, Survey and Mapping Department, and all individuals whom have been directly or indirectly involved in this research.

7. References

[1] Othman F 2014 GIS-based Measures of Burglary and Connectivity to Housing Layout. *International Real Estate Research Symposium*, 7th IRERS. Selangor: INSPEN, Valuation and Property Services Department, Ministry of Finance Malaysia.

[2] Benedikt M L 1979 To take hold of space: isovists and isovist fields. *Environment and Planning* **P** 47-65.

[3] Asami Y, Kubat A S, Kitagaw, K and IIda S 2002 Introducing the third dimension on space syntax: application on the historical Istanbul. (p. Paper No 44). *Tokyo: Center for Spatial Information Science.*

[4] Schroeder T and Saboya R T 2015 Configurational characteristics of sociospatial segregation in Brazilian cities. Brazil.

[5] Sullivan W C and Kuo F E 2016 Environment and Crime in The Inner City - Does Vegetation reduce crime? *SAGE Publication Inc.*, 343-367.

[6] Du Y and Law J 2016 How Do Vegetation Density and Transportation Network Density Affect Crime across an Urban Central-Peripheral Gradient? A Case Study in Kitchener—Waterloo, Ontario *Journal of Geo-Information* **118**
[7] Donovan G H and Prestemon J P 2012 The Effect of Trees on Crime in Portland, Oregon. *Environment and Behaviour* 3-30.
[8] Hillier B and Sahbaz O 2012 Safety in Numbers: High-resolution Analysis of Crime in Street Networks. London, *UK: Springer*.
[9] Bolton T, Francis N and Froy F 2017 The Impact of Space Syntax on Urban Policy Making: Linking research into UK Policy. *11th International Space Syntax Symposium*. Lisbon.
[10] Rosser G and Davies T 2017 Predictive Crime Mapping: Arbitrary Grids or Street Networks? *Journal of Quantitative Criminology* **33** 569–594.
[11] Nes A V and Rueb L 2013 Spatial Behaviour in Dutch Dwelling Areas- How Housing Layouts Affects the Behaviour of its Users. *7th International Space Syntax Symposium*. Stockholm.
[12] Turner A 2007 From axial to road-centre lines: a new representation for space syntax and a new model of route choice for transport network analysis *Environment and Planning* **34** 539–555.
[13] Weitkamp G 2011 Mapping Landscape openness with Isovists *Research in Urban Series* **2** 205-223.
[14] Turner A, Doxa M, O'sullivan D and Penn A 2001 From Isovists to visibility graphs: a methodology for the analysis of architectural space *Environmental and Planning* **28** 103-121.
[15] Wiener J M and Franz G 2004 Isovists as a means to predict spaial experience and behaviour. *International Conference on Spatial Cognition* 42-57
[16] Varoudis T and Penn A 2015 Visibility, accessibility and beyond: Next generation visibility graph analysis. *Proceedings of the 10th International Space Syntax Symposium*, (p. 152). London.
[17] Lee S and Ha M 2015 The Duality of Visibility: Does Visibility Increase or Decrease the Fear of Crime in Schools' Exterior Environments? *Journal of Asian Architecture and Building Engineering* **14** 145-152.
[18] Yang P P, Putra S Y and Li W 2007 Viewsphere: A GIS-Based 3D Visibility Analysis for Urban Design Evaluation *Environment and Planning*
[19] Gonzalez-Jorge H, Diaz-Vilarino L, Lorenzo H and Arias P 2016 Evaluation of driver visibility from mobile lidar data and weather condition *International Archieves of the Photogrammetry, Remote Sensing & Spatial Information Science*, 41.
[20] Guth P L 2012 Probabilistic Line-of-Sight with Lidar Point Clouds. *ASPRS 12 Annual Conference*. California, USA.
[21] Murgoition J J, Shrestha R, Glenn N F and Spaete L P 2013 Improved Visibility Calculation with Tree Trunk Obstruction Modelling from Aerial LiDAR. *International Journal of Geographical Information Science* **27**1865-1883.
[22] Peter R, Ledoux H and Biljecki F 2015 Visibility Analysis in a Point Cloud based on The Medial Axis Transform. UDMV.
[23] Diaz-Vilarino L, Boguslawski P, Khoshelham K, Lorenzo K and Mahdjoubi L 2016 Indoor navigation from point clouds: 3D modelling and obstacle detection. *International Society for Photogrammetry and Remote Sensing*.
[24] Spreiregen P D 1965 Urban Design: The Architecture of Towns and Cities. *USA: Mcgraw-Hill*.
[25] Jacobs J 1961 The Death and Life of Great American Cities. *New York: Random House*.
[26] Al-Sayed K, Turner A, Lida B and Penn A 2014 Space Syntax Methodology. London, UK: Barlett School of Architecture, UCL.
[27] Jacoby K 2006 What is Space Syntax? Does the urban form of the city affect the level of burglary and crime? Stockholm: Royal Institute of Architecture.
[28] Schueren M 2017 New Visibility Analysis Tools in CityEngine 2017.1. Retrieved from esri city engine: http://www.esri.com
[29] McElhinney S 2018 isovist_2.1: a basic user guide, v1.0. London: isovist. Retrieved from isovist.

IOP Conf. Series: Earth and Environmental Science **385** (2019) 012025 doi:10.1088/1755-1315/385/1/012025

A triangulation for passive design strategies in public hospital

J Muhamad[1], H Ahmad[1] and A Abd Aziz[1]
[1]Faculty of Architecture, Planning and Surveying, Universiti Teknologi MARA, Seri Iskandar Campus, Perak Branch, 32610 Perak, Malaysia.

Email: 53a.jmaarchitects@gmail.com

Absract: This study touches on passive design elements used at the preparatory design stage of public hospital in Malaysia. Passive design is an aspect of sustainable architecture that needs to be understood and adopted in an environmentally sensitive architectural design. The most common problem is that the designers do not have proper guidelines for passive design to help them in designing a public hospital. Currently there is no comprehensive guidelines on passive design or sustainable architecture that may help designers. This consequently affects the increase of energy usage in hospitals. The intent of this study is to furnish evidence that the initial process of systematic design and guidance is essential in planning for a sustainable public hospital environment. In this study, the researcher uses a mixed method to obtain outcome for guideline. Through this method, a combination of qualitative and quantitative metadata to foster effective interventions to improve the quality of environmental friendly hospital design. The conclusions obtained from this study are to assist in the process of more efficient public hospital design. In addition to making these passive design guidelines as helpful in providing explanations and working to ensure that the guideline produced has credibility and meet the criteria.

1. Introduction

In social science research, there are several ways to collect and analyze data onward formalization guideline. It is included simulation methods, expert interviews, case studies and triangulation or mixed methods. Greene [5] states this data is often referred to as dialectal hearing bridging post positivist and social constructivist worldview, pragmatic perspective, and transformative perspective [4]. The production of guidelines using this mixed method has been successfully developed by the World Health Organization and National Institute For Health & Care Excellence. Therefore, it is a guide to the researcher to use the method as it has proven successful.

The combination between the questionnaire (quantitative) and the interview (qualitative) session was aimed to gain the impact value of the implementation of building designs, especially hospitals to consumers. The use of this method also has three sequences to follow: a) convergence, (b) sequential explanation, and also (c) sequence of exploration [1]. It encompasses the characteristics of data collection, analysis, interpretation and verification of research. Therefore, the combination will provide a more in-depth understanding of the research's problems.

2. Methodology

In research design, the use of mixed methods as an implication to the impact investigated by researchers. The findings that will be conducted in select hospitals in Malaysia are dependent on the limited area of the patient's ward area. Qualitative interpretation study with exploration of users using the hospital environment. The start of the study is based on the topic selection as well as the paradigm that comes from the theoretical framework and methodology which then understand the impact of the factors investigated and the structure of variables [6].

In data collection, interviews with expert panels are one of the methods in qualitative. The Interview is an active interaction between two or more people leading to negotiation decisions and based on the context and topic chosen. According to Yin [15], bridging is a must-do for researchers to prove case studies. In this context, the use of partial structural propagation adapted to which the question has been determined, but the order and the sentence can be changed to match the atmosphere of the conversation or to be added during the interview.
Through a field survey (observation) of the hospital environment, it is a very challenging psychological or physical task [11]. The purpose of this study is to help researchers understand the function and routine of medical care routine. Researchers also need to understand the privacy of patients as well as do not interfere with nursing duties to carry out their daily routines as it is a requirement to be complied with by the authorities without any compromise [12]. To achieve the goal, researchers need more creativity in information gathering in hospitals.
During the field survey, a questionnaire was also conducted on hospital users i.e. staff and patients. The study conducted by Kothari [8] found that this method is often used by researchers in the built environment and is an effective tool in research for architecture. This is because it involves socio-cultural, environmental and management interactions as an inquiry tool known as Post Occupancy Evaluation (POE) to assess the performance of buildings. This method is one of the examples of questionnaires on the most effective building design and population response to their environment [10].

2.1 Nested Methodology Approach
The nested research methodology proposed by Robson [16] through his study aims to test the theory and reflect research questions. In addition, it also serves as a guide to researchers to understand the complexities and also to provide an overview of the current research. The combination found in this nested research encompasses: the overall philosophy of research, research approaches and research techniques.
This method is also referred to as an integrated research as it creates a framework for guiding the researcher's approach and consequently leads to a more appropriate selection of techniques [13]. Therefore, through this study mapping the methodology of passive design and its effect to the public hospital in Malaysia. The implication is that the use of active energy increases from time to time. Preparation of the pre-design of the physical environment of the public hospital, especially in the ward should be taken into account in order to contain this energy use from continuing to increase.

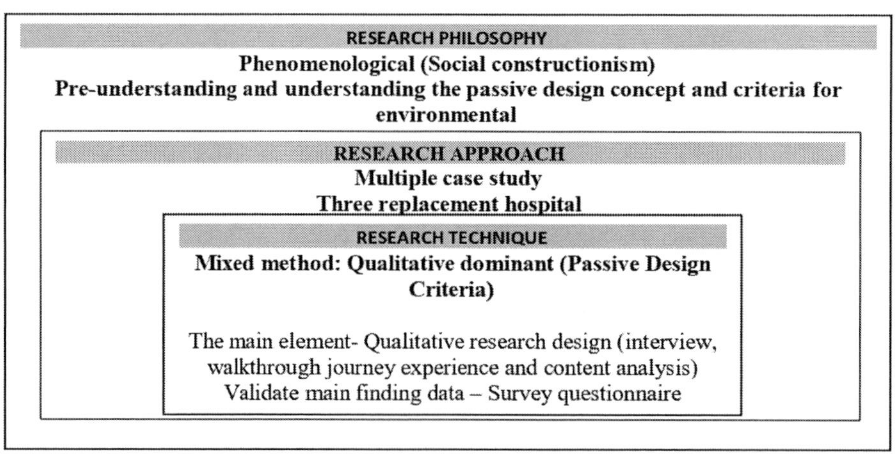

Figure 1. [7] Nested Methodology Approach Model and Its Application in the Main Study

2.1.1 Research Philosophy: Phenomenological. Philosophy of phenomenology is a philosophy associated with empirical communities and human cultures. Through this philosophy reflection on passive designs based on cultural anthropology and space studies based on social construction needs and comforts (interpretisvm) [9]. The emphasis on research and exploration is the principle of knowledge assessment that supports the theoretical and practical developments in the design process.

The use of philosophy in this study is to see the implications of the active energy use and its effect on hospital design. In addition, wastage and increased use of active energy for health care. The selection of philosophy is a reflection of the objective and adapted evaluation framework during the design process. Catalytic studies and case studies with relevant information and guides for designers in producing more efficient and practical design.

2.1.2 Research Approach: Multiple Case Studies
According to Yin [15] and J. W. Creswell [3], the philosophy associated with social constructs and paradigm constraints, environmental preservation of users and managing organizations can advance the theoretical perspective in a more complex and practical form. The problem statement through cross-analyzes arising from research questions will provide the corresponding pattern of equality as well as the differences in case studies [9]. This statement is supported by Yin [15], which states that the relatively used sample affects the establishment of parameters that are case studies aimed at guiding the building designers.
Creswell [2], in his study noted that the case study research approaches were carried out covering a comprehensive method of data collection during exploration of each case. Hence, the findings from this diverse case study are challenging in implementation to researchers [15] and strategies involving the analysis need to be deepened so that issues raised can be achieved and resolved more efficiently.

2.1.3 Research Technique: Mixed Method with the Qualitative-Dominant
The topic selection and paradigm associated with the design are the first step in building the theoretical framework and methodology that need to be understood as it is a guide to the investigation of the construction of the design process [2]. Thus, this qualitative methodology is a priority for the entire research design and is further supported by quantitative method [1] [2] to get results implications of design impacts to users and their experience. The two methods carried out in this research are an approach called 'dominant and less dominant'. This less dominant approach was during the collection of data through a questionnaire (quantitative) and validated by expert panels (dominant) through interviews (qualitative) on research conducted through participants' beliefs, views, understanding and experience [3] [9].
Therefore, qualitative research is an inductive procedure that is relevant to interviews, expert panel about the production of hospital designs produced in a sustainable environment. Subsequently the data is supported through quantitative data verified through surveys obtained and questionnaires distributed to consumers (staff and patients) aimed at strengthening qualitative data. Therefore, the mixed method approach has many advantages related to the exploration to complement the development and development of the triangulated research.

3. Result And Discussion Of Triangulation Of Data
Researchers need to understand the quality requirements for data validation obtained as stated by Kothari [8], and Lucas [9]. Verification of key finding data must be transparent as well as to ensure the credibility of this qualitative study and the development of knowledge gained is true and accurate [6] [1]. Questionnaire data obtained from the respondents will be analyzed to support the findings of qualitative research results that respond from the start of both data [2]. Therefore, the re-authentication of the data obtained after the discovery by users of this public hospital is known as cross-validation [1]. For the validation of the data, it aims to establish the final interpretation of the data obtained from the study as well as to minimize any biased findings.
Therefore, this triangulation method will benefit, the researchers in producing guidelines later. This method is closely related to one another and can also cross check the data obtained. This framework relates to the assessment of passive design criteria as dimensions. This conceptual and practical assessment framework supports the clarity of the conclusions discussed.

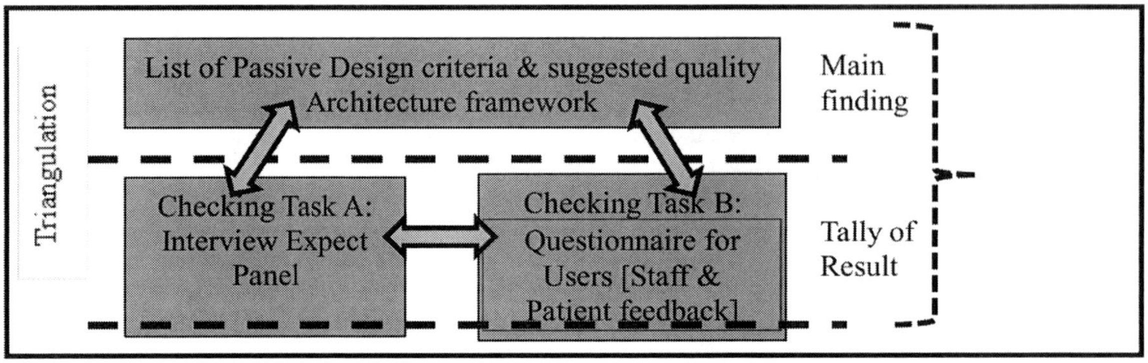

Figure 2. The Validation process of the Main Study Results

4. Conclusion

Through the findings of the study, it is clear that the cross-checking of data through triangulation is important so that the continuation of the design of the public hospital is more efficient. The main case study is the continuation of pioneering hospital studies with some issues that need to be expanded in relation to the limitations and scope of the study on the issue of sustainable environmental hospital. Therefore, in making guidelines, it is necessary to study the aspects from the initial design process and implications to end users based on their experience.

5. References

[1] Creswell J, Klassen A C, Plano V and Smith K C 2011 Best Practices for Mixed Methods Research in the Health Sciences *Methods* **29** 1–39

[2] Creswell J W 2012 Educational research: Planning, conducting, and evaluating quantitative and qualitative research *Educational Research* **4**

[3] Creswell J W 2014 Research Design: Qualitative, Quantitative and Mixed Methods Approaches. (V. Knight, Ed.).

[4] Creswell J W 2014 Research Design [Fouth Editon]: Qualitative, Quantitative and mixed Methods Approaches *Animal Genetics* **39**

[5] Greene J C 2006 Toward a Methodology of Mixed Methods Social Inquiry *Research in the Schools: Mid-South Educational Research Association* **13** 93–98

[6] Ivankova N V, Creswell J W and Stick S L 2006 Using Mixed-Methods Sequential Explanatory Design: From Theory to Practice *Field Methods* **18** 3–20

[7] Kagioglou M, Cooper R, Aouad G and Sexton M 2006 Rethinking construction: the Generic Design and Construction Process Protocol.

[8] Kothari C R 2004 Research Methodology: Methods and Techniques (Second Rev). New Age International.

[9] Lucas R 2016 Research Methods for Architecture. United Kingdom: Laurence King Publishing.

[10] Mohammad I S 2011 Post Occupancy Evaluation of Building Performance In Malaysia.

[11] Rina S and Zakaria R 2014 Green Assessment Criteria for Public Hospital Building Development in Malaysia *Procedia Environmental Sciences* **20** 106–115

[12] Smith F 2014 Sustainable Healthcare & ebola, (June).

[13] Turner S F, Cardinal L B and Burton R M 2017 Research Design for Mixed Methods: A Triangulation-based Framework and Roadmap *Organizational Research Methods* **20** 243–267

[14] Yin R K 2006 Case Study Reserach - Design and Methods *Clinical Research* **2** 8–13

[15] Yin R K 2009 Case Study Research Design and Methods. Applied social research methods (Fourth Edi).

[16] Zhao Y and Mourshed M 2012 Design indicators for better accommodation environments in hospitals: Inpatients' perceptions *Intelligent Buildings International* **4** 199–215

IOP Conf. Series: Earth and Environmental Science **385** (2019) 012026 doi:10.1088/1755-1315/385/1/012026

The Effectiveness of Training Programmes Towards the Level of Knowledge of the Personnel Involved in Stamp Duty Valuation

S H Rosman[1], A H Nawawi[1], S N F Ab Latif[1] and M I Abdullah[1]
[1]Faculty of Architecture, Planning and Surveying, Universiti Teknologi MARA, 40000 Shah Alam, Selangor, Malaysia

Email: sitihasniza@uitm.edu.my

Abstract. Effective training programmes that is underpinned by integrated content holds the role to ensure that effective delivery of knowledge can be absorbed appropriately by the personnel of Valuation and Property Services Department (JPPH) that have undergone the training programmes being organized. This study is conducted to identify the effect of training programmes towards the level of knowledge of the personnel involved in stamp duty valuation. Due to its importance of the provision of the training programmes given to the personnel of an organization, hence, an investigation related to the effect of the content of the training programmes towards the knowledge of the personnel of Valuation and Property Services Department (JPPH) in stamp duty valuation has been conducted. This is to identify if the content of the training programmes conducted by the National Institute of Valuation (INSPEN) towards the personnel of Valuation and Property Services Department (JPPH) has ensured effectiveness and has ensured the fulfillment of the requirement of the personnel in executing their tasks effectively. This study has been conducted focusing on stamp duty involving six (6) offices of the Valuation and Property Services Department (JPPH) within Klang Valley that has included the District Valuer, Valuation Officer and Assistant Valuation Officer, who are engaged in the valuation process of stamp duty, as respondents. Interview approach, pilot study, content analysis, application of Analytical Hierarchy Process (AHP) has been employed to achieve the main objective of this study. Analytical Hierarchy Process (AHP) which has been used in this study, is to determine the relative importance of the level of knowledge of the personnel in Valuation and Property Services Department (JPPH) and also acts as the instrument that can be applied as a guide in the effort to improve the performance in the content of training programmes to fulfill the expectation and requirement of the personnel during the valuation process of the stamp duty. In summary, this study has proven that there is strong relationship between knowledge contained in the training programmes content with the level of knowledge of the personnel in stamp duty valuation.

1. Introduction

Nowadays, the stamp duty is one of the main sources of government revenue. The rate used is dependent on the market value of a property during when the transfer is completed. As provided under the First Schedule to the Stamp Act 1949 (Act 378), this duty is based on a consideration of monetary value or the market value of the property, whichever is higher. The Valuation and Property Services Department (JPPH) has been appointed to conduct valuation for the purposes of stamp duty. Almost in every state in

Peninsular Malaysia, the valuation for the purposes of stamp duty is carried out by the Valuation and Property Services Department (JPPH). Measurement of property that has been reported is for the purpose of stamp duty to be performed by Valuation Officers in Valuation and Property Services Department (JPPH).

Interest in the improvement of knowledge can also not to be denied because individuals and organizations begin to understand and appreciate that knowledge as the most valuable asset in a competitive environment [1]. It is beyond doubt that knowledge is crucial in carrying out valuation of stamp duty, starting from inspection activities of the property, analysing comparisons, making opinion of value, and eventually producing a detailed report on the value of the property that involved stamp duty. Each officer involved in conducting valuation of stamp duty has different level of knowledge about stamp duty, despite the fact that it is known that knowledge is something very complex and controversial and can be interpreted in various ways [2]. The role of Valuation and Property Services Department (JPPH) in providing quality service to customers is dependent on the performance level of personnels consisting of Valuation Officers and Assistant Valuation Officers in performing a given task effectively and upholding good level of quality.

This paper will highlight the assessment on knowledge level of personnel involved in stamp duty valuation using Analytical Hierarchy Process (AHP) types of knowledge in stamp duty valuation, in which there are different types of knowledge in the handling process of the stamp duty for which it is not noticed by the valuation personnel.

This paper will discuss regarding the three (3) objectives have been structured as follows; i) to identify the content of training programmes attended by personnel associated with the types of knowledge in stamp duty valuation, ii) determine the relative importance of the level of knowledge of the personnel involved in stamp duty valuation and; iii) to measure the types of knowledge that has been acquired by the personnel in training programmes related to stamp duty valuation. This is intended to see to what extent the level of knowledge possessed by the personnel assessment based on the types of knowledge involved in stamp duty valuation in accordance with (3) three stages during their inspection, comparable analysis and opinion of value.

2. Content Analysis to Identify the Types of Knowledge in the Stamp Duty Valuation

According to [3], content analysis is a method that is widely used in research covering various fields [4]. [3] also stated that, among the researchers, content analysis has several advantages including the level of sensitivity on the textual content of a document or report. Content analysis method is also widely accepted in the field of communication and information science, as a method for the detection of media images as well as text content [5] and was also applied in the field of environmental research [6].

The content analysis has been used based on previous journals, articles and books related to the issue of this study. This is identified intellectually through assessment of previous research; the types of knowledge that form the assessment. The researcher for this study has listed the many types of knowledge that have been found as a result of the content analysis of previous literature for the purpose of the study. After the researcher has listed several types of knowledge which are likely to be involved in stamp duty valuation, so the researcher will use these data for the purpose of examining the validity.

This data is collected through interview with respondents that comprised of District Valuer, Valuation Officers and Assistant Valuation Officers who are experienced in stamp duty valuation in particular for types of knowledge necessary for the task. Results from the interview held in meeting were recorded and the data found will be interpreted in text forms for the purpose of qualitative content analysis of the data types that will be used systematically.
In this paper, the researcher has divided the types of knowledge involved in stamp duty valuation into three (3) main stages of the valuation process namely inspection, comparable analysis and opinion of value. For each stage of the evaluation process, the researcher has established seven (7) types of knowledge involved. The results of the content analysis of the previous literature review and interviews were conducted in order to achieve the first objective. The researcher has found the findings are shown in Figure 1.

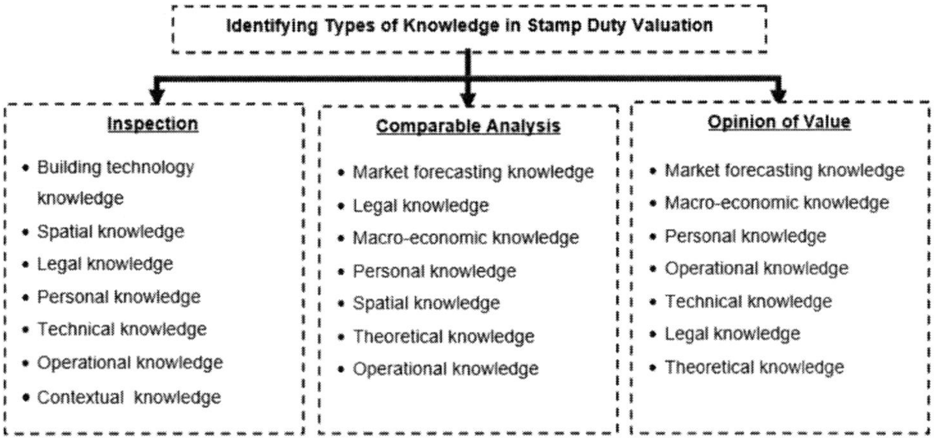

Figure 1. Types of Knowledge Involved in Stamp Duty Valuation

Each type of information that has been identified is the types of information needed by the personnel of the Department of Valuation and Property Services Department (JPPH) while they conduct the work of valuation.

3. Content Analysis to Identify the Training Programmes Attended by Personnel Associated with the Types of Knowledge in Stamp Duty Valuation

This paper has identified the contents of the training programme which has been attended by personnel. It is associated with the types of knowledge in stamp duty valuation. In this study, researcher have used the method of content analysis using an inductive approach that involves coding [7] to analyse the content of each training programme which was attended by personnel of Valuation and Property Services Department (JPPH). The researcher used the book documents of the training programme published by the National Institute of Valuation (INSPEN) as an instrument for content identification of training programme attended by personnel associated with the types of knowledge in stamp duty valuation.

Training programme books published from year 2011 to 2013 will be used as reference material for the purpose of inductive content analysis in which researchers will use the coding of words or sentences that give meaning associated with the types of knowledge that has been gained from the first objective.

The researcher has compiled a list of all types of training programmes conducted by the National Institute of Valuation (INSPEN) from 2011 until 2013 which was attended by respondents involved in this research, namely by the Valuation Officer and Assistant Valuation Officer involved in the work of valuation of stamp duty. Figure 2 has shown a summary of the analysis conducted by year and type of categories involved as well as the overall respondents.

It can be observed that "personal knowledge" recorded the highest frequency for all levels of stamp duty valuation process. This shows that "personal knowledge" is the type of knowledge that is commonly applied to all types of training programmes conducted by the National Institute of Valuation (INSPEN). The importance of "personal knowledge" in improving an organization's performance in carrying out valuation work is crucial for stamp duty exercise. This statement was supported by [8] in which he stated that the quality of human resource performance often depends on the implementation of the competence skills and knowledge of an individual or personnel.

This suggests that the ability of personnel performance depends on the skills and knowledge they have in order to achieve the main goal of the organisation they represent. According to [9], training programme act as a tool for performance improvement in organisations with low skills and knowledge. This shows that there is interest in training to improve the skills and knowledge that is required on every member of personnel to ensure organisational excellence. Therefore, every individual believes that by adopting the skills and knowledge learned during the training exercise will improve their performance [9].

In the findings of this study, it is evident "personal knowledge" is the most dominant and achieve the highest frequency in which each training programme organised by the National Institute of Valuation (INSPEN) implement a "personal knowledge" which is especially important in improving performance work and knowledge to ensure that the work undertaken can be executed with excellence. Figure 2 also shows that the "operational knowledge" and "technical knowledge" has a relatively high frequency value compared with other types of knowledge to another.

Good knowledge of skills for work refers to the knowledge of an personnel to perform work effectively and to achieve effectiveness and best workmanship [10]. With this, it can be reinforced that "personal knowledge" is a medium to enhance other types of knowledge involved in the valuation of stamp duty. The results of these research findings have been able to achieve the second objective of this study for which the data were obtained prove to be useful information for researcher to continue obtains the results for the next objective.

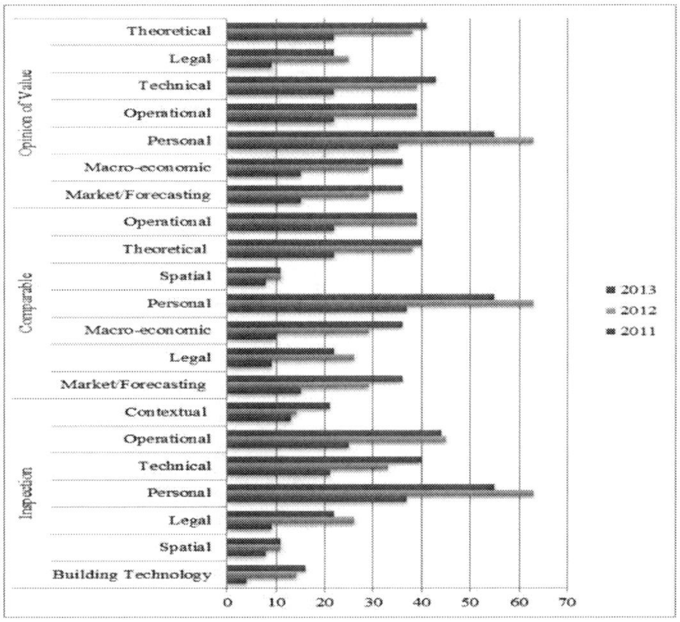

Figure 2. Total Frequencies of Contents of Training Programmes Attended by Personnel Related to Types of Knowledge in the Stamp Duty Valuation by Categories of Respondents

4. Methodology

This section will discuss the research methodology that has been applied to recover information that is needed to complete this research as depicted in Figure 3. In identifying the types of knowledge in stamp duty valuation, the researcher uses literature review of previous methods to identify the types of knowledge related to the valuation of stamp duty. To ensure the accuracy of information on the types of knowledge that have been classified as closely related to the valuation, an interview with the Valuation Officer has been carried out. This is to confirm the identification of the types of knowledge is valid and precise to be studied in-depth.

For the third objective, the method of Analytical Hierarchy Process (AHP) is used for the purpose of measuring the level of knowledge of the personnel involved in stamp duty valuation. Questionnaires were used as the main medium in obtaining the required information effectively for the purpose of obtaining a measure of the level of knowledge of each personnel member involved as respondents. The personnel selected as respondent were those involved in the work of stamp duty valuation comprising the District Valuer, Valuation Officer and Assistant Valuation Officer of six (6) selected Valuation and Property Services Department (JPPH) based in Klang Valley.

The Analytical Hierarchy Process (AHP) methods are used to determine the stages of the valuation process as well as the weightage for each type of knowledge involved in the training programme attended by each personnel of Valuation and Property Services Department (JPPH).

The purpose of this rule is to learn about more information on score weightage that was found in the results of the questionnaire that can identify the level of knowledge of the types of knowledge dominant in their opinion. Analysis using the AHP also involves a perception survey which ranked the relative importance of the criteria in the evaluation process and sub-criteria of which types of knowledge in stamp duty valuation is using pairwise comparison method in Analytical Hierarchy Process (AHP).

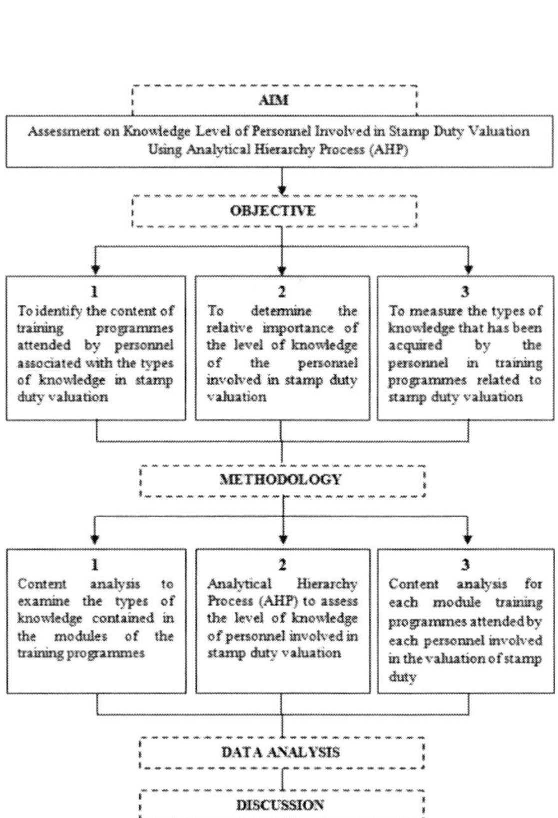

Figure 3. Research Methodology Process

5. Data Analysis using Analytical Hierarchy Process (AHP)

Analysis Analytical Hierarchy Process (AHP) will produce findings individually according to each individual or tendency of each personnel involved in this study and to identify the overall level of personnel in the Valuation and Property Services Department (JPPH). The evaluation process of different kinds of knowledge in stamp duty valuation requires measuring the weight of each criterion whether more value or par value each other. The scale used in the evaluation of these weightage is to use a scale of Saaty's as shown in the previous table on Analytical Hierarchy Process (AHP) [11].

5.1 Analytical Hierarchy Process (AHP) Analysis

Saaty's scale uses pairwise comparison method in which it is a method of comparison of each other for each criterion based on one hundred and fifty (150) of the respondents to determine the relative importance and weightage of these criteria to the criteria Level 1. With this, the objective of which is targeted to achieve the third objective was successfully implemented and detailed results are acquired. The findings have been analysed in detail once again to provide an easy conclusion to explain the level of knowledge of personnel in Valuation and Property Services Department (JPPH) for all respondents involved in this study, which can be referred to in Table 2.

Table 1. Results on Eigenvector Calculation Criteria – Overall Process Level 1

				Total	Weighting
Inspection	53.6254	66.5863	40.4266	160.6383	0.3570
Comparable Analysis	45.2070	27.3662	53.6445	126.2177	0.2805
Opinion of Value	51.1508	56.0275	55.8513	163.0296	0.3623
				449.8856	1

Based on Table 3, it shows the final stage of pairwise comparison method for making comparisons in determining the relative importance of the criteria in the valuation process stage in the study area. As a result, the criteria are those to obtain the highest weightage is the criteria "opinion of value" that is 0.3623, while the criteria to obtain the lowest weightage is the criteria "comparable analysis" of 0.2805 followed by the criteria "inspection" of 0.3571.

The criterion "opinion of value" got the highest weightage following the criterion respondents which has concerned about this than the other criteria in ranking of stamp duty valuation process by the Valuation and Property Services Department (JPPH) of Klang Valley area as criterion for "opinion of value". The elements of different types of knowledge are very important for the respondent to conduct the valuation process involving different types of knowledge such as market/forecasting knowledge, macro-economic knowledge, personal knowledge, operational knowledge, operational knowledge, technical knowledge, legal knowledge and theoretical knowledge.

In the "opinion of value", we can state that based on the results obtained, this criterion is very important compared to the other criteria of personnel in the Department of Valuation and Property Services Department (JPPH) involved in the valuation of stamp duty, for one's knowledge of valuer to the "opinion of value" is very high. These findings are supported by a statement, from [12], which states that the value of property acquired from a real estate valuer must also be accepted by the court and if there arises any disagreement or dispute of any valuation or substantially the same, able to defend their decision in court. It has been shown that the "opinion of value" is very important for valuer in the performance of their valuation work.

In addition, the criterion "inspection" got the second highest weightage criteria due to the importance of knowledge as a kind of building technology knowledge, spatial knowledge, legal knowledge, personal knowledge, technical knowledge, operational knowledge and contextual knowledge as an important asset to ensure that the valuation is done correctly and according to the latest market developments in the specification. At the same time, the valuation by Valuation and Property Services Department (JPPH) can reduce the number of appeals in revaluation and can improve personnel effectiveness.

The results obtained are of interest to respondents' perceptions of the level of stamp duty valuation process, in determining the criteria for evaluating the level of knowledge of the personnel involved in stamp duty valuation. Figure 4 shows clearly the position or ranking each of these criteria for Level 1, Criteria.

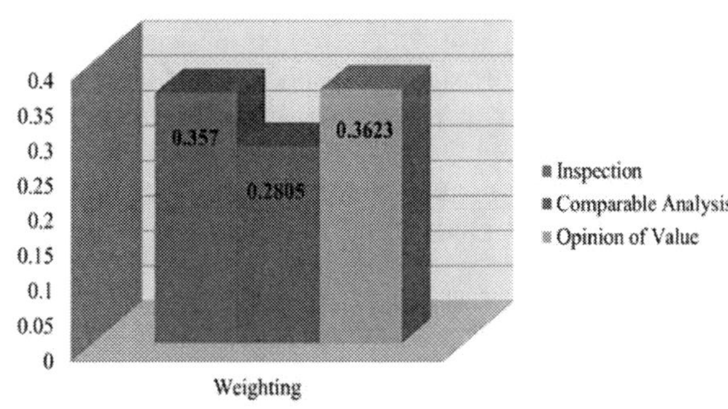

Figure 4. Overall Ranking of Weighting Criteria Level 1

Through this pairwise comparison method, the analysis continues using the same method for the next stage of comparing the relative importance of each criterion in Level 2. These criteria which are in Level 2 are involved in sub-criteria for Level 1. This process is based on hierarchy, namely by using Analytical Hierarchy Process (AHP). Thus, each of the criteria and sub-criteria necessary to obtain the weightage according to the relative importance of these rules, in the hierarchical method of Analytical Hierarchy Process (AHP).

The pairwise comparison process is similar to that done in Level 1. The results of the criteria analysis and calculations are made. Table 3 shows the ranking of each sub-criterion with the relative importance of each sub-criterion in Level 2 (Inspection).

Table 2. Results on Eigenvector Calculation Sub-Criteria Level 2 Overall – Inspection

								Total	Weighting
Building Technology Knowledge	61.2326	90.9770	52.8768	37.5674	29.3555	24.4973	22.2489	**318.7555**	**0.3036**
Spatial Knowledge	14.7589	16.9829	52.8275	37.7474	29.3580	24.4981	22.2495	**198.4223**	**0.1890**
Legal Knowledge	14.7616	8.5243	10.9183	37.8650	29.5795	24.4584	22.2255	**148.3326**	**0.1413**
Personal Knowledge	14.7826	8.2723	8.1769	9.0427	30.2812	24.8992	22.5615	**118.0164**	**0.1124**
Technical Knowledge	14.8268	8.3973	8.2472	9.2105	7.8167	24.8750	22.6927	**96.0662**	**0.0915**
Operational Knowledge	14.7299	8.3309	8.2823	9.2105	11.7825	8.0425	22.5188	**82.8974**	**0.0789**
Contextual Knowledge	14.8703	8.4559	8.6281	9.3105	11.7825	18.6841	15.4528	**87.1842**	**0.0830**
								1049.6750	**1**

Table 4 and Figure 6 show the results of the calculation of pairwise comparisons in Level 2 (Inspection) for six (6) Valuation and Property Services Department (JPPH) offices in the Klang Valley. The table shows the criterion for "building technology knowledge" which is having the highest ranking of importance that weights 0.3036, while the criteria for "operational knowledge" is the lowest importance weight of 0.0789. The criteria "spatial knowledge" has the second highest weightage of 0.1890.

The criteria "legal knowledge" gained the third highest weightage of 0.1413, while the criterion "personal knowledge" of 0.1124 obtained the fourth highest weightage criteria followed by "technical knowledge" and "contextual knowledge" in fifth and sixth places respectively with a total weight of 0.0915 and 0.0830.

The criterion "building technology knowledge" has got the highest weightage and this has proven that this type of knowledge is the most dominant and has been given priority in researching the content of the training programme conducted by the National Institute of Valuation (INSPEN) to personnel of the Valuation and Property Services Department (JPPH), while a stamp duty valuation process is in the "inspection".

These findings are reinforced by statements of [13] that relate to building knowledge which involves valuation with excellent knowledge of all aspects of construction technology, materials used in construction of buildings and other concerns. Often this knowledge is one of building technologies related materials and technologies as well as knowledge of health and safety aspects of a building.

Knowledge relating to "technical knowledge", "contextual knowledge", and "operational knowledge" are to be given appropriate concern by the National Institute of Valuation (INSPEN) in controlling the content of the training programme to enhance knowledge to ensure that the knowledge acquired by personnel of Valuation and Property Services Department (JPPH) can be balanced, and this eventually can produce something useful for the future.

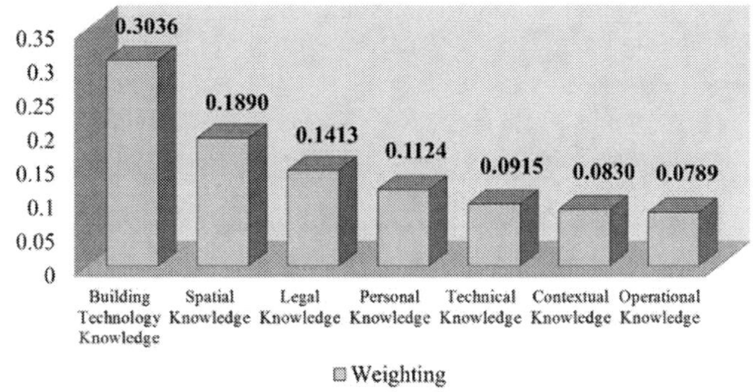

Figure 5. Overall Ranking Sub-Criteria Weightage Level 2 – Inspection

The results of pairwise comparison analysis for sub-criteria Level 2 (Comparable Analysis) is also shown in Table 4. This ranking is assessed based on the relative importance of each sub-criterion in the types of knowledge involved rated that is "comparable analysis" where it is very important in stamp duty valuation process.

Table 3. Results of Eigenvector Calculation of Sub-Criteria Level 2 Overall – Comparable Analysis

								Total	Weighting
Market/Forecasting Knowledge	60.1153	88.5735	51.6327	36.9265	29.6284	25.4102	22.6846	314.9712	0.3000
Legal Knowledge	15.2855	17.0878	52.7597	37.6531	28.6568	24.0375	21.8184	197.2988	0.1879
Macro-economic Knowledge	15.2882	8.6108	11.1181	38.0692	29.0073	24.1950	21.9120	148.2006	0.1411
Personal Knowledge	15.1402	8.3588	8.2237	9.1175	30.5541	24.7767	22.3657	118.5367	0.1129
Spatial Knowledge	14.8317	9.1948	8.6936	9.2897	8.0263	24.4708	22.8931	97.4000	0.0927
Theoretical Knowledge	14.5940	9.1284	8.5917	9.3992	12.1261	8.2744	22.7192	84.8330	0.0808
Operational Knowledge	14.7062	8.9867	8.9375	9.4992	11.9569	18.7898	15.5576	88.4339	0.0842
								1049.6740	1

Table 5 shows the results of pairwise comparisons for the evaluation of the relative importance weightage of sub-criteria in Level 2 (Comparable Analysis) for six (6) Valuation and Property Services Department (JPPH) offices in the Klang Valley. Results show that the criterion of "market/forecasting knowledge" gained the highest weightage in the evaluation process whereby 0.3000 is at the "comparable analysis", this kind of knowledge is dominant where it has been proven that any stamp duty valuation made by the officers have a very good information on this type of knowledge. [14] also states that "the market/forecasting knowledge" is very important in the evaluation process as it is generally more in touch with the market and can provide assistance to Valuer in making market forecasts.

The criterion of "legal knowledge" gained the second highest weightage of 0.1879, while the criteria for "macro-economic knowledge" and "personal knowledge", each has a weightage of 0.1411 and 0.1129. The criteria "spatial knowledge", "operational knowledge" and "theoretical knowledge" has obtained the lowest score of 0.0927, 0.0842 and 0.0808.

Therefore, it can be seen here that three (3) types of knowledge with the lowest score weightage should be given greater importance in determining the content of the training programme in the future and can improve the performance of personnel in conducting the evaluation process at this stage. Figure 6 shows a further explanation on the ranking weightage for each sub-criterion in Level 2 (Comparable Analysis).

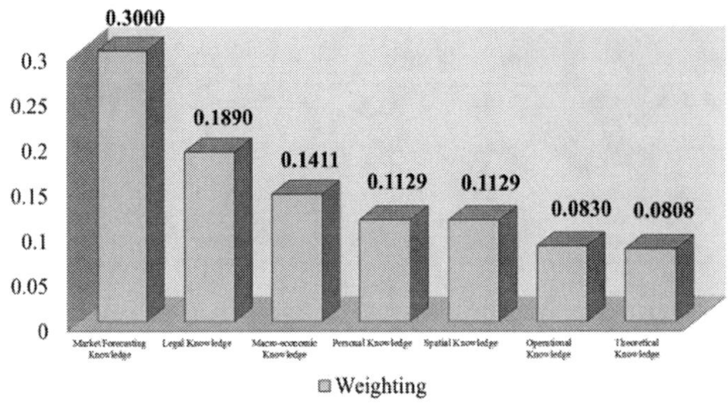

Figure 6. Ranking of Weightage on Sub-Criteria Level 2 Overall Comparable Analysis

Analysis of the relative importance of weightage of sub-criteria in Level 2 (Opinion of Value) also involves sub-criteria comprising market/forecasting knowledge, macro-economic knowledge, personal knowledge, operational knowledge, technical knowledge, theoretical knowledge and legal knowledge.

Assessment methods using pairwise comparisons to obtain the relative importance of each sub-criteria are shown in Table 4, based on respondents' perceptions of the types of knowledge involved sub-criteria for the level of stamp duty valuation process for Level 2 (Opinion of Value).

Table 4. Results of Overall Eigenvector Calculation Sub-criteria Level 2 – Opinion of Value

								Total	Weighting
Market/Forecasting Knowledge	60.1965	89.8114	52.2497	36.9046	28.8767	24.0282	21.7266	313.7957	0.2989
Macro-economic Knowledge	14.4251	16.6347	51.9402	37.7193	28.8801	24.0574	21.7272	195.3840	0.1861
Personal Knowledge	14.6608	8.6588	10.8329	37.2484	29.5169	24.2359	22.1365	147.3302	0.1403
Operational Knowledge	15.1921	8.5597	9.2784	9.2036	30.3842	25.6093	22.5347	120.7620	0.1150
Technical Knowledge	15.2106	8.8191	8.5586	10.7135	8.7760	25.5487	23.2538	100.8803	0.0961
Legal Knowledge	15.0211	8.5727	8.5413	9.0822	11.7603	8.0803	23.2538	84.2917	0.0803
Theoretical Knowledge	15.2541	8.8447	8.5566	9.0822	11.7603	18.4167	15.3181	87.2327	0.0831
								1049.677	1

Table 6 shows the results of analysis of the relative importance of weightage of sub-criteria for Level 2 (Opinion of Value). Perception of respondents indicated the criteria "market/forecasting knowledge" has the highest weightage over other criteria, that is 0.2989 followed by criterion "macro-economic knowledge" that has the second-highest weightage that is 0.1816. "Market/forecasting knowledge" had the highest scores in which knowledge of this type is dominated by personnel at the Valuation and Property Services Department (JPPH) in the process valuation of stamp duty.

"Macro-economic knowledge" was the second highest of the value associated with the development of the world economy in the present example, the current interest rate charged by the Central Bank, recent policy and other relevant impact on the value of the property. [15] have stated that the "macro-economic knowledge" is the most important knowledge in the field of economics.

The criteria that has the third highest weightage and is the fourth highest are "personal knowledge" and "operational knowledge" with a weightage of 0.1403 and 0.1150. In this stage, the criteria for "technical knowledge", "theoretical knowledge" and "legal knowledge" is the most important criteria that must be addressed by the National Institute of Valuation (INSPEN) in handling the content of the training programme for the personnel of Valuation and Property Services Department (JPPH). The lowest weightage is "legal knowledge" of 0.0803 followed by the "theoretical knowledge" of 0.0831 and "technical knowledge" at 0.0961. Figure 7 below shows clearly the ranking sub-criteria for Level 2 (Opinion of Value).

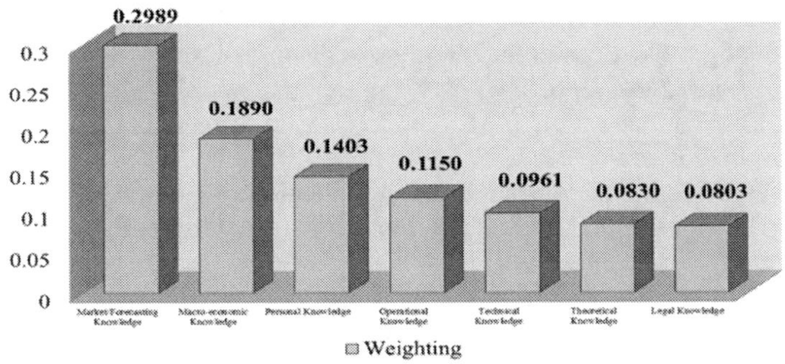

Figure 7. Ranking of Overall Sub-Criteria Weightage Level 2 – Opinion of Value

6. Conclusion

The findings of the study have proven all the objectives in this study and hence, have been achieved. Each of the criteria has been identified and determined on the relativity of the criteria weightage for each. The score is given for each type of knowledge involved and based on the value of the index of the quality of the ratings process which is obtained in the form of percentage.

All the features and criteria for the types of knowledge involved in the stages of the valuation process has to be adapted according to the needs of personnels of Valuation and Property Services Department (JPPH) in the Klang Valley area in improving the quality of the achievement of better knowledge in accordance with the requirements of the country, Malaysia, to generate international public service at the international arena.

References

[1] Syed Ikhsan S O S and Rowland F 2004 Benchmarking Knowledge Management in a Public Organisation in Malaysia Benchmarking An International Journal.

[2] Funk J B, Baldacci H B, Pasold T and Baumgardner J 2004 Violence exposure in real-life, video games, television, movies, and the Internet: is there desensitization? *Journal of Adolescence* **27** 23-39

[3] Krippendorff K 2013 Content Analysis. An Introduction to Its Methodology, 3rd ed., SAGE Publication.

[4] Duriau V J, Reger R K, Pfarrer M D 2007 A Content Analysis of the Content Analysis Literature in Organization *Organizational Research Methods* **10** 5-32

[5] D'Agostino, Louis A, Sovacool, Benjamin K, Trott, Kirsten, Ramos, Catherine Regalado, Saleem, Saleena and Ong Y 2011 What's The State of Energy Studies Research? A Content Analysis of Three Leading Journals From 1999 to 2008 *Energy* **36** 508-519

[6] Altaweel M and Bone C 2012 Applying content analysis for investigating the reporting of water issues *Journal of Computers, Environment and Urban Systems* **36** 599-613.

[7] Hsieh H F and Shannon S 2005 Three Approaches to Qualitative Content Analysis *Qualitative Health Research* **15** 1277-1288.

[8] Lyons M 2007 Place-shaping: A Shared Ambition For The Future of Local Government

[9] Bates R A 2001 Public sector training participation: An empirical investigation *International Journal of Training and Development* **5** 134-150.

[10] Yahaya A, Yahaya N, Ismail J, Sharif Z, Saud M S, Mohd Kosinin A, Yahaya N and Abbas F 2011 Effects of personality trait, motivation factors on performance of customer service personnel (CSP): A case of MPH bookstore *African Journal of Business Management* **5** 4519-4530.

[11] Saaty T L 2001 Decision Making with Dependence and Feedback: The Analytic Network Process (Second ed.), Pittsburgh, USA: RWS Publications, 4922 Ellsworth Avenue, Pittsburgh, PA 15213 USA.
[12] Ismail and Buyong 1998 Residential Property Valuation Using Geographic Information System *Buletin Geoinformasi* **2** 249-266.
[13] INSPEN 2014 Latihan dan Akademik, Pusat Penilaian & Perkhidmatan Harta, http://www.inspen.gov.my/
[14] Nawawi A H 2005 Computer Assisted Rating Valuation of Commercial and Industrial Properties in Malaysia: Developing and Expert System from a Multiple Experts Knowledge Elicitation Methodology *Built Environment Journal* **2** 17-26.
[15] Blanchard O 2011 Macroeconomics Updated (5th ed.). Englewood Cliffs: Prentice Hall. ISBN 978-0-13-215986-9. A Concise Introduction to Logic by Patrick J. Hurley. 2007. Cengage learning. Entry on "Theoretical Definitions" may even be available through google books.

Acknowledgement
The authors would like to acknowledge the support given by the LESTARI grant, Universiti Teknologi MARA, File Number: 600-IRMI/MyRA 5/3 LESTARI (K) (235/2017) during the carrying out of this research.

IOP Conf. Series: Earth and Environmental Science **385** (2019) 012027 doi:10.1088/1755-1315/385/1/012027

Temporal Changes of Mangrove Distribution In Mukim Kuala Selangor Using GIS Approach

N I Hashim[1], N F Zaharin[1], N H Hashim[1] and M Abdul Halim[1]
[1]Centre of Studies for Park and Amenity Management, Faculty of Architecture, Planning and Surveying, Universiti Teknologi MARA, 40450 Shah Alam, Selangor, Malaysia
[2]Centre of Studies for Surveying Science and Geomatics, Faculty of Architecture, Planning and Surveying, Universiti Teknologi MARA, 40450 Shah Alam, Selangor, Malaysia

Email: nuridzhaineehashim@gmail.com

Abstract. Mangrove forests are commonly found along coastal line where they grow in saline soil and brackish water. With the advantage of new technology, mangrove distribution information can be studied by using Geographic Information System (GIS) technique. This study is to analyse the mangrove distribution changes using GIS over Mukim Kuala Selangor, Malaysia of years 2014, 2016 and 2018. Through this study, area selection technique from Landsat-8 OLI/TIRS data was used in image processing. Supervised and unsupervised classification were applied in image processing along with the maximum likelihood classification. The findings of this study show that there were changes occured essentially on mangrove area at the study area. Therefore, it is hoped that this study can help stakeholders for the management of mangrove distribution as well as the other land cover which is vital to provide ecological balance and essential services for the community.

1. Introduction

Mangrove is a species of trees and shrubs that are salt tolerant and commonly found in tropical climate country along intertidal zones and coastal area [1]. Mangrove gives huge contribution to the environment especially towards the marine life and it also gives protection towards land from the flood, erosion due to harsh tidal currents and sea waves like tsunami [2]. Mangrove grow lavishly in the brackish water and saline soil together with water inundation. They are typically found along the coastal area where plants are able to adapt with saline water [3].

In order to face with increasing of human population that persistently changes the earth, many parties like researchers, planners and land managers are striving to monitor the continuous loss of living organisms and ecosystems [4]. Therefore in Mukim Kuala Selangor, GIS technique was used by the researcher to monitor and manage the mangrove ecosystem as well as the mangrove distribution changes. This study used high resolution satellite imagery to detect latest mangrove area changes from the year of 2014, 2016 and 2018.

Mangrove ecosystems have a remarkable significance in terms of ecological and economical. Hence, there is a vital need for conservation and restoration measures. Therefore, gaining up-to-date data with regard to the range and condition of mangrove ecosystems is fundamental to management and decision-making processes. In Malaysia, there is a need in monitoring mangrove using GIS since the data obtained from this technique in Malaysia is still limited [5]. This paper analyses and discuss the process of using satellite images and the details that can be extracted such as detection of land cover changes and vegetation density.

2. Methodology

2.1 Study area

The precinct of Kuala Selangor is one of the areas in Kuala Selangor District and is located 64 km from the city of Kuala Lumpur. It is located between 3° 10′N to 3° 34′ N and between 101° 6′ E to 101° 30′ E (refer Figure 1). It covers an area of 21.6 km² [6]. Natural resources in Selangor are impending under pressure since the development and growth of Selangor has widen into its surrounding districts. However, Mukim Kuala Selangor still has several patches of mangroves. One of the famous mangrove conservation area in Mukim Kuala Selangor is Kuala Selangor Nature Park (KSNP).

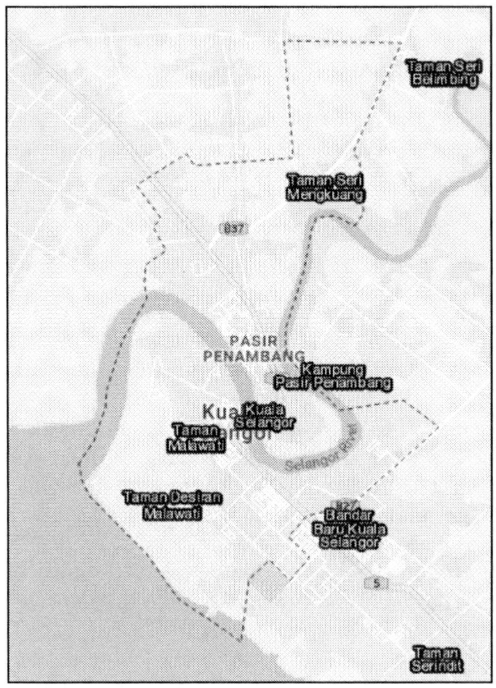

Figure 1. Study area. Source: JPBD Kuala Selangor [7]

According to [8], Kuala Selangor Nature Park (KSNP) is said to be the first site in Selangor to be designated as a Ramsar site (Convention on Wetlands) as the society analysed the criteria set by Ramsar and found that the park could qualify for the recognition. Hence, there is a need for hard data aimed at the area and its surrounding area in order to strengthen the argument of the analysis.
Agriculture especially oil palm plantation is the other major land use in the study area. The river in Mukim Kuala Selangor is Selangor River that supplies water for domestic, agriculture and fishing industries. Most areas in Selangor River especially in its upstream reaches is still in natural state. It also serves for many recreational activities such as river tubing, whitewater rafting, kayaking and one of the famous activities is watching fireflies along the Selangor River. With the benefits it gives, mangrove forests in Kuala Selangor are highly valuable for wildlife conservation and ecotourism.

2.2 Sources of imagery

The Landsat-8 OLI/TIRS data (path/row 127/58) which was taken in March 2014, March 2016 and May 2018 respectively was aqcuired for this study. Satellite images has panchromatic and multi-spectral band sensor but to be specific, there are seven multi-spectral bands for Landsat [5]. Landsat data which has resolution of 30 m for multi-spectral bands contained in path 127, row 58 which have 7 multi-spectral bands (blue, green, red, infrared, two short-wave infrared and thermal) was

IOP Conf. Series: Earth and Environmental Science **385** (2019) 012027 doi:10.1088/1755-1315/385/1/012027

downloaded from United State Geological Survey (USGS) [5]. Landsat-8 OLI/TIRS was used due to the quality of satellite data which is medium to high spatial resolution. The satellite data was chosen because of the 3 images have <10% cloud coverage which is almost cloud-free over the study site. Table 1 shows the properties of satellite data used in the study in terms of the name of the satellite, spatial resolution and date of acquisition.

Table 1. Properties of satellite data used in the study

Satellite	Spatial Resolution	Date of Acquisition
Landsat-8 Operational Land Imager (OLI) and Thermal Infrared Sensor (TIRS)	30m multispectral 15m panchromatic	24/03/2014
Landsat-8 Operational Land Imager (OLI) and Thermal Infrared Sensor (TIRS)	30m multispectral 15m panchromatic	29/03/2016
Landsat-8 Operational Land Imager (OLI) and Thermal Infrared Sensor (TIRS)	30m multispectral 15m panchromatic	06/05/2018

2.3 Image processing

Image processing was started from geometric correction and it is important for georeferenced setting of the satellite image to ensure that the coordinate of the places is at the correct position. The importance can be seen when the data is transferred to different software such as from ArcGIS to ERDAS.

When the geometric correction was done, doing a radiometric correction was the next step. This type of correction is vital as to enhance the satellite image visualisation and for the satellite image to become easier to interpret.

2.4 Area selection

The elements of the visual interpretation have been used in the process of selecting the training area. One of the example is mangrove area has been selected through its habitat located nearer to the water and shows same density pattern.

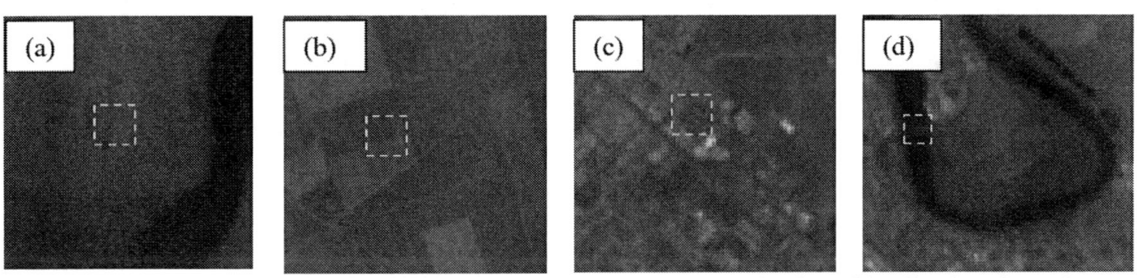

Figure 2. Selection for different types of area. (a) Mangrove selection. (b) Oil palm plantation selection. (c) Urban and associated areas selection. (d) Water selection.

Additionally, another criteria that was looked for the study is the texture of the mangrove. Mangrove texture is seen to be rougher than texture of the water and the soil. The tone which is also included in the combination of mangrove pattern is different from other elements whereby its tone colour is darker than other types of vegetation and makes it clear to analyse the differences as well as to identify the mangrove area. Apart from mangrove plantation, oil palm plantations were also detected and usually located far from the water bodies such as the river and the sea.

2.5 Image classification

Image classification is important to classify the features on the satellite image. In this study, supervised classification process were conducted where the classification interpret familiar element with the training area or location of the satellite image.

Supervised classification requires the pixel to be grouped into classes whereby the interpreter need to do the selection for each of the training area. To do that, hundreds of polygons need to be drawn and the greater the number of polygon drawn, the higher the accuracy of the group classification. However, the pixel group can result to false information due to errors. For example, some of the oil palm plantation classes can be accidentally included in the mangrove area. Other example, the presence of cloud can make the cloud class becomes urban class or other types of class because the pixel value from the cloud has same pixel value with the building in urban area. The value also might be same with other vegetation and caused pixel group belongs to other classes that is not related [9]. Figure 3 shows how the polygon is being drawn in order to classify the image.

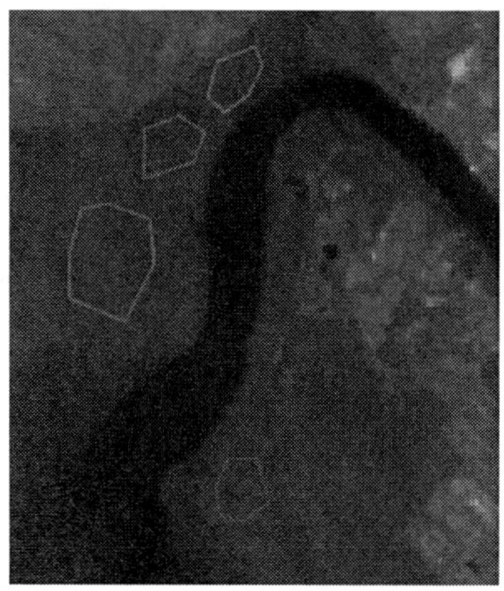

Figure 3. Drawing polygon to classify image.

In this study, maximum likelihood supervised algorithm has been use as type of classification. Maximum likelihood supervised algorithm is an algorithm that assigned the pixel to a classification that has probability own by some classes and through this process, the probability of the density function is determine by computing the probability of the pixel that belonging to each class [9]. In the meantime, the unclassified pixel during the classification will be assigned to the group based in relative likelihood of the pixel occurring within each class probability density function [9].

This paper has been described extensively on research methodology that are relevant to the study of temporal changes of mangrove forests in Kuala Selangor. Therefore, this research has clarified the area and method used in order to identify the changes on the mangrove distribution. These methods represent description of study area, sources of imagery, image processing, area subset and image classification. The interaction between elements in the aspect forms its basis and paves the way to produce land use or land cover map which shows mangrove distribution changes. Figure 4 shows the flow chart of the study which describe the methodology used during the process of this research.

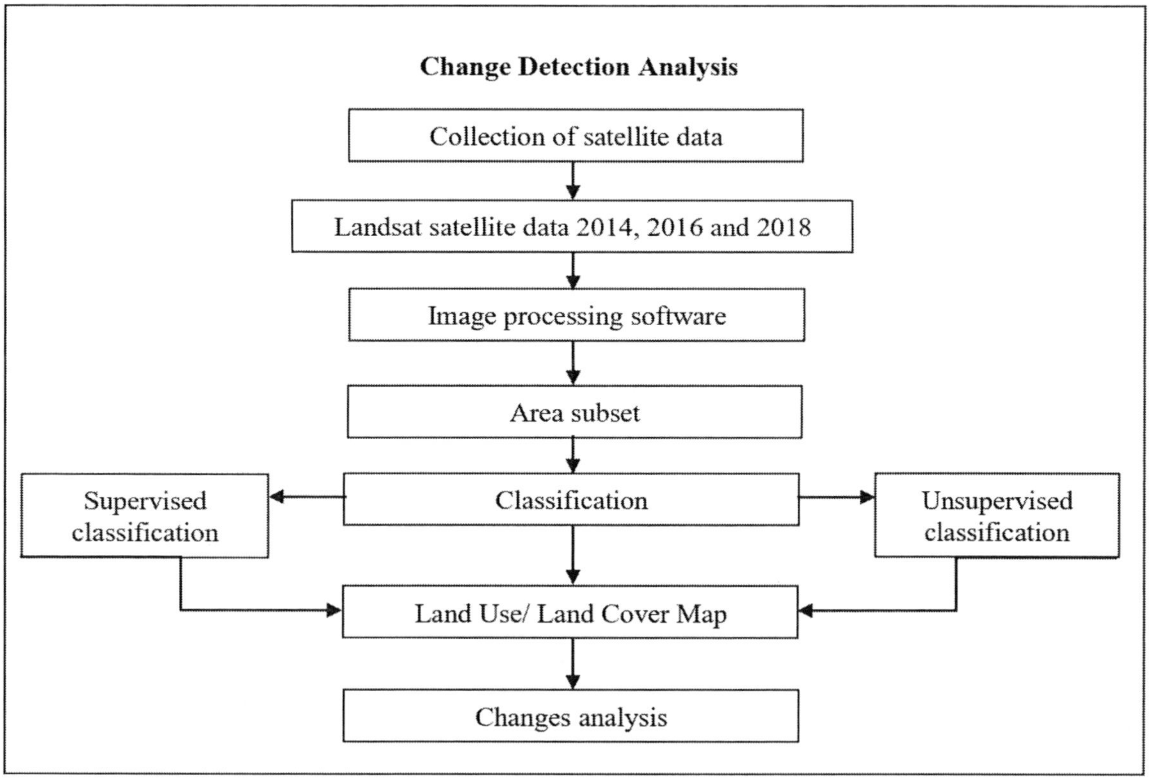

Figure 4. Flow chart of the study

3. Result and Discussion

Mangroves are usually darker in colour on any combination of spectral bands of multispectral image due to the natural ecosystem of mangroves, which is covered by swamps and sometimes by tidal water. In addition, the chlorophyll content of the mangrove leaves is higher than those of trees and crops which make them appear darker on satellite images as shown in Figure 5.

Figure 5. Images showing combination of bands 5, 6 and 4 of Landsat-8 OLI/TIRS

IOP Conf. Series: Earth and Environmental Science **385** (2019) 012027 doi:10.1088/1755-1315/385/1/012027

In order to understand how a mangrove ecosystem changes, information is needed on what changes occured, where and when they occured as well as the rates at which they occured. Despite ongoing research efforts on ecosystem change patterns, there is a need for research on gaining datasets providing quantitative, spatial land-cover information [10].

The method applied to estimate the mangrove ecosystem changes was by conducting image processing on satellite image maps to detect changes occurred on land cover and land use in 2014, 2016 and 2018. An assessment for accuracy was carried out and the result for the assessment is 80%. This assessment result meets the 80% accuracy standard of the National Park Service (NPS) in the United States that showed the percentage is a good accuracy value for land cover classification [3].

Figure 6 shows the final data that indicates the land use classification and distribution including 4 types of elements which is mangrove forest, oil palm plantation, water bodies, urban and associated areas. The colour used in the final data for each element was referred from the Classification and Land Use Colour Code According to Category 2013 which was published by The Department of Survey and Mapping Malaysia (JUPEM).

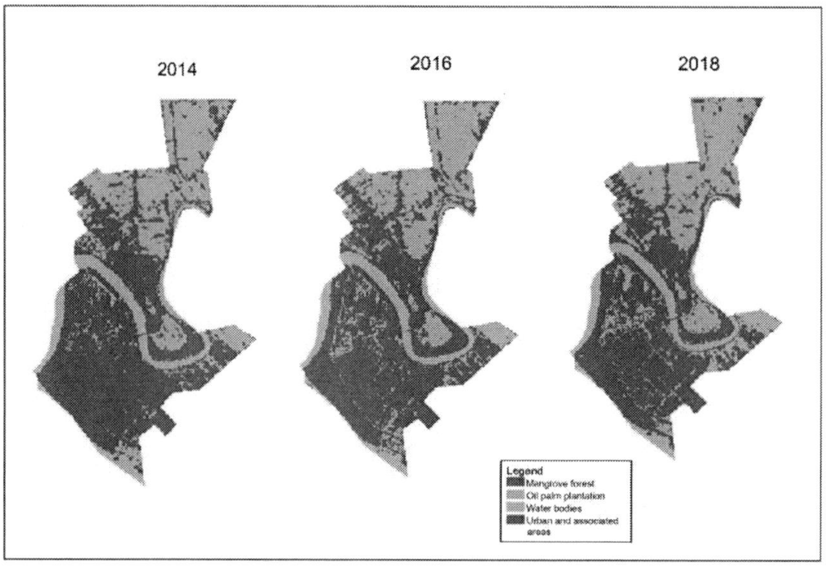

Figure 6. Landuse classification in Mukim Kuala Selangor for year 2014, 2016 and 2018

Based on the image analysis, the values of rate of change in land use were found to range from -2.3% to 5.2%. The most affected land use type was mangrove forest and oil palm plantation, which decreased from 5.1 km^2 in 2014 to 4.4 km^2 in 2018, and increased from 6.5 km^2 in 2014 to 8.0 km^2 in 2018, respectively. The decreasing of mangrove area is due to illegal land clearing mainly for oil palm plantation purpose and it is also a strong reason for the increasing of oil palm plantation area. The other land use types also experienced changes during the 5-year period. Water bodies has no significant changes but increased only from 2.2 km^2 to 2.3 km^2 for 2014 to 2016 and decreased from 2.3 km^2 back to 2.2 km^2 in 2018. Urban and associated area decreased from 7.7 to 7.5 and from 7.5 to 7.0. The changes is caused by the presence and absence of cloud and some error caused from the image classification process. Table 2 shows the statistical analysis of the land cover changes for the year of 2014, 2016 and 2018 which indicates data of area cover and image percentage according to each class element.

Table 2. Statistical analysis of the land cover changes for the year of 2014, 2016 and 2018

	2014 Landsat map		2016 Landsat map		2018 Landsat map	
Class Name	Area cover (km^2)	Image (%)	Area cover (km^2)	Image (%)	Area cover (km^2)	Image (%)
Mangrove forest	5.1	23.7	4.9	22.7	4.4	20.4
Oil palm plantation	6.5	30.2	6.9	31.8	8.0	37.0
Water bodies	2.2	10.4	2.3	10.6	2.2	10.0
Urban and associated areas	7.7	35.7	7.5	34.8	7.0	32.6

4. Conclusion

This paper gives an overview of the mangrove distribution changes study which has been done by using the GIS approach. Methodology of research which is a basis of strategy to gain final data has been used to monitor the area during the year of 2014, 2016 and 2018. The information about land use has become essential for Malaysia's plan in order to overcome problems such as uncontrolled development, destruction of important wetlands and loss of wildlife habitats. Based on other previous studies, the GIS has proven to be powerful tools for landuse analysis based on the existing natural resources.

Other than that, the study has determined that the mangrove distribution changes can be analysed by using landsat sources of imagery and area selection technique for image processing. These techniques were popular and commonly used for land use especially for conservation purpose. Kuala Selangor can be further planned for activities related to its unique mangrove forests as the area has high potential value for wildlife conservation and ecotourism.

Acknowledgement

We would like to express our deepest appreciation to Univesiti Teknologi MARA (UiTM) Malaysia for the financial support for both research and publication through the internal funding of LESTARI Grant Schem (600/-IRMI 5/3/LESTARI (014/2018). A great thanks to all staffs and members of Research Management Institute (RMI) UiTM Shah Alam, Selangor, Malaysia.

References

[1] Jensen, R., Mausel, P., Dias, N., Gonser, R., Yang, C., Everitt, J. & Fletcher, R. (2007). Spectral analysis of coastal vegetation and land cover using AISA+ hyperspectral data. Geocarto Int., Vol. 22. Pg. 17-28.
[2] Mcivor, A., Möller, I., Spencer, T., & Spalding, M., (2012). Reduction of Wind and Swell Waves by Mangroves. Nat. Coast. Prot. Ser., Pg. 1–27.
[3] Ibrahim, N. A., Mustapha, M. A., Lihan, T., & Mazlan, A. G. (2015). Mapping mangrove changes in the Matang Mangrove Forest using multi temporal satellite imageries. Ocean and Coastal Management, Vol. 114, Pg. 64-76.

[4] Rubino, M.J. & Hess, G.R. (2003) Planning open spaces for wildlife 2: modeling and verifying focal species habitat. Landscape and Urban Planning. Vol. 64. Pg. 89–104.

[5] Beh, B.C., MatJafri, M.Z. & Lim, H.S. (2012). Temporal Change Monitoring of Mangrove Distribution in Penang Island from 2002-2010 by Remote Sensing Approach. Journal of Applied Sciences. Vol. 12 (19). Pg. 2044-2051.

[6] Laporan Profil Daerah (n.d.). Jabatan Pengairan dan Saliran Daerah Kuala Selangor.

[7] JPBD Kuala Selangor (n.d.). Sistem Maklumat Perancangan Negeri Selangor. Available online at http://sismaps.jpbdselangor.gov.my/sismapsv2/PenggunaAwam/RTKualaSelangor/viewer/

[8] Chen, G. (2018). Kuala Selangor out to get Ramsar listing for wetlands. The Star Online. Available online at https://www.thestar.com.my/metro/metro-news/2018/10/26/kuala-selangor-out-to-get-ramsar-listing-for-wetlands

[9] Samad, A.M., Iliyas, N.S., Sahriman, N. & Ruslan, F.A. (2017). Mangrove Area Detection by Using High Resolution Satellite Imagery. 2017 IEEE 13th International Colloquium on Signal Processing & its Applications (CSPA 2017. Pg. 293-298.

[10] Xavier, B. & Szejwach, G. (1998). LUCC Data Requirements Workshop: survey of needs, gaps and priorities on data for land-use/land-cover change research (Barcelona, Spain: Institut Cartogra. c de Catalunya).

Malaysian Urban Residents' Readiness to Reduce Car Usage and their Perception on Non-Motorised Facilities.

N Nasrudin[1], M A Marzukhi[1], Y A Abdullah[1], N S Khalid[1] and M Abdul Wahab[2]
[1]Centre of Studies for Town and Regional Planning, Faculty of Architecture, Planning and Surveying.Universiti Teknologi MARA, 40450 Shah Alam, Selangor, Malaysia.
[2]Centre of Studies for Estate Management, Faculty of Architecture, Planning and Surveying.Universiti Teknologi MARA, 40450 Shah Alam, Malaysia.

Email: naasah76@gmail.com

Abstract. The research in this paper tries to establish the readiness of urban residents' in Malaysia to minimize the car usage and their inclination to use non-motorised transport, such as cycling and walking, as alternatives to using a car for commuting. This study also tries to measure the perception of the residents regarding the existing scenario of walking and cycling infrastructure facilities provided in their cities. The data used in the study was collected through a survey in three cities, namely-Shah Alam, Batu Pahat, and Putrajaya, as a representation of the main cities in Malaysia. The results produced suggested- the majority of the population was not ready to give up travelling via car and take up cycling and walking in lieu of it. The causes listed for "refusal" of reducing the car usage included hot weather, surrounding safety concerns and undesirable pedestrian and cycling infrastructure conditions. The research explored that attitude, lifestyle and high level of dependency on cars are crucial factors dictating the travel behaviour, which is the reason why residents refuse to shift to sustainable and environment friendly options of travel. The research also suggests that these three cities need to improve and uplift their walking and cycling infrastructure facilities, especially in terms of provision of cover and shade, connectivity, design and maintenance and cleanliness to motivate the urban residents to minimize usage of the cars and take up walking and cycling instead.

1. Introduction

Being completely dependent on the car, as a mode of travel, has become the current nature of commutation because of the lifestyle and stature of the society, which inculcates the need for more comfortable travelling choices. The flexibility offered by cars in accessing multiple destinations, that too right at the doorstep, makes them more reliable, convenient, and safe in comparison to public transport. These factors influence the desire of residents to own purchase and a car [1]. In 2012, Malaysia, a country of 5 million households, had a registration of more than 20 million vehicles, which posed a serious challenge and a threat for the country to tackle. [2]. According to a 2014 press release by Nielsen (market research agency), Malaysia is a country which has the third highest car ownership in the world where 93 percent of the households possess a car. Increased dependency on cars, and other vehicles on the road, contributes to rising traffic congestion and constant degradation of the environment because of high pollution levels emitted by them.

2. Car Dependency

Car dependency occurs when the city and its inhabitants decide to take up commuting through a car, as a priority over public transportation, infrastructure, and land use. The factors influencing people's dependency upon car are the high levels of per capita automobile travel, automobile-oriented land use. on cars instigates the traffic congestion and degradation of environmental quality caused by continuous emission of pollutants from the rapid increase of vehicles on the road. Numerous studies have proven that some people like to drive out of their own will and choice but not need. [3] Cars affixed with various features provide humans with psychosocial value, which affects the human psychology to choose a car over other modes of transportation. To encourage the commutation via more sustainable transport modes, reduction in the pattern of car dependency is needed. To outstrip the -unrestrained reliance on motorised vehicles, the Malaysian government traversed the five-pronged transportation policy with contemporary laws. This policy allows the Malaysian government to spend road tariff, as a means to fund other issues and abolition of fuel subsidies will lessen the government's expenses. Fuel, cars, congestion, and road taxes were proposed, in anticipation of decreasing motorised vehicles, resolving congestion and pollution issues [4].

The previous studies on sustainable transportation have identified numerous elements that influence the degree of non-motorised travel in the city as demographic and socio-economic characteristics, trip characteristics, environmental factors, and their viewpoint and perceptions.[5] From an emotional viewpoint, which alternate should one choose to travel can be perceived as either an intentional activity, or begin from behavioural habits.[6] Habit has an automated link between an individual's goal and their behaviour. It is believed that a strong habit can supposedly block intentional processing; even before behaviour. Before a habit evolves, the travel behaviour is most likely to be aligned with an individual's intentions and personal preferences. However, once a strong habit to commute via car has developed, then the motivation to reduce it (e.g. following awareness campaigns to understand the problems linked with motorised travel) will not influence travel pattern and behaviour much, because the individuals are accustomed to behave in alliance with their habits and not their motivation. Thus, an existing strong habit of using cars to travel can be a hurdle to motivate people for reducing car usage [7].

3. Methodology

The three cities, namely Shah Alam (the first planned city in Malaysia), Putrajaya (the new planned administrative city) and Batu Pahat (old city) were selected for this research and survey, to represent urban residents in Malaysia, by using stratified random sampling technique. The questionnaires were prepared and taken to numerous selected houses along with approaching the respondents at various picked recreation and shopping centres in each of the three cities mentioned above. The respondents were provided with a survey form, which had various tests to respond to, to help us understand the extent of car usage by the selected respondents. Numerous factors determining the degree of car dependency of the respondents were identified in the form of the number of cars owned; the emotion and perception of owning and driving a car; the level of readiness to reduce car usage and shift to non-motorised modes and the evaluation of walking and cycling infrastructure facilities.

4. Results

4.1 Number of cars owned in a family

Table 1 indicates that the average number of cars owned by each family was two in Shah Alam and Batu Pahat, and three in Putrajaya. These figures depict that virtually every family, in these cities, has a car to commute and every family has more than one member with a legal driver's license.

Table 1. Total Number of cars owned by families in each city

Number of cars owned in the family	Shah Alam	Putrajaya	Batu Pahat
1	40.4%	1.7%	48.1%
2	41.4%	21.4%	40.4%
3	9.4%	42.8%	10.9%
4	5.7%	9.8%	0.6%
5	1.8%	5.6%	-
More than 5	1.3%	1.1%	-

4.2 Car Dependency Ratio

As per the calculations derived from the Littman's formula, Shah Alam and Batu Pahat cities indicate a mediocre dependency on cars, whereas on the other hand, Putrajaya City has a high car dependency. Table 2 depicts the comparison of car dependency in these three cities (Shah Alam, Putrajaya, Batu Pahat), compared to Malaysia's average ratio. The growth in the number of vehicles in the country has been 3.3 times faster than the growth in the population. [8]The traffic on the road has increased tremendously over the past few years because most of the households today have access to two or more cars.

Table 2. Car Dependency Ratio for Shah Alam, Putrajaya and Batu Pahat Cities

Car Dependency Level	Shah Alam	Putrajaya	Batu Pahat
Car : Population	1:2.2	1:1.5	1:2.6

Note: low dependency- 1:4 or higher, medium dependency- 1:4-1:2.2, high dependency-1:2.2 or lower

4.3 Emotions and perceptions towards cars

In the questionnaires, when respondents from these cities were asked about their experience while driving cars, the majority of them responded with a positive feedback. Most of them considered driving a car as "relaxing" and "safe". The Majority of them also stated that driving a car was practical and gave them liberty.Hiscock et al., established a positive car perception from his research, quoting that a car is seen as something that offers convenience, reliability and capability in providing easy access to multiple destinations as compared to public transport and can provide security from unwanted people, events and happenings [9].

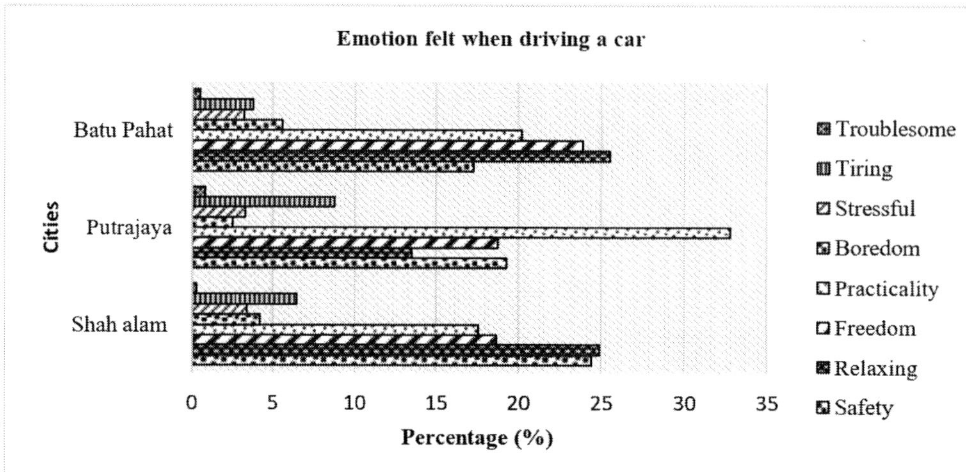

Figure 1. Emotion felt while driving a car

4.4 Willingness to reduce car usage

Upon being asked regarding the willingness of these respondents to reduce the car usage, the majority of them from these three cities gave a negative feedback with the highest percentage of people being 'unready'. However, some people who responded with a positive feedback for being 'ready' to reduce car usage, gives a ray of hope for these cities (Shah Alam and Putrajaya) to switch to more sustainable modes of transport.

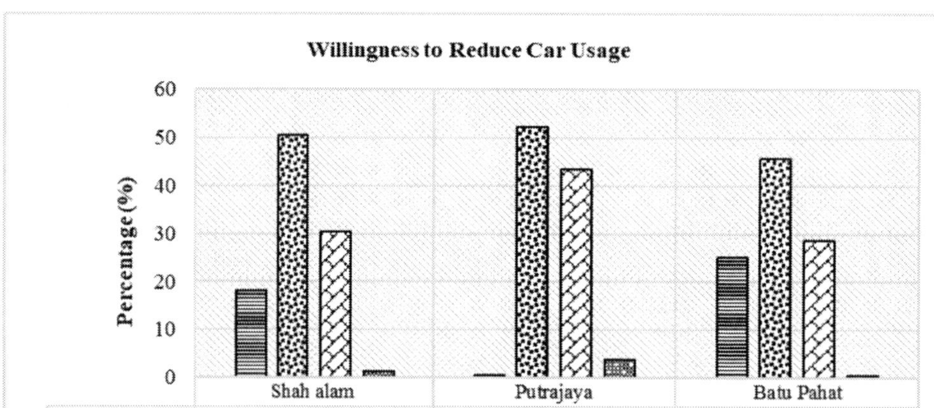

	Shah alam	Putrajaya	Batu Pahat
Very Unready	18	0.55	25
Unready	50.5	52.2	45.8
Ready	30.3	43.45	28.6
Very Ready	1.2	3.8	0.6

Figure 2. Willingness to reduce car usage

4.5 Reason for being unwilling to reduce car usage

The majority of the respondents from the above-mentioned three cities stated that they are not ready to reduce car usage because they did not prefer to walk or cycle in the hot weather. They also considered walking as "exhausting, tiring" and "not safe." According to them, driving a car was more convenient than walking. "No proper pedestrian walkways" contributed to being one of the reasons why people refused to walk. These findings are similar to Rose and Marfurt (2007), who revealed that distances and other aspects such as weather conditions, physical abilities and safety issues often influence an individual's perception [10]. McDonald's, in his research, believed that humans normally follow their norms in terms of travelling pattern [11].

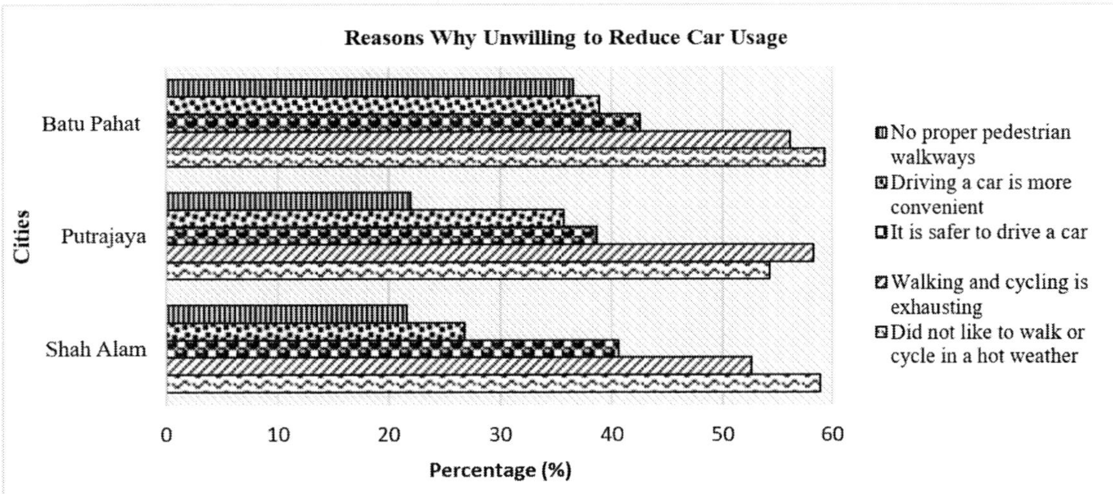

Figure 3. Reason why not willing to reduce car usage

4.6 Readiness to walk and cycle

This research also investigates and evaluates if Malaysian urban residents are ready to use non-motorised transport. Through this study it is found, that the majority of the people are not ready to cycle or walk in lieu of using motor vehicles. However, in case of Putrajaya and Shah Alam cities, the majority gave a positive feedback with a higher percentage of respondents willing to walk and cycle in comparison to Batu Pahat city respondents. This finding could be related to Cao et al. (2006), who suggested that residents of a "New Urbanist" area are more willing and able to commute via public transport, walk, or cycle to their destinations, owing to their personal principles and philosophies on transport and environment [12]. Lensink also concluded that achieving a more sustainable transportation system needs to pay more focus on the inter-relations between the infrastructure planning, traveller's decision behaviour and the transportation energy used in government transportation policies [13].

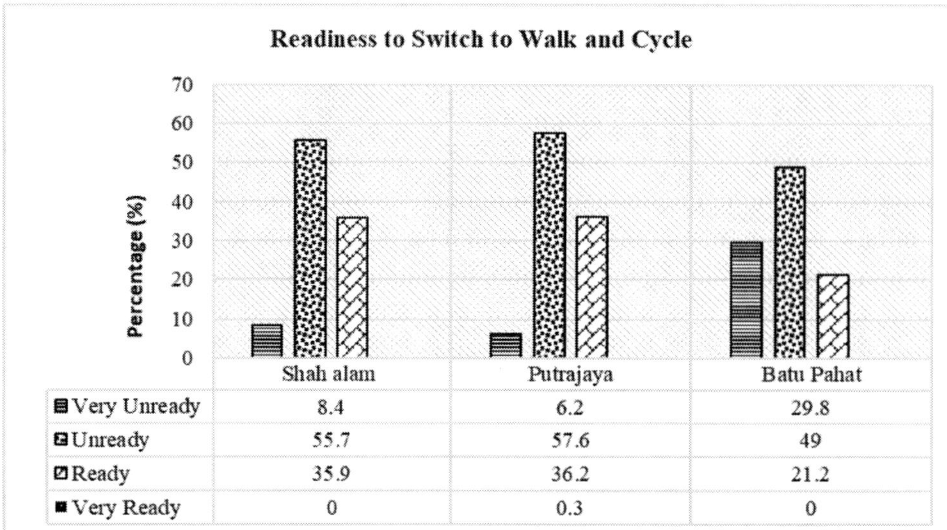

Figure 4. Readiness to switch to walking and cycling

4.7 Existing Condition of Walking and Cycling Facilities

Table 3 below shows residents' satisfaction level about the existing condition of walking and cycling infrastructure facilities available in their cities. The responses received from urban residents ranged from poor to moderate level. This indicates that these three cities have to improve their walking and cycling infrastructure, especially in terms of covers and shades, connectivity, design, and maintenance and cleanliness to motivate urban residents to walk and cycle in order to reduce car usage.

Table 3. Existing Condition of Walking and Cycling Facilities

	Existing Condition of Waking Facilities	Mean Score for Each Cities		
		Shah Alam	**Putrajaya**	**Batu Pahat**
WALKING	Design	2.41	2.75	2.94
	Width	2.85	3.12	2.96
	Design Material	2.33	3.13	2.91
	Separation of lane	2.89	3.24	2.75
	Landscape	2.45	3.35	2.83
	Covered and shades	1.91	2.33	2.78
	Trash bin	2.64	3.28	3.12
	Street light	2.53	3.51	3.04
	Benches	2.55	2.84	2.97
	Connectivity	2.37	2.32	2.61
	Maintenance & cleanliness	2.82	3.01	2.77

	Existing Condition of Cycling Facilities	Mean Score for Each Cities		
		Shah Alam	Putrajaya	Batu Pahat
CYCLING	Design	2.34	2.91	2.55
	Width	2.58	2.54	2.52
	Design Material	2.12	2.72	2.50
	Separation of lane	2.27	2.46	2.44
	Landscape	2.33	2.91	2.79

Covered and shades	2.18	2.38	2.41
Trash bin	2.52	2.72	2.64
Street light	2.75	2.62	2.55
Benches	3.31	3.69	2.63
Bicycle parking	2.31	2.54	2.45
Connectivity	2.72	2.31	2.41
Maintenance & cleanliness	3.11	3.56	2.54

Note : 1.00-1.49 =Very poor, 1.50-2.49 = Poor, 2.50-3.49 =Moderate, 3.50-4.00= Good 4.50-5.00= Excellent

5. Summary

This research indicates that the majority of people in these three cities are not ready to reduce their car usage and to consider cycling or walking as alternatives to using a car. The reasons provided with for "refusal" to reduce car usage were more associated with their attitudes and perceptions like- they did not like to walk or cycle in a hot weather, they had safety concerns and unsatisfactory conditions of pedestrian and cycling infrastructure facilities. According to them, driving a car was more convenient than walking and cycling, which was a reflection of their habits. This study eventually reveals that attitudes and lifestyle of people in these cities are important determinants of travel behaviour and high level of car dependency is the key factor why they refused to shift to more sustainable modes of travel. According to Stern, attitudinal factors (such as values, personal norms and attitudes), habits, personal resources, and contextual factors are important for environmentally significant behaviours [14]. Thus, it is not unanticipated that car habits, attitudinal factors such as moral motivation analysed in this study, explicated only a small amount of willingness in reducing car use. In order to reduce car dependency, personal resources, the time and knowledge to transform travel behaviour as well as contextual factors, such as the availability of alternate travel modes, supporting social norms and policy strategies to influence the sustainable transformation may be critical.

This study also suggested that these three cities have to improve their walking and cycling infrastructure facilities, especially in term of covers and shades, connectivity, design, and maintenance and cleanliness to encourage urban residents to walk and cycle to discourage the existing habit of car usage. Advertising campaigns, with the aim of increasing more sustainable transport usage, needs to focus on the environmental and health benefits attained by using the sustainable modes such as tagging walking, cycling and public transport as environmental symbols; Hence, opposing the status symbol of cars.

6. Recommendations

Malaysian citizens should have mental and physical readiness aligned with the goal of implementing an environment friendly sustainable transportation system. The importance of protecting the environment by using sustainable vehicles should be strengthened through various programs, campaigns, advertisements and promotions. A well-coordinated, joint campaign by local authorities and non-government institutions such as public transport corporations, health organizations, and environmental lobby groups could help in influencing the attitude of citizens towards car ownership and its usage amongst the next generation of potential drivers and may contribute towards restraining the demand for car travel.

Acknowledgments

The authors would like to thank the Ministry of Higher Education, Malaysia for funding this research through the FRGS grant 600-IRMI/FRGS 5/3 (46/2016) and Universiti Teknologi MARA (UiTM) for supporting the research.

IOP Conf. Series: Earth and Environmental Science **385** (2019) 012028 doi:10.1088/1755-1315/385/1/012028

References

[1] Anable J 2005 Complacent car addicts' or 'aspiring environmentalists'? Identifying travel behaviour segments using attitude theory *Transport Policy* **12** 65–78
[2] Ministry of Transport Malaysia 2012 Transport Statistics Malaysia 2012 Kuala Lumpur
[3] Handy S, Weston L and Mokhtarian P L 2005 Driving By Choice Or Necessity? *Transportation Research Part A* **39** 183–203
[4] Kasipillai J and Chan P Travel demand management: Lessons for Malaysia *Journal of Public Transportation* **11** 41-55
[5] Chandra R B, Jessica Y G and Rupali S 2005 Non-motorised travel in San Fransisco Bay Area, Texas *The University Texas, Austin*
[6] Garling T and Axhausen, K W 2003 Introduction: Habitual travel choice *Transportation* **30** 1–11
[7] Erisson L, Garvill J and Nordlund A M 2008 Interrupting habitual car use: The importance of car habit strength and moral motivation for personal car use reduction *Transportation Research Part F* **11** 10-23
[8] Ministry of Federal Territories and Urban Wellbeing, 2011.
[9] Hiscock R, Macintyre S, Kearns A and Ellaway A 2002 Means of transport and ontological security: do cars provide psycho-social benefits to their users? *Transportation Research Part D: Transport and Environment* **7** 119–135
[10] Rose G and Marfurt H 2007 Travel behaviour change of major ride to work day event *Transportation Research Part A* **41** 351-364
[11] McDonald N C Critical factors for active transportation to school among low-income and minority students' evidence from the 2001 National Household Travel Survey *American Journal of Preventive Medicine* **34** 341-344
[12] Cao X, Handy S L and Mokhtarian P L 2006 The influences of the built environment and residential self-selection on pedestrian behavior: evidence from austin *Transportation* **33** 1 - 20
[13] Lensink S M 2005 Capacity building for sustainable transport, optimising the energy use of traffic and infrastructure *University of Groningen*
[14] Stern P C 2000 Toward a coherent theory of environmentally significant behavior *Journal of Social Issues* **56** 407–42

IOP Conf. Series: Earth and Environmental Science **385** (2019) 012029 doi:10.1088/1755-1315/385/1/012029

Seasonal Windstorm Pattern and Damages in Peninsular Malaysia 2018

N H Zakaria[1] , S A Salleh[1,2*], Arnis Asmat[3] , and M A Islam[4]

[1]Centre of Studies of Surveying Science and Geomatics, Faculty of Architecture, Planning and Surveying, Universiti Teknologi MARA, 40450 Shah Alam, Selangor, Malaysia.

[2]Applied Remote Sensing and Geospatial Research Group, Faculty of Architecture, Planning and Surveying, Universiti Teknologi MARA, Shah Alam, Selangor, Malaysia.

[3]Faculty of Applied Sciences, Universiti Teknologi MARA (UiTM), 40450, Shah Alam, Selangor, Malaysia.

[4]Faculty of Engineering, Department of Civil Engineering, University of Nottingham, Malaysia Campus, Jalan Broga, Semenyih, Selangor, Malaysia, 43500.

Email: aekbal@salam.uitm.edu.my & nur1812@gmail.com

Abstract Natural disasters cause damages, losses and social problems. Windstorm has been a major contributor to the losses and damages to the sustainability of human lives and built environment. The formation of seasonality in Southeast Asian countries are affected by the monsoon which brings large scale seasonal interchange of wind pattern. The trend of windstrom events is not properly recorded in any recent or previous studies.This paper analyses the windstorm pattern in Peninsular Malaysia. NCEP FNL Operational Model Global Tropospheric Analyses and Meteorological Operational (MeTOP) datasets were used to retrieve the wind speed and the direction during the study period. Both datasets show approximate value of wind pattern based on reported events. Southeast-monsoon (SM), Northeast-monsoon (NM) and Inter-monsoon (IM) impacts are considered in this study. The result exhibited the highest windstorm event was Southeast-monsoon, and the highest wind speed is from NS that is 5.96 m/s to 6.23 m/s. While the highest wind speed outside of peninsular that blows and impacts the land is from SM and the value calibrates in between 9.20 m/s to 10.96 m/s.

1. Background

Windstorm classified as tropical cyclone, thunderstorm, tornadoes, monsoon and gale [1]. The increase of windstorm event leads to loss of public and private properties and causes death. Centre for Research on the Epidemiology of Disasters (CRED) in Emergency Events Database (EM-DAT) reported that there is increased of total damage by storm events in 2017 as compared to the previous year [2]. Seasonal monsoon event takes place in existence with windstorm events. Malaysia experiences four types of monsoon known as Northeast Monsoon (NM) occurs from December to March, Southeast Monsoon (SM)occurs from June till September, Inter-monsoon (IM) from April to May and also from October to November [3]. The highest wind speed value during NM is recorded 15 knots during SM it is 10-20 knots [3].

Malaysian Meteorological Department (MMD) is the organization in Malaysia that responsible to deliver and alert the citizen about upcoming weather condition including windstorm events through news, social media and website. Malaysia currently does not have any complete database of windstorm occurrence and resulting damage [4]. The complete information about windstorm will benefits the society especially in planning the emergency response. as well as for future planning and development.

The windstorm event impact has cause big impact to the urban metropolitan society. 80% damages of roofing system in Peninsular Malaysia caused by windstorm event. Study in Pulau Pinang on 2010 to 2013 shows, the damages breakdown is about 47% damages targeted the steel sheet roofing, 30% the truss system, 13% the roof tiles and 20% other related component [5]. Figure 1 shows the number of houses damaged during windstorm events from 2010 to 2013 in five districts of Pulau Pinang. The district was divided to island portion and mainland. Northeast Penang Island (TL) and Southwest Penang Island (BD) in the island portion and three (Northern Seberang Perai (SPU), Central Seberang Perai (SPT) and Southeast Seberang Perai (SPS) on the mainland. During the period, the highest number of houses damaged recorded in SPU with 538 cases (47%) and the lowest number houses damaged in BD with 111 cases (10%) respectively. [5]

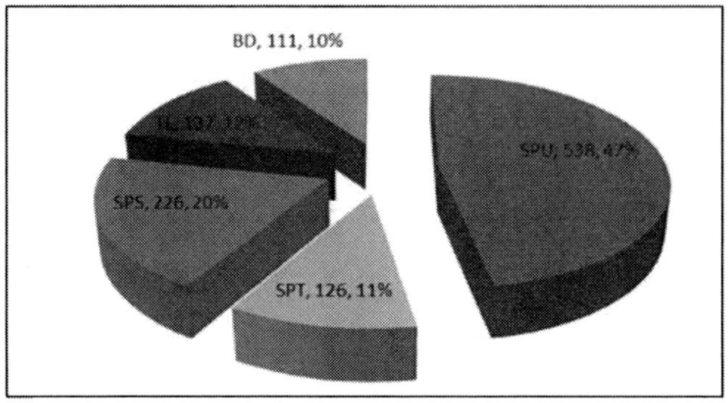

Figure 1. Total of house damage and percentage in Pulau Pinang from year 2010 to 2013

With the emergence of geomatics technology such as Meteorological Operational (MetOp), Geographical Information System (GIS) and numerical simulation Advanced Research Weather Research and Forecasting Model (WRF-ARW), the analysis about windstorm event has become reliable. The limitation of ground data to recording windstorm event in Malaysia can be resolve with the used of geomatics technology as mentioned before. MetOp data give wind speed and direction reading as well as WRF-ARW which is the data is approximately as the ground data. For analysis and illustrate the data, the use of GIS was implementing to mapping the wind speed and direction associates with the windstorm event data.

The wind speed and direction of wind was retrieve using NCEP FNL Operational Model Global Tropospheric Analyses (1° x 1°) that is under Global Data Assimilation System (GDAS). GDAS continuously collects observational data for required analyses [6] and parameters include surface pressure, relative humidity, geopotential height, temperature, sea level pressure, sea surface temperature, soil values, ice cover, u component (East-West) and v component (North-South) of winds, vertical motion, ozone and vorticity.

MetOp satellite carries the Advanced Scatterometer (ASCAT) onboard as its sensor is the lastest satelitte that provides wind data. MetOp is a satellite from the European Organization for the Exploitation of Meteorological Satellites (EUMETSAT). ASCAT is a C band radar, which is operated by measuring the wind field over global ocean [7]. The wind speed and wind direction are collected using C band radar in ASCAT on board of the MetOp platform [8].

Hence, this paper aim collecting and compile the data of windstorm event in Peninsular Malaysia on 2018. The data was compile based on the MetOp data was used to explore the wind speed and wind direction pattern from ocean into the land of Peninsular Malaysia. The use of numerical simulation was implementing to see the wind speed and direction in Peninsular Malaysia. These three results were analyzing to see the wind speed and direction compare to real event reported.

2. Methods

This study integrates MetOp data with numerical simulation to analyze the windstorm event in Peninsular Malaysia on 2018. Hence, the method and data use are WRF-ARW Simulation, ASCAT MetOp , Remote Sensing data extraction and collection of past windstorm event.

2.1 WRF-ARW Simulation

Table 1. Data LBC specification

Specification / Data	**NCEP FNL Operational Model Global Tropospheric Analyses, continuing from July 1999**
Data type	Grid
Data Format	WMO_GRIB1 WMO_GRIB2
Temporal Resolution	6 hours
Spatial Grid Resolution	1° × 1°
Data Coverage	From 0E to 359E and 90N to 90S

The dataset NCEP FNL Operational Model for Global Tropospheric Analyses which provides data from 1999 until now. Table 1 shows the specification of the lateral boundary condition (LBC) of the simulation settings. The datasets are employed due to their common applications in representing the wind speed and direction for simulation modelling [10][11][12]. The dataset used are based on monsoon event on NM (December to March), SM (June to September) and IM (April to May and October to November) for year 2018. The data is simulated for 1 km resolution to cover peninsular Malaysia.

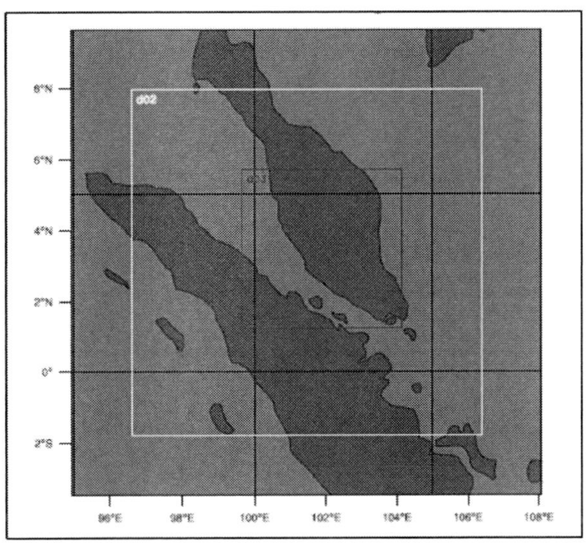

Figure 2. Domain design and coverage area.

The regional dynamic downscaling is conducted using Advanced Research Weather Research and Forecasting Model (WRF-ARW) version 3.8 nested domains are used with the grid size approximately 30 km, 10 km and 1 km the near surface wind. The domains are focused on Peninsular Malaysia. The largest domain cover Southeast Asia region and the smallest domain covers the Peninsular Malaysia. Figure 2 shows the domain covering the study area. WRF-ARW default physics scheme is used for this study.

2.2 ASCAT MetOp

This study uses datasets from MetOp satellite for wind speed and direction retrieval. The data is requested from EUMESAT official webpage. The spatial resolution MetOp datasets provide is 12.5 km x 12.5 km.

IOP Conf. Series: Earth and Environmental Science **385** (2019) 012029 doi:10.1088/1755-1315/385/1/012029

ASCAT is an active microwave sensor and its main operational objective is designed to detect surface wind speed and wind direction over global ocean with the consideration of measured winds speed with accuracy less than 2 m/s [7]. Besides that, the data is available in real-time with a correctness retrieval enhanced within 3 hours and below for assessment.

Moreover, ASCAT data gives information about eastward and northward wind, wind speed, surface downward, eastward and northward stress, wind stress, sampling length and land ice mask. The characteristics of ASCAT, MetOp satellite is shown in Table 2. MetOp-A stopped functioning after October 2006 and later replaced with MetOp-B launched in 2012 working till now.

The ASCAT data is provided with multiple variables in the form of Network Common Data Form (netCDF) data type. The variables are added as fields in feature attribute table. For this study, wind speed data was used to retrieve the average wind speed. Meanwhile, eastward wind and northward wind data was used to retrieve wind direction. The data was processing to retrieve wind speed and wind direction. Theorem Pythagoras's equation stated in later part of this article is applied to produce the final output.

Table 2. MetOp data specification

Satellite	**ASCAT satellite MetOp** **a) MetOp-A- launched 19 October 2006** **b) MetOp-B launched 17 September 2012 till now**
Product Name	Wind speed data
Data Format	Grid
Data Type	NetCDF-CF
Accuracy	RMS error = > 2 m/s Bias = >0.5 m/s
Variables	a) Longitude (degree) b) Latitude (degree) c) Wind Speed (meter per second) d) Northward wind e) Eastward wind
Sampling (Resolution)	a) 50km resolution product with 25km cell spacing. b) 25km resolution product with 12.5km cell spacing

2.3 Data Extraction for Remote Sensing

The equation 1 to 5, study from Grange on 2014 were used in this study to calculate the wind speed and direction from the MetOp satellite image [12] [13]. The wind speed can be directly obtained from equation 1, where $\vec{u}$ is scalar average wind speed and u_i is defined as wind speed. This is the standard arithmetic mean for wind speed. Wind is represented in vector and specified as speed and direction. Wind speed is defined as the air motion of kinetic energy with respect to the earth surface. The equation 1 shows the average value of the wind speed. The monthly average wind speed data are calculated according to the monsoon seasonal periods.

$$\vec{u} = -u_i \times sin\left[2\pi \times \frac{\theta_i}{360}\right] \quad (1)$$

To retrieve the velocity amplitude or resultant vector, Pythagoras's theorem and trigonometry formula are used to obtain wind direction. The meridional and zonal (u and v respectively) represent the two wind components namely the east-west and north-south to be calculated using the formula to measure the average vectors for horizontal wind [14].

The two wind components (u and v) in equation 2 and equation 3 need to be calculated first before the average value is calculated [12]. The negative sign is calculated along with wind speed (u_i). Wind direction, by meteorological convention, is defined as from where the wind is blowing from, while the vectors are defined as the direction where the flow is heading to. Wind direction (θ_i) measured in degrees and the units for the components in radians [13].

$$\vec{u} = -u_i \times sin\left[2\pi \times \frac{\theta_i}{360}\right] \quad (2) \qquad \vec{v} = -u_i \times cos\left[2\pi \times \frac{\theta_i}{360}\right] \quad (3)$$

To calculate the resultant vector average wind speed ($\bar{U}_{RV}$) and direction (θ_{RV}) the following equation 4 and equation 5 are used:

$$\bar{U}_{RV} = (\vec{u}^2 + \vec{v}^2)^{\frac{1}{2}} \quad (4)$$

$$\bar{\theta}_{RV} = arctan\left(\frac{u}{v}\right) + flow \quad (5)$$

2.4 Past Windstorm event and damages

For this study, data of windstrom event were collected based on reports and social media[15][16]. The used of big analytic data has give a lot of information for windstrom event from side of human that faced the situation. The picture andvideo collected from social media proven the case of windstorm event even the case not reported by related state authority [17]. Due to a very limited resources of storm data, this study opted to use these methods and same study by Majid [18] in Penang have also used this method for windstorm event data collection.

The collection of the data need be done manually since a reliable windstrom database in Malaysia still unvailable because lack of funding, recent advance monitoring technologies, and unrecorded data [4]. The development of a database is crucial to provide information on windstorm distribution and the damaged. The use of big analytic data was utilized to gather the windstorm event and damages. The data was collected based on related keyword that refer to event. In Malaysia, the study about damages by windstorm focused on roof structure [19].For this study, the considered damage was falling trees, building damage, car damage, accidents and human death. Those damages were selected based on cases damages reported.

3. Result and analysis

3.1 Windstorm event and damages on 2018

Figure 3 shows the total windstorm cases reported and recorded in newspaper and article [15], [16]. Total cases recorded in this study is 53. The bar chart shows the highest windstorm event reported in Southeast Monsoon with 17 cases, followed by Northeast monsoon 15 cases, Inter-monsoon 10 and 2 cases each. Figure 4 shows the total cases based on state in Peninsular Malaysia. It is shown that the highest windstorm impacted Kedah with 21 cases, follow by Selangor with 14 cases and other state less than 10 cases.

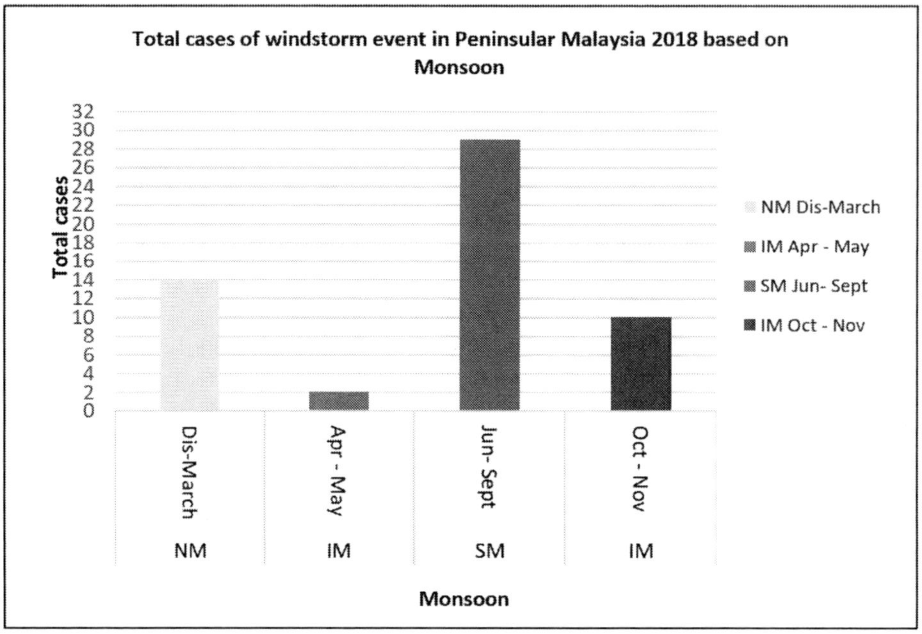

Figure 3. Statistics of Windstorm event cases based on seasonal monsoon in 2018

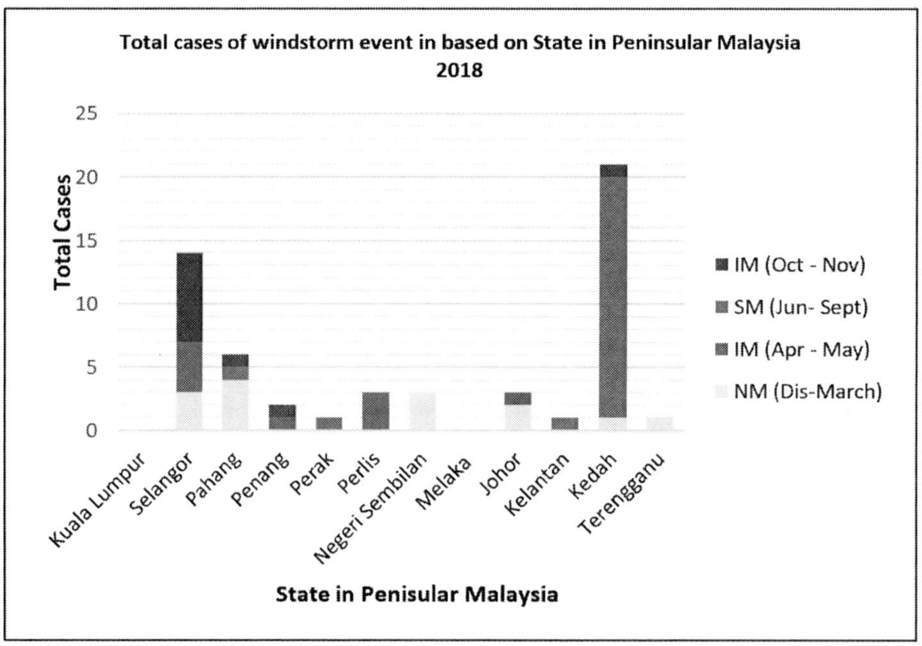

Figure 4. Statistics of windstorm event based on State of Peninsular Malaysia in 2018

Figure 5 shows the total damages by windstorm event in 2018. The highest damage was caused by tree falling with 30 cases, building damage of 19 cases, car damage, accident and death 1 cases each. Kedah shows 21 cases of damages by windstorm followed by Selangor with 17 cases Perlis with five cases, Pahang and Penang three cases, while Perak, Johor and Kelantan one cases and no cases in Kuala Lumpur, Negeri Sembilan and Terengganu. In Perlis, the windstorm event has causes car accident and death reported. Figure 6 shows the map of windstorm event reported in 2018.

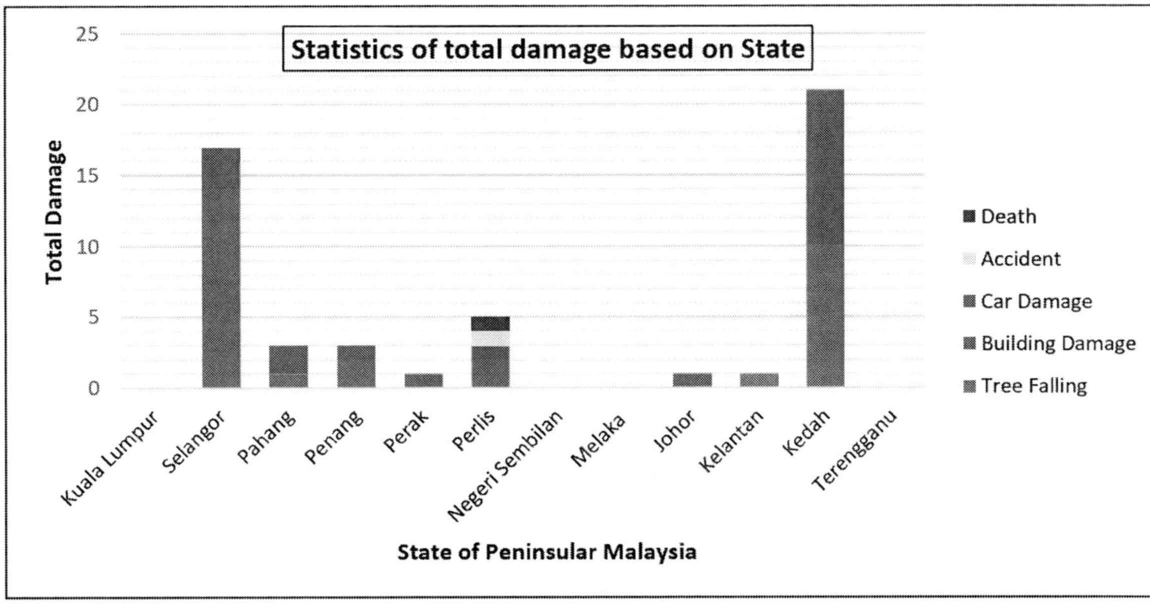

Figure 5. Statistics of total damage based on State

Figure 6. Maps of windstorm event in Peninsular Malaysia on 2018

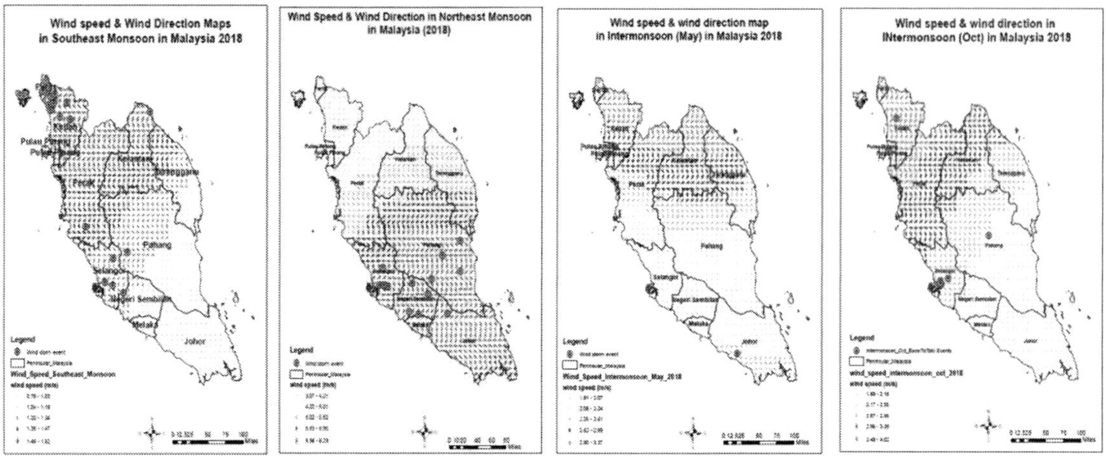

(a) (b) (c) (d)

Figure 7. Maps of wind speed and wind direction simulate from WRF-ARW simulation (a) Southeast Monsoon (NM) , (b) Northeast Monsoon (SM) , (c) Inter-monsoon April to May (IM) and (d) Inter-Monsoon October to November (IM).

3.2 WRF simulation for wind speed and wind direction

Figure 7 show the maps of wind speed and direction stimulated by WRF-ARW for seasonal monsoon in Malaysia in 2018. The highest wind speed occurred on NM that is 6.23 m/s and the lowest wind speed occurred in SM 0.78 m/s.

Figure 7 (a) show the wind speed and direction for SM. The wind speed recorded between 0.7 m/s to 1.7 m/s. The highest wind speed focused on northern-region which is in Perlis, Kedah, Pulau Pinang and Perak. While in centre Peninsular of Malaysia shows the wind speed reading was between 1.2 m/s to 1.4 m/s impact the Selangor and Pahang state. The reason Selangor is impacted by windstorm event hazard because the increases of urban area and the orientation of building in Selangor is very typical which causes damage to the building in this area during windstorm event. Hence the intensity of vertical and orientation of building has affecting the formation of wind in the urban area. Building orientation has strong correlation with wind direction [20] .

In NM as shown in figure 7 (b), the highest wind speed at 5.96 m/s to 6.23 m/s. The lowest value lies in between 3.07 m/s to 4.21 m/s. The high wind speed occurs at the centre of Peninsular Malaysia [21] and causes the windstorm event that brings damages. The area impacted by windstorm and damages are Selangor, Pahang, Negeri Sembilan and Johor. In the northern-region in Perlis, Kedah and Pulau Pinang the wind speed is less than in the centre of Peninsular and there is no event of windstorm and damages reported from the collected data of windstorm event.

While in the IM, less cases were reported. Figure 7 (c) show the IM wind speed on April to May 2018. The highest and lowest wind speed recorded are 2.90 m/s to 3.37 m/s and 1.81 m/s to 2.07 m/s. The highest wind speed at north-region and the lowest wind speed at central region of Peninsular Malaysia. Only 2 cases were reported and recorded by newspaper and article.

Figure 7 (d) shows the IM wind speed on October to November 2018. The highest wind speed is 3.04 m/s to 4.02 m/s while the lowest is 1.69 m/s to 2.16 m/s. For this IM, the highest wind speed recorded in north-region while the lowest wind speed at Southeast region. In Selangor, the wind speed range between 2.57 m/s to 2.95 m/s, which is higher comparatively with the other states. The impact of unplanned urban expansion caused Selangor to become prone area for windstorm impact although the wind speed is medium and is not supposed to cause any damage.

3.3 Wind speed and direction pattern from MetOp satellite image

The transboundary wind circulation flows from one place to another has caused the winds fluctuation one the land surface. Figure 8 shows the wind speed and direction map from MetOps satellite images. The highest wind was on SM at 9.20 m/s to 10.96 m/s and the lowest wind speed was 0.99 m/s to 2.01 m/s in IM in October to November.

The high SM wind speed was 9.20 m/s to 10.96 m/s and blows from outside of Johor as shown in Figure 8 (a). But no windstorm cases are reported even though the wind speed in the land surface is high. It is because type of landuse in Johor 73% of total area was green area and only 27% was developed area [22].Different cases are observed in northern region and Perlis where the wind speed from outside areas are 6.84 m/s to 8.29 m/s which forms the windstorm event and causes death reported by online Astro Awani on 18 September 2018 [23].

Figure 8 (b) shows the wind speed of NM from MetOps. The highest wind speed is 7.3 m/s to 8.93 m/s that comes from Straits of Melaka and circulates into the land surface. The highest wind speed also comes from outside next to Selangor state and intensifies the windstorm event. While the other impacted areas are in Pahang where the outer wind comes into Pahang at 5.57 m/s to 6.38 m/s.

In IM as shown in figure 8 (c) and 8 (d), the highest reading for wind speed on April-May and October-November are approximately same. The lowest wind speed for IM April-May was 1.76 m/s to 3.38 m/s and for IM October-November was and 0.99 m/s to 2.01 m/s._For IM in April-May the highest wind speed blows from outside of western region (Terengganu and Kelantan) and from outside of Selangor, while for IM in October- November the highest wind speed blows from southeast region (Johor). The IM in October-November causes bigger windstorm impact at Selangor state.

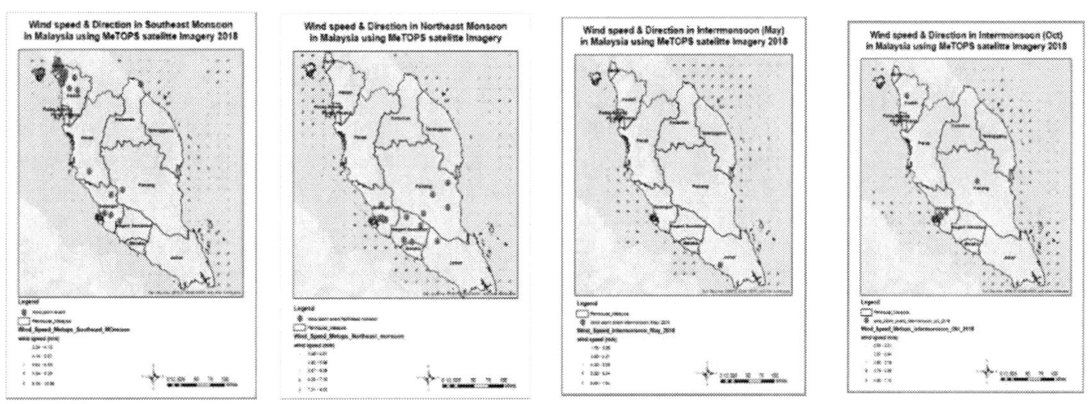

(a) (b) (c) (d)

Figure 8. Maps of wind speed and wind direction simulate from MetOp satellite image (a) Southeast Monsoon (NM), (b) Northeast Monsoon (SM), (c) Inter-monsoon April to May (IM) and (d) Inter-Monsoon October to November (IM).

4. Conclusion

The wind pattern changes during windstorm events and causes damages to the public and private property. This study analyzed the wind speed and direction using WRF-ARW simulation and MetOp satellites images. It has been determined, that the highest impact by windstorm is spotted on SM and NM. The highest wind speed occurs on NM. The circulation of wind from ocean has causes the stronger or weaker circulation of wind patterns inside the Peninsular Malaysia. The type of land use also impacts the development of windstorm events. The numbers of events do not cover all Peninsular Malaysia because weak windstorm detection network. The limitations of ground stations to retrieve wind data spatially leads to the use of WRF-ARW and MetOp in this study. Future study needs to be conducted to examine the impact of wind coming from outside to a certain boundary layer and change the impact of wind pattern locally causing windstorms. As the study in Penang on 2016, the impact of windstorm event has been recorded. The total cases show the windstorm event, destruction, and total loss for the study area. But there is no study show the causes of windstorm event that might be associate with the other related parameter such as rain, humidity, temperature and orientation of building.

Acknowledgements
The authors would like to express their gratitude to the Universiti Teknologi MARA (UiTM) for funding this project under Geran Bestari Perdana 600-IRMI/DANA 5/3 BESTARI (P) (074/2018).

References

[1] D. Henderson and J. Ginger, "Role of building codes and construction standards in windstorm disaster mitigation," *Aust. J. Emerg. Manag.*, vol. 23, no. 2, pp. 40–46, 2008.

[2] N. Kishore *et al.*, "Mortality in Puerto Rico after Hurricane Maria," *N. Engl. J. Med.*, vol. 379, no. 2, pp. 162–170, 2018.

[3] M. O. F. Science, "General Information Natural Hazard in Malaysia Dissemination of weather information / forecast / warning Additional Utilisation of Weather / Climate Information."

[4] F. A. Wan Chik, C. D. S. N. A, N. I. Ramli, M. K. A. Muhammad, T. A. Majid, and N. Z. Zulkarnain, "Development of Windstorm Database System for Wind Damages in Malaysia," *J. Civ. Eng. Res.*, vol. 4, pp. 214–217, 2014.

[5] T. A. Majid, S. A. S. Zakaria, F. A. Wan Chik, S. N. C. Deraman, and M. K. A. Muhammad, "Past windstorm occurrence trend, damage, and losses in Penang, Malaysia," *J. Eng. Sci. Technol.*, vol. 11, no. 3, pp. 397–406, 2016.

[6] Y. Dong, G. Li, M. Yuan, and X. Xie, "Evaluation of five grid datasets against radiosonde data over the eastern and downstream regions of the tibetan plateau in summer," *Atmosphere (Basel).*, vol. 8, no. 3, pp. 1–19, 2017.

[7] A. Bentamy, S. A. Grodsky, J. A. Carton, D. Croizé-Fillon, and B. Chapron, "Matching ASCAT and QuikSCAT winds," *J. Geophys. Res. Ocean.*, vol. 117, no. 2, 2012.

[8] A. S. J. Figa-saldaña,J. j. w. Wilson, Evert Attema, R. Gelsthorpe , Mark R Drinkwater, "The Advanced Scatterometer (ASCAT) on the Meteorological Operational (MetOp) platform: A follow on for European wind scatterometers," *Curr. Sci.*, vol. 107, no. 4, pp. 589–595, 2014.

[9] H. Zhang, Z. Pu, X. Zhang, H. Zhang, Z. Pu, and X. Zhang, "Examination of Errors in Near-Surface Temperature and Wind from WRF Numerical Simulations in Regions of Complex Terrain," *Weather Forecast.*, vol. 28, no. 3, pp. 893–914, Jun. 2013.

[10] Y. Takeyama, T. Ohsawa, K. Kozai, C. B. Hasager, and M. Badger, "Effectiveness of WRF wind direction for retrieving coastal sea surface wind from synthetic aperture radar," *Wind Energy*, vol. 16, no. 6, pp. 865–878, Sep. 2013.

[11] D. Carvalho, A. Rocha, M. Gómez-Gesteira, and C. Silva Santos, "WRF wind simulation and wind energy production estimates forced by different reanalyses: Comparison with observed data for Portugal," *Appl. Energy*, vol. 117, pp. 116–126, Mar. 2014.

[12] S. K. Grange, "Technical note : Averaging wind speeds and directions Technical note : Averaging wind speeds and directions," no. June 2014, 2015.

[13] S. K. Grange, "Averaging wind speeds and directions," no. October, p. 12, 2014.

[14] S. K. Grange, "Technical note : Averaging wind speeds and directions Technical note :," no. June 2014, p. 12, 2014.

[15] "Plaza Tol Bukit Raja gempar dilanda ribut | Kes | Berita Harian." [Online]. Available: https://www.bharian.com.my/berita/kes/2018/08/465683/plaza-tol-bukit-raja-gempar-dilanda-ribut. [Accessed: 09-Jan-2019].

[16] "Tiga lelaki nyaris maut, ribut landa Bachok | Astro Awani." [Online]. Available: http://www.astroawani.com/berita-malaysia/tiga-lelaki-nyaris-maut-ribut-landa-bachok-182838. [Accessed: 09-Jan-2019].

[17] "Malaysia Country Report | GardaWorld." [Online]. Available: https://www.garda.com/crisis24/country-reports/malaysia. [Accessed: 09-Jan-2019].

[18] T. A. Majid, S. A. S. Zakaria, F. A. Wan Chik, S. N. C. Deraman, and M. K. A. Muhammad, "Past windstorm occurrence trend, damage, and losses in Penang, Malaysia," *J. Eng. Sci. Technol.*, vol. 11, no. 3, pp. 397–406, 2016.

[19] N. I. Ramli, M. I. Ali, T. A. Majid, W. C. F. A, D. S. N.C., and M. M. K. A, "Wind-related Disasters in Malaysia and Changes in Regulations and Practice," *Wind Eng. JAWE*, vol. 40, no. 3, pp. 290–293, 2015.

[20] E. Prianto, F. Bonneaud, P. Depecker, and J.-P. Peneau, "TROPICAL-HUMID ARCHITECTURE IN NATURAL VENTILATION EFFICIENT POINT OF VIEW A Reference of Traditional Architecture in

Indonesia," 2000.
[21] P. Malaysia, A. M. Razali, M. S. Sapuan, K. Ibrahim, A. Zaharim, and K. Sopian, *Mapping of Annual Extreme Wind Speed Analysis from 12 Stations in*. .
[22] Jabatan Perancangan Bandar dan Desa Negeri Johor (JPBD), "Draf Rancangan Struktur Negeri Johor 2030," 2016.
[23] Salim Aziz (ASTRO AWANI), "Tiga maut, kereta bertembung ketika ribut di Padang Besar | Astro Awani," 2018. [Online]. Available: http://www.astroawani.com/berita-malaysia/tiga-maut-kereta-bertembung-ketika-ribut-di-padang-besar-185845. [Accessed: 04-Mar-2019].

Modular Construction System (MCS) in Malaysia: Mass Customization Through Combinatorial

S Aziz[1], S N Che Mohd Nasir[2], R Hatrom[3], L Ahmad Bazuli[4] and M R Abdullah[1]
[1]Faculty of Architecture, Planning and Surveying, Universiti Teknologi MARA, 32610 Seri Iskandar, Perak, Malaysia
[2]Faculty of Architecture & Ekistics, Universiti Malaysia Kelantan, 16100 Pengkalan Chepa, Kelantan, Malaysia
[3]Razak Faculty of Technology and Informatics, Universiti Teknogi Malaysia, Jalan Sultan Yahya Petra, 54100 Kuala Lumpur, Malaysia
[4]Jabatan Kejuruteraan Awam, Politeknik Ungku Omar, Jalan Raja Musa Mahadi, 31400 Ipoh, Perak, Malaysia

Email: myyaarc@gmail.com

Abstract. Combinatorial is another sub-topic and sub ordinate of Golden Mean theory. This paper is part of on-going research of modular construction system (MCS) in Malaysia. Objectives of the research were to propose an appropriate scheme and concept for sustainable design for modular home based on research findings.The researcher conducted case study (Personnel Expert Interviews) to collect data of MCS in Malaysia. Proposed combinatorial design for spatial planning of home design has been carried out by the researcher at the end of analysis as design experimental yet need validation from computer science experts. Combinatorial in modular architecture is the theory of conceptual design for future building design process of MCS. The researcher discuss in details and to clarify the proportion combinatorial for MCS in aspect of building forms and arrangements through proposed combinatorial formula. In Malaysia, design technique should relate to local culture, weather, and available materials sources as mentioned by the experts. Combinatorial design of modular architecture is an alternative design for compact affordable home. This combinatorial geometry (3D module) repeated, duplicate and can be oriented in a few ways to forming combinatorial models of building design. Pattern A and B are considered to be the best in economic aspects and practicality in term of shape of green. Proposed concept possibly can improve user decision making (digital design process) e.g. building players to plan their building lifecycle projects in term of economical design, green shape, reduced construction time and cost. Thus, it may be able to promote user friendly design, fast track construction, and quality based product, less wastage and varied of design selection can be chosen of the user.

1. Introduction

The basic aspects of architectural design were the buildings must conveniently and functional to use, durable and delight [1]. Alexander stated pattern was important part of architectural design. Pattern defined as ordered and logical structure [2]. Pattern can be found in natural environment, human, art, space, time, and music. Successful buildings obey the same system law as a complex organism and an efficient computer program. The user used the building to suite their own needs of common and daily life while the contractor tried to minimize cost and to standardize the building parts. In Malaysia, the construction industry is highly aware of the need to improve the integration, planning and control of its design and production process [3]. Consequently, the reality of construction is that most of the problems encountered in the field such as reworks, delay, low in quality or productivity and even legal entanglement and claims often compounded by inherent design flaws that generated in the design phase [3]. Architecture is designed to satisfy the different representational, functional, aesthetic, and emotional needs of organisations and people who live or work in structures [4]. Affordable home price normally in between RM 250, 000 to RM400, 000 [5]. The type of affordable home can come in different range of types and sizes such as for singles, couples, families and flat shares as well as for seniors.

2. Modular Architecture in Malaysia

In Malaysia, MCS or Offsite Manufacturing included in Industrialised Building System (IBS). The definition of offsite is the manufacture and pre-assembly of components, elements or modules, before installation into their final location [6]. As mentioned by Zaikhan (2015) from the author case studies, the problems of modular construction in Malaysia were about limited durable materials and higher cost. The improvement of constructability rarely been mention as the benefit of offsite but in theory offsite can improve constructability by providing designers with the fresh perspective and outlook on the concept of repetition, preassembly and standardization [3].

2.1 Analysis of MCS in Malaysia

The trends of MCS in Malaysia are creating big waves among building players but to risk and use this option of construction system is a bit scarier to them. Majority of the experts agreed that MCS is getting well known in the Malaysian construction industry as an option over conventional system as it is more cost and time saving. The experts are positive and optimistic of the MCS industry in Malaysia because there is high demand among Malaysians.

Based on case studies, it has been established that MCS is defined as a 3D module which is produced, manfactured and assembled in a factory (control environment). As for the construction of MCS module, it ses various tools and machinery and the MCS has a good quality of construction. There are many advantages of MCS compared to the conventional system. The experts agreed that MCS product are of high quality. MCS has a high quality product because approcimately 80% of the process is conducted in a factory. 14% of the expert suggest that MCS is more suitable for future construction in Malaysia. In terms of green shape, rectangularity is a practical shape to use in design. The experts claimed that square or rectangularity is a commonly use shape in construction. They suggested that one of the factors in which rectangular is considered as a green shape is because this shape is economical and has less wastage of materials compared to other shape. Beside that, another factor that rectangular is considered as a green shape in architectural design is because it has been proven that this design has been in architectural history for a long time. Moreover, the flexibility of square and rectangular shapes makes it the best shapes to use in architectural design and can make the space functional. These factors were agreed by all the experts. In addition, to promote MCS in Malaysia, the building players should not follow exactly what other cuntries such as in the UK,Australia, and USA, the cost of labour is expensive. Therefore, these countries result in using prefab and MCS as a solution to overcoe the high cost of labour. Malaysians are looking for versatility in home design. They want something new but cost saving and can be constructed in a shorter period of time. MCS is one of the solutions for this home design. The experts agreed that the future of MCS in Malaysia is bright. There is a demand among Malaysians in using MCS for construction. Howevver, there are still is a few issues that need to be tackled, such as home price of MCS within home of Conventional System type should be fair. As a result of design sustainability in terms of green shape, the researcher discovered that pattern B was the most practical shape of affordabe home design (high-rise type). The second most practical shape to be used in house design was Pattern A. All these pattern have been proposed using block arrangements and Combicatorial Concept (Mathematical Way of Design). These two pattern are considered as the most economical and practical shape in MCS. Other patterns were not favorable because of certain issues such as ess sconomical, uses a lot of footprint and poor ventilation in terms of arrangements and orientation. Therefore, the experts agreed that MCS in Malaysian construction industry but the government should play a big role in supporting building players in the IBS field itself. The suitability of MCS to be a new contruction system for affordable homes could still be achieved if the MCS house is priced at a fair rate within conventional house design.

3. Combinatorial for Modular Architecture

Modularity in general is part or series module of something whether it became appliances, furniture, machinery, automotive, aerospace, or buildings. Constructed a modules or unit packaging schemes. Standard parts of modular approaches offer gains in time, cost and quality [7]. For building design, modular refer to parts of a building define by 3D compartments such as LEGO. As combinatorial is arrangements and combination of geometric objects and wiht discrete properties of these objects. Mentioned by Christopher Alexander houses were generated by patterns. Spatial configuration is concerned wih finding feasible locations and dimensions for a set of interrelated objects that meet all design requirements and maximize design quality in term of design preference [8]. Spatial configuration is relevant to all physical design problems, so it is an important area inquiry [8]. Reported attempts to automate the process of layout design started over 35 years ago [9]. Geometry decompostion is one of the ideas to relate modeular construction and cobinatorial things. Combinatorial geometry is the study of combinations and arrangements of geometric objects and with discrete properties of these objects. Historically, modularity has play an important role in the study of combinatorial geometries [10]. Theory development of combinatorial design has brought incredible success in the application is not expected, in connection with basic math, and the desire to produce order out of chaos [11]. Moreover, he questioned about the future of application of combinatorial geometry. He explained that new mathematical truths will be found and that unanticipated application will arise. Christopher Alexander in his book, Pattern Language defines that pattern in a solution of space. In an aspects of design, combinatoric as mathematical approach can be found in that research. The building blocks can be combined in an infinite number of ways. The question is for a given fixed area which shape will create the greatest feeling of spaciousness [12].

4. Research Methodology

The aim of this study is to investigate current practises and issues of IBS and modular construction. In addition, it opens a prospect to test the new method of modular construction using 'Combinatorial Concept' during design phase. Therefore, a collection of data is obtained through literature review in the aspects of IBS and MCS development to achieve innovation in green construction and the implementation of combinatorial geometry. The data is supported using case studies and questionnaires. Both method are described as follows.

A) Case studies

Case studies performed in data collection are x Intrinsic and x Instrumental [13]. The reason is x intrinsic offer better understanding in cases carried out while x Instrumental help to examined and refine the theory proposed. Thus, case studies is carried out to acquire in-depth knowledge about registered MCS companies under the Malaysia Construction Industry Development Board (CIDB) to define controlled environment. The area selected is located in three cities highly involved in MCS, i) Kulala Lumpur; ii) Johor Bahru and iii) Sarawak involving four registered companies in 2015. It is limited because the implementation of modular construction in Malaysia are low [14]. It is essential to select registered companies as it has reliable track record and gradually monitored by professional bodies. Data observation is validated through content analysis with the outcome of all variables in theoretical model tested showed significant relationship. The data collection are categorized in two. First is validated documentations related to MCS and the other is on site data. Data collection is limited to photos, size and dimensions, specifications, cost estimation, materials and location.

B) Questionnaires

The supporting data is run through structured questionnaires among target group experts in design, engineering, costing and manufacturing in MCS. Specifically the MCS project is focusing on affordable housing scheme in Malaysia. Professionals selected are architects and contractors directly involved in MCS. In specific, nine issues were discussed and evaluated. The outcome is described in Table 1.

Table 1. Data collection by Different Company on MCS in Malaysia

Name /Company	Definitions of MCS	Trend of MCS in Malaysia	Advantages of MCS Vs. Site Built	Factors of Rectangularity is more practical shape	Agreement of Rectangular Shape is green shape of design (Yes/No)	Future MCS in Malaysia	Best option of shape design for MCS	Generalization of MCS and Affordable (Suitability)
Company A (10 years experienced in MCS)	-Standard size & form -expandable -pre-assamble in factory -fabricated on site -Flexible / variety of shapes (complex shape lead to difficult joinery system) -Conventional form (square) most practical & economical	-Increased, getting popular.	-Fast -Easy installation -Smarter -Green -Controlled environment (quality and clean logistic) -Easy for transportation	-Most of furniture, transportation , house floor plan and machinery were square (design logic of daily life) -Hexagonal was too difficult to organized things in the house (only customized the furniture)	Yes. Same with No. 4 reasons.	-House price need to be fair within using conventional system or MCS - There is hope for MCS in Malaysia -Don't compare MCS in Malaysia with other countries because different issues such as labor in Australia and USA were too expensive.	-B -Shared wall -Cost saving -Easy for transportation -Lack of aesthetic value	-House price need to be fair within using conventional system or MCS (suitable cost/price of the house) -Gov. have to create the plan to educate Malaysian on green & sustainability of MCS.
Company B (15 years experienced in MCS)	-LEGO -IBS;Prefab -Produced in factory -Repeated -Duplicate -Volumetric -Extendable/ Expandable -ISBU	-Existed demand among Malaysian -Inquiries; 10 questions/day -Inevitable option	- Containner (ISBU); Durability 2.5 to 108 times strength than site built.	-Trend timeline; history, square to organic to square (contemporay) -Golden Section by Le Corbusier -Economist agreement on rectagularity can minimize materials waste & economical	Yes. To avoid wastage of materials.	-Structure replacement -ID Modularize -Optimist of future of MCS	-B - Squarish > Rectangular	-Inevitable option

Company C **(15 years experienced in MCS)**	-Everything should be done at factory (controlled env.) -Accurate -Very fast -Quality	-Going up but very slow because of the limitation of manufacture interest to do MCS projects; all projects were different size MC; too rigid, inflexible -Past & current problems faced in IBS. MC was too rigid (cannot generalized and apply to other projects)	-process in factory -prepare very fast -quality control -all machinery	-Most practical (e.g; furniture, cabinert etc) -Other shapes; Round/ Circle unused space	Yes. Practicality can be rectangularity. Depends on your creativity to make it nicer.	We can do over here.	-A -E -It gives the best cost in construc tion - Increase d of number of shared wall, cost of footprint and ventilati on	-Needs to have market - Past & current problems faced in IBS
Company D **(10 years experienced in MCS)**	-Build in control env. (factory). -fully fitted out include finishes. - transport & delivery -ready to use within the day	-Infancy -Lack interest among developers -Labor cheaper cost	-Build in controlled env. (easy to control on time management/ schedule & quality. -Cost controlled -Use machinery & tools (less labor)	-Cheaper to construct -Easier to plan & fit out. -Lesser need for very skilled workers.	Yes. Maximum use of our climate factor.	Excellent. Building players should work together to tap the benefits of MCS.	B	Should be strongly pursued.

The analysis attempts to review the definitions, characteristics, issues, trends and sustainable design of modular construction and scientific way of architectural design (in aspect modular architecture).

5. Findings

The results showed that the combinatorial concept for modular design has a potential to widened the opportunity for affordable housing in Malaysia as it offers alternatives design layout, cost effective and time saving compared to the conventional system. Moreover, the product of MCS is highly reliable in quality as the manufacturing process is under a controlled environment involving tools and machineries. MCS is considered as high quality product because 80% of the process is conducted in the factory supported by 14% experts recommended the MCS for fast track project which is high in demand. However, a thorough research to adapt MCS in architectural heritage is yet to be define (Ar. Mastor, 2015)

The objective of promoting combinatorial concept to develop effective design strategies, standard guidelines, specifications and procedures aiming for cost reduction and efficiency is achievable through equipment modularization and standardization with process simplification. The products of combinatorial providing a repetition of identical units, both in the building dimensions and building layout.

As suggested by experts, Pattern A and B were the best shape for house pattern arrangement. Therefore, the researcher proposed five (5) combinatorial way of most practical pattern for arrangement. The pattern proposed is based on a single unit of high rise home. The method proposed by the researcher entitled as Combinatorial Concept of design. The combinatorial tiling is defined as a process or technique to figure out the floor plan and pattern composition by repeating the modules in the given square area.

Figure 1 indicates the best shape for MCS of public housing design. 75% experts agreed pattern Bis the most practical shape of affordable home design. Meanwhile, only 25% selected pattern A as the second best shape design for MSC in Malaysia. Pattern A and B are considered to be the best in economic aspects and practicality. Other patterns are not favourable as it is less economical, has larger footprint and offer poor ventilation when it comes to space arrangement and orientation.

In order to relate the concept of green shape, the researcher discovered, that Pattern B was the most practical shape of affordable home design for high-rise. The second most practical shape is Pattern A. The experts claimed that square or rectangularity is a commonly used shape in construction. They suggested rectangular is considered as a green shape because it is economical and has less wastage of materials compared to other shapes. Besides that, rectangular shape is synonyms in architectural design as it is proven that this design is functional and flexible, and has been used and fully utilized in architectural history for a long time. All these patterns have been proposed using block arrangements and Combinatorial Concept. However, the experts agreed that MCS in Malaysia needs to improve IBS issues according to building players in the construction industry and the end user.

In addition, to promote MCS in Malaysia, it is not recommend to follow exactly what other countries use for modular construction. Experts claimed that in other countries such as in the United Kingdom, Australia, and United States of America, the cost of labour is expensive. Therefore, these countries resulted in using prefab and MCS as a solution to overcome the high cost of labour. Malaysians are looking for versatility in home design. They want something unique but cost saving and can be constructed in a shorter period of time. MCS is one of the solutions for this home design. The experts agreed that the future of MCS in Malaysia is bright. There is a demand among Malaysians in using MCS for construction. However, there are still is a few issues that need to be tackled such as home price of MCS within home of Conventional System.

The experts also remarked that it is possible to implement MCS in the Malaysian construction industry but the government should play a big role in supporting building players in the IBS field itself. The suitability of MCS to be a new construction system for affordable homes could still be achieved if the MCS house is priced at a fair rate compare to conventional house design.

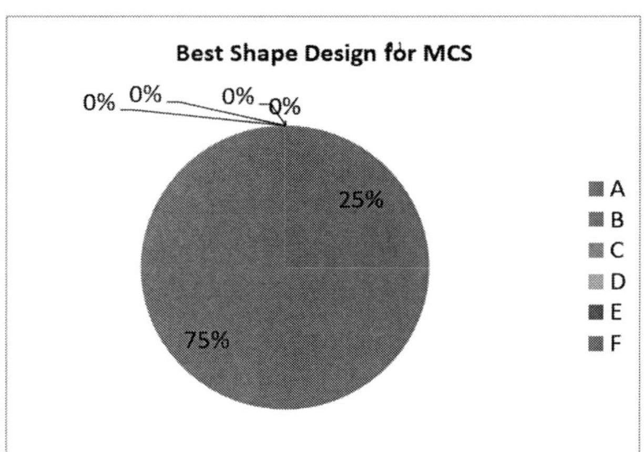

Figure 1. Best Shape for MCS of Public Housing Design

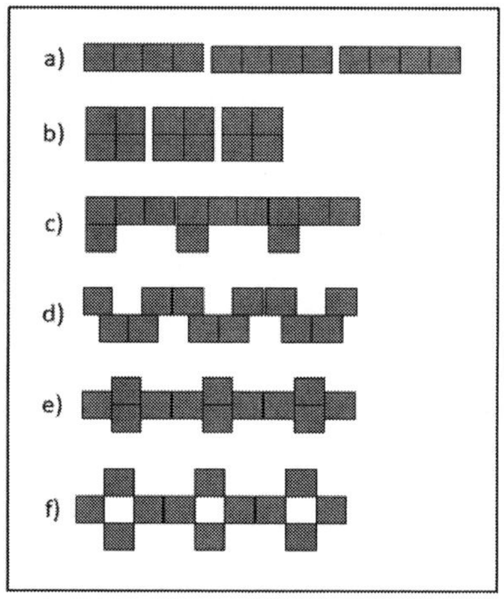

Figure 2. Modular Home Pattern

6. Recommendations and Conclusions

Suggested of the Combinatorial Concept for MCS might be able to cover issue product modularity in sustainable shape design especially. As Pattern A and B most practical and economical shape for home pattern, experts agreed rectangular is considered as a green shape because it is economical and has less wastage of materials compared to other shapes. This concept will improve user decision making e.g. building players in order to look for appropriate methods before to proceed with any decision of prototype, testing and hands-on construction at site. Thus, with the concept proposed by the researcher it may be able to promote user friendly design, fast track construction, and quality based product, less wastage and varied of design selection can be chosen of the user. The Combinatorial Concept was developed based of economization of product and risk management approach (based on Literature Review) with direct engagement from practitioner in construction industry to choose best shape of design to look for in MCS. Despite this Combinatorial Concept not been testing yet unto real world consideration, should future research to look into account this method for further research and assessment. Area to be considered in order to sustain of the concept development in future such as computational software based. Due to the formulation of affordable home based on combinatorial probability and combinatory code. As combinatory code has been used in field of Artificial Intelligence (AI) and Internet of Things (IOT) The objective was to develop to necessary strategies, standards, designs, specifications, and procedures for cost reduction through equipment modularization, equipment standardization and process simplification. The products of combinatorial providing a repetition of identical units, both in the dimensioning building spaces, as well as in the layout of the building. With this theory, it might be provided more idea in producing affordable home in the near future. Through the reuse of formwork, sustainable and inexpensive designs can be attained. The system allows for the customization of designs, while maintaining the benefits of a regular prefabrication product, so their cost and performance can be improved over time [15]. Challenge of MCS in Malaysia is vastly complex but to apply this alternative construction system in near future is inevitable action. So, the right track and steps of conceptual and construction technique certainly can help narrow down suitable solution of current problem faced by building players to apply MCS in construction industry. In short, the combinatorial design was an experimental design to find out the sustainable shape and economic way of design.

Acknowledgement

The Author wishes thanks to En. Nazrol (Staff of CIDB Center, Kuala Lumpur, Malaysia) provided the Author of CIDB registered's MCS companies in Malaysia and Henry Losoi (MSc Tech in Applied Mathematics and Physics, Department of Mathematics and System Analysis, Alto University, Finland) guided the Author detailed and narrow down the Combinatorial Concept to suite the context of architectural design. Besides that, thanks and full of appreciation of MCS and IBS companies in Malaysia for all their kind help towards this research. All of their cooperation in helped given information for the research to the Author was valuable and highly appreciated.

IOP Conf. Series: Earth and Environmental Science **385** (2019) 012030 doi:10.1088/1755-1315/385/1/012030

References

[1] Salama A M 2007 Nikos A. Salingaros: A New Vitruvius for 21st–Century Architecture And Urbanism? *International Journal of Architectural Research* **1** 114-131
[2] Salingaros N A 1999 Architecture, Patterns and Mathematics *Nexus Network Journal* **1** 75-85
[3] Mohd Nawi M N, Baluch N and Bahauddin A Y 2014 Impact of Fragmentation Issue in Construction Industry: An Overview *MATEC Web of Conferences* **15** 1-8
[4] Werner S and Long P 2003 Cognition Meets Le Corbusier – Cognitive Principles of Architectural Design *International Conference on Spatial Cognition* **2685** 112-126
[5] Wan Abd Aziz W N A, Hanif N R Singaravello K 2011 A Study on Affordable Housing Within The Middle-Income Households In The Major Cities And Towns In Malaysia *Australian Journal of Basic and Applied Science* **5** 258-267
[6] Gibb A G 1999 Off-Site Fabrication United Kingdom: Whittles Publishing.
[7] Bertram 2005 Modularization of Ships, Report within the Framework of Project "Intermodul" s/03/G IntermareC, Available from: https://www.scribd.com/document/377274256/Modularization-of-Ships
[8] Michalek J J Choudhary R, Papalambros P Y 2002 Architectural Layout Design Optimization *Engineering Optimisation* **34** 461-484
[9] Ebadi A 2014 Computer as the Designer: Form and Space Layout Allocation KTH Royal Institute of Technology, November 2011, Available from: https://www.academia.edu/9091807/Computer_as_the_Designer_Form_and_Space_Layout_Allocation
[10] Brylawski T 1975 Modular Constructions for Combinatorial Geometries *Transactions of the American Mathematical Society* **203** 1-44
[11] Colbourn C J, Dinitz J H and Stinson D R 2003 Applications of Combinatorial Designs to Communications, Cryptography, and Networking, Available from: https://pdfs.semanticscholar.org/352f/daddd85be594fa5690a0b98cc5c03e93aad1.pdf
[12] Dawes M J and Ostwald M J 2017 Christopher Alexander's A Pattern Language: analysing, mapping and classifying the critical response *City, Territory and Architecture* **4** 1-14
[13] Stake R 1995 The art of case study research *Thousand Oaks, CA: Sage* Available from: https://legacy.oise.utoronto.ca/research/field-centres/ross/ctl1014/Stake1995.pdf
[14] Musa M F, Yusof M R, Samsudin N S, Muhammad Halil F 2017 The industrialised building system modular system (IBSMS) Framework *AicQoL2017Bangkok* pp.105-113
[15] Xue H, Zhang S J, Su Y K and Wu Z Z 2014 Factors Affecting the Capital Cost of Prefabrication—A Case Study of China *Sustainability* **9** 1-22

IOP Conf. Series: Earth and Environmental Science **385** (2019) 012031 doi:10.1088/1755-1315/385/1/012031

Evolution of Investments through Sale and Leaseback Investment: A Qualitative Investigation from Malaysian REITs

N L F Johaimi Ling[1], M H H Aliasak[1] and K Alauddin[1]
[1] Faculty of Architecture, Planning and Surveying, Universiti Teknologi MARA, Seri Iskandar, 32610 Seri Iskandar, Perak, Malaysia

Email: nurle187@perak.uitm.edu.my

Abstract. Sale and leaseback is known as a voluntary investment by owners to free up their equities for other uses while maintaining the use of the facility. In Malaysia, Real Estate Investment Trust (REIT) is one of the institutional investors adopting the sale and leaseback approach for the wide range of new investment technologies to enhance the value of the investment. The aim is to revolutionize the property industry by switching the design of physical asset investments from direct investments to indirect investments influenced by factors determining the property investment yield by analysing and responding to the information received. The research uses a qualitative method due to the limitation in the number of REITs involved in sale and leaseback investments in Malaysia. Using semi structured interviews, the data was recorded, transcribed and analyzed using thematic analyses for multiple interviews. Results indicated three factors that influence the yield which are location, tenants and property type. The result of the study would be a useful guide to Real Estate managers in developing countries to use sale and leaseback as one of the alternative ways to increase the revenue from the property investments.

1. Introduction

The traditional form of transaction, which is the direct investment, has transformed into a new generation of indirect investments in the form of sale and leaseback arrangements. These investments are currently enjoying a new wave of demand among Real Estate Investment Trusts in Malaysia intending to acquire new properties in their portfolios.

The institutional investor sector is largely dominated by listed property companies (or known as property shares) and real estate investment trusts (REITs; formerly known as listed property trusts (LPTs)) [12]. Thus, sustainable efforts have been undertaken by the Government to improve the structure of REITs and accelerate their growth. These efforts have renewed the vigour of institutional investors towards the Malaysian securitized real estate market. The total number of Malaysian REITs listed on Bursa Malaysia has increased rapidly from 3 in 2004 to 18 in 2016. According to records of listed Real Estate Investment Trusts as at 31 December 2018, the total net asset value was at RM34.5 million and the total market capitalization was RM41.4 million [2]. Unfortunately, previous studies on Malaysian REITs only emphasized on the performance analysis in their asset portfolios. To date, there is still a lack of studies conducted specifically on the sale and leaseback investments in Malaysian Real Estate Investment Trusts.

1.1 Purpose of the study
The purpose of this study was to investigate the Malaysian REIT's preference in purchasing property through the arrangement of sale and leaseback investments. Results of the investigation will lead towards the formulation of factors influencing the yield which aids in the investor's decision-making process when expanding their property portfolio.

1.2 Objective of the study
In line with the above purpose, the objective of the study was to forecast the factors affecting the Yield from the perspective of Malaysian REITs in the arrangement of sale and leaseback investments. The ability among Malaysian REIT Managers to make useful forecasts of yield is crucial especially in investment decisions and portfolio construction. Similarly, the ability to identify factors affecting yield is valuable for academic research.

Brief overview of sale and leaseback investment by Real Estate Investment Trusts (REITs)
Real estate is traditionally considered a lumpy and illiquid investment. Purchase of real estate investment involves the transfer in ownership of land rights, usually with buildings. It involves long-term investment and commitment due to the nature of its returns. It commonly involves substantial capital expenditure which, in most cases, is beyond the reach of small investors. Therefore, indirect property investment becomes the closest alternative where small investors can cater for property cycles and raise economic benefits by investing in stocks of listed property companies. In this study the listed property companies will become the Real Estate Investment Trusts. Indirect investments are now seen by institutional investors as means of improving real estate performance in the medium term by creating a diversified market for property investment. Sale and leaseback investments are examples of such investments made in listed real estate companies.
Sale and leaseback is a special financing technique used in real estate market transactions. It is often used among large real estate investors such as Real Estate Investment Trusts and Operator Company firms. The sale and leaseback approach is a valuable way to release the capital needed to invest in lieu of constructing a new building. It is an alternative extension where it needs a continuous supply of buildings by finding new sites and the onus of obtaining planning permission lies in the future occupants or developer's partners.
Historically, many large REIT companies in Malaysia have amassed large portfolios of real estate. These properties were either acquired through organic growth over a considerable time or through the amalgamation and takeover of existing businesses such as sale and leaseback investments [13]. The REITs industry has been diversified through segmentation which depends on the functional category of property such as hotels, offices, industrial, residential, healthcare, retail and diversified REITs [8]. The purpose is to allow the investors to be vigilant on the performance of every REIT sector.
Investors such as the REIT companies need to recognize the importance of acquiring quality property in their property portfolio. Hence, their professional advisors need to identify the greatest factors that influence their yields in order to maximize the returns on their property investments in dealing with sale and leaseback arrangements. The ratios between income and capital value or cost in the property market are reflected in the terminologies used; such as rate, return and yield [15]. There are a few factors that may affect the performance of yields in property investments.

2. Research Methods
This research adopts a descriptive and exploratory method, with 'what' and 'how' questions as a guide [6]. The explanatory approach was employed to have a better understanding on the phenomenon. Therefore taking all these into account, a qualitative approach with case study design approach was adopted for this research.
Besides that, Malaysian REITs hold great potential in raising local financial markets as a whole [5]. This is also supported by the Malaysian Securities Commission Annual Report 2013 which concluded that the Malaysian REITs market has matured over the years particularly in the period of post-Global

Financial Crisis of 2008/2009 (GFC) [7]. Taking all these into account, this research survey was thus conducted only on Malaysian REIT companies especially those involved in sale and leaseback investments. Data was obtained through semi structured interviews conducted directly with persons in charge of the process of arrangement of sale and leaseback investments.

Only 6 out of 9 REIT companies involved in sale and leaseback investments participated in the survey out of a total of 18 REIT companies in Malaysia. The use of multiple case studies is the best choice to be explored in this research as these case studies act as intervention reflecting the real-life context in which it occurred. Besides that, similarities and differences within and between cases can be better explored through multiple case studies [1]. As limited conclusive views have been given on the factors influencing yields within the context of sale and leaseback in Malaysian REITs, qualitative triangulation of multiple case studies were conducted to establish the factors through expertise interviews survey.

Multiple-case studies can produce results with a higher degree of certainty for external validity when theoretical replications and analyses are conducted in similar contexts [11]. Cross case analyses provide benefits to this research through multiple case studies. The results of the cross-case analyses can be holistically achieved between cases [10].

3. Methods

3.1 Data Collection

The data collection procedure implemented by the researcher in this study is divided into two phases which are:

3.1.1 Phase 1: Literature Review. This research commenced with a review of a range of literature on the research topic itself. In particular, it draws upon a limited amount of academic literature either in Malaysia or from world-wide studies including theoretical and conceptual framework studies. This research deals only with

i) Real Estate Investment Trusts (REITs)
ii) Sale and leaseback transactions in other countries
iii) Factors most influencing the yields.

3.1.2 Phase 2: The Main Survey. The main survey for this study is through semi structured interviews since the number of participants is limited and data was gathered through field research technique.

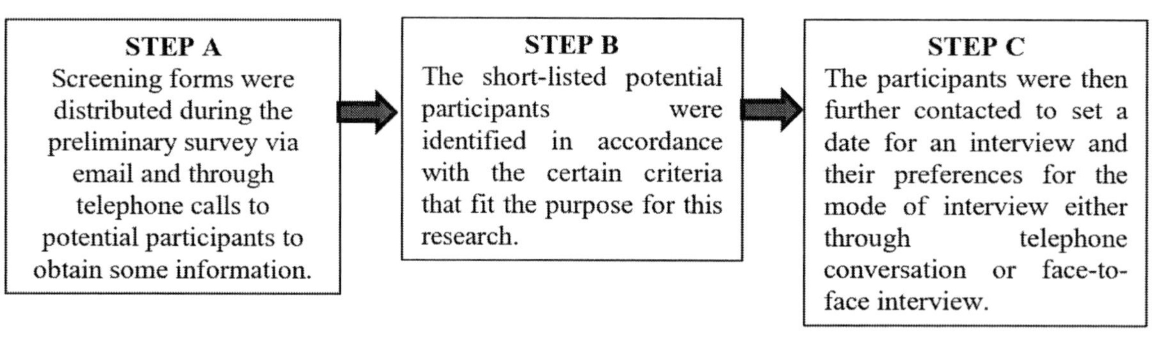

Figure 1. Before the Process of Main Survey

Figure 1 shows the process of main survey conducted in this research. In step A, the questions prepared were relatively straightforward, aimed at identifying any of the property portfolio among the 18 REIT companies that are involved in the arrangement of sale and leaseback investments.

After the preliminary survey was conducted, the next step is to shortlist the potential participants to be selected in the main survey whereby they fulfil the criteria that have been determined before the preliminary survey was completed. The main selection criterion of a participant is: only REIT companies experienced in dealing of sale and leaseback arrangements were selected. It transpired that only 9 out of 18 REIT companies have their properties under the arrangement of sale and leaseback investments.

Once the potential participants were selected, the researcher then approaches the participants to request for interview appointments by either telephone calls or email communication. The main survey of this research uses semi-structured interviews with the participants who are the professionals in charge at the case studies. This is done to allow the participants to generate their own perspectives that act as guide questions leading to the topics and issues about which the interviewer aims to explore. As a consequence of the short-listing of the participants, only 6 out of 9 REIT companies agreed to participate in the main survey. 4 REIT companies agreed to participate through face-to-face interviews while 2 REIT companies were interviewed by telephone. The interviews were conducted in October 2018. The number of interviews was considered sufficient and tallies with the minimum sample size ranging from 5 to 25 for semi-structured or in-depth interviews [3].

3.2 Data Analysis

After the interview surveys recorded individually for each of the participants were completed, the data was subsequently transcribed into a paper script. The interview transcripts were analysed using content analysis and was imported into QSR NVIVO Version 12 for the process of coding and quotations that have been identified are as illustrated in Figure 2.

Through NVIVO the data coding that have been labelled were converted into codes known as nodes which refer to the answers or responses from the interviewees. These codes were then grouped into categories, and themes were then developed.

The themes were identified from a pattern or trend for these nodes using thematic and content analyses. The thematic analysis began with familiarization of data, data coding, searching for a theme, and recognizing relationships; and ended with refining themes and testing propositions [3].

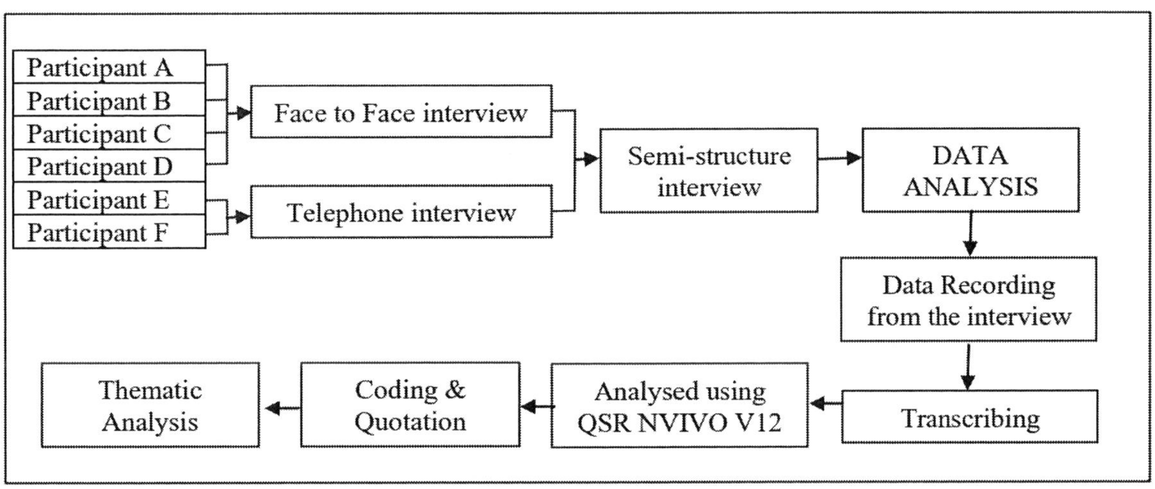

Figure 2. Data Collection and Data Analysis

4. Result and Findings

This research highlights the main factors that influence the yields in order to maximize the returns on property investments dealing with the sale and leaseback arrangements by Real Estate Investment Trusts. Even though the current revenue in investments cannot be calculated until the transaction is made, an investor "may set a target yield for an investment prior to or during ownership". Historical rates obtained from comparable transactions may be relevant, albeit they do not embrace the future. Therefore, the focus should be on the yield forecast. Predicted rates are estimated either through market participants' interviews and their views on future market expectations; or through indirect evaluations of factors affecting yields. There are 3 factors that have been identified in this survey to be the critical considerations by all investors prior to purchasing the property under the sale and leaseback investment agreement. They are:

4.1 Location

Location of the property has been identified as the main factor influencing the yield. Whether the property is located in the urban area or the suburbs will produce a different rate of yield. The REIT may consider a property located in the suburban area but they will demand for a higher yield to ensure that it is more secure for the shareholder returns. Properties that are located in the urban areas are more favoured by REITs. This preference is supported by research findings that location in main town centres and prosperous market towns were favoured and considered to be more likely to retain their business over time [4]. Each of the REIT companies have their own target rate.

4.2 Tenant

Selecting an owner or also known as potential tenant is important for the REIT because it may influence the yield of the building itself, as it is dependent on the ability of the tenant to secure income from the business in the building for a long-term tenure in order to pay the rental to the REIT companies. The financial background of the tenants is a determining factor to guarantee their incomes are stable enough to sustain the business. The number of tenants in the building whether single tenant or multiple tenants will influence the returns of the building in future. Simultaneously, it will also affect the distribution of income by the REITs to their shareholders.

4.3 Property Type

Each type of building produces a different type of yield due to the risks that the building face. Some of the REIT companies prefer to invest in a different type of properties, for example mixed type of development or it can be single type of building such as industrial, office or commercial buildings. For instance, Participant 1 prefers to invest his property in the industrial sector in his company's property portfolio because the yield is more secure compared to Participant 2 who has different preferences such as to have a mixed property portfolio because the yield received is more reliable. In support of these findings, it has been shown that factors having the most influence on yields have a clear emerging theme in the form of property type and each of it may have different forms of variables [4][14].

5. Conclusion

A sale and leaseback investment is a surrogate method of investment transaction in Malaysia. It has been adopted by Malaysian REITs since the establishment of REITs in Malaysia in 2005. This kind of investment is beneficial to both parties; the seller (operational companies) and the buyer (REIT Companies). It is an opportunity for the operation companies to expand their business in the long term by receiving a bunch of capital through selling off their properties but maintaining their businesses and activities in the same building.

Concurrently, REIT companies could increase their property portfolio with low risks due to long term guaranteed rental income. This research paper has identified three factors that influence the yield in sale and leaseback arrangements for Real Estate Investment Trust Companies in Malaysia which are location, tenant and property type. With these factors, it becomes possible to maximize returns by professional practitioners and investors in predicting high yields for portfolio building purposes. Nevertheless, REITs are calling for those sale and leaseback arrangements from the strong operation companies to remain high.

This research focuses on Real Estate Investment Trusts as investors involved in the arrangement of sale and leaseback investments. Further research may also be expanded to cover the other types of investors such as individuals or organizations that are involved in sale and leaseback investments.

Acknowledgments
The authors wish to acknowledge the support funds for this research under the Geran Khas Insentif Penyeliaan Perak (GKIPP) from Universiti Teknologi MARA, Perak Branch.

References

[1] Carrie A W M and Kelly L W 2018 *Making Research Relevant: Applied Research Designs for the Mental Health Practitioner.* https://books.google.com.my/books?id=UWNgDwAAQBAJ&printsec=frontcover#v=onepage&q&f=false
[2] SC 2018 *List of listed real estate investment trusts (As at 31 Devember 2018)* https://www.sc.com.my/api/documentms/download.ashx?id=bf905fdd-b9c5-4a95-bdf3-6c953d7d43c0
[3] Saunders 2016 *Research Methods for Business Students.* Harlow-Munich: Pearson
[4] Lam T Y M & Tipping M 2016 *A case study of the investment yields of high street banks. Journal of Property Investment & Finance, 34*(5), 521–534. https://doi.org/10.1108/JPIF-03-2016-0019
[5] Wong Y M 2015 *Malaysia REIT: First Decade Development and Returns Characteristics.* Ssrn, (January). https://doi.org/10.2139/ssrn.2555174.
[6] Creswell J W 2013 *Research Design: Qualitative, Quantitative, and Mixed Methods Approaches.* SAGE Publications. https://doi.org/10.1007/s13398-014-0173-7.2
[7] SC 2013 *Securities Commission Malaysia Annual Report 2013*: Securities Commission Malaysia.
[8] Ong T S, Teh B H, and Chong M P 2011 A Study on The Performance of Malaysia Real Estate Investment Trusts From 2005-2010 by Using Net Asset Value Approach. *International Journal of Economic and Research*, **2** 1–15.
[9] Yin R K 2011 *Qualitative research From Start to Finish.* New York London: The Guilford Press.
[10] Thomas G 2011 '*How to do Your Case Study - A guide for students and researchers*', Cornwall: SAGE Publications.
[11] Yin R K 2009 *Case Study Research: Design and Methods.* Sage Publications, California.
[12] Ting K H and Lee C L 2009 The role of Malaysian securitised real estate in a mixed-asset portfolio. *Journal of Financial Management of Property and Construction* **14** 208–230 https://doi.org/http://dx.doi.org/10.1108/13664380911000440.
[13] Tipping M and Bullard R 2007 *Sale and leaseback as a British real estate model.* Journal of Corporate Real Estate. Retrieved from http://www.emeraldinsight.com/doi/abs/10.1108/14630010710848458
[14] Evans A W 1995 The property market: ninety per cent efficient? *Urban Studies* **32** 5-30
[15] Estates Gazette 1993 *Yields and Rates of Interest. In: Mainly for students.* http://www.egi.co.uk/egarchive_detail.asp?fprint=1&auto=1&multi=1

The Implementation of Green Neighbourhood Initiatives in Urban Living: Methodology of Research

R Ramli[1], D Omar[1], P Ahmad[1] and O L H Leh[1]
[1]Faculty of Architecture, Planning and Surveying, Universiti Teknologi MARA, 40450 Shah Alam, Selangor, Malaysia.

Email: rohanaramliuitm@gmail.com

Abstract. Green neighborhood is a neighborhood that practices a way of life to meet the needs of people's activities. Community practices in this green neighborhood also have the potential to regulate pollution, save energy, accessibility, safeguard the environment, increase employment opportunities and reduce crime rates. Green neighborhoods can form friendships, practice agricultural areas, use renewable energy methods, and support sensitive nature. This practice also makes easy access to home, workplace, infrastructure, public utilities, transit facilities and easily accessible from one neighborhood. Conflicts between rapid development and local communities required a new model to understand about the relationships between green neighbourhood initiatives from local communities. This paper discusses how to doing research method on the effectiveness for implementation of green neighbourhood initiative in urban living and part on my ongoing PhD study. It starts by discussing how the theoretical background of green neighbourhood initiatives in urban living, then how the initiatives contribute to sustainable environmental and quality of life in urban living. Subsequently, it discusses the method of the study to conduct research on green neighborhood practices with a special emphasis to seek a clear understanding of acceptance from society on new actions, problems arising, community perceptions and community proposals. It followed by data collection methods, how data analysis conducted and results. It ends with preparation of questionnaires to look more in detail on the perceptions of society in terms of economic, social and environment aspects. Results and proposals designed to help communities, local authorities and agencies to carry out five selected green neighborhood initiatives.

1. Introduction

In recent years, sustainability of urban development has characterized as a development that can cater to the needs of today's development without affecting the capabilities of development and future generations to meet their own needs. Since the United Nations Conference on the global agenda and the Human Environment in 1972, Malaysia has taken serious steps in addressing environmental issues and issues. In 1992, Malaysia also demonstrated its commitment to the Rio Summit afterwards the Malaysian Nature Environment. Sustainable and Environmental issues are undoubtedly the challenge of the 21st century. The current global climate change brings the need to serious in designing urban development. Cities now find themselves in the middle of the 'Green Revolution' as one of the most important development components to do sustainability. Some of the innovations and city initiatives meant to make city more sustainable, greener, healthier and environmentally friendly. One of the concepts that has introduced for sustainable city formation is the Green Neighbourhood Initiative. This initiative is the first attempt to join the principles of urban sustainability with the level of community development at a micro level. Green neighborhood designated using various indicators that go beyond traditional variables such as plant protection, convenience, energy conservation and park size. In the

IOP Conf. Series: Earth and Environmental Science **385** (2019) 012032 doi:10.1088/1755-1315/385/1/012032

Tenth Malaysia Plan, emphasis was on formation of a vibrant, compressible and vibrant city, green technology development and green building, which is part of the green neighbourhood.

1.1 Background

Green neighbourhood is a relationship that people have with nature resources. Green Neighbourhood is an application and a practice for sustainable in term of economic, social and physical. The challenge of neighbourhoods regards how involving sustainable practices attached with forms, technologies and process of 21st sustainable cities. This is not a cultural changes made necessary by the negative impacts of global warming. It is an opportunity to enhance the living environment. It is also one of afford to practice the green neighbourhood initiatives for environment. Green neighbourhoods and sustainable is an area with mixed-use with community had a good relationship with each other, livable communities for all ages and a place to live, work and play, facilities that can meet the needs of various populations in the present and the future. This problem is also due to global warming and global climate change has affected the lives of the world's people in various stages. [1] mentioned the global population has risen by about 80 million people each year. In 1990, only 10 percent of the world's population lived in the city and by 2008, urban population has risen to more than 50 percent. The increase expected to 60 percent by 2030. The city's estimated growth rate is four times in rural areas, and it expected that urban populations would increase to 5 billion by 2030.

Lack of awareness about the important of preserving environment from stakeholders and communities. [2] mentions around the world, waste generation rates are rising. In 2012, people in the world's cities produce 1.3 billion tons of solid waste annually with an estimated 1.2 kilograms per day. By 2025, with the growing population growth and rapid urbanization, municipal waste generation expected to increase to 2.2 billion tonnes. The waste managed will contribute as a breeding ground for disease vectors, contributing to global climate change through methane generation and promoting urban violence.

Lack of awareness and public participant of community in green neighbourhood initiatives program by local authority. According to [3] the level of understanding and awareness of community to participation in green neighbourhood initiatives program are very low. [4] mentioned local agenda in Malaysia has introduced since 1999 and today, in local authorities in Malaysia still in the process of implementation.

1.2 Aim and Objectives

The purpose of this paper is to carry out a research method on the effectiveness of implementing green neighbourhood initiatives in urban living. The aim is specifically to find the implementing research methods by discussing how the concept of green neighbourhood initiatives in urban living and how it can give to the environment and sustainable quality of life. This paper discusses the basics of method to conduct research on green neighbourhood practices, investigate the effectiveness of current practices, acceptance of new concepts and perceptions of society, local authorities and agencies. This paper will then take into account how the proposed data collection and analysis is to see more detail for the perceptions of the economic, social and environment aspects. It also discusses how other steps helped develop this questionnaire. From the survey, it can see results, potentials and suggestions.

2. Theoretical Background

Refer to [5], green neighbourhood guideline as guidance to developers and Local Authorities on the elements that should include in developing green neighbourhood initiatives. It is initiatives include walkability and connectivity, creating a transit facility in the near distance, green network, mixed-use development, high density, neighbourhoods safe environment, application of green technology in building construction and infrastructure facilities, selection and planning for neighbourhoods community network.

2.1 Differences between Conmon Neighbourhoods and Green Neighborhood Initiatives

Green Neighbourhood initiatives are gaining popularity in the develop area, the concept, initially derived out of concern for community urban living and environmental.

Table 1. The Differences between common neighbourhoods and green neighbourhoods in Malaysia

Common Neighbourhood	Green Neighbourhood
Not apply environmental friendly in building, infrastructure and utilities construction	Apply green elements in construction for buildings, infrastructure, utility and built environment.
Depending on all natural resources and no energy savings.	Apply and always practice energy savings and good resource management.
Communities use private vehicles, no efficient public transport facilities and no transportation facilities integration	Apply the concept of transit-oriented development, integration public transportation mode. More concept of connectivity, promotes walkability and narrower streets.
Uncontrolled development, no green element and encouraging sprawl.	Development practices conserve land and energy-saving and resources saving.

Source: PLANMalaysia (2012) [5]

Table 1 indicating the differences between common neighbourhoods and green neighbourhoods in Malaysia. Green neighbourhood apply the green approaches in buildings, infrastructures, practice efficient, transit-oriented development and to conserve land and energy resources.

2.2 Malaysian Green Neighbourhood Development Action Plan

According to [6], for the realization and success of this action plan must requires support from stakeholder, local authorities, developers, urban planners, engineers, landscapes, environmental officers, architects and related agencies in the public and private sectors. Base on [6], the guidance aimed at assisting local authorities and agencies to carry out five (5) selected green neighbourhood initiatives as pedestrian walkway, bicycle lane, rainwater harvesting system, waste composting and community farming. These five actions to prioritize because it can be implemented with a quick and easy time and a less expensive.

3. Conceptual Framework

This research is to study the implementation of green neighbourhood initiatives in urban living. Conceptual framework can adapted to explain the achievement from implementation of green neighbourhood initiatives to get aims of study from the social, economic and environmental aspect.

A research hypothesis for this study come from the main research question. According to the PLANMalaysia 2019, there is still a state in Malaysia, which has not yet adopted these initiatives. While according to Town Planning Department, Subang Jaya Municipal Council (2019), the level of understanding and awareness of community to participation in green neighbourhood initiatives program are very low. There is a significant implementation of green neighbourhood effect to social, economic and environment in urban living.

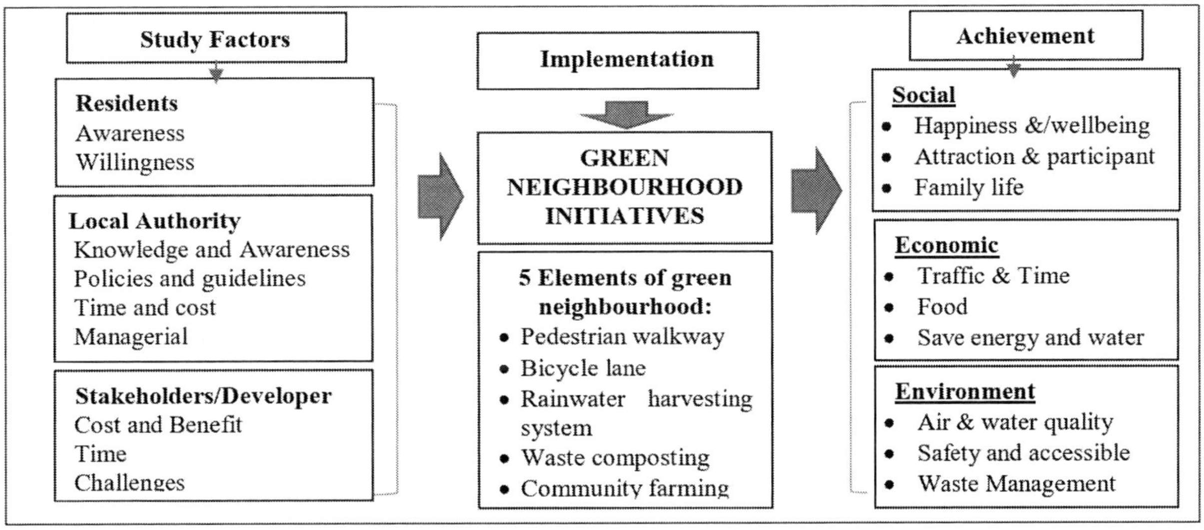

Figure 1. Conceptual Framework

3. Methodology

According to [7], the methodology is a theoretical and systematic analysis of the methods used to conduct a study. It includes theoretical analysis of the methods and principles related to the study. Methodology includes concepts such as paradigms, theoretical models, phases and techniques of obtaining data. [8] note that this methodology is a guideline for problem solving, with specific components such as methods and tools for planning, reviewing and conducting projects.
The case study area chooses located at Subang Jaya Municipal Council, located 25 km south of Kuala Lumpur. While Shah Alam is, the state capital of Selangor, Malaysia situated within the Petaling District.

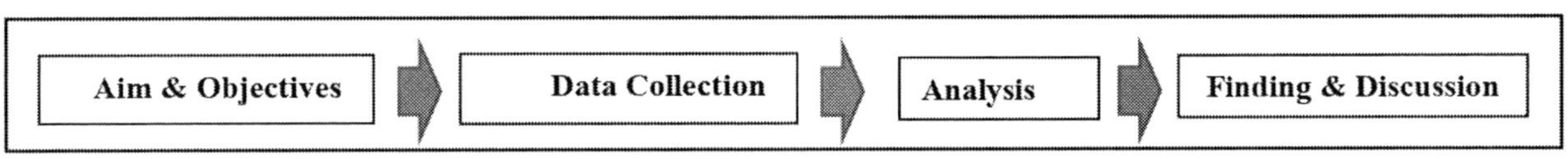

Figure 2. Methodological Framework

Figure 2 indicating the proses of research method which include aim, objectives, data collection, analysis, finding and discussion on implementation of green neighbourhood in urban living.

3.1 Aim and Objectives

These research objectives are to investigate the current practices of green neighbourhood initiatives, to study the criteria of green Neighbourhood in urban living, to investigate the green neighbourhood initiatives that influence involvement and willingness among community and to develop strategies of green neighbourhood initiatives in urban living.

3.2 Data Collection

Data collection using mixed method integrates both forms of data, various sources and uses different designs. The primary assumptions of this inquiry forms are integration of qualitative and quantitative results to produce a wealth of accurate information with data on population questionnaires and interview. [9] describe various methods, including collecting data through surveys, observation methods and using existing secondary data.

3.2.1 Quantitative Method. This method most often carried out by collecting resident questionnaires. This questionnaire is important to decide the exact number of respondents required. For this paper, the questionnaire survey method will carried out with 400 questionnaires from the communities of Subang Jaya Municipal Council.

Table 2. Respondents for Questionnaire Method

Parties	Respondents	Reasons
Communities/Residents	• Residents • Jawatankuasa Penduduk JKP (24 JKP)	• Issues of communities, willingness • Public responsibility, knowledge, awareness • Public relationship, advantage, disadvantage • Community's needs, comments, suggestions

Table 2 indicates the research method conducted by the questionnaire survey that involves collecting quantitative data from communities or residents.

3.2.3 Qualitative Method. The qualitative method is to hold interview sessions, the basis of respondents about experience and those who manage green neighbourhood initiatives. Samples selected are people from Subang Jaya Municipal Council and other agencies. Questions provided to respondents for data gathering related, policies and guidelines, initiatives, management, planning strategies to solve implementation problems and programs conducted to resolve these issues. All interviews will conducted face-to-face and will need important skills in conducting research to know perceptions, knowledge and behaviour.

Table 3. Respondents for Interview Method

Parties/Agencies	Name of Departments	Reasons
Local Authority	• Town Planning Department • Environmental Department • Engineering Department • Landscape Department	• Issues of local authority and community • Support green neighbourhood. • Implementation of green neighbourhood. • Managing the problems, expertise, ability • Future planning, comments, suggestions
PLANMalaysia Developer	• Research and Development Department	• Issues and problem, guide line, policies • Knowledge, willingness, expertise, ability • Future planning, comments, Suggestions

Table 3 indicates the research method conducted by the interviews that involves collecting qualitative data from local authorities, agency and stakeholders.

3.3 Analysis

3.3.1 Quantitative Analysis. Analytical techniques for quantitative data can use descriptive statistical analysis for frequencies, cross tabulation and descriptive ratio. These techniques associated with numbers as ideal tool for various data analysis tasks. Quantitative data will analyze using respond to codes on Likert scale from 1 (lowest) to 5 (highest). The main nodes used are elements of green neighborhood initiatives, challenges and benefits that affect the green neighborhoods implemented, factors that influence the involvement and willingness of the community in urban life and the effectiveness of communities using green neighborhood initiatives within the urban living.

3.3.2 Qualitative Analysis. Technical analysis to analyze qualitative data is computer qualitative data analysis with coding question technique and performed by identifying all the nodes associated with this study. Data will organize, store and retrieve. The use of key nodes is issues and problems, planning, managing, comments and suggestions to carry out green neighbourhoods initiatives.

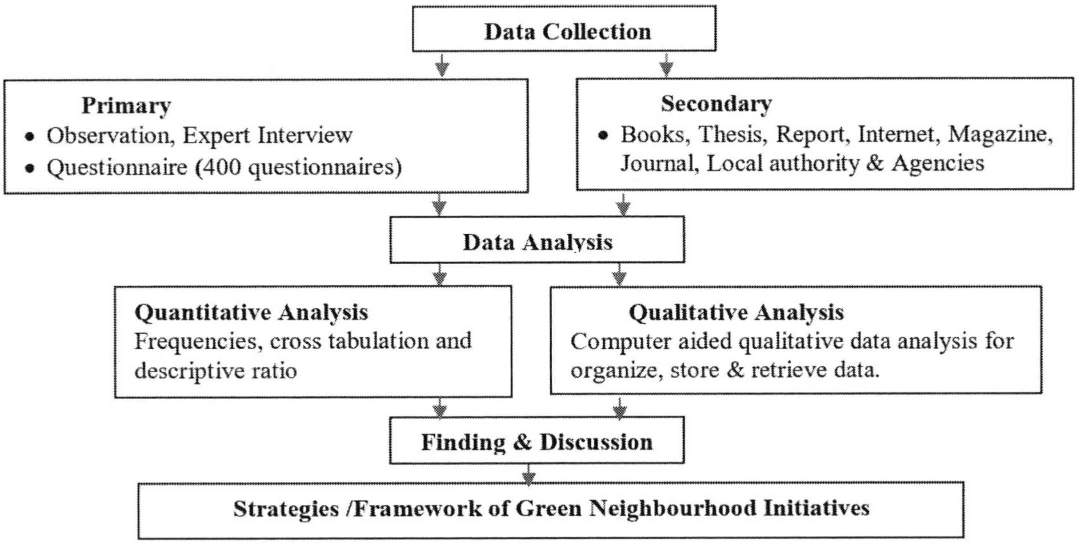

Figure 3. Relationships between data collection and data analysis

3.4 Finding and Discussion
Result from observation, interview and resident's questionnaire that conducted from difference neighbourhood, organisation and backgrounds. The findings consist of various data and information's from perceptions among the neighbourhood and current practise from local authority. All the data and information's about implementation of green neighbourhood initiatives will have reported. The finding will include the underlying perceptual categories of planning for green neighbourhood initiatives, trend and demand of green neighbourhood initiatives, implementation from local authority and stakeholders, perception of communities, problems, community needs and correlations between factors.

4. Conclusion
This paper concludes that method is a very important part of the research. Research method is a specific procedure used to find, select, process, and analyse information about a research topic. The method can check the validity and reliability of the entire study conducted in order. This research aim to gather the feedback from communities, stakeholders and local authority on how implementation of these initiatives in urban living and affect to social, economic and environment. The most suitable type of community activities in urban living will make sure the high level of development in urban living. Finally, this study will bring a lot of benefit to resident, developer and local authority and the area itself. Approach from stakeholders, local authorities and residents with implementation of green neighbourhood initiatives will increase quality for communities, economic and environment.

Acknowledgement
The authors would like to thank Universiti Teknologi MARA (UiTM) for the support and partly funding the study and publication through the BESTARI research grant (600-IRMI/MYRA 5/3/BESTARI (008/2017)). The authors are also thankful to all the departments, organisations, and person who had contributed to this study, specially PLAN Malaysia, and Subang Jaya Municipal Council.

References
[1] Wahi N, Mohamad I, Mohamad Zin R, Munikanan V, Junaini S 2017 *The High Rise Low Cost Housing : Sustainable Neighbourhood Elements (Green Elements) in Malaysia*. (Malaysia: Universiti Teknologi Malaysia)
[2] United Nation- Habitat 2018 *The World Urban Forum (WUF)*
[3] The World Bank 2018 *Solid Waste Management. Understanding Poverty*

[4] Sustainable and Local Agenda 21 Unit 2019 *Town Planning Department, Subang Jaya Municipal Council.* (Malaysia : Selangor)

[5] Mat Nurudin S, Hashim R, Hamik S A, Rahman S, Zulkifli N, Mohamed A S P 2015 *Public Participation in Local Agenda 21 Programs Implemented by Seremban Municipal Council* (Malaysia)

[6] PLANMalaysia 2012 *Garis Panduan Perancangan Kejiranan Hijau: Bahagian Penyelidikan dan Pembangunan.* (Malaysia : PLANMalaysia)

[7] Rosly D, Mohd Puzi N Z, Arshad M R 2013 *Planning Strategies, Guidelines And Action Plan For Green Neighbourhoods In Malaysia.* (Malaysia : PLANMalaysia)

[8] Igwenagu C 2016 *Fundamentals of Research Methodology and Data Collection* (Nigeria: LAP Lambert Academic Publishing) p1

[9] Ishak I S and Alias R A 2005 *Designing A Strategic Information Systems Planning Methodology For Malaysian Institutes Of Higher Learning (Isp-IPTA)* (Malaysia :Universiti Industri Selangor, Universiti Teknologi Malaysia)

[10] Smith, Thorpe and Jackson 2012 *Management Research: 4th Edition* (Great Britain: Sage Publication Ltd)

IOP Conf. Series: Earth and Environmental Science **385** (2019) 012033 doi:10.1088/1755-1315/385/1/012033

Approaching BIM-GIS Integration for 3D Evacuation Planning Requirement Using Multipatch Geometry Data Format

S A F Syed Abdul Rahman[1], K N Abdul Maulud[1,2]
[1]Earth Observation Centre, Institute of Climate Change (IPI), Universiti Kebangsaan Malaysia, 43600 Bangi, Selangor, Malaysia.
[2]Smart and Sustainable Township Research Centre, Faculty of Engineering & Built Environment, Universiti Kebangsaan Malaysia, 43600 Bangi, Selangor, Malaysia.

Email: p94717@siswa.ukm.edu.my

Abstract. Building Information Modelling (BIM) is known as a new technology that developed specifically to serve the architecture designing activities. It is used as a tool for researchers in Architecture, Engineering and Construction (AEC) fields to collaborate between different parties in order to manage the building information. Currently, there are plenty of researches that discussed on the concept of translating data between BIM into Geographical Information System (GIS) for various purposes. This trend leads this paper to focus on the method and findings of integrated 3D technologies between GIS and BIM. The implementation of integration between two different formats requires deep understanding about the core of both platforms such as differences and similarities between both formats. It is important to identify all related aspects of evacuation such as least risk paths, path simplicity, navigable network improvement, more detailed routes, and human behavior. To fulfil the paper research topic of Approaching BIM-GIS Integration for 3D Evacuation Planning Requirement Using Multipatch Geometry Data Format, four major concepts were the main focus namely; BIM-GIS Integration Method, Integration Data Level, Integrated Data Consideration and Requirement Analysis. Digital component involves was Revit Autodesk as BIM 3D development model software, the final integrated data format was 3D multipatch geometry and ArcGIS Pro v2.2 software used for 3D model visualization to conduct evacuation planning. The requirement for data integration and indoor evacuation planning analysis were also itemized where plenty of specific decision to be taken on based on the capabilities of current technology. Future research will focus on access-to-road strategy to test the connection of indoor spatial networks with outdoor spatial networks that based on the different data structure (BIM/vector and DEM/raster) for detailed 3D evacuation planning.

1. Introduction

In the era of Forth Industrial Revolution technology development at a high pace, man-made ecosystem is becoming more complex and sophisticated. These intricate collections of infrastructure, materials, machinery and people, require progressively more sophisticated tools to design and manage them with countless spatial and temporal relationships and dependencies [1]. To maximize the potential of those fields, some studies were conducted which related to existing systems for evacuation activities. Currently, most of the building evacuation system is build and analyzed through a 2D platform where the elevation aspects are not emphasized. Due to the rapid growth of elevated building all around the world, these aspects need to be featured in 3D visualization as it can offer plenty of advantages as compared to the 2D platform. With the technically advanced of BIM in 3D, it surely can be a catalyst for the process of evolution in building information technologies.

BIM was introduced in this era of elevated building development as a new technology which technically used to serve the architecture designing activities. BIM offers a flexible approach that covers all important element structure for a stage in designing a certain model. BIM also has capabilities in creating, managing, derive and share building information with another platform that also involves in the construction process to conduct collaboration of information sharing between them [2].One of the BIM platforms which is Revit Autodesk was used in this research and the development of 3D BIM model was using Revit Autodesk (.rvt) data format. Revit Autodesk is a building information modelling software for AEC. The software is capable in designing a building structure and components in 3D. Since there are plenty of potentials on BIM, it became research pit for researchers in AEC fields to conduct experimentation in order to collaborate with different parties especially GIS where it carries a pile of information storing widely used around the world.

On the other hand, GIS is an advanced technological tool which specialized in acquainting geography and provide intelligent decisions [3]. GIS basically works by categorizing the geographic data to enable user understanding on the map and select necessary data for a specific task. A thematic map contains a table of contents that give space to the reader to overlay layers of information to a real-world base map of locations. BIM and GIS have their core in different knowledge areas [4]. BIM is being used by providing detailed 3D building models that could be used for a construction project, operation, and so on [5]. Meanwhile, GIS is used to analyze and examine location-related information in geospatial science, by integrating attribute and spatial besides deriving information through various spatial analysis tools and visualization approaches. This integration promotes the application of geospatial analysis and visualization in business processes that occur inside buildings by making it possible.

The integration between BIM and GIS is being discussed from time to time. With rapid development of technologies, the technique of translation data model between BIM and GIS is evolving fast. For example, the capabilities of GIS software in reading Revit Autodesk (.rvt) format is recently developed. This makes the visualization of BIM 3D models are spread widely among GIS software especially ArcGIS Pro v2.2. Unfortunately, there are still limitations on this specifically in exploiting the data into the 3D analysis. In this paper, the authors focus on the necessary portion of translating BIM data into multipatch geometry (.shp) format through several procedures in order to meet indoor navigation for evacuation planning framework. To be clear, there is a supplementary GIS specific requirement needed for indoor evacuation planning analysis based on the purpose of the project. This is because different type of functionalities requires a different type of information supplied from data translation.

Evacuation route planning for emergency response is heavily bonded with indoor routing information. It is relying on geospatial data and network structure model to utilize its potential. Unfortunately, indoor geospatial data are complicated and time consuming as each structure have specific characteristics especially semantic information. Currently, AEC has been used widely in the generation of indoor network models. These models usually focus on visualization but lack in semantics information. To achieve the data interoperability in emergency response models, both geometry and semantic information must be connected to each other, hence BIM and GIS can be used to give a balance for indoor integrated analysis application.

This paper is organized as follows: Section 2 will present briefly about interrelation between BIM and GIS and the differences between both platforms. This is important to give a big picture about BIM-GIS relationship when comes to data structure and visualization. Section 3 will involve integration activities in which three methods at highpoint to achieve the multipatch geometry as an integrated data format. Section 4 will focus on the integrated data level of multipatch geometry from the perspective of BIM and GIS platform. Section 5 will highlight the integrated data consideration by measuring with selected active 3D analysis which can profit the evacuation planning quality. Section 6 will present the requirement needed for indoor evacuation planning analysis where was conducted based on the specific GIS software perspective, ArcGIS Pro v2.2.

2. Building Information Modelling (BIM) and Geographic Information System (GIS)

2.1. Interrelating between BIM and GIS

As mentioned before, BIM and GIS are two different platforms. Each of them has a specific objective in their own type. BIM system is basically built on building information platform that does not exist physically. On the other hand, GIS comes in totally different concept as it is established to represent objects that already exist around us. For the geospatial information models, it is definitely for representing the objects in the most intellectual way. The representation of semantic information has constantly been vital for geospatial information models. In the simplest way, BIM offers indoor spatial information but it lacks spatial functionality [6] while GIS delivers spatial information related to outdoor [7].

The detailed information on 3D models designed in BIM help GIS to extend its scope by applying spatial analysis at a better scale. It is also can be exploited into several contexts such as several active research about 3D analyses that were used in this study to measure the effectiveness of integrated data. The selected 3D analysis are as follows:

a. Indoor network analysis such as evacuation navigation network [7]. The consume building models is used to create indoor networks for emergency response as information of definite area measurements of hallways and leave entryways is an essential spatial information to build up evacuation path during the ongoing crisis. This can be accomplished adequately with the integration of 3D GIS and BIM.
b. Location based services analysis such as asset or facility management [8]. The potential for enhancing evacuation route decisions by integrating BIM and GIS frameworks and set up a two-way connection between two spatial information's which is 3D geometry and asset position. This analysis also helps in identifying a high percentage of areas to engage with any emergency matter.
c. Layout planning such as wireless sensor coverage [9]. Through combination between BIM detail building information and GIS precise spatial information, the layout planning can maximize its potential by a collaborating with the real time wireless sensor coverage (smoke or fire sensor), emergency events can be detected more quickly. Besides that, the option for evacuation navigation network also can be narrowed down by specifically on which route is safe to be used.
d. Environmental analysis such as view coverage analysis [10]. The integrated model can be linked with its surrounding information. For example, while planning a partition in an existing environment, information from BIM models enable us to integrate with environmental information to carry out view coverage analysis for rescuing and evacuation process.
e. Dynamic factor analysis such as fire or smoke analysis [11]. The building models need to consider the computation and simulation of indoor emergency cases such as smoke, fire and pedestrian flow. Smoke and fire signify hazards during emergency circumstances and change over time, the occupant evacuation needs to be planned for the pedestrian flow.

The selected 3D analysis above can be combined under one simulation and have capabilities to improve the decision making in identifying the evacuation path. The most imperative involvement of BIM to GIS is providing comprehensive 3D building models also their rich building information. These reasons would be tough to comprehend GIS without the details level of BIM. Another reason in which BIM may profit from GIS ability is because it needs some of the rich spatial analysis purposes from GIS to extend its ability, such as identifying the safer route for safety and evacuation in precise measurement. The specific concept for BIM and GIS are highlighted in Table 1. BIM and GIS is complementing each other. GIS is helping BIM to associate to the outdoor, and BIM consultants are presenting GIS to the indoor environment. They could attain much more in combination rather than by working in isolation.

Table 1. The specific concept for BIM and GIS.

Concept	BIM	GIS
Field Industry	Architecture	Geospatial/ Planning
Object Representation	Represent non-exist physical object	Represent exist physical object
Spatial Information	Indoor information	Outdoor information
Semantic Information Model	Detail information	General Information
Spatial Functionality	Low level	High level
3D Geometry Model	High level visualization	Medium level visualization
Visualization Specification	LOD100 - LOD500	None
Analysis Capability	Require third party software	Widely used for analysis

2.2. 3D Structure in BIM and GIS

In order to integrate BIM and GIS, the understanding of the current 3D structure availability of both platforms is important. For BIM, the 3D structure is built in details scale. Based on the definition used by The American Institute of Architects [12] there are five basic levels of development. All structure is designed based on each function and separated into five main levels same as IFC scheme Level of Development (LOD) which are:

a. LOD 100 – It is fundamentally on the corresponding of conceptual design, the model contains total building mass. Based on this type of LOD, it can be used to accomplish mass building types of analysis such as building orientation, volume, cost per square foot, and so on.
b. LOD 200 – It is developed same as the schematic design. The model assemblies with approximate quantities, location, shape and orientation. The use enable an analysis of particular systems by application of comprehensive performance standards to be standards.
c. LOD 300 – This model is fit for the creation of construction of documents and shop design. It is capable to run a simulation and analysis for complete elements and systems.
d. LOD 400 - This model is suitable for manufacturing and assembly. The details of this LOD is most possible to be fitted usually outside the scope of the architect's or engineers. It usually integrates with sub-surface utilities mapping for GIS fields.
e. LOD 500 - This level of development signifies the models as it has been manufactured the as-built conditions. This model is suitable for the operations and maintenance of the facility.

In short, it can be explained that as the type of LOD increased, more particulars are contained in the model. On LOD 100, one solid structure is representing by the model, but on LOD 500, it is built as a complex model with numerous components, including enlargement joints and so on [13]. Examples of the BIM model LOD are shown at Figure 1. From the viewpoint of GIS, spatial analysis was focused on 2D data, and its ability in 3D data has been incomplete. Usually, the 3D models were extruded from 2D drawings, and some of the simple editing functions are provided such as rotate, move, scale, merge, split and union. The models created could reach LOD 100 according to the standard of BIM, which are the most basic block models [13].

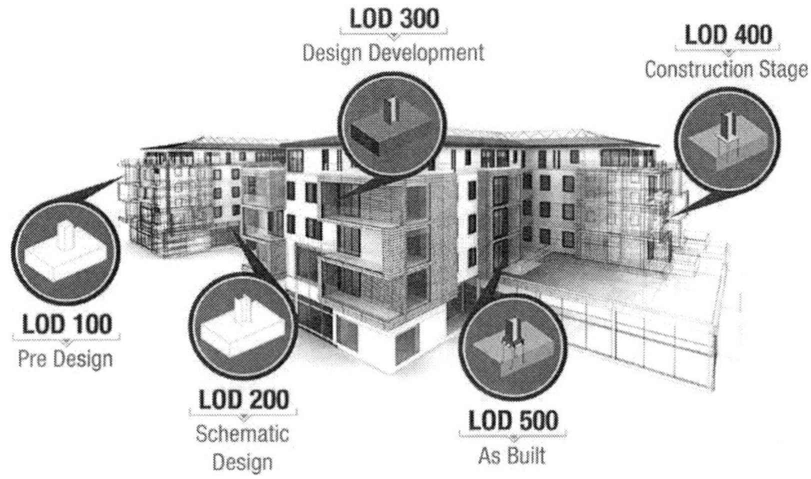

Figure 1. Example BIM model LOD [12].

2.3. Integration Method

The BIM-GIS integration is important as the integrated product can be manipulated into the multipurpose analysis. A 3D spatial query is essential in the GIS field and it can be utilized with the information extracted from BIM data. There are some options which interrelating BIM and GIS. For example, Boyes et al. [8] proposed a combination use of BIM and GIS for asset management. It is based on the capabilities of GIS in location-based query and asset information from BIM. Kang and Hong [15] also preferred the conversion from BIM data to GIS for extraction of information and to be utilized for facility management. The authors prefer to integrate BIM data into a GIS platform.

There were many papers published in the field of BIM-GIS integration. As mentioned before, this paper focused on two main phases which were data integration for indoor mapping generation and indoor evacuation planning analysis framework. In this section, the main objective is to integrate BIM data into a GIS format for 3D evacuation planning. The designed BIM model is in .rvt format and the GIS indoor mapping format is in multipatch geometry (.shp). There were three methods being used for data integration which are:

2.3.1. Method 1: Revit (.rvt) to IFC (.ifc) to multipatch geometry (.shp)

For this direct integration method through IFC exporter, two elements were translated; boundary representation (B-rep) and sweep solid (SS) to represent 3D Geometry [16]. B-rep represents a 3D object using its bounding surfaces. It is used for composite objects, such as doors (IfcDoor) and windows (IfcWindow). In SS, a 2D surface organized with a path is used to state a solid, the path outlines the route over which the surface is extruded. But it requires data interoperability extension on ArcGIS Pro v2.2 for converting the .ifc format into multipatch geometry. This method doesn't have the capabilities to generate room volume based on room boundaries. This become the limitation for the user who seeks for 3D analysis that relates to space and volume such as fire or smoke simulation. By integrating the BIM data through this method, it is appropriate for LOD 300 and above as LOD 100 and LOD 200 does not attain any impact other than data visualization.

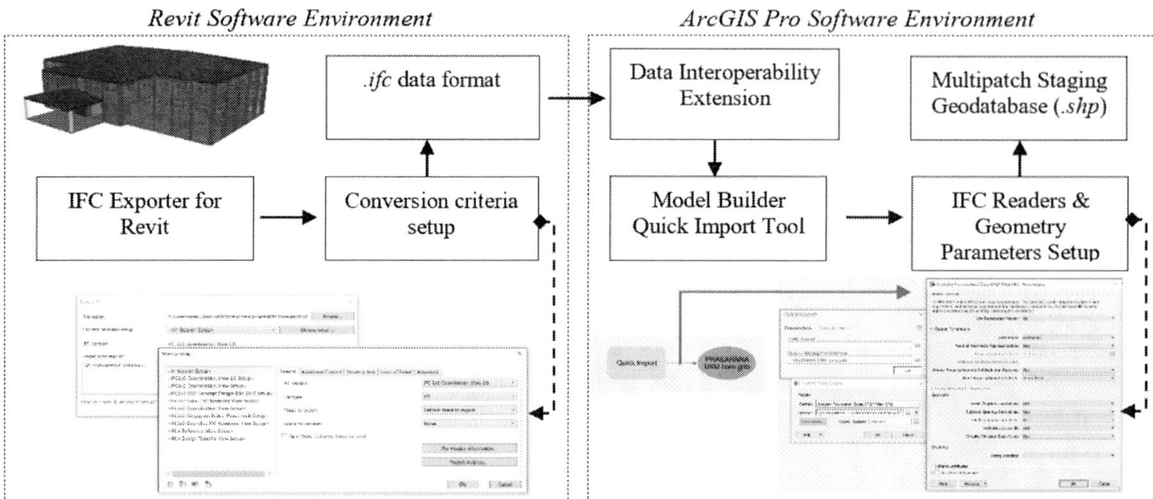

Figure 2. Method 1 Integration Method

IFC exporter is a build in extension within Revit Autodesk. The conversion process from .rvt data format in Method 1 is using .ifc data format due to the limitation of GIS platform in reading directly .rvt data format. Several criteria needed to be specified such as level of detail (low, medium, height) and property set (base quantity, common property set) before the conversion take place. After converted in .ifc data format, a data interoperability extension is needed in ArcGIS Pro enabling the software to activate the Quick Import tool for migrating the data into a multipatch data format (.shp) as shown in Figure 2. Through the Quick Import tool, IFC readers and Geometry parameter can be set to make sure the converted data follow the user requirement.

2.3.2. Method 2: Revit (.rvt) to multipatch geometry (.shp)
This is second method that integrates BIM data directly into multipatch geometry and just currently developed in July 2018. This method can be used in ArcGIS Pro v2.2 as it has capabilities to reading the .rvt format. The Revit file is translated on the fly to be represented in a data model, like CAD data, so that it is used as a direct input to many of the geoprocessing tools but also displayed on a map. Unfortunately, to translate those data into the multipatch geometry format, it requires the user to export the data directly from the software. The advantages of this method is it's practicality in data visualization as the data arranged in the default layer group consist of Exterior Shell, Architectural, Structural, Electrical, Mechanical and Pipping. This method has limited capabilities as only B-rep of 3D geometry is integrated from the BIM model. It unable to generate the room volume based on room boundaries but have enough content to run simple 3D analysis such as indoor pedestrian route network. This gives advantages for the user who seeks for 3D analysis that relates with visualization and spatial analyst. Other than that, the integrated data also can be used to preserve the asset database to monitor the energy consumption of the building as suggested by Fosu et al. [17]. By integrating the BIM data through this method, it is appropriate for LOD 300 and above for advanced 3D analysis as LOD 100 and LOD 200 does not attain any impact other than data visualization.

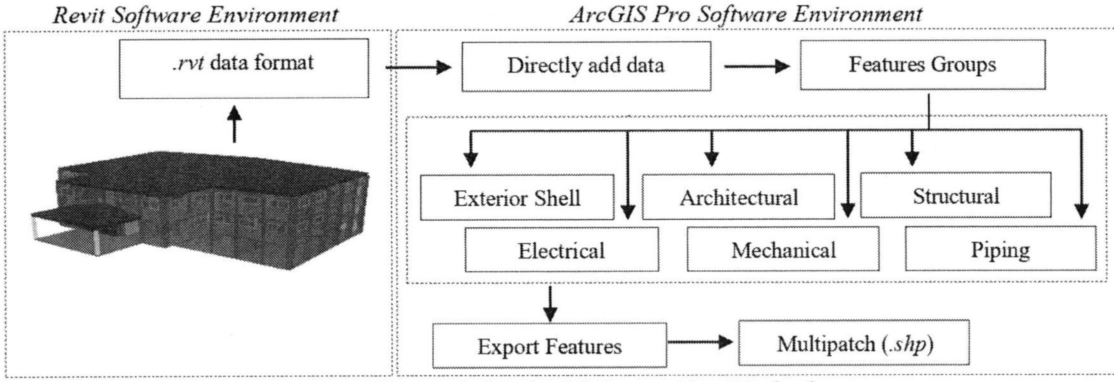

Figure 3. Method 2 Integration Method

By adding the .rvt data into ArcGIS Pro, the data will be organized into several categories such as Architectural, Structural and so on. These data sourced Revit properties in feature classes and the parameter is defined as ArcGIS features attribute. In order to migrate those data into 3D multipatch geometry data format, the user can use the export features tool as shown in Figure 3. The data products sourced Revit properties in feature classes but the parameter is defined based on ArcGIS features attribute.

2.3.3. Method 3: Revit (.rvt) to FME Workbench (.rvz) to multipatch geometry (.shp)

For this indirect integration method, it is selected to be used in the integration phase because of its flexibilities in designing the semantic information which requires further analysis. The semantic information is important as there are plenty of information stored in the BIM model. The selection of data information based on the 3D analysis requirement for the project can help in reducing the unnecessary data besides can reducing the size of integrated data. The 3D Geometry elements such as B-rep, CSG, and SS can be extracted from this method. For CSG, it used to represent the post-processing object for the operation of simpler objects such as the intersection of the wall in shape of cones, cylinders and so on. Besides that, there are numbers of geometry that can be translated from BIM models such as segments, path, simple area, simple surface, composite surface, simple solid, composite solid and so on. This method has the ability to generate room volume and gives advantages for the user who seeks for 3D analysis that relates to space and volume such as fire or smoke simulation. By integrating the BIM data through this method, it is suitable for LOD 300 and above as LOD 100 and LOD 200 does not attain any impact other than data visualization.

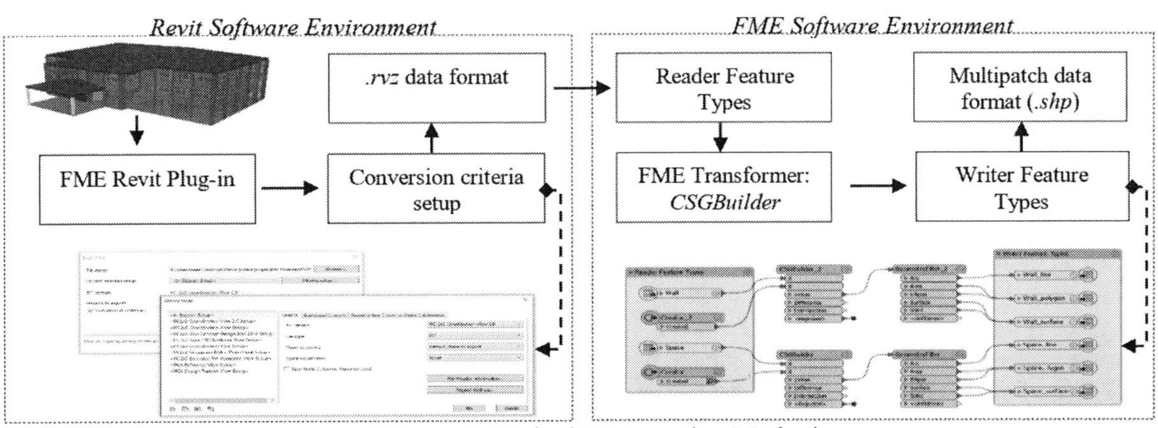

Figure 4. Method 3 Integration Method

In order to integrate BIM 3D model from Revit using Method 3, the FME Revit Plug-in extension is needed. By enabling the extension, the conversion process from .rvt data format can be easily converted into .rvz data format by specifying several details information for conversions such as level of detail (low, medium, height) and property set (base quantity, common property set). After the model converted into .rvz, a workspace needs to be set (reader feature types, writer feature types). In order to make sure the converted data in CSG geometry form, CSGBuilder in FME Transformers must be used as shown in Figure 4. If not, the converted data will act in b-rep geometry form. Since the target data format is in Multipatch Geometry, the writer format need to be set ESRI Shapefile while the reader data format is in Autodesk Revit. Run the Translation tool when the workspace has been confirmed.

Table 2. Integration of data output from three method.

Factor		**Integration Method**		
		Method 1	**Method 2**	**Method 3**
3D Geometry	**B-rep**	Supported	Supported	Supported
	CSG	Not Supported	Not Supported	Supported
	SS	Supported	Supported	Supported
LOD	**LOD 100 – LOD 200**	Visual	Visual	Visual
	LOD 300 – LOD 500	3D Analysis	3D Analysis	3D Analysis
Integration Level	**Geometry Level**	Median	Median	Details
	Semantic Level	Median	Median	Details

Based on the method listed in Table 2, each of them starts (.rvt) and finish (.shp) with the same data format but it produces different characteristics of the final product. Each final product has its own capabilities to be utilized into a 3D analyst. It shows the importance of selection of data integration method to define the suitability of data preparation for 3D evacuation planning.

3. Integration Data Level

3.1. Geometry Level

At the geometry level of the integration process, the focus is more towards the translation of information between BIM and GIS. There are two major issues that can be seen in the translation phase namely; 3D Geometry and Geometry Classification.

For 3D geometry, GIS and BIM are using different approaches. For example, Method 3 integration BIM through IFC format contains a combination of CSG, SS and B-rep to represent 3D geometry while Method 1 and Method 2 integration do not translate the CSG. This is because generally, the GIS platform is only supporting B-rep elements. Due to the integration conducted on a GIS platform, the 3D geometry elements will fulfill the requirement based on its current platform. It is important to understand this matter because the geometry data level is playing a major role in 3D indoor mapping.

The different in nature becomes a barrier for both platforms to be integrated automatically. The only solution for solving this 3D geometry problem is to understand the requirement for integration and identify the 3D analysis research to be carried out. For example, to conduct dynamic factor analysis such as smoke simulation, CSG is important so the integration method must be able to produce the CSG element on the output integrated product for volume usage. There is extension builds in GIS software environments such as Data Interoperability in the ArcGIS environment and IFC explorer in BIM environment to minimize the effect of data integration in geometry level. However, it still requires data identification to fulfill the objectives of model integration.

For Geometry Classification, the physical detailing in the 3D model by all three methods are different. For example, 23 classes BIM data such as ceilings, column, and door, wrap into "Architecture" layer in ArcGIS Pro v2.2 for the integrated data from Method 1. Another examples are the roof, floor and stairs from integrated data from Method 3 wrap into Slab_surface layer. Table 3 show some examples of integrated geometry data level. There are some commercial software packages such as Data Interoperability and Feature Manipulation Engine (FME) which build to support the data migration between both platforms but none of them can fully transfer the geometry and semantics between BIM and GIS. The semantic information loss for data visualization is the biggest disadvantage which mostly happen to a semantic incongruity among the two domains. So, the solution for integrated geometry data classification is the layer group for integrated geometry data need to be reclassified by the user.

Table 3. Example of integrated geometry data level.

BIM Features	Integration Method	Integrated Multipatch Geometry Features	Requires Actions
Roof Floor Stairs	Method 1	Slab_surface	Features Extraction
Roof Floor Stairs	Method 2	Architecture	Features Extraction
Roof Floor Stairs	Method 3	Roof Floor Stairs	Maintain Features

3.2. Semantic Level
In the semantic level, the information transferring is required to make sure the integration become successful. There are two mismatch issues related to that which are; Domain Definition and Semantic Classification. Firstly, is when the integrated BIM data defined some component with a specific domain and GIS database was define with another name. For example, when integrated BIM data define a window as IfcWindow while GIS define as Window in the database. This situation does not bring a major problem as the user can identify the type of features through the name. The second issue is Semantic Classification when the attribute component of integrated BIM data is generalized with a single specific name in the attribute. For example, BIM data defines a column, beam, stair, and so on, while integrated BIM data defines as BuildingInstallation. This situation gives the foremost problems for applications. In the case of generating an indoor evacuation network, the lack of matching connection of information between stairs makes it difficult to link the geometric between them and obstructs the capability to navigate between floors [18]. The semantic level of data integration becomes complicated when the BIM data contains plenty of information than GIS data. The main effort in semantic level data interoperability is then connecting the hole between the two representations [19], which means an alteration to current representations is needed. Table 4 below shows the integration factors and its effect on the data level.

Table 4. Integration factors of data level.

Factors	Geometry level	Semantic level
Level of difficulty	Median	High
Purpose	Visualization	Visualization & Analysis
Richness of information	Low	High
Semantic Loss	Yes	No

4. Integrated Data Consideration
The main objective of BIM-GIS integration for this paper is to give a review on the integrated multipatch geometry data based on a different method for appropriateness of 3D evacuation planning. In order to measure the effectiveness of integrated data, several active research about 3D analyses

were selected as the benchmark for assessment as explained in Sub-Section 2.1. Selected 3D analysis were listed as follows:

a. Indoor network analysis such as evacuation navigation network [7].
b. Location based services analysis such as asset or facility management [8].
c. Layout planning such as wireless sensor coverage [9].
d. Environmental analysis such as view coverage analysis [10].
e. Dynamic factor analysis such as fire or smoke analysis [11].

These 3D analyses were selected based on similarity in 3D mapping analysis aspect and the potential to improve the decision making of 3D evacuation planning. In the future, the results of each analysis above can be combined under one generated model and improve the evacuation planning. Based on the integrated data from the given method, type of LOD and data level as mentioned before, the data consideration can be concluded as Table 5 below:

Table 5. Integrated data consideration.

Capable Analysis	Consideration					Convenient Integration Method
	LOD	Data Level				
		Geometry Level		Semantic Level		
		3D Geometry	Geometry Classification	Domain Definition	Semantic Classification	
Indoor Navigation Analysis	LOD300 LOD400 LOD500	B-rep SS	Details	For every Features	Details	Method 1 Method 2 Method 3
Location Based Services	LOD300 LOD400 LOD500	B-rep SS	Median	For every Features	Median	Method 1 Method 2 Method 3
Layout Planning Analysis	LOD300 LOD400 LOD500	B-rep CSG SS	Median	For every Features	Median	Method 3
Environmental Analysis	LOD300 LOD400 LOD500	B-rep CSG SS	Details	For every Features	Details	Method 3
Dynamic Factor Analysis	LOD300 LOD400 LOD500	B-rep CSG SS	Details	For every Features	Details	Method 3

Based on the integrated data consideration, authors conclude that the most suitable method for integration BIM to GIS in context of 3D evacuation planning is Method 3: Revit (.rvt) to FME Workbench (.rvz) to multipatch geometry (.shp). The authors evaluate from two standpoints. First is through a literature review of 3D analysis that can improve the quality of evacuation planning where integrated data from Method 3 support most of the criteria. The second is based on the data testing, multipatch geometry integrated from Method 3 shows a minimum of geometry and semantical error than other method.

5. Requirement Analysis

This section deliberates the requirement analysis for developing 3D evacuation planning considering the factors on BIM-GIS integration method is deliberate. The requirement covers two main topics which are: 1) generation of 3D indoor mapping and 2) indoor evacuation planning analysis based on BIM-GIS integration method.

5.1. Generation of 3D Indoor Mapping

The indoor map in GIS platform is the fundamental of this research. The generation of the 3D indoor map from integrated 3D BIM data offers a 3D model of building with precise dimension. Based on integration Method 3, the integrated multipatch geometry can provide multiple 3D analysis factors in considering the evacuation planning with 3D indoor map integrated capabilities. This 3D indoor map can be utilized into several functions such as indoor navigation. Unfortunately, some part of 3D models is in disadvantage such as stairs. This is because, in current technologies, an existing method to identify network navigation automatically generates from the flat network in the single floor [11]. This requires detail attention to generate nodes on each stair to connect with another floor. The integrated model must semantically rich and able to keep useful route data. The changes (inaccessible halls, exits, and so on.) in the environment because of smoke, water, or different kinds of harms are represented just on the network for navigation in 3D indoor mapping.

In order to generate a 3D indoor mapping, the components of integrated product geometry and semantic data must be separated based on the type of features as per explained in Section 3. This is important because the types of features integrated from BIM have specific classification and each of them brings different role in evacuation network simulation that affects the 3D indoor mapping. The integrated features from Method 1 and Method 2 requires further action in order to extract features and it will consume time to get satisfactory results. The 3D GIS integrated model needs to be checked and validate before being used for 3D evacuation planning. This is because the 3D geometry in BIM is manually designed before integrated into a GIS platform. It is possible to involve with a geometric error. Diakite and Zlatanova [20] has provided a test for geometric data validation for indoor navigation and concluded that the geometry needs to be validated before the generation of indoor navigation is derived. The structure of semantic information is important in the generation of 3D indoor mapping. This is because it can become an option to identify the evacuation network as the information of structure that can be broken down for access during evacuation simulation based on its structural properties such as wall partition. Therefore, the requirements for the generation of 3D indoor mapping can be concluded as follows:

a. Features semantically rich. In most of emergency situations, the most interesting question to evacuees is "where are the exits?" [21].
b. The components of integrated product geometry and semantic data must be separated based on the type of features.
c. The 3D GIS integrated model need to be checked and validate before being used for 3D evacuation planning.
d. Considering the structure semantic information that can be broken down for access during evacuation based on its structural properties such as wall partition.

5.2. Indoor Evacuate Planning Analysis

Based on all explanation before, it shows that all the method of BIM-GIS integration, type of LOD, 3D indoor mapping and 3D analysis are bonded to each other. The method selection for data integration is important as each method produces its specialty output to translate the 3D geometry element for every type of LOD. Every type of LOD is playing a huge role in the designation of 3D indoor mapping and influenced 3D analysis based on 3D geometry elements capabilities.

The existing BIM data is used for holding infrastructure without the ability for data analysis. Thus, the 3D indoor framework can make use of the spatial information from BIM integrated data for indoor evacuation planning analysis. As for the different with a traditional 2D spatial analyst on maps, the 3D analysis function requires a generic data structure to a certain extend. By confederation geometry elements, 3D concept and GIS capabilities, the indoor evacuation planning can be derived from the analysis that carried out for the best results [19]. For example, through the indoor navigation analysis with basic 3D geometry structure, it will provide the suitable route for evacuating. On top of that, with a combination of volume in dynamic factor analysis, the evacuation route can be improved with the consideration from fire or smoke analysis. By the combination of sensors, 3D spatial analyst and the quality of integrated BIM data under single simulation, the superiority of result surely will increase. The evacuation planning can be regarded as an optimization problem which optimizes the computation of multiple criteria [11]. An interface with multiple criteria can be designed to obtain different evacuation planning results. On the other hand, with the capability of integrated data in providing volume information, the consideration of space optimization can be used to planning the evacuation planning with the real surrounding and user quantity for evacuation network. This is important because those factors are reflecting the ability of route to be served during emergency situation. This also becomes a barrier that differentiates between the functionalities of 2D and 3D spatial information. Therefore, the requirements for indoor evacuation planning analysis are presented as follows:

a. Incorporate the evacuation planning functionality in the framework with the collaboration of multiple additional analysis criteria.
b. Generate the 3D indoor framework for evacuation planning to incorporate the sensor data for fire or smoke simulation.
c. Include the consideration for space optimization in the evacuation process especially on stairs pedestrian area.

6. Conclusion
In this paper, it can be concluded that there are plenty of decisions to be taken on as it interacts with two different platforms of BIM and GIS. BIM being used in the AEC field to concentrate on the description of properties and visualization of the building, while GIS is use in conducting 3D spatial analyst for multipurpose factor decision making. By combining both subjects, it can produce an effective method for achieving more in decision making and properties description model. To achieve that target, the current method of integration was being researched and the multipatch geometry data format has been selected because of its potential in designing 3D analyst for the 3D model. At present, the priority for integration data is to achieve effective data interoperability between both platforms through several methods. Integration Method 3 were selected by identifying the potential of integrated data through current active 3D analyst requirement from multipatch geometry perspective. The selection of the most effective method is important because it is the indicator for data interoperability and for further 3D analysis especially for evacuation planning in the future. To consider the 3D evacuation planning for data interoperability, future research will extend an access-to-road strategy to link indoor spatial networks (based on integrated BIM data/vector) with outdoor spatial networks (based on DEM/raster) for detailed evacuation planning.

Acknowledgements
The author would like to give gratitude to the Institute of Climate Change, Universiti Kebangsaan Malaysia for the funding (KRA-2018-019) and guidance during conducting this research.

References
[1] Rich S and Davis K H 2010 Geographic Information Systems (GIS) for Facility Management *IFMA Foundation.*
[2] Hijazi I, Ehlers M, Zlatanova S and Becker T 2011 Initial Investigations for Modeling Interior Utilities Within 3D Geo Context: Transforming IFC-Interior Utility to CityGML / UtilityNetworkADE **95**–113.
[3] Esri 2009 Space Utilization Optimization *An Esri White Paper* **J-9780**.
[4] Azhar S 2011 Building Information Modeling (BIM): Trends, Benefits, Risks, and Challenges For The AEC Industry. *Leadership Management in Engineering* **11(3)** 241–252.
[5] Volk R, Stengel J and Schultmann F 2014 Automation in Construction Building Information Modeling (BIM) For Existing Buildings — Literature Review and Future Needs *Automation in Construction* **38** 109–127.

[6] Daum S and Borrmann A 2014 Advanced Engineering Informatics Processing of Topological BIM Queries Using Boundary Representation Based Methods *Advanced Engineering Informatics* **28(4)** 272–86.

[7] Teo T A and Cho K H 2016 BIM-Oriented Indoor Network Model For Indoor and Outdoor Combined Route Planning *Advanced Engineering Informatics* **30(3)** 268-282.

[8] Boyes G A, Ellul C and Irwin D 2017 Exploring BIM For Operational Integrated Asset Management – A Preliminary Study Utilising Real-World Infrastructure Data *ISPRS Annals of the Photogrammetry, Remote Sensing and Spatial Information Sciences* **4** 26–7.

[9] Tomasi R, Sottile F, Pastrone C, Mozumdar M M, Osello A and Lavagno L 2015. Leveraging BIM Interoperability For UWB-Based WSN Planning *IEEE Sensors Journal* **15(10)** 5988–5996.

[10] Rafiee A, Dias E, Fruijtier S and Scholten H 2014 From BIM To Geo-Analysis: View Coverage and Shadow Analysis By BIM/GIS Integration *Procedia Environmental Sciences* **22** 397–402.

[11] Liu L, Li B, Zlatanova S and Liu H 2018 The Path From BIM to a 3D Indoor Framework – A Requirement Analysis *The International Archives of the Photogrammetry, Remote Sensing and Spatial Information Sciences* **42** 1–5.

[12] Brodie M 2010 Building Information Modelling: The UK Legal Context *International Journal of Law in the Built Environment* **2(3)** 246-259.

[13] Zhu J, Wright G, Wang J and Wang X 2018 A Critical Review of the Integration of Geographic Information System and Building Information Modelling at the Data Level *ISPRS International Journal of Geo-Information* 1–16.

[14] Taher M 2018 BIM in Academia [Internet] Available from: https://www.thebig5constructegypt.com/media/4978/taher-mohamed-updated-final.pdf

[15] Kang T W and Hong C H 2015 A Study On Software Architecture For Effective BIM / GIS-Based Facility Management Data Integration *Automation in Construction* **54** 25–38.

[16] Donkers S, Ledoux Ã H, Zhao J and Stoter J 2015 Automatic Conversion of IFC Datasets To Geometrically and Semantically Correct CityGML LOD3 Buildings *Transactions In GIS* **41201379** 1–23.

[17] Fosu R and Cory C 2015 Integration of Building Information Modeling (BIM) and Geographic Information Systems (GIS) – A Literature Review And Future Needs *Proceedings of the 32nd CIB W78 Conference, Eindhoven, The Netherlands.*

[18] Xu M, Hijazi I, Mebarki A and Meouche R 2016 Indoor Guided Evacuation : TIN For Graph Generation And Crowd Evacuation *Natural Hazards and Risk* **7** 47-56.

[19] Yuan L I and Zizhang H E 2008 3D Indoor Navigation : A Framework of Combining BIM With 3D GIS *44th ISOCARP Congress* 1–10.

[20] Diakité A and Zlatanova S 2016 Valid Space Description in BIM for 3D Indoor Navigation *International Journal of 3-D Information Modeling* **5(3)** 1-17.

[21] Liu L and Zlatanova S 2011 Towards a 3D Network Model For Indoor Navigation *Proceedings of the Urban Data Management Society Symposium* 79-92

IOP Conf. Series: Earth and Environmental Science **385** (2019) 012034 doi:10.1088/1755-1315/385/1/012034

Determining Indoor Air Quality in Gymnasium for Indoor Recreation Activity at University Teknologi MARA, Shah Alam

N H Mohd Hashim[1,2], W H Wan Ismail[3], F R Sulaiman[4] and M A S Mokhtar[3]
[1]Institute of Quality and Knowledge Advancement, Universiti Teknologi MARA, Shah Alam, Selangor, Malaysia.
[2]Faculty of Architecture, Planning and Surveying, Universiti Teknologi MARA, Shah Alam, Selangor, Malaysia.
[3]Institute of Postgraduate Studies, Universiti Teknologi MARA, Shah Alam, Selangor, Malaysia.
[4]Faculty of Applied Sciences, Universiti Teknologi MARA Cawangan Pahang, Bandar Jengka, Pahang, Malaysia.

Email: norhanisah@salam.uitm.edu.my, w.hazwatiamani@gmail.com

Abstract. This study is conducted to determine the indoor air quality in the Faculty Sport and Recreation's gymnasium, UiTM Shah Alam. The objective of the study is to determine possible type and cause of pollutant, determining the health and performance of occupants and assess whether the IAQ parameters comply with the standards set by DOSH. The study consists of questionnaire survey, walkthrough assessment and reading measurement. A total of 350 questionnaires were distributed to the gymnasium users and a walkthrough assessment were conducted at the surrounding area of the gym in order to identify the possible sources of pollutants as well as the reading of selected parameter of IAQ. Several criteria of indoor air quality parameter are identified and measured. Result indicated that concentration of carbon dioxide increases in relation with the number of the occupant. As a conclusion, indoor air quality in the gymnasium is merely complying with DOSH standards. Thus, relevant actions should be taken pertaining to the quality of indoor air in a gymnasium for the betterment of healthy lifestyle.

1. Introduction

In these new millennia, people live in a desirable and comfortable condition and are more conscious about their healthy lifestyle. [1] stated that about 70% to 90% of time spent by people is in indoor environment. Some of them prefer to do their recreation and leisure activity in indoor environment due to time constraint and appropriate facilities provided in the gymnasium rather than doing outdoor activities which they feel expose to unhealthy environment [2]. With proper ventilation system and air conditioner in the gymnasium, it resulted to a pleasant place for occupants doing their activity and also improves their health. They perceived exercising in indoor fitness centre is safer compared to outdoor due to deteriorating of air quality. According to [3], in modern indoor environments, sources of polluted air are identified coming from outdoor particles, emissions produced by pollutants inside the building and also from the ventilation system.

A research on the parameters of indoor air quality at 11 fitness across Lisbon show that indoor air quality of the fitness centres are not really in a good condition, high concentration of carbon dioxide, volatile compound and formaldehyde identified in the fitness centre [4]. According to Parliament of the United Kingdom (2010) indoor air pollutants is a cocktail of pollutants where it is the combination of pollutants produced from inside the building, outside the building and natural radon gas that can go inside the buildings from the ground. Indoor air quality (IAQ) in any building can be compromised by microbial contaminants (mold, bacteria), chemicals (such as carbon monoxide, formaldehyde), allergens, or any mass or energy stressor that can induce health effects.

In that case, the equipment and the furnishing inside the fitness centre or gymnasium can possibly becoming the source of contaminants. With humid and dry environment of Malaysia, it is inevitable to acquire a proper ventilation system or air conditioner for a comfortable place for recreation. But this might be a double edge sword. [5] stated that without adequate ventilation and cooling, an excessive increase of efficient use of given spaces may reduce the quality of its indoor climate to the extent that affect the users. It means that poor indoor air quality may risk the health of the occupants while risking the performance. The aim of this study is to identify the indoor air quality in fitness centre. Thus, the objective of this study is to determine the effects of occupant with current condition of indoor air quality and to analyse the current parameter with the standard by Division of Occupational Safety and Health (DOSH).

2. Literature review

Indoor air quality is defined as the concentrations of incoming chemical and biological substances and the comfort-related factors which affect people [3]. Concentration of the indoor air pollutant is crucial to be taken care of, therefore, the level of the indoor air quality at the gym can be controlled.

2.1 Indoor air quality parameter

2.1.1. Carbon dioxide. According to ASHRAE [6], the recommendation level of the carbon dioxide in indoor environment should not be exceeded the outdoor carbon dioxide concentration level by more than 650 ppm. While for the good standard quality, the level of carbon dioxide in indoor air should be about 1030 ppm.

2.1.2. Temperature. The optimum temperature range between 20 to 30°C but depend on the relative humidity, seasons, clothing and activity done [3].

2.1.3. Relative humidity. The acceptable range of air temperature inside a building is 23-26 degree Celsius while the relative humidity should be at the level of 40-70% [7].

2.1.4. Particulate matter. The standard set by DOSH Malaysia for the particulate matters is 0.15 mg m^{-3}.

2.2 Cause of poor indoor air quality

2.2.1 External cause. The cause of the microbial sources is usually the water problem such as stagnant water or the humidity is too high. The polluted air can easily infiltrate indoor area if there are openings such as crack, windows and doors. The polluted air can be as a result of natural based pollution and human induced activity [8]. Biological pollutants are one of the causes of poor indoor quality, these microbial included bacteria, molds, viruses, and mites.

2.2.2 Internal cause. Buildings with the lower amount of outdoor air coming in may have higher pollutant levels [9]. The improper cleaning and organizing processes such as storage and disposal of the garbage, the sweeping, dusting and vacuuming during the presence of the occupants will contribute to indoor air quality problem [8]. Materials such as building materials and furnishing such as paint will emit certain chemicals that are harmful to the health.

2.2.3 Heating, ventilation and air-conditioning (HVAC). A HVAC system operation that are poorly design or had an interrupted flow in response to thermal conditioning needs may cause the increase of indoor contaminant levels [10]. A split unit can contribute to the increase concentration level of carbon dioxide compared to centralized air-conditioning system [13]. If HVAC systems are not maintained, the wheel can severely become dirty or damaged which will cause metal dust to be generated and transmitted to the indoor air [8].

2.3 Impact of poor indoor air quality

2.3.1 Health and well-being. The health impacts due to poor IAQ can be determined as acute or chronic illness for examples asthma, infections in respiratory systems, allergy and even lung cancer. These health problems are referred to as Sick Building Syndrome and Building Related Illnesses [11]. In indoor environment such as inside the fitness centre and gyms, the presence of high concentration of chemicals such as formaldehyde was alarming because it can cause respiratory problems and asthma [4].

2.3.2 Performance and productivity. Health problem will lead to poor performance and productivity among the users inside the gym, making them felt uncomfortable while exercising and felt weak. High level of carbon dioxide and volatile organic compounds would harm the human cognitive function and decision-making skills [12].

3. Methodology

3.1. Study area

The study was conducted at the gyms in Faculty of Sport and Recreation, UiTM Shah Alam. The building had operated for about 57 years and was converted into gym in 1996 which make the gym operated for almost 22 years. The building is generally sealed airtight and covered well. The targeted respondents for this research are among UiTM Shah Alam students and staff that using the gyms for their leisure and recreation purposes.

3.2. Research design and data collection

In order to complete the research, the research approach that has been selected is both qualitative and quantitative method.

3.2.1 Physical measurement. The parameters that have been chosen were the thermal comfort (temperature and the relative humidity); and some IAQ parameters which were carbon dioxide and particulate matters (PM). All measurement was done using a Multi-parameter Environmental Monitoring (EVM-7).

3.2.2 Physical assessment. The forms covered; (1) the symptom of Sick Building Syndrome (SBS), (2) ventilation system, (3) related possible cause and (4) problem of indoor air quality presence, and inside plan/sketch of the gyms and fitness centre.

3.2.3 Questionnaire survey. A set of questions were prepared and given to a total of 350 respondents. The questionnaire was designed in a Likert-scale construction way. This survey is important to determine the possible effect and the opinion of the respondents. The mean of 3-4 shows the respondent is agree, and the mean below 3 indicates the respondent disagree with the statements given in the survey.

3.3. Data analysis

In order to analyse the data, a descriptive analysis and content/inferential analysis were conducted.

3.3.1 Physical measurement. The numbers were compared to the standard for ventilation rate.

3.3.2 Physical assessment. Content analysis was done, and the results were compared to the standard set by DOSH Malaysia and ASHRAE.

3.3.3 Questionnaire survey. A descriptive analysis using SPSS was applied.

4. Result and discussion

4.1 Physical measurement

4.1.1 Carbon dioxide. There are about 12 people inside the gyms and the highest CO_2 recorded was 510 ppm. This level of CO_2 is still within the acceptable standard recommended by DOSH [9].

4.1.2 Relative humidity. The range of relative humidity is 55.2% – 68.9% and it is complied with the standard set by the DOSH which is 40 – 70%. Relative humidity will likely to change when the temperature change, this due to warmer air able to retain more water vapour than cooler air.

4.1.3 Temperature. The range of temperature recorded is between 22.3 – 27.4 °C. This temperature range values are slightly above the guidelines range set by the DOSH which is 23 – 26 °C but comply with ASHRAE which is 24 – 27 °C. The temperature of the gyms decreased significantly when all the six air conditioners inside the gyms were turn on.

4.1.4 Particulate matters. The average reading of particulate matter recorded is 61.5 ug m^{-3}. This indicates that the particulate matter level complies with the DOSH standard for indoor environment (0.15 mg m^{-3}). The highest reading determined (125 ug m^{-3}) when the workers cleaning the floor.

4.2 Physical assessment

4.2.1 General description. Cracks were found inside the building. There are present of mould inside the building. It will lead to allergic reactions towards the building occupants. The flooring has not been changed for quite a long time which about 10 years.

4.2.2 HVAC Systems. By far, the air conditioners work and perform well without any malfunctioning. There are no leakages found such as the leakage of the duct and the supply fan works great.

4.2.3 Facilities and equipment. The condition of the equipment is quite poor in the appearance aspect, the leather of the seat already torn out and cracked as well as the soft leather already worn out, the metal part of the equipment's are rusty. The equipments are not cleaned properly, there were dust everywhere surrounding the place.

4.2.4 Alteration and activities conducted in the building. The building is altered by flooring covered with rubber like mat used in the exercise centre. The interior of the building also painted. Paint is one of the contributors of volatile organic compound that is one of the causes of poor indoor air quality.

4.2.5 Cleaning and sanitizing process. Cleaning and sanitizing process are a must and the workers done the cleaning and sanitizing process twice a day.

4.3 Questionnaire survey

4.3.1 Demographic profile. The respondents are 58.3% male and 41.7% female. Most of the respondents are in the age of 21-30 years old by the percentage of 78.0%. The total numbers of users that often come to the gym are 79.3% while those who do not often come are 20.7%.

4.3.2 Gym physical condition. For the indoor ventilation system inside the gym, the mean is 3.99 which mean it more towards the likert scale of 4 which is agree (Table 1). During the assessment period, the gyms is properly ventilated, and the air conditioner is at peak and good condition. While for the physical condition of the gym, the mean is also 3.99 which make it more towards the likert scale of 4 which is agree. The furnishing and the equipment are still great in condition because they were maintained about twice a year. The occupants agree with the mean of 4.01 that the interior of the gym is properly design and it is easy to move around inside the gym without being blocked or disturb by the gym's equipment positioning. The mean for the statement that stated that the occupant is comfortable with temperature and humidity inside the gym is 3.93. The temperature and humidity of the gym in fact is within an ambient condition and according to the recommended standard set by DOSH Malaysia. The occupants disagree (the mean value of 2.47) that there is an unusual chemical odour presence inside the gym while they were exercising and working out. This may due to the air that has a proper circulation and ventilation inside the building.

Table 1. Respondent perception on gym physical condition

Items	Mean	Std. Deviation
Indoor is properly ventilated	3.99	0.848
Gym physical condition is great	3.99	0.821
Gym layout is properly design	4.01	0.878
Comfortable with temperature	3.93	0.927
Gym atmosphere is fresh	3.93	0.915
Unusual chemical odour	2.47	1.255
Presence of strong sweat odour	2.42	1.234
Equipment and place are dusty	2.60	1.258

4.3.3 Performance, health and productivity. Table 2 shows the mean for the question *I perform well* during inside the gym is 4.01. Overall occupants are agreed that they do perform well while doing activities and working out inside the gym. The mean of 2.39 is recorded for the statement that the occupants experiencing the symptom of poor indoor air quality such as dizziness, coughing and difficulty to breathe. Overall, the occupants are disagreed that they are experiencing the symptom of poor IAQ.

Table 2. Satisfaction of respondent on performance, health and productivity

Statements	Mean	Std. Deviation
I perform well	4.01	0.759
I have symptom of poor IAQ	2.39	1.162
I have sudden allergy	2.12	1.089
Easy to concentrate	3.57	1.197
I don't bother spending more time inside the gym	3.78	1.081
My reflexes are good	3.94	0.824
I can make good decision making	3.74	1.035
My psychomotor performance decrease	2.41	1.181

5. Conclusion

Indoor air quality inside the gymnasium of Faculty of Sport and Recreation UiTM Shah Alam can be recognized as good and still on par with the recommendation and standards by DOSH and ASHRAE. Majority of occupants that working out and exercising inside the gymnasium perceived and showing great performance and productivity as well as showing good sign of health. However, not all the occupants do agree with the indoor air quality inside the gym and this may the exceptional cases due to several reasons. Therefore, further details investigation is needed.

6. Recommendation

6.1 Recommendation to gym management

- Controlling and managing the facilities and the building maintenance inside the gym.
- Enforce proper guidelines and regulation inside the gym that will likely to contribute towards the improvement of the indoor air quality inside the gym.
- Invest more on the consultation involving indoor air so that the indoor air quality inside the gymnasium really comply with the rules and standards.

6.2 Recommendation on parameters of indoor air quality

- The temperature and the relative humidity play important roles for the comfort inside the gymnasium.
- The humidity and temperature are the factors that interrelated with each other.

6.3 Recommendation for the future research

- Continuous monitoring should be done and focus on the dust collection especially from the equipment inside the gym.
- Analysis of dust composition using the appropriate laboratory analysis.
- Study on the bacteria count in the air inside the gymnasium.
- Conduct a health risk assessment based on United States Environmental Protection Agency approach.

Acknowledgement
This study is supported by Bestari Perdana Research Grant provided by the Ministry of Education Malaysia, and Innovation & Research Management Institute (IRMI), Universiti Teknologi MARA (600-IRMI/PERDANA 5/3 BESTARI (075/2018)).

References

[1] Sharpe M 2004 Safe as houses? Indoor air pollution and health. *Journal of Environmental Monitoring.* **6** 46 – 49.
[2] Hajian M and Mohaghegh 2015 Indoor air pollution in exercise centers. *International Journal of Medical Toxicology and Forensic Medicine.* **5** 22-32.
[3] Camfil 2015 A Brief Guide to indoor air quality for elite athletes. *Indoor Air Quality.* Retrieved from http://www.camfil.com.my/FileArchive/Brochures/Clean%20Air%20Knowledge/IAQ%20handbook_EN.pdf
[4] Ramos C A and Wolterbeek H T 2014 Exposure to indoor air pollutants during physical activity in fitness centers. *Building and Environment* **82** 349–360.
[5] Saari A, Tissari T, Valkama E and Seppanen O 2006 The effect of a redesigned floor plan, occupant density and the quality of indoor climate on the cost of space, productivity and sick leave in an office building. *Building and Environment* **41** 1961-1972
[6] ASHRAE 2013 Thermal environmental conditions for human occupancy. *ANSI/ASHRAE Standard 55.*
[7] Department of Occupational Safety and Health, DOSH 2010 *Industry Code of Practice on Indoor Air Quality.* Ministry of Human Resouces, Malaysia.
[8] Rajkumar G 2003 Causes and effects of poor indoor air quality. Retreived from http://www.bvsde.paho.org/bvsacd/cef-2/GRindairqu.pdf
[9] United States Environmental Protection Agency 2017 Introduction to indoor air quality. Retreived from https://www.epa.gov/indoor-air-quality-iaq/introduction-indoor-air-quality
[10] United State Environmental Protection Agency 1990 Ventilation and air quality in offices. *Fact Sheet.*
[11] Ambu S, Chu W L, Mak J W, Wong S F, Chan L L and Wong S T 2008 Environmental health and building related illnesses. *International E-Journal of Science, Medicine & Education*, **2** 11–18.
[12] Loki R 2015 *Environment Alternet.* Retrieved from http://www.alternet.org/environment/indoor-air-pollution-negatively-impact-human-cognition

Mapping Informal Settlements Using Geospatial Method

F K Maula[1], D N Choerunnisa[1] and R Akbar[1]
[1]Urban Design and Planning Research Group, Department of Regional and City Planning, School of Architecture, Planning and Policy Development, Institut Teknologi Bandung, Ganesha Street 10, Bandung City, West Java, Indonesia

Email: faridamaula@sappk.itb.ac.id

Abstract. As one of the largest city in Indonesia, Bandung faces informal settlement, or *"Permukiman Kumuh"* problems. Based on the criteria of informal settlements used in KOTAKU Program, Bandung contains 8,7 percent of slums within it. During its development, these informal settlements experiences phase based on the development of the city, which are infancy, consolidation and saturation. Each developmental phase needs differing solutions. This study formulates the spatial development phases from the informal settlements in Bandung through Geospatial methods to give new perspective in settling informal settlement problems.
Keywords: informal settlements, spatial phases, models

1. Introduction to Informal Settlements in Indonesia

Informal settlement is a serious and common problem in Third World Countries (Nassar & Elsayed, 2017). UN-Habitat (2008) defined informal settlements in three types: (1) slum: a neighborhood of housing that was once in good condition but has since deteriorated or been subdivided into a state of high crowding and rented out to low-income groups; (2) squatter: an area of poor quality housing built on illegally occupied land; and (3) An irregular subdivision, in which the legal owner subdivides the land into sub-standard plots and sells or rents them out without following all relevant building bylaws. UN-Habitat (2008) also defines the informal settlements as a group of household who live under the same "roof" in urban areas which lack one or more of these conditions: durable housing, sufficient living area, access to clean water, access to proper sanitation and security of tenure.

Informal settlements grow along with the development of the city (Fakade, 2000). There are three phases of informal settlements growth (Abebe, 2011): infancy, consolidation and saturation. The phases of informal settlements showing the formation and level of the informal settlements. These phases are analyzed based on spatial characteristics of the informal settlements.

In Indonesia, informal settlements have become one of the problems that get special attention from the government. In Indonesia, informal settlements are also known as *"Permukiman Kumuh"*. *Permukiman Kumuh* is inhabitable settlements because of the high density of the buildings and the quality of buildings and facilities and also infrastructures that do not meet the requirements (Minister of Public Works and Public Housing Regulation Number 02/PRT/M/2016). One of the efforts by the government to resolve informal settlements' problems is through a program, called KOTAKU program. KOTAKU or *Kota Tanpa Kumuh* (City without Slums) program is a Minister of Public Works and Public Housing owned program that accelerates the handling of informality in Indonesia and also support the "100-0-100 movement", namely 100% universal access to drinking water, 0% of slums (*permukiman kumuh*), and 100% of access to proper sanitation.

KOTAKU program handles informality by building collaboration platforms through increasing the role of local government and community participation. Almost every city in Indonesia has informality which belongs to the KOTAKU program. An informality belongs to the KOTAKU program based on the criteria: buildings, local streets, drinking water supply, community drainage, wastewater management, garbage management, and fire protection. In handling the informality which belongs to KOTAKU program, the government have several handling patterns, they are restoration, rejuvenation and re-settlement.

In handling the informality, it should be known at what phases the informality settlements occur because the phases can help to determine which handling patterns are possible to do. Hence, this study aims to formulate the model of informal settlements phases based on their spatial characteristics as one of the informal settlements criteria. This study will use GIS as a tool of analysis. Taking Bandung as a case, which is the third most inhabited cities with a population around 2.5 million citizens and the second highest population living in slums (Anindito, 2017 citing Tarigan et al, 2015), this study should suggest the possible model of informality.

2. Study Area

Bandung City is one of the largest cities in Indonesia, having the total area of 167,31 km^2 with 2.497.938 people (BPS,2018) (approximately 14.930 people/km^2). Based on the Mayor's Decree number 648/Kep.286-DisTarCip/2015 regarding the establishment of informal settlements and housing environment locations, there are 454 slums areas in Bandung with the total area of 1,457.45 Ha (8.7 percent of Bandung total area). Based on this decree, the slums were divided into three planning target areas, which are:

1. Accelerated Slum Areas, including 15 areas with the total area of 448.9 Ha.
2. Slum Areas, including 106 areas with the total area of 963.91 Ha.
3. Slum Prevention Areas, including 30 areas with the total area of 44.64 Ha.

The following map shows the spread of slums in Bandung based on Mayor's Decree number 648/Kep.286-DisTarCip/2015 regarding the establishment of informal settlements and housing environment locations.

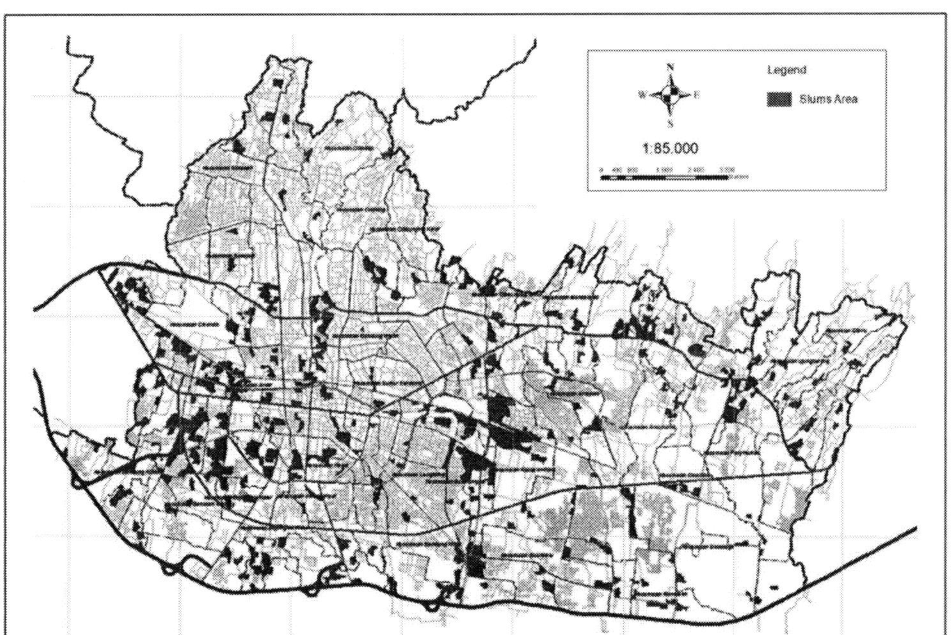

Figure 1. Slums Area in Bandung
Source: DPKP3 Bandung City, 2017

IOP Conf. Series: Earth and Environmental Science **385** (2019) 012035 doi:10.1088/1755-1315/385/1/012035

3. Methodology

3.1. Criteria, Indicator and Data

The criteria that are used for this study are based on the phases of informality development (Abebe, 2011; Sori, 2012).

1. Infancy/ starting stage: the initial occupancy stage where patches of the vacant land, like on step sided land, swamps, near river banks, hazardous areas, conserved areas, etc, become available to the slum dwellers. They can also occur in suitable lands. In this phase, the vacant land consist of at least 50 percent of the land is used for housing.

2. Consolidation stage: the intermediate stage between infancy and saturation. There is fast outward expansion and the available land will be reduced by filling up additional buildings. About 80 percent of the land would be used for housing.

3. Saturation stage: this stage whereby additional construction is primarily through vertical development because of the vacant lands gets filled up. At this stage, overcrowding is highest and lowers the living standards of slum dwellers.

From the definition of the phases, the keyword of staging the phases are vacant land and residential density. Therefore, spatial data that can be used to formulate the model of the development phase of informality are the geographic and the built-up area map (land use and buildings) of Bandung City. Because of Bandung City does not have a standard of building density per area, comparisons between areas in the city are used to categorise the phasing in each area. For example, if an area has the building density (d_1) which is larger than the density in other areas (d_2), d_1 can be categorised as having a higher level of informality (consolidation or saturation).

3.2. Model Formulation

The following indicators shown on the table below will be used to formulate the classification model based on the informality growth phase.

Table 1. Data and Indicators

Datum	Indicator
Percentage of built areas (%)	The higher the percentage of built areas, the higher the phase
Percentage of undeveloped areas (vacant land) (%)	Vacant land > 50% = Infancy 50% > Vacant land < 80% = Consolidation Vacant land <20% = Saturation
Areas used for housing (m^2)	The more areas used for housing, the higher the phase
Building Density (units./ Ha)	The higher the building density, the higher the phase

For mapping the informal settlement using geospatial methods, carried out in the following stages:

IOP Conf. Series: Earth and Environmental Science **385** (2019) 012035 doi:10.1088/1755-1315/385/1/012035

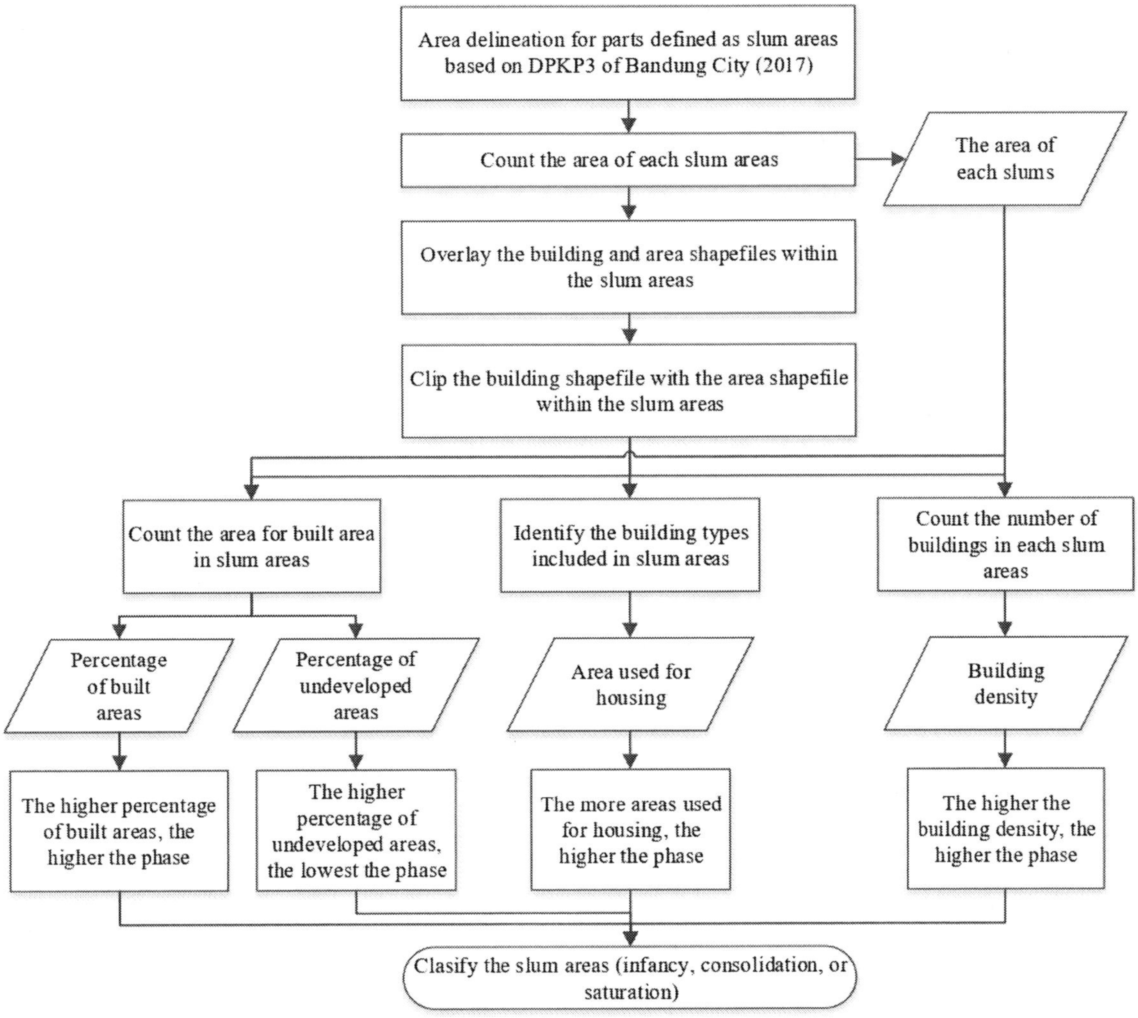

Figure 2. Staging of Geospatial Method

4. Result

Based on existing data of buildings in Bandung, in each slums area is more than 70 percent of the buildings are residential. From the **Figure 3**. Almost all areas in each slum areas are covered by buildings. The density of buildings in each slum area is more than 50 units/ hectare and residential land use are dominates in almost all areas.

Based on the results of building density and land uses, 70 percent of slums area in Bandung is classified as the saturation phase. This phase is the last phase of informal settlement development. Based in this saturation phase, the condition of informal settlements in Bandung are in bad condition because they have the highest density of informal settlement. The density will trigger a decrease in quality of facilities and infrastructures. Also will causes the decrease of environmental quality in that area.

In area on the consolidation phase, the vacant lands (undeveloped area) are mostly form of abandoned land or shrubs. That condition will trigger people in that area to build the land become residential area. So that, the slums area in the consolidation phase also have potential to become saturation phase.

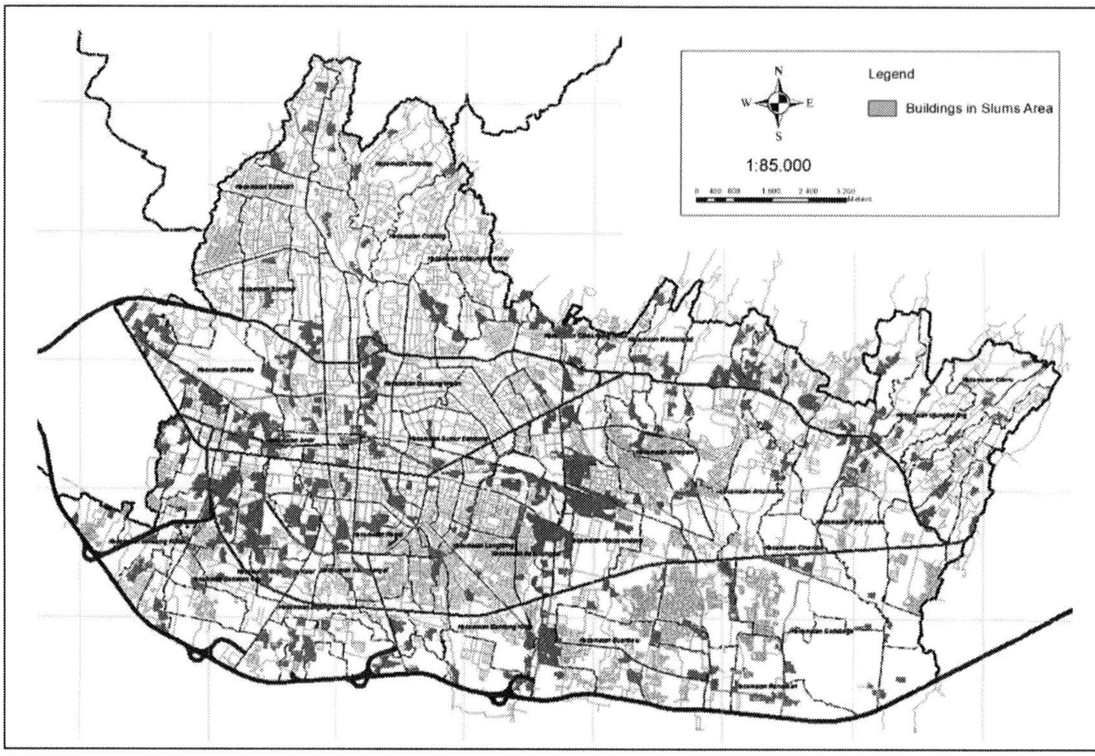

Figure 3. Buildings Map in Slums Area

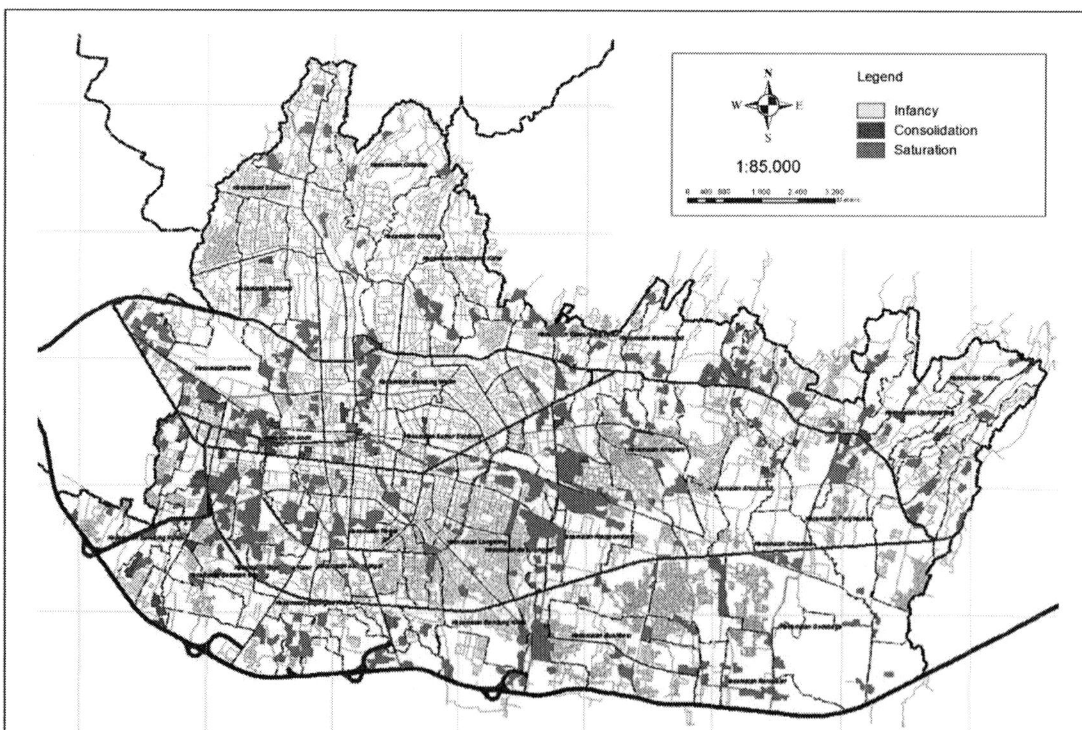

Figure 4. Phases of Slums Area in Bandung

5. Conclusion and Recommendations

More than 70 percent of informal settlements in Bandung are in the saturation phase. This phase represents that informal settlement conditions in Bandung are in poor condition. Judging from its phase, informal settlement in Bandung has developed from long time ago. The high density in informal settlement has triggered the decrease of facilities qualities. In informal settlement which is in the consolidation phase, there is potential to develop into saturation phase.

To prevent the growing number and phase of informal settlements in Bandung to the worse condition, Bandung government needs to determine the priority of handling informality based on their level. In informal settlement in the latest phase, it may takes action form the government to re-settlement the slums. So that the people in that area can get better quality of environment. Whereas for informal settlement which is in the infancy and consolidation phase, it is necessary to monitor and anticipate from the government they do not grow to higher phase, especially monitoring in land use. In both phases, it is still possible to use restoration and rejuvenation handling patterns to organize the informal settlement. This research can provide a new knowledge that there are staging of informality development and the development in each area is different. So, the government cannot always provide the same solution in handling the informality. City characteristics should be considered in determining the standards for informality phase category for that peculiar city

Acknowledgement

This study is conducted by the research grant from Institut Teknologi Bandung entitled *"Program Penelitian, Pengabdian Kepada Masyarakat dan Inovasi (P3MI)"*.

References

[1] Abebe Fikreselassie Kassahun 2011 *Modelling Informal Settlement Growth in Dar es Salaam* Tanzania Thesis for Master of Science in Geo-Information Science and Earth Observation (Netherlands: University of Twente)

[2] Anindito D B et al 2018 *Modelling the Kampungkota: A Quantitative Approach in Defining Indonesia Informal Settlements* ICRMBEE (IOP Conf. Series: Earth and Environmental Science 117) 012005 DOI: 10.1088/1755-1315/117/1/012005

[3] Fekade Wubalem 2000 *Deficits of Formal Urban Land Management and Informal Responses Under Rapid Urban Growth, an International Perspective* (Habitat International 24) p 127-150

[4] Nassar, D M and Elsayed H G 2017 *From Informal Settlements to Sustainable Communities* (Alexandria Engineering Journal)

[5] Sirueri Felix Orina 2015 *Comparing Spatial Patterns of Informal Settlements Between Nairobi and Dar Es Salaam* Thesis for Master of Science in Geo-Information Science and Earth Observation (Netherlands: University of Twente)

[6] Sori Negera Dinsa 2012 *Identifying and Classifying Slum Development Stages from Spatial Data* Thesis for Master of Science in Geo-Information Science and Earth Observation (Netherlands: University of Twente)

[7] UN-Habitat 2008 *Quick Guide for Policy Makers 2, Low-Income Housing* (Thailand: United Nations Economic and Social Commission for Asia and the Pacific (UNESCAP))

IOP Conf. Series: Earth and Environmental Science **385** (2019) 012036 doi:10.1088/1755-1315/385/1/012036

Importance of lightning detection network development in Peninsular Malaysia

M.A Islam[1], A. Chan[1], S. K. K. Khaw [1], S. A. Salleh [2,3], N. H. Zakaria[2], N.A. Isa[3], M.O.C. Gee[4] and M. Azari[1]

[1]Department of Civil Engineering, Faculty of Engineering, University of Nottingham, Malaysia Campus,Semenyih 43500, Selangor, Malaysia.

[2]Centre of Studies of Surveying Science and Geomatics, Faculty of Architecture, Planning and Surveying, Universiti Teknologi MARA, Shah Alam 40450, Selangor, Malaysia.

[3]Applied Remote Sensing and Geospatial Research Group, Faculty of Architecture, Planning and Surveying, Universiti Teknologi MARA, Shah Alam, Selangor, Malaysia

[4] Department of Atmospheric Sciences, National Central University, Chung Li 32001, Taiwan.

Email: islamccs@gmail.com (M. A. Islam)

Abstract. Lightning is a major climatic phenomenon in Southeast Asia over Indian Ocean and Malaysia is the most affected area in the region. Damages in the different industries, real estates, defense and human fatalities are caused by lightning strikes and it is very common in Malaysia. Existing non-effective lightning detection network in Malaysia now has become an urge to modify the system and improve it according to the need of the Malaysian Meteorological Department. Research facilities in the Malaysian universities and related research organizations need to give substantial effort in developing lightning science and technology to solve this problem. In this article, current status of lightning detection network in Malaysia and its weaknesses have been discussed briefly. Also, suggestion have been proposed to improve it.

1. Background

Maritime Continent in Southeast-Asia is one of the most important areas of attention for the atmospheric scientists [1]. What plays important role in this area is the Inter-tropical Convergence Zone (ITCZ) as it affects the global circulation system of the atmosphere [2]. ITCZ is also familiar as Equatorial Convergence Zone or Inter-tropical Front. This low-pressure belt encircles the earth near the equator where the trade winds meet and converge [3]. According to Köppen's classification of the climate, Malaysia and Indonesia are in the tropical rain-forest climate [4]. Thunderstorm, and lightning to be specific, is a major natural hazard in this area not only to the humans but also to many industries, real estates, power producers, wildlife managing governmental organizations, defense sector etc. [5]. According to Gomes et al. (2011) every year globally over 20, 000 human lives are affected and thousands of them get direct injured by lightning which left them with life time injuries, temporary or permanent disabilities or trauma, and even death [6].

Malaysian Meteorological Department characterizes Malaysia as a high lightning activity zone and claimed that it has over 200 days in a year lightning of different intensities occur in this area [7].The biggest electric utility company in Peninsular Malaysia - Tenaga Nasional Berhad (TNB) claimed that, 70% power supply failure and transient tripping in their transmission and distribution network occur due to lightning problem and the effects account for damages of the electrical equipments, downtime, data loss and the controlling automated system they use. These losses costs millions of dollars and human fatalities [8]. The defense sector of Malaysia also suffers due to lightning related accidents (e.g. radar and other equipments malfunctioning) which is a big threat to the national security [9].

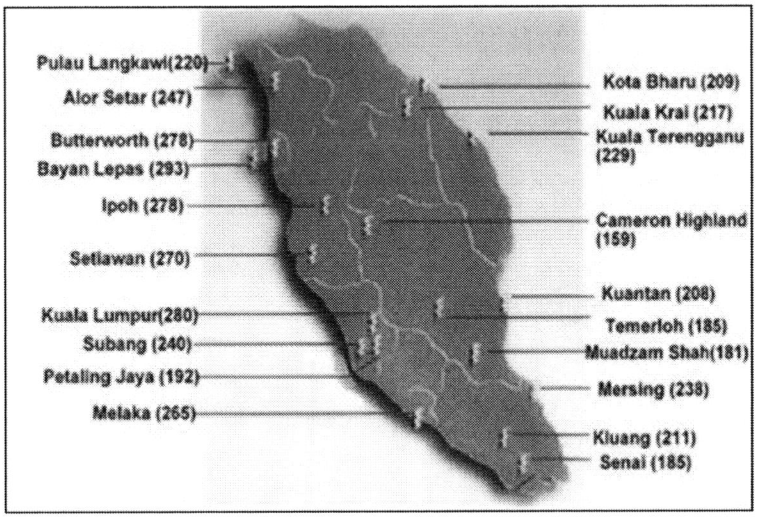

Figure 1: Average numbers of lightning strikes in Peninsular Malaysia [10]

1.2 Lightning in Malaysian Climate

There are intracloud – IC, cloud to cloud – CC and cloud to ground – CG lightning types and occurrence follows the local weather pattern [11]. Malaysian wind pattern is characterized by 2 monsoon wind patterns – Southeast Monsoon (April-September) which makes the area drier and Northeast Monsoon (October-March) that brings rainfall to this area [12]. Most lightning in Malaysia is observed during the inter-monsoon period. In Malaysia lightning type CG makes up around 23% of total lightning and during the daytime lightning strikes are the most. During the night and day times lightning changes its frequency. It is more frequent over the land during the daytime and less over sea and opposite during the night time [9].

1.3 Lightning caused fatalities in Malaysia

In terms of lightning density, Malaysia is ranked among top 3 in the world which is more than any other country in the Asian region [13]. The citizen of this area also lacks proper education about lightning protection which increases the fatality rate. It is estimated that 1 in 10 people is a possible victim of lightning strike [13]. Table 1 below is an indicator of fatalities and injuries due to lightning strikes recorded in Malaysia during 2008-2015 [13].

Table 1: Fatalities and injuries during 2008-2015 due to lightning in Malaysia [13]

Year	JAN	FEB	MAR	APR	MAY	JUN	JUL	AUG	SEP	OCT	NOV	DEC	Total
2008	3	0	0	3	3	2	0	0	0	0	9	0	20
2009	0	0	0	0	13	6	0	0	0	8	0	5	32
2010	0	0	0	2	0	0	0	0	9	2	0	0	13
2011	1	2	1	0	0	3	3	12	2	1	4	1	30

IOP Conf. Series: Earth and Environmental Science **385** (2019) 012036 doi:10.1088/1755-1315/385/1/012036

2012	0	5	1	56	7	1	5	5	1	0	5	0	86
2013	0	12	1	2	0	0	2	1	2	0	0	0	20
2014	0	0	4	1	2	0	0	0	0	2	2	0	11
2015	0	0	0	12	4	0	1	1	5	0	0	0	23
Total	4	19	7	76	29	12	11	11	19	13	20	6	235

2. Lightning distribution and density in Peninsular Malaysia

According to *Ibrahim and Ghazali (2012)* [14] in Peninsular Malaysia the range of lightning intensity is mostly in between 5kA to 25kA, and sometimes with exception even up to 32kA is recorded. More lightning strikes are observed during the Southeast Monsoon period rather than Northeast Monsoon. Despite less lightning during Northeast Monsoon, the intensity is stronger than during the Southeast Monsoon and lies between 15kA-25kA. The Klang Valley area is found to have most lightning**.** Because of the influence of sea breeze, land area along the mountain used to have more lightning. Strait of Malacca during the Southeast Monsoon receives more lightning than average. During Northeast Monsoon lightning becomes weaker in the east coast and during Southeast Monsoon east coast sees more lightning.

3. Effective factors for improved lightning detection

Before working on the lightning protection system (LPS), lightning detection system (LDS) and therefore lightning prediction, it is very important to understand the background lightning physics, its related engineering and technical matters [15]. Lightning protection system works by providing the electric charge produced by the clouds a path of least resistance to the ground. It is a passive means of preventing damages of property and lives from the effects of lightning strikes. Lightning detecting system is a group of equipments that detects lightning produced by thunderstorms and they are ground based, mobile systems (usually by aircraft) based and space based (satellites). Lightning prediction is determined by the equipments which work on the detection system and according to the measured and validated data, observing the weather pattern they can predict the possible outcome of lightning [16].

Lightning itself is a very complex atmospheric phenomenon and its proper detection depends on some basic important factors. Factors which should be systematically maintained to establish effective lightning detection network are:

- configuration of the network
- strength of the waveform
- unique character of lightning
- distance of the lightning from the network
- sensor status
- local sensor interference
- geography of the network [17]

4. Current lightning detection network in Malaysia, its weaknesses and necessary work for improvement

Jabatan Meteorologi Malaysia (Malaysian Meteorological Department - MMD) operates 2 subsystems:

- wide area lightning localization sub-system (SAFIR) and
- precision lightning warning sub-system (PLWS).

SAFIR station consists of 8 interferometry sensors, 2 central processors and 2 main user terminals. They provide per second lightning data and status of the detection station per second. This system is also capable of auto data testing every 5 minutes. The other subsystem (PLWS) involves 7 remote alarm displays, 3 electrostatic field mills and 2 consoles for precision lightning warning. But MMD claims that these detection systems are weak and are unable to provide correct and enough data to understand, detect and predict lightning effectively. MMD also claims that, the number of available sensors is not enough to cover the area of Peninsular Malaysia, the installation sites are not suitable for smooth functioning as they are located in noisy areas and get frequent obstructions and the spacing of the sensors are also uneven and fails to provide information with maximum efficiency [18]. Lightning detection sensors need to be upscaled and sensor replacement with modern equipments has become crucial at this stage. For the relocation of the new sensors better sites need to be chosen using proper scientific methods and calculations.

5. Conclusion

Malaysia is a very important area in the Asian region for its special geographical location and its climate. It requires to develop a generation of scientific researchers to work on developing an advanced lightning detection, protection and prediction system for Malaysia. With available funding it is of course possible to collaborate with organizations who are specialized in this field and can help with lightning science and technology, but it still requires own skilled people to learn from the skilled collective, maintain it up-to the mark and working on further improvement.

Acknowledgments

This paper is the introductory report of a collaborative study between the University of Nottingham, Malaysia campus and Universiti Teknologi MARA, Shah Alam, Malaysia. Authors would like to acknowledge the assistance and encouragement of the internal research funding authorities of their respective universities.

References

[1] Islam, M.A.; Chan, A.; Ashfold, M.J.; Ooi, C.G.; Azari, M. Effects of El-Niño, Indian Ocean Dipole, and Madden-Julian Oscillation on Surface Air Temperature and Rainfall Anomalies over Southeast Asia in 2015. *Atmosphere* 2018, 9, 352.

[2] Marshall J.; Donohoe A.; Ferreira D.; McGee D. The ocean's role in setting the mean position of the Inter-Tropical Convergence Zone. *Clim Dyn* (2014) 42: 1967

[3] Krider, E. P.; Noggle, R. C.; Uman, M. A. A gated, wideband magnetic direction finder for lightning return strokes. *Journal of Applied Meteorology*, vol. 15, Mar. 1976, pp. 301-306

[4] Köppen, W. The thermal zones of the Earth according to the duration of hot, moderate and cold periods and to the impact of heat on the organic world. *Meteorologische Zeitschrift* vol. 20 No. 3 (2011), pp. 351 – 360.

[5] Colin G. P. Lightning Applications in Weather and Climate Research. *Surveys in Geophysics*, November 2013, vol 34, Issue 6, pp. 755–767.

[6] Gomes C. and Ab Kadir M.Z.A. A theoretical approach to estimate the annual lightning hazards on human beings. *Atmospheric Research* vol 101, Issue 3, August 2011, pp.719-725.

[7] Abdullah N; Yahaya M.P.; Hudi N.S. Implementation and use of lightning detection network in Malaysia. *IEEE 2nd International Power and Energy Conference, 2008*

[8] Wooi C.L.; Abdul-Malek Z.; Ahmad N.A.; Mokhtari M.; Khavari A.H. Cloud-to-Ground Lightning in Malaysia: A Review Study. *Applied Mechanics and Materials, 2016* vol. 818, pp. 140-145

[9] Ab-Kadir M.Z.A. Lightning severity in Malaysia and some parameters of interest for engineering applications. *Thermal Science* 2016, vol. 20, suppl. 2, pp. 437-450.

[10] Abidin H.Z.; Ibrahim R. Thunderstorm day and ground flash density in Malaysia. *National Power Engineering Conference, IEEE, 2003.*

[11] Tan Y.; Tao S.; Liang Z.; Zhu B. Numerical study on relationship between lightning types and distribution of space charge and electric potential. *J. Geophys.Res. Atmos.,*119, pp.1003–1014

IOP Conf. Series: Earth and Environmental Science **385** (2019) 012036 doi:10.1088/1755-1315/385/1/012036

[12] Salimun E.; Tangang F.; Juneng L.; Behera S.K.; Yu W. Differential impacts of conventional El Niño versus El Niño Modoki on Malaysian rainfall anomaly during winter monsoon. *Int. J. Climatol.* 2014, 34: pp.2763 – 2774

[13] Ab Kadir M.Z.A.; Misbah N.R.; Gomes C.; Jasni J.; Wan Ahmad W.F.; Hassan M.K. Recent statistics on lightning fatalities in Malaysia. *International Conference on Lightning Protection (ICLP), IEEE, 2012.*

[14] Ibrahim W.I.; Ghazali M.R. An overview of Pekan Lightning Detection System (PLDS). *IEEE International Conference on Power and Energy (PECon), 2012.*

[15] Dwyer J.R.; Uman M.A. The physics of lightning. *Physics Reports* Volume 534, Issue 4, 30 January 2014, pp.147-241

[16] Betz H.D.; Schmidt K.; Oettinger P.; Wirz M. Lightning detection with 3-D discrimination of intracloud and cloud-to-ground discharges. Geophysical Research Letters 2004, vol. 31, L11108

[17] Peterson D.; Wang J.; Ichoku C.; Remer L.A. Effects of lightning and other meteorological factors on fire activity in the North American boreal forest: implications for fire weather forecasting. *Atmos. Chem. Phys.*, 10, pp.6873-6888, 2010

[18] Wooi C.L. Cloud-to-Ground Lightning in Malaysia: A Review Study. *Applied Mechanics and Materials*, vol. 818, pp. 140-145, 2016.

IOP Conf. Series: Earth and Environmental Science **385** (2019) 012037 doi:10.1088/1755-1315/385/1/012037

Spatial and Statistics for Profiling Risk Factors of Diseases: A Case Study of Tuberculosis in Malaysia

A R Abdul Rasam[1*], N Mohd Shariff[2], J F Dony[3] F Othman[1]
[1]Faculty of Architecture Planning and Surveying, Universiti Teknologi MARA Shah Alam, Selangor, Malaysia
[2]Geography Programme, School of Distance Education, Universiti Sains Malaysia, Penang, Malaysia
[3]Sabah State Health Department, Ministry of Health Malaysia

Email: rauf@uitm.edu.my

Abstract. Understanding concepts of a proper disease transmission risk is not a straightforward process. In the context of tuberculosis (TB) dynamics, the concepts require the exploration of two meticulous criteria to produce an accurate epidemic modelling of the risk areas of the disease. The criteria include interpreting the biological transmission of the disease and applying multidisciplinary approaches. Spatial statistics were used to evaluate the preferences of risk factors in Shah Alam, Malaysia. GIS-multicriteria decision making (MCDM) method and logistic regression method were specifically integrated to select the local risk factors and seven influential factors were ranked accordingly i.e. human mobility, high risk group, socio-economic status (SES), population, type of house, distance of factory and urbanisation. Each has relative risk rate that affects the cases and the combination of them will even impact more on the overall risk concentration of TB. Human–based factors are identified as dominant effects to the risk than biophysical factors, for example, a location of TB risk will be increased by four times if individuals are living together with people who have TB disease for a particular time period. This geospatial method is expected to predict a better factor prediction in identifying hotspot areas of the disease.

1. Introduction

The Ministry of Health (MOH), Malaysia has established several guidelines to control tuberculosis systematically. Although the guidelines seem to be focusing on human or biomedical approach (TB screening and x-ray methods) to control the disease as stated by WHO in 2014, the main aim of the guidelines or agendas is still the same which is to emphasise on strengthening control, prevention and elimination of the disease as suggested by the TB and Leprosy Sector, MOH Malaysia [15].

However, the findings from the national technical report on TB in Malaysia (TB and Leprosy Sector, MOH, Malaysia, 2015) asserted that the existing methods for TB screening among high-risk groups (such as missing and undiagnosed TB cases) need to be strengthened in order to increase TB cases detection rate (CDR). Similarly, [22] of WHO also agreed that the current method still fails to address inequitable distribution of disease and does not diagnose many TB patients especially among poor and vulnerable communities, and marginalized people.

This situation is caused by several factors especially the inefficiency of the existing method to comprehensively detect TB cases. For example, the risk factors influencing the TB cases is multi-sources from geographical or environmental factors human-based factors (such as high-risk group and socio-economic status). Consequently, this method needs to be combined with other techniques [2][14][21] in order to improve management of cases and analytical power. As such, this study examines the capabilities of spatial statistics to determine the risk factors contributing to TB cases in Shah Alam. This

IOP Conf. Series: Earth and Environmental Science **385** (2019) 012037 doi:10.1088/1755-1315/385/1/012037

geospatial method can be used as a tool to characterize potential high-risk factors and potential risk areas of TB.

Spatial epidemiology (SE) is empirically applied to provide integrated spatial frameworks for enhancing current technical issues on case detection of a disease (such as TB). The core framework includes the spatial analysis (spatial pattern and spatial risk correlation) and spatial modelling as a local disease surveillance management tool. A spatial statistics method can perform a better spatial decision-making system for public health management [6][28] to estimate potential high-TB risk areas [12][16[18][29] especially in using spatial statistical approach [30][31][32].

2. A review of spatial and statistics for risk disease assessment

2.1. Spatial epidemiology of tuberculosis transmission risk

At present, TB is still a global concern and it can occur in any part of the world. [4] estimated that one-third of the global community is infected with M. tuberculosis. In 2010, the World Health Organization (WHO) stated that there were approximately 8.5–9.2 million cases and 1.2–1.5 million deaths from TB worldwide. The global distribution of TB cases in 2000 is skewed heavily toward the low-income and emerging economies. [33] reported that the majority of the TB cases come from African and Asian regions especially in South Africa and India. Other countries of high TB prevalence in Asia include China, Bangladesh, Indonesia, and Pakistan.

There are notable differences in the contributing risk factors between the developing and developed countries. TB cases in developing countries are clustered in populated density, low socioeconomic status and poor housing conditions or environments, whereas in the developed countries the cases are driven by foreign-born immigrants and animal-borne TB factors. However, an interesting aspect is that most of the foreign-born immigrants in developed countries have similar characteristics found in developing countries that are mainly caused by crowded and poor environments.

In Malaysia, the trends of TB cases are slightly increasing [17]. Sabah, Sarawak and Selangor recorded the highest prevalence of TB cases in 2011 in the country. The general spatial clustering of the recent TB distribution patterns in Peninsular Malaysia showed that there is no specific clustering in the study area, but the overall spatial pattern of TB cases concentrated in urban areas especially those with a high number of the population, a low socioeconomic status, and high urbanisation. However, it is insignificantly related with and non-forested area of all the states [1]. Specific analysis at smaller areas needs to be conducted in Malaysia to determine the precise risk factors contributing to particular regions.

2.2. Spatial statistics for tuberculosis dynamics

An appropriate understanding on local TB transmission and its epidemic stages are fundamental steps to develop a better predictive model of diseases. A sufficient concept of TB transmission can simplify the dynamic representation of real TB phenomena or system according to a local actual situation and dynamics. Dynamics is the interaction process which is related to clear changes or finding the causality underlying the type and speed of the observed evolution. The concept of dynamics can provide a simplified map, statistical information, explanatory model based on a known fundamental mechanism, and towards the application of a multidisciplinary model [7][10].

Tuberculosis causation and transmission process is not a straightforward matter and the accepted models of disease causation require the precise interaction of factors and conditions before a disease occurs. However, the traditional model (epidemiologic triangle, epi-triad) or cycle of infection illustrates that infectious diseases result from the interaction of three components including external agent resources (bacterium in human, environments), a susceptible host (droplet, airborne) and an environment that brings the host and agent together. Specifically, transmission occurs when the agent or pathogen leaves its reservoir or host through a portal of exit,and is conveyed by a mode of transmission and enters through an appropriate portal of entry to infect a susceptible host.

Thus, the factors that determine the probability of TB transmission need to be studied in-depth and they can be divided into several epidemic stages from human body (intrinsic or exogenous factor) to human environment (extrinsic or endogenous factor). According to CDC USA (2011) and Narasimhan

IOP Conf. Series: Earth and Environmental Science **385** (2019) 012037 doi:10.1088/1755-1315/385/1/012037

et al. (2013), internal or human level conditions are dominant or well-established factors of susceptibility and infectiousness, while external factors include environment and exposure. Determining risk factors is important to identify the characteristics of high-risk population and areas according to global perspective. Thus, efforts should also be made to collect risk factors data in routine surveillance for TB disease [20].

Techniques in spatial epidemiology (SE) are applied in certain fields to quantify pattern, correlation and assessment of a situation. For example, in the context of health geography, the effect of a certain risk factor on disease outbreaks might be determined by estimating its correlation and risk areas using a certain model. In this study, the combination of statistical method and spatial method is applied in estimating TB risk areas and their correlation with influential risk factors such as socio-economic status, human and physical environments in Shah Alam. These methods can help in explaining complex and often difficult aspects of system behaviour to 'non-experts', especially if the models can produce graphical visual, even animated, outputs [27]. Besides this, [7] also stated that the methods can be a powerful tool to assist animal health policy development and disease prevention and control.
SE introduced and applied a spatial framework for epidemiological studies [8][9][23][25] for spatial pattern, cluster, and prediction of TB disease risk.

GIS-MCMD and statistical methods are combined to produce a risk map of the TB incidence and population as conducted in local studies [34]. These methods also emphasise on the population or community health assessment rather than individual decisions (clinicians are concerned with the health of an individual). Lena Sanders (2007) stated that spatial dimension plays a key role in modelling for phenomena including social and human population and ecology physic environments phenomena besides knowledge from various fields and statistician model, while statistical framework is a common method to set out a hypothesis in the thematic model using dependant/to be explained (dependent variable), exploratory of independent variables (IV) both of quantitative and/or qualitative data. This combination would be enhancing the technical aspects in spatial epidemic models and draw attention to the lack of models that explore the relationship between communicable [5][26].

3. Data and materials

The main secondary data used in this study are TB cases (all types) from 2013 to 2015 obtained from MyTB, Selangor State Health Departments. Spatial and statistical methods were integrated to determine the potential risk factors of TB in Shah Alam. This spatial-statistical method is relevant to cover a larger scope of the epidemic model. The technical framework used in this geospatial-based method is referred to as GIS modelling and spatial epidemiological (SE). A binary logistic statistical method has been selected because the data did not meet the assumption of the normality, while spatial method has the capabilities in data mapping and analysis of disease risk assessment. Shah Alam is a suitable location to be selected as a study area due to its geographical condition, dynamic population growth and human movement. Four main steps applied in the study were i) risk factor selection, ii) spatial data collection, iii) spatial data processing, and iv) spatial statistical analysis.

3.1. Selection of risk factors

Establishing a spatial concept of local TB transmission and data collection of risk factors are essential tasks to identify the transmission components involved in the study and. Scholarly sources (a systematic review) from global TB studies, exclusive reports from WHO and MOH were used in
selecting local risk factors based on a multi-criteria analysis, including human-factors and biophysical indicators.

Biophysical environment is defined as the biological environment as well as the physical activity within it, including land use/urbanisation, type of house, proximity of healthcare facilities and industrial zone. Human-population based environments or human risk factors refer to human and individual characteristics, which influence behaviour at work in ways which affect health such as density of population, social economic status, concentration of risk group and population mobility. The combination of these indicators of risk factors is important to develop a standard and holistic guideline for local levels of external risk factors and risk areas priorities in Malaysia.

Several technical methods are then used such as GIS-MCDM, and regression model. This main analytical framework of research is innovatively combined from GIS operations [4], spatial epidemiological methodology [25]. Table 1 presents a list of selected risk factors and their variables proposed in this study. These predictors were included in the MyTB databases for risk assessment as individual-level spatial information.

Table 1. Global perspectives on potential risk Spatial data collection

Indicators	Independent variables/Factors (Code) (Scale /No_Class) Data Model
Natural environment of landscape (Biophysical and ecological environment)	Urbanisation-land use (Score_Urb) (Continuous scale from 1 to 5) Surface data from point inputs
	Types of house-settlement (Score_Hou) (Continuous scale from 1 to 5) Surface data from point inputs
	Distance to healthcare facilities (Score_Hea) (Continuous scale from 1 to 5) Surface data from point inputs
	Distance to industrial factory locations (Score_Fac) (Continuous scale from 1 to 5) Surface data from point inputs
Anthropogenic (Human and population)	Number of people in a house (Score_Peo) (Continuous scale from 1 to 5) Surface data from point inputs
	High risk groups (Score_Gro) (Continuous scale from 1 to 5) Surface data from point inputs
	Household income-Socio economic status, SES (Score_Soc) (Continuous scale from 1 to 5) Surface data from point inputs
	Concentration of Patient Mobility Score_Mob) (Continuous scale from 1 to 5) Surface data from point inputs

Data and instruments used in the study are the disease cases, spatial-environmental data for risk factors, and spatial-statistical packages. Two approaches were applied to collect the data, which are primary and secondary data collection as displayed in Table 2. Primary research consists of a collection of original primary data collected by the researcher especially from expert opinions (Health staff from TB Unit of Pejabat Kesihatan Daerah, PKD, Petaling) and site visit at high risk areas of the study area. Systematic reviews and questionnaires were conducted to define local risk factors and their risk rates. Secondary data refers to the data that was collected by others such as TB cases from MyTB System.

Table 2. Main data sets used in the study.

Data	Data structures and model
TB Cases (Secondary data)	Non-spatial (attribute Data) of all types of TB cases from 2013 to 2015.
Risk Factors of TB (Primary and Secondary data)	Combination of spatial and non-spatial data. They include 8 influential risk factors of local TB in Shah Alam as shown in Table 1.
Base maps of TB mapping (Primary and Secondary data)	Spatial data.

Reviews, questionnaires and site visit of TB risk factor and study areas (Primary data)	Non-spatial data and image data.

3.2. Spatial data processing and weightage calculation

Data processing and analysis were employed with ArcGIS and SPSS Statistics 21 for descriptive and scientific analysis. All non-spatial data required in this study were converted into spatial data using ArcGIS to become a geodatabase of TB. After selecting the risk factors, standardisation and classification of TB Risk Scale were also conducted. The attributes/tables were added with eight risk factors (as independent variables, IV of risk factor of TB), and one for total score of risk factors (as a dependent variable, DV) with their risk values. Each table has a standardised value of risk ranking from 1 to 5 since some original data were formed as nominal values.

A ranking procedure is based on local expertise and government standard guidelines. For household income, the Economic Planning Unit of the Prime Minister's Department Malaysia (2012) has defined poverty as referring to people who have average monthly household income at or below RM 830.00. Thus, this study assumed that people who earn average incomes of less than RM 830.00 to be included as risk part of groups (value 5), while people earning more than RM 2490.00 or low-risk groups were categorised as value 1 and value 2. The values of 1-5 were added in the database for each IV and then the average was calculated to obtain a Total_Risk Score (DV). If a classification scale of risk factor could not be clearly decided using the judgement from the expert opinion or theoretical context, the statistical evaluation (mean value) is also alternatively used in ranking the risk.

Then, these scaling values were added into the existing databases of risk factors (IV). Total_Risk Score (DV) can be replaced with a new value as 1 (rank 4 and 5) and 0 (rank 1 to 3) for logistic regression modelling (Logistic_Outcome, DV).

Furthermore, Total_Risk Rank (DV) was created to develop a GIS–MCDM model according to expert opinion or knowledge driven values. These DV values can be used in calculating the risk level of TB factors. In general, MCDM method comprises of three main stages: i) selection of risk factors, ii) calculation of risk factor weights and iii) eliciting local risk factors. MCDM method [4][13] analysis was implemented to rank the TB infection risk rate or weight. Each criterion or risk factor was directly ranked (from 1 to 8) by the selected experts and then the values or weight were standardised using rank sum techniques as showed in equation (1) from 0 to 1. The final stage is eliciting the influential risk factors and risk weights of local TB according to expert's rank or MCDM method. The values of 1-5 were added to the TB geodatabase. The scale values were also placed into logistic outcome using 1 (Yes Risk) and 0 (No Risk) for calculating the risk level of TB.

$$W_j = \frac{n - r_j + 1}{\sum (n - r_k + 1)} \qquad (1)$$

Where:
Wj = the normalized weight for the jth criterion
n = the number of criteria under consideration (k=1,2,3...n) and
r_j = the rank position of the criterion
each of the criterion is weighted $(n - r_k + 1)$ and then normalized by the sum of all weights and that is $\Sigma (n - r_k + 1)$.

3.3. Spatial statistics analysis

A logistic regression analysis was used in characterising and prioritising the risks of TB after the data passes data assumptions. The model is an inferential statistical technique used to predict the presence or absence of a characteristic or outcome values (DV) of a set of predictor variables. The regression coefficients can be used to estimate the odds ratios (OR) for each of the independent variables in the model. A logistic regression model used to predict the potential location of TB disease by estimating the value of $\beta_0, \beta_1 ... \beta_m$ (individual characteristics) and outcome probability of disease (Y= a chance or ratio of an outcome)is shown in equation (2). The result can be used to understand whether TB risk area/locality can be predicted based on environmental risk factors (i.e., "Yes Risk" or "No Risk". The following prediction equation (2) was used to calculate a predicted probability of having disease or to estimate the risk of disease [11][15]. In selecting the suitable modelling of risk assessment, a comparative study was conducted using 3 different assessments entering the risk factors. To determine which model is suitable for the study area, the tables in block report were mainly referred to as the statistical tests to determine the model goodness.

$$Y = \frac{1}{1 + \exp^{-(\beta_o + \beta_1 X_1 ... \beta m X m)}} \quad (2)$$

For the results of MCDM, logistic regression model was used to estimate the unmeasured points of TB risk in the study area using GIS interpolation or geostatistical method. These points have all the characteristics and scale of TB risk factor (IV) and total TB risk factor (DV). In Geostatistical Analyst Wizard in ArcGIS, the input data, method and attribute must be correctly selected. The desired parameters should also be set on the IDW dialog box to provide the risk points, including Optimize Power value=2, Symbol size=standard, Neighbour including 15 with at least 10 points, Sector types=standard ellipse, Anisotropy factors=1, and coordinate estimate points. In addition, the box of weight, value menu and preview (neighbourhood) were selected.

4. Result and discussion on profiling spatial effect of TB risk factors

Spatial statistics is quantitative assessment is applied to support the spatial qualitative result in identifying the extent of the risk factors that have affected the incidence of TB, and ultimately use them to control and prevent the spread of TB. It was found that 7 out of 8 common risk factors are influential factors in the incidence according to multi-criteria and regression analysis. It includes factory distance, urban areas, house type, high-risk group, income status, mobility and population. The number of populations in a house is not a statistically significant factor, but according to local experts, this factor should be considered in defining the real effect of TB on the local study. While the distance of healthcare facilities is not statistically significant even though it is a common indicator of TB cases based on global and local perspective.

Table 3. Justification of risk rate of selected TB risk factors in Shah Alam

Risk Factors of TB (Risk Value/Rate)	Theory and Previous Studies
Mobility (3.654)	Theory yes, especially for patient with an active TB. They will be dangerous and infect others through closed space, short distance and exposure in long duration of time (8 hours).
Risk group (2.841)	This is a well-established factor as occurred in the world TB.
SES (2.395)	This is a well-established factor as occurred in the world TB, especially in Malaysia.
Population (2.061)	This is a well-established factor as occurred in the world TB, either in the house or outside.
House (1.715)	This is a potential risk factor of the local TB. The types of houses reflect the condition and space of the building. A poor, low and small/limited house will influence TB occurrences.

Factory (1.598)	This is a potential risk factor of the local TB. There is still no strong evidence to show its correlation with local TB occurrences, but it can be related with air pollution and the risk group in the industrial factories.
Urban (1.161)	The factor has been proven from previous studies. Most of the cases were located at urban areas with poor housing areas.

The selection of the factors is the basis for the formation of the proposed model of the impact risk assessment and mapping potential areas using GIS-based logistics regression technique. Risk assessment using a logistic technique shows that statistical test indicates a strong relationship between prediction and grouping ($p=0.797$), and the classification was generally and correctly completed at 93.2%. In addition, human-based indicators, especially human mobility, made a significant contribution to prediction instead of biophysical indicators.

Overall risk factors used in the proposed geospatial model is as shown in Table 3. According to the influential risk ranking and values, human indicators are more dominant factors than biophysical indicators as found in previous studies in Malaysia [1][17] and local expert opinion. Regression was even clearer by illustrating that the Coefficient (ß) of risk factors according to ranking includes Score_Mob (3.654), Score_Gro (2.841), Score_Soc (2.395) and Score_Peo (2.061). These factors are included as human indicators. While biophysical factors only recorded minimum Coefficient (ß) such as Score_Hou (1.715), Score_Fac (1.598), and Score_Urb (1.161), urbanisation is evidently not significant, but in terms of expert opinion it should be one of the major risk factors in local condition. The distance of healthcare facilities is not included as a factor due to the fact that it is statistically insignificant, as well as supported by experts' opinion.

It can, therefore, be said that the combination of these seven factors can produce relevant TB risk factors in Shah Alam. Four risk factors based on human indicators are riskier than three biophysical indicators, but the combined results of these indicators can make the area more significant to TB risk. The results are also not only important for local health staff to fully understand the risk factor but also to use them in developing the proposed geospatial epidemic model [1].

In summary, spatial-logistic regression is based on real data and predictive analysis. The results suggest that the logistics may have minimal risk, but according to existing data and local experts' view, this area should only have a medium burden of TB. This study suggests several recommendations to further strengthen the findings. It includes the logistic regression technique used as a direct (binary) technique that provides outcome to two choices Yes (1) or No (0) only and therefore, it is suggested that a comparison is made with other techniques such as spatial multivariate and ordinal regression models.

The techniques used will also take into consideration the possible variation effects on the environment and different distribution patterns of local TB because the variation can affect the accuracy of the results. This proposed spatial logistic regression model could help the health staff to predict the prevalence of TB in Selangor according to a population or areal level and combined with a GIS mapping for early identification of TB progression.

5. Conclusion

In Malaysia, the presence of spatial heterogeneousness of tuberculosis (TB) incidence may lead to the complexities of risk factors identification because each area has its own unique risk factors and dynamics. This paper is aimed to quantitatively assess the selected eight risk factors in a locality by employing an integration of GIS-MCDM and logistic regression method. Findings from MCDM suggests that 8 risk factors have contributed to the local TB cases, but quantitative risk assessment with a GIS-logistic method found that all seven risks are significant ($p < 0.05$), including urbanisation, distance to factory, SES, high risk group, human mobility, house type, distance to healthcare centres, and the number of populations in a house. Each relative risk rate that affects TB occurrences and their combination will give more impact on the overall occurrences. Overall, the combination of spatial/GIS and statistical method can enhance the classical method to determine the risk factors contributing to local TB occurrences. Furthermore, the proposed risk factors in this study can provide a holistic platform in identifying targeted spots for TB risk concentration in the study area.

Acknowledgement
The authors would like to express appreciation to the TB and Leprosy Sector of Ministry of Health Malaysia and Selangor States Health Department for providing TB datasets in this research paper. The research has been registered in the National Medical Research Register, Malaysia (ID: NMR R -15-2499-24207). The authors would also like to sincerely thank Universiti Teknologi MARA Selangor for providingLESTARI Research Fund (600-IRMI5/3/LESTARI (005/2018).

References
[1] Abdul Rasam, A.R.., Shariff, N.M., Dony, J.F and Misni, A. (2018). Socio-environmental factors and tuberculosis: an exploratory spatial analysis in Peninsular Malaysia, *International Journal of Engineering & Technology, 7* (3.11), 187-192.
[2] Burgos, M. V., & Pym, A. S. (2002). Molecular epidemiology of tuberculosis. *European Respiratory Journal, 20*(Supplement 36), 54S–65s. doi:10.1183/09031936.02.00400702
[3] CDC USA. (2013). *To View or Order the Core Curriculum on Tuberculosis*. USA. Retrieved from https://www.cdc.gov/tb/education/corecurr/pdf/corecurr_all.pdf
[4] Chang, K. (2011). *Introduction to Geographic Information Systems*, New York, USA.McGraw Hill, 389-399.
[5] Childs, C. S., Florida, S., Health, P., & College, M. (2013). *Evaluating the Spatial Relationship Between Remotely-Sensed Urban Vegetation and Tuberculosis in Guayaquil , Ecuador* (Vol. 5).
[6] Dhanaraj B, Papanna MK, Adinarayanan S, Vedachalam C, Sundaram V, Shanmugam S, et al. (2015). Prevalence and Risk Factors for Adult Pulmonary Tuberculosis in a Metropolitan City of South India. PLoS ONE 10(4): e0124260. https://doi.org/10.1371/journal.pone.0124260
[7] Fuller, D. O., Troyo, A., Alimi, T. O., & Beier, J. C. (2014). Participatory Risk Mapping of Malaria Vector Exposure in Northern South America using Environmental and Population Data. *Applied Geography (Sevenoaks, England), 48*, 1–7. doi:10.1016/j.apgeog.2014.01.002
[8] Garner M.G. and Hamilton S.A. (2011). Principles of Epidemiological Modelling. *Review Scientific et Technique, 30(2)*, 407-416.
[9] Lai, P.; Wong, C.M; Hedley, A.J; Leung, G.M. (2009). Hong Kong Medical Journal, 15 Suppl 9, 17-19.
[10] Lawson, A.B, & Leimich, P. (1999). Approaches to the space time modelling of infectious disease behaviour. IMA Journal of Mathematics Applied in Medicine and Biology, 17, 1-13.
[11] Lena Sanders (2007). *Models in Spatial Analysis*. (L. Sanders, Ed.) (First., pp. 1–26). Great Britain: ISTE Ltd.
[12] Leung, P., Tran, L. ., & Fast, A. . (2000). A logistic regression of risk factors for disease occurrence on Asian shrimp farms. *Dis Aquat Org, 41*, 65–76.
[13] Lin, H., Shin, S. S., Contreras, C., Asencios, L., Paciorek, C. J., & Cohen, T. (2012). Use of Spatial Information to Predict Multidrug Resistance in. *Emerging Infectious Diseases, 18*(5), 5–7. doi:DOI: http://dx.doi.org/10.3201/eid1805.111467
[14] Malczewski, J. (2000). On the use of weighted linear combination method in GIS:Common and best practice approcahes.*Transanctions in GIS* 4:50-22.
[15] Mathema, B., Kurepina, N. E., Bifani, P. J., & Kreiswirth, B. N. (2006). Molecular epidemiology of tuberculosis: current insights. *Clinical Microbiology Reviews, 19*(4), 658–85. doi:10.1128/CMR.00061-05
[16] Mcnutt, L., Holcomb, J. ., & Carlson, B. . (2000). 088626000015010003 #.pdf. *Journal of Interpersonal Violence, 15*(10), 1050–1059.
[17] Middelkoop, K., Bekker, L., Morrow, C., Lee, N., & Wood, R. (2014). Decreasing household contribution to TB transmission with age : a retrospective geographic analysis of young people in a South African township. *BMC Infectious Diseases, 14*(221), 1–7. doi:10.1186/1471-2334-14-

221

[18] Ministry of Health Malaysia [MOH]. (2015). *The National TB strategic plan (2010-2015) and NTLP*. TB/Leprosy Sector. Putrajaya

[19] Moonan, P. K., Oppong, J., Sahbazian, B., Singh, K. P., Sandhu, R., Drewyer, G., … Weis, S. E. (2006). What Is the Outcome of Targeted Tuberculosis Screening Based on Universal Genotyping and Location ? *Am J Respir Crit Care Med*, *174*, 599–604.

[20] Musenge, E., Vounatsou, P., Collinson, M., Tollman, S., & Kahn, K. (2013). The contribution of spatial analysis to understanding HIV/TB mortality in children: a structural equation modelling approach. *Glob Health Action 2013*, *19266*(1).

[21] Narasimhan, P., Wood, J., Macintyre, C. R., & Mathai, D. (2013). Risk factors for tuberculosis. *Pulmonary Medicine*, *2013*, 1–11. doi:10.1155/2013/828939

[22] Narayanan, S. (2004). Molecular epidemiology of tuberculosis. *Indian J Med Res*, (October), 233–247.

[23] Nishikiori, N. (2011). *Tuberculosis control among high risk and vulnerable populations*. Retrieved from http://health.gov.fj/PDFs/Tuberculosis control among high risk.pdf

[24] Peterson. (2014). *Mapping Disease Transmission Risk. Enriching Models Using Biogeography and Ecology*. USA, John Hopkins University Press.

[25] Pfeiffer, D. U., & Stevens, K. (2011). *Using Data - and Knowledge - driven Approaches for Spatial Modelling* (pp. 1–44).

[26] Pfeiffer, D. U., Robinson, T. P., Stevenson, M., Stevens, K. B., & Rogers, D. J. (Techniques). (2008). *Spatial Analysis in Epidemiology United Kingdom*. Oxford University Press, USA.

[27] Ribeiro, F.K.C., Pan, W., Bertolde, A., Vinhas, S.A., Peres, R.L., Riley,L., Palaci, M., Maciel, E.L.(2015). Genotypic and Spatial Analysis of *Mycobacterium tuberculosis* Transmission in a High-Incidence Urban Setting, *Clinical Infectious Diseases*, 61, 5, 1 September 2015, Pages 758–766.

[28] Roberts, M., Andreasen, V., Lloyd, A., Pellis, L., Sciences, M., Mail, N. S., … Epidemiology, D. (2016). Nine challenges for deterministic epidemic models. *Epidemics*, *10*, 49–53.

[29] Sun, W., Gong, J., Zhou, J., Zhao, Y., Tan, J., Ibrahim, A.N., Zhou, Y. A (2015). Spatial, Social and Environmental Study of Tuberculosis in China Using Statistical and GIS Technology. *Int. J. Environ. Res. Public Health*, *12*, 1425-1448

[30] Taylor, N. (2003). *Review of the use of models in informing disease control policy development and adjustment* (pp. 1–80). Reading, United Kingdom. Retrieved from http://www.veeru.rdg.ac.uk/documents/UseofModelsinDiseaseControlPolicy.pdf

[31] Torre, A., Iglesias, I., Carballo, M., Ramírez, P., & Muñoz, M. J. (2012). An approach for mapping the vulnerability of European Union soils to antibiotic contamination. *The Science of the Total Environment*, *414*, 672–9. Wei, W., Wei-sheng, Z., Ahan, A., Ci, Y., Wei-wen, Z., & Ming-qin, C. (2016). The Characteristics of TB Epidemic and TB / HIV Co-Infection Epidemic : A 2007 – 2013 Retrospective Study in Urumqi , Xinjiang Province , China. *PloS One*, *11*(10), 1–12.

[32] WHO, W. H. O. (2014). Drug-Resistant Tb Surveillance & Response;Supplement Global Tuberculosis Report 2014 (pp. 1–20). 20 Avenue Appia, 1211 Geneva 27, Switzerland.

[33] Abdul Rasam, A.R., Mohd Shariff, N., Dony, J.F, & Maheswaran, P (2017). Mapping Risk Areas of Tuberculosis Using Knowledge-Driven GIS Model in Shah Alam, Malaysia, *Pertanika Journal of Social Sciences &Humanities*, *25 (S)*,135-144.

IOP Conf. Series: Earth and Environmental Science **385** (2019) 012038 doi:10.1088/1755-1315/385/1/012038

Online Mapping and GIS Mobile Healthcare Systems for Intensifying Low-and-Middle-Income Community Accessibility

A H Zain Rashid[1], A R Abdul Rasam[1,2*], A N Idris[1], N A Adnan[1] and F Othman[1]
[1] Centre of Studies of Surveying Science and Geomatics, Faculty of Architecture, Planning and Surveying, Universiti Teknologi MARA, 40450 Shah Alam, Selangor,
[2]Environmental and Social Health Research Group,
[3]Faculty of Architecture Planning and Surveying, Universiti Teknologi MARA Shah Alam, Selangor, Malaysia

Email : rauf@uitm.edu.my

Abstract. Systematic healthcare facilities and accessibility system in developing countries are mostly located in urban areas instead of rural areas, causing inconvenience to low-and-middle-income residents to reach healthcare centres. Therefore, an innovative mapping and geomobile application or app is developed for the modernization of healthcare systems in Pulau Indah, Klang Malaysia. This study aimed to provide relevant geoinformation of the healthcare centres and the optimum route to the nearby healthcare facilities. Data collection comprised of road access, lists of healthcare facilities, the type of healthcare facilities, travelling time, cost, as well as base map from Google Earth. The framework of geospatial functions and System Development Life Cycle (SDLC) were implemented to create the proposed system. Survey on 45 respondents revealed that 80% of them are displeased with the existing system performances. Thus, the proposed healthcare accessibility system has created eminent geospatial modules and cartographic elements to intensify the existing system. This personal app system also offers several spatial network analyses to assist the local community in finding the optimum accessibility to reach ideal healthcare centres. Inclusively, seven selected local communities agreed with the functions proposed in this system but there are several parts can be further improved.

1. Introduction

Mobile applications or Mobile apps are software programs developed for mobile devices such as smartphones, mobile books and tablets. They turned mobile devices into miniature powerhouses of function and fun [13]. [13] also added that the purposes of these apps run the gamut, with purposes not limited to utility, productivity, and navigation but also to entertainment, sports, fitness, and just about any other imaginable daily activities [13]. The common mobile apps are usually available to be downloaded via mobile operating platforms, either Apple App Store, Google Play, Windows Phone Store or Blackberry App World which can be free downloaded or in-apps purchases. They also generally offer productivity and information retrieval, such as email, calendar, contacts, a stock market and weather information.

However, the rising demand by the public and the development of the tools by the developers has become the main factors in the apps into other categories such as mobile games, factory automation, GPS and location-based services, banking, order tracking, ticket purchase, accessibility and medical services. The services offered by the mobile apps has attracted many users from time to time to use mobile applications in their daily life to make their lives easier.

Health and well-being are the most important element in human life for longevity. The combination of a healthy diet and a well-practiced of exercise in daily routine are not limited to maintaining a good health, but also to prevent any complex diseases infection and to reduce high medical cost. Although healthcare facility has rapidly grown from time to time, the handling capacity is still insufficient for people around the area. Therefore, the solution for finding healthcare services must be taken seriously from different perspectives.

This study focuses on the development of a mobile application with a function to provide users with information on healthcare facilities. Today, smartphones and mobile electronic devices such as tablets and Ipads have transformed our daily communication lifestyles [10]. It is recorded that three-quarters of world population expected the mobile apps to be free, and their willingness to pay for apps are varied by several categories including technology, finance, travel, social, entertainment, and retail [10]. Hence, the development of a mobile application will be able to provide information on direction, location, and cost-estimation of the chosen healthcare facility thus making it easily accessible to the public.

Pulau Indah, Klang, Malaysia was selected to be the study area on several factors. This area is located remotely from Klang City and has a limited number of clinics. Most of the healthcare centres are located at Port Klang where local clinics are insufficient to accommodate many patients at a one time. Thus, a good information system of healthcare accessibility is proposed in this study area. There are three objectives of this study which are to carry out the study of user requirement on the proposed mobile applications system, to design and develop the proposed geomobile application according to user requirement, and lastly to evaluate the performance of the system functionality and application.

2. A review on geospatial mobile apps for healthcare community services

2.1. Healthcare network accessibility using a GIS approach

Healthcare accessibility is recognized as an important component of the overall health system due to its capability to promote socio-economic growth and development for the nation and reduce the burden of diseases by improving the quality mental and physical of all well-being (Surage et al., 2017). The study on the accessibility of healthcare is aimed to investigate the level of satisfaction from the public ~~on~~ whether they have an easy, simple and user-friendly access to healthcare services or otherwise. There has been a number of existing healthcare services that does not offer people the guarantee to have a good healthy life especially for people who live in a rural area as there is limited access for them. According to [7][15][16] which conducted a study on accessibility for healthcare centres the best way to find optimum access to reach ideal healthcare centres especially in peak time by focusing on the element of global accessibility. By focusing globally, the availability and suitability of the healthcare centre can be identified and determined.

As the fundamental principle for the health geography is even distribution of the healthcare services across space and time and easy access by the populace [11]. Often, hospital and other healthcare facilities are built in the area that has good access to public transportation. Most people in the urban area prefer to commute using cars as there are more stable financially but it is a different situation for people in rural areas. Most of them commute using motorcycles which is within their income, but it is very unsafe and impractical for long distance journey especially if they travel for the purpose of seeking health treatment. Besides that, the distribution of healthcare facilities also impacts people's health. This is because the location of healthcare facilities can affect people's ability to acquire and use the services. A great example would be for people who live in an area that has no healthcare facility and the one nearby is quite distant from their home hence there are high possibilities they will face problem in order to get help during an emergency.

The measurement of physical accessibility is well established and has been applied in many areas with the use of Geographical Information System (GIS). This includes transport analysis, retail site analysis, emergency service and healthcare planning. GIS is well suited to measuring spatial accessibility to

IOP Conf. Series: Earth and Environmental Science **385** (2019) 012038 doi:10.1088/1755-1315/385/1/012038

healthcare as they contain the core components needed for such analysis namely data capture storage, management and manipulation tools for both spatial and attribute (textual) data with the following capabilities:

- Programming environments to customize and extend existing algorithms and create new analysis tools.
- Core analysis algorithms such as buffering, overlay, proximity analysis, shortest path and raster cost-distance analysis.
- Mapping and visualization tools to communicate the results of the analysis title.

2.2. Online mapping and GIS mobile applications for healthcare information system

System navigation has been used widely nowadays and gradually it has evolved to a simpler and better method and it has become a part of people's everyday life. System navigation is a tool that helps people move or navigate from one location to another. This system is a combination between the modern mobile technology, satellite navigation system and cartography capability which makes this function a vital application in this era. There are many applications that have been developed to provide the same function which comprises navigation. Google maps are one of the best navigation applications available for free. Other than showing directions, it also shows real-time traffic information whether it is congested or not and provide alternative routes to avoid the issue. Consequently, users will be able to identify which roads can be chosen to avoid traffic congestion. This is beneficial for users in making their travel plans and selecting the best routes to reach their destinations.

Another mapping technology in cartography is GIS mobile apps that can be developed in various platforms and brands such as Lenovo, Smart Phone mobile operating system (Table 1). A mobile operating system is a software system designed to control computer hardware while the operating system that operates on mobile devices is called a mobile operating system [9]. This operating system operates on a smartphone, PDA, cellular phone, tablet or another digital mobile device. The system has the combined features of a personal computer operating system with including the features of the touch screen, cellular, Bluetooth, Wi-Fi, GPS mobile navigation, camera, video camera, speech recognition, voice recorder, music player, Near field communication and Infrared Blaster [8].

Today, there are many accessibility apps that are available to the public, such as Google Maps, Waze, Navigation – GPS Maps Traffic, Life Navigation and many other apps related to the navigation concept. However, for the healthcare facility accessibility apps, they are not as popular compared to other healthcare apps. This is probably due to the existing apps that are related to navigation which also provide accessibility assistance to the users for monitoring patient, personal advice to aware the patient, education or training to the nurses or doctor regarding new available resources, treatment and decision support for health professional applications in health care accessing with the apps [2][3][5][6][12][14].

3. Data and materials

The methodology utilized in the study consisted of several stages to achieve the main objectives as presented in Figure 1. These stages used the framework of System Development Life Cycle or (SDLC), including user requirement, system design and development, system testing and implementation. SDLC assisted the developer in establishing a proper step in system development with an organized procedure at the beginning of the process.

3.1. Preliminary study and user need assessment

From the comparative review of the existing android apps for healthcare facility services, this study has decided to use MIT Apps Inventor to develop the Androids Apps. MIT Inventor is an open source web application originally provided by Google and today it is operated by the Massachusetts Institute of Technology. It uses a graphical interface, and quite like Stratch and the StarLogo TNG user interface. This MIT Inventor allows a user to drag and drop the visual objects with the aim to create an application

IOP Conf. Series: Earth and Environmental Science **385** (2019) 012038 doi:10.1088/1755-1315/385/1/012038

that can run on Android devices. In creating this mobile app, Google drew upon significant prior research in educational computing, and the work is done within Google on online development environments. For the user requirement, the online questionnaire via Facebook and Whatsapp were distributed to the community in Pulau Indah, Klang in getting the user feedbacks on the proposed apps. From the findings, there are several components used in the analysis for the mobile apps, from the operating systems of the mobile phones to the agreement from the users to use the function for the development of researcher mobile apps. The questions are also focused on the functions of healthcare facility, information of healthcare, information of health, comment for users review, and costing information i.e. travel cost to healthcare facilities. The questionnaires received feedback from 45 respondents from the community.

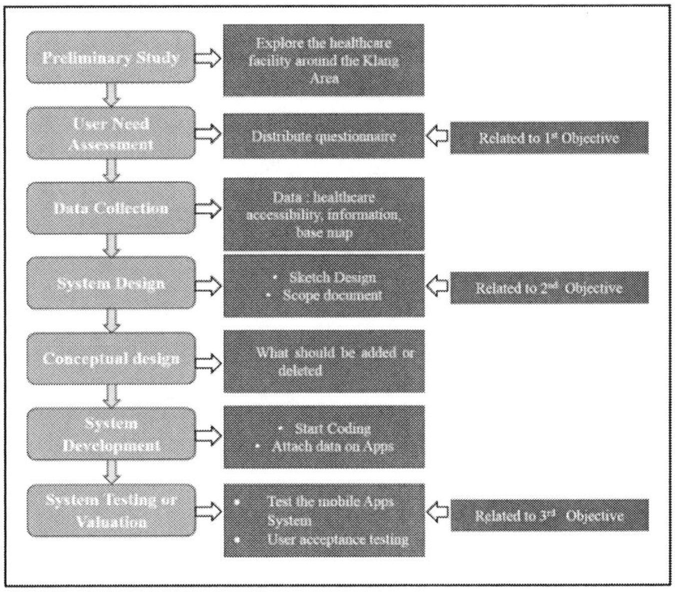

Figure 1. Methodology of GIS mobile healthcare apps system

3.2 Data collection

There are several data that were combined in the mobile application to produce the application that will meet the need of the community of Pulau Indah. The data were integrated into the application as maps, health information, healthcare information and transportation. The data can help the community to obtain information, navigation then provides users with the ability to solve their problems on the healthcare facility. The maps and the information on healthcare data can be obtained from open sources such as Google maps while the data for the transportation were collected through the survey of the transportation services provided by the people of Pulau Indah.

3.3 System design, development and testing

There are five interfaces in the application which are the home page, healthcare page, health info page, driver contact page and complaint page. For example, healthcare page provides information about health care facilities services listed in the app and the navigation assistance to the desired healthcare. There are two buttons provided in this page; a button for navigation and a back button. Maps are provided in this page together with a list of the clinics or hospital that is shown in the page on the chosen health care facility's distance. Additional function is added in the app, which is useful for cost travel calculation and vehicle information services when using Grab car and My Car. For system development, the tasks were app coding system and app data attachment. Meanwhile, in the last stage, it was all about system testing which conducted among 11 selected communities in Pulau Indah, Klang.

4. Results and Discussion

4.1. *User requirement for the proposed mobile apps*

There are three sections of a questionnaire that have been answered by residents in Pulau Indah, Klang. The first section focuses on demographic, while the second section emphasizes on the level of technology savvy of the residents and the last section is on the introduction of the apps and the ways to improve it. In terms of demography, seven residential communities involved in this questionnaire are Kampung Sungai Kembung, Kampung Perigi Nenas, Pulau Indah's Customs Resident, Kampong Teluk Nipah, Bandar Armada Putra, Kampong Sungai Pinang, and Kampung Orang Asli Pulau Indah. 75.5 % of the respondents have been living in the areas of more than 5 years with a majority range of household incomes are between RM 1000 to RM 2000 (low-and-middle-income). Therefore, it is not surprising if most of the respondents in these communities are using motorcycles (45.5%) and personal cars (54.5) only as main transportation, with difficulties using or having access to public transports (0%).
Majority of the respondents (80%) seem not to be satisfied with existing mobile applications as shown in Figure 2. One of the reasons is that 61.4 % of the respondents do not have any experience using the specific type of mobile apps while another 38.6 % have experiences in using mobile apps that provide navigation to healthcare facilities.

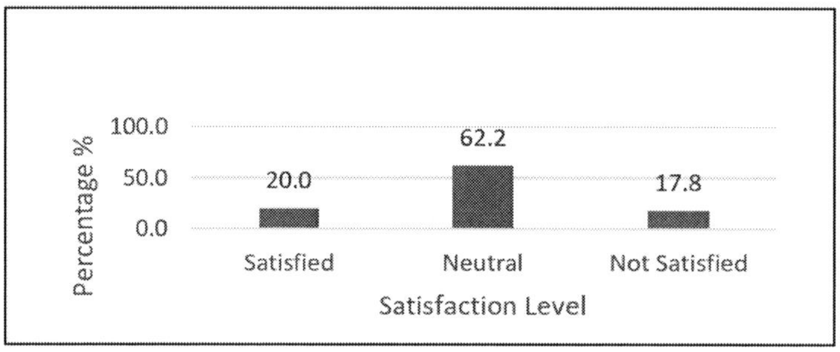

Figure 2. Respondent satisfaction on the existing mobile applications

Therefore, Figure 3 reveals that most of the respondents agreed with the proposal to develop a new mobile application for healthcare services in Pulau Indah community. 55.6 % from 45 respondents agreed with the proposal, while 28.9 % were neutral with the proposal and another 15.6 % disagreed with the idea to develop the proposed mobile application. The respondents also expressed strong agreement if the proposed apps do not only provide a directory map but also offers some important menu such as healthcare database, healthcare information, transportation cost estimation and transportation report. Some of the respondents are disagree with the proposal due to having general knowledge on the local accessibility to healthcare centres and some of them have also own transportations.

4.2. *System interface and elements*

The interface is designed based on the requirement that was proposed by the respondents. The interface was developed using the MIT Inventors. There are three main layouts of the interface designed for these apps as shown in Figure 4. The first layout is the Home page. It contains three buttons which are Healthcare button, Healthcare Information and User Manual button. Each of the button in this page is the door to get through to another interface. The second interface is the main page. This page gives the users the option to choose their healthcare treatment. The government and private buttons are for the user's option to choose different types of healthcare facilities listed in the app.

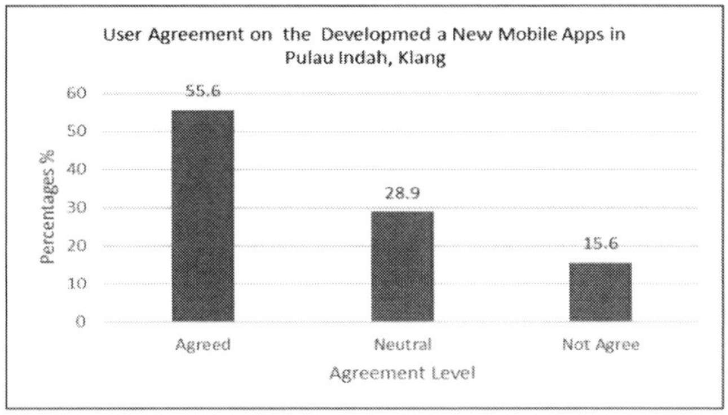

Figure 3. Respondents' responses on the proposed mobile applications

Besides this, this page also provides a button (*Aduan* button) for users to lodge a report or comment regarding their feelings toward the healthcare facilities. The third layout is the next page after users choose the option on the healthcare facility, either government or private healthcare facility. In this layout, the users will be provided with the information on the healthcare facility i.e. its address, operation hours and specialization. This page also provides users the option maps to get the best routes for their destination healthcare centre.

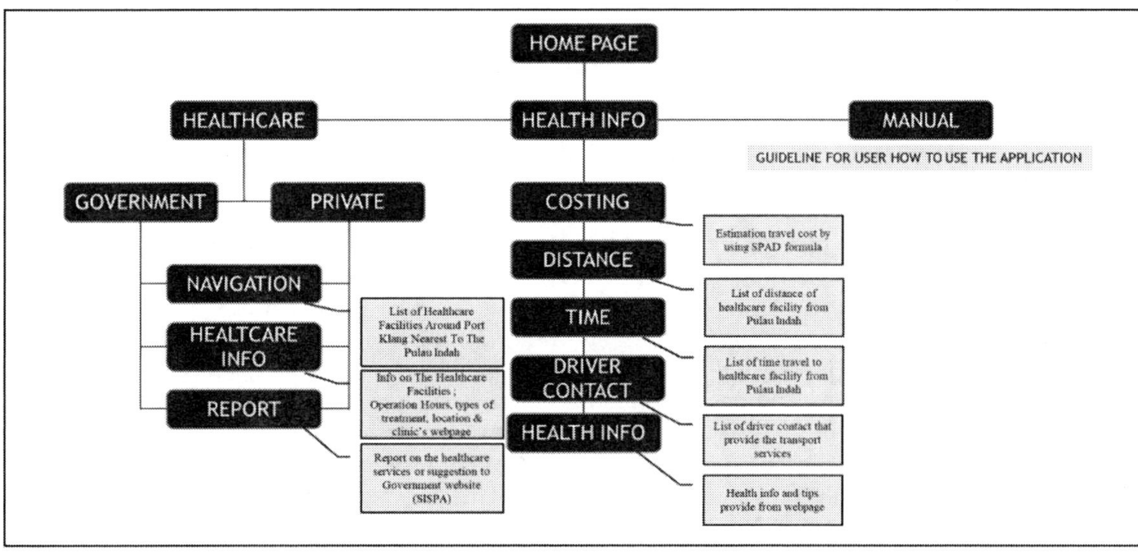

Figure 4. Structure of system design and elements

4.3. System functions and capabilities

The proposed app has three main menus as displayed in the home page i.e. Healthcare Accessibility, Healthcare Information and Manual for user's guideline as shown in Figure 5. The Main page is divided into two parts which are government healthcare facilities and private healthcare facilities. This page will list all the healthcare nearest to Pulau Indah and users will be guided by Google maps to get to the healthcare facility of their choice. In addition, the page also provides some information related to the chosen healthcare facility such as its address, operation hours and the capability of the healthcare to perform the treatment.

The Map page is also one the most important function in this app due to the navigation function it provides. It is generated from the government or private healthcare button in healthcare accessibility page. When users choose healthcare facility, a map is shown with the direction to the destination. This map is linked to the Google maps application from the user's smartphone. The app's system also has the following functions; for querying, locating, sending a report (comment) of the local healthcare information. Its functions also include Sending Message of the public driver, travel cost estimation and health info.

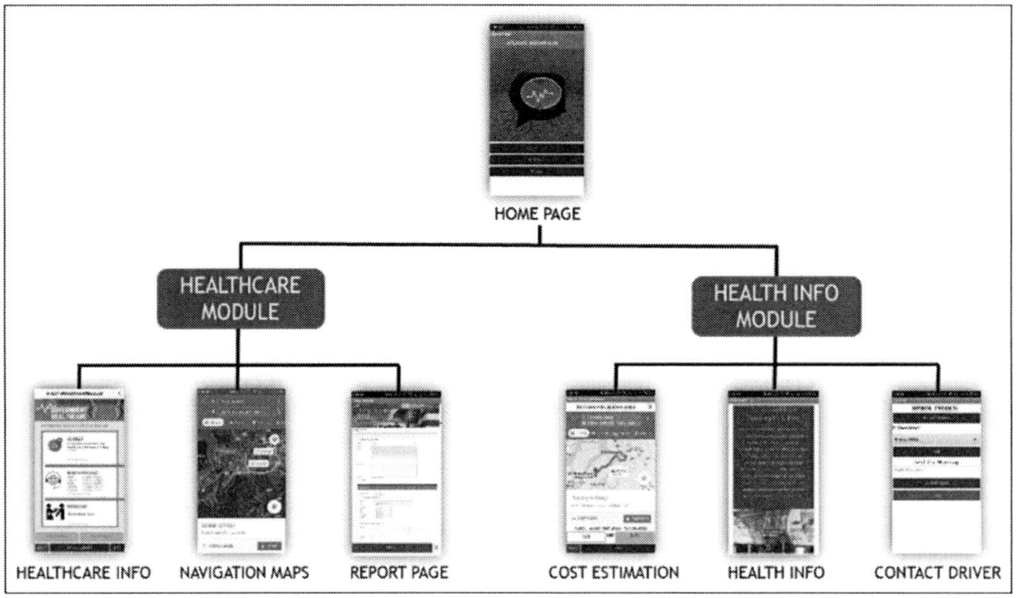

Figure 5. Main functions of the proposed mobile applications system

4.4. System functions and capabilities

According to the respondents' satisfaction level on the mobile app organization as illustrated in Figure 6, 37.5% of the respondents feel that this app is well organized while another 62.5% feel that this app is moderate. Notably, no respondent thinks that this app is not well organized. An interesting fact is that more than 80% of the selected respondents were satisfied with the functions and performances developed in the system, including Maps Function, Healthcare Information Function, Health Information Function, Cost Estimation Function, and Complaints/ Report Function.

In terms of the practicality and possibilities of the proposed apps to be applied in the community routine, a majority of the respondents agreed that the apps are easy to understand, and it could be adapted in the local daily activities. Figure 7 illustrates that 50% of the respondents agreed that the application is easy to understand while the remaining 50% disagreed. This may be due to the fact that some of them are not well versed with such mobile technology.

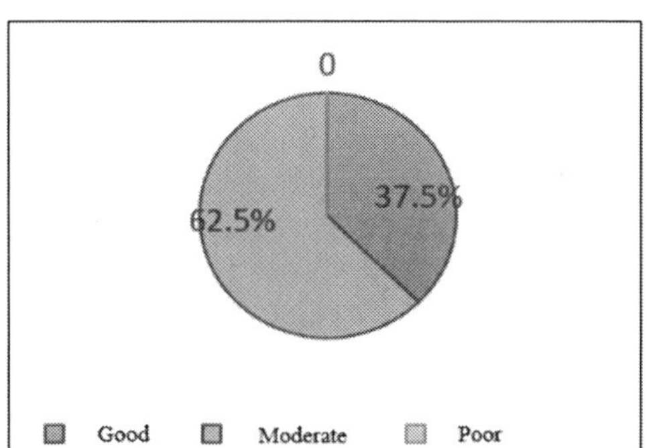

Figure 6. Respondents' satisfaction on the proposed mobile application structure.

With regards to respondents' agreement for apps adaption into daily applications, Figure 7 shows that 75% of the respondents agreed to recommend the app to the public while 25% expressed their disagreement. This may be due to the fact that they are not well versed with the usage of such technology.

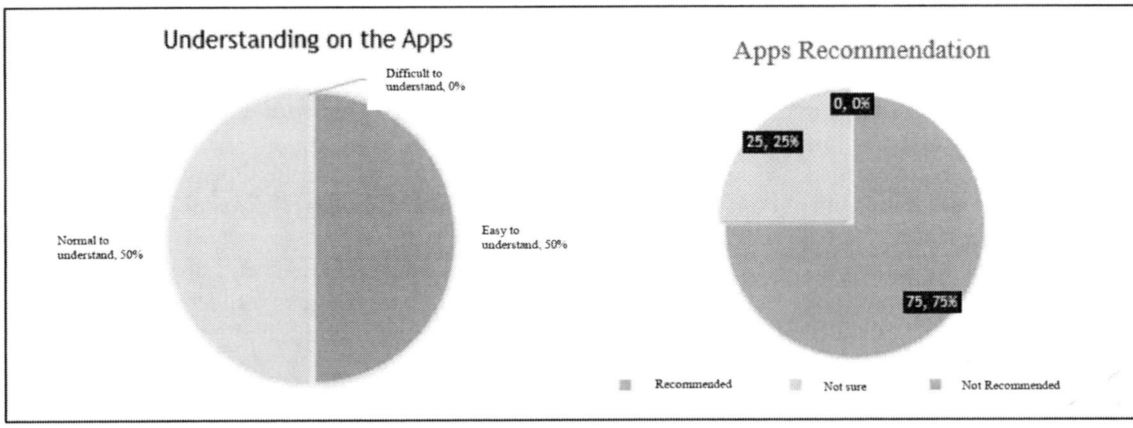

Figure 7. Respondents' understanding and recommendation on the proposed mobile app

5. Conclusion

From the site visit conducted at Pulau Indah, the communities faced several problems to reach the healthcare facilities. For a community of senior citizens, in general, their knowledge of technology is relatively weak and most of them do not know how to use the existing apps on the market and some do not use smartphones. Besides this, services limitation at available local healthcare centre, transportation problem and cost constraint causing the communities to get the appropriate health services, thus they need a practical information system for healthcare accessibility and information services. A mobile healthcare app has been developed in this study with some benefits and special features. The app functions to help residents obtain information on the nearest healthcare services for treatment and supplementary information of health clinic. The app also assists users in the selection of healthcare centres that suit their needs. Users could also get information on operating hours, the level of treatment provided, estimated travel costs, health info information, address and navigation assistance via the mobile app. By using the information provided in this app, it will make it easier for users to decide on a clinic closest to them.

Acknowledgement

The authors would like to express their appreciation to the local communities in Pulau Indah, Selangor for sharing their inputs in this research paper and sincerely thank Universiti Teknologi MARA, Selangor for providing LESTARI Research Fund (600-IRMI5/3/LESTARI (006/2018).

References

[1] Beal, V. (4 January, 2011). *Webopedia.* Retrieved from mobile operating systems: http://www.webopedia.com/DidYouKnow/Hardware_Software/mobile-operating-systems-mobile-os-explained.html

[2] Devlin, A. S. (2014). Wayfinding in Healtcare Facilities: Contribution from Environmental Psychology. *Behaviour Science*, 4(4), 423-436.

[3] Doughty, K. (2012). Low Cost Connected Healthcare Systems in Developing Countries. *Journal of Assistive Technologies*, 216-220.

[4] Friedman, L., Gyr, H., & Gyr, A. (2010). The Changing Patient in the Digital Era: A Typology for Guiding Innovation in Healthcare. *International Journal of Innovation Science*, 239(4), 39-

46.
[5] Higgs, G. (2004). A Literature Review of the use of GIS-based measures of access to health care services. *Health Services and Outcome Research Methodology*, 5:119.
[6] Iyer, S. (13 September, 2016). *How healtcare mobile apps help doctors and patients.* Retrieved from softwebsolution: http://www.softwaresolution.com/resources/healtcare-mobile-app-development-services.html
[7] Jalil, I. A., Abdul Rasam, A. R., Adnan, N. A., Saraf, N. A., & Idris, A. N. (2018). Geospatial Network Analysis for Healthcare Facilities Accessibility in Semiurban Area. *2018 IEEE 14th International Colloqium on Signal Processing & Its Application (CSPA)* (pp. 255-260.doi:10.1109/CSPA.2018.8368722). Batu Feringghi: CSPA.
[8] Okediran, O., Arulogun, O., Ganiyu, R., & Oyeleye, C. (2014). Mobile Operating Systems and Application Development Platforms: A Survey. *International Journal Advanced Networking and Applications*, 2195-2201.
[9] Renner, T. (15 April, 2014). *Mobile os feature renner.* Retrieved from snet: http://www.sneet.tu-berlin.de/fileadmin/fg220/courses/SS11/snet-project/mobile-os-features_renner.pdf
[10] Shen, S. T. (2016). People and their smartphones - Mapping mobile interaction in the modern connected world. *Enginerring Computation*, 0153.
[11] Surage, J., Tawariah, R., & Twumasi-Mensah, T. (2017). Geographical perspective of modelling primary healtcare accessibility. *International Journal of Human Riights in Healtcare*, 10(1), 56-67.
[12] Ventola, C. L. (2014). Mobile Devices and Apps for Health Care Professional: Uses and Benefits. *Journal for managed care and hospital formulary management*, 39(5): 356-364.
[13] Viswanathan, P. (16 October, 2017). *What's a Mobile Apps?* Retrieved from lifewire: http://www.lifewire.com/what-is-a-mobile-application-2373354
[14] West, D. M. (3 October, 2013). *Improving Health Care through Mobile Medical Devices and Sensors.* Retrieved from brooking: http://www.brookings.edu/wp-content/uploads/2016/06/West_Mobile-Medical-Devices_v06.pdf.
[15] Lokhman, M.T., Abdul Rasam A. R. & Mohd Noor, A.M (2012). Reforming health care facility using geographical information system, *2012 IEEE Control and System Graduate Research Colloquium*, Shah Alam, Selangor, 267-270.
[16] Abdul Rasam, A.R., Azlin. A.H. & Saraf, N.M. Mobile Apps and Web GIS-Based Accessible Health and Social Care System for People with Disabilities, *2018 IEEE 8th International Conference on System Engineering and Technology (ICSET)*, Bandung, 2018, 85-90.

The Elderly (Senior) Housing Options Preferences Among Malaysian Generations

Hafiszah Ismail[1], Fatin Aziz[1], Maszuwita Abd. Wahab[1]
[1]Faculty of Architecture, Planning and Surveying, Universiti Teknologi MARA, 40450 Shah Alam, Selangor, Malaysia

Email: hafiszah@salam.uitm.edu.my

Abstract. Trends of changing demographics would greatly grant impact on the supply and demand for housing. Different types of populations or generations equipped with different needs and preferences which award basis for more studies mainly on the relationship between population and housing. Relevantly, the numbers of senior (elderly) population currently increasing and projected to persistently increase due to the enhancement of the health quality and lifestyle. Despite the soaring fraction of the elderly moving to a nursing home or retirement community with increasing age, the majority of the elderly age 85 and over still want to stay in their own homes (Ngarmyarm and Panichpathom, 2012). Thus, studies on senior (elderly) housing options preferences were significant to guarantee enhanced future housing provisions for this aged population. The main objective of this paper is to give an overview on the concept of aging-in-place and the preferred elderly (senior) housing options for the Malaysian generations (Baby Boomers, Generation X, Generation Y, and Generation Z). This study employs mixed methods approaches with Selangor as a case study. A sum of 1,067 respondents was contacted through drop-off, face-to-face interview, postal interview and internet survey (web-survey). The findings show that the Malaysian generations accept the concept of Retirement/Age-Restricted Community with health-care services.

1. Introduction

By the year 2030, Malaysia will be obtaining the position of an aging nation due to sturdy raises numbers of elderly (senior) populations. The numbers of elderly (senior) population in Malaysia has increased significantly from 0.5 million in the 1970s to about 2.3million in 2010 and continues to rise annually. Statistics forecasts show that by 2030, the elderly (senior) will boost to 15 percent of the total population in Malaysia (Hamid et al., 2013; Sulaiman, 2011). The escalating numbers of the elderly (senior) population will intensely influence the housing market and give effect to the Malaysian housing demand in particular. Nevertheless, as compared to other developed countries and despite the continuous increase of elderly (senior) populations currently less attention is being given on housing provisions which specifically tailored according to the needs and preferences of this specific elderly (senior) generations in Malaysia. Various types of elderly (senior) housing options should be made available and affordable for the benefit of the Malaysian elderly (senior) generations.

2. Generations and Housing Mobility

Each generation is unique as they were equipped with diverse characteristics alongside with different needs and preferences. The generations are different types of the population which is identified according to their age group or year born. The elderly (senior) which is also known as the Baby Boomers generation is a group of generation of 60 years and above of age. Basically, they are of those who were born between the year of 1946 to 1961. Next, was Generation X (Gen-X) who were born between 1962-1976 followed by Generation Y (Gen-Y) born between 1977-1999 and Generation Z (Gen-Z) born after 1992.

Each generation accompanied by different needs and preferences due to different generational characteristics. Thus, reasons or motives of the generations would also vary. In addition, household motives for moving would differ with age or by generations. Mobility on the housing market plunges with age. Conversely to younger age groups, older adults were less mobile and display some trend to 'stay put' or to age-in-place. The elderly (senior) or households more than 50 years tend to move less often than the younger households and when they do move or decides to be mobile, they move for diverse reasons. The continually improved health conditions and life expectancy drives the aged generation to become more active (mobile) and independent. Bamzar (2018) found that households in the 50-64 range of age be likely to move for better jobs or better housing while households over 75 years old move to gain more affordable housing or for health reasons (Bamzar, 2018). According to Mathews (2009), the desire to be closer to family and friends escalate in significance as the household ages. The rising quantity of elderly within the whole population and discussion about lifestyle changes among the young old, in particular, the Baby boomers, induce questions about shifting patterns of residential mobility (Kramer and Pfaffenbach, 2009). Thus, it is clear that aging will not stop the elderly (senior) to be mobile. Therefore, due to the continually increasing numbers of these generations in Malaysia proper planning for future elderly (senior) housing provisions are vital to guarantee better welfare of this particular age group. As compared to other developed countries including Singapore as Malaysia neighbouring country, the elderly (senior) housing provisions were well-taken care, especially by the government. Each housing development is required to include basic features of senior (elderly) in line with the needs and made mandatory by the governing body or Housing Development Board (HDB) in Singapore. In the Malaysian context, comprehensive and detail information on the Malaysian elderly (senior) housing needs and preferences need to be sought for proper planning and best delivery of the concept. As a start, the concept of elderly (senior) housing needs to be introduced to the Malaysian generations together with the indication of acceptance on this concept. This study aim is to provide an overview of the Malaysian preferred elderly (senior) housing options by generations. The main objective of this study is to determine the acceptance of the Age-Restricted Community concept for the aged generations by comparing familiarity with preferences on the introduced elderly (senior) housing concept.

3. Methodology

Combinations of qualitative and quantitative strategies or the mixed method research were used in this study of elderly (senior) housing options preferences. A research problem can better be understood by the use of mixed method strategy (qualitative and quantitative) an amalgamation of data collection and analysis in a single study or cycle of studies (Creswell, 2006; Creswell and Plano, 2007). Study on consumers' preferences categorises this study as a consumer behaviour type of study and this provides an explanation for the use of mixed methods. Cresswell (2009) stressed on the advantage of using mixed method strategies in research that would provide better insight into research problems than a single unaided approach. The significance that mixed methods add up to the study can be accomplished through numerous basis (Cresswell, 2009). Three (3) main method of data collection in survey research involve interviews, questionnaire administration and observation on people or phenomena (Sekaran and Bougie, 2010). Demographics (population) as the prime factor of property development especially for housing provisions that strongly relate with demand and supply for housing were proved via qualitative data. This data obtained through interviews with the local authorities and housing developers in Selangor. Additionally, determinations of the preferred elderly (senior) housing options among Malaysian generations were obtained via quantitative data through survey questionnaires with a total sum of 1,067 respondents as samples. The quantitative respondents for this study were contacted through several

IOP Conf. Series: Earth and Environmental Science **385** (2019) 012039 doi:10.1088/1755-1315/385/1/012039

survey strategies that include drop-off, face-to-face interview, postal interview, and internet survey (web-survey).

4. Findings and Discussions

The generational distributions on the elderly (senior) housing choice preferences are shown in the tables below (Table 1, Table 2, Table 3 and Table 4). The respondents were asked to rank their preferred choices criteria of elderly (senior) housing from 1 to 5 ranks of importance.

Location near shopping centres and distance from family and relative were ranked as the 1st and 2nd preferred choice of elderly (senior) housing by three (3) generations namely the Baby Boomers, Gen-Y and Gen-Z. Location near shopping centres was ranked as the 1st choice by the majority of Baby Boomers (59.3 percent) and Gen-Z (54.5 percent). This preferred choice of criteria was ranked as the 2nd choice by Gen-Y (25 percent). In contrast, these criteria were ranked as 4th by Gen-Z (40 percent). Whereas, distance from family and relative were ranked as the 1st choice by Gen-X (34 percent) and Gen-Y (40.2 percent). The same criteria were ranked as the 2nd preferred choice by Baby Boomers (50 percent) and Gen-Z (33.3 percent).

The other most important choice of the elderly (senior) housing selected by the generation is access to medical services/hospitals. These criteria were ranked as; 1st rank by Gen-Y (38.9 percent), 2nd rank by Gen-X (32.9 percent) and ranked as 3rd rank by both Baby Boomers (31.3 percent) and Gen-Z (30.6 percent).

In addition, access to planned social activities was ranked as the 4th and 5th choice by the majority of the generations. The criteria were ranked 4th choice by Gen-X (26.4 percent). The other three (3) generations ranked access to planned social activities as their 5th choice; Baby Boomers (32.3 percent), Gen-Y (22.4 percent) and Gen-Z (34.4 percent).

Table 1. Elderly (Senior) Housing Choice Preferences - Baby Boomers

Rank	Considerations	Percentage	Number responded
First	• Location near shopping centers	59.3	16
Second	• Distance from friends and relative	50.0	16
Third	• *Access planned social activities (3&5)*		
	• Access to medical services/hospitals	31.3	10
	• Home or personal security	29.6	8
Fourth	• Access to personal & home-care services	37.9	11
Fifth	• Access to public transportation	29.6	8
	• *Access to planned social activities (3&5)*	*32.3*	10

Table 2. Elderly (Senior) Housing Choice Preferences - Generation X

Rank	Considerations	Percentage	Number responded
First	• Distance from friends and relative	34.0	50
	• Home or personal security	35.5	54
Second	• Access to medical services/hospitals	32.9	51
	• Access to public transportation	25.6	23
Third	• Access to personal & home-care services	28.2	29
Fourth	• Access to planned social activities	26.4	38
Fifth	• Location near shopping centers	27.5	22

Table 3. Elderly (Senior) Housing Choice Preferences - Generation Y

Rank	Considerations	Percentage	Number responded
First	• Distance from friends and relative	40.2	88
	• Access to medical services/hospitals	38.9	88
	• Home or personal security	34.1	71
	• Access to public transportation	24.1	34
Second	• *Location near shopping centers (2&5)*	*25.0*	37
	• Access to personal & home-care services	27.7	46
Third			
Fourth			
Fifth	• Access to planned social activities	22.4	44
	• *Location near shopping centers (2&5)*		

Table 4. Elderly (Senior) Housing Choice Preferences - Generation Z

Rank	Considerations	Percentage	Number responded
First	• *Location near shopping centers (1&5)*	54.5	12
Second	• Distance from friends and relative	33.3	12
Third	• Access to medical services/hospitals	30.6	11
Fourth	• *Location near shopping centers (1&5)*		
	• Access to personal & home-care services	40.0	8
	• *Home or personal security (4&5)*	*24.2*	8
	• Access to public transportation	33.3	8
Fifth	• Access to planned social activities	34.4	11
	• *Home or personal security (4&5)*		

The major difference between generations was evidenced here referring to the criteria of home or personal security. Two (2) generations namely Gen-X (35.5 percent) and Gen-Y (34.1 percent) regard

IOP Conf. Series: Earth and Environmental Science **385** (2019) 012039 doi:10.1088/1755-1315/385/1/012039

this choice of preferred criteria as among the 1st rank of importance. In comparison, Baby Boomers (29.6 percent) and Gen-Z (24.2 percent) perceived these criteria as less important and ranked it as their 3rd and 4th ranked of preferred choice. The other dissimilarity between generations were found on the criteria of access to public transportation. Gen-Y (24.1 percent) ranked these criteria as among their 1st preferred choice for elderly (senior) housing followed by Gen-X (25.6 percent) as their 2nd choice. In contrast, Baby Boomers (29.6 percent) and Gen-Z (33.3 percent) both regard these criteria as less important with 4th and 5th given rank.

In continuance, the following tables below (Table 5, Table 6 and Table 7) show the results for housing options preferences at retirement age. Table 5 shows that most of the Baby Boomers (52.3 percent) and Gen-X (43.4 percent) wishes to 'age in place' during the elderly (senior) age. According to Mathew and Turnbull (2007), the sense of ageing in place falls under two (2) types of living senses; (i) The sense of living in a specific home as long as probable, or; (ii) The sense of living in a series of home in a specific community. Hence, the result of excessive preference to age in place by the Baby Boomers and Gen-X indicates two (2) probabilities; (i) the preferences to be living in a specific home as long as probable indicates the sense of belonging to a certain type of home or known as 'home attachment'; and (ii) the preferences to be living in a series of home with specific community shows the sense of belonging to particular community or recognised as 'community attachment'. This consequence also due to previous findings (Ismail, 2016) that show Baby Boomers and Gen-X prefer to be living in similar housing location for their current, future and senior (elderly) housing (i.e: the same location of same local authorities. Likewise, choices of preference of focal attraction factors for future housing between location, house and neighborhood also uncover Baby Boomers and Gen-X who select neighborhood as their prime consideration give a signal sense of community attachment among these generations.

Table 5. The Elderly (Senior) Housing Options Preferences

Generation	Baby Boomer		Gen-X		Gen-Y		Gen-Z	
Elderly (Senior) Housing options	**Number of respondents**	**Percent**	**Number of respondents**	**Percent**	**Number of respondents**	**Percent**	**Number of respondents**	**Percent**
• **Aging in place**	23	52.3	75	43.4	97	33.9	15	34.1
• **Move to other location & live on your own**	8	18.2	68	39.3	107	37.4	20	45.5
• **Move to other location & live with family members**	11	25.3	18	10.4	73	25.5	8	18.2
• **Move to other types of housing especially for the elderly**	2	4.5	12	6.9	9	3.1	1	2.3

The generational preference variances were found on their subsequent preferred options of elderly (senior) housing. Here, the Baby Boomers as well as their prime wish to 'age in place' (52.3 percent)

at the same time highly willing 'to move to other location and live with family members' (25.3 percent) during their elderly (senior) age. Likewise, Gen-X also highly prefer to 'age in place' however they are more willing to 'move to other location and live on their own' at their senior age.

The younger generations of Gen-Y and Gen-Z display the clear difference of preferences on the elderly (senior) housing options as compared to the older generations (Baby Boomers and Gen-X). Most of Gen-Y (37.4 percent) and Gen-Z (45.5 percent) placed their greatest priority and choose to 'move to other locations and live on their own' throughout their elder (senior) age. Additionally, the generations following preference Gen-Y (37.4 percent) and Gen-Z (34.1 percent) were to 'age in place' at the elderly (senior) age. The single similarity found between the generations were on the preference to 'move to other types of housing especially for the elderly/senior citizen *(i.e: Single Family housing, Retirement/Age-Restricted Community, Nursing home)*. The generations regard this as their least preferred elderly (senior) housing options.

Living alone in old age is the strongest predictor of isolation (Wu and Chan, 2012). The most important indication from the above findings demonstrates that most of the generations (Baby Boomers, Gen-X, and Gen-Z) prefer to 'age in place' throughout their elderly (senior) age. There were dissimilarities of preferences between generations on the other elderly (senior) options. It was found that Baby Boomers willing to be mobile and willing 'to move and live with family members' during the elderly (senior) age if have to and this was stated as their second preferred option. The willingness to be mobile and live with family members during retirement age by the Baby Boomers show that this particular generation does not agree with 'isolation' to age. In contrast, the results shown by Gen-X, Gen-Y and Gen-Z who be willing to 'move to other location and live on their own' during their elderly (senior) age as the subsequent option show that these generations despite their willingness to be mobile during retirement age also prefer to be living independently on their own with some tendency to isolate themselves with others.

Table 6. Types of Elderly (Senior) Housing by Generations

Generation	Baby Boomer		Gen-X		Gen-Y		Gen-Z	
Types of Elderly (Senior) Housing	**Number responded**	**Percent**	**Number responded**	**Percent**	**Number responded**	**Percent**	**Number responded**	**Percent**
Single-family housing	28	65.1	111	63.8	195	69.1	28	65.1
Retirement	12	27.9	60	34.5	80	28.4	15	34.9
Nursing home	3	7.0	3	1.7	5	1.8		
Others (*)					2	.7		
Total	43		174		282		43	

In addition, Table 6 depicts the preferred type of senior housing by generations. Majority of the generations prefer single-family housing; Baby Boomers (65.1 percent), Gen-X (63.8 percent), Gen-Y (69.1 percent) and Gen-Z (65.1 percent). Next preferred type of senior housing is the Retirement/Age-restricted Community; Baby Boomers (27.9 percent), Gen-X (34.5 percent), Gen-Y (28.4 percent) and Gen-Z (34.9 percent).

Furthermore, referring to the Retirement/Age-restricted Community in particular (refer Table 7), most of the generations; Baby Boomers (87.1 percent), Gen-X (75 percent), Gen-Y (91.5 percent) and Gen-Z (20 percent) required this type of senior housing to be provided with health-care services. As for the nursing home, it was regarded as the least preferred type of senior housing by all generation; Baby Boomers (7.0 percent), Gen-X (1.7 percent), Gen-Y (0.7 percent) and Gen-Z (0 percent.

Table 7. Preference for Retirement/Age-Restricted Community – With or without health-care services

	Overall		Baby Boomer		Gen-X		Gen-Y		Gen-Z	
Options	**Number responded**	**Percent**	**Number responded**	**Percent**	**Number responded**	**Percent**	**Number responded**	**Percent**	**Number responded**	**Percent**
With health-care services	128	87.1	6	75.0	47	85.5	65	91.5	9	20.0
Without health-care services	19	12.9	2	25.0	8	14.5	6	8.5	3	6.7
Total	147		8		55		71		12	

Most importantly although the generations prefer to be staying in 'single-family housing' (conventional housing) during elderly (senior) age their decision of choosing the 'Retirement/Age-Restricted Community' (with health-care services) as their subsequent preferred option indicates sign of acceptance to the introduction of elderly (senior) housing concept in Malaysia.

Table 8. Familiarity and Preference for the types of Elderly (Senior) housing option Retirement/Age-Restricted Community

Retirement/Age-Restricted Community	**Familiarity***	**Preference***
• **Type 1: Assisted Community** ***(Assisted Living Residence)***	36.6	45.2
• **Type 2: Unassisted Community**	51.5	39.1
• **Shared home/accommodation** ***(Age-Restricted Apartment)***	60.5	50.2
• **Supported home/accommodation** ***(Disabled or having mental health problems)***	64.3	54.2

Familiarity* refers to '*Familiar*' or '*Very familiar*'
Preference* refers to '*Preferable*' or '*Most Preferable*'

Additionally, the comparison between familiarity and preferences of the generations for the type of elderly (senior) housing is particularly focussing on the Retirement/Age-Restricted Community housing option presented in Table 8. Generally, the result illustrates the majority of the samples were familiar with Supported home/accommodation (64.3 percent), Shared home/accommodation (60.5 percent) and Type 2: Unassisted (51.5 percent). On the contrary, the respondents demonstrate low (36.6 percent) level of familiarity with Type 1: Assisted Community housing. Logically, familiarity will shape preference at some point. The above-presented findings clearly show that a great percentage of familiarity appears to be aligned with the great percentage of preference. The only difference is on familiarity and preference for Type 1: Assisted Community and Type 2: Unassisted Community. Despite a low level of familiarity (36.6%), the generations show a higher preference for Type 1: Assisted Community (45.2%). In addition, notwithstanding of great familiarity (51.2%) with Type 2: Unassisted Community concept the generations show a low level of preference (39.1%) with this type of elderly (senior) housing option.

The preferences between Type 1: Assisted Community and Type 2: Unassisted community was further validated by examining the correlation between the two (2) variables of familiarity and preferences among generations (Table 9). The results evidenced that most of the Baby Boomers were highly familiar (60 percent) with Supported-home/accommodation. Nevertheless, high preference marked for Type 1: Assisted Community (Assisted Living Residence) (64.7 percent). While, the majority of Gen-X and Gen-Z show that they were both highly familiar and highly prefer with Supported-home (Gen-X: Familiarity - 70.4 percent and Preference - 60.2 percent, and; Generation Z: Familiarity – 72.2 and Preference – 46.1 percent). As for Generation Y, the familiarity and preference were different as they were greatly familiar (60.8 percent) with the Shared home but highly prefers (53.7 percent) Supported home.

Table 9. Distribution of Familiarity and Preference for Retirement/ Age-Restricted Community

Retirement/ Age-Restricted Community housing	Baby Boomer		Gen-X		Gen-Y		Gen-Z	
	Familiarity*	Preference*	Familiarity*	Preference*	Familiarity*	Preference*	Familiarity*	Preference*
Type 1: Assisted Community	43.2	64.7	28.0	35.4	38.4	49.4	53.5	43.2
Type 2: Unassisted Community	43.6	35.1	54.3	44.1	49.6	37.8	58.1	30.6
Shared home	54.8	42.1	61.1	52.2	60.8	52.1	63.8	40.0
Supported home	60.0	41.3	70.4	60.2	59.6	53.7	72.2	46.1

Familiarity* refers to '*Familiar*' or '*Very familiar*'
Preference* refers to '*Preferable*' or '*Most Preferable*'

Table 10. Correlation Analysis for Retirement/Age-Restricted Community

Retirement/Age-Restricted Community housing	Overall	Baby Boomer	Gen-X	Gen-Y	Gen-Z
Type 1: Assisted Community	0.331*	.237	.311*	.369*	.123
Type 2: Unassisted Community	0.317*	.346*	.336*	.329*	.075
Shared home	0.372*	.221	.392*	.416*	.260
Supported home	0.436*	.379*	.453*	.447*	.216

Finally, Table 10 demonstrates the correlation analysis finding on the types of Elderly (Senior) housing options of Retirement/Age-Restricted Community by generations. The results clearly indicate the Malaysian generations' level of acceptance on this specific type of housing concept for the Malaysian Elderly (Senior) populations. In detail, the Baby Boomers prefer Type 1: Assisted Community (Assisted Living Residence) an independent type of senior (elderly) living arrangement with assisted living facilities features in the senior (elderly) housing environment. This type of elderly (senior) housing is an age-restricted neighbourhood with individual residence each person. Disregard with great familiarity on Supported home/accommodation (60 percent) the Baby Boomers show a signal of acceptance on Type 1: Assisted Community (Assisted Living Residence) with a high level of preference (64.7 percent). This indications of acceptance further support the wish of Malaysian generations to age in place and live independently which mean living in an individual housing unit under an Age-restricted community housing with some level of assistance. Generally, the key characteristics for the Type 1: Assisted Community (Assisted Living Residence) of Age-restricted community housing concept can be divided into two (2) types; (i) Independent Living with Assisted Living Facilities (ALFs) which offer supervision or assistance with basic daily living activities, and (ii) Assistance Living that may provide the administration or supervision of medication or personal care services by a trained staff.

On the other hand, Gen-X and Gen-Z highly prefer Supported home/accommodation (disabled or having mental health problems). The key criteria offer by Supported home/accommodation concept were an independent living with support workers like security and warden to assist the housing consumer. The high level of familiarity and preference of Gen-X and Gen-Z with Supported home/accommodation (disabled or having mental health problems) show that these generations expect that elderly (senior) might be faced with some kind of physiological condition during the senior age (might become disabled or have a mental health problem). Thus, this would require 24-hours care from someone (guardian) with regard to an independent living arrangement.

Lastly, the results show that Gen-Y greatly prefers Supported home/accommodation (disabled or having mental health problems) and Shared home/accommodation (Age-Restricted Apartment). This is an independent living concept with elderly (senior) housing assisted living facilities/arrangements: shared residence (seniors housing blocks/apartments/complexes type of residence). Although Gen-Y is very familiar with the Shared home/accommodation concept, similarly with Gen-X and Gen-Z, Gen-Y also highly prefers Supported home/accommodation (53.7 percent). Nevertheless, the next preferred type of Retirement/Age-Restricted Community by Gen-Y is Shared home/accommodation (Age-Restricted Apartment) (52.1 percent). Several basic criteria for Shared home/accommodation (Age-Restricted Apartment) concept consist of an active senior apartments housing offering meals, services, transportation and planned activities that may generally not include levels of life seniors' services within the same community. For example, the 'Congregate Living' housing concept an elderly (senior) housing concept that offers independent living in separate apartments and opportunities of sharing activities of living with other residents.

5. Conclusion

Aging is certain. Statistical data shows that the numbers of the elderly (senior) population currently and will continue to increase each year. The increase of this population will give effect to all aspect including the property market, especially on the housing provisions. Additionally, this specific population chose to age-in-place and at the same time prefer to be living independently. However, at some point, the elderly (senior) the housing mobility do occur to this 'silver hair' population. This study found that the Malaysian generations of housing consumers (Baby Boomers, Gen-X, Gen-Y and Gen-Z) would accept the initiation for the senior (elderly) housing provision. There is also an indication of approval on the idea of Retirement/Age-Restricted Community housing concept for the Malaysian elderly (senior) housing consumers. The generational preferences of the Malaysian generations on diverse types of Retirement/Age-Restricted Community housing such as the Assisted, Unassisted, Shared and Supported home housing despite their familiarity on the concepts indicate the need of specifically tailored elderly (senior) housing concept for Malaysian context. Thus, it is time for Malaysia to start focussing on affordable elderly (senior) housing provision due to the demand and most importantly to guarantee the welfare of the population with other countries as examples for a start.

Acknowledgments

The authors would like to thank The Institute of Research, Management and Innovation (IRMI), Universiti Teknologi MARA (UiTM) for the award of funding for this study under the Lestari Grant 2018 (600-IRMI 5/3/LESTARI (023/2018)).

References

[1] Bamzar, R. (2018). Assessing the quality of the indoor environment of senior housing for better mobility: A Swedish case study. Journal of Housing and the Built Environment.

[2] Creswell, J. W., & Plano, Clark, V. L. (2007). Designing and conducting mixed methods research. Thousand Oaks, CA: Sage

[3] Creswell, J.W. (2009). *Research Design Qualitative, Quantitative and Mixed Methods Approaches*. California: SAGE Publications, Inc.

[4] Duflot, L. (2012). Urbanisation Policy in Malaysia and its Impacts.Graphite Publications. Retrieved from http://graphitepublications.com/urbanisation -policy-in-Malaysia-and-its-impacts/

[5] Hamid, T. A., & Tyng, C. S. (2013). Meeting the Needs of Older Malaysians: Expansion, Diversification and Multi-sector Collaboration. Malaysian Journal of Economic Studies, 50(2), 157.

[6] Ismail, H. (2016). The Influence of Generational Differences on the Malaysian Housing Preferences. Unpublished Ph.D. Thesis. Institute of Graduate Studies, Universiti Teknologi MARA (UiTM) Shah Alam, Selangor.

[7] Kramer, C. and Pfaffenbach, C. (2009). Persistence preferred – on future residential (in)mobility among the generation 50plus. *Erdkunde*, 63, 2, 161–72

[8] Mathews, Thomas, J., and Hamilton, Brady (2009). Delayed Child Bearing: More Women Are Having Their First Child Later in Life. National Center for Health Statistics. Data Brief No. 21

[9] Narayan, A. and Panichpathom, S. (2012). Preferences for Retirement Housing: A Case Study of Thai Senior Citizens. World Journal of Social Sciences Vol. 2. No. 7. November 2012 Issue. Pp. 64 – 78

[10] Sekaran, U. and Bougie, R. (2010), "Research Methods for Business A Skill Building Approach 5th Edition", John Wiley & Sons Ltd., United Kingdom

[11] Sulaiman, N. (2011). Opportunities for the transfer of United Kingdom best practices for the provision of public residential care facilities for the elderly to Malaysia (Doctoral dissertation, Salford: University of Salford).

[12] Wu, T. and Chan A. 'Families, Friends, and the Neighborhood of Older Adults: Evidence from Public Housing in Singapore'. Journal of Aging Research. Vol. 2012

Residential Satisfaction: *Literature Review and A Conceptual Framework*

Nurzafira Zainul Abidin[1]*, Mohamad Isa Abdullah[1], Noraliza Basrah[1] and Muhammad Nazim Alias[1]
[1]Centre of Studies for Estate Management, Faculty of Architecture, Planning and Survey, Universiti Teknologi MARA, Shah Alam, Malaysia

Email: nurzafira@uitm.edu.my

Abstract. Residential satisfaction, defined as the feeling of contentment when one has or achieves what one needs or desires in a house, is an important indicator for the planners, architects, developers and policy makers use it in several ways. There are three theories related to residential satisfaction which are housing needs theory, housing deficit theory and psychological construct theory, and most empirical studies have used these theories or a combination of these theories in their research design. Several variables representing housing and neighbourhood characteristics, socio-demographic attributes as well as their perceptions of housing and neighbourhood conditions have been analysed to indicate that further studies are required until a general theory of residential satisfaction or dissatisfaction occurs. Variables which are significant that influences on the level of residential satisfaction or dissatisfaction are vary according to the objectives of the study. This indicates that further studies on residential satisfaction can be carrying out on case specific context in guiding the housing policies.

1. Introduction

Measuring residential satisfaction is very complicated as it is very subjective to the particular place, time, purpose of evaluation and involvement various range of people [1]. There are a lot of studies on residential satisfaction which have came out with varies of result. Studies on residential satisfaction have become an important indicator for various people such as architects, policy makers and developers. The concept of residential satisfaction has been used in at least four different ways [2]. The first is that it has been used as a key predictor of individuals' perceptions of general which is quality of life. Second, it has been used as an ad hoc evaluative measure for judging the success of housing developments constructed by the private sector and the public sector. The third is, it has been used as an indicator of initial state of residential mobility and, hence, altered housing demands and neighbourhood change. And the last is, it has been used to assess residents' perceptions of insufficiencies in their current housing environment. Thus, it is very important to understand the concept of residential satisfaction in order to gain the proper information. The aim of this paper is to provide an overview of residential satisfaction as well as a conceptual framework for residential satisfaction. This paper addresses two main objectives to create a conceptual framework.

a. To explore the concept of residential satisfaction
b. To determine the variables involved in residential satisfaction study

2. Theories related to Residential Satisfaction

The term of residential satisfaction is very wide used in research all around the globe. This is very significant indicator used by the people in the development industries as well as researchers. There are several theories which are related to residential satisfaction such as housing needs theories [3], housing deficit theories [4] and psychological construct theory [2]. The first theory, housing need theories was first introduced by [3] which stated that there are different levels of life cycle and the evolution through these levels makes households become uncomfortable. Households showed this discomfort through migration to other places. According to Maslow's hierarchy of needs, there are five stages. These stages are physiological needs, safety needs, belongingness and love need, esteem needs, and the need for self-actualization. It is very important to satisfy a person's basic need for developing a person's potentiality and capability in a society.

Housing deficit is generally defined as lack of houses accommodated for the population. But according to the [4], conception of the housing deficit theory is the lack of house condition itself. People assess their own housing by compare to others. Differences in the housing conditions makes their house became housing deficit. Resulted from this situation, people keen to do adjustment on their house. As for the psychological theory, this theory is more to the physical aspect of the houses. Situation of the residential influence people to compare with others and people will have the feeling of satisfaction or dissatisfaction [2]. Dissatisfaction of the physical condition of the residential may lead to house modification. These show that the residential satisfaction research must be studied further as the result for each residence may vary.

3. Residential Satisfaction

In the previous researches, the term of residential satisfaction or housing satisfaction should be separated the word as Housing and Satisfaction and define them apart. Housing is a composite of the total physical and social parts that makeup the housing system. Housing also is a multidimensional phenomenon that includes structural type, tenure, location and political jurisdiction [5]. As for the word Satisfaction, it is a process of evaluation between what was gained or received and what was expected. It also can be accurately elaborate as the perceived discrepancy between objective and accomplishment. Satisfaction was not only about the physical aspect but also by the ability to form social networks and it is subjective reaction to an objective environment [5]. The housing's character that is multifaceted makes assessment of the housing more complicated. Nevertheless, the single-facet can't provide adequate information related to other areas of housing to housing quality [6].

Residential satisfaction is defined as the feeling of contentment when one has or achieved what one needs or desires in a house [5]. Residential satisfaction can also be defined as indicator of homeowners' view of the general quality of their life and it can be mean that an individual's expectation of housing is met [7]. Residential satisfaction is the descriptions of the quality of life of the inhabitants of a determinate residential environment and act as trigger factor affecting residential mobility [8].

4. Empirical Studies of Residential Satisfaction

There are various researches of empirical studies on the residential satisfaction that have been done by the professionals and experts especially research on the characteristic of the users. To achieve the aims of the different studies, there are several instruments that will be in the questionnaire on residential satisfaction [8]. The general characteristics are divided into four sections which are to obtain perceived environmental quality indices, to determine the subjects' satisfaction with their residential environment, to collect the social demographic and personal characteristics and the last part is about behavioural aspect and questions on individuals in their residential environment. The factors as neighbourhood and housing conditions, physical and mental health, and socioeconomic conditions such as economic outcomes and social contact are included in research [9]. On the other hand, to measure residential satisfaction through factors concerning the placement of the neighbourhood environment surrounding housing, along with surrounding conditions such as "Safety and walkability," "Access to destinations," "Social network," and "Travel network [10].

A research which summarized the empirical studies on residential satisfaction at cross-cultural level [5] was based on the research [8] which consists of four parts which are socio-demographic characteristics, housing characteristics, neighbourhood characteristic and behavioural characteristics. There are also other researchers which defined the socio-demographic factors which is relevant to the study. Another study shows that building quality, owners' culture of maintenance, social, neighbourhood, management, and dwelling unit features significantly predict students' satisfaction in public halls of residence, using Takoradi Technical University as a case study [11]. Which they add one more of variables that is owners' culture of maintenance in their study to relate its significance to the residential satisfaction.

4.1 Social Demographic Characteristic

Education appears to be insignificant effects in housing satisfaction [12]. However, other studies stated that the education level of the affecting the residential satisfaction level [13,14]. This statement also [supported by the statement of another study [15] where there is positive relationship has been found between housing satisfaction and age, income, education and job status. Important factors also belong to residents' characteristic such as age, income, duration of residence and ownership of house [12]. Some researchers [12,16] stated that age give a positive affect on residential satisfaction where older people tend to be more satisfaction with their dwelling than younger people.

Previous study reported that higher income enables household to move to a suitable house in a attractive neighbourhood which may give higher level of satisfaction [17]. This statement supported by the works by other study [12] where the higher income households are generally satisfied with their housing. Employment sector and household size significance to the residential satisfaction level [14].

In this part, it was stated that the finding is mainly about the residents' characteristics including age, income, duration of residence, house ownership, household type, and the impact of residents' satisfaction variously, positively or negatively [5]. But the finding between residential satisfaction and residents' characteristics is not conclusive. Analysis done by other study shows that socio-demographic characteristic which are size of household, duration of residence, type of tenure is positively correlated to overall residential satisfaction [18].

4.2 Housing Characteristic

Housing characteristics were more important than demographic characteristic was mentioned in a study [19]. It shows that building features such as number of bedrooms, size and location of kitchen and quality of housing units are strongly related to residential satisfaction [20]. Structural indicator of housing is a significant factor affecting housing satisfaction was found in previous researches [21]. The indicators involved include physical characteristic of housing such as laundry and washing area, kitchen space, size of living area and dining area, number and level of sockets, number of bedrooms and bathrooms, other aspect of housing such as housing quality, privacy and housing services provided by developers such as safety and ventilation of the house.

Physical characteristic of the housing as comfort, quality of building, housing plan and size of the house are significance in measuring the higher level of residential satisfaction [22]. This is supported by other study which summarized that the housing characteristic such as number of bed rooms and toilets, size and location of kitchen, living room, quality of housing unit affect the residential satisfaction differently at cross-cultural levels [5].

4.3 Neighbourhood Characteristic

A study was stated that neighbourhood satisfaction is an important predictor of residential satisfaction [5]. Most of the neighbourhood dissatisfied because of the higher distance travelled for school, work, shopping and medical centres. The elements of the neighbourhood such as safety from crimes and accidents are positively linked with residential satisfaction. It is concluded that the residential satisfaction does not only rely on the housing unit only but also the neighbourhood.

Previously, a study reported that a family evaluates a neighbourhood based on the normative criteria which are area should be predominately residential, accessibility to quality school, quality of streets and roads and homogeneity regarding social class, race and ethic group [23]. Therefore, it is emphasized that neighbourhood satisfaction has been shown to be an essential predictor to housing satisfaction [12]. Accessibility to the public transportation, community and shopping facilities and physical environment variables has also been recorded as factors of neighbourhood satisfactions [24].

A study found that the issue of safety from accidents as a factor of housing satisfaction was considered [25]. This is supported by another study where it is stated that level of crime or lack of amenity or industrial development or location of workplace, are probably to be contributors of neighbourhood dissatisfaction [26]. Therefore, it was concluded that residential satisfaction does not only rely on the housing units itself, but neighbourhood plays an important role in residential satisfaction [21].

4.4 Behavioral Characteristic of Residents

Behavioural characteristic of residents or 'Housing adjustment and adaptation' as conceptualized in a study are the family's effort to redress the discrepancies between the housing it has and the housing it and others feel they should have [27]. Housing adjustment is a process that may occur when a family experiences a normative deficit that causes a significant reduction in housing satisfaction. In their study, they also pointed out that residential alterations and additions consists of two main situation which are increases in the amount of space or number of rooms in housing unit and improvements in the quality of the housing unit.

Behavioural characteristic of the residents reflects their feelings about their residential satisfaction and dissatisfaction [5]. The residents react differently with their housing dissatisfaction. They may be adapting with the situation of their housing unit, or they may be done some modification at the certain part of their housing unit, or it may lead to migration. It is all depend on the level of their dissatisfaction or the ability of relocation.

The new attribute has been included in a study which is owners' culture of maintenance. It was found that this attribute to be significant together with existing residential attributes. Significant of attributes in the study according to the rank is start with building quality feature and followed by owners' culture maintenance, social feature, neighbourhood feature, management feature and dwelling unit feature [11].

There are four (4) characteristic from the finding of residential satisfaction which are socio-demographic characteristics of residents, housing characteristics, neighbourhood characteristics and behavioural characteristics of residents [5]. The table below concluded the variables involved in residential satisfaction studies.

Table 1. Components of Residential Satisfactions.

Components of Residential Satisfaction	Determinants	Elements (varies according to objective of study)
Socio-demographic characteristic	Demographic of residents Socioeconomic of residents	Age, gender, marital status income, duration of residence, tenure status, race, education, household size, job status etc.
Housing and ancillary characteristic	Physical characteristic Support services Management factors Building quality features	Number of rooms and sockets, size of rooms, dining, kitchen, and living, dry area, quality of house, toilet/bathroom, ventilation, parking, corridor, staircase etc.
Neighbourhood characteristic	Neighbourhood facilities Social environment Public facilities	Accessibility, safety and security, infrastructure, privacy, urban activity and noise, neighbours' relationship etc.
Behavioural characteristic	Residents behavioural towards residential	Mobility, modification, adjustment, adaptation, maintenance culture

It shows that, the studies which involve different types of housing, tenures, case study, and culture have different effects on the level of residential satisfaction and the significant of the variables in the study. The use of elements is not fixed to a certain number, but it is may be adjustable according to the objectives of the studies. Case specific situations is suitable in indicating that further studies are required to determine residential satisfaction to guide public policies [28].

5. Conceptual Framework of Residential Satisfaction

Indicator framework of residential satisfaction has multi-faceted [5]. This multi-faceted involved six (6) components which are socio-demographic characteristics, physical features of the house, housing support services, public facilities, neighbourhood facilities and social environment.

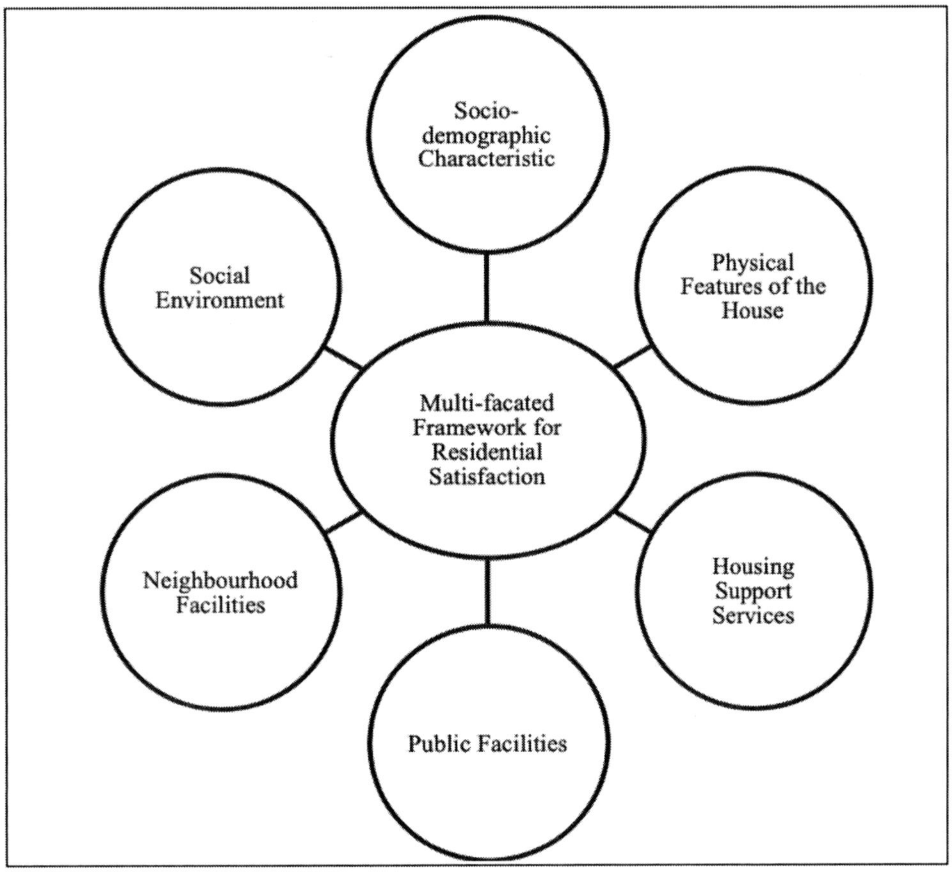

Figure 1. Multi-faceted Framework for Residential Satisfaction. [5]

The determinants and variables involved in residential satisfaction have been discussed earlier. Based on the exploration on the current studies on residential satisfaction, this study is proposing a conceptual framework of residential satisfaction.

IOP Conf. Series: Earth and Environmental Science **385** (2019) 012040 doi:10.1088/1755-1315/385/1/012040

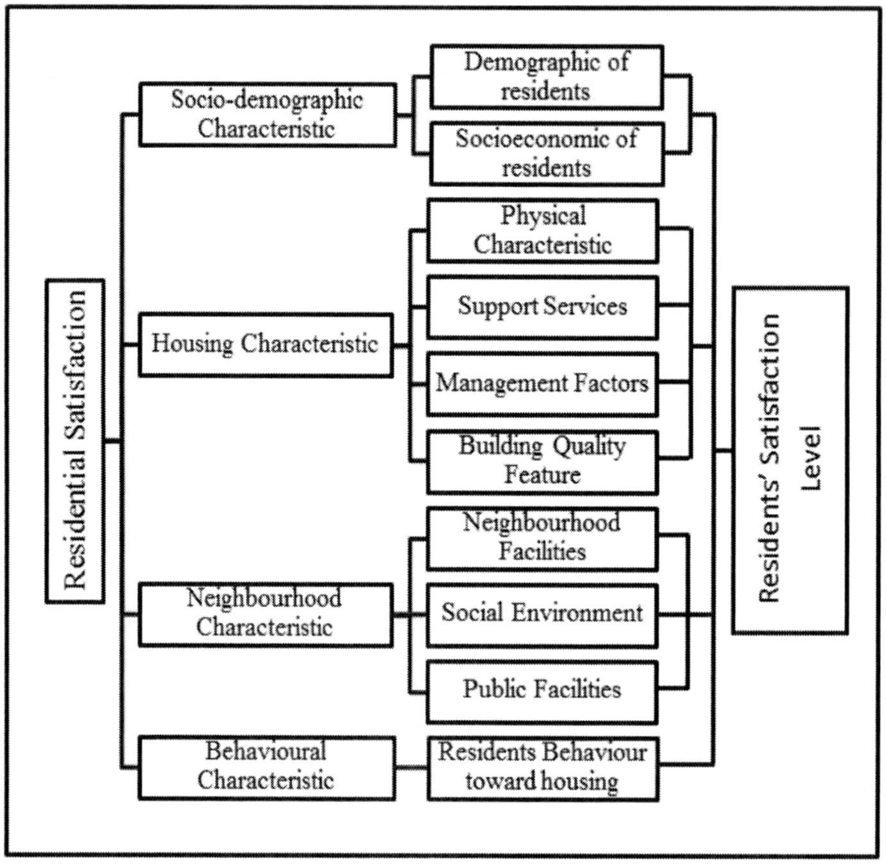

Figure 2. Components and Determinants of Residential Satisfaction. (Researchers, 2019)

Figure above is the conceptual framework of residential satisfaction. This conceptual framework was developed by adapting the indicator framework developed in previous study by [5]. There are six (6) components in their studies. But, in this study, it has been reduced to four (4) and included the determinants of the components that refer to the previous studies.

6. Conclusion

Residential satisfaction is a very wide study where it can be adapted in various cases. Study on residential satisfaction may be varies in terms of places, people, culture and most important is the housing itself. All this study will be resulted in different level of satisfaction according to the objective of the study. Nevertheless, residential satisfaction is very complicated construct because of three reasons [3]. The first is because it has two terms which are residential and satisfaction. Secondly is because it involved three (3) different processes which are cognitive, affective and behavioural. And the last reason is because the measurement of the residential satisfaction is based on those three (3) processes. This study indicates that further studies on residential satisfaction can be carrying out on case specific context in guiding the housing policies.

Acknowledgement
The authors wish to acknowledge the support grant in performing this study under the LESTARI Grant, Universiti Teknologi MARA, File Number: 600-IRMI 5/3/LESTARI (026/2018). The authors also wish to thank the Centre of Studies for Estate Management of Universiti Teknologi MARA and all our colleagues who kindly provided valuable and helpful comments of this paper.

References
[1] Bardo J.W and V. Dokmeci 1992, 'Modernization, tradionalization and the changing structure of community satisfaction in two-sub-communities in Istanbul, Turkey: A Procrustean', Genetic, Social and General Psychology Monographs, 118(3)
[2] Galster, G. C., 1985. Evaluating indicators for housing policy: Residential satisfaction VS marginal improvement priorities. Social Indicator Research. Vol.**16**. pp.415- 448.
[3] Rossi, P.H. (1955) Why Families Move: A Study in the Social Psychology of Urban Residential Mobility. The Free Press, Glencoe.
[4] Morris, E. & Winter, M. (1975). A Theory of Family Housing Adjustment. Journal of Marriage and Family, 37(1), 79-88.
[5] Mohit, M. A., & Raja, A M M A. 2014. Residential satisfaction e concepts, theories and empirical studies. Planning Malaysia *e Journal of Malaysian Institute of Planners*, **3**, 47-66
[6] Weideman S. and Anderson J. 1985 A conceptual framework for residential satisfaction. In: Altman, I. and Werner, C. (eds) *Human Behavior and Environment*, pp. (Plenum Press, New York)
[7] Tan, T. H., 2016,"Residential satisfaction in gated communities", *Property Management*, Vol. **34** Iss 2 pp. 84 - 99
[8] Amerigo,M. and Aragones.J 1997. A Theoretical and Methodological Approach to the Study of Residential Satisfaction. *Journal of Environmental Psychology* .Vol.**17**.pp.47 -57
[9] Fauth, R. C., Leventhal, T., & Brooks-Gunn, J. 2004. Short-term effects of moving from public housing in poor to middle-class neighborhoods on low-income, minority adults' outcomes. *Soc Sci Med*, 59(11), 2271-2284.
[10] Leslie E. and Cerin E. 2008, Are perceptions of the local environment related to neighbourhood satisfaction and mental health in adults?, *Preventive Medicine*. Vol **47**, Issue 3, pp. 273-278.
[11] Somiah, M.K., Aidoo, I., and Braimah, A. 2017, An Empirical Enquiry into the Attributes of Residential Satisfaction that Predict Students' Satisfaction in Public Halls of Residence in Technical Universities in Ghana. *International Journal of African and Asian Studies*. Vol.**39**, pp 54-61
[12] Lu, M. 1999. Determinants of residential satisfaction: ordered logit vs . regression models. *Journal of growth and change* .Vol.**30**.pp 264- 287.
[13] Vera-Toscano, E. & Ateca-Amestoy, V. 2008. The Relevance of Social Interactions on Housing Satisfaction. *Soc Indic Res*, **86**, 257-274.
[14] Ibem, E. O., & Aduwo, E. B. 2013. Assessment of residential satisfaction in public housing in Ogun State, Nigeria. *Habitat International*, **40**, 163–175.
[15] Campbell, A.et.al 1976. The quality of American life. perceptions, evaluation and satisfaction. Russell Sage foundation. .U.S.A.
[16] Chapman, D., & Lombard, D. 2006. Determinants of neighborhood satisfaction in fee based gated and non-gated communities. *Urban Affairs Review*, 41(6), 769-799
[17] Frank B. and Enkawa T., 2009 Economic drivers of dwelling satisfaction: Evidence from Germany, *International Journal of Housing Markets and Analysis*, Vol. 2 Issue: *1, pp.6-20*
[18] Mohit, M. A, & Azim, M. 2012. Assessment of residential satisfaction with public housing in Hulhumale', Maldives. *Procedia - Social and Behavioral Sciences*, **50**, 756–770.

[19] Lane, S., & Kinsey, J. 1980. Housing tenure and housing satisfaction. *Journal of Consumer Affairs*, **14**, 341–365.
[20] Raja Noriza Ariffin, et ,a.l 2010. Residential satisfaction in private low – cost housing in the urban area: A case study of the Klang Valley, Malaysia. International Housing Research Conference. Istanbul, Turkey, 4 -7 July 2010.
[21] Parkes,A. et.al. 2002.What makes people dissatisfied with their neighbourhoods? *Journal of urban studies*. Vol.**39**. No.(13). pp . 2413- 2438.
[22] Türkoğlu, H. 1997. Residents' satisfaction of housing environments: the case of İstanbul, Turkey, Landscape and Urban Planning (39) 55-67.
[23] Morris, E. W., Crull, S. R., & Winter, M. 1976. Housing Norms, Housing Satisfaction and the Propensity to Move. *Journal of Marriage and Family*, 38(2), 309-320.
[24] Ozo, A. O. 1990 Low cost urban housing strategies in Nigeria. *Habitat International* 14(1), 41-54.
[25] Francescato, G., Weidemann, S., Anderson, J. R., and Chenoweth, R. Residents' satisfaction in HUD-assisted housing: Design and management factors. Washington, D. C.: U.S. De-partment of Housing and Urban Development, 1979.
[26] Mulligan, G., Carruthers, J. & Cahill, M. 2004. Urban quality of life and public policy: A survey. In R. Capello & P. Nijkamp (Eds.), Advances in urban economics (729–802). Amsterdam: Elsevier Science B.
[27] Morris, E. W., Crull, S. R., & Winter, M. 1976. Housing Norms, Housing Satisfaction and the Propensity to Move. *Journal of Marriage and Family*, 38(2), 309-320.
[28] Mohit, M.A. et al 2010. Assessment of residential satisfaction in newly designed public low- cost housing in Kuala Lumpur, Malaysia, *Habitat International*, Vol.**34**. pp.18 -27.

IOP Conf. Series: Earth and Environmental Science **385** (2019) 012041 doi:10.1088/1755-1315/385/1/012041

Preserving the School Landscape and its Relationship with the Learning Process of Secondary School Students

Salina Mohamed Ali[1], Abd Hair Awang[2], Faridatul Akma Abdul Latif[1], Noriah Othman[1], Rostam Yaman[1]
[1]Faculty of Architecture, Planning and Surveying, UiTM Malaysia
[2]School of Social, Development & Environment, UKM Malaysia

Email: salina77775@yahoo.com

Abstract. A peaceful and tranquil environment can influence the school landscape by assisting the learning process either directly or indirectly, through various methods, such as providing an area or place for a variety of activities, to socialize, to interact and to do revisions outside the classroom. The purpose of this research is to determine the role or function of the school's landscape in preserving the environment at the school level. For this reason, primary data were collected from the surrounding schools located in Klang Valley, Malaysia that were chosen randomly. Secondary data were obtained from various relevant agencies. Structured interviews were conducted with the higher management of the schools concerned with regards to the development, management of landscape and the contribution of soft and hard landscapes in the learning process. The outcome of the research indicates that the landscape functions in assisting the learning process and nourishes the awareness of the students for their affection towards natural environment. This study has proven that landscape plays an important role in the learning process and nourishes the students' awareness in appreciating the environment beginning at the school level. Thus, the school has to take the necessary initiatives by way of increasing soft landscapes such as planting more shaded trees; hard landscapes such as gazebos, garden chairs and tables as well as man-made landscapes such as fish ponds, waterfalls and fountains which could be used by the students and teachers in the learning process and to encourage them to appreciate the environment.

1. Introduction

The landscaping function in primary and secondary learning institutions is not only for the surrounding aesthetic values and to offer shade, besides it is also to provide and create learning process and to encourage environmental appreciations of students. Recognizing the significance of landscaping in these learning institutions, this study endeavors to promote key factors in regards of knowledge functions of school's landscape in education understanding process [1, 2] and also in encouraging students to value the surrounding environment [3]. This is due to schools as a social institutions context and it play an important function in molding the end-users means. Primary and secondary learning institutions command changes to end-users mentality and their intellectuality level, an institution which delineates the nation and determines the direction or path which a country is on. Schools are also the institution which determines future generation and provides the basis for the development of a country. A country will not experience any development in the absence of an effective education system. An effective education system could be implemented much effectively if the functions of the surroundings of the school assist in achieving the said purpose [4]

Based on the preamble above, the intent of this study is to develop the affection towards the environmental issues from the early schooling stage. Apart from above justification, the information gathered from this research can be used as a guideline in assisting the education process, thus it parallel according to the National Education Philosophy. Landscaping in schools are portion of the learning component which need to be highly addressed.

The role or function of the school's landscape in preserving the environment should be from the schools level. We need to have a clear way of preserving the schools environment. Recognizing the circumstances that there are problems in developing landscaping in schools, the government has its plans to permeate these problems [5]. Several programs are introduced in liven up the surrounding and environment in schools were held at municipality, district and also at state levels. Programs are introduced and organized via the Ministry of Education, Malaysia in ensuring every school performs its position in developing their own landscaping. Ministry of Education introduced and launched its 3K Programme in year 1993 and the programme stresses on the safety, cleanliness and beautifying the schools [6].

The key intention in this study is to determine the role or function of the school's landscape in preserving the environment from the school level. The specific study objectives are to distinguish the contribution of soft and hard landscapes in the learning process and to encourage students and teachers to appreciate the environment.

2. Literature Review

2.1 The Natural School Environment (Outdoor)

It is indicated where there are inadequacies in highly managed and risk free play-area essentially found in school ('play-area' is terminology employed locally to portray the playground in school) and that supplementary advantages could be resultant from playing in natural outdoor environment in schools. According to Stephenson (2003), by giving an example, relates adolescent and children physical risk-taking in natural environment outdoor with children potentials in developing both self-confidence and disposition in managing the risk well [7]. Correspondingly, taking natural environment risk were correlated to the children's learning routes and also dispositions development [8].

Bell and Dayment (2008) explored the potential and capacity of the natural school environments in promoting well-being and health, and to transpire an integrated element of multidimensional, learning-institution-based promotion health strategies [2]. Particularly, latest research in measuring natural school environments as places for interests to the teachers and children's well-being advocates can meet-up, be informed and to support each other. Furthermore, Bell and Dayment (2008) highlighted that there are bodily growth of verification that natural school environments, as of school's setting, may impact to children's mental, physical, social and inner well-being [2].

A growing literature body suggests that individuals need connection with natural environment for selves' well-being. Conversely, at this junction, the young children are increasingly becoming more separated from their natural surrounding as their access to the surroundings are gradually diminishing [3]. The significance of schooling and prior-to-school settings in correlating children's with natural surroundings has been recognized. Furthermore, the yielded outcomes suggested that natural outdoor environments furtherance children's imaginative act, the developmental of pro-behavioral relationships and permits natural environment as a place of play-learning. Dowdell et. al., (2011) concluded that in order to make use of an effective outdoors play-learn settings, early childhood learning centers ought to provide children with an access to the natural outdoor environment and educators who are supportive towards children developmental relationship with the nature [3].

According to the past studies, provision for children to move freely within the space of the outdoor environment is one of the main advantages of psycho-behavioral context factor [9]: physical movement, together with play-act, has been portrayed as one of the most powerful and natural learning approaches for the youth [10]. As noted by numerous researchers [9]-[11], when learning or playing outdoors children can develop on a larger scale context, explore and discover the world first hand and experiencing natural phenomenon such as the weathering, the changing climatic and casting of shadows.

In relationships with their peers, children can easily back away from any confrontation when at outdoors and so likely less to indicate signs of self-frustration and lack of mutual cooperation [11]. Furthermore, adults apparently to relate more differently towards children in the outdoor environment. Rivkin (1998) noted that when inside, the children are expected to be quiet and sit still; meanwhile when outside children are expected to make a noise and run freely [12]. This suggested that children can further push their boundaries of who they were and what they may do without being fear of admonishing for being too overexcited, messy or loud [11][12]

3. Methodology

This research was carried out at public schools sited in urbanized locales in the Klang Valley, of the state of Selangor. The chosen case studies are at Shah Alam and Subang Jaya in the Petaling district in order to represent the learning institution in the urbanized area. The grouping of schools in the urbanized area were classified according to criteria set by the Education Department of Selangor and the Federal Territory Education Department (2011). The justification for choosing Klang Valley as the case study area due of the these factors: firstly, it is strategically located within prime region in Malaysia where urbanization and sprawl is at rapid pace; secondly, schools within this area are experiencing modernization due to the positioning of nation's centre of innovation, particularly the capital Kuala Lumpur; thirdly, the extent of landscaped area at primary and secondary institution in Klang Valley which is decreasing due the shortage of land-plot and price hike of the land-plot as a results of rapid development and impact to its environment [13][14].

This study employed both gathering of primary and secondary data. The secondary data were documents, records and statistics gathered by diverse agencies or departments such as the Selangor and Federal Territory Education Department and also from Ministry of Education, Malaysia. General data obtained from Statistics Department, Malaysia were also utilized. The similar method were applied to specifics students and educators were obtained from chosen schools in this study. Information related to National Education Philosophy and National Landscape Policy which concerns the learning institutions planning and development, social economy, population density, local infrastructure and planning are also gathered for the objective of this study. Data were also obtained through the websites of government agencies namely the Economic Planning Unit of the Prime Minister's Department, Ministry of Trade and Industry, Economic Planning Unit of Selangor and Department of Statistics, Malaysia. Other tertiary sources of information included books, journals, technical report, thesis and dissertation. Prior to the introduction of the research form, researchers have conducted specific reference procedures namely research by experts in the field of landscape plants, landscape architects at the National Landscape Department, experts in the education field, teachers who are experienced and other individuals (landscape consultants) with extensive experience in the related field.

The quality of the schools landscape function in the Klang-Langat Valley were valued based on the aspect of landscape function by co-relating its relevancy in aiding the process of learning and nurturing the awareness among students towards environmental appreciation [15]. Evaluation on the purpose of landscaping in schools were based on landscape expert's opinion from National Landscape Department, Putrajaya Holdings Sdn. Bhd., Putrajaya Development Corporation, and Ministry of Education Malaysia and also from school headmasters.

IOP Conf. Series: Earth and Environmental Science **385** (2019) 012041 doi:10.1088/1755-1315/385/1/012041

4. Result & Discussion

4.1 Planning for Selection of Landscape Themes and Concepts

Table 1. shows nine major landscape concepts selected by school management in the Klang-Langat Valley. The most popular school landscape concept is the type of artificial landscapes involving 53 schools (51%), followed by natural landscapes at 49 (47.1%) and Nusantara landscapes at 21 (22%). However, almost 50 percent of respondents stated that they did not have a specific landscape concept that could be linked to the school environment. This may be due to the lack of exposure of managers and school administrators on the concept of school landscape. Planning to choose the theme and concept of the school landscape solely depends on the creativity of the school management.

Table 1. Theme and landscape concepts selected by school management

Concept	Total	
	Frequency	%
Natural Landscape	49	25.8
Artificial Landscape	53	27.9
Islamic Concept	3	1.6
English Concept	1	0.5
Chinese Concept	3	1.6
Mediterranean Concept	1	0.5
Japan Concept	4	2.1
Bali Concept	10	5.3
Nusantara Concept	21	11.1
No Concept	45	23.7
Total	190	100.0

Source: Field Review 2013

4.2 Relationship and Association between Quality of Landscape Function and School Achievement in Academic Field, Knowledge and Environmental Awareness

For the purpose of testing the relationship and association between the qualities of landscape function with school achievement in academics, character and environmental awareness; the research used the Pearson correlation test and Chi-square. The purpose of the analyses were to determine whether there are significant correlation concerning the functions of landscaping or otherwise within learning process (independent variable) and with students in academic achievement, personality and activities, fostering awareness of the environment for five years (dependent variable).

Table 2. shows the mean of dependent variables, which are landscape functions in assisting learning, characterizing students and promoting environmental awareness. For the function of the landscape that helps in cognitive learning, the highest mean score is for the component of the facility (54.8), and followed by flora component mean score(43.59), then the mean score for the component landscaping (33.94). Meanwhile, for landscape function to form the character of the students, the highest mean score was achieved by the component of the facility (66.46), followed by the flora component (47.12), and the components of the landscape (29.52). This proves that the components of the facility consisting of elements such as pedestrian paths, bicycles parking, lights, badminton court, field and others have a high contribution compared to other components in its function to help the learning process and shape students either directly or indirect.

With reference to the total score, overall schools tend to have landscapes that can be attributed to environmental awareness with mean 218.71. This shows that the landscaping elements in school surroundings have a role in fostering the love of the environment since at the school level. Assessment made on landscaping functions by professionals comprising of officers from Education Ministry and Selangor State Department of Education Office; National Landscape Department, Perbadanan Putrajaya and Putrajaya Holding Sdn. Bhd. may cause landscape function to foster environmental awareness of the school earning a high mean.

Table 2. Mean and standard deviation for variable score (dependent), landscape function (cognition, personality and environmental awareness)

Landscape component	Cognitive Function		Personality Function		Environmental Awareness Function	
	Mean	Std D	Mean	Std D	Mean	Std D
Hard Landscape	33.94	6.270	29.52	5.420	34.15	6.164
Soft Landscape	21.83	3.230	21.02	3.034	29.97	3.114
Artificial Landscapes	4.47	1.954	2.99	1.248	4.48	1.889
Facility Element	54.80	2.120	66.46	2.828	51.00	1.936
Fauna	23.04	4.435	19.01	3.390	28.53	5.086
Flora	43.59	8.402	47.12	8.544	72.00	13.538
Facility Element Total Score	180.20	18.270	185.17	17.006	218.71	22.980

Source: Field Research 2013

Table 3. Chi-squared^2 Relationship between Landscape (learning) Function with School Academic Achievement (Total)

School Academic Achievement	Landscape Function Score (Learning)					
	High		Moderate		Low	
	No	%	No	%	No	%
High	39	79.6	23	51.1	1	10.0
Moderate	10	20.4	20	44.4	5	50.0
Low	0	0.0	2	4.4	4	40.0
Total	49	100.0	45	100.0	10	100.0

Significant at $p < 0.01$; df = 4; Chi-squared^2 = 35.993

Using the Chi-squared^2 (Pearson correlation) statistics method, landscape function variables in helping learning have been tested against the segment of academic achievement variables. Table 3 presents what is aimed at detecting the relationship between landscape functioning helps with learning with academic achievement of the school. The study found that the landscape function score helped the learning to be high (79.6%). A significant linkage between landscaping function helps the process of learning with higher achievement of academic at 0.01 level with the value of Chi-squared$^{2\,a}$ at 35.993.

A more detailed test is performed between the scores of the component of the landscape function that helps with the school achievement score segments (Table 4). The results showed that there was significant correlation between weak and simple (Guilford, 1956) among the overall result for function in landscaping in assisting the process of learning with the achievement of academic in school ($p < 0.01$, $r = 0.504$). Generally, this test proves there is correlation between school landscape and process of learning at school within Klang-Langat Valley. Results of the analysis shows that there are significance between landscape function helps learning process against achievement in Malaysia Examination Certificate ($p < 0.01$, $r = 0.355$); with Middle Lower Examination achievement ($p < 0.01$, $r = 0.370$,); with the achievement awards of quality ($p < 0.01$, $r = 0.243$). Hence, it shows that landscaping function helps process of learning by signifying relationships all analyzed segments in academic achievement.

Table 4. Pearson Correlation on the Relationship between Landscape Function in Assisting Learning Process and Academic Achievement

Relationship	r	Sig	Level of Relationship
Landscape function (helping learning) with SPM achievement	0.355**	0.000	Weak
Landscape functions (helping to learn) with the achievement of PMR	0.370**	0.000	Weak
Landscape function (helping learning) with achievement of quality awards	0.243*	0.013	Weak
Landscape function (helping learning) with academic achievement (overall)	0.504**	0.000	Moderate

Note: ** Correlation relationship is significant at 0.01 (two-tailed) levels;
* Significant correlation relationship at 0.05 (two-tailed)

Overall, there is a significant relationship though at a moderate and weak levels between the landscape function in assisting the learning process with the academic achievement of the school. At this point, the relationship involves the landscape function of helping learning to cover the entire landscape components of the school consisting of elements such as pavilion, pergola, bench, garden table, signboard, flowerpot, and soft elements such as shade, shrubs, fruits, cover of earth, herbs and others.

5. Conclusion

In conclusion, the importance of landscape in school not only benefits students, teachers, but the parents as well. Commentary on the importance of school landscapes can impact student psychology, student behavior, physical activity and the school environment. Issues and challenges of the school landscape have a profound impact on the development of school landscapes in Malaysia. School landscape management is an important topic because effective landscape management can affect the academic achievement of the school and the quality of the school environment in general. Challenges of the school's landscape development have slightly affected the landscape design of the school.

The characteristics of the school landscape are important aspects of the study highlights as each school has different landscape features influenced by the location namely urban and rural schools. Some of the factors identified have become the barriers to the development of the school landscape can be solved based on observations on the school area. With an interview with the school management, some problems with regard to the landscape can be identified. In addition, observations conducted by researchers on the school area can be portray the issues and problems of the school landscape. Finally, article writing that refers to earlier studies can provide some guidance towards improving this present article.

Acknowledgement
The author would like to extend the appreciation and gratitude to Universiti Teknologi MARA, for granting Lestari Grant to complete this article. RMI file no : 600-IRMI 5/3/LESTARI (043/2018).

References

[1] Anne C. Bell, Janet E. Dyment & Lucas, J.A (2009). The relationship between school ground design and intensity of physical activity, Journal Children's Geographies Volume 7, 2009 - Issue 3.
[2] Anne C. Bell & Janet E. Dyment (2008). Grounds for health: the intersection of green school grounds and health-promoting schools, Journal Environmental Education Research Volume 14, 2008 - Issue 1
[3] Dowdell, K. Gray, T. & Malone, K. (2011). Nature and its Influence on Children's Outdoor Play, Australian Journal of Outdoor Education, 15(2), 24-35, 2011
[4] Matsuoka, R. H. (2008). *High school landscapes and student performance* (Doctoral dissertation, University of Michigan).
[5] Salina, S. M. A., Katiman, K. R., & Hair, A. H. A. A. (2014). Characteristics of Landscape and Surrounding Qualities of Schools in Klang Valley. *Asian Journal of Multidisciplinary Studies*, *2*(6).
[6] Rohhayati, N. (2008). Student perception of school landscape and its influence on learning experience at the award winning secondary school in Kelantan, Malaysia. *Unpublished Master Thesis. University of Putra Malaysia.*
[7] Stephenson, A. (2003) Physical risk-taking: dangerous or endangered?, Early Years, 23(1), 35–43.
[8] Waller, T. (2005) 'This is the way we go to the park!' Recording and evaluating young children's knowledge and perspectives of geography, paper presented at the British Educational Research Association Annual Conference (BERA), September.
[9] Rivkin, M. (1998) Happy play in grassy places: the importance of the outdoor environment in Dewey's educational ideal, Early Childhood Education Journal, 25(3), 199–202.
[10] Bilton, H. (2002) Outdoor play in the early years (London, David Fulton).
[11] Ouvry, M. (2003) Exercising muscles and minds: outdoor play and the early years curriculum (London, National Early Years Network).
[12] Rivkin, M. (1995) The great outdoors: restoring children's right to play outside (Washington, DC, NAEYC).
[13] Katiman Rostam. (2006). Migration to the suburbs Metropolitan Region Lembah Klang. *Akademika*, 68: 3-27.
[14] Katiman Rostam, Mochamad Rosul, Er Ah Choy, Abdul Rahim Mohd Nor, Zaini Sakawi, Norazuan Md Hashim & Aishah@Esah Hj Muhammad. (2010). Urbanization and the spread of the city on the outskirts of Klang-Langat Metropolitan Region. *Geografia : Malaysian Journal of Society and Space*, 6 (2): 37 – 50.
[15] Salina, S. M. A., Katiman, K. R., & Hair, A. H. A. A. (2015). School Landscape Environments in Assisting the Learning Process and in Appreciating the Natural Environment. *Procedia-Social and Behavioral Sciences*, *202*, 189-198.

IOP Conf. Series: Earth and Environmental Science **385** (2019) 012042 doi:10.1088/1755-1315/385/1/012042

Reliability of Terrestrial Laser Scanner Measurement in Slope Monitoring

M A Abbas[1,2,*], N A Fuad[2], K M Idris[2], Y D Opaluwa[3], N M Hashim[1], Z Majid[2] and S A Sulaiman[1]
[1]Faculty of Architecture, Planning & Surveying, Universiti Teknologi MARA, 40450 Shah Alam, Malaysia
[2]Faculty of Built Environment & Surveying, Universiti Teknologi Malaysia, 81310 Skudai, Malaysia
[3]Department of Surveying & Geoinformatics, School of Environmental Technology, Federal University of Technology Minna, Nigeria

Email: mohdazwan@perlis.uitm.edu.my, mohdazwanabbas@gmail.com

Abstract.. Dense 3D point clouds provided by terrestrial laser scanner (TLS) has demonstrated significant reliability of TLS in landslide monitoring. However, existence of errors in measurement is inevitable which eventually has decreased the quality of TLS data. To concretely measure the capability of TLS in landslide monitoring, this study has performed two epoch measurements using tacheometry (for benchmarking) and TLS (Topcon GLS-2000) at Kulim Techno City, Kedah, Malaysia. Sixteen (16) artificial targets were well-distributed on the slope to determine the accuracy of the employed TLS. Results obtained revealed that Topcon GLS-2000 provides 0.006m of accuracy. However, the presence of high incidence angles in TLS measurement has limited the capability to identify the significant displacement of the targets.

1. Introduction

Landslide is becoming common issue in worldwide and it can be defined as the deformation of a mass of rock, debris, or earth down a slope. Landslide occurs when the strength of the earth material that composes the slope has less forces than the gravity effect which eventually causes the down-slope movement of soil and rock [1]. Currently, there are two approaches available for land slope monitoring (figure 2), which are using geodetic or geotechnical [2] methods.

Tacheometry and global navigation satellite system (GNSS) methods can be considered as established geodetic approach in deformation measurement and landslide monitoring. Taking into account the accuracy in measurement, reliability of both methods have long been proven [3]-[6]. However, the precision (i.e. point density) provided from tacheometry and GNSS measurement is quite limited. To resolve that limitation, several non-contact measurement techniques were introduced, which are remote sensing [7], photogrammetry [8]-[9] and LiDAR [10]. With the capability to provide dense three-dimensional (3D) data (according to resolution of sensor), these measurement techniques have enabled surface to surface comparison assessment or often known as surface deviation analysis. This kind of analysis can homogenously measure and identify the trend of movement that happened from two differ measurement epochs.

Taking accuracy as first priority among those three approaches that able to provide dense 3D data (i.e. remote sensing, photogrammetry and LiDAR), terrestrial laser scanner (static LiDAR) can be considered as the best. With significantly less errors propagation in measurement and processing procedures, terrestrial laser scanner (TLS) is also not dependent upon the lighting conditions and surface texture as well as able to provide direct, rapid and high-density 3D data compared to other approaches [11]. However, TLS instruments are complex tools with many moving parts whose relative positions can change over time depending on use, handling frequency and care. Furthermore, similar to other optical sensors, observations from TLS is also can be impaired by numerous of errors especially from local environment [11]. Thus, reliability of TLSs in slope monitoring is still remain questionable. For that purpose, this study has robustly examined the accuracy of TLS data in landslide monitoring with the aid of tacheometry measurement.

2. Terrestrial laser scanner measurement

In contrast to traditional 3D data acquisition approaches (i.e. tacheometry and photogrammetry), terrestrial laser scanner is capable to rapidly acquire data without any requirement of direct contact to the object and extensive processing procedure. Similar to reflectorless tacheometry, TLS did measure three main components: i) Range (r); ii) Horizontal direction (ϕ); and iii) Vertical angle (θ). According to Abbas et al. [12], there are three options employed by TLS in order to measure range: i) time-of-flight; ii) phase shift; and iii) triangulation. While the other components were obtained based on the scanner pre-determined axes.

To ensure TLS data are significant for further processing, most of the TLSs on-board software have automatically convert raw TLSs data (i.e. spherical coordinate system) into Cartesian coordinate system. However, raw data in spherical coordinate system is essential for this study to conduct least square adjustment for TLS data. Conversion from Cartesian (X_A, Y_A and Z_A) and spherical (range, horizontal direction and vertical angle) coordinates system can be expressed as follows [12]:

$$\text{Range, } r = \sqrt{x_A^2 + y_A^2 + z_A^2}$$
$$\text{Horizontal direction, } \varphi = tan^{-1}\left(\frac{x_A}{y_A}\right) \tag{1}$$
$$\text{Vertical angle, } \theta = tan^{-1}\left(\frac{z_A}{\sqrt{x_A^2 + y_A^2}}\right)$$

3. Experiments

Due to the active slope movement that occurred as reported by Kedah state authorities, this study was conducted at Kulim Techno City, Kedah, Malaysia. For benchmarking reason, as illustrated in figure 1, sixteen (16) artificial targets were well-distributed over the land slope area. Other than TLS measurement, this study also exploited conventional approach (i.e. reflectorless tacheometry) to establish 3D known coordinates for each target. Based on the reference coordinates obtained from tacheometry measurement, accuracy of TLS data can be mathematically determined.

For the tacheometry measurement, Topcon ES-105 with accuracy of 0.0014° and 2mm for angular and range measurement, respectively was utilised to measure all the artificial targets. To secure the accuracy yielded, triangulation observation technique was employed and least squares adjustment has been entrusted to compute the most probable values of all targets along with their quality (i.e. precision).

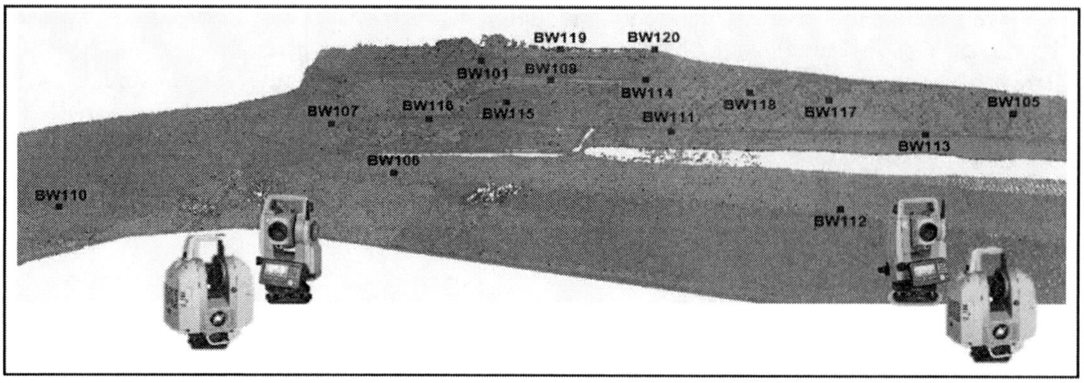

Figure 1. Research site located at Kulim Techno City, Kedah, Malaysia.

As depicted in figure 1, Topcon GLS-2000 scanner was used to scan all targets with the surface of the land slope. According to the instrument specification sheet, this time-of-flight scanner able to provide accuracies of 3.5mm and 0.0017° for range and angular measurements, respectively. It should be noted that the accuracies mentioned above are based on single point measurement, if involves with determination of target centroid procedure, the accuracy is theoretically decrease due to the error propagation with other uncertainties during measurement phase.

4. Results and analyses

Assigned 0.0014° as precision for tacheometry angular measurement, least squares adjustment manage to converge at third iteration. Reference standard deviation obtained from tacheometry observation is 0.0027°, which is equal to 2mm data quality for 50m range measurement. To evaluate the accuracy of Topcon GLS-2000 scanner data acquisition, fifteen independent vectors were established from sixteen (16) targets. Employing target BW111 as reference point, fifteen (15) vectors were obtained from both tacheometry and scanner data as shown in table 1.

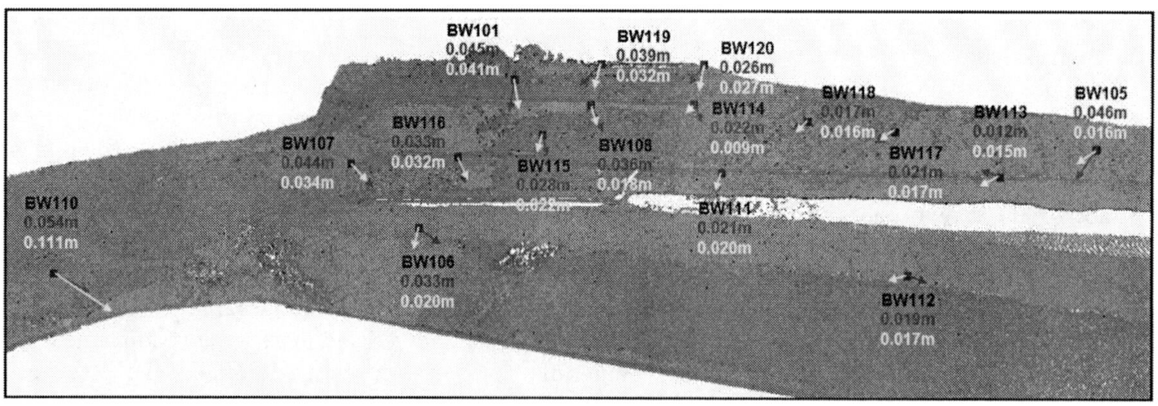

Figure 2. Magnitude and direction of computed displacement vectors for both tacheometry and TLS data.

Based on the computed vectors and with the aid of law of propagation of variance algorithm, it was found that the accuracy of Topcon GLS-2000 scanner is 6mm. As expected, under 95% confidence interval (two sigma), accuracy of the scanner as stated by manufacturer is 7mm. Depicted in figure 2 is magnitudes and directions for all displacement vectors obtained from tacheometry and TLS measurements. Through graphic evaluation, all displacement vectors have shown similar trend for both measurement approaches except for target BW112. The uncertainty that occurred may be due to the position of target which is quite close to the sensors location (refer figure 1). Thus, high incidence angle was expected to reduce the quality of measurement for targets that are near to the occupied sensors (e.g. BW112, BW106 and BW110).

Table 1. Independent vectors for TLS accuracy assessment.

Vector	Tacheometry (m)	TLS (m)	Discrepancies (m)
BW111 - BW101	23.427	23.424	0.003
BW111 - BW105	24.727	24.729	0.001
BW111 - BW106	21.174	21.177	0.003
BW111 - BW107	27.793	27.792	0.001
BW111 - BW108	14.607	14.610	0.003
BW111 - BW110	41.379	41.359	0.020
BW111 - BW101	20.624	20.622	0.002
BW111 - BW105	17.927	17.927	0.000
BW111 - BW112	9.783	9.786	0.003
BW111 - BW113	15.371	15.371	0.000
BW111 - BW114	20.113	20.110	0.003
BW111 - BW115	13.186	13.194	0.009
BW111 - BW116	9.013	9.012	0.002
BW111 - BW117	19.920	19.917	0.003
BW111 - BW118	16.417	16.415	0.002

5. Conclusions

The aim of this study is to examine the reliability of TLS measurement in slope monitoring. Robust experiments were performed through vector deviation analysis to eventually conclude the capability of TLS measurement. Sixteen (16) artificial targets were well-distributed on the land slope and measured using tacheometry (for benchmarking) and TLS approaches. As claimed by manufacturer, with 95% confidence interval, the experiment demonstrates that accuracy of Topcon GLS-2000 scanner is 0.006m. However, the accuracy has been constrained by incidence angle when later experiment (deformation analysis) has indicated that only 50% of the findings are similar to tacheometry output. It is essential to reduce the existence of high incidence angle by properly selecting the position of the scanner.

Acknowledgement

Authors would like to thanks the Universiti Teknologi MARA for the funding of this research. Our special thanks also goes to Photogrammetry and Laser Scanning Research Group, Universiti Teknologi Malaysia for providing instruments and experimental site.

References

[1] Lynn M H and Peter B 2008 *The Landslide Handbook - A Guide to Understanding Landslides USGS Science for a Changing World*, U.S. Geological Survey, Reston, Virginia: 2008

[2] Chowdhury R and Flentje P 2010 Geotechnical Analysis of Slopes and Landslides - Achievements and Challenges *Geologically Active, Proceedings of the 11th IAEG Congress of the International Association of Engineering Geology and the Environment*, Auckland, New Zealand, 2010

[3] Josep A G Jordi C and Joan R 2000 Using Global Positioning System Techniques in LandslideMonitoring *Engineering Geology*, Vol. 55, Issue 3: 167-192

[4] Malet J P Maquaire O and Calais E 2002 The use of Global Positioning System techniques for the continuous monitoring of landslides: application to the Super-Sauze earthflow (Alpes-de-

Haute-Provence, France) *Geomorphology*, Vol. 43, Issues 1–2: 33-54

[5] In T Y Jae K P and Dong M K 2007 Monitoring the Symptoms of Landslide Using the Non-Prism Total Station *KSCE Journal of Civil Engineering*, Vol. 11, Issue 6: 293–301

[6] Andreas W 2017 A New Approach for Geo-Monitoring Using Modern Total Stations and RGB D Images *Measurement*, Vol. 82: 64-74

[7] Mirzaee S Motagh M and Akbari B 2017 Landslide Monitoring Using Insar Time-Series and Gps Observations, Case Study: Shabkola Landslide in Northern Iran. *The International Archives of the Photogrammetry, Remote Sensing and Spatial Information Sciences, Volume XLII-1/W1*, 2017, ISPRS Hannover Workshop: HRIGI 17 – CMRT 17 – ISA 17 – EuroCOW 17, 6–9 June 2017, Hannover, Germany

[8] Stumpf A Malet J P Allemand P Pierrot-Deseilligny M and Skupinski G 2015 Ground-Based Multi-View Photogrammetry for The Monitoring of Landslide Deformation and Erosion *Geomorphology*, Vol. 231: 130-145

[9] Peppa M V Mills J P Moore P Miller P E and Chambers J E 2016 Accuracy Assessment of A UAV-Based Landslide Monitoring System. *ISPRS - International Archives of the Photogrammetry, Remote Sensing and Spatial Information Sciences*, XLI-B5: 895-902

[10] Malet J P Ferhat G Ulrich P Boetzlé P and Travelletti J 2016 The French National Landslide Observatory OMIV – Monitoring Surface Displacement Using Permanent GNSS, Photogrammetric Cameras and Terrestrial Lidar for Understanding the Landslide Mechanisms *Joint International Symposium on Deformation Monitoring (JISDM)*, Vienna University of Technology, Vienna, Austria

[11] Reshetyuk Y 2009 *Self-Calibration and Direct Georeferencing in Terrestrial Laser Scanning* Doctoral Thesis in Infrastructure, Royal Institute of Technology (KTH), Stockholm, Sweden

[12] Abbas M A Luh L C Setan H Majid Z Chong A K Aspuri A Idris K M and Mohd Ariff M F 2014 Terrestrial Laser Scanners Pre-Processing: Registration and Georeferencing. *Jurnal Teknologi (Sciences & Engineering)* 71:4 (2014) 115–122

IOP Conf. Series: Earth and Environmental Science **385** (2019) 012043 doi:10.1088/1755-1315/385/1/012043

Scale Factor Effect in Terrestrial Laser Scanner Datum Transformation

M A Abbas[1,2,*] Z Majid[2], M A A M Azmi[1], A K Chong[3], L C Luh[2], K M Idris[2] and M A Mustafar[1]
[1]Faculty of Architecture, Planning & Surveying, Universiti Teknologi MARA, 40450 Shah Alam, Malaysia.
[2]Faculty of Built Environment & Surveying, Universiti Teknologi Malaysia, 81310 Skudai, Malaysia.
[3]School of Civil Engineering & Surveying, University of Southern Queensland, Australia.

Email:mohdazwanabbas@gmail.com, mohdazwan@perlis.uitm.edu.my

Abstract. Datum transformation is an inevitable procedure in terrestrial laser scanners (TLSs) pre-processing. By taking into account the uncertainties in multi distances measurement, neglecting the scale factor in registration and georeferencing phases should be reconsidered especially when involved with applications that demand high accuracy data. With the motivation to robustly examine the effect of scale factor in the issue raised, this study has designed experiment which consider the element of multi distances. Two types of TLSs utilised in this study, which are Faro Focus 3D and Leica ScanStation C10 to represent phase and pulse based scanners, respectively. Exploiting robust statistical analysis, all seven (7) parameters were computed from both experiments and scale factors were extracted for the further assessment. Significant analysis was performed by comparing the ideal value of scale factor (i.e. 1.000 or no scale effect) and values obtained from the experiments. The results show that, the null hypotheses for the experiment is accepted with a 95% confidence level. This result also indicates that scale factor can be neglected in datum transformation process for both type of terrestrial laser scanners.

1. Introduction

Investigation of uncertainties in TLSs measurement have been massively studied and most researchers still continue searching any possibilities to simplify and improve current approaches and results [1]-[4]. In contrast, very few research performed were related to uncertainties in TLSs processing. Due to the dependency on the algorithm employed to yield final products, the errors produced in post-processing phase are significantly variant and difficult to standardise the investigation. Differing in implementation, pre-processing use standard algorithm in performing registration and georeferencing. Both procedures employ seven-parameter similarity transformation involving seven parameters, which are three rotations, three translations and one scale factor. However, in TLSs datum transformation procedure, Gordon and Lichti [5] did claim that the scale factor can be neglected because it has shown to be irrelevant in the transformation. Thus, the six parameters that are also known as the rigid-body-transformation parameters (without scale factor) or six degrees of freedom are as follows: i) Translation for the 3 coordinate axes (Δ_X, Δ_Y, Δ_Z); and ii) Rotation around the 3 coordinate axes (ω, φ, κ).

The ability to acquire direct and automated millions of 3D points with considerably rapid and accurate measurement, TLSs have been widely employed in many purposes. Most of the yielded information processed from TLSs data were used for documentation, investigation, preservation and

IOP Conf. Series: Earth and Environmental Science **385** (2019) 012043 doi:10.1088/1755-1315/385/1/012043

decision making. Thus, data quality assurance is a mandatory factor. Regarding the scale issue in TLSs pre-processing phase, Rueger [6] did mention that sources of errors that can affect the scale factor are very numerous. Other than instrument causes, there are also external influences that can affect the results of transformation parameters, such as derivation of transformation parameters using poor network can cause a weak solution [7], false determination of target centroid due to the less resolution [8] and algorithm employed for form fitting has wrongly identify the object [9]. Furthermore, dependency of other measurement approaches (e.g. tacheometry and global navigation satellite system) in establishing control points for georeferencing procedure also need to be taken into account. Orienting measurement from various sensors (which utilise different mechanism in range acquisition) into one global coordinate system will definitely cause errors in scale factor. Those arguments have indicated that neglecting scale factor especially in applications that demand high accuracy data is questionable. Further investigation is necessary to robustly prove that contribution of scale factor in TLSs pre-processing phase is insignificant. For such purpose, this study focused on quantitative investigation of scale factor for phase-based (i.e. Faro Focus 3D) and time of flight (i.e. Leica ScanStation C10) scanners. To ensure the evaluation was robustly performed, experiment was performed using multi distances configuration. Statistical analysis was employed to mathematically verify the results obtained from the experiments.

2. Terrestrial laser scanner pre-processing

To ensure TLSs data are significant for further processing, all data need to be oriented from several local coordinates system into one global coordinate system. For mapping and database purpose, registered data are required to be georeferenced or transformed into ground or national coordinate system. Due to the similarity of algorithm employed to derive transformation parameters for both pre-processing procedures [2], thus, discussion on the mathematics were combined using registration example.

Assuming that only two scan stations involves in the registration procedure, coordinates of two scan systems are as following: i) Reference station (X_i, Y_i, Z_i); and ii) Subsidiary station (x_i, y_i, z_i). All seven parameters were computed using resection method, hence, minimum three well-distributed non-collinear common targets need to be measured from both scanner positions (e.g. reference and subsidiary stations). Based on the seven-parameter similarity transformation, relation of the reference target scanned from the subsidiary scanner station can be derived using the equations as follows [10]:

$$X_i = T + S.R.x_i \tag{1}$$

where:

X_i = Coordinates of the ith target in the reference (or global) scanner coordinate system (X_i, Y_i and Z_i).
S = Scale factor.
R = Components of rotation matrix between the two coordinate systems (ω, φ, κ).
x_i = Coordinates of the ith target in the subsidiary coordinate system (x_i, y_i and z_i).
T = Translations of the subsidiary scanner station in the global coordinate system.

However, as stated by Gordon and Lichti [5], the scale factor can be neglected in TLSs datum transformation, thus, only six (6) parameters were used to relate between two coordinate systems. Thus, the relation between the two coordinate system using seven (6) parameters are as follows:

$$X_i = T + R.x_i \tag{2}$$

In order to investigate the significance of scale factor in TLSs datum transformation for registration and georeference procedures, this study has employed equation 1.

3. Experiments

To ensure the robustness of TLSs measurement with respect to multi ranges observation, this study has established indoor calibration site at Faculty of Built Environment and Surveying, Universiti Teknologi

Malaysia, Malaysia. Six (6) points were marked as visualised in figure 1 with interval 10m, range from wall where targets distribute (right side of figure 3) into last point is 60m. For registration purpose, four (4) plane and two (2) sphere targets were utilised. Both phase-based (i.e. Faro Focus 3D) and pulse-based scanners were positioned in every mark to scan all six (6) targets.

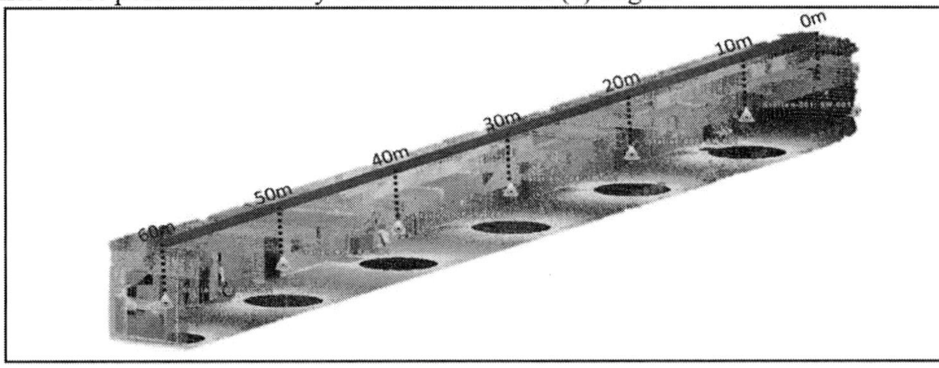

Figure 1. Calibration site for multi-distances experiment.

Taking into consideration that the closest scan station (with range 10m from targets) will contribute less errors in determining target centroids, pairwise registration for other stations was performed by exploiting this closest station as reference. This has result five (5) registration pairs yielded for multi-distances investigation. To concretely support the results obtained, this study has consumed statistical analysis to verify the significant of scale factor in TLSs registration processing.

4. Results and Analyses

The experiment was conducted by exploiting the multi ranges of scan stations from distributed targets. Both scanners were move gradually every 10m until 60m. Considering 10m position able to well-determine the targets centroid among others, thus, this occupied station was selected as reference for datum transformation computation. The scale errors obtained for 20m until 60m ranges were illustrated in figure 2. The largest scale error (i.e. 0.001) was contributed 40m range by pulse based, while the largest for phase based scanner is 0.0004. The largest uncertainty occurred in pulse based scanner may be due to the 40m position has high exposure of light compare to others [12]. This argument is supported by similar trend demonstrated by phase based scanner. Nevertheless, the conducted statistical test to determine whether the calculated scale factors are similar to the ideal value (i.e. one) at a 0.005 level of significance has yielded fascinating results. As presented in table 1 and table 2, in all conditions computed values of t are smaller than the tabulated or critical t, thus, the null hypothesis can be accepted at 95% level of confidence. In other words, multi distances experiment has mathematically proved that scale factor is insignificant in TLSs datum transformation.

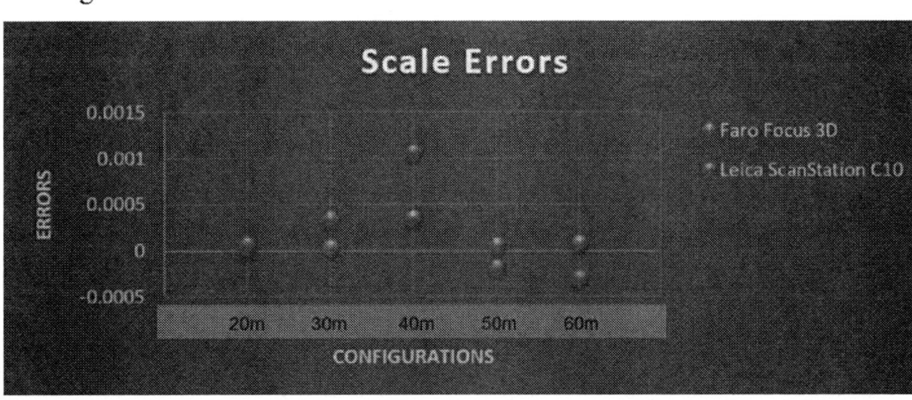

Figure 2. Multi sensors experiment using phase based (a) and pulse based (b) scanners.

Table 1. Multi distances statistical test for Leica ScanStation C10 (pulse based scanner).

Configuration	Scale	DoF	Calculated T	< / >	Critical T
20m	1.000065	11	0.228666	<	2.201
30m	1.000338	11	0.513983	<	2.201
40m	1.001061	11	0.385331	<	2.201
50m	0.999815	11	0.181687	<	2.201
60m	0.999697	11	0.272587	<	2.201

Table 2. Multi distances statistical test for Faro Focus 3D (phase based scanner).

Name	RSO Geocentric			Orthometric Height	
	North (m)	Easting (m)	Horizontal RMS (m)	Height (m)	Vertical RMS (m)
GNSS01	172129.333	628106.384	0.002	17.645	0.004
GNSS02	172148.751	628444.568	0.001	24.488	0.002

5. Conclusions

With the aim to quantitatively examine the significant of scale factor in TLSs (i.e. phase and pulse based scanners) datum transformation, this study has performed robust experiment. The experiment which employed multi distances configuration has shown the effect of light in TLS measurement that can affect the quality of data. However, statistical analyses have indicated that null hypotheses of the experiment showed that the scale factor yielded from both scanners can be neglected. In other words, the scale factor is not significant in TLSs datum transformation. This conclusion is applicable for the phase and pulse based scanners with medium range measurement. Further significant study on scale factor is necessary when measurement involved with long range measurement.

Acknowledgements

Authors would like to acknowledge Universiti Teknologi MARA for financial funding. Special thanks goes to Photogrammetry and Laser Scanning Research Group, Universiti Teknologi Malaysia for providing instruments and experiment site.

References

[1] Lichti D D 2007 *Error Modelling, Calibration and Analysis of an AM-CW Terrestrial Laser Scanner System* ISPRS Journal of Photogrammetry & Remote Sensing 61 (2007): 307-324

[2] Reshetyuk Y 2009 *Self-Calibration and Direct Georeferencing in Terrestrial Laser Scanning* Doctoral Thesis in Infrastructure, Royal Institute of Technology (KTH), Stockholm, Sweden

[3] Chow J C K 2014 *Multi-Sensor Integration for Indoor 3D Reconstruction* Thesis submitted for the Degree of Philosophy Geomatic Engineering, The University of Calgary, Calgary, Alberta

[4] Abbas M A Lichti D D Chong A K, Setan H Majid Z Luh L C Idris K M and Mohd Ariff M F 2017 *Improvements to the Accuracy of Prototype Ship Models Measured By Terrestrial Laser Scanner* Measurement 100 (2017): 301-310

[5] Gordon S J and Lichti D 2004 *Error Propogation in Directly Georeferenced Terrestrial Laser Scanner Point Clouds for Cultural Heritage Recording* FIG Working Weeks, Athens, Greece, 2004

[6] Rueger J M 1996 *Electronic Distance Measurement: An Introduction* Fourth Edition, Springer-Verlag Berlin Heidelberg, Germany

[7] Fraser C S 1996 *Network Design In Close Photogrammetry and Machine Vision* Edited by K. B. Atkinson, Whittles Publishing, Roseleigh House, Latheronwheel, Scotland, United Kingdom

[8] Abbas M A Setan H Majid Z Chong A K Idris K M and Aspuri A 2012 *Calibration and Accuracy*

Assessment of Leica ScanStation C10 Terrestrial Laser Scanner Development in Multidimensional Spatial Data Models, Springer Lecture Notes in Geoinformation and Cartography (LNG&C), March 2013

[9] Rabbani T 2006 *Automatic Reconstruction of Industrial Installations Using Point Clouds and Images* A thesis for the degree of Doctor of Philosophy at TU Delft

[10] Wolf P R Dewitt B A and Wilkinson B E 2014 *Elements of Photogrammetry with Applications Applications in GIS* Fourth Edition, McGraw-Hill Education: New York

[11] Ghilani C D 2010 *Adjustment Computations: Spatial Data Analysis* Fifth Edition, Copyright © 2010 by John Wiley & Sons, Inc., Hoboken, New Jersey

[12] Lemeš S and Zaimović-Uzunović N 2009 *Study Of Ambient Light Influence On Laser 3D Scanning* Seventh International Conference on Industrial Tools and Material Processing Technologies, Slovenia, Ljubljana, 04-07 October 2009

IOP Conf. Series: Earth and Environmental Science **385** (2019) 012044 doi:10.1088/1755-1315/385/1/012044

A Study on Current Dynamics in Adjudication Implementation in Malaysian Construction through Law Cases Analysis

Ahmad Arzlee Hassan[1], Abdul 'Izz Mohammad Kamil[1] and Ruth Lua Ejau[1]
[1]Faculty of Architecture, Planning & Surveying, 40450 Universiti Teknologi MARA, Shah Alam, Selangor MALAYSIA

Email: arzleehassan@gmail.com

Abstract. Adjudication is currently in its green age, where many lessons and experience are yet to be learnt. The implementation since its first introduction in 2012 has shown a progressive dynamic as it is widely applied, with many claims resolved, despite numerous being referred to arbitration and litigation. Challenges made against the adjudication decision under various bases and the court only allows for limited grounds for challenges. The paper aims to establish the valid approaches to challenge adjudication decision within the Construction Industry Payment and Adjudication (CIPA) Act 2012 implementation. The objectives include underlining the common problems occurring within the implementation through the behaviour of adjudication practice and studying the basis for challenges in adjudication decision. Law case analysis was carried out using thematic analysis by extracting court judgements from several law cases and categorizing them within a particular theme. Once the data with similar theme was filtered, deductive reasoning was used to extract the conclusion. Limited grounds for admissibility of challenges included a claim against adjudicator acting in excess of his given jurisdiction and/or adjudicator who is in breach of natural justice. Limited ground for challenges are important to ensure the relevance of the adjudication institution while maintaining the role of the judiciary in conserving the fairness and justice within its implementation.

1. Introduction

Adjudication implementation in Malaysia through the introduction of Construction Industry and Adjudication Act (CIPA Act) 2012 is currently still at its green age, where lessons are still being learnt and there are still much to gain in terms of experience regarding its execution. It is opined to be the solution for many payment problems happening in the industry. Azam and Beh is in the opinion that speedy and cheap recovery of payment is pinnacle [1][2].

Hence, an authorized enforcement channel such as adjudication is introduced. This supports the claim that Adjudication through CIPA Act 2012 is what the industry actually needs to resolve pending payment issues.The early years of its implementation have not revealed much of the subsequent problems. However, the recent developments prominently exhibit through the judgement of *View Esteem V Bina Puri Holdings [2018] 1 CIDB - CLR 38* and other concurrent law cases have finally exposed the problems. It is detected that the same repetitive theme emerges through challenges against adjudicators' decision in legislation.

Adjudication is becoming more and more prevalent these days. The awareness towards CIPA Acts 2012 and the rights the construction players have under the Act leads to numerous claims being referred to adjudication under the Act. As it is important to shine lights on the proper procedural rules, it is as pertinent to acknowledge the rights of parties within the quasi-judicial adjudication proceedings and subsequent judicial proceeding, which comes within the rules. To avoid misleading claims, themes for challenges in the decision of adjudicators must be established.

1.1 Aim and Objectives

The study aims to establish the valid approaches to challenge adjudication decision within the CIPA Act 2012 implementation. Few objectives are highlighted in order to achieve this aim, i.e.;

a) To underline the common problems occurring within the implementation through the behaviour of adjudication practice.
b) To study the basis for challenges in adjudication decision.

1.2 Problem Statement

As reported in the Malaysian Reserve [3], the then Director of Asian International Arbitration Centre (AIAC) Datuk Professor Sundra Rajoo shared that AIAC records revealed a steady rise in the number of adjudication cases led year-on-year, with over 700 cases recorded in 2017 of up to RM1.38 billion worth of claims resolved through adjudication. The rise signified awareness towards the CIPA Act and the rights construction player has under them. With several drawbacks and comments as being fast, rough-and ready justice, temporarily binding and merely interim in nature adjudication is open to the vulnerability for challenges [4][5]. Seah ,observed that 54.35% of adjudication decisions are re-arbitrated or re-litigated, with almost half of the adjudication decisions were referred to the High Court for setting aside orders and some numbers were contributed by the question of jurisdictional issues [6]. Although the Act allows for such challenges to be made especially if there are; elements of fraud or bribery, when there are denial of natural justice, where adjudicator has not acted independently, or when he has acted in excess of his jurisdiction [4], which will subsequently set aside the adjudicators decision, some claims were dismissed for not being valid for challenges. To avoid inconsequential claim, which only cause parties to spend money for futile claims in court, a study to understand the ground for challenges against adjudication decision is highly pertinent.

2. CIPA Act 2012: The story so far

2.1 The Purpose of CIPA Act 2012 and Implementation

CIPAA 2012 and its statutory adjudication rules addresses effective, speedy and cheap dispute resolution to ensure better cash flow [7], through sensible payment regime provided by a speedy dispute resolution, i.e. statutory adjudication. Recovery for payment mechanism [6] is guaranteed for aggrieved claimant. Through its implementation over these past few years, adjudication managed to successfully secure parties' payment for work done, eventually compels mandatory payment of an amount due to the contractor [8][9] and also ensures fairer risk allocations between the employers and the contractors. In addition, the confidential nature is attractive to parties who wish to resolve construction disputes quickly and quietly [4][5][6][7][8][9].

The loopholes and questions were only raised and detected after years of implementation. The occurrences of parties unhappy with the decision of adjudicators were as expected. Hence the allowance for challenges to be made in arbitration and litigation. However, it was a concern that court and arbitration might be treated as complain counters for unsatisfied aggrieved parties. It is clearly stated in the Act that only under certain limited grounds allow for challenges to be made. Some rules were vague in their definitions and court clarification was deemed required.

3. Methodology

Legal research is unpopular in the construction industry due to its nature being unfamiliar to the technical drawn industry. Maidin and Sulaiman and Quapp and Holshemacher agreed that although the construction (namely surveyors and also civil and structural engineers, in their research) have little interest in legal researches and studies due to the terms, language, nature, the theoretical content and basically not prepared for the legal profession, the importance of legal research in construction cannot be dismissed as it is prominent to understand the contract-centric environment[10][11]. The knowledge may also circumvent parties from engaging into expensive legal lawsuit and issues. Hence, this research encompasses comprehensive legal research framework structured to give a better view from a construction perspective.

Due to its applied legal research nature, doctrinal legal research was used for this research [12]. Legal doctrinal research examines principles and law obtained from known principles and established case law. The black letter research was used to analyse impact of a particular law on a community, in this research scope; the construction community. Cases related to adjudication were gathered and analysed to demonstrate the operation and efficacy of a particular law [13]. Therefore, all relevant cases enlisted are defined, categorized and analysed.

3.1 Law Case Analysis

Law cases relating to adjudication and implementation under CIPA Act 2012 were collected. These primary data was gathered from the CIDB Construction Law Report 2015 to 2017 [14][15][16]. Cases filtered and enlisted are as follows:

a) ACFM Engineering & Construction Sdn Bhd V Esstar Vision Sdn Bhd [2017] 1 CIDB-CLR 205
b) Bina Puri Construction Sdn Bhd V Hing Nyit Enterprise Sdn Bhd [2016] 1 CIDB-CLR 180-182
c) WRP Asia Pecific Sdn Bhd v NS Bluescope Lysagth Malaysia Sdn Bhd & Other Case [2016] 1 CIDB-CLR 140
d) Innovative Engineering(M) Sdn Bhd V Nomad Engineering Sdn Bhd [2017] 1 CIDB-CLR 254
e) Naim Engineering Sdn Bhd & Anor v Pembinaan Kuantiti Sdn Bhd [2017] 1 CIDB-CLR 284
f) Ranchan Heavy Engineering Sdn Bhd V Pelabuhan Tanjung Pelepas Sdn Bhd [2017] 1 CIDB-CLR 284
g) Rimbunan Raya Sdn Bhd V Wong Brithers Buidling Construction [2017] 1 CIDB-CLR 151
h) View Esteem Sdn Bhd V Bina Puri Holdings Sdn Bhd [2018] 1 CIDB-CLR 38

Most of the above cases were tried in the High Court and Court of Appeal, with the exception of View Esteem V Bina Puri [2018], which was appealed in the Court of Appeal (reported CIDB-CLR 2016) and finally decided in the Federal Court of Malaysia (reported CIDB-CLR 2018). This turn of event has promoted the case to be the benchmark adjudication case.

3.2 Thematic Analysis

Thematic analysis was the qualitative analysis approach applied in this research. The qualitative analysis software Atlas.t.i.8 was used to process the data. The themes were predetermined through a pattern revealed from the frequency of used terms extracted from the cases. Judgements from each of the cases were then individually dissected, labelled according to themes or codes, and linked to produce a diagram of network, which will be encrypted in the next stage of the research. A report was also produced enlisting judgement according to the theme preconceived at the earlier process. With the data thematically extracted and structured, deductive reasoning can be easily conducted.

3.3 Deductive reasoning

The structured data were analysed using deductive reasoning. The process includes defining the principle, identifying a factual event and lastly, making a relation between the two to determine whether the principle applies or otherwise [12]. The research mainly focused on the black letter law, thus the implementation and their effects through judgements were taken into consideration. Concluding the observation involves deducing the facts from the judgement into several key focuses before arriving to a framework, which will eventually summarize the outcome. The research design is as follows:

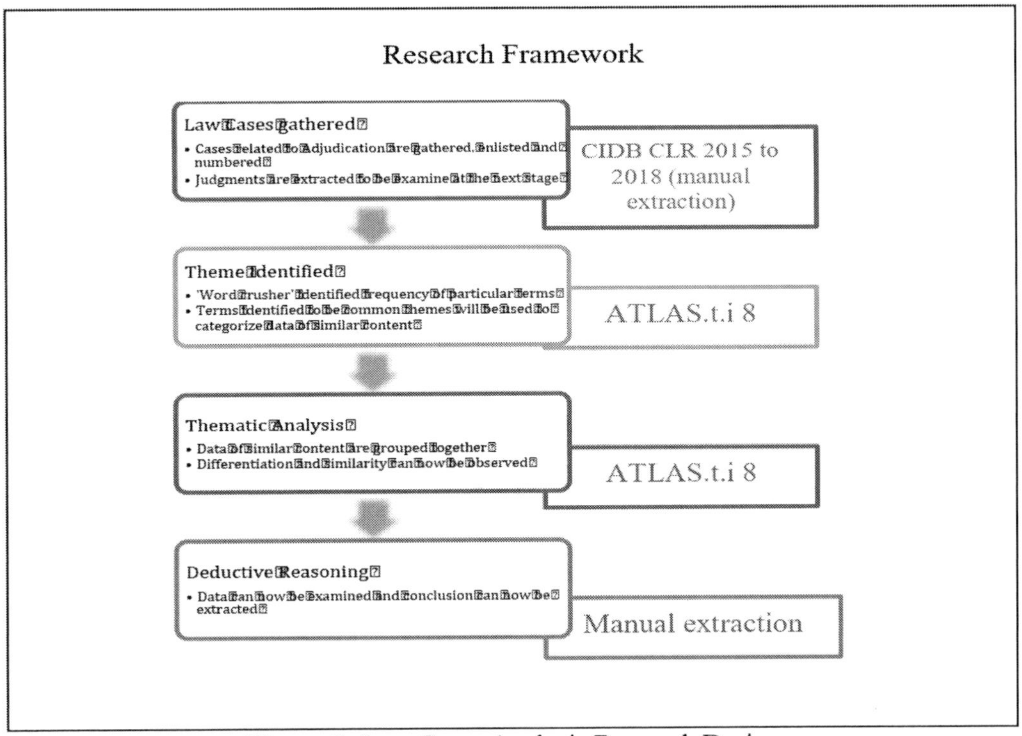

Figure 1. Law Cases Analysis Research Design

3.4 Limitations

The law cases examined were limited to the ones captured in the CIDB Construction Law Report from 2015 to 2018. The timeline for the research were also not inclusive of cases being currently on trial or not being published in any law report yet. In addition, as adjudication being strictly private and confidential, adjudication claims not referred to litigation and resolved through either arbitration or other forms of quasi-judicial means were excluded from this analysis. However, arbitration awards and other decision from other alternative dispute resolution (ADR) were arbitrary and can be quite trivial to the outcome affecting adjudication provisions as only the court judgements were seen as precedence and impactful to the implementation of any particular rules.

4. Data Analysis

4.1 The Outcome of an Adjudicator's Decision

According to section 28 of the Act, a party satisfied with the decision of an adjudicator may have them enforced in High Court. Under section 26 (7), section 31 and section 15 also recognized that the court might also grant a stay of adjudication decision, if the decision is challenged in arbitration or the court through certain procedural rules. Hence dictate the permissibility of any challenges to an adjudication decision. This also signifies the non-finality of an adjudication jurisdiction. However, challenges to adjudication decision must adhere to the limited grounds admissible by the court and Act under section 15, which includes; the possible influenced by bribery or fraud, breach of natural justice, biased and partiality in the conduct and judgement, and lastly in such situation the adjudicator has exceeded his jurisdiction. It is observed from the law cases that the common themes used to challenge the decision are as follows:

a) The adjudicator has committed a breach of natural justice by being unfair and partial to either party
b) The adjudicator has exceeded his jurisdiction allowed in the claims and as sanctioned by the Act
c) Other reasons (such as non-existence of construction contract and challenge against the competency of the adjudicator)

4.2 The Grounds to Challenge the Adjudicator's Decision

The data gathered from the judgements in each of the law cases were encoded and the following diagram was produced:

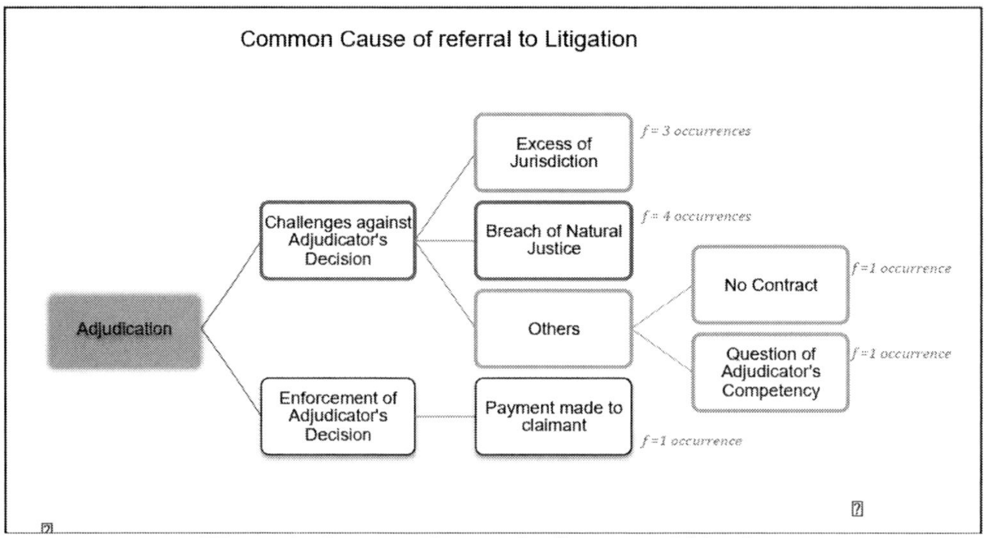

Figure 2. Common Cause of reference to Litigation

It can be concluded that only the said challenges are permissible to set aside adjudication decisions and allowed for challenges in the court. Other challenges do not have enough ground for challenges and therefore were dismissed. The following Table 1 shows the main events relating to such admission;

Table 1. Valid Ground for Challenges and their Relevant Events

Grounds	Relevant Events
Excess of jurisdiction	• Adjudicator tried events not relating to payment disputes • Adjudicator addressed and decided on issues relating to other than disputed claims
Breach of natural justice	• Adjudicator not giving enough space for either party to put forward their argument • Adjudicator communicated to a party without the knowledge of the other party • Being bias and partial
Others	• No known written contract between the parties • Adjudicator's competency in question • Bribery, fraud and other misconduct

5. Conclusion

The implementation of statutory adjudication since its introduction in 2012 has seen moving to a particular dynamic. They are used widely and the industry's maturity on statutory adjudication is improving. Although it is well accepted, there were some drawbacks, which require support from a higher legal institution, i.e. the judiciary. Further exploration on the rules and their effect could only be made through judiciary.

Lastly, as challenges are allowed, there are some checklists claimants need to adhere to before forwarding the challenges to courts to avoid treating the court like some complain counters and at the same time undermining the authority and integrity of the adjudication institution.

References

[1] Azman, M. N. A., Dzulkalnine, N., & Abd, Z. (2013). Payment Scenario In The Malaysian Construction Industry Prior To CIPAA. *Law and Dispute Resolution*, 56

[2] Beh, L. C. (2017). Construction Industry And Payment Adjudication Act (CIPAA 2012) - Messrs Ravindran. Retrieved September 27, 2018 from https://ravindran.com.my/construction-industry-and-payment-adjudication-act-cipaa-2012/

[3] Malaysian Reserve (2018). Construction Sector Sees A Higher Number Of Dispute Cases. Retrieved September 27, 2018 from https://themalaysianreserve.com/2018/04/25/construction-sector-sees-a-higher-number-of dispute-cases/

[4] Mah, R. (2016). Construction Adjudication in Malaysia - Faster and Cheaper Dispute Resolution -. Retrieved September 27, 2018 from http://www.mahwengkwai.com/construction-adjudication-in-malaysia-faster-and-cheaper

[5] Paulraj, J.J (2016.). Adjudication Decisions Pursuant To The Construction Industry Payment And Adjudication Act 2012 (CIPAA 2012); Open To Abuse? *TKP Law*, Retrieved September 27, 2018, from http://www.tkplaw.co/articles_7.html

[6] Seah, K. (2018). CIPAA: Adjudication Leading the Way? Retrieved from https://themalaysianlawyer.com/2018/09/05/cipaa-adjudication-leading-the-way/

[7] Nasir, N. I. I. M., Ismail, Z., & Ng, N. K. M. F. (2018). Comparative Analysis On Construction Adjudication Systems Towards Effective Implementation Of Statutory Adjudication In Malaysia. *Malaysian Journal of Civil Engineering*, *30*(2)

[8] Raji, B. A., Ali Mohamed, A. A., & Mohammed, N. S. (2017). Statutory Adjudication: A Global Trend For Resolution Of Payment Problems In Construction Industry. In *International Conference on Dispute Resolution* (p. 418). IIUM

[9] Choon, B. L. (2017). Construction Industry And Payment Adjudication Act (CIPAA 2012) - Messrs Ravindran. Retrieved 27 September 2018 from https://ravindran.com.my/construction-industry-and-payment-adjudication-act-cipaa-2012/

[10] Maidin, A. J., & Sulaiman, S. S. (2011). Importance of legal education for quantity surveying professionals: A proposal for developing a legal studies module for Malaysian system. *Journal of Applied Sciences Research*, *7*(SI), 2249-2256.

[11] Quapp, Ulrike & Holschemacher, Klaus. (2014). The Importance of Construction Law Teaching in Civil and Structural Engineering Education, Conference Paper, Penza, Russia : Retrieved from https://www.researchgate.net/publication/280931761_The_Importance_of_Construction_La w_Teaching_in_Civil_and_Structural_Engineering_Education

[12] Chynoweth, P. (2008), *Legal Research* in Knight, A. & Ruddock, L (2008) ed.,*Advance Research Method in the Building Environment*, Oxford; Wiley & Blackwell Corbin, J. & Strauss, A. (2008) *Basic of Qualitative Research* 3rd Edition, California: Sage Publicatio

[13] McConville, M & Wing, HC (2007), *Research Method for Law*: Edinburgh University Pres

[14] Construction Industry Development Board. (2015). *CIDB Construction Law Report* (002344785-W).Kuala Lumpur, Malaysia: ARK Knowledge Solution.

IOP Conf. Series: Earth and Environmental Science **385** (2019) 012044 doi:10.1088/1755-1315/385/1/012044

[15] Construction Industry Development Board. (2016). *CIDB Construction Law Report*. Kuala Lumpur, Malaysia: ARK Knowledge Solution
[16] Construction Industry Payment and Adjudication Act (2012), *Law of Malaysia Act 746,* Percetakan Nasional Malaysia Berhad

IOP Conf. Series: Earth and Environmental Science **385** (2019) 012045 doi:10.1088/1755-1315/385/1/012045

Analyse the Barriers that Influencing Green Design Implementation among the Contractors in Construction Industry

Heng Yong Ming[1], Zul Zakiyuddin Ahmad Rashid[2]*, Hamimah Adnan[2]
[1]Universiti Sains Malaysia, Penang, Malaysia,
[2]Faculty of Architecture, Planning and Surveying, Universiti Teknologi MARA, Shah Alam, Selangor.

Email: zakiphd@gmail.com

Abstract. This paper aims to find out the vital barriers that affecting the implementation of Green Design practice, which is part of the Green Supply Chain Management (GSCM) application in construction industry. First of all, the objectives of this study are to determine the barriers that preventing Green Design related activities among contractors and also analyse the critical barriers for Green Design application in construction industry. Quantitative research method with survey questionnaire is employed in this study. Around 450 sets of questionnaire are distributed with 21.8% response rate. The independent variables in this study are the barriers of Green Design implementation in construction industry while the dependent variable is the level of adoption of Green Design practice in construction industry. In short, four (4) barriers are identified, which are Government Supports, Company Resources, Knowledge and Information, and Financial issue barriers. Results of the study show that Government Supports and Company Resources barriers are significant for Green Design practice. Significance of this research is to provide a better understanding of green practices, such as GSCM, to deal with current environmental issues and to realise the recent problems and obstacles faced by all construction players, so that further actions are required for a successful GSCM implementation in order to move towards a sustainable environment in the future.

1. Introduction

Construction industry is very significant in socio-economic development of every countries around the world. It plays an active role in contributing in Gross Domestic Product (GDP) and also providing job opportunities specifically in developing countries. Around 3% to 5% of national GDP of Malaysia is steadily contributed by the construction industry since 2000s [1]. Despite of the contribution of the construction industry, it has caused destructive effects to the environment. Those harms are attributed from massive material waste generated by the industry, resources exhaustion such as depletion of water, and etcetera [2].

Green Supply Chain Management (GSCM) is found to be an integrated supply chain-wide management approach with environmental management which is likely to mitigate the environmental damages from the firms and achieve the operational performance concurrently [3, 4 & 5]. After revising some previous studies about the definition of GSCM, it can be defined as an innovative strategy/concept that integrates environmental and social considerations with the involvement of all parties in product (building) design stage, procurement, materials sourcing and selection, completion and handover to the ultimate users without overlooking the end-of-life management of the product in order to improve both short and long-term competitiveness and profitability of the organization.

The Malaysian government has been promoting sustainability initiatives since year 2000 [6]. Regrettably, not more than 2% of qualified construction projects are evaluated with sustainable building rating systems like Green Building Index (GBI) in Malaysia according to CIDB statistics 2014-2015 [7]. For those assessed projects, not more than 50% have been rated with satisfied GBI.

In a nutshell, GSCM application in construction is still at infancy stage and there is a need to fill up the knowledge gap in construction field [8]. Thus, the main objective of this study is to find out the significant barriers that influencing the adoption of Green Design practice among contractors in the Malaysian construction industry.

2. Background of Research

The Green Design approach is discussed as well as the barriers of green initiatives adoption are covered in this section.

2.1 Green Design Approach

Green design is a strategy to encourage a company's environmental performance by practising internal inter-functional cooperation all the way through the whole company while having external collaboration with other partners such as suppliers along the supply chain [9]. Green Design process can be categorized into three (3) key stages as follows: impacts assessment stage, design strategies formulation stage and design implementation stage [10]. Lifecycle Impact Tool (LIT) is useful to provide a structured means to discover the product or service's impacts at each stage in its lifecycle.

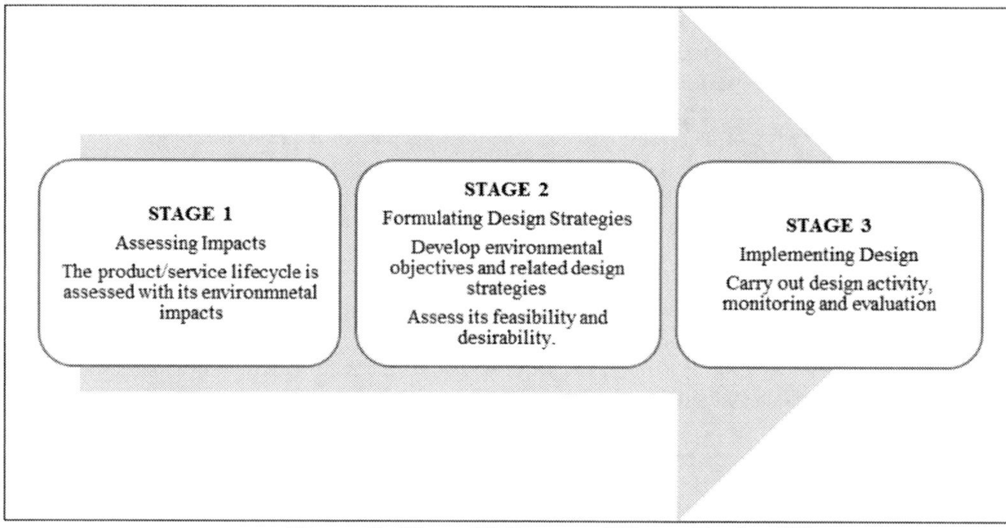

Figure 1. Three (3) Stages of Green Design [9]

2.2 Barriers of Implementation

There are four (4) barriers in this study, which are government supports barrier, company resources issue barrier, knowledge and information barrier and lastly, financial issues barrier.

2.2.1 Government Supports Barrier

Government influential policies play a vital role in promoting green initiatives [11]. Regulations set by the government is the major enabler in implementing GSCM. Yet it is also one of the barriers for the sustainable supply chain application. Occasionally, the stipulated techniques and requirements from the environmental regulations might inhibit the innovation and creativity [12]. But if there is no government initiatives system for green supply chain practitioners, which means that special paybacks are not given to the GSCM practitioners from the environmental friendly policies [13].

Porter and van der Linde (1995) illustrated that strict environmental regulations is the primary driver for sustainable supply chain management which can encourage the companies to become more resourceful and effective [12]. Friedman (1992) finalized that environmental management was integrated to become a foremost attention within the organisations after 1970s due to the financially significant regulation [14].

2.2.2 Company Resources Issue Barrier

In addition, GSCM's multiple complexities and uncertainties has caused the firms are less favourable to undertake GSCM practices in their operations. The complexity and uncertainty of GSCM may be due to the inter-organisational and cross-functional integration of environmental, production, engineering,

IOP Conf. Series: Earth and Environmental Science **385** (2019) 012045 doi:10.1088/1755-1315/385/1/012045

marketing, and logistics personnel [9]. Lack of experienced and well-trained green experts in construction industry is the hindrance of the adoption of GSCM [14]. The examples of experts are green architects, contractors, professionals and developers. Sufficient green experts are required in order to well implement green supply chain [15]. The firm is likely to decrease the expenditures, increase competences and designate social and environmental responsibility in a capable way with the existence of skilled and experienced green experts.

Dashore and Sohani (2013) think that the incorporation of IT system into the green supply chain approach is an essential requirement for this new concept to be implemented properly [13]. There are a lot of computer based applications, programmes, IT enabled procedures and software which are able to enhance the GSCM practices with advanced data analysis method and information interchange process [13].

2.2.3 Knowledge and Information Barrier

Furthermore, certain level of knowledge is required to perform green practices, however, the lack of professional knowledge and information among managerial personnel, staffs and suppliers resulted that GSCM practices cannot be implemented successfully and this becomes the biggest barrier in GSCM adoption [16]. Besides, insufficient information causes the importance of GSCM hard to be delivered to the employees and causes them do not have adequate awareness about the benefits of GSCM.

Lack of information resources or expertise in dealing with environmental matters and the size of the organisation are the major factors for GSCM implementation [17]. Thus, larger firms are more eager to be involved in and practise green supply chain initiative when compared to smaller firms.

2.2.4 Financial Issues Barrier

Ambec and Lanoie (2008) identified the most critical barriers to implement environmental practices are the economic issues and costs related matters [18]. Price is an important element that consumers always consider for. Most of the time, the consumers used to ask for lower prices and thus actions are required to lower the cost incurred in order to offer lower price deals. Unfortunately, numerous studies have discovered that integration of sustainability practices into the supply chain is more expensive and it requires large amount of money particularly for SMEs [14]. Study also found that SMEs have limited available resources and therefore incurring costs are particularly substantial for them with vulnerable inventories [19]. More than half of the SMEs think that the main difficulty for them to practise green supply chain practices is the high costs to adopt such practices [20].

Hence, most of the SMEs do not implement GSCM as they face scarcity of information about the environmental benefits when greening their products, weak environmental legislation as well as requirements and also, they do not recognize their social responsibility in greening the environment [21]. Wycherley (1999) stated that implementation of GSCM will change the current investment, information system and habits which are too expensive and overpriced [22].

3. Methodology

Quantitative research design approach is applied in this research with closed-ended questionnaire survey method. With quantitative strategy, a relative large number of data can be collected in shorter time if compared with qualitative approach. Moreover, quantitative approach can be engaged to observe and measure information as well as employ statistical procedures.

The construction companies that hold a Grade 7 (G7) certificate of registration issued by the Construction Industry Development Board ("CIDB") and located in Penang, Wilayah Persekutuan and Johor are the population of the study. The reason that G7 class is selected is because the tendering sum exceed 10 million or even more in the construction projects, the projects involve a lot of parties such as sub-contractors and sub-suppliers either locally or internationally and advanced construction technologies and practices can also be discovered.

Two (2) types of variable are covered in this study, which are independent variable and dependent variable. The independent variables in this study are the barriers of Green Design adoption, while the

IOP Conf. Series: Earth and Environmental Science **385** (2019) 012045 doi:10.1088/1755-1315/385/1/012045

dependent variable is the level of adoption of Green Design in the construction industry. The variables and items of the questionnaire are shown as Table 1.

Table 1. Variables and Items of Questionnaire Survey

Variable	Description of Items
Green Design	• Design of building or structure for reduced consumption of material • Design of building or structure for reduced consumption of energy • Design of building or structure for reuse, recycle, recovery of material
Government Supports Barrier	• Lack of government support to adopt environmental friendly policies. • Lack of government incentives for practitioners in greening the supply chain • Inadequate enforcement of environmental regulations by government officers • Lack of government incentives and tax reduction for green practitioners
Company Resources Issue Barrier	• Lack of new technology, materials and processes • Lack of human resources (eg. green architects, consultants, green developers and contractors) • Lack of Information Technology (IT) system such as environmental monitoring system • Lack of technical expertise • Current practices lack of flexibility to switch over to new system
Knowledge and Information Barrier	• Lack of professionals exposed to green system • Lack of environmental knowledge • Perception of "out-of- responsibility" zone • Lack of awareness about reverse logistics adoption • Disbelief about environmental benefits
Financial Issues Barrier	• High cost of hazardous waste disposal • Non-availability of bank loans to encourage green products/processes • High investment and less Return- On-Investment (ROI) • Financial constraints

4. Data Analysis and Results: Quantitative Study

450 questionnaires were distributed through mail and email in December 2016. Follow up process was made by phone calls and reminder emails. A total of 98 responses were received with different level of completeness after 3 months. 92 useable responses are used and loaded into SPSS software. Although 21.8% (98 out of 450) response rate is quite low, however it is not uncommon and tolerable as Dulami, Ling and Bajracharya (2003) stated that the normal response rate in the construction industry is in the range of 20-30% for postal questionnaire [23].

The data is examined with multiple linear regression analysis with IBM SPSS Statistics software, version 22.0. The results of the analysis is displayed as Table 2 with the Beta coefficients value (β) of four (4) adoption drivers as independent variables and Green Design practice as dependent variable.

Table 2. Results of Coefficients for Adoption Barriers and Green Design

Model		Standardized Coefficients (Beta)	t	Sig.
1	(Constant)		3.745	0.000
	Government Supports	0.225	1.956	0.054*
	Company Resources Issue	-0.236	-1.739	0.086*

Knowledge and Information	-0.186	-1.585	0.117
Financial Issues	0.098	0.778	0.439

* significant at 0.10 **significant at 0.05 ***significant at 0.01

Table 2 depicts the results of the coefficients for Green Design practice and its barriers of adoption. When linking the independent variables (adoption barriers) with the dependent variable (Green Design), the Government Supports barrier (β=0.225, p=0.054) and Company Resources barrier (β=-0.236, p=0.086) indicate substantial characteristics to the Green Design variable. This shows that both barriers are the influencing variables among all four barriers, which exert positive and negative impacts onto the Green Design respectively as the other barriers do not prove any statistical significance with the Green Design variable. Even though Knowledge and Information barrier has a β-value of -0.186 and is categorised as third influence barrier with Green Design practice with negative impact, but it does not show any significance as the p value is more than 0.10. While Financial Issues barrier has low β-value of 0.098 and shows no significant relationship with Green Design.

5. Discussion

This study found that Government Supports barrier (β=0.225) and Company Resources barrier (β=-0.236) are two barriers that influence the adoption of Green Design practice.

Green Design practice is positively affected by Government Supports barrier (β=0.225). In other words, the implementation of Green Design practice can be stimulated by encouraging the Government Supports barrier. Nevertheless, this does not seem sensible as barrier is a kind of hindrance that prevents or limits some actions or some activities to continue developing and it is supposed to limit the implementation of Green Design practice. Nevertheless, this can be explained in alternative way, which shows that supports from government can encourage the implementation of Green Design practice as government agencies are the influential groups that may affect the movements of an organisation. Therefore, encouragement could be made from government agencies in stimulating green initiatives.

Secondly, negative impact of Company Resources barrier (β=-0.236) indicates that increase of Company Resources barrier decreases the implementation of Green Design practice. Modern technology, materials, processes and human resources, suchlike architects, contractors and consultants are the examples of company resources. Lacking of human resources such as green consultants and professionals might possibly reduce the interest of the clients to implement the green practices as there may have uncertainties in completing the project successfully. Insufficiency of all these resources might result a failure of green design implementation. Besides that, Information Technology (IT) system such as environmental monitoring system and technical expertise might also make the Green Design practice fail to be performed. In Malaysia, construction players are less likely to take initiatives in switching to contemporary construction methods as this may affect the inventories of the company and thus causes inadequacy of new technologies and machineries to support the implementation of Green Design practice. In short, Company Resources barrier is the major barrier to the adoption of GSCM initiatives. This finding matches with the study from Wooi and Zailani (2010) [24].

6. Conclusion and Recommendation

This study provides a platform for construction players to realise the key barriers that slow down the popularity of green initiatives in construction industry and eventually let the players find out pertinent ways to overcome these barriers. More importantly, the effects of unsustainable construction activities can be reduced with green initiatives while the benefits could significantly be reducing greenhouse gases emission, minimising consumption of natural resources and energy, cutting down the waste disposal from construction sites and so on. With this, we can only ensure the preservation of natural resources for our future generations. Recommendation for future studies is that more barriers could be identified and analysed which are related to the adoption of GSCM application. Additionally, studies can be conducted on different GSCM practices such as Green Purchasing and Investment Recovery approaches with same barrier variables.

Acknowledgements
This research is financially supported by the Fundamental Research Grant Scheme (FRGS) for grant title "Developing Policy Framework for Successful Green Supply Chain Management (GSCM) Application in Malaysian Public Infrastructure Projects" (PPBGN 6711556).

References

[1] Johnston P Everard M Santillo D and Robèrt K 2007 Reclaiming the definition of sustainability. *Environ Sci Pollut Res Int*, **14**(1), 60–66.
[2] Aigbavboa Ohiomah and Zwane 2017 Sustainable construction practices: "a lazy view" of construction professionals in the South Africa construction industry. *Energy Procedia* **105** pp 3003–10
[3] Zhu Q Sarkis J Cordeiro J and Lai K 2008 Firm level correlates of emergent green supply chain management practices in the Chinese context. Omega – *Int J Manage S.*, **36**(4) pp 577–91.
[4] Zhu Q Sarkis J and Lai K 2012 Environmental management innovation diffusion and its relationships to organizational improvement: an ecological modernization perspective. J. *Eng & Tech Manage* **29**(1) pp 168–85
[5] Svensson G 2007 Aspects of sustainable supply chain management (SSCM): conceptual framework and empirical example. *SCM–An Int J.* **12**(4) pp 262–66
[6] Suliman L and Omran A 2009 Sustainable development and construction industry in Malaysia. Manager Journal: *University of Bucharest, Faculty of Business & Administration.*
[7] CIDB 2015 Construction industry transformation programme Retrieved 9th May 2018 from www.cidb.gov.my/cidbv4/index.php?option=com_content&view=article&layou t=edit&id=1191&lang=en [Online]
[8] Khan M Hussain M and Ajmal M 2016 Green supply chain management for sustainable business practice. *IGI Global*
[9] Zhu Q and Sarkis J 2006 An inter-sectoral comparison of green supply chain management in China: Drivers and practices. *J.Cleaner Production* **14**(5) pp 472–86
[10] Charter M Keiller S and Clements V 2013 A guide for SMEs on eco-design for the Construction industry. *Southampton, UK: EDECON*
[11] Vachon S and Klassen R 2006 Green project partnership in the supply chain: the case of the package printing industry. *J.Cleaner Production* **14**(6– 7) pp 661–71
[12] Porter M and van der Linde C 1995 Green and competitive: ending the stalemate. *Harvard Business* Review **73**(5) pp 120–34
[13] Dashore K and Sohani N 2013 Green supply chain management - barriers & drivers: a review. *Int J. Eng* 2 pp 2021–30
[14] Alzawawi M 2013 Drivers and obstacles for creating sustainable supply chain management and operations
[15] Holt D and Ghobadian A 2009 An empirical study of green supply chain management practices amongst UK manufacturers. *J. M Tech Manage* **20**(7) pp 933–56
[16] Al Zaabi S Al Dhaheri N and Diabat A 2013 Analysis of interaction between the barriers for the implementation of sustainable supply chain management. *Int J Adv Manuf Tech* **68**(1) pp 895–905
[17] Lee S 2008 Drivers for the participation of small and medium-sized suppliers in green supply chain initiatives. *SCM–An Int J.* **13**(3) pp 185–98
[18] Ambec S and Lanoie P 2008 Does it pay to be green? A systematic overview. *Academy of Management Perspective*, **22**(4) pp 45–62
[19] Hervani A Helms M and Sarkis J 2005 Performance measurement for green supply chain management. *Benchmarking: An Int J.* **12**(4) pp 330–53
[20] Revell A Stokes D and Chen H 2010 Small businesses and the environment: turning over a new leaf. *J Business Strategy and the Env* **19**(5) pp 273–88
[21] Perron G 2005 Barriers to environmental performance improvements in Canadian SMEs.

Dalhousie University, Canada
[22] Wycherley I 1999 Greening the supply chain: the case of the body shop international. *Business Strategy and the Environment* **8**(2) pp 120–27
[23] Dulami M Ling F and Bajracharya 2003 Organisational motivation and inter-organisational interaction in construction innovation in Singapore *Construct Manag Econ* **21**(3) pp 307–18
[24] Wooi G and Zailani S 2010 Green supply chain initiatives: Investigation on the barriers in the context of SMEs in Malaysia. *Int Business Management* **4**(1) pp20–27

A BIM ORIENTED MODEL TO A 3D INDOOR GIS FOR SPACE MANAGEMENT– A REQUIREMENT ANALYSIS

S N A Syed Mustorpha[1] and W M N Wan Mohd[1]
[1]Centre of Studies for Surveying Science and Geomatics, Faculty of Architecture, Planning, and Surveying, Universiti Teknologi MARA, 40450 Shah Alam, Selangor, Malaysia

Email :wmn@salam.uitm.edu.my

Abstract. Building Information Modeling (BIM) is a process concerning the formation of digital representations of the physical and functional features of built spaces. A significant increase of the BIM utilization in the Architecture, Engineering, and Construction (AEC) sector are becoming more accessible and functional. The BIM methodology was developed in order to provide an effective system for building management as well as to organize the collaboration between different parties. Industry Foundation Classes (IFC) is maintained by buildingSMART; enable an interoperable format for exchanging BIM data between different software platform which is capable of restoring both geometric information and rich semantic information of building components to support lifecycle data sharing. BIM provides a detailed 3D geometrical model with rich semantic data. In contrast, Geography Information Science (GIS) offers powerful spatial analytical tools. As the benefits brought by the integration of Building Information Modelling (BIM) and Geographic Information Systems (GIS) are being proved by many researchers, further spatial analysis can be applied that can benefit indoor-outdoor integrated applications. Recent research has mainly discussed the process required for data conversion from BIM to GIS. Thus, more specifics requirements on the functionalities of a 3D indoor applications need to be considered. In this study, the use of BIM IFC for 3D indoor GIS was reviewed, especially on the data conversation, data management, and 3D spatial analysis. The aim of this paper is to reveal the 3D indoor GIS requirements for space management. Therefore, the requirements were analyzed in four aspects: 1) data level integration; 2) data management; 3) 3D indoor GIS analysis; and 4) 3D space management. 3D space management was developed according to the requirement. Future research will be testing the conversion data in a GIS environment, the data structure for the 3D indoor GIS will be designed and a prototype will be developed to demonstrate the spatial analysis functionalities for 3D space management.

1. Introduction

Building Information Modelling (BIM) is one of the major technological advances in the Architecture, Engineering, and Construction (AEC) domains by providing detailed 3D building models, as it is letting stakeholders capture, manipulate, update and exchange information during the lifecycle of a building construction project [1]. This platform serves in bridging the gap of interoperability among industry players thus, boosting its popularity between stakeholders [2]. Specifically, Industry Foundation Classes (IFC) is the primary open data scheme developed by buildingSMART is used as an interoperable format for exchanging BIM data between different software platform [3][4]. The need for sophisticated analysis tools arises because different stakeholders use BIM for diverse purposes. Therefore, some attempts were made to design tools and systems to integrate building models into a GIS environment [5][6].

GIS is a powerful tool that can visualize and analyses location for archival and management of geospatial data by integrating different spatial and attribute data as well as allowing for spatial analysis and modeling operations [7]. BIM IFC contains many specific geometrical and semantic information on building components and can be treated as an ideal source of indoor spatial information and interpreted in GIS field studies, such as indoor modeling and mapping [8]. In precise, BIM IFC offers a standard data model, which can be selectively used for various purposes on GIS platforms such as location-based services (LBS) and navigation location [9][10], asset management [11][12], cultural heritage management [13][14] or hazard simulations [15]. In a building management relation, facility managers should play their roles for the translation of building models into geospatial context.

BIM and GIS interpret 3D modeling from two points of view which are BIM stresses detailed building components and project information from the architectural and construction point of view, while GIS is specific on the geographical information of buildings and their components from a geographical point of view [16]. Although traditionally, 2D GIS suffice to analyze data over a large area, the requirement of handling the insides of building drives to the development of a more detailed 3D GIS. Since IFC is one of the standard building formats, it is one of the interoperable tools for exchanging information from BIM to 3D GIS, enabling more complex analysis, visualization and documentation. BIM 's most important contribution to GIS would be the provision of detailed 3D building models for both geometry and semantic information, but it does not contain the surrounding information. In contrast, GIS provides a concise way of managing data through a geodatabase and offers plenty of tools for spatial analysis [17].

There are many researchers investigating the requirement of translation from BIM to GIS. Majority of the current studies were more focused on buildings, although tunnels, bridges, and other constructions are also vital. Isikdag et al. had proposed related conceptual requirements on a building model [18]. Although the integration of the IFC-based information model was investigated using various approaches, there were practical problems in solving the integration problem. For example, during the exchange of information between heterogeneous systems, which means more than two different software or commercial modeling software using IFCs, some information loss or changes have been reported [19]. In addition, a related study identified several issues relating to the integration of BIM and GIS [20].

A major barrier to achieve the integration of BIM-GIS is the nuts and bolts process for BIM data integration, often in the form of a variety of the proprietary file, web, and database-based information stores with GIS. Frequently, information from BIM is reused by engineers and facility managers in GIS analytical and visualization tools. The first step of this type of integration is getting BIM data into ArcGIS. By using ArcGIS Data Interoperability Extension, individual BIM elements are converted to GIS features depends on the level of details (LOD) of the building models.

However, specific requirements for GIS applications with data from BIM IFC should be clarified, such as how data converted from BIM IFC can be used, such as how BIM IFC converted data can be used, the level of detail I am looking for and the functionality of spatial analysis. For this application, a generic 3D indoor GIS was considered to provide modeling and analytical functionality for 3D space management. Although this research is in its early stages, this paper seeks to disclose the requirements of 3D indoor GIS for space management. Generally, this procedure was designed to incorporate a 3D building model with 3D spatial analysis. Therefore, BIM IFC is able to support this procedure in the following aspects: 1) data level integration; 2) data management; 3) 3D indoor GIS analysis, and 4) 3D space management such as space utilization optimization. Currently, there is no generic solution that can address the above topics completely. In this paper, a requirement analysis was carried out focusing on the four aspects and potential application scenarios explained with an example. The following sections will examine the four aspects of requirement analysis. Subsection 2.1 introduces the data level integration of BIM GIS; subsection 2.2 discusses the requirements on data management; subsection 2.3 presents the need for 3D indoor GIS analysis, and subsection 2.4 provides the requirements on 3D space management. Section 3 explains two case studies of large-scale facilities management and section 4 concludes this paper with some future work.

2. Requirement Analysis

This section presents an analysis of requirements regarding the development of a 3D building space management by using BIM IFC data. These requirements cover four main topics: 1) data level integration; 2) data management; 3) building management analysis, and 4) 3D space management based on BIM IFC data.

2.1. Data level integration

Data level integration of BIM and GIS was divided into two paths; geometry and semantic level [21]. The geometry level is the transformation of information related to geometry. Semantic level concentrates on full attribute information transfer. There are three approaches for 3D geometry representation of an object, which are boundary representation (b-rep), Sweep Solid (SS) and Constrictive Solid Geometry (CSG) [22][23]. IFC file uses one of b-rep, SS and or CGS or takes their combination to represent a 3D geometry. For complex shapes in IFC files, such as rectangular shapes, CSG is an efficient way to represent them. For example, IfcSpace represents an area or volume bounded with certain functions within a building, where space is often associated with a building storey and can potentially be used to manage space utilization inside buildings. For modeling physical parts of buildings, "IfcBuildingElement" entity is specialized into several subclass entities such as IfcDoor, IfcWall, IfcSlab and so on [24].

GIS adopt the use of only b-rep, like multipatch. Multipatch is the feature type used for storing 3D solid or surface features in ESRI systems [21]. Every space in the 3D building model has a unique ID which is used to specify the geometry location. While the requirements for information and level of detail (LOD) of the BIM objects in GIS environments depends on in which it is applied and should thus be taken into consideration in the integration.

Hence, Figure 1 and Figure 2 show the geometric information exported from Autodesk Revit software through IFC into multipatch features by using ESRI Data Interoperability Extention for developing ESRI geodatabase. Then, ESRI ArcGIS Pro was used to check the quality of the transformation. The quality of the transformation was evaluated by counting the number of features lost, by visual comparison, by checking that each transformed element could still be identified and then finally by testing whether the features were topologically closed. The ArcGIS Pro 3D scene was used to validate that a spatial query could be performed to find the 3D space where an object existed. Data level integration shows in Figure 1.

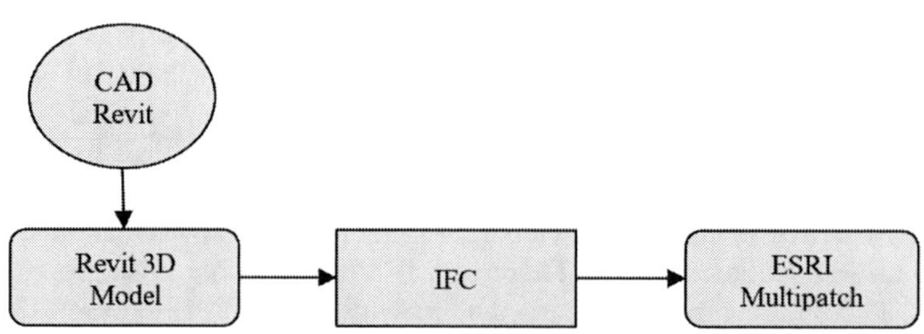

Figure 1. Data level integration

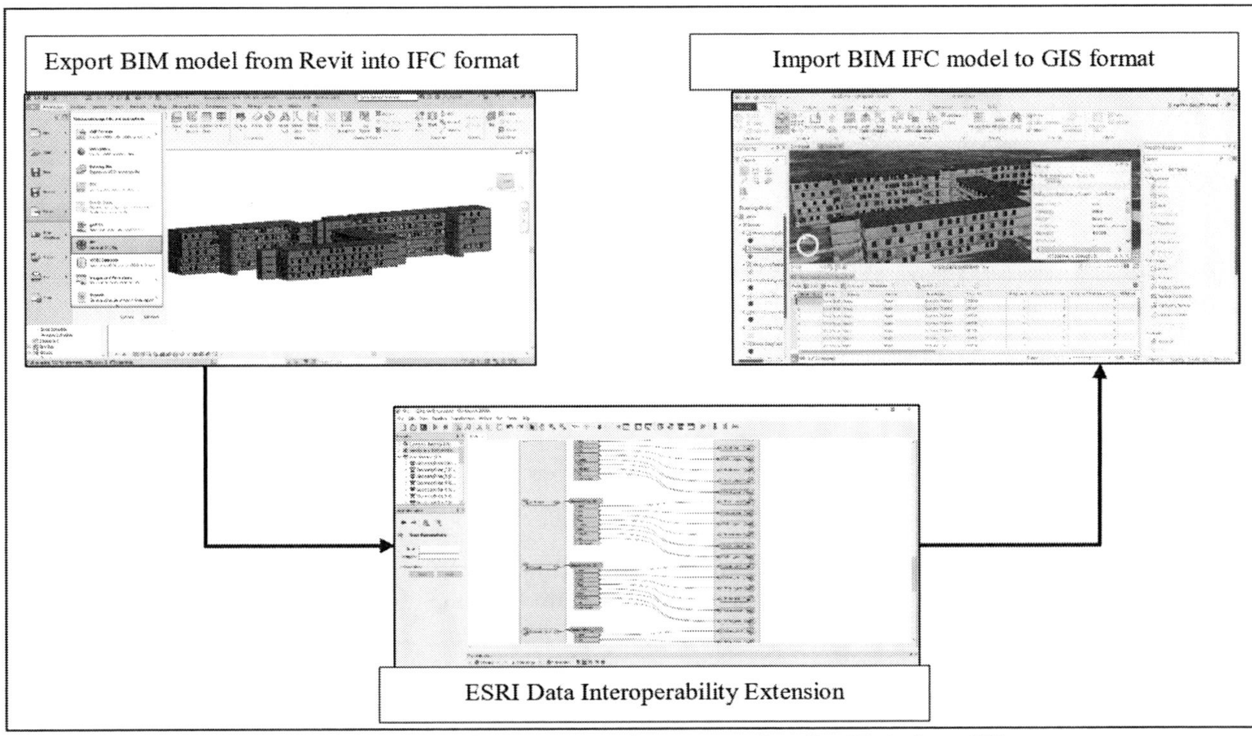

Figure 2. Example of BIM to GIS data conversion for 3D indoor GIS

Therefore, the requirements in regard to the integration of data levels for 3D indoor GIS are transforming CSG and Sweeping representations of IFC BIM into geometric representations in GIS which is fundamental for volume features in space management. Besides, the semantic data of the geometry must be preserved [18]. Thus, integrated geometry and semantic data must be separated based on specific standalone features. The transformations of BIM IFC data are validated in the GIS platform.

2.2. Data Management

The requirement to perform 3D indoor GIS data management is the geometric and semantic data need to be adequately organized in an effective way. Based on the functionalities of 3D indoor GIS, the enriches of data is an initiative to build focused GIS solutions for the facilities domain, enabling users to get up and running quickly and to be able to do their work more efficiently. The extension procedures can be added to absorb new information types from BIM IFC for the current implemented data models in GIS.

However, it may be considered insufficient as the building facilities are not reflected in the aspects of facility management (FM) in order to reduce the probability of failure because it is too extensive to build an entire FM system all at once in 3D. All data converted from BIM IFC must be managed properly. The data needed based on the functionalities queried to make better decision making includes detailed spatial data on the facility that satisfies 3D visualization and space search. Therefore, the geometry and semantic data were focused on space management and removed unnecessary information, making it easier to downsize files.

For instance, several BIM elements in IFC format that can be transformed into the 3D indoor GIS shows in Table 1. Therefore, the information from IFC data needs to be properly organized in 3D space management to keep their source correctly.

Table 1. Semantic data in IFC organized in GIS environment

BIM elements	GIS elements
IfcWall	Wall
IfcWindow	Window
IfcDoor	Door
IfcCovering	CeilingSurface
IfcSpace	Room

A developed 3D indoor GIS for space management facilitate the general users including FM managers to control the related work on the timetabling and space allocation problems, and so on because of object-oriented BIM modeling data. Besides, database is constructed according to space management and data would be managed by space, floor, and building unit.

All geometry and semantic data for spatial analysis stored in the 3D indoor GIS. The generic data model can be designed for different applications to maintain data integrity and independence. For example, 3D space with class schedule data can be stored and the room schedule can be analyzed and viewed in the 3D indoor GIS.

Figure 3 shows the conceptual data management of BIM information to the 3D indoor GIS. Four IFC data components, geometry, semantics, relationship classes, and properties of building elements are converted into the 3D indoor GIS [23]. The IFC geometry has been converted into boundary representation (b-rep), Sweep Solid (SS) and Constrictive Solid Geometry (CSG). The semantics of building elements and relationships between IFC classes could be transformed into the semantics in the GIS environment. Moreover, properties are converted to the attribution of geometric forms. At last, in GIS DBMS, geometry, associated semantics and attributes could be managed. In the implementation of the ESRI Geodatabase, all classes of the model are made up of polygons. The multipatch data type is used to populate objects in the implementation of Geodatabase.

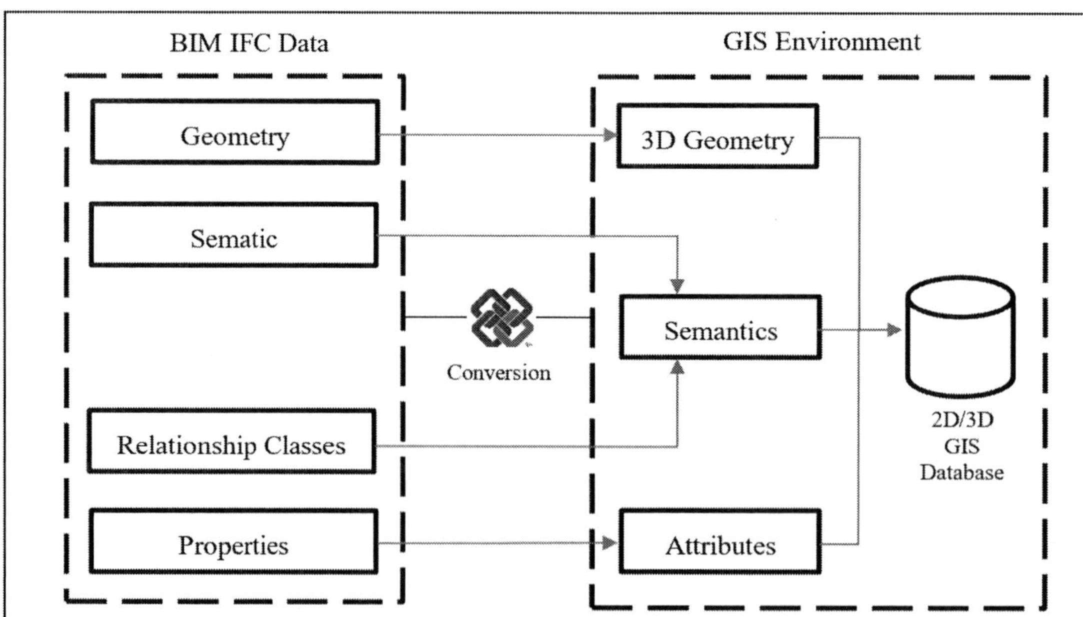

Figure 3. Conceptual data management of BIM IFC into a GIS environment

Therefore, the requirements on data management for BIM IFC information in the 3D indoor GIS are the original source from IFC models need to be kept and well organized with the model data of the generic 3D indoor GIS. Besides, four components of IFC data need to be linked to the corresponding parts in 3D indoor GIS to make the model more compact, and easy to find and search.

2.3. 3D indoor GIS analysis

BIM addresses details of individual buildings without the contextual geospatial information support. It will be insufficient and difficult to meet FM needs of a geographically dispersed, for instance, a wide range of building facilities such as campus buildings. Consequently, GIS is a technology that can offer FM scalable solutions far beyond the building perimeter if it is integrated with BIM and the myriad facilities management technologies. By integrating BIM and geospatial information, decision-makers can significantly build and evaluate various scenarios for moving planning, evaluate the current allocation of organizational space and determine overcrowded or underused facilities in real time. Moreover, coupled with the optimization algorithm, this procedure enabled the decision to optimize space utilization, can be quickly evaluated.

Building information that extracted from IFC provides highly structured and detailed information in a 3D or multidimensional format which is a lack in GIS. Thus, the data for space management incorporated with 3D building information that can use the spatial analysis for innovative, data-driven and performance-based solutions for strategic planning and decision-making. It should not only provide queries concerning single building elements, but also the whole building management environment involves. For example, room utilization can be analyzed and the room availability for an event can be checked using three-dimensional manage spaces that were formerly two-dimensionally managed with a spreadsheet and 2D CAD.

The blocks within the building that extracted from IFC represent the space of each room. Again, size is a proportional and relative position to the real-world location is preserved, where multiple floors of buildings delineated with shadow lines. There are many areas in a building, such as circulation areas, stairwells, bathrooms, and mechanical equipment rooms, which may not be germane to a space allocation problem thus, do not need to be included for analysis via the visualization tool.

The campus currently updates its drawing and data information on separate formats which requires a manual update in creating duplication of workload. For example, DWG floor plans which come in 2D representations and an MS Excel database. On the other hand, photographs scanned elevations and sections were also used to verify specific details from the original drawings' sheets. With the rapid development and changes in building utilization occurring year-round, it is dragging the full-time attention of a CAD technician. The lack of comprehensive, up-to-date, accurate, reliable and accessible information can seriously inhibit the improvement of space planning and management. The new transformation model has a rich set of spatial features and attributes. Thus, the data need to well-structured to be used for spatial analysis [25].

By properly designing data structure and modeling the building in 3D GIS model, it allows the function to automatically update the required schedules, elevations, 3D visuals, produce instant sections and generate drawing sheets, from a single program through the creation of geometric information and inclusion of specific building information [26]. This provides information that is currently not available to the facility management team and promotes instant efficiency gains as well as providing interactive information.

In summary, without losing the necessary details, the generic data structure is used to ensure that the geometries of the new 3D model are easily recognized and visualized. Besides, 3D geometric and semantic representation of a building must be validated and should have a structure that supports the data analysis requirement for the application. Thus, the data structure must be designed properly to meet the needs of the implemented spatial analysis.

2.4. 3D space management

The generic 3D space management needs to consider the computation of the space utilization optimization cases such as learning space usage rate based on the teaching and learning schedule (timetable). Timetabling represents room usage over time. BIM IFC data provides a 3D building model that can be used to visualize the room occupancy and exploring opportunities to use available data regarding room usage to match student need with available spaces. By utilizing this information, the simulation of space utilization can be performed and based on the optimization computation, it can support implicit cost saving.

Although there are studies on space utilization optimization [27], the computational functionalities can also be introduced to 3D space management. The core of the computational of space utilization optimization is the mathematical models that able to solve the complex space utilization optimization in an automatic and efficient manner. In addition, IFC data such as IfcSpace contains IfcLengthMeasure for room dimension. The computational of space utilization optimization can take the function of IfcLengthMeasurement into account. In this case, the 3D space management can provide computational and visualization of space utilization. It can also estimate the expenditure cost.

As mentioned before, up-to-date data collection in this 3D space management should be considered. In this instance, the up-to-date data can support the visualization of room usage over time. Space optimization calculation can be viewed as a variable which consists of several criteria. These criteria include used space, capacity, dimensions and so on. The visualization can be used to increase the efficiency of transition planning and management.

Therefore, to fulfill the requirement of 3D space management in the 3D indoor GIS, 3D building model development need to be incorporated with the timetables to schedule the room usage by assigning them to timeslots and places in a way that makes optimal use of the available resources. Besides, the 3D building model also needs to be integrated with the space utilization optimization computational functionality which is used to calculate the space usage and at the same time to save costs for the operational costs of the building.

3. Case Study

In recent years, various types of building information mostly CAD file has been integrated with GIS for asset and facilities management. Most efforts were focused on the management of facilities, in particular, space management. However, studies on these cases will gain some insight into the integrated GIS and BIM from a general perspective. This section provides an example of the case study in these implementations.

3.1. NASA, USA

Langley Research Center (LaRC) of the National Aeronautics and Space Administration (NASA) was implemented integrated spatial data management and decision-making system based on GIS [27]. The general scope of LaRC has been 800 acres and 400 buildings comprising 6,700 rooms totaling 3.7million square feet, with assets valued at approximately $3 billion. The facilities were designed to accommodate more than 4,000 employees.

LaRC has developed a move planning tools that allow space utilization managers to construct and evaluate various personnel move scenarios for both within buildings and between buildings. By consolidating space, LaRC able to reduce building dependence with uneffective operating costs while maintaining and improving the coherence of personnel distribution within each organization.

Web tools have been developed to make the analysis capabilities available to LaRC personnel with standard Web browsers. This approach will allow the user to access the data representations and minimizes the processing and requirements for plugins. The capabilities allow users to reuse additional conventional map views and building layouts with room details, increased analysis capability, uses any symbolization and labeling available, and easier integration with CAD. However, there are some useful capabilities include metadata access or linking, area measurement and statistical analysis has not been addressed in this system [27].

3.2. Korea Institute of Construction Technology (KICT)

Korea Institute of Construction Technology (KICT) was implemented a prototype system developed by utilizing the BIM/GIS-based for facility management. The databases were constructed based on existing managed documents and drawing. Thus, many information missing due to illegible handwriting. Furthermore, facilities management history data in terms of BIM objects were difficult to obtain. Therefore, historical data of facilities management were constructed based on space. The databases were integrated to the current system for information interoperability of the main building at the Korea Institute of Construction Technology (KICT), were Excel- based structures and the BIM objects were modeled using the Revit software.

The databases were divided into structural and facilities history data for space and managed in Excel files. According to each work based on field survey, databases were constructed, and FM data were managed by space, floor and building unit. This can systematically search and display data using the hierarchical order of the space-floor-building-area depending on the user's objective. Based on the existing KICT portal system, FM manager able to search, read, edit, and check with fundamental basic functions. FM manager can control a room schedule, history management of joint equipment, site navigation, remodeling plan of the institute, and so on because of object-oriented BIM modeling data [25]. The developed FM system was integrated with the KICT main system. Figure 5 shows the prototype system.

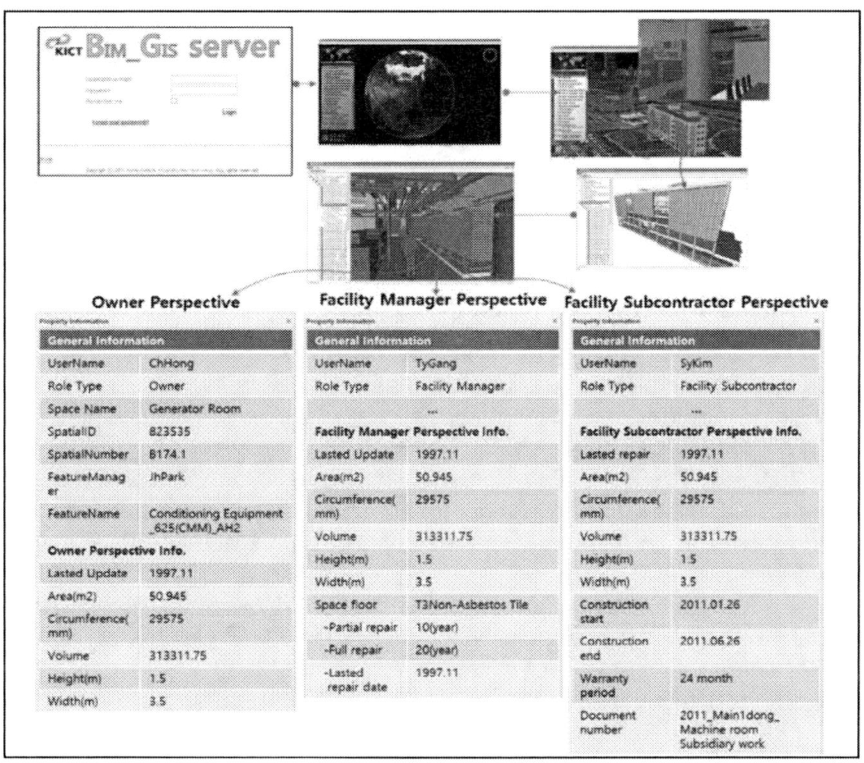

Figure 4. Prototype System for BIM/GIS-based FM software

4. Conclusion

A new approach to the integration of BIM and GIS allows managers to visualize the 3D model of the campus effectively. BIM IFC data provides information that can be refined to be used for spatial analysis in 3D indoor GIS. The aim of this paper is to reveal the 3D indoor GIS requirements for space management. The motivation behind this work is to facilitate the translation from BIM to the 3D indoor

GIS and to apply more suitable spatial analysis functionalities from the extracted 3D geometries and semantic information of BIM IFC.

Although this research is in the early stages, an analysis has been presented on the requirement for the possible development of 3D indoor GIS from BIM IFC. In specific, the requirements were analyzed in four aspects: 1) data level integration; 2) data management; 3) 3D indoor GIS, and 4) 3D space management based on BIM IFC data. Accordingly, the requirement for conducting 3D space management in 3D GIS environment are summarized as follows:

1. The conversion method to transform 3D geometry and semantics data in IFC and validation of the data.
2. The translation of the BIM IFC data component into the GIS environment and data storage management.
3. A generic data structure to facilitate the new 3D building model and data structure design that suit the spatial analysis implementation.
4. Timetabling data that relate to space usage and space utilization optimization computational functionality.

In the future, the feasibility integration of the BIM IFC into the 3D indoor GIS will be tested; the structure data model will be designed and implemented in GIS DBMS. Based on the structure data designed, the functionalities of spatial analysis for 3D space management will be established. After that, a demonstration prototype will be developed and presented.

Acknowledgments
The authors would like to express their gratitude to the Universiti Teknologi MARA (UiTM) and Ministry of Higher Education (MOHE) for Ph.D. funding through MyBrain15 scholarship scheme.

References

[1] Fosu R, Suprabhas K, Rathore Z and Cory C. 2015 Integration of Building Information Modelling (BIM) and Geographic Information Systems (GIS)–A Literature Review and Future Needs *Proceedings of the 32nd CIB W78 Conference, Eindhoven, The Netherlands* 27-29

[2] Azhar S, Khalfan M and Maqsood T. 2015 Building Information Modelling (BIM): Now and Beyond *Construction Economics and Building* **12(4)** 15-28

[3] Laakso M and Kiviniemi AO 2012 The IFC Standard: A Review of History, Development, And Standardization, Information Technology *Journal of Information Technology in Construction (ITcon)* **17(9)** 134-61

[4] Song Y, Wang X, Tan Y, Wu P, Sutrisna M, Cheng J, and Hampson K. 2017 Trends and Opportunities Of BIM-GIS Integration in The Architecture, Engineering, and Construction Industry: A Review from A Spatio-Temporal Statistical Perspective *ISPRS International Journal of Geo-Information* **6(12)** 397.

[5] Liu X, Wang X, Wright G, Cheng J, Li X and Liu R. 2017 A State-Of-The-Art Review on The Integration of Building Information Modelling (BIM) And Geographic Information System (GIS) *ISPRS International Journal of Geo-Information* **6(2)** 53

[6] Boguslawski P, Mahdjoubi L, Zverovich V, Fadli F and Barki H. 2015 BIM-GIS Modelling in Support of Emergency Response Applications *Building Information Modelling (BIM) in Design, Construction And Operations* 381-391

[7] Zhu J, Wright G, Wang J, and Wang X. 2018 A Critical Review of the Integration of Geographic Information System and Building Information Modelling at the Data Level *ISPRS International Journal of Geo-Information* **7(2)** 66.

[8] Pispidikis I, Tsiliakou E, Kitsakis D, Athanasiou K, Kalogianni E, Labropoulos T and Dimopoulou E 2018 Combining Methodological Tools for The Optimum 3D Modelling of Ntua Campus *ISPRS Annals of Photogrammetry, Remote Sensing and Spatial Information Sciences* **4** 57–63

[9] Li Y. 2011 Building Information Model for 3D Indoor Navigation in Emergency Response.

InAdvanced Materials Research **368** 3837–3840
[10] Teo TA and Cho KH 2016 BIM-oriented Indoor Network Model for Indoor and Outdoor Combined Route Planning *Advanced Engineering Informatics* **30(3)** 268–282
[11] Costin ATeizer J 2014 Utilizing BIM for Real-Time Visualization and Indoor Localization of Resources *Computing in Civil and Building Engineering* 649–56
[12] Park T, Kang T, Lee Y and Seo K Project Cost Estimation of National Road in Preliminary Feasibility Stage Using BIM/GIS Platform *Computing in Civil and Building Engineering* 423–430.
[13] SanJosé-Alonso JI, Finat J, Pérez-Moneo JD, Fernández-Martín J andMartínez-Rubio J 2009 Information and Knowledge Systems for Integrated Models in Cultural Heritage *Proceedings of the 3rd ISPRS International Workshop 3D-ARCH* **38** 5
[14] Dore C and Murphy M 2012 Integration of Historic Building Information Modeling (HBIM) and 3D GIS for Recording and Managing Cultural Heritage Sites *Virtual Systems and Multimedia (VSMM), 2012 18th International Conference on IEEE* 369-376
[15] Amirebrahimi S, Rajabifard A, Mendis P and Ngo T 2015 A Data Model for Integrating GIS and BIM for Assessment and 3D Visualisation of Flood Damage to Building *CEUR Workshop Proceeding* 78–89
[16] Liu X, Wang X, Wright G, Cheng J, Li X and Liu R 2017 A State-of-the-Art Review on the Integration of Building Information Modeling (BIM) and Geographic Information System (GIS) *ISPRS International Journal of Geo-Information* **6(2)** 53
[17] Liu L, Li B, Zlatanova S and Liu H 2018 The Path From Bim to a 3D Indoor Framework–A Requirement Analysis *International Archives of the Photogrammetry, Remote Sensing & Spatial Information Sciences*
[18] Isikdag U, Zlatanova S and Underwood J 2013 A BIM-Oriented Model for Supporting Indoor Navigation Requirements. *Computers, Environment and Urban Systems* 41 112-123
[19] Boyes G, Thomson C and Ellul C 2015 Integrating BIM and GIS: Exploring the Use of IFC Space Objects and Boundaries *Proceedings of the GISRUK*
[20] El Meouche R, Rezoug M and Hijazi I 2013 Integrating and Managing BIM in GIS, Software Review *International Archives of the Photogrammetry, Remote Sensing and Spatial Information Sciences* **2**
[21] Tobiáš P 2015 An Investigation into the Possibilities of BIM and GIS Cooperation and Utilization of GIS in the BIM Process *Geoinformatics FCE CTU* **14(1)** 65-78
[22] Cheng J, Deng Y and Du Q 2013 Mapping Between BIM Models And 3D GIS City Models of Different Levels of Detail *13th International Conference on Construction Applications of Virtual Reality, London* 30-31
[23] Deng Y, Cheng JC and Anumba C. 2016 Mapping Between BIM and 3D GIS In Different Levels of Detail Using Schema Mediation and Instance Comparison *Automation in Construction* **67** 1-21
[24] Lin YH, Liu YS, Gao G, Han XG, Lai CY and Gu M. 2013 The IFC-Based Path Planning for 3D Indoor Spaces *Advanced Engineering Informatics* **27(2)** 189-205
[25] Kang TW and Hong CH. 2015 A Study on Software Architecture for Effective BIM/GIS-Based Facility Management Data Integration *Automation in Construction* **54** 25-38
[26] Kassem M, Kelly G, Dawood N, Serginson M and Lockley S. 2015 BIM in Facilities Management Applications: A Case Study of a Large University Complex *Built Environment Project and Asset Management* **5(3)** 261-277
[27] Ball WB. 2012 Esri News To Optimize Space Utilization And Cut Costs *NASA Langley Research Center*

IOP Conf. Series: Earth and Environmental Science **385** (2019) 012047 doi:10.1088/1755-1315/385/1/012047

The Potential Benefits of Building Information Modelling (BIM) in Construction Industry

Hilfie Safwan Ibrahim[1], Norfashiha Hashim1 and Khairool Aizat Ahmad Jamal[2]
[1] Faculty of Architecture, Planning & Surveying, Universiti Teknologi MARA, 40450 Shah Alam, Malaysia,
[2] School of Architecture, Building and Design, Taylor's University, 47500, Subang Jaya, Selangor, Malaysia

Email: norfashiha@uitm.edu.my

Abstract. Construction industry known as complicated, risky and uncertain that certainly contribute to project delay, cost overrun, low product quality and others. Building Information Modelling (BIM) knows as a promising platform that can solve those problems. BIM has been developed at a rapid pace to cope with the increasing complexity and difficulty of project management. The construction players especially contractors still unclear about what BIM truly is, how and when to use it. The purpose of this paper is to explore the extent of which use of BIM producing potential benefits across construction project life phase. The exploration is done by collecting secondary data from 56 contractors (G7) across Klang Valley. The data of were analysed using descriptive analysis which are the frequency analysis, average mean and cross-tab analysis. Based on contractors Level of Development (LOD), most benefits can be obtained were from the Design Phase rather from Construction Phase and Facilities, Operations and Maintenance Phase.

1. Introduction

BIM has been developed at a rapid pace to catalyst on the increase in the aspect of complexity and the difficultness of project management. Despite that, the genuine information still makes the usage of BIM falling low of its capabilities [23]. The development of BIM adaption in Malaysia is considerably very little and stagnant. This is far behind if make comparison with developed countries for instance Hong Kong, Australia, Denmark, Finland, Singapore and Norway [18]. In addition, education and awareness are the third thrust of Malaysian BIM roadmap 2014-2020. Some criteria and conceptual of BIM adaption were reviewed and conclude in most study of BIM technology [4]. The number of studies that focus on the BIM awareness among Malaysian contractors are still low [8].

At present, success level and the BIM progression are still not in a satisfying level. BIM is not fully entrenched when it comes to construction industry [5]. BIM consist of major advancement especially in technical and design aspects. Despite of that, the implementation of BIM yet still not being fully utilized by construction industry users [12]. Apart from that, BIM is lacking of widespread uptake and it is linking to the challenges and risks that are potential impending its effectiveness [6]. The interacting complex system in BIM for instance building, services machine floor layout, all these multiple layers requiring modelling to be done in careful manner [13]. BIM adaption among Malaysian construction player are significantly low because of insufficient knowledge on BIM and mostly they do not know how, when and what to start [25]. Hence, it necessary to study the current implementation of BIM and together with determining the potential benefits to increase BIM awareness among contractors.

The construction industry players that in contacts with the project of 100 million sum and above are required to obtain a minimum score of 40% of stage 2 BIM maturity by 2020 [11]. In the report produced by CIDB at 2016 had justify that the willingness of construction industry to change for BIM only at 17% adoption which is extremely low. The previous work minister Datuk Seri Fadillah Yusof urged that BIM adoption in the country relatively lower only at 17%. It is very high difference as compared to other developed country such as United Kingdom, Singapore and United States which at 54%, 65% and 71% respectively [10].

There is an iniative to develop technology by the construction experts in a way to serve the demands in construction projects now to BIM technology [2]. Thus, it is compulsory to follow up to date technology and it is the moment we need to change the way of construction in Malaysia that has come. Nevertheless, the implementation of BIM is still limited and processing slowly [1].

The business of construction development made item orientated building item demonstrating displaying since early 90's. A few market divisions for example the structural steel used the parametric 3D modelling. These days, BIM apparatuses has turned out to be promptly accessible in the development business. It has turned into the reward of development industry's devotion to BIM for as long as 20 years [26]. What's more, development industry had meet to the degree of to acknowledge and mindful of the genuine advantages of mechanical headway. The hole in the work profitability and effectiveness can be cure with the assistance of Building Information Modeling idea. Consequently, it is the need to consider more on BIM and it tools to decide the best possible benefits and misfortunes it has toward the construction management [15].

2. Literature Review

The construction sector could be one crucial and proactive sector when it comes to Malaysia economy. Asian nation had been noticed that as a developed nation, the crucial function of the construction industry not just the aspects of economics process however additionally in enhancing the standard of life and expectations for everyday comforts of Malaysian people [3]. In the course of the most recent twenty years it's been contributing between 3 to 5 % of the blend economy GDP. The improvement blast in Asian country started in mid Nineties, basically once the launch of Vision 2020. Nowadays Malaysian construction industry keeps on developing altogether inside the local moreover as worldwide market [22].

Base on its size, the business of construction industry might be a mammoth customer of every item and administrations [7]. For sample, the construction industry ingest an essential extent of Malaysia's producing sector, and additionally fundamental amounts of basic metals, ceramic, cements and distinctive building materials. The improvement of construction section go about as an outsized customer of higher esteem included hardware together with machinery also. Separated of been as the purchaser of made products, the construction business go about as a key shopper of a scope of administrations. Every last bit of it incorporates the learning driven consultancy and designing, other than go about as a wide base of money related administrations [10].

There are three major lifecycle phase in construction projects that has been induced by construction industry [24]. Design (D), Construction (C) and Operation (O) are the three major focus area. Somehow, there are more division in project phase. The division are namely; Feasibility (F), Program (P), Conceptual Design (CD), Detail Design (DD), Construction Planning (CP), Construction (C) and Operation (O) [14]. It is included that, Tendering and adjudication is one among the project lifecycle which coming after Detail Design and before Construction.

The definition given by above authors nonetheless, give off an impression of being conventional to incorporate any computerized innovation and some specialist examined it a bit much as long as it could add to frame office data. Some researcher underlined that, BIM go about as building simulation that gives am vastly improved procedure of correspondence toward all construction participants at different phase of the building life cycle [17]. Building Information Model allude to the spread sheet of exceptional information drawing in the use of alphanumeric information to reenact a genuine business process in the building life cycle [14]. Not just that, it additionally includes the multifaceted nature and an interrelationship of association.

BIM development are categorized in five main stages which are called Level of Development (LOD) 100, 200, 300, 400 and 500. These LOD referring to the determination of project phase from early design process towards the construction process ended. Basically, LOD 100-200 focus on the traditional process involving 2D approach and for LOD 400 and LOD 500 are the exact BIM process. The LOD known as the phases that Model Progression Specification had been establish by BIM application and information input [20]. The application of each LOD divided into the pre-design, schematic design, construction design, construction documentations, fabrication and as built.

In light of a review contemplated, BIM selection level had been filled in as basic standard for the progress inside development industry [11]. The Malaysian development industry has been perceived as critical division that drives the economy of the nation [4]. In any case, regardless of being one of critical part, this industry perceive as among the business with most minimal profitability level. In this manner, several benefits of BIM application shall be listed with the goal that the construction players know about it.

The presentation of BIM benefits, will improve the design to be which can control the task cost life cycle and the project procedure too [21]. The potential benefits are can be sorted in a wide range of perspectives. The benefits can be classified in five viewpoints which are: Design aspects, Scheduling perspectives, Budget viewpoint, Communication viewpoints, and Documentation perspectives [19]. In addition, a research underlined that, six fundamental concern on the benefits of BIM adaption mainly due to; project client, system of innovation, project speed, cost vs time, competition and efficiency [16].

3. Research Methodology

To identify the level of BIM implementation and to perceived BIM benefits during the implementation, a quantity approach was used. Quantity approach provides comprehensive finding on the benefits of BIM from various perspectives [27]. The method approach includes a questionnaire survey among others which enables the generalizability of the results.

The set of questionnaire survey derived based on the collection of previous research. The questionnaire contains of 56 respondents demonstrating the potential benefits and being measured in the means of Likert Scale, with the score of one to five signifying 'strongly disagree' to 'strongly agree' respectively. BIM benefits section was derived as the benefits or value that respondent receive during each project life cycles. The potential benefits is divided into three project phase as in the questionnaire section where the first 24 variables are in the Design Phase, the following 19 variables in the Construction Phase and the last 13 variable in the Facilities, Operations and Maintenance Phase

Questionnaire was randomly distributed to 220 G7 contractors in Klang Valley. Klang Valley is an area in Malaysia which is centered in Kuala Lumpur and includes adjoining cities and towns in the state of Selangor. The limit of the area of geographical are mainly because of the time constraint and the accessibility of the researcher to carry the survey in Klang valley area. These set of questionnaires were hand out by the process of drop-and-collect technique and web survey to all selected area. G7 contractor are chosen because they are more experienced in the construction industry in Malaysia. The respondent contractors will be chosen based on types of projects being done and the contractor's profile.

The data were quantitatively analysed by using the Statistical Package for Social Science (SPSS). Descriptive analysis such as average mean and cross tab analysis were used in order to achieve the research objectives.

4. Results and Discussions

A total of 220 sets questionnaire were distributed to G7 contractor companies in Klang Valley. The questionnaire forms were distributed by hand and via online survey tool – Survey Face. A total of 56 questionnaires were collected, Table 1 summarized the socio – demographic of the respondents. Analysis shows that the male dominant the response rate which at 43 whereas the female respondent at

IOP Conf. Series: Earth and Environmental Science **385** (2019) 012047 doi:10.1088/1755-1315/385/1/012047

13 only with the percentage rate for male and female are 76.80 % and 23.30 % respectively. The majority of the respondent are from the Drafter or Designer position which is at 17 respondents (30.4%) from total respondents. Next, Site Engineer having 21.4% and the number of respondents is 12. There 11 respondents having current job that is not mentioning in the categorization. They are currently working as BIM modeler and Planning Engineer.

Table 1: Socio – Respondents Demographic

Description		No	%
Gender	Male	43	76.80
	Female	13	23.30
Position	Project Manager	5	8.90
	BIM Manager	3	5.40
	Construction Manager	1	1.80
	Managing Director	4	7.10
	Site Engineer	12	21.40
	Quantity Surveyor	3	5.40
	Drafter / Designer	17	30.40
	Others	11	19.60
Highest Academic Qualification	SPM	1	1.80
	Diploma	4	7.10
	Bachelor's Degree	44	78.60
	Master's Degree	5	8.90
	Others	2	3.60
Years of Experiences in Construction Industry	Less than 1 year	6	10.70
	1 – 5 years	31	55.40
	6 – 10 years	14	25.00
	11 – 15 years	5	8.90
	16 – 20 years	0	0.00
	More than 21 years	0	0.00

The age of the respondents are divided into seven categories which are the interval of 18-25 years old, 26-30 years old, 31-35 years old, 36-40 years old, 41-45 years old, 46-50 years old and lastly 51 years old and above. The response rate are only coming from 18-25 years old, 26-30 years old, 31-35 years old, and 36-40 years old. More than half of the respondents (51.80%) are in the age 26-30 years old which is 29 numbers of respondent. There is a fair of amount for respondents in between 18-25 years old and 31-35 years old which are 12 and 11 respectively. The remaining four numbers of respondents at the age of between 36-40 years old.

Academic qualification of the respondents are important factor that need to be consider in this research by taking in considering their expertise, knowledge, thinking and understanding. Bachelor's degree has domain at 44 numbers of respondents out of 56. Five person having master degree and it seems like SPM on have one number of frequency. The data in table above showed that most of the respondents are in the construction industry in the range between 1-5 years at 31 numbers of frequency and dominant up to 55.4% of overall. Coming next, half of it conquer by the range of age in between 6-10 years which at 14 numbers of respondents.

Although the contractors were fully aware of BIM, and its application, the extent of its usage is not so much on the BIM features in the scope of level of development (LOD). Table 2 shows most of the contractors are using the BIM features up to LOD 3 which is at 37.4%. This indicates, the contractors have use the capability of estimating construction, monitoring the clash detections and visualizations. Somehow, the usage of LOD 5 adaptability is only at 3.3%. This showed that the facility management system cannot fully adapt to the building models. This will affect the operation and maintenance of building systems.

Table 2: Extent of BIM Usage

		Responses	
		N	Percent
LOD Level	Level of Development (LOD) 1	16	13.0%
	Level of Development (LOD) 2	13	10.6%
	Level of Development (LOD) 3	46	37.4%
	Level of Development (LOD) 4	39	31.7%
	Level of Development (LOD) 5	5	4.06%
	None	4	3.3%
Total		123	100.0%

Table 3 showed the results based on the survey answer by contractors by listing the possible potentials of BIM in the design phase. Results showed that the average mean for all 24 variables in this question are 4.32, indicate that the contractors mostly agreed with the benefits that the respondents received at design phase stages. The average index of this findings is in between 3.80 (σ = .961) to 4.59 (σ = .596). BIM detected potential clashes in the early design stage before construction process starts is the highest mean with the average mean is 4.59 (σ = 0.596). Clash avoidance is actually the key part of the design and construction process. Most of the time, during the design process, every BIM modules or users shall assess design decision and clashes to see if the problem can be resolve internally by the contractor themselves. BIM also will benefits contractors in support decision making regarding the design (*M*= 4.54, σ =0.503). In actual BIM process, the viability of the end products can be detected earlier with the help of BIM. Hence, any clash can be well proven by BIM implementation. Decision making is compulsory throughout the design stage of any projects. Thus, with BIM the decision can help to solve all design related problem. In addition, BIM benefits in clearing the concepts and project conceptualization easier (M= 4.50, σ =0.539). It can be generalize that with all the design making with BIM, the models become an ultimate concept tools to convey the project scope, step, outcomes and requirement. Among contractors, there is quite lacking in project conceptuality, they are more focus on constructability because they depends more on the consultants. This is the reason this potentials variable becoming the top three potential benefits.

However, respondents feels that BIM will not be able to reduce the knowledge gap of senior and junior staff (*M*= 3.80, σ =0.961). Mostly, in contractors' firm, they were reluctant to appoint the personnel related to BIM. There were still many organization does not have a BIM manager. Thus, there is not much gap identification of senior and junior staff. Next with BIM, it is not necessary improving the organization itself (M= 3.93, σ =0.892). All respondents agreed that it contributes to organization improvement. Yet, the improvement of the organization can be get through many medium not only the implementation of BIM. Respondents were as well less agree of the practicing current innovation as the main potential benefits (M= 4.00, σ =0.972). Digital technology such as BIM is the key to improve construction industry productivity. Somehow, contractors does not quite agreed with this variable as it seems that BIM has been developed earlier and the main concern is not be fully implemented.

Table 3: Benefits of BIM at Design Phase

Benefits of BIM	Mean	Std. Deviation
Concepts becomes clearer and project conceptualization easier	4.50	.539
Earlier and more accurate visualizations of a design to the owner	4.50	.539
Support decision making regarding the design	4.54	.503
Improve feasibility studies	4.29	.624
Improve simulations and coordination	4.30	.658
Improve design quality	4.43	.628
Design and installation services coordination improved	4.41	.565
Building proposal capability analysis increase	4.32	.716
Improved lean in design	4.16	.708
Improve design effectiveness	4.45	.570
Save design time and cost	4.39	.731
Improve mistake identification thus reducing conflict errors	4.48	.572
Utilize the coordination resolution in pre-construction	4.36	.672
Ease communication	4.12	.875
Organization improvement	3.93	.892
Practicing current innovation	4.00	.972
Reducing the knowledge gap of senior and junior staff	3.80	.961
Improve collaboration between construction players	4.18	.765
Enhance and improve planning process	4.43	.657
Enhance accuracy of existing documentation	4.41	.682
Risk reduction	4.27	.726
Simultaneous access	4.34	.721
Increase attention to the selection of the components of construction at early stages	4.41	.708
BIM detected potential clashes in the early design stage before construction process starts	4.59	.596

Table 4 below shows the benefits of BIM in construction phase. Majority of the respondents agreed BIM benefits in construction phase is clash detections (*M*= 4.25, σ =0.640). Respondents agreed that BIM helps to detect clash in construction phase due to the fact that with all of the building design completed on a capture and alteration of existing reality, the model will be the ultimate references to the ongoing particulars in a project. Hence, with the creating of design completed with information, clash can be detected earlier, and contractors able to reduce time, cost and improve quality. Respondents also agreed that constructability issues still arising during filed operations and once in the project environment, contractors can still improve constructability benefits with BIM implementation. With BIM, the many potential issues can be resolved by the 3D modelling (*M*= 4.41, σ =0.626). Implementation of BIM in construction phase will be able to reduce error, rework and waste for better sustainability for construction (*M*= 4.41, σ =0.626). With the availability of shared model in BIM, there is less rework need and duplicating drawings for the different requirement of a buildings. Since it is connected to the database, during construction the sustainability of construction will be enhance by less error and waste.

However, respondents feels that BIM will least help in the improvement of communication among stakeholders (M= 4.41, σ =0.626). When having the better handling process in clash detection and coordination, it helps in handling the keeping of sub-contractors work to be predictable. Hence, the communication will be improvised. Somehow, the reason most of contractor less agreed with the points is that, not all sub-contractors are of BIM implementation, and base more on the technical experienced technical point of view on site rather that follow up to the proposed design by BIM.

Table 4: Benefits of BIM in Construction Phase

Benefits of BIM	Mean	Std. Deviation
The improvement in understanding the sequence of construction activities and duration	4.36	.616
The improvement in visualization of construction details	4.27	.618
The improvement in synchronization of design and construction planning	4.34	.640
The improvement in constructability	4.41	.626
The improvement in risk identification and the suitability of risk management to made	4.13	.788
The improvement of safety features	4.14	.724
The improvement of quality and efficiency in construction management	4.20	.796
The improvement of project quality and BIM digital data quality gain	4.20	.796
The enhancement of contractors ability to make informed decisions, by estimation, coordination and scheduling the construction process	4.32	.636
The enhancement of productivity through time and cost saving	4.34	.640
The accuracy of scheduling	4.16	.757
Error reducer, rework and waste for better sustainability for construction	4.41	.565
The improvement of communication among stakeholders	4.13	.740
Quick reaction to design changes	4.29	.731
Optimizing the clients experienced and satisfaction	4.30	.685
The increase in workforce effectiveness	4.23	.763
Dispute resolution	4.27	.674
Reducing whole project's life cycle cost	4.32	.690
Clash Detection	4.54	.602

Table 5 showed the findings of BIM in the facilities, operations and maintenance phase. BIM is very broad in terms of the process creating and managing digital information of a project. It creates available concurrent information on the use or the performance of the building. Contractors strongly agreed that the information on a building life cycle can be share more easily through the BIM as a medium ($M= 4.25$, $\sigma =0.640$). Next, BIM will be able to reduce the risk of losing project information ($M= 4.11$, $\sigma =0.731$). BIM as digital recorder of project coordination able to accelerates the adaption of standard building prototypes. Thus, the BIM provides the physical information about the building such as furniture, types of finishing and necessary equipment. The risk of losing the project information can be reduce through this means when it comes to management phase.

Somehow, respondents feels that BIM will least help in increasing the staff personal efficiency ($M= 2.82$, $\sigma =0.596$). This variables in connected to design phase by including the staff in design process, the construction team will be more capable of making decision accurately reflect to the design intent. Contractors may not aware of this potential benefits. In addition, Most of the contractors do not agreed to BIM as enabling review of maintenance history ($M= 3.00$, $\sigma =0.640$) as the BIM conducted is currently below LOD 500. Facility maintenance management (FMM) is not much being conducted by respondents. Hence, most of them were having neutral answer on this variable. BIM helps in FMM by usage of geometry 3D oriented data to support the maintenance services which including the basic descriptions, parameter-related data together with records and reports of maintenance.

Table 5: Benefits of BIM in Facilities, Operations and Maintenance Phase

Benefits of BIM	Mean	Std. Deviation
The information on a building life cycle can be shared more easily	4.25	.640
The improvement in collaboration	4.11	.652
The enhancement of environmental sustainability	4.11	.652
The enhancement of the control of the whole life costs	4.09	.640
The improvement in emergency management	3.93	.657
Enhancing project closeout	3.95	.699
Reducing risk of losing project information	4.11	.731
Increasing the staff personal efficiency	3.82	.855
Keep track of built property	4.07	.828
Manage the property proactively	4.02	.556
Enable schedule maintenance	3.98	.674
Enable review of maintenance history	3.00	.745
Assist in controlling the subcontractors	3.91	.815

Table 6: Benefits in each LOD Level

LOD Level		Benefits in Design Phase	Benefits in Construction Phase	Benefits in in Facilities, Operations And Maintenance Phase
LOD 200	Mean	4.4583	4.5789	4.8462
	Std. Deviation	.	.	.
LOD 300	Mean	4.2652	4.3828	4.0559
	Std. Deviation	.51142	.43087	.46928
LOD 400	Mean	4.4278	4.3912	4.0744
	Std. Deviation	.35422	.42143	.49088
LOD 500	Mean	4.1574	3.9825	3.8632
	Std. Deviation	.47268	.60182	.55971

For the perceived benefits for each LOD (Table 6), respondents who implement the LOD 200 contributes highest means on the potential benefits given in the Facilities, Operations and Maintenance Phase with the mean of 4.8462. However, the validity of the information cannot be guarantee due to the fact that only one contractor's organization tied to LOD 200. Next, for LOD 300 the highest mean goes to the potential benefits of BIM in Construction Phase with the means of 4.3828 and the standard deviation of 0.43087. This information is gained by eleven number of respondent who established BIM at LOD 300. There are total of 30 numbers of respondents applied BIM LOD 400. With the means of 4.4278 and the standard deviation of 0.35422, most contractors agreed that potential benefits of BIM can be seen in Design Phase. Last but not least, total organization which applied BIM up to LOD 500 is not much; nine numbers. The highest mean goes to Design Phase with the mean of 4.1574 and the standard deviation of 0.47268.

5. Conclusions

BIM is definitely bring its own potential used at all stages of the project life-cycle, especially towards all the stage mentioned above. These all good to be practiced widely by the contractors to cartelist the project requirements. Apart from that, the potential has been applied in by the design team to analyze and design together with develop the particular projects. Next, it leads to contractors by process of managing the construction of the project and during the facilities, operations and maintenance Phase managing by the facility manager.

Given the findings of the result, it had shown that there has not been a greater take up on BIM when it comes to construction projects. And of course, there are many barriers identified thus experts should know about these difficulties so as to guarantee the potential benefits of BIM in development industry been connected and figured it out.

There is a need for every means to enhance the BIM adoption level in Malaysia. It is necessary to have marketing and BIM selling which supported by rigorous cost or benefits analysis. Although the initial cost of BIM is quite higher than CAD platform, however over the long haul the expanded profitability conceivably accomplished by BIM. Perhaps, based on this dissertation it is proven that it is more difficult to achieve the BIM objectives when it comes to people perspectives. Hence, it is a need for senior management in contractors firm to infuse in BIM education and training.

All in all, contractors needed to see that potential benefits of BIM can lead in bigger context. BIM definitely has its very own potential use at all phase in construction industry; benefits of BIM in Design Phase, benefits of BIM in Construction Stage and benefits of BIM in Facilities, Operations and Maintenance Phase. Through the survey made, contractors have been exposed with list of potential benefits. It can conclude that, most of respondents agreed that BIM provides many benefits. In order to drive and enhance the adoption level of BIM in construction industry, contractors needed to be well expose of the benefits of BIM can lead to. Transformation towards the new technology; BIM require strategic plan. In order to achieve BIM fully adoption, enforcement by government is needed. At early stage of implementation, the enforcement by government towards BIM as a driver for construction project is necessary.

References

[1] Abdulfattah, N. M., Khalafallah, A. M., & Kartam, N. a. (2017). Lack of BIM training: Investigating practical solutions for the state of Kuwait. *World Academy of Science, Engineering and Technology, International Journal of Civil, Environmental, Structural, Construction and Architectural Engineering*, *11*(8), 1050–1056. Retrieved from http://www.waset.org/publications/10007742

[2] Abdullah, S. A., Sulaiman, N., Latiffi, A. A., & Baldry, D. (2014). Building Information Modeling (BIM) from the perspective of Facilities Management (FM) in Malaysia. *International Real Estate Research Symposium*, (April)

[3] Alias, Z., Zawawi, E., Yusof, K., & Aris, N. M. (2014). Determining Critical Success Factors of Project Management Practice: A Conceptual Framework. *Procedia Social and Behavioral Sciences*, *153*, 61–69.

[4] Ali, K. N., Mustaffa, N. E., & Enegbuma, W. I. (2016). Building Information Modelling (BIM) educational framework for quantity surveying students. Journal of Information Technology in Construction, (November)

[5] Alwan, Z., Jones, P., & Holgate, P. (2017). Strategic sustainable development in the UK construction industry, through the Framework of Strategic Sustainable Development, using Building Information Modelling. Journal of Cleaner Production, 140(140 (1)), 349–358.

[6] Alzahrani, J. I., & Emsley, M. W. (2013). The impact of contractors' attributes on construction project success: A post construction evaluation. *International Journal of Project Management*, *31*(2), 313–322

[7] Arayici, Y., Egbu, C., & Coates, P. (2012). Building information modelling (BIM) implementation and remote construction projects: Issues, challenges, and critiques. *Electronic Journal of Information Technology in Construction*, *17*(May), 75–92.

[8] Chong, H., Lee, C., & Wang, X. (2016). A mixed review of the adoption of Building Information Modelling (BIM) for sustainability. Journal of Cleaner Production, 121, 64–71.

[9] Chovichien, V., & Nguyen, T. A. (2013). List of indicators and criteria for evaluating construction project success and their weight assignment. *Proceedings of the 4th International Conference on Engineering, Project, and Production Management (EPPM 2013)*, (Eppm), 130–150.

[10] CIDB. (2017). 22 nd AsiaConstruct Conference Seoul, Korea. Malaysia Country Report, (45), 5–6. Retrieved from http://www.cidb.gov.my/images/content/international/Malaysia----Country-Report-2017--22nd-Asia-Construct.pdf
[11] Construction Industry Development Board (CIDB) Malaysia. (2016). Malaysia Building Information Modeling Report 2016.
[12] Ghaffarianhoseini, A., Tookey, J., Azhar, S., & Efimova, O. (2017). Building Information Modelling (BIM) uptake: Clear benefits, understanding its implementation, risks and challenges. Renewable and Sustainable Energy Reviews, vol. 75, issue C, 1046-1053.
[13] Gourlis, G., & Kovacic, I. (2017). Building Information Modelling for analysis of energy efficient. Renewable and Sustainable Energy Reviews, 953-963.
[14] Haron, A. T. (2013). Organisational Readiness to Implement Building Information Modelling : A Framework for Design Consultants in Ahmad Tarmizi Haron PhD. Thesis Building Information Modelling
[15] Hergunsel, M. F. (2012). Benefits of Building Information Modeling. *Design*, (May), 1136–1145.
[16] Ismail, N. A. A., Owen, R., & Drogemuller, R. (2015). Cost estimating practice incorporating building information modelling (BIM): Malaysian quantity surveyors perspectives. *The 7th International Conference of Sustainable Development in Building and Environment (SuDBE) 2015.*
[17] Kushwaha, V. (2016). Contribution of Building Information Modeling (BIM) To Solve Problems in Architecture, Engineering and Construction (AEC) Industry and Addressing Barriers to Implementation of BIM. *International Research Journal of Engineering and Technology (IRJET)*, *3*(1), 100 – 105.
[18] Latiffi, A. A., Brahim, J., & Fathi, M. S. (2016). Transformation of Malaysian Construction Industry with Building Information Modelling (BIM). Ibcc, 00022.
[19] Latiffi, A. A., Mohd, S., Kasim, N., & Fathi, M. S. (2013). Building Information Modeling (BIM) application in Malaysian construction industry. *International Journal of Construction Engineering and Management*, *2* (March 2016), 1–6.
[20] Monteiro, A., & Poças Martins, J. (2012). SIGABIM: a framework for BIM application. *38th IAHS World Congress Visions for the Future of Housing, Magacities*, 422–427.
[21] Newton, K., & Chileshe, N. (2012). Awareness, Usage and Benefits of Building Information Modelling (BIM) Adoption – The Case of the South Australian Construction Organisations. 28th Annual Association of Researchers in Construction Management (ARCOM) Conference, At Edinburgh, Scotland.
[22] Ofori, G. (2015). Nature of the Construction Industry, Its Needs and Its Development : A Review of Four Decades of Research, *20*(2), 115–135.
[23] Qinghua, H., Wang, G., Luo, L., Shi, Q., Xie, J., & Meng, X. (2017). Mapping the managerial areas of Building Information Modeling (BIM) using scientometric analysis. International Journal of Project Management, 35(4), 670–685.
[24] Succar, B. (2012). Building information modelling framework: A research and delivery foundation for industry stakeholders. *Automation in Construction*, *18*(3), 357–375.
[25] Zakaria, Z., Nasly, M. A., Tarmizi, H., Marshall-Ponting, A., & Zuhairi, A. H. (2013). Exploring the adoption of Building Information Modelling (BIM) in the Malaysian construction industry: A qualitative approach. IJERT: International Journal of Research in Engineering and Technology, 2(8), 384–395.
[26] Zhang, S., Teizer, J., Lee, J. K., Eastman, C. M., & Venugopal, M. (2013). Building Information Modeling (BIM) and Safety: Automatic Safety Checking of Construction Models and Schedules. *Automation in Construction*, *29*, 183–195.
[27] Zikmund, W. G., Babin, B. J., Carr, J. C., and Graffin, M., (2013) Business Research Methods, Mason

IOP Conf. Series: Earth and Environmental Science **385** (2019) 012048 doi:10.1088/1755-1315/385/1/012048

Tendering Issues and Improvement in Public Infrastructure Project Through Public-Private Partnerships (PPP)/ Private Finance Initiative (PFI)

Shamsida Saidan Khaderi[1] , Anis Sazira Bakri[1] , Ani Saifuza Abd Shukor[1] , Abdul 'Izz Mohamad Kamil[1] and Rohana Mahbub[1]

[1]Centre of Studies for Quantity Surveying, Faculty of Architecture, Planning & Surveying, Universiti Teknologi Mara, Shah Alam, Selangor, Malaysia.

Email: shamsida72@yahoo.com

Abstract. In recent year, there have been an increasing market of Public-Private Partnerships (PPP)/Private Finance Initiative (PFI) for the development and operation of public infrastructure projects in Malaysia. The aim of PPP/PFI is to encourage greater participation of private sector in delivering public services. However, due to lack of knowledge on PPP/PFI procurement bring it being criticized especially in tendering process. Some PPP/PFI projects are the subject of lengthy political debate before they go to tender, causing further delay in their execution. Though, very little empirical studies have been conducted in the area of its overall implementation and tendering process. Therefore, the objectives of the research are to identify and understand the PPP/PFI implementation and practices in public infrastructure project in Malaysia; to determine issues and challenges in PPP/PFI tendering stage; and to propose PPP/PFI tendering improvement. The overall research process and methodology is divided into three phases: The first phase starts with extensive literature search on the current state and issues regarding to tendering in PPP/PFI public infrastructure projects and analysed the relevant documents of PPP/PFI. Based on these searches appropriate problem statement, research aim and objectives are established. The second phase deals with questionnaire survey. The study mainly focuses on concessionaires who are involved in PPP/PFI projects in Malaysia. The third phase is to synthesize the data collected from the literature review, document analysis and questionnaire survey. Finally, is to report the findings, conclusion, recommendation and further research based on the final outcomes. Expected outcomes from the research project are to provide new insight on the theoretical perspectives on tendering for public infrastructure project through PPP/PFI implementation.

1. Introduction

During the 10th Malaysian Plan, the Government plans to expand the implementation of PPP-based projects to increase the opportunity for the private sector to participate in infrastructure and utility development. This measure will also contribute to the growth of the construction industry, which has contracted in the past eight years. PPP/PFI is now first introduced in Malaysia, but in the United Kingdom (UK) it has been practiced for nearly 22 years. The PPP / PFI concept was introduced in the United Kingdom in 1995 and was used as a mechanism to procure public services from administrative bodies and to exploit the private sector 's expertise in financing, design, construction and operation during the provision of public services [1]. The PPP / PFI vehicle is different from the privatization in which the private sector is responsible for service provision. Instead, this vehicle primarily involves a structure in which the government buys services such as the construction and operation of public facilities in the private sector and provides public administrative services [2].

The PPP / PFI market for the development and operation of infrastructure projects in Malaysia has increased over the past year. However, few empirical studies have been carried out in the field of overall implementation and tendering. Some PPP / PFI projects are subject to long political debate before tendering, causing further delays in their implementation [3]. In New South Wales, Australia, the planning and decision- making process for a motorway project has gone on for almost a decade and half. However, despite many projects to date, the effectiveness of the PPP / PFI is still under considerable debate.

Several studies have suggested that PPP / PFI has benefited from conventional procurement in the public sector [4]. For instance, a survey of PPP/PFI projects by the National Audit Office revealed that only 22% of projects experienced increasing construction costs after the contract had been awarded [1]. According to [5][4][6], many of the constraints that occur in PPP/PFI projects in Malaysia are due to the failure in the tendering process. It is crucial to adapt the open competitive tendering process to select the most qualified contractor based on their expertise and experience. This problem also arises in other countries like the UK and Australia.

A study conducted by [1][2][7][8][9][5] has shown that the most critical stage in the implementation of PPP/PFI projects is at tendering stage. Among the things that have identified, a lengthy negotiation period for a decision and the high cost involved in the process and the tender price. The very high value of the bidding process is not only a waste of resources that could invest in buildings and services, but also a barrier to the entry of all but a small number of important players into the PFI market. This has led to a reduction in competition in the sector, which many specialists, niches and innovative organizations are now turning away from PFI projects. Furthermore, given the increasing complexity of bids, the time required to undertake the bidding process is now excessively long, with subsequent project delivery delays. The PPP / PFI procurement process requires a complete systematic approach compared to the traditional procurement method.

2. Literature Review

2.1Issues in PPP/PFI Tendering Process

The tendering process is crucial for the public customer to obtain the value for money in PPP / PFI venture procurement. It relies upon the existence of a competitive bidding process. The guideline of the significance of value for money and whole life costing would require the government to emphasize the best value, and not necessarily the lowest initial price. In this way, the bidders must guarantee that their costs are the lowest for the whole life of the concession, not just at the initial of construction. There is an issue where the contractors often said that high bid costs prevent them from bidding for more PFI projects. There is a healthy number of contractors in PPP/PFI market, but the number of bidders for each project is falling according to Great Britain Parliament: House of Lords [10].

The Construction Industry Council [11] stressed that the bidding for PFI projects is intrinsically complicated and initially expensive. Once the bidder has the experience, the following bids should be cheaper. [12] mentioned that the PFI projects had awarded through competitive bidding. It is a necessity to process in extracting value for money from the PFI. It is significant for both public and private sector partners to recognize the importance of an effective procurement competition as bidding costs for PFI projects. PFI/PPP explicitly designed for large-scale, high-value projects such as road/rail infrastructure networks or hospitals. It is a result of the shift toward privatization. A construction company looking for a long-term involvement needs to formulate a strategy to deliberate over the type of PFI projects to be a focus on.

PPP/PFI tendering procedure is more complicated and costlier compared to traditional process [13]. As an example, in UK's PFI project tender costs, Design and Build and traditional method reveal high total project cost as compared to PPP/PFI (higher range between 0.48% to 0.62%). The surveys by [9] showed that the bidding cost is a significant problem in PFI. This review additionally upheld by the Contract Journal and Knowles survey in the same year showing that the respondents rated the price and the bidding process in the scale of 8 from 10 for PFI as being unattractive [8] stated that high bidding costs represent opportunities and a high risk of entering the market for companies, act as barriers to market participation and reduce market competitiveness. [14] expect that the bid cost for a PFI project could be as much as £0.5m, contingent upon the project size with even the biggest contractors discouraged. Due to tendering cost risk, most of the contractors have refused to become engaged with PPP/PFI work.

The lengthy procurement period and a high-cost impact on the negotiation of PPP/PFI deals. The average procurement time for PPP / PFI companies in the UK is about 22 months and in Australia between 12 and 18 months [15]. An acceptable transfer of risks, payments and other terms has taken a long time for both sectors. In some cases, the main reasons for the delay were either preventable or could have been mitigated by the public sector without jeopardizing the value for money, such as insufficient marketing specifications, changes in the scope and design of projects and poor management processes [16]. In most PFI projects, for example in the case of the road project in Scotland, the risk of procurement, time and cost overruns was part of. Due to a large meeting with bidders, the procurement period had to be revised between 17 and 27 months [17] cited that the National Audit Office [18] showed that, due to this and other unforeseen events, the procurement costs for the public sector procurer from £ 1.3 million to £ 2.85 million.

2.2PPP/PFI Tendering Process Improvement

Many authors have proposed an alternate arrangement of policy to enhance PPP bidding process. All phases of a project need to be overseen well, including the tendering stage. Many authors have reviewed several proposals to strengthen the skills and support of the public sector and to reduce procurement times and costs for both the public and private sectors. These incorporated a more prominent accentuation on the readiness of projects before they go to the market, increased monitoring and scrutiny of projects, including a new stage of investigation before selection of the preferred bidder. PPP procurement development takes long-time than numerous traditional types of procurement imported into the construction industry previously. To reduce the bidding cost which will not affect the value for money of the PPP/PFI projects, there are few strategies recommended for the government.

[19] had classified three strategies as to avoid a premature projects announcement and allowing enough time for pre-tender phase arrangement as a following; a proper PPP market study is required before any project plan is issued. Next, release the demand for a structured proposal once all the necessary preliminary works have completed, so that it can lower the need for addendum and re-bids. Another strategy to be adopted is to reduce the amount of bid design work required by bidders instead of relying more on project requirements.

The government should also assure more financial institutions to participate in PPP/PFI to facilitate the issue of adequate finance facilities faced by the concessionaires. It may divert the scepticism in providing financial support for PFI projects. An education programmed on the mechanism of PPP/PFI should be held to all parties to make it more attractive. After so much constraints derived respecting PPP/PFI matters, the best way forward is to improve the PPP/PFI procurement commonly and tendering procedure undoubtedly in the future. According to Reeves et al. [20], the history of PPP procurement in the United Kingdom recommends that efforts to improve the execution of the tendering process be continued since the beginning of the PFI. The official report of the PFI, HM Treasury [16] has, over time, emphasized the objective of strengthening the procurement procedure.

3. Research methodology

3.1Method of Data Collection

The importance of research methodology is to explain the research process and tools used to analyse its objectives. This research consists, in general, of two mixed approaches, PPP / PFI studies from previous studies and quantitative research. First, the investigation began with a review of relevant literature covering materials from journals, magazines, government agencies, seminar reports, conference proceedings, and web trawl. The purpose of the early literature review was to identify the current status and tendering issues in public infrastructure projects in PPP / PFI. Then a quantitative research approach was conducted.

Quantitative research usually involves the collection and conversion of data into a numerical form, so that a statistical calculation can be carried out and the conclusion drawn. Questionnaires are the most popular method for obtaining the data required for research. The distribution of questionnaires is an example of primary data to strengthen the report findings. This quantitative method will identify problems and challenges in the PPP / PFI tendering phase and propose improvements in the PPP tendering. A flow chart in Figure 1 shows the research methodology process.

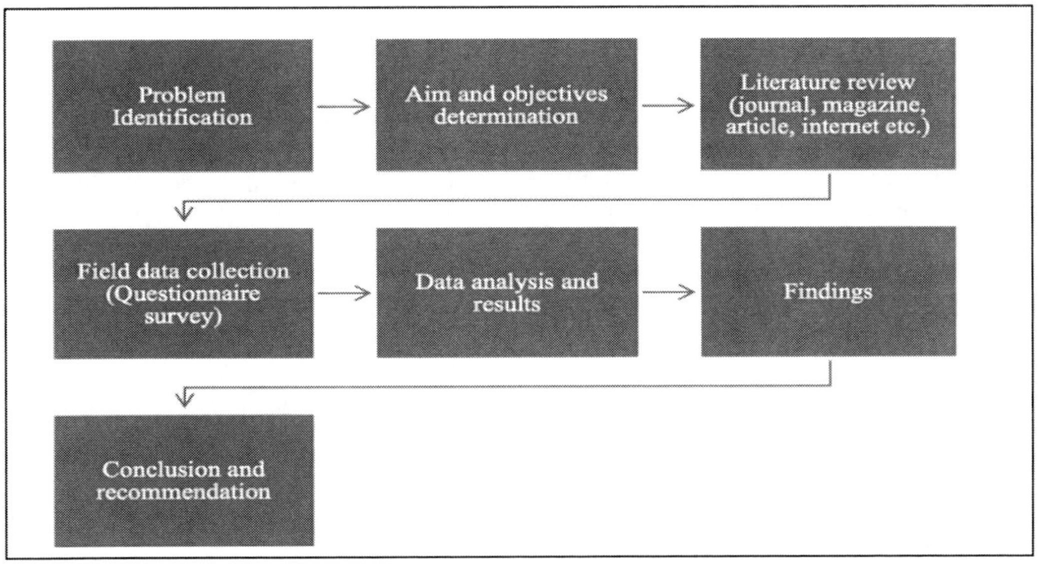

Figure 1. Research Methodology Process

3.2 Questionnaire Survey

The objective of the survey is to collect primary, relevant and up-to-date information on the issues and challenges associated with the implementation of the PPP / PFI tendering stage, to support and validate the knowledge gained from the literary review. A questionnaire survey was carried out to gather information from a construction company (concessionaires and contractors), which involve in PPP/PFI projects in Malaysia. They were selected because they have reasonable capacity in both financially and expertise to participate in PPP/PFI projects. In this study, the survey based on four section names Sections A, B C and D. In section A, the questionnaire describes the information of the respondent. Section B requires information related to PPP / PFI projects, knowledge level, and PPP / PFI mechanism. Section C required respondents to identify the issues and challenges reported in tendering for PPP/PFI, common procedures used in tendering for PPP/PFI, criteria and reasons for withdrawal from competition. Section D aims to identify the solution to the problems associated with the tendering procedures for PPP / PFI. Respondents must provide some advice to alleviate the issues.

The Likert Scale is used in Sections C and D to ask respondents about their perception of the questions listed in the questionnaire. The researcher uses scale 1 with strong disagreement and scale 5 with strong agreement. Typically, there are five categories of response, from (for example) 1=mighty disagreement to a 5=mighty agreement, although there are arguments in favor of scales with 7 or even numbers of response categories. A total of 100 sets of questionnaires have been distributed to the correspondents. All the questionnaire was spread via email and manually send to concessionaire and contractors office. Among 100 set that had been distributed, only 30 sets were duly answered and given back. One of the limitations of this research is there is a small scale of respondents that involved in PPP/PFI projects which not all of them were exposed with the implementation of PPP/PFI.

The software used in the data analysis is version 22.0 of the Statistical Package for Social Science (SPSS). Some results derived from the SPSS software questionnaire analysis, such as descriptive statistics; bivariate statistics; prediction for numerical results; and prediction for group identification. In this research, even many features can access from the method, only frequencies statistics, cross-tabulation and descriptive have been used to obtain the frequency of problems and improve the tendering stage for public infrastructure projects in PPP / PFI.

4. Research Findings

4.1Issues and Challenges in PPP/PFI Tendering Process

Although all the respondents aware of the benefits and advantages of adopting PPP/PFI in managing public infrastructure project tendering, still PPP/PFI must face various challenges in its use. The primary data showed that there is a high degree of agreement on issues such as high tendering costs due to its process, the complexity of the negotiations because it took a long time to conclude a deal, the length of the tendering periods and the high bidding costs. Practically, projects taking the PPP/PFI route are often more involved in public infrastructure projects. It, therefore, requires a more complex form of procurement, technology, and specification which leads to a much more extended period of negotiation before the work begins. Most of the respondents (mean score of 4.51) 85 percent strongly agree that the crucial issues relating to PPP / PFI are high tendering costs and complicated negotiation. This medium is because PPP / PFI is more complicated than traditional procurement and the conclusion of an agreement took a long time. This situation can lead to the unattractive participation of PPP / PFI in the private sector and can lead to less market competitiveness.

PPP / PFI took longer than other types of traditional procurement due to many reasons, including the need for the government not only to identify and articulate a long- term requirement for a short term but also to consider it before calling for tenders. It can also say that the customer must list bids before requesting detailed proposals. Not all bidders will invest in the cost of preparing a comprehensive plan if the field of potential bidders is not reduced. Most of the types used in the Malaysian PPP / PFI tender are invited to tender in comparison with PPP / PFI in the United Kingdom, which use open competition. One respondent indicates that the reason why the Malaysian PPP / PFI tender has adopted such procedures is to reflect transparency and good governance while ensuring that only capable and financially stable firms are involved. It is essential for a concessionaire to have the most appropriate criterion for winning the PPP / PFI tender to meet the specification, to provide a secure financing term and to be a coherent consortium, all of which have been rated by the respondent as strongly agreed.

4.2 Suggestions to Respond to the Issues Discussed

After so many problems and challenges arise about the PPP / PFI issue, this is the best way to improve the PPP / PFI procurement in general and the tendering process in public infrastructure projects. All respondents strongly agreed that enough time and resources should be allocated to procure authorities to develop and prepare their plans before formal commitments are made to attract more private sector participation in PPP / PFI projects. The reasons for this are that the structure of PPP / PFI projects involves various parties carrying out a different project task and risks. It also shows that most respondents also agreed that standardized contractual guidance needs to be developed and enforced. This medium is essential to ensure that every building expertise has clear concepts of PPP / PFI and attracts the PPP / PFI market. A standard guideline should be followed to ensure that all procedures are carried out accordingly so that the products and services provided represent a good value. It will also increase the total number of private sector participants in the provision of public infrastructure. The government should even persuade more financial institutions to participate in the PPP / PFI to alleviate the problem of the concessionaires ' inadequate financing facilities. This factor can probably change the way of being sceptical about financing the PPP / PFI project.

5. Conclusion

The research methodology was established to achieve the research objectives. A set of questionnaires developed in the quantitative method based on the findings from the literature review. The survey was designed using the Likert scale of five ordinal measures of agreement. The results analysed from the collection of data could then be used to justify and validate the facts of the literature reviews. The findings and recommendations are finally made to conclude the investigation.

The methodology approach showed that most of the problems and challenges associated with the implementation of PPP / PFI procurement in Malaysia is concessionaires who are not trained and may not be familiar with the process, detailed and tedious. The tendering procedures faced high tendering costs and problems that are more complicated than conventional procurement, complexity in negotiations because it took a long time to conclude an agreement and high tender costs. From this study, it can find that the procedures of PPP/PFI tendering process for public infrastructure project might differ from one to other countries that implement it. The method used in awarding the private sector (concessionaire) also can vary due to the complexity of the PPP/PFI projects. Thus, contractors must have an effective plan and mechanism in place to produce a winning bid for PPP/PFI contract. Overall this paper provides the explanation and sound reasoning to the selection of the research methodology. Subsequently, the description as to how the quantitative approach will use in the context of this study through literature review and questionnaire method presented.

References

[1] NAO, Improving the PFI tendering process. *Report by the Controller and Auditor General*. HC **149** Session 2006-2007 (2007)
[2] Merna A, & Owen M. The Private Finance Initiative. *Centre for research into the Management of Projects,* UMIST, 1999
[3] The Edge. Untested model for Malaysian companies, *The Edge Malaysia*, pp 75, 2006
[4] Abdul Rahman I, Aftab H M, and Nora Sheda M Z. Public private Partnerships procurement methods in Malaysia construction industry. *Journal of American Science*, **10(7)** : pp 91-97, 2014
[5] Shamsida S K. Private Finance Initiative (PFI) Implementation in pre-construction stage. *Unpublished Ph.D. Thesis*. University Sains Malaysia, 2011
[6] Syuhaida I, & Aminah M D. The provision of infrastructure via PFI – Theoritical and Empirical. *Research in Urban Management.* Special Number **15**/April, 2009
[7] Broadbent J, and Laughlin R C, Public Private Partnerships introduction. *Accounting Auditing & Accountibility Journal*, **16(3)** :pp332-341, August, 2003
[8] Dixon T, Pottinger G and Jordon A. Lesson from the PFI in the UK : Benefits, Problems and Critical Success Factors. *Journal of Property Investment and Finance.* **23(5)** :412-423 Oct. 2005
[9] Akintoye A, Beck M, and Hardcastle C. Public Private Partnerships – Managing Risks and Opportunities. *Blackwell Publishing* (2003)
[10] Great Britain Parliament : House of Commons. Private Finance Initiative (PFI). *Seventeenth Report of Session 2010-12* Vol. **1**, 2010
[11] The Construction Industry Council. *Contractors' Key Guide to PFI,* 1998
[12] Bain R, (2006), "Private finance rates of return: evidence from the UK'S PFI roads sector", available at: www.robbain.com/ETC%20Paper%20Bain%202008.pdf
[13] Birnie J, Risk Allocation to the Construction Firm within a Private Finance Initiative (PFI) Project. *ARCOM Conference Proceedings*, pp. 527-534 (1997)
[14] Akintoye A, Hardcastle C, Beck M, Chinyio E, and Asenova D. Achieving best value in private finance initiative project procurement'. *Construction Management and Economics*, Vol. **21**, pp. 461-470 (2003)
[15] Grimsey D. and Lewis M K. (2005), "Are public private partnerships value for money?", *Accounting Forum*, Vol. **29** No. 4, pp. 345-378
[16] HM Treasury, PFI: Strengthening long-term partnerships. *HMSO*, London, 2006 [Online] Abvailable. http://www.hm-treaury.gov.uk/media/IEI/33/bud06_pfi_618.pdf [Accessed December 10, 2006]
[17] Ahadzi M, Bowles G, The Private Finance Initiative: The Procurement Process in Perspective". *Proceedings of the 17th Annual Conference of ARCOM*, Salford, pp. 971-980 (2001)
[18] NAO. Private Finance Initiative – Construction Performances. *National Audit Office*, 1999.
[19] Nurfarhana S. Tendering Strategies in Winning PFI Contract. *Unpublished Degree Thesis.* University Teknologi MARA, Shah Alam, 2015
[20] Reeves E, Palcic D and Flannery D. PPP Procurement in Ireland : An Analysis of Tendering Periods. *Local Government Studies* **41(3)** :379-400 March 2015

IOP Conf. Series: Earth and Environmental Science **385** (2019) 012049 doi:10.1088/1755-1315/385/1/012049

Deriving Urban Design Principles for Jeddah Corniche Developments: User-Preferences Approach

Reem Fuad Oudah[1], Tarek Ragab[1] and Mohammed Shokry[1]
[1]Collage of Architecture and Design, EFFAT University, An Nazlah Al Yamaniyyah, Jeddah 22332, Saudi Arabia

Email:tragab@effatuniversity.edu.sa

Abstract. Today, many recreational waterfront developments are only superficially pleasant as they lack significant design qualities needed by users. Notions on waterfront qualities create communication gaps between a designer vision and end-user perception. Designers might formulate objectives and ideas on waterfront qualities without adequately understanding the true needs and preferences of users, which encompass aspects deeper than merely visual features. This study evaluated the design quality of the recent development of Jeddah north Corniche. It aimed to analyse the relationship between physical patterns and users activity patterns, which will assert in understanding how seafronts physical pattern produces different social interactions; and how these activity–physical pattern relationships make this public realm work or not. This research employs questionnaires data collection method to interpret the activity patterns that appear to relate with particular use of design features. Analysing this relationship will conclude seafront socio-physical design principles, which take into account users preferences and the unique characteristics of Jeddah location and its Islamic culture, and encourage excellence in the design quality of urban waterfronts; focusing on creating prosperous, sustainable and liveable social environments. In addition, it will add insights and complement the application of urban design theories and practices, which could lead to further studies aiming to improve recreational seafronts design quality.

1. Introduction

Dong [1] defined the word waterfront as "the part of a town or city adjoining a river, lake, harbour, etc.". The waterfront is defined generally as the area of interaction between an urban development and the water. The urban waterfront could include any urban area that overlooks or adjacent to a river, sea, lake, canal or an artificial water body. At the city or macro level, the waterfront is the window reflecting the quality of the city [2]. The waterfront contributes to the formation of the city's image and identity. Many cities are well known for their scenic waterfront landscapes. Some of these are very traditional, such as London, Paris, and Venice, while some are modern, such as Boston, Sydney, and Hong Kong. Recently, many waterfront cities have tried to implement measures by which to stimulate economic growth, improve environmental condition, and enhance competitiveness and vitality by taking advantage of their unique waterfronts and utilizing their respective cultural and historical resources [3].

Identifying the physical form of waterfronts is essential in finding out whether designed places are adaptable and able to accommodate people's activities [4][5][6]. According to Rasouli [2], investigating the physical dimension will lead to understand whether designed spaces are determined by the types of activities. Nonetheless, the local officials and urban designers lack of appropriate knowledge about the strong integration of social activities and public spaces use with the quality of physical design features in the process of designing quality waterfronts. The importance of such appropriate knowledge proved by argument of many scholars[6][7][8][9].

Designers and end-users may have different notions on waterfront qualities, and communication gaps possibly exist between the two groups. Designers might formulate objectives and ideas on waterfront qualities without adequately understanding the true needs and preferences of users, which encompass aspects deeper than merely visual features. Hence, perceptions on waterfront physical and social attributes need to be analysed. The designers should create more human-oriented waterfronts by gaining a better understanding of real user needs. User's perceptions regarding waterfront and the key aspects associated with those perceptions should be studied. Therefore, the main objective of this research is to explore an alternative approach to waterfront design through the eyes of users. Four key attributes of successful places: accessibility, comfort, socialization, and activity are accepted as quality indicators of today's urban spaces and they became an outline for the study methodology.

2. Methodology

The chosen waterfront to be studied is Jeddah's Corniche. In year 2012, Jeddah's Corniche project was awarded the best urban planning award in the Middle East. Thus, it was of great interest to evaluate this development. Furthermore, The Corniche has been indicated as the most popular public recreational site in the city due to the amount of open space it offers and because of its location on the Red Sea. Due to that, visitors come from various areas and nearby cities. Hence, their opinions reflect, to some extent, that of the total population of Saudi Arabia. Since the Corniche is very wide, this study has been narrowed down to the northern part of the Corniche only as shown in Figure 1. This selected site presents a typical example of various public recreational areas all over different parts of Saudi Arabia. Various social and physical problems had been noted by the author such as privacy intrusion, behavioural conflicts, harassment faced by the female visitors, social tension and conflicts between visitors and automobiles, between pedestrians and those sitting on the pavements.

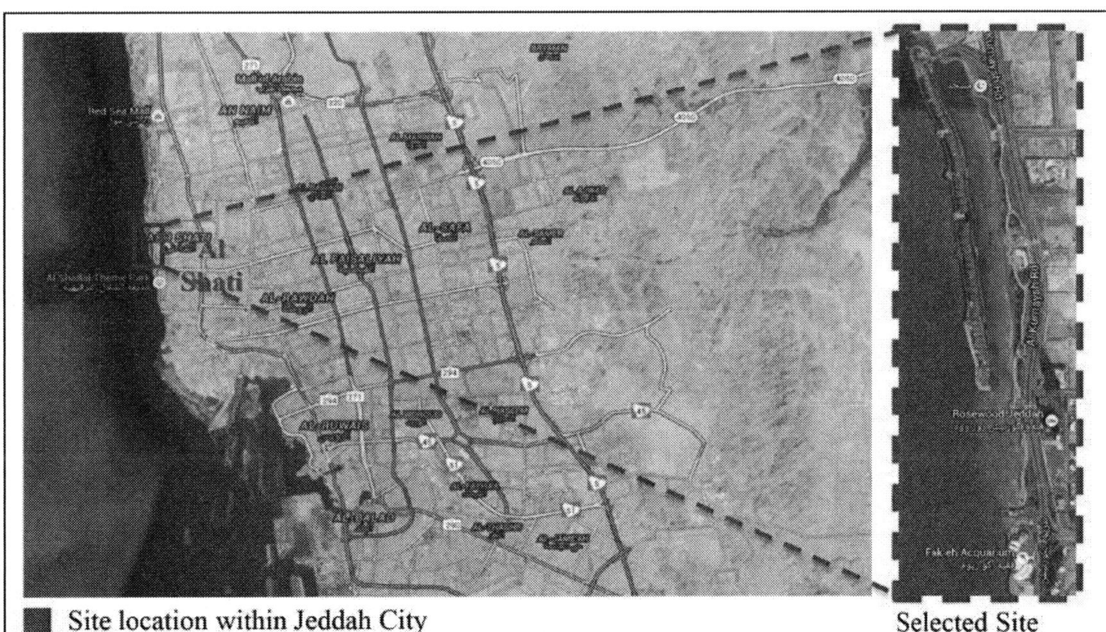

Figure 1. Study site (North side of Jeddah Corniche)

The data for this study is gathered using questionnaire whereby the 100 participant samples are randomly selected. The design of the questionnaire is intended to assess user's perception, which is associated with how users form a mental image of their socio-physical experience and accordingly prefer a specific place rather than another. These perceptions sometimes affect the cultural background and social profile of the user. The questionnaire aims to explore the relationship between users' satisfaction and the seafront recreational provision in order to investigate how that might influence or determine the Corniche public realm design. Moreover, it focuses on finding out how people behave in relation to the leisure site.

3. Results and Discussion

3.1 Parking

To ease congestions and avoid disturbance of parking lots, a question was asked to test how far people are willing to walk if their cars were parked far away from the Corniche. Results were positive with 29% of people willing to walk 10-50 meters, 26% did not mind walking 50-100 meters while 36% preferred not to walk at all (Figure 2). Li et al. reported that tourist like to choose cheaper and more convenient parking solution [10]. Nonetheless, they are willing to spend more time and money on the journey with scenic roadway and recreational characteristics. In addition, Wang et al stated that parking space availability, parking charge, and walking distance have significant effects on holiday parking choice of visitors.[11].

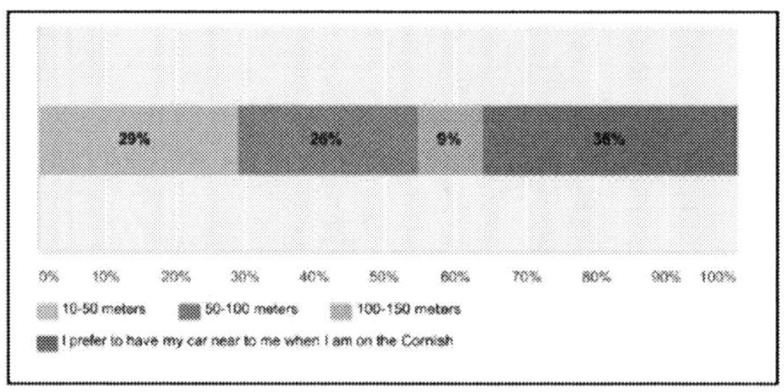

Figure 2. Walkability distance from parking lots to corniche

3.2 Visiting Frequency

The majority of responses confirmed that peaks on the Corniche happen in weekends and vacations with 45% of respondents; 25% go every month and 20% go every week (Figure 3). As for the time of these visits during the day, 57% usually go at night, 33% in the morning while only 10% go in the afternoon (Figure 4). During these visits, 48% would spend 1-2 hours and 38% would spend 3-4 hours.

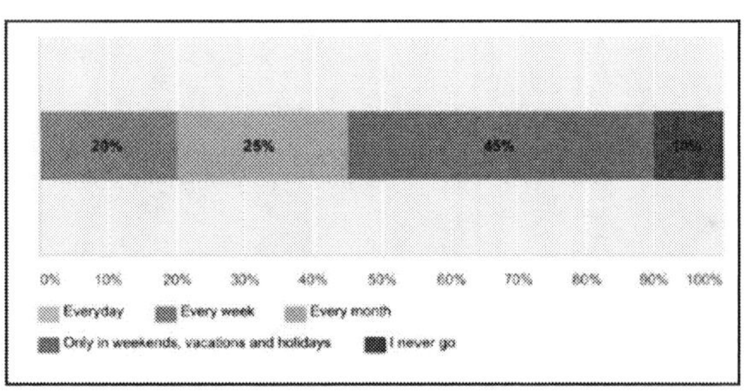

Figure 3. Frequency of visiting Corniche

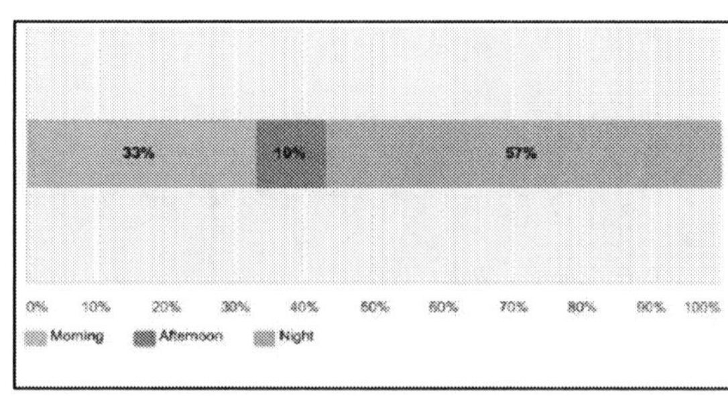

Figure 4. Preference of visiting time in a day

3.3 Reasons for Attraction or Leaving

When the respondents were given options on attraction to the Corniche, the top five preferences (refer Figure 5) were to see the sea and the natural view (98 votes), relaxation and meditation (81 votes), to meet other family members or friends (61 votes), spaces for family and children to play (59 votes), lastly casual trip (47 votes). Others highlighted reasons like walking, eating, nearby house and the only place to go to. Figure 6 presented the outcome of the answers on why do most of the visitors decide to leave Corniche. 88 of them stated that the place is being overcrowded, 69 mentioned that the location is very sunny and hot, and 33 of the respondents stated that they left to find a better place. These were the three main factors selected whereas cleanliness, poor lighting at night, not designed for different age groups, lack of services and no comfort while sitting and the poor mixing between activities without proper definition were some of the highlighted reasons as well.

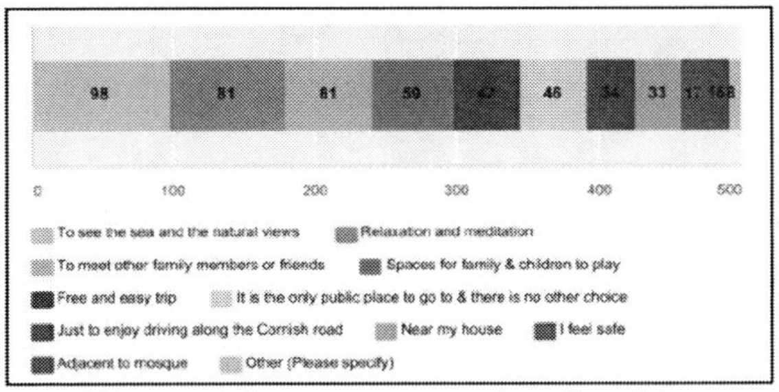

Figure 5. Reasons attracted to Corniche

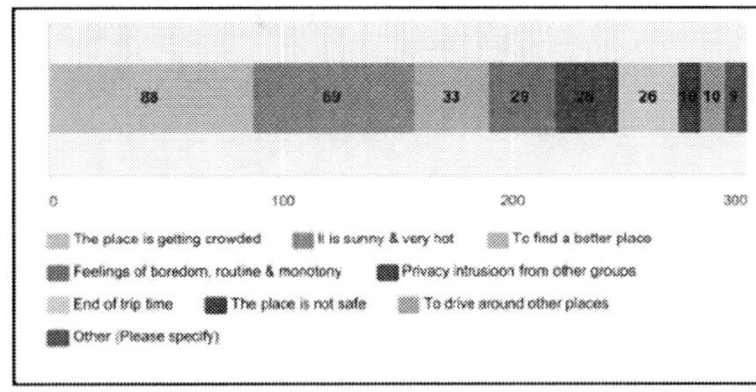

Figure 6. Reasons to leave Corniche

3.4 Perception and Satisfaction of Physical Features

Aspects such as thermal, noise and lighting, along with functionality need to be considered in order to obtain global comfort of visitors [12]. Four critical design features were questioned. 94% of respondents stated that the number of seats is not enough. 87% stressed on the importance of having shaded areas. 76% complained from noise on the Corniche. As for the lighting at night, 31% were satisfied, 65% said to add more lighting as it is not enough at night (Figure 7).

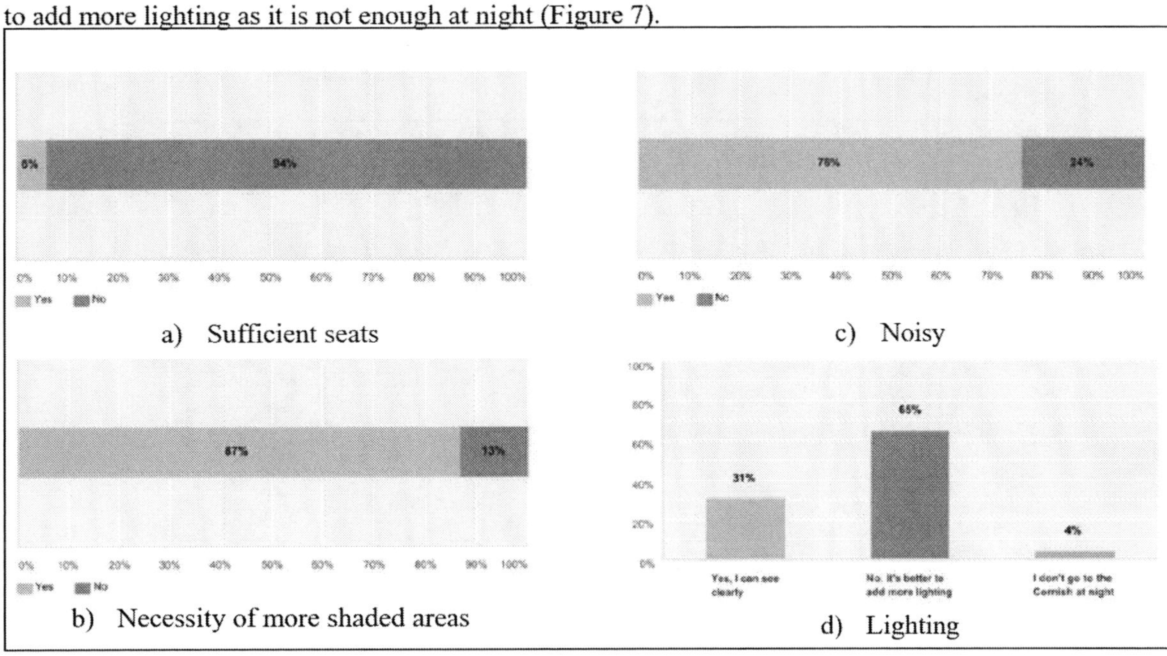

Figure 7. Perception on physical features at the Corniche

3.5 Patterns of Use

Several questions were asked to investigate levels of comfort in relation to the physical feature. Figure 8 showed the responses obtained regarding the seating comfort at the Corniche. As for seats, results were almost equal since 30% preferred to sit on the grass, 25% use available seating's while 31% would rather bring their own seats. With reference to Figure 9, for safety, 90% of respondents were unhappy to be away from their children while they are playing. To measure accessibility (Figure 11), 54% of respondents were familiar with main entrances while 46% did not recognize it. Similarly, 78% of respondents did not notice handicapped access as figure 12. As for signage and way finding, 61% stated that they need more signage that is clearly distributed around the Corniche while 25% highlighted that they could not find clear directional signage (Figure 10).

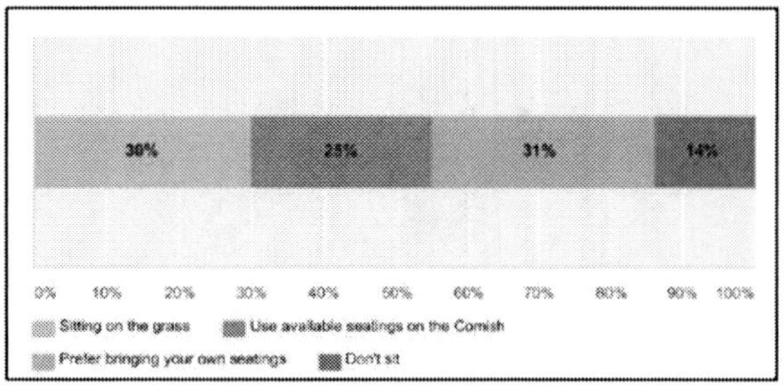

Figure 8. Comfort in seating

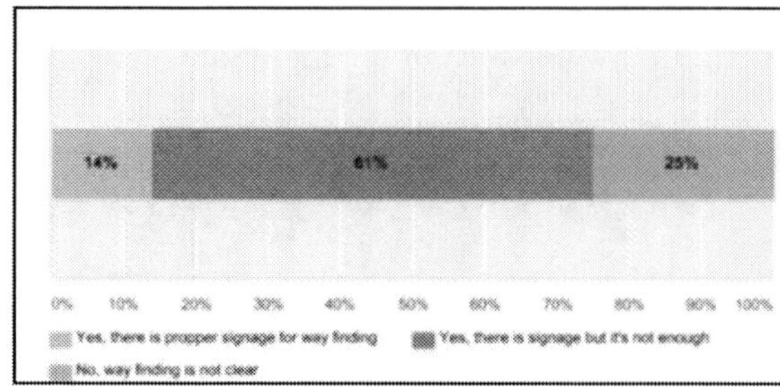

Figure 10. Locating places at Corniche

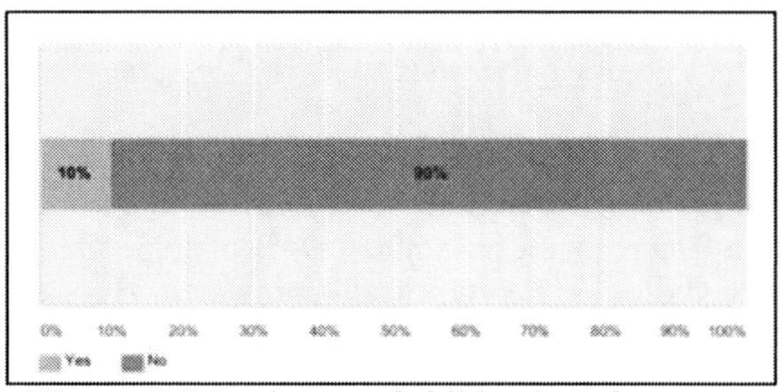

Figure 9. Worriment of children's safety

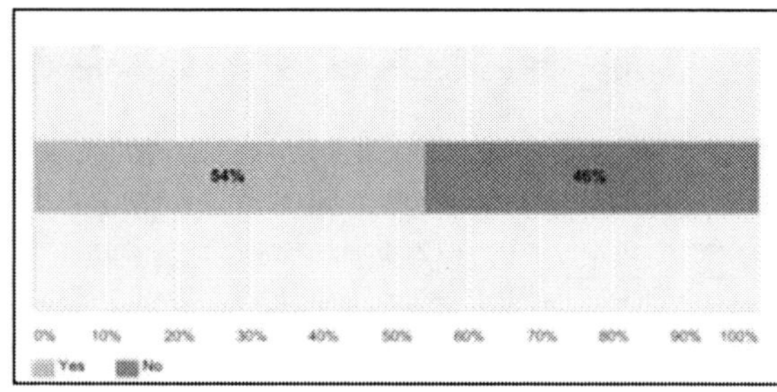

Figure 11. Ease in finding main entrances to an area in Corniche

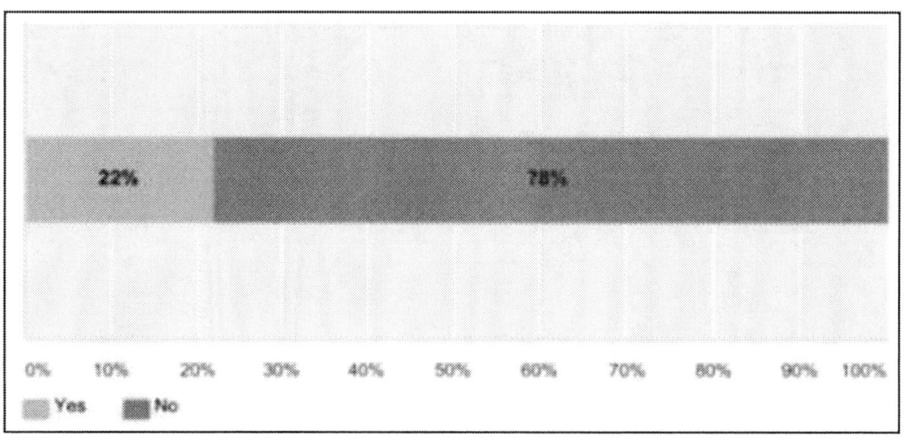

Figure 12. Handicap special parking areas and ramped entrances provided

3.6 User's Preferences

There are many activities carried out around the Corniche area. In Figure 13, the question intends to classify the top five activities according to user's preferences. 91% prefer sitting and relaxing, 69% prefer eating, 68% like walking and exercising, 61% prefer barbeque and finally 48% like taking photos. Whereas activity such as fishing and swimming were the least done with only 22% and 6% of the population respectively. Deenihan et al [13] found that a tourist is willing to increase their cycling time by approximately 100% in order to cycle upon a fully segregated from traffic cycling facility rather than along a road without cycling infrastructure. While they are willing to increase their time by 40–50% to be able to cycle along a road with a cycle lane rather than a road without cycling facilities. Silva [14] suggested that a comfortable microclimate for the pedestrian could be achieved through urban planning and design, considering street orientation and aspect ratio.

In terms of spatial functions (Figure 14), 87% of respondents preferred shaded areas and green open spaces as their number one favourite space. Second, 57% needed to have clean toilets. Third, 50% required easy and adjacent car parking areas. Fourth, 47% preferred safe children playgrounds. Fifth, 41% would like to have designated BBQ areas.

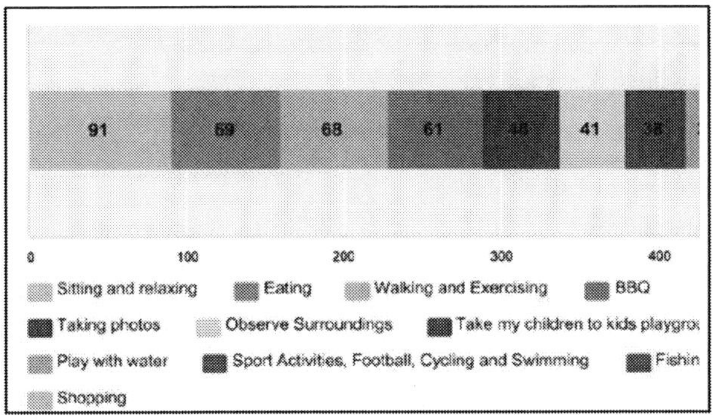

Figure 13. Activities frequently done

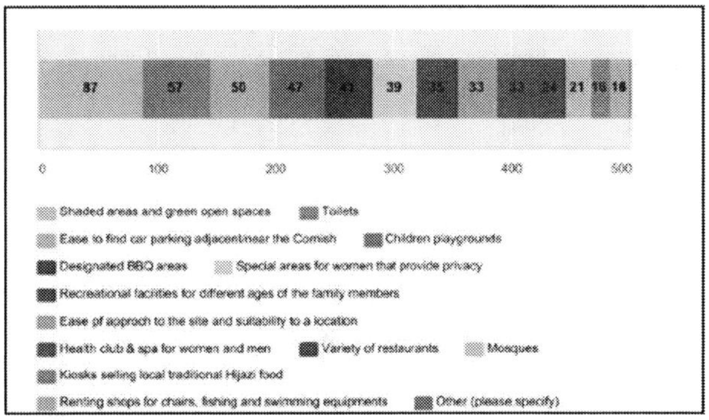

Figure 14. Preferred facilities at Corniche

4. Conclusion

Waterfronts play a vital role in the revitalization of cities economically, environmentally and socially. Waterfronts design qualities are an important topic have and continue to be research until today. Some of these studies aim is to fill the gap between urban designers and users perception of waterfront developments. Therefore, involving users in the creation and management of their built and natural environment will provide insights about their needs and preferences. This "Participatory Design" approach is a well-known trend these days as it helps in involving citizens in the planning and decision-making, which consequently improve the quality of designed projects. Following the application of the methodology, design principles for Jeddah Cornish development are concluded. These principles consider the two main variables of this study: 1) spatial "physical" principles and 2) social "activity" principles in order to improve Jeddah Cornish design quality. Results from the questionnaire revealed that the development of waterfronts depends on the existence of social interaction. Therefore, waterfront design should increase sense of community in users which is able satisfy the needs of comfort, relaxation, passive and active engagement. Some of the recommendations provided by the respondents are as follow:

1. Provide well-defined, wide and welcoming accesses: physical, visual and symbolic, with special accommodations for the disable.
2. Provide enough parking areas with special consideration for disable: nearby compact parking areas or located within 10-50m walking distance.
3. Create a variety of climate environments to facilitate activity in different seasons and weather conditions: include sunny, wind-protected areas for winter and shaded zones for summer.
4. Provide lights on sidewalks and activity areas to extend opportunities for activity into the evening.
5. Create multi-functional and mixed uses on the waterfront.
6. Regulate activities according to the dependency on water: water dependent activities should be located near the water. However, co-locating activities is recommended between sitting and walking.
7. Facilities and services should be within a reasonable price and support the waterfront activities like shaded chairs rental shops, local foods markets, outdoor gym equipment's, and a variety of affordable to high-end restaurants.

References

[1] Dong L 2004 *Waterfront development: a case study of Dalian, China* thesis

[2] Rasouli M 2013 *Analysis of Activity Patterns and Design Features Relationships in Urban Public Spaces Using Direct Field Observations, Activity Maps and GIS Analysis. Mel Lastman Square in Toronto as a Case Study* thesis

[3] Shang W 2011 *Role of waterfront in shaping city center landscape: perception of Tianjin Haihe riverfront landscape* thesis

[4] Bornberg R 2008 Identity by spatial design: Topos, a principle derived from historic and vernacular cultures *URBAN DESIGN International* **13** 182–200

[5] Stephenson J 2010 People and Place *Planning Theory & Practice* **11** 9–21

[6] Carmona M 2010 *Public places - urban spaces: the dimensions of urban design* (Amsterdam: Architectural Press, Elsevier)

[7] Varna G and Tiesdell S 2010 Assessing the Publicness of Public Space:The Star Model of Publicness *Journal of Urban Design* **15** 575–598

[8] Francis J, Giles-Corti B, Wood L and Knuiman M 2012 Creating sense of community: The role of public space *Journal of Environmental Psychology* **32** 401–409

[9] Ansari F A 2009 *Public open space on the transforming urban waterfronts of Bahrain: the case of Manama City* thesis

[10] Li J, Ye J, He Q and Shao C 2016 A Novel Scheme to Relieve Parking Pressure at Tourist Attractions on Holidays *Sustainability* **8** 164

[11] Wang B, Shao C, Li J, Zhao D and Meng M 2015 Investigating the interaction between the parking choice and holiday travel behavior *Advances in Mechanical Engineering* **7** 168781401558949

[12] Rossi F, Anderini E, Castellani B, Nicolini A and Morini E 2015 Integrated improvement of occupants comfort in urban areas during outdoor events *Building and Environment* **93** 285–292

[13] Deenihan G and Caulfield B 2015 Do tourists value different levels of cycling infrastructure? *Tourism Management* **46** 92–101

[14] Silva J P 2017 Solar radiation and street temperature as function of street orientation. An analysis of the status quo and simulation of future scenarios towards sustainability in Bahrain *E3S Web of Conferences* **23** 02002

IOP Conf. Series: Earth and Environmental Science **385** (2019) 012050 doi:10.1088/1755-1315/385/1/012050

Survey Based Sustainable Socio-Cultural Guidelines for Neighbourhood Design in Jeddah

Khadijah A. Fatani, Mady Mohamed[1]* and Samah Al-Khateeb[1]
[1]Collage of Architecture and Design, EFFAT University, An Nazlah Al Yamaniyyah, Jeddah 22332, Saudi Arabia

Email: momohamed@effatuniversity.edu.sa

Abstract. Today, many neighbourhoods around the world are suffering from several problems. These problems affected the culture, social interaction, economic aspects as well as environmental impacts. Accordingly, urban development should address social and cultural aspects of each particular community within the environmental and economical context. The current research aims at developing a set of design guidelines that promotes the concept of socially sustainable neighbourhoods in Saudi Arabia. Several rating systems and urban design codes around the world set guidelines for urban design development and particularly in neighbourhood design. LEED (Leadership of Environment and Energy design) for neighbourhood design, and QSAS (Qatar Sustainability Assessment System); are examples of these rating system. Nonetheless, these rating systems are not applicable in the context of Saudi Arabia without considering its local social and cultural aspects. Therefore, this research resulted as a reaction to the abbreviation of socio-cultural aspects in neighbourhood design approaches. This research depends on an analytical review of different neighbourhood design approaches and case studies. In addition, subjective data is collected from experts and professionals about their points of view and their recommendations regarding neighbourhood design for Jeddah city. The result of the research is a set of design guidelines that consists of design principles adjusted to fit the context of Saudi Arabia (Jeddah City). The formulated guidelines are adapted and tested in an evaluation application of two residential neighbourhoods in Jeddah city. This application validated the proposed design guidelines. Considering the proposed design guidelines; in the design process of Jeddah's neighbourhoods, this will promote a leading approach to the concept of socially sustainable neighbourhoods in the kingdom.

1. Introduction

Rating systems and evaluation assessments have emerged to target the ultimate goal of achieving sustainability in the built environment. Sustainable rating systems and evaluation assessments are defined as "*systems typically designed to evaluate performance of construction projects based on various assessment parameters addressed through credits, which are classified under set of categories. Each credit is assigned a weight reflecting the presumed saved impact achieved by implementing such a credit*" [1]. There are numerous attempts in creating evaluation assessments or rating systems targeting sustainability. Most of these attempts are concerned with the building level and the concept of green building design [2-3].

The concept of designing sustainable neighbourhoods as a practice is relatively new. The initial rating system to pursuit the concept of sustainable neighbourhood development is LEED for Neighborhood Development in 2009. However, LEED rating systems accused to be more environmentally oriented abbreviating socio-cultural aspects in neighbourhood design [4]. Rating systems and evaluation assessments have abbreviated the social dimension of sustainability and become more environmentally oriented. Thus, rating systems and evaluation assessments are lacking the strong bond with social and cultural aspects [5-8]. As a result, other approaches and rating systems are trying to address these lacks according to their own distinctive socio-cultural aspirations and needs. Two regional official approaches are QSAS – Qatar Sustainability Assessment System as well as, Estidama Pearl Rating System by Abu Dhabi Urban Planning Council [9] have been introduced. Both QSAS and Estidama Pearl Rating System aim to address sustainability for their countries. QSAS implemented credits targeting local environment and culture of Qatar [10] whereas Pearl Rating System has added a new principle to sustainability principles, which is culture [9]. Both of these regional rating systems have added the two dimensions that are culture and climate. Table 1 tabulated the differences between LEED, QSAS and PEARL Rating System.

Table 1. Comparisons between LEED, QSAS and PEARL [11-13]

	International	**Regional**	
	LEED	QSAS	PEARL
Location	United States of America	Qatar	Abu Dhabi, United Emirates
Design principles	1.Compact, 2.Walkable, 3.Mixed-use, 4.Connectivity to nearby communities, 5.Neighborhood morphology, 6.Pedestrian scale, 7. Safe Travel.	1.Heritage and cultural identity. 2.Pedestrian pathways. 3.Mixed-use. 4.Regional Materials 5.Thermal Comfort 6.Natural Ventilation	1.Neighborhood connectivity. 2.Open space network. 3.Accessible community facilities. 4.Housing diversity. 5.Community walk-ability. 6.Improved outdoor thermal comfort. 7.Safe and secure community. 8.Showcase of regional and cultural practices. 9.Cluster arrangement.

Until today, Saudi design approaches are following the same approaches used in the West. Therefore, Jeddah's neighbourhoods are not socially sustainable due to the adaptation of foreign design approaches. These foreign design approaches have neglected the socio-cultural aspects in the design process. Hence, the current research proposed a set of design guidelines governed with bio-climatic considerations and cultural aspects to stimulate social sustainability in Jeddah's neighbourhoods. Main urban design principles combined with the cultural aspects and bio-climatic considerations to propose a set of design guidelines. The proposed sustainable socio-cultural guidelines for neighbourhood design in Jeddah aim to stimulate social sustainability for Jeddah's neighbourhoods. These guidelines are validated according to professional perspectives in the field.

2. Methodology

In order to utilize the proposed socio-cultural guidelines for neighbourhood design in Jeddah, the authors of the current research believed that it should be validate through the perspective of professionals and experts in the field of architecture and urban design. This is done through subjective assessment of self-directed/closed ended questionnaires for experts in the field. The questionnaire included 14 questions with time limit of five days starting from 1st to 5th of May 2016. The total number of experts participated in the questionnaire are 62 experts and professionals. The questionnaire covers their opinion regarding the different proposed guidelines through closed ended questions in a rating scale of five points from 1 to 5, starting with strongly agree, agree, neutral, disagree and strongly disagree. These questions include design factors such as walk-able distance, street width, and building heights. In addition, this category of questions include the effect of the climatically factors on the performance of the open space such as the effect of the ventilation, shading, orientation and vegetation. The third and last category is an open-ended question asking about the expert opinion regarding additional important aspects that can affect the performance of the open space of Jeddah's neighbourhood.

3. Results and Discussion

3.1 Problems faced in the neighborhood of Jeddah

The analysis presented some problems facing neighbourhoods of Jeddah from the perspective of professionals. These problems are taken to be the main factors affecting negatively social interaction in Jeddah's residential neighbourhoods. Jeddah's residential neighbourhoods are not walk-able from the perspective of 40.3 % of experts (Figure 1). Moreover, the adequate walking distance comfortable for Saudis taking into consideration socio-cultural aspects and weather conditions range from 200 – 300 m according to 45 % of the respondents (Figure 2).

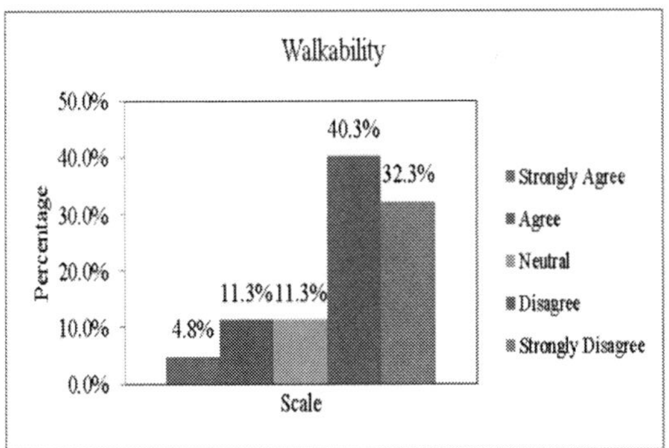

Figure 1. Walkability of Jeddah's neighbourhoods

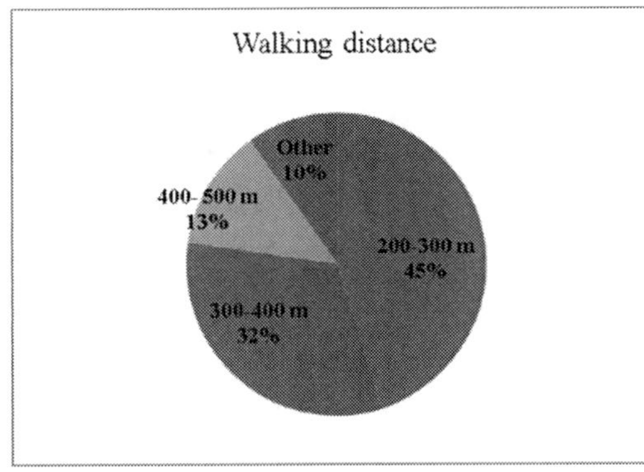

Figure 2. Walking Distance in Jeddah

29 % of experts found that the weather conditions of Jeddah is the main discouragement for utilizing open gathering spaces which validate the point of view of the author regarding the relationships between the bio-climatic aspect and the socio-cultural aspect of Jeddah's residential neighbourhood. Whereas another 29 % of experts found that both weather conditions and cultural restrictions are the main discouragements for utilizing open gathering spaces (Figure 3). 41.9 % of experts believed that 15 square meter is the proper green area per capita suitable for Jeddah's neighbourhoods. 32.3 % believed that adequate green area per capita in Jeddah is 12.5 square meter while 25.8 % of experts found it better to be 10 meter square per capita (Figure 4). 46.8% of field professional strongly agreed that cultural identity and character are neither well reflected nor considered in the design of Jeddah's neighbourhoods (Figure 5).

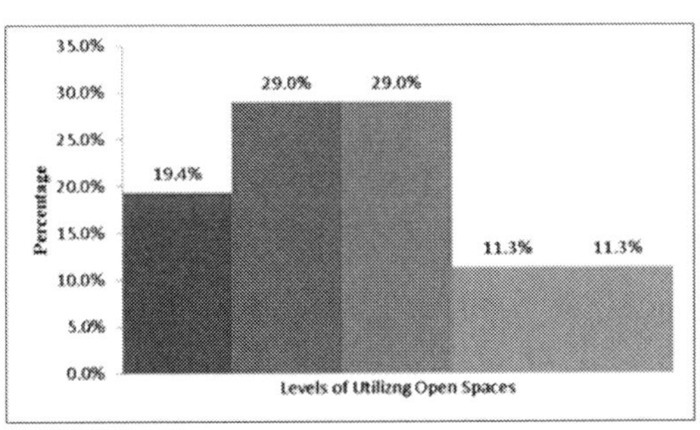

Figure 3. Levels of Utilizing Open Spaces In Jeddah (from left weather conditions and towards the right is cultural restrictions)

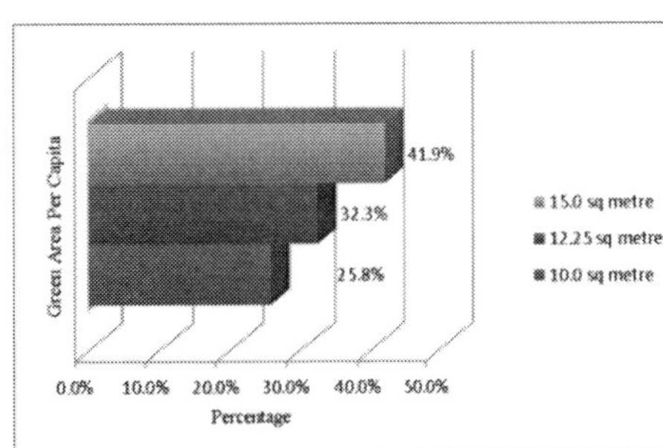

Figure 4. Green Area Per Capita

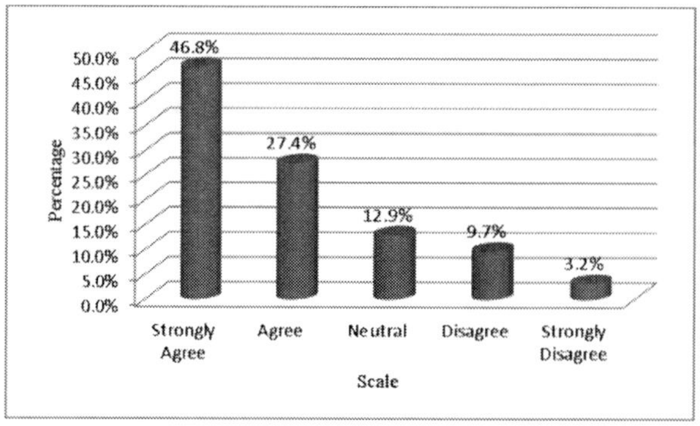

Figure 5. Reflection of cultural identity in Jeddah's neighbourhoods

3.2 Validating Proposed Sustainable Socio-Cultural Guidelines for Neighbourhood Design in Jeddah

Validating sustainable socio-cultural guidelines for neighbourhood design in Jeddah through the perspective of professionals is the main concern of this section. The results from the survey are tabulated in Table 2. Social sustainability (socio-cultural aspects) in Jeddah's neighbourhoods is achieved through the consideration of Bio-Climatic conditions in respect to cultural aspects. 22.6 % strongly agreed to this statement while the other 27.4 % responded neutrally. From the perspective of 51.6 % of experts; orientation of buildings, streets and open spaces is a main aspect to achieve ventilation. 56.5 % found ventilation as a main consideration for urban designs in the context of Jeddah due to its weather conditions. Shading elements are major motivations for utilizing gathering spaces and walkways according to 61.3 % of professionals. Dadamouny [14] revealed that although according to Hagen [15] ventilation is considered the most important aspect that could enhance the thermal performance of open spaces in the hot humid climate; the shading comes at the first priority. Ventilation is being created by the difference in the air pressure, which comes from the difference of air temperature. Difference of air temperature could be found when sunny and shaded area found. From that end, creating shaded areas to lower the air temperature than sunny areas forces the air to move and create the ventilation. 38.7 % of experts find that achieving privacy in residential neighbourhoods, along with its associated residential units is essential in the design of Jeddah's neighbourhoods. Linking the neighbourhood to adjacent ones will increase both social interaction and economic base in accordance with 46.8 % experts (Figure 11). In response to 37.1 %; economic facilities located in the neighbourhood will benefit both the residents of the neighbourhood and adjacent ones. The role of community and their involvement along with neighbourhood council will increase social interactions and relationships as stated by 46.8 % of professionals.

Table 2. Survey results on socio-culture

Aspects	**Responses (%)**				
	Strongly agree	Agree	Neutral	Disagree	Strongly disagree
Achieving social sustainability for Jeddah's neighbourhood	22.6	27.4	27.4	14.5	8.1
Importance of Orientation	51.6	30.6	8.1	1.6	8.1
Importance of Ventilation	56.5	29.0	0	9.7	4.8
Importance of Shading	61.3	22.6	6.5	3.2	6.5
Privacy levels in Jeddah's neighbourhoods	38.7	33.9	16.1	1.6	9.7
Locating economic facilities in the neighbourhood	37.1	37.1	11.3	9.7	4.8
Community and neighbourhood council role importance to	46.8	33.9	11.3	3.2	4.8

increase social interactions and relationships	

As a result of the conducted survey, the primitive proposed design guidelines were updated according to the gained perspectives. The results assure that both Bio-Climatic conditions and cultural aspects are main factors affecting social sustainability in Jeddah's neighbourhoods. The Bio-Climatic considerations and Cultural aspects are arranged according to its importance rates from the perspective of professional in the field as shown in Figure 6. The elements of the Bio-Climatic conditions factor were arranged such way that having shading as the first priority. After that comes ventilation and then orientation. Next are density and the use of local materials. On the other hand, elements of the cultural aspects factor are also arranged according to importance. First is the economic base then community role and the neighbourhood council. Subsequently, privacy and lastly religious aspects.

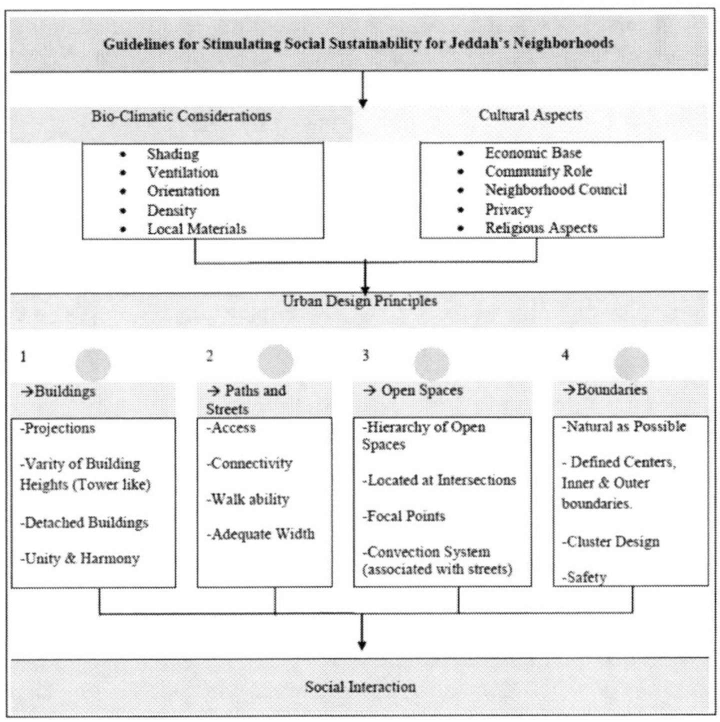

Figure 6. Guidelines according to professional's perspectives

4. Conclusion

The community way of thinking need to be reform through affective design approaches to bring back social interactions into Jeddah's neighbourhoods. In accordance with 29 % of experts, both bio-climatic considerations in respect to cultural aspects are major factors for stimulating social interactions and social sustainability in Jeddah's neighbourhoods. The study proposed Sustainable Socio-Cultural Guidelines for Neighbourhood Design in Jeddah. These guidelines can utilize by urban designers in the process of designing neighbourhoods in the context of Jeddah city. These guidelines are summarised in the following points:

1. The first factor of the proposed design guidelines is Bio-Climatic considerations. This factor aims to design considering shading, ventilation, orientation, local materials and proper densities. Urban designers should consider the mentioned factors in order to achieve natural thermal comforts in urban spaces.
2. Cultural aspects are considered the second factor of the proposed design guidelines. To design urban spaces that invite users and inhabitants, their culture and life style should be address.

The proposed design guidelines have set important cultural aspects (economic base, community role, neighbourhood council, privacy and religious) should be taking into consideration when design for the context of Jeddah.

References

[1] Attallah S, Kandil A, Senoucy A, Alderham H and Elwakil E 2014 Modeling Impact of Sustainability Policies in Qatar using Agent Based Approach and Life Cycle Analysis *Computing in Civil and Building Engineering (2014)*
[2] Haapio A 2012 Towards sustainable urban communities *Environmental Impact Assessment Review* **32** 165–169
[3] The Business Case for Green Building *U.S. Green Building Council.*
[4] Black E 2008 Green Neighborhood Standards from a Planning Perspective: The LEED for Neighborhood Development (LEED-ND) *Focus* **5**
[5] Ameen R F M, Mourshed M and Li H 2015 A critical review of environmental assessment tools for sustainable urban design *Environmental Impact Assessment Review* **55** 110–125
[6] Emmanuel R 2012 Assessment of sustainability in the built environment: possible directions for developing countries *Built-Environment Sri Lanka* **11** 1
[7] Opoku A 2015 The Role of Culture in a Sustainable Built Environment *Sustainable Operations Management Measuring Operations Performance* 37–52
[8] Masri M, Yunus R M and Ahmad S S 2016 Underlying Concerns of Socio-cultural Aspects in Green Building Rating Systems towards Improving Quality of Life *Procedia - Social and Behavioral Sciences* **222** 710–719
[9] Banani R, Vahdati M and Elmualim A 2013 Demonstrating the importance of criteria and sub-criteria in building assessment methods *Sustainable Development and Planning VI*
[10] Lewis E Estidama Pearl Rating System *Sustainaspeak* 107–108
[11] Attallah S O, Senouci A, Kandil A and Al-Derham H 2013 Utilization of life-cycle analysis to evaluate sustainability rating systems for construction projects with a case study on Qatar Sustainability Assessment System (QSAS) *Smart and Sustainable Built Environment* **2** 272–287
[12] Alawadi K and Benkraouda O 2017 What happened to Abu Dhabi's urbanism? The question of regional integration *Journal of Urban Design* **23** 367–394
[13] Ameen R *A framework for the sustainability assessment of urban design and development in Iraqi cities.* Thesis
[14] Dadamouny M A and Schnittler M 2015 Trends of climate with rapid change in Sinai, Egypt *Journal of Water and Climate Change*
[15] Hagen K and Stiles R 2010 Contribution of Landscape Design to Changing Urban Climate Conditions *Urban Biodiversity and Design* 572–592

Synthesizing the *Maqasid al-Syariah* for the waqf property development

Ahmad Shazrin Mohamed Azmi[1,3], Noor Rosly Hanif[1], Siti Mashitoh Mahamood[2], Siti Nadiah Mohd Ali[4]

[1] Faculty of Built Environment, University of Malaya, 50603 Kuala Lumpur, Malaysia

[2] Syariah and Law Department, Academy of Islamic Studies, University of Malaya, 50603 Kuala Lumpur, Malaysia

[3] Centre of Studies for Estate Management, Faculty of Architecture, Planning and Surveying, Universiti Teknologi MARA, 40450 Shah Alam, Selangor, Malaysia

[4]Department of Estate Management, Faculty of Architecture, Planning and Surveying, Universiti Teknologi MARA, Seri Iskandar Campus, 32610 Bandar Seri Iskandar, Perak, Malaysia

Email: ahmadshazrin@salam.uitm.edu.my

Abstract. The development of Islamic religious related property under the waqf (religious endowment) has shown some progress in many countries. Waqf has propagated its significant towards enhancing the public interest through the supply of many beneficial developments regardless the religious or non-religious purposes. The practice of waqf is maintained under the Shariah rulings and fiqh principles that form the overall Islamic Laws. In relation to *Maqasid al-shariah*, it has become a fundamental element in seeking for the higher objectives and intents of Islamic Law. It is also known as the wisdom behind the Shariah rulings that the Almighty has imposed on the Muslims. *Maqasid al-shariah* has three hierarchy of purposes and five essentials to preserve under the Shariah rulings. Examined under the lens of *Maqasid al-shariah*, waqf is indeed matched with the purpose of *Maqasid* itself. Moving forward, this study is intended to propose *maqasid* for waqf property development, which is the extension from the existing studies of *maqasid* for waqf. This study was using content analysis to search for suitable essentials/components to add into the proposed *maqasid* of waqf property development. The proposed *maqasid* of waqf development still maintain the preservations of religion, property and life from the original *Maqasid* essentials as the primary objectives. But being reinforced with six additional essentials/components to be regarded as the secondary objectives. Through the lens of *maqasid*, it provides an extension and insight to acknowledge the objective of developing the waqf properties.

1. Introduction

In the context of built environment studies, lands that fall under the Islamic religious purposes are still not fully explored although those lands are carrying significant function as the factor of production. There are several types of land under the religious scope, but this paper will only focus on the waqf land. Waqf land becomes unique due to its permanence, inalienability and irrevocability characteristics that have been preserved by the Islamic jurisdiction. Data in 2013 from Malaysia has revealed that there were 14,356 lots of waqf land with 30,888.89 hectares in size [1]. Whereby 11.35% of it has been developed through out Malaysia with the cost of more than RM949.84 million. Meanwhile waqf in Indonesia as of March 2016 recorded 435,768 lots with the size of 4,359,443,170 sq.m and Singapore as of year 2017 has 156 properties with worth approximately S$769 million [2, 3]. In the Middle Eastern region, the value of waqf assets has also accummulated, for instance Morocco has approximately $600 million, Algeria ($220 million), Libya ($250 million) and Qatar ($1.3 billion), which will continue to grow [4]. These figures have surely become evidence to gauge the latent potential of waqf if the structure of waqf development is properly established.

Depending on the deeds by waqf donors, the development of waqf land may not solely for religious purposes but may broadly serve the social needs across different social layers. It becomes one of the instruments in the third sector that able to facilitate the well-being of the society at large although mainly to cater the Muslims. Despite of its clear benevolent actions and benefits, people usually oversights to see the wisdoms or objectives of why Islam regards waqf as an important instrument through those dedications and sacrifices. This is where the matter should be observed through the lens of *maqasid al-shariah* that this paper would explore in the latter parts. The *maqasid* for waqf development will be proposed at the end of this paper.

2. Bridging the context of shariah, *maqasid al-shariah*, waqf and property development

Shariah principles can be described as the formal code of Islam that derived from the Quran and hadith [5]. Shariah has been explained as Islamic laws that are broad and wide-ranging to all aspects of human life by the Creator with purposes and objectives [6]. Moving forward, there is the *Maqasid al-shariah*, that is literally bring the meaning of intent, objective, purpose, aim and end-goal in any implemented Islamic law that people should understand [7]. Referred as the higher objectives of Islamic law, *Maqasid al-shariah* entails the principles that putting wisdoms behind rulings, thus could provide answers to various questions regarding the practice of Islam and to preserve people's minds and souls [8]. It becomes a framework that outlines the purpose for Islam to become a system of life. It will rationalize all the needed values embedded upon divine revelation and Islamic epistemology to solve various issues and guide the Muslim's life [5]. Also, it provides guidance and framework for the process of *ijtihad* (independent reasoning) by the Islamic jurists in solving the issues at hand while maintaining the compliance with the Islamic laws [9]. In short, *maqasid al-shariah* is to safeguard and protect the essential of human rights under the shariah framework during the lifetime in this world and facilitates the journey in the hereafter.

Waqf on the other hand is one of the optional contributions in Islam precedent from the Sunnah of Prophet Muhammad (ﷺ). Waqf is not explicitly spells out in the Quran, but it conforms with many Quranic verses that encourage the Muslims to do charity, generous help, benevolent and good deeds such as from the verse 261 of Al-Baqarah as below.

> *"The likeness of those who spend their wealth in the way of Allah, is as the likeness of a grain which grows seven ears, in every ear has a hundred grains. Allah gives manifold increase to whom He pleases. Allah is All-Embracing, All-Knowing."*
>
> (Surah Al-Baqarah, 2:261)

After the arrival of Islam, the starting of waqf practice could be inferred by the action of building up the Quba Mosque by the Prophet himself soon after His arrival in Medina during the event of migration from Makkah. It then has been practiced under the purview of Islamic jurisprudence, whereby the waqf doctrinal legitimacy was stemmed from several hadith that recorded the Prophet's advices or encouragements to His companions and family to bequeath their assets for the long-lasting benefits to others [10–14]. As the time goes by after the demise of the Prophet, waqf has expanded tremendously and require for the additional rules to handle the emergence issues. That included the addition of new categorisation of waqf assets and dedications, the validity of the waqf, the condition of the waqf pillars, the appointment of mutawalli and contemporary waqf instruments, just to name a few. All of that require for Islamic jurists to formulate the rules through the process of *ijtihad* (reasoning) based on their understanding from the text of Quran and Sunnah [15, 16]. As of today, the Shariah rules for waqf are fairly well-developed and would open the subject to progress in line with the needs of the Muslim community [17]. The development of Shariah principles or rules for waqf that mostly furnished by the process of *ijtihad* entails greater relationship to *maqasid al-shariah*.

Property development, which this paper confines the scope within the waqf land entails the effort of providing the improvement on the land. The type of the development and the expected benefits must in accordance to the waqf deeds of the donors. It is not necessarily for religious purposes, but could be any development that the Shariah would allow. The donors could outline the intended development in specific or general in their deeds for the *mutawalli* (trustee/manager) to materialize. Overall, the development must be beneficial to the specified beneficiaries and conform with the regulated laws in the country to ensure the public's harmony and safety. The idea of *maqasid* could be instilled in the waqf property development so that everything is aligned with the purpose of waqf.

3. Methodological Approach

The study intended to synthesis the components of *maqasid* as specified in the Islamic shariah framework into the application for waqf development. Although the context of *maqasid al-shariah* has been well-developed to accommodate every aspect of shariah ruling, but this study felt that waqf needs for a specific *maqasid* to attain the purpose of developing the waqf properties in more detail. Extending the works from several scholars who had studied the *maqasid* for waqf, this study was scrutinizing the *maqasid* within the context of waqf property development.

The process to identify the *maqasid* for waqf property development required the study to conduct the content analysis from several text, articles and references related to *maqasid al-shariah*. It includes the area of muamalat because waqf also categorised under the same scope. The consideration was mainly to extend the original essentials of preservation under the Shariah rulings and create the secondary layer of essentials which tailored to the waqf development context.

4. Analysing the *Maqasid*

4.1. Maqasid al-shariah under its traditional classifications

Maqasid al-shariah was developed after the Prophet Companions' era in between the third and fifth Islamic century but only clearly developed in between the fifth and eight Islamic century by many Islamic scholars such as Al-Tirmidhi al-Hakim, Abu Zayd al-Balkhi, Abd al-Malik al-Juwayni, Abu Hamid al-Ghazali, Abu Ishaq al-Shatibi and many others [8, 18]. The level of necessities under the traditional concept of *Maqasid al-shariah* is illustrated in Figure 1.

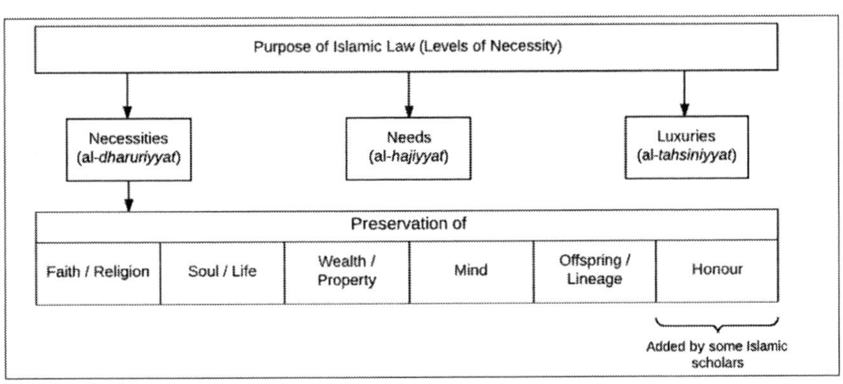

Figure 1: Traditional Classifications of *Maqasid* – Three levels of necessity [8]

Maqasid al-shariah has three levels of necessity known as the *dharuriyyah*, *hajiyyah* and *tahsiniyyat* that translated as the necessities, needs and luxuries (embellishment/complementarities). *Al-Dharuriyyat* (necessities) is the essential to achieve spiritual and material well-being, whereby the missing of these essentials would bring imbalance, corruption, disturbance and destruction in people's lives [9, 15]. *Al-*

hajiyyat (needs) functions to alleviate or relieve hardship and difficulty, and to create ease and comfort in life [7, 19]. Meanwhile *Al-tahnisiyyah* (luxuries) is for 'beautifying purposes', facilitate the quality of life and would make human well-being more comfortable [8, 9].

Although the classification is notable, but these levels are actually overlap on each other, where each level serves and protects the lower level [8]. Inversely, the subordinate level will function to consolidate and advance the realizations and preservations of the main purpose of the Shariah [9]. However, in order to address the contemporary socio-cultural realities, the contemporary scholars, especially in the twentieth century have searched for the improvisations, thus witness the *maqasid* to evolve over the centuries [8]. Under the modern environment, the *maqasid* should get enhancement in its dimension and becomes more versatile in the modern context without suppressing its fundamentals.

4.2. Studies on maqasid of waqf

Waqf under the *maqasid* of religion preservation connects with doing, mobilize and spread the good deeds to people regardless the Muslim or Non-Muslim, animals and environment, which need to be carried out in any way at all times and places [19]. It proposes that through the good deeds, waqf may fulfills all the five necessities (*dharuriyyat*) of preserving faith, soul, mind, wealth and lineage while maintaining the needs (*hajiyyat*) and luxuries (*tahsiniyyat*). Having said that, waqf according to Ibn `Ashur is the *hajiyyat* because it provides ease to the people's hardships and bring comfort to the people's necessities [7]. It is reasonable because waqf functions as an optional contribution that become secondary after the *dharuriyyat* but may become vital supports in preserving the necessities at the *dharuriyyat* level.

Regardless the level of necessities of *dharuriyyat*, *hajiyyat* and *tahsiniyyat*, scholars still referring the earlier mentioned five necessities as the component of fulfillment. For instance, waqf is viewed to be more specific into preserving the religion and property. But in some countries, the management of waqf is quite lacking. Thus, calling for the establishment of *maqasid* for waqf management [21]. It is consistent with the notions from many scholars who urging for better management structure for waqf as mechanism to uphold the overall *maqasid* [22, 23].

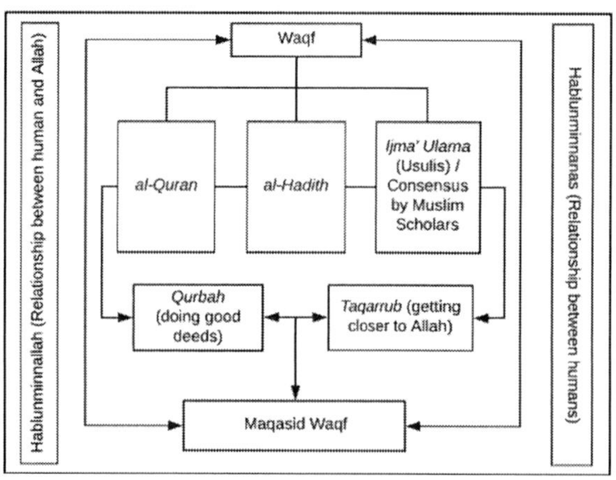

Figure 2: The *Maqasid* of Waqf [24]

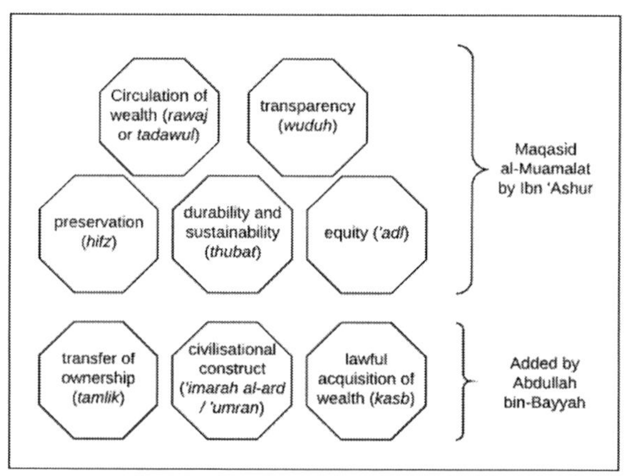

Figure 3: *Maqasid al-Muamalat* [22]

The *maqasid* of waqf also has been proposed by Razali Othman [24] as illustrated in

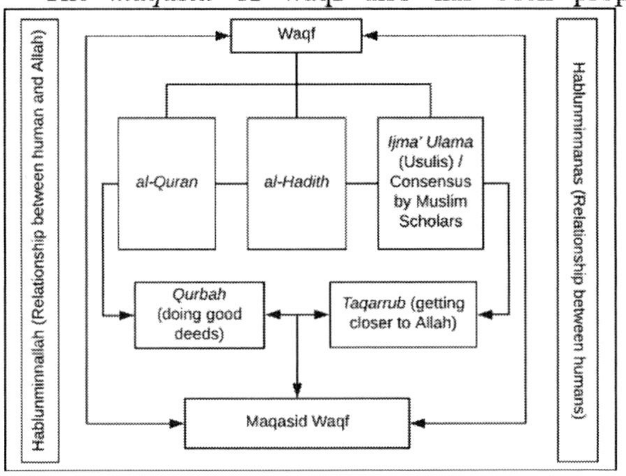

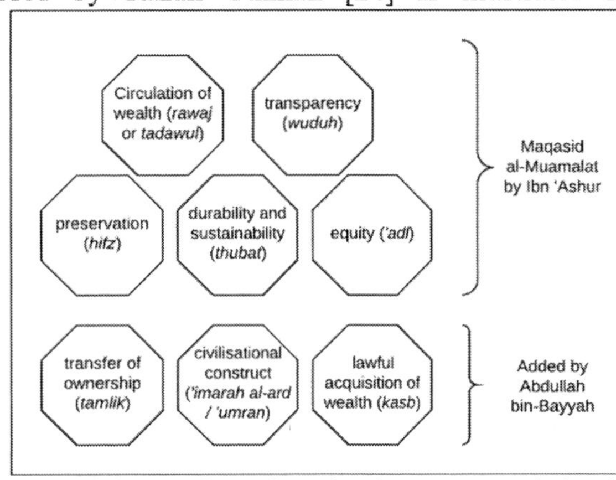

. He has firmly stated that the philosophy of waqf was originated from the Islamic *tasawwur* (Islamic worldview) that stands from the Quran, hadith and ijma' ulama (Islamic scholars' consensus). Thus, rejecting the notion that waqf was influenced or inspired by the practice of religious givings in the society before the arrival of Islam or from the Near Eastern civilization territories like Rome. He then outlined the objectives of waqf under the context of preservation of property (*hifz al-mal*) and coupled with the intention to do good deeds and to get nearer to Allah [24]. It establishes the vertical and horizontal relationship through the relationship between human and Allah as well as the relationship between humans in order to induce harmony in the whole society.

Taking the perspectives broader than the traditional *maqasid al-shariah*, waqf could be viewed under the context of *maqasid al-muamalat*, that refers to dealings [22]. Al-Mubarak had included five Ibn 'Ashur's *maqasid* for financial transaction and contracts with three inclusions from Abdullah bin Bayyah to associate with waqf (Refer **Error! Reference source not found.**). In his elaborations, waqf is an instrument to circulate wealth (*rawaj/tadawul*) from the donor, that involves ownership transfers (*tamlik*) to Allah for the enjoyment of the public. Its perpetuity characteristic requires for preservation of its property/corpus (*hifz al-mal*) and involves in activities that can generate sustainable benefits or incomes, thus serve the objective of permanence and sustainability (*thubat*). Developing waqf assets would usually involve construction projects, thus meets the objective of constructing the civilization (*i'mar al-ard*). Transparency (*wuduh*) and good governance would ensure the trustworthy of mutawalli in protecting people's equity (*'adl/'adalah*) and build up the social inclusion in the waqf sector.

5. Proposing the *maqasid* of waqf development

Synthesizing the input from other studies, this study would like to propose the *maqasid* of waqf property development from the context of preservation of religion (*hifz al-din*), property (*hifz al-mal*) and life (*hifz al-nafs*) as the primary objectives towards getting the blessings of Allah and realizing the development. Then reinforced by six essentials that form as the secondary objectives (Refer Figure 4).

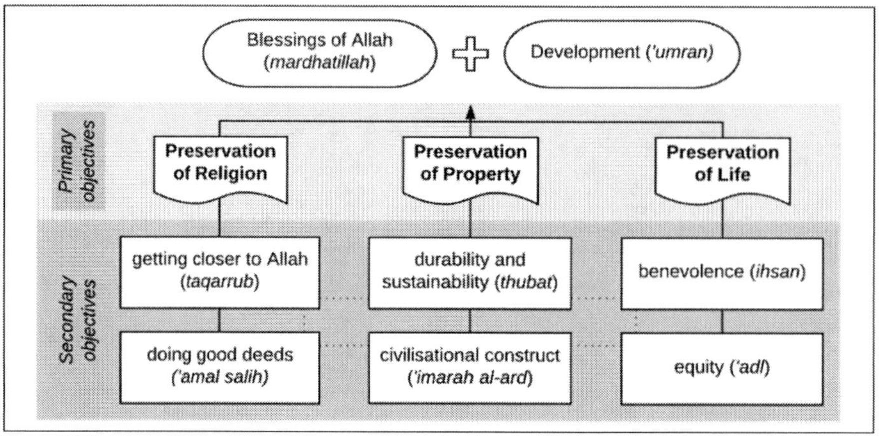

Figure 4: The *maqasid* of waqf development

The preservations of religion, property and life are regarded as vital in maqasid of waqf development. The preservation of religion is core considering waqf exists due to the Islamic religion encouragements on good virtues, which in exchange with tawheedic spiritual enrichment and manifold rewards as promised by Allah in verse 261 of Surah Al-Baqarah and verse 7-8 of Surah Al-Bayyinah. It entails the essentials of getting closer to Allah and take up the encouragement of doing good deeds.

The preservation of property has become the central objective considering that the number of undeveloped waqf lands is huge and need for urgent but careful attention. It entails the objectives of maintaining the sustainable aspects of waqf properties and to help for the construction of civilization.

Meanwhile the preservation of life outlining the benevolence and equity or justice objectives. The benevolence actions together with equality or justice would internally enrich people towards achieving a greater and blessed civilisation. Moreover, waqf development projects would bring comfort and increase the quality of life.

6. Conclusion

Shariah and its *maqasid* can evolve within the allowable parameters to meet the challenges of contemporary environment but of course it cannot deny the needs of certain verifications. The effort to develop waqf properties would automatically uphold the spirit of Islamic components and at the same time would either provide necessities in peoples' life, lifting difficulties or providing fulfilments through various benefits from waqf properties. But waqf development need to be guided under specific *maqasid*. It should have its own *maqasid* or objectives that can portray a better reasoning of the practice. Especially when dealing with new expectation or the contemporary needs in the society. This study successfully proposing the maqasid for waqf development, but it stills open for expansion and contention through future studies.

References

[1] Yayasan Waqaf Malaysia. *Waqf Property Development Transformation Master Plan Study in Malaysia*. Master Plan, Putrajaya, Malaysia: Yayasan Waqaf Malaysia, 2016.

[2] Indonesian Waqf Board. Data Tanah Wakaf Seluruh Indonesia. *Badan Wakaf Indonesia*, http://bwi.or.id/index.php/in/tentang-wakaf/data-wakaf/data-wakaf-tanah.html (2018, accessed 9 January 2019).

[3] Warees Investments Pte. Ltd. Official Website. Warees Investments Pte. Ltd, http://www.warees.sg/ (2018, accessed 9 January 2019).

[4] Al Salahat S. Awqaf Landmarks in the Arab World.

[5] Azhar Mohamad, Imtiaz Mohammad Sifat. The Fiat Fiasco: The Erosion of Maqasid of Shariáh and Practical Alternatives. In: *Proceedings of IDMAC2015*. University Conference Hall,

Universiti Sains Malaysia, Penang, Malaysia: Centre for Islamic Development Management Studies (ISDEV) Universiti Sains Malaysia, 2015, pp. 577–590.
[6] Borham AJ. *Pengantar Perundangan Islam*. 1st ed. Skudai, Johor: Penerbit Universiti Teknologi Malaysia, 2002.
[7] Mek Wok Mahmud, Syed Sikandar Shah. Optimisation of Philanthropic Waqf: The Need for Maqasid-Based Legislative Strategies. *Shariah Law Rep* 2010; 2: 45–61.
[8] Auda J. *Maqasid Al-Shariah as Philosophy of Islamic Law: A Systems Approach*. South East Asia edition. Kuala Lumpur: Islamic Book Trust & The International Institute of Islamic Thought (Joint Publication), 2010.
[9] Abdulaziz A. Al-Dharā'i' and Maqāṣid al-Sharī'ah: A case study of Islamic insurance. *Intellect Discourse* 2010; 18: 261–281.
[10] Abu Zahrah M. *Wakaf Menurut Agama & Undang-Undang*. Puchong, Selangor, Malaysia: Berlian Publication Sdn. Bhd., 2007.
[11] Hisham Yaacob. Waqf History and Legislation in Malaysia: A Contemporary Perspective. *J Islam Hum Adv Res* 2013; 3: 387–402.
[12] Lita HN. The Construction of Commercial Building on Waqf Land Based on Indonesia Law. *IOP Conf Ser Earth Environ Sci* 2018; 175: 012199.
[13] Muhammad As-Sayyid Sabiq. *Fiqh Sunnah*. Kuala Lumpur: Al-Hidayah Publication, 2009.
[14] Mustofa al-Khin, Mustofa al-Bugho, Ali Asy-Syarbaji. *Kitab Fikah Mazhab Syafie (Al-Fiqh Al-Manhaji Mazhab Al-Imam Al-Syafie*. Kuala Lumpur: Pustaka Salam Sdn. Bhd., 2009.
[15] Al-Raysuni A. *Imam Al-Shatibi's Theory of the Higher Objectives and Intents of Islamic Law*. South East Asia edition 2006. Petaling Jaya, Selangor, Malaysia: The International Institute of Islamic Thought (IIIT) & Islamic Book Trust (IBT), 2006.
[16] Kamali MH. *Equity and Fairness in Islam*. Kuala Lumpur: Ilmiah Publishers Sdn. Bhd., 2006.
[17] Kamali MH. Shariah Regulations and Contemporary Reforms of Waqf. In: *Proceedings of the SC-OCIS Roundtable 2014 - Harnessing Waqf into a Bankable Social Financing and Investment Asset Class*. Kuala Lumpur: Securities Commission Malaysia, 2014, pp. 22–26.
[18] Auda J. *Memahami Maqasid Syariah: Peranan Maqasid dalam Pembaharuan Islam Kontemporari*. Edisi Kemas Kini. Selangor, Malaysia: PTS Islamika Sdn. Bhd., 2014.
[19] Mohamad Akram Laldin, Mek Wok Mahmud, Mohd. Fuad Sawari. Maqasid Syariah dalam Pelaksanaan Wakaf. In: *Prosiding Seminar Konvensyen Wakaf Kebangsaan Jabatan Wakaf Zakat dan Haji*. Kuala Lumpur: Jabatan Wakaf Zakat dan Haji, Jabatan Perdana Menteri (JPM), http://irep.iium.edu.my/28549/1/Maqasid_Waqaf.pdf (2006).
[20] Kamali MH. Law and ethics in Islam- The roles of the maqasid. In: Vogt K, Larsen L, Moe C (eds) *New Directions in Islamic Thought: Exploring Reform and Muslim Tradition*. United Kingdom: I.B.Tauris & Co. Ltd., 2008, pp. 23–46.
[21] Md. Ibrahim AA, Khan SH. Waqf management in Bangladesh: An analysis from Maqasid al-Shar'ah perspective. IAIS Malaysia, Kuala Lumpur: International Institute of Advanced Islamic Studies (IAIS) Malaysia, http://www.iais.org.my/e/attach/2015/maqasid2015/pdf/s3s3_paper.pdf (2015).
[22] Al-Mubarak T. The Maqasid of Zakah and Awqaf and Their Roles in Inclusive Finance. *Islam Civilisational Renew ICR*; 7, http://www.icrjournal.org/icr/index.php/icr/article/view/548 (2016, accessed 20 January 2017).
[23] Habib Ahmed. *Role of Zakah and Awqaf in Poverty Alleviation*. First edition. Jeddah, Kingdom of Saudi Arabia: Islamic Development Bank (IDB) and Islamic Research & Training Institute (IRTI), 2004.
[24] Razali Othman. *Wakaf Tunai: Sejarah, Amalan dan Cabaran Masa Kini*. First. Kuala Lumpur: Dewan Bahasa dan Pustaka, 2015.

Assessment of complexity factors in briefing stage of refurbishment projects in Malaysia

M Mokariantabari[1], H Adnan[1] M A J Hussin[1], Z Z Abidin[1], H E A Baharuddin[1] and W N W Ismail[2]
[1]Faculty of Architecture, Planning and Surveying, Universiti Teknologi MARA, 40450 Shah Alam, Selangor, Malaysia
[2]Faculty of Architecture, Planning and Surveying, Universiti Teknologi MARA Perak Branch, Seri Iskandar, Perak, Malaysia

Email: mimad856@gmail.com

Abstract. In Malaysia, building refurbishment is fast becoming an important sector in the construction industry. Refurbishment projects are generally characterised to be of higher risk, more complex and needs greater coordination than new build. As briefing stage is one of the main source of design information, meeting the clients' requirements in briefing stage and working within the complexities of existing building may pose a difficult task to produce the required information for designers to start design. Quality of required design information in refurbishment projects is affected by level of complexity factors in briefing stage. Affected design information is one of the considerable factors, which leads to refurbishment projects performance failure. Therefor this research aim is to assess the level of potential complexity factors in briefing stage of refurbishment projects that hinder designers to acquire the required information to start design. This research employs the quantitative approach, that is, a questionnaire survey targeted at experienced architects and interior designers involved in refurbishment projects within Klang valley area to provide the data in investigating the level of complexities in briefing stage of refurbishment projects. The result indicated that the key complexities in briefing stage of refurbishment projects can be divided into three main categories; Lack of document, Project characteristic and Client's characteristic.
Keywords: Briefing stage, complexity, refurbishment

1. Introduction

In recent years, there has been an increasing interest in building refurbishment works due to the change in economic conditions and the emphasis on sustainable development [1-2]. Buildings owners still need to refurbish their property, despite the economic slowdown [3]. In Malaysia, building refurbishment is fast becoming an important sector in the construction industry. The key factors that contribute to the growth of building refurbishment works in Malaysia are; economic recession, building deterioration, obsolescence and growing demand for green buildings [4]. Hence, building refurbishment of existing buildings plays a significant role in achieving sustainable environment in the urban area [2]. In Malaysia, the building refurbishment sector contributes to 17 percent in 2005 and increased to 22 percent in 2014 [5]. This shows the increasing importance of building refurbishment sector in Malaysia and It should be highlighted that many building refurbishment projects are not reported. The main reason of performance failure of building refurbishment projects is the high level of complexity in these projects [6-7-8]. Quality of required design information in refurbishment projects is affected by level of complexity and uncertainty factors in these projects. Affected design information is one of the considerable factors, which leads to refurbishment projects performance failure. [8]. Ali [8], also indicated briefing is one of the most important information sources for designer in refurbishment projects. In refurbishment projects, there is a need to improve the briefing process through capturing the client's needs, information and requirements to acquire the required information for design. Despite of conducting several researches on complexity and uncertainty factors in refurbishment projects, none of them focused on briefing stage. Due to the increasing importance of briefing stage of refurbishment projects, more research needs to be done to resolve some of the pertinent issues of designers at early stage and it is only possible to rightly identify and assess the complexities in briefing. The purpose of this study is to identify and assess the level of complexity factors faced by designers in briefing stage of refurbishment projects in Malaysia.

2. Literature Review

2.1 Briefing stage of refurbishment projects
The briefing process in construction projects involves the documentation and communication of clients' requirements and represents a cornerstone for achieving client satisfaction. The inefficiency of the briefing process has long been recognised as an area which needs improvement. Refurbishment projects are generally characterised to be of higher risk, more complex and needs greater coordination than new build, thus meeting the clients' requirements and working within the constraints of existing building structures may pose a difficult task in briefing stage of refurbishment projects hence It is crucial to the success of projects [9].

2.2 Complexities in briefing stage

2.2.1 Lack of archive documents
Lack of archive documents in building refurbishment projects is one of the main factors that causing complexity [10]. Archived document is an accumulation of historical records. Archived documents and files can serve as records of results of the decision made [11]. Moreover, from the archive documents, the architects obtained a relatively high portion of their design information [8].

2.2.2 Incomplete building site survey result
Watt et al., [12] defined Site surveys as detailed studies done to verify and supplement site information provided by the client. The literature review also reveals that incomplete and inadequate building site survey is a factor that contributes to complexity in building refurbishment projects [4].

2.2.3 insufficient building inspection results
A building inspection has to be carried out to obtain accurate, sufficient, and clear information about the building conditions before any building refurbishment project started [13]. Also, complete and accurate building inspection results are considered ideal for building refurbishment buildings [14].

2.2.4 Lack of services information of the existing building
The literature review also reveals that apart from archive documents, lacking services information is a factor that contributes to uncertainty [15]. Services information of the existing building is information about the services, structural elements and the systems installed in buildings and in building refurbishment projects, services information is often critical, especially in old concrete buildings [16].

2.2.5 Difficulty in accessing to the existing building
Doran et al. [17] revealed that access to the existing building site is one of the factors that contributes to uncertainty in building refurbishment projects. Access to the site is a fundamental problem in most of the building refurbishment projects which make operations very slow and challenging due to an occupied building and existing conditions [18].

2.2.6 Unforeseen Site Condition
Bernstein et al. [10] observed that all participants involved in building refurbishment projects know unforeseen site conditions as one of the main factors which cause complexity on the projects. Similarly, Acharya et al. [19] noted that unforeseen site conditions cause the complexity in early design stage.

2.2.7 Unclear scope of the work

A building refurbishment works often suffer from delays and escalating cost due to a high degree of complexity in defining the scope of work during briefing and design stage [20-21]. However, Rahmat [22] noted that the unclear scope of works caused the complexity of tasks in building refurbishment projects. When the scope of work is not clear or well defined in briefing stage, clients tend to add, delete, or revise the existing scope of work within the current project by submitting written changes order [23].

2.2.8 Lack of client's knowledge in refurbishment project

Knowledge of the clients is a very important factor in decision making through the briefing and planning of the building refurbishment projects [7]. Haapasalo et al. [24] observed that because clients are rarely knowledgeable in construction and building, client's lack of skill and knowledge contributes to the complexity in early stage of building refurbishment projects.

2.2.9 Lack of client skill and experience in briefing stage of refurbishment project

The client must be able to communicate the information and at the right time in briefing stage. Having enough experience in the construction process especially in briefing stage, will enable clients to choose the right material and equipment options [25]. In refurbishment projects, most of the clients don't have enough skill and knowledge about the construction process, and they do not know about the complexity of the construction procedure.

2.2.10 Ambiguity of client's needs

Ali et al. [26] observed that the client's needs are the main cause of complexity in building refurbishment projects. Meeting client's needs in building refurbishment projects can pose a challenging task for designers. There is a need to improve the briefing process through capturing the client's needs. It is only possible to rightly identify the client's needs through a team of knowledgeable participants involved in briefing stage of refurbishment [9].

3. Research Methodology

A quantitative method research approach is employed in this study to provide the basis for primary data in investigating the level of complexities in briefing stage of refurbishment projects. The main research method involved the distribution of the questionnaires to experienced architects and interior designers within Klang valley area. A total of 700 questionnaires were distributed to architects and interior designers involved in refurbishment projects. From those questionnaires distributed, 143 were found to be valid enough to form a database for data analysis.

3.1 Data Collection

The data collection method was accomplished through distributing questionnaire survey. The research instrument used in this study was a questionnaire containing open – ended and closed – ended type questions. A 5-point Likert scale was used in gathering some of the data, using a measurement scale of 1-5 in which number 1 stands for very complex factor and number 5 stands for not complex factor.

3.2 Method of Data Analysis (Quantitative Method)

Quantitative approaches can help to provide large representative samples of communities reliably assert cause and effect relationships among constructs as well as confirm or disconfirm theoretical hypotheses and summarize numerical data [27]. The collected data were analyzed using the Statistical Package for Social Science (SPSS) computer software. Descriptive statistics such as frequency tables were used to analyze the data. The mean scores computed from the descriptive statistics of SPSS were used to formulate the ranking of potential complexities in briefing stage of refurbishment projects. Basically, the mean is used to describe the average score or central tendency of variables. The formula of mean is derived as:

$$\text{Mean} = \Sigma\, X/N \tag{1}$$

where ΣX is the sum of the X scores; N the total number of scores.

4. Results and Findings

Seven-hundred questionnaires were distributed and collected, with a response rate of 20.4%. The results for the Mann-Whitney U test for the objective of the research is as shown in Table 1.

4.1 Complexity factors in briefing stage of refurbishment projects

Table 1. The mean and ranking of complexity factors in briefing stage of building refurbishment

Construct	Items	Rank	Mean
Complexity	*Lack of documents factors*		
	• Lack of Archive document in briefing stage	6	2.7794
	• Incompleteness of service information of the existing building	5	2.6912
	• Unavailability of building inspection	3	2.5662
	• Unavailability of building site survey result	7	2.7794
	Client's characteristics factors		
	• Lack of client skill and experience in briefing stage of refurbishment project	4	2.6765
	• Lack of client's knowledge in refurbishment project	1	2.5147
	• Ambiguity of client's needs		
	Project characteristics factors	2	2.5273
	• Difficulty in accessing to the existing building	9	3.1985
	• Unforeseen Site Condition	8	3.0000
	• Unclear scope of the work in refurbished projects	10	3.5074
		Overall	***2.8240***

From Table 1, it can be concluded that all of the chosen complexity factors have been assumed as a considerable complexity by architects and interior designers. The result shows the highest rank of complexity in briefing stage of building refurbishment is the Lack of client's knowledge in refurbishment project followed by the Ambiguity of client's needs in briefing stage of refurbishment project. Surprisingly the lowest considerable factors in this survey is Unclear scope of the work followed by Difficulty in accessing to the existing building.

Table 2. The mean and ranking of the main three categories of complexity factors in briefing stage of building refurbishment

Construct	Items	Rank	Mean
Complexity	Client's characteristics factors	1	2.57
	Lack of documents factors	2	2.70
	Project characteristics factors	3	3.23

Table 2 shows that among the three main categories of complexity factors in briefing stage of building refurbishment, Client's characteristics factors is the highest concern for designers to challenge with to get the required information to start design. Lack of documents factors ranked as the second more complex factor from the respondents' point of view and Project characteristics factors is chosen as less complex factors comparing to the first two mentioned categories.

4.2 Discussion on findings

From the results, it can be shown that the Client's characteristics factors' impact in briefing stage of building refurbishment is more critical compared to the Lack of documents factors and Project characteristics factors due to the vital role of clients in briefing stage to prepare and sharing the required information for designers to start design.

Based on the findings, the highest rank is the Lack of client's knowledge in refurbishment project followed by Ambiguity of client's needs. This is supported by Ali [26] that observed the client's needs are the main cause of complexity in building refurbishment projects and understanding client's needs in building refurbishment projects can pose a challenging task for designers. Ali [26] also indicated that poor briefing and breakdown of communication always happens when the client always changed their needs. Client's input is one of the main source of information and according to the findings of this research, unclear client's needs and their poor knowledge about refurbishment projects are the biggest concern for designers in early stage of building refurbishment. This is supported by Shen [28] that mentioned inexperienced and not knowledgeable client is unable to provide an appropriate and adequate brief information and this always contributes to additional work and design variations.

It can be concluded that the quality of required information that designer needs to start design is partly dependent upon the knowledgeable and experienced client's involvement in the briefing stage and similarly, a strong design team that plays effective roles can better ensure clients' active participation to communicate more and acquire the needed information.

5. Conclusion

This study investigated the level of complexity factors faced by architects and interior designers in briefing stage of refurbishment projects in Malaysia. The results indicate that lack of client's knowledge in refurbishment projects and lack of client's skill and experience in managing briefing stage of refurbishment projects are the main concern for the designers of refurbishment projects to acquire the required information to start design. These results can be a useful basis for developing a framework to assist construction practitioners to get awareness about the main complexity factors in briefing stage of refurbishment projects that they may encounter. Furthermore, understanding the level of complexity in briefing stage of building refurbishment can be valuable in further researches when considering and applying some initiatives for mitigating such complexity factors, which may lead to successful briefing by generating the desired level of information for designers to start design. The outcomes of this study can help researchers and participants involved in briefing stage of building refurbishment to increase their knowledge on the levels of complexity factors and prepare proper initiative actions to overcome these factors.

Acknowledgements

The authors would like to extend their utmost appreciation to all participants for sharing their valuable knowledge, time and support. The authors also acknowledge the Universiti Teknologi MARA Lestari grant No. 600 IRMI/MyRA5/3/Lestari (K) (23/2017) for funding the research.

References:

[1] Kohler, N., & Hassler, U. (2002). The building stock as a research object. *Building Research & Information*, *30*(4), 226–236.
[2] Riley, M., & Cotgrave, A. (2011). *Construction Technology 3 The Technology of Refurbishment and Maintenance.* Palgrave Macmillan.
[3] Chan, A. P. C., & Hon, C. K. H. (2016). *Safety of repair, maintenance, minor alteration, and addition (RMAA) works: a new focus of construction safety*. Routledge.
[4] Zolkafli, U., Zakaria, N., & Yahya, Z. (2012). Risks In Conservation Projects. *Jurnal Of Design*.
[5] CIDB. (2015). Construction Industry Development Board, Malaysia Official Portal.
[6] Egbu, C. O. (1997). Refurbishment management: challenges and opportunities. *Building Research & Information*, *25*(6), 338–347.
[7] Rahmat, I. (2008). *Managing Refurbishment Projects*. Shah Alam, Malaysia: Pusat Penerbitan Universiti, Universiti Teknologi MARA.
[8] Ali, A. (2010). Design information in managing refurbishment projects in Malaysia. *The International Journal of Physical Sciences*, 768–773.
[9] Lee, C. C. (2015). Identifying knowledge gap to meet client project requirements in refurbishment projects. *Journal of Information & Knowledge Management*, *14*(02), 1550016.
[10] Bernstein, H. M., Jones, S. A., Russo, M. A., & Laquidara-Carr, D. (2014). *Managing Uncertainty and Expectations in Building Design and Construction. SmartMarket Report*. (H. M. Bernstein, S. A. Jones, M. A. Russo, & D. Laquidara-Carr, Eds.). Bedford: McGraw-Hill Construction.

[11] Shu, E. (2008). Retaining and archiving records. In *The Architect's Handbook of Professional Practice* (pp. 445–457).
[12] Watt, S., Simpson, C., McKillop, C., & Nunn, V. (2010). Electronic Course Surveys: Does automating feedback and reporting give better results? *Assessment & Evaluation in Higher Education*, *27*(4), 325–337.
[13] Douglas, J., & Ransom, B. (2007). *Understanding Building Failures*. Routledge.
[14] Kohler, N., König, H., Kreissig, J., & Lützkendorf, T. (2010). *A life cycle approach to buildings: Principles - Calculations - Design tools*. Walter de Gruyter.
[15] Volk, R., Stengel, J., & Schultmann, F. (2014). Building Information Modeling (BIM) for existing buildings - Literature review and future needs. *Automation in Construction*, *38*(July 2015), 109–127.
[16] Martin, A. J., & Gold, C. A. (1999). Refurbishment of Concrete Buildings: The decision to refurbish. *Guidance Notes*, *GN 7/99*.
[17] Doran, D., Douglas, J., & Pratley, R. (2009). *Refurbishment and repair in construction*.
[18] Okoroh, M. I. (1992). *Knowledge based decision support system for the selection and appointment of sub-contractors for building refurbishment contracts*. © Michael Iheoma Okoroh.
[19] Acharya, N. K., Lee, Y. D., & Kim, J. K. (2006). Critical construction conflicting factors identification using analytical hierarchy process. *KSCE Journal of Civil Engineering*, *10*(3), 165–174.
[20] Ali, A. S., Rahmat, I., & Noorhashimah, A. (2005). The design process of building refurbishment on project performance. *Built Environment Journal*, *2*(2), 1–13.
[21] Daoud, O. (1997). The architect/engineer's role in rehabilitation work. *Journal of Construction Engineering and Management*, (March), 1–5.
[22] Rahmat, I., & Ali, A. (2010). The involvement of the key participants in the production of project plans and the planning performance of refurbishment projects. *Journal of Building Appraisal*, *5*(3), 273–288.
[23] Okeil, A. M., Ph, D., & Asce, M. (2013). Enforceability of Limitation of Liability Clauses in Engineering Contracts, (August), 128–135.
[24] Haapasalo, H., Aapaoja, A., Söderström, P., Aapaoja, A., Haapasalo, H. (2013). Early Stakeholder Involvement in the Project Definition Phase: Case Renovation. *ISRN Industrial Engineering*, *2013*
[25] Ahmad, N., Jaapar, A., Adnan, H., Bari, N. A. A., & Rashid, R. A. (2012). Important qualities of design teams in managing client's brief and their impact on delivering quality buildings. *Procedia-Social and Behavioral Sciences*, *35*, 728-738.
[26] Ali, A. S., Kamaruzzaman, S. N., & Salleh, H. (2009). The characteristics of refurbishment projects in Malaysia. *Facilities*, *27*(1/2), 56–65.
[27] Fassinger, R., & Morrow, S. L. (2013). Toward Best Practices in Quantitative, Qualitative, and Mixed-Method Research: A Social Justice Perspective. *Journal for Social Action in Counseling & Psychology*, *5*(2).
[28] Shen, Q., Li H., Chung, J. & Hui, P. Y. (2004). A framework for identification and representation of client requirements in the briefing process". *Journal of Construction Management and Economics,* Vol. 22, 213-221.

IOP Conf. Series: Earth and Environmental Science **385** (2019) 012053 doi:10.1088/1755-1315/385/1/012053

SEM-PLS Analysis Approach in Identifying the Effective Residential Bikeway Design Factors

Mohd Zahid Mohd Salleh[1], Nurhayati Abdul Malek[2], Noriah Othman[1], Siti Zabeda Mohammad[2]

[1] Centre of Studies for Postgraduate Studies, Faculty of Architecture, Planning & Surveying, Universiti Teknologi MARA, 40450 Shah Alam, Malaysia.

[2] Centre of Studies for Landscape Architecture Faculty of Architecture, Planning & Surveying Universiti Teknologi MARA, Kampus Puncak Alam, 42300 Bandar Puncak Alam, Selangor, Malaysia.

Email: nurhayati.abdmalek@gmail.com

Abstract. The study was conducted to investigate the impact of residential-bike environment influences cycling participation among resident. The underpinning theories used in the study were the Ecological Model of Active Living (EMAF) and the Theory of Routine Mode Choice Decision (TRMCD). A structured questionnaire online survey was employed by engaging two resident associations (RA) in City of Elmina that contributed, 242 respondents. The data was analysed using Smart PLS 3.2.6 to test a developed model through measurement model analysis and followed by a structural model analysis. This approach offers comprehensive, quantifiable and flexible causal-modelling capabilities in a multi-regression analysis for non-parametric data that run simultaneously. The measurement model analysis was archived the threshold values for all criteria such as reliability and validity test. The structural model analysis result showed that perception, preference and satisfaction have significant positive relationship with the bike-built environment directly. Besides, the bike-built environment and cycling participation among resident has significant positive relationship. The Coefficient Determination, R^2 of the bike-built environment was substantial value with 33 percent while the cycling participation was 7.3 percent. The study confirmed that the residential-bike built environment has impact on the cycling participation among resident.

1. Introduction

Presently, the township development in Malaysia is accelerating transform towards wellness living concept that offers more green spaces and green infrastructure to pushed carbon emission remediation as well [1], [2]. An example, sustainable mobility like cycling is well promoted in keeping resident active in the urban area, and it is incarnating among Malaysian nowadays [3][4]. The support system such as an appropriate infrastructure endowment, integrated land use planning and robust education promotion are the key elements that need to be concentrated [5][6]. Unfortunately, we face some restriction such as infancy bike built environment that affects ridership in Malaysia [7]. Consequently, the understanding of the environmental behaviour towards cycling activity among Malaysian is a focus in the study. The study will develop a model to understand the relationship between the perceived physical environment that influence the bike-built environment as well as cycling participation promotion.

The model was fabricated from the underpinning theories such as the Ecological Model of Active Living (EMAF) and the Theory of Routine Mode Choice Decision (TRMCD). The EMAF is illustrated that when people are motivated (perceived environment) and setting support (residential-bike built environment) would lead the high cycling participation [8][9]. Furthermore, three steps evaluation of perceived environment will enhance the understanding of cycling behaviour through the perception, preference and satisfaction that introduced by TRMCD. When the three essential elements (1. Safety/security; 2. Convenience & 3. Enjoyment) are archived, the people will be adopted the cycling mode practically and regularly in daily life [10]. Thus, the developed model as exhibited in Figure 1 is showed to understand the relationship between perceived perception (Safety-TRMCD), perceived preference (Convenience-TRMCD) and perceived satisfaction (Enjoyment-TRMCD) on the bike-built environment (Setting-TEMAF) towards cycling participation (Behaviour-TEMAF/ TRMCD).

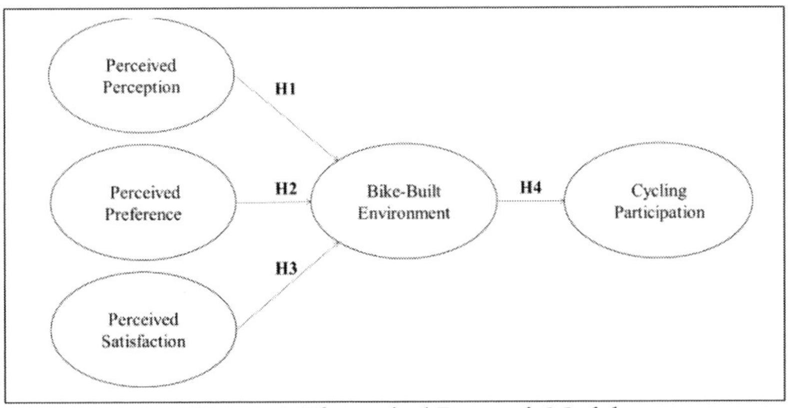

Figure 1. Theoretical Research Model

The study will evaluate the impact of residential-bike built environmental that force to environmental intervention (effective bikeway design) towards cycling promotion [11]–[14]. The perceived perception residential environment revealed by filtering the environmental information of surrounding to expose the critical factors that appropriate towards the behaviour. The perception of personal safety, crime, traffic, infrastructure availability, network system, public facilities, park and open spaces, climate and natural features have been significantly positive relationship towards cycling [16]–[18]. Therefore, the first hypothesis was established as follows: *H1- The perceived perception of residential environment has a positive effect on the bike-built environment.*

Moreover, the perceived preference of route choice is prioritising or choosing the specific elements of condition to fit their insight towards behaviour [18]. It is vital to explore due to understanding the interaction way of the individual with a set of distinctive values of the specific environment. The previous studies was focused on the characteristics of the road, traffic ambience, environment setting, trip purpose and the route character that has been affected the cycling activity [19]–[21]. Thus, the second hypothesis is *H2- The perceived preference of route choice has a positive effect on the bike-built environment.*

The cycling habit also affected by perceived satisfaction when all expectation and needs had been fulfilled. The evaluation operationalised with the circumstance on the roadway corridor environment that emphasised the width, traffic volume, activities, pavement and stoppages [22]–[24]. It shows that the satisfaction consideration will enhance the behaviour with some intervene effort from the evaluation. Besides, the maintenance, environment, network, design and personal satisfaction are the factors that need to be determined further [25]. Hence, the third hypothesis comes out such as *H3- The perceived satisfaction of travel environment has a positive effect on the bike-built environment.*

The bike built environment is the behavioural setting that serves people to do cycling in proper manner equivalently with other types of modes that need a specific system. The cycling-friendly cities have been proper layout planning, facilities, design methods, and detailed design that can be contemplated with some adjustment to suit Malaysia context [26]. Besides, the key aspect that has been stressed in the bike-built environment enhancement are coherence, directness, attractiveness, traffic safety, comfort, and spatial integration [27]. All these elements will lead to the cycling participation among people when they assist with the good ambience of the cycling environment. Therefore, the final hypothesis developed is as follow: *H4- The bike-built environment has a positive effect on cycling participation.*

2. Methodology

The City of Elmina, Selangor was the chosen site for the study conducted. We engaged with the residential association (RA) to distribute the survey link on the resident web pages. The minimum sample size of 92 respondents has been calculated at 80 per cent with effect size 0.15 using Daniel Soper Calculator [28]. However, the actual sample size should be exceeded 92 and the actual sample size will be discussed in 3.1 afterward. A structured questionnaire online survey is used with six segments including demographic profile (6 items), perception (5 items), preference (5 items), satisfaction (5 items), bike-built environment (6 items) and participation (3 items). The measurement level is comprised continues (a five-level Likert-scale with "Strongly Disagree" until Strongly Agree") and definite measurement. Eventually, the analysis operationalised by using IBM SPSS version 23 and Smart PLS 3.2.6 to extract the significant statistical result. The rules of thumb consideration to applying Smart PLS can examine both reflective and formative measurement models and works precisely with small sample sizes, it does not require normal-distributed data, and importantly the study aims for theory confirmation proposition [29]. The process of analysis comprises two-stage procedures which are measurement model analysis (Reliability And Validity Test) and followed by a structural model analysis (Hypothesis Testing) [30]. Besides, a bootstrapping method subsequent with path coefficients and significant loadings value. This approach offers comprehensive, quantifiable and flexible causal-modelling capabilities in the multi-regression analysis for non-parametric data that run simultaneously [31].

3. Result

The PLS rules of thumb were occupied accordingly in the study. The small size data (n=242) was confirmed, and non-normal distributed data was proved by Mardia's multivariate skewness and kurtosis test stated that ($\beta = 6.665$, $p < 0.01$) and ($\beta = 44.876$, $p < 0.01$) respectively above the cut-off value of skewness ± 3 and kurtosis ± 20 [30]. Thus, the non-parametric analysis software can be used for analysis purpose like Smart PLS.

3.1. Demographic Profile

It is confirmed that 242 respondents were employed. The study obtained information on gender is dominated by male (59.5%) compared to 40.5% are female. The respondent age ranged between 25 and 44 years old (47.6%) is prominent. Among them, the study identified 59.5% was led by Malay and followed by Chinese and Indian which were 21.4% and 19% respectively. This group of respondent entailed of diploma (26.2%), bachelor (40.5%) and higher bachelor degree (31%) holders. We found that the respondent had preferred to use private transport with 71.4% (Car) and 26.2% (Motorcycle). Surprisingly, 2.4% has been used as a bicycle as a transport mode in daily life.

3.2. Measurement Model Analysis

3.2.1. Construct and Convergent Validity

The measurement model analysis was examined the construct and convergent validity with threshold values (factor loadings, > 0.70; Cronbach Alpha (CA), > 0.70; Average Variance Extracted (AVE), > 0.50; and Composite Reliability (CR), > 0.70) [32] for reflective measurement model. As shown in Table 1, it was found that the individual items loadings were 0.496 – 0.948. Although two items stated as 0.496 and 0.634 it still useable due to the measurement threshold (AVE and CR) had met, thus, both indicators have remained. Consequently, CA reached above 0.70 threshold value with ranged between 0.717 to 0.864. The AVE values were above 0.50 (0.554 – 0.888) while CR were ranged from (0.827 – 0.936), above 0.70. For the formative measurement model does not have CR and AVE but it measured by weight loading with significant level t-values and collinearity assessment by evaluating variance inflation factors (VIF) with values less than 5.0 [30]. All the data were met the threshold value as per stated in Table 1 and Figure 2 (in Appendix A). So the measurement model analysis has met the requirement.

Table 1. The Convergent Validity of measurement Model Analysis Result

Measure	Construct	Items	Loadings/Weight	Cronbach Alpha	CR/VIF	AVE/T-Value
Reflective	Perception	PER4	0.750	0.717	0.827	0.554
		PER5	0.828			
		PER8	0.496			
		PER9	0.850			
	Preference	PRE2	0.716	0.846	0.890	0.620
		PRE3	0.832			
		PRE4	0.807			
		PRE5	0.854			
		PRE6	0.716			
	Satisfaction	SAT2	0.634	0.743	0.840	0.570
		SAT3	0.796			
		SAT4	0.737			
		SAT5	0.838			
	Participation	PART6	0.927	0.864	0.936	0.880
		PART7	0.948			
Formative	Bike Environment	BDF10	0.522		1.667	3.203
		BDF11	0.447		2.228	2.334
		BDF2	0.286		1.900	2.003
		BDF5	-0.634		2.569	3.285
		BDF7	0.454		1.482	2.586
		BDF9	-0.342		1.053	2.933

3.2.2. Discriminant Validity
Next, there are two approaches to assess the discriminate validity by using Fornell and Larcker's (1981) criterion and Heterotrait-Monotrait Ratio of correlations (HTMT) by Henseler et al. (2015). For the study, we used the HTMT ratio as per Table 2; it stated that the constructs were confirmed the threshold value less the 0.90 [33].

Table 2. The Discriminant Validity of Measurement Model Analysis Result (HTMT Ratio)

	1	2	3	4
1. Participation				
2. Perception	0.470			
3. Preference	0.225	0.137		
4. Satisfaction	0.080	0.567	0.257	

3.3. Structural Model Analysis
Before we proceed with the structural model analysis, a model fit assessment need to be carried out by statistical inference bootstrapping called Standardized Root Mean Square Residual (SRMR). It used to escape the model misspecification with implying the values less than 0.10 [34], and the value of the study is 0.097 whereby it passed the threshold value.

3.3.1. Hypothesis Testing Result

The next stage is to proceed with the hypothesis testing by looking at path coefficient (β), standard error, t-values, collinearity, coefficient determination (R^2), effect size (f^2) and predictive relevance (Q^2) via performing a resample of 5,000 bootstrapping procedure [30]. Currently, the combination of statistical (p-value) and substantive significant (f^2) are vital information to be reported due to effect size is not revealed by p-value alone [35]. Subsequently, the quantitative reporting result should be included all elements that have been mentioned above with additional items comprises confidence interval and decision-theoretical modelling [36] as per Table 3.

Environmental Perception ($\beta = 0.193$, $t = 3.023$, $p< 0.01$, $f2 = 0.078$), Environmental Preference ($\beta = 0.405$, $t = 6.843$, $p< 0.01$, $f2 = 0.232$) and Environmental Satisfaction ($\beta = 0.215$, $t = 3.376$, $p< 0.01$, $f2 = 0.056$) had influenced bike-built environment positively whereby 33.3% of R^2 (bike-built environment predictors) indicates substantial model with above 0.26 threshold value [37]. Thus, the hypothesis H1, H2 and H3 were supported. For instance, the predictive effects between bike-built environment and cycling participation were looked up next. Bike-built environment ($\beta = 0.270$, $t = 3.023$, $p < 0.01$, $f2 = 0.078$) was significant predictor of cycling participation with only 7.2% of R^2 (cycling participation predictor) indicates weak model with above 0.02 threshold value [37]. However, the hypothesis H4 accepted to support. Afterward, the predictive relevance value also should have explained by using blindfolding procedure to obtain Q^2 value. It can be assessed for the reflective measurement model only and the value for the study was 0.057, more than zero value [30] as shown in the Table 3 and Figure 3 (in Appendix A).

Table 3. The Result of Hypothesis Testing (Structural Model Analysis)

	Std. Beta	Std. Error	t-value	Decision	2.50%	97.50%	VIF	R^2	Q^2	f^2
H1-Bike Environment -> Participation	0.270	0.089	3.023**	Supported	-0.139	0.388	1.000	0.072	0.057	0.078
H2 -Perception -> Bike Environment	0.193	0.084	2.296**	Supported	-0.114	0.323	1.188			0.051
H3-Preference -> Bike Environment	0.405	0.059	6.843**	Supported	0.271	0.509	1.042			0.232
H3-Satisfaction -> Bike Environment	0.215	0.064	3.376**	Supported	0.083	0.333	1.225	0.333	-	0.056

** p-value, <0.01, one-tailed, >1.960 of t-value
*b p-value, <0.05, one-tailed, >1.645 of t-value

4. Discussion and conclusion

The study was found the effective bikeway design factors via SEM-PLS analysis approach (statistical and substantive significant) through built environment evaluation. Initially, it is uncommonly used in the social, built environment study. However, the study tries to explore the significant application towards understanding the environmental behaviour by using developed model testing to obtain significant predictors. The environmental perception implicates that resident's perception of the convenience infrastructure, good street network system, acceptable climatic condition and pleasant natural setting have been influenced by the behavioural setting (bike-built environment). These items have similar findings from the previous study using the Neighbourhood Environment Walkability Scale (NEWS) [13]. It shows that perception is a fundamental awareness of the environment at a macro-scale level.

Moreover, the study reveals pavement types, dedicated space, security, proximate and scenic are the environmental preferences that need to emphasise in cycling ambience intervention. This prediction confirmed that preference has a most robust positive relationship among the other predictors. Thus, the preference takes action as a micro-scale level of evaluation before conducting cycling [19]–[21], [38], [39]. Environmental satisfaction also has a positive effect on the bike-built environment. The study associated with the fulfilment of expectation towards cycling in the residential neighbourhood. The elements of linkages, maintenance, design and personal satisfaction have been influenced towards the cycling environment whereby it parallels with other studies' findings [22]–[24]. This is the complementation of evaluation for both macro and micro-scales. For instance, the high value of R^2 indicates that perception, preference and satisfaction have a significant to predict bike-built environment.

Lastly, the effective bikeway design factors can be revealed such as lighting provision, continues cycling network, shaded trees incorporated, greens spaces integration, sufficient width, proper surface and material selection are the key elements for the enhancing cycling participation among residents [27]. Although the f^2 value is weak it still has a positive effect on cycling participation. It also has supported the theoretical predictors when the people awoke (perceived environment) and the setting support provided (bike-built environment) would lead the high cycling participation (behaviour) [8], [9]. In conclusion, the impact of residential-bike built environment was confirmed with significant influenced on cycling participation among residents.

Acknowledgments
High appreciation is given to IRMI, Universiti Teknologi MARA for providing the funding support through the Research Grant (LESTARI) 600-IRMI/DANA 5/3/LESTARI (0152/2016).

References

[1] N. A. Salleh, R. Zainol, F. Ahmad, and N. M. Noor, "Factors that Contribute to Green Neighbourhood : A Review," *Int. J. Prop. Sci.*, vol. 4, no. 1, pp. 1–10, 2014.
[2] T. Lokman, "PM : Malaysia on course to reduce carbon emissions by 40 pct by 2020," *New Straits Times*, Kuala Lumpur, pp. 3–7, Dec-2017.
[3] N. Ibrahim, "Pembuka tirai 2018 Shah Alam Hari Tanpa Kenderaan sedia pelbagai acara menarik," *Selangorkini*, Shah Alam, pp. 1–3, Jan-2018.
[4] J. Chan and Y. Nathan, "Cycling culture picking up speed," 2017.
[5] J. Pucher and R. Buehler, "Cycling towards a more sustainable transport future," *Transp. Rev.*, 2017.
[6] R. Ewing and R. Cervero, "' Does Compact Development Make People Drive Less ?' The Answer Is Yes," vol. 4363, 2017.
[7] R. Shokoohi and A. Nikitas, "Urban growth, and transportation in Kuala Lumpur: Can cycling be incorporated into Kuala Lumpur's transportation system?," *Case Stud. Transp. Policy*, 2017.
[8] T. L. Burton, "The environment and physical activity," *Phys. Act. Hum. Exp. Interdiscip. Perspect.*, pp. 231–265, 1997.
[9] J. F. Sallis, R. B. Cervero, W. Ascher, K. A. Henderson, M. K. Kraft, and J. Kerr, "an Ecological Approach To Creating Active Living Communities," *Annu. Rev. Public Health*, vol. 27, no. 1, pp. 297–322, 2006.
[10] R. J. Schneider, "Theory of routine mode choice decisions: An operational framework to increase sustainable transportation," *Transp. Policy*, vol. 25, pp. 128–137, 2013.
[11] L. Mertens *et al.*, "Perceived environmental correlates of cycling for transport among adults in five regions of Europe," *Obes. Rev.*, 2016.
[12] K. C. Heesch, B. Giles-Corti, and G. Turrell, "Cycling for transport and recreation: Associations with the socio-economic, natural and built environment," *Heal. Place*, vol. 36, pp. 152–161, 2015.
[13] J. F. Sallis *et al.*, "Environmental and demographic correlates of bicycling," *Prev. Med. (Baltim).*, vol. 57, pp. 456–460, 2013.
[14] C. Zhao, T. A. S. Nielsen, A. S. Olafsson, T. A. Carstensen, and C. Fertner, "Cycling environmental perception in Beijing – A study of residents' attitudes towards future cycling and car purchasing," *Transp. Policy*, vol. 66, no. February, pp. 96–106, 2018.
[15] J. Kerr *et al.*, "Perceived neighborhood environmental attributes associated with walking and cycling for transport among adult residents of 17 cities in 12 countries: The IPEN study,"

Environ. Health Perspect., vol. 124, no. 3, pp. 290–298, 2016.
[16] A. Ghekiere *et al.*, "Critical environmental factors for transportation cycling in children: A qualitative study using bike-along interviews," *PLoS One*, vol. 9, no. 9, 2014.
[17] D. Van Dyck *et al.*, "Perceived neighborhood environmental attributes associated with adults' transport-related walking and cycling: Findings from the USA, Australia and Belgium," *Int. J. Behav. Nutr. Phys. Act.*, vol. 9, pp. 1–14, 2012.
[18] M. Haybatollahi, M. Czepkiewicz, T. Laatikainen, and M. Kyttä, "Neighbourhood preferences, active travel behaviour, and built environment: An exploratory study," *Transp. Res. Part F Traffic Psychol. Behav.*, vol. 29, pp. 57–69, 2015.
[19] N. Waintrub, C. Peña, M. Niehaus, R. Vega, and P. Galilea, "Understanding cyclist traffic behaviour: Contrasting cycle path designs in Santiago de Chile," *Res. Transp. Econ.*, 2016.
[20] L. Mertens *et al.*, "Differences in environmental preferences towards cycling for transport among adults: A latent class analysis," *BMC Public Health*, vol. 16, no. 1, pp. 1–10, 2016.
[21] S. E. Vedel, J. B. Jacobsen, and H. Skov-Petersen, "Bicyclists' preferences for route characteristics and crowding in Copenhagen – A choice experiment study of commuters," *Transp. Res. Part A Policy Pract.*, 2017.
[22] S. K. Beura and P. K. Bhuyan, "Development of a bicycle level of service model for urban street segments in mid-sized cities carrying heterogeneous traffic: A functional networks approach," *J. Traffic Transp. Eng. (English Ed.*, vol. 4, no. 6, pp. 503–521, 2017.
[23] S. K. Beura, H. Chellapilla, and P. K. Bhuyan, "Urban road segment level of service based on bicycle users' perception under mixed traffic conditions," *J. Mod. Transp.*, vol. 25, no. 2, pp. 90–105, 2017.
[24] A. Ilie, C. Oprea, D. Costescu, E. Roşca, O. Dinu, and F. Ghionea, "The use of the bicycle compatibility index in identifying gaps and deficiencies in bicycle networks," *IOP Conf. Ser. Mater. Sci. Eng.*, vol. 161, no. 1, 2016.
[25] J. C. Calvey, J. P. Shackleton, M. D. Taylor, and R. Llewellyn, "Engineering condition assessment of cycling infrastructure: Cyclists' perceptions of satisfaction and comfort," *Transp. Res. Part A Policy Pract.*, 2015.
[26] A. Forsyth and K. Krizek, "Urban design: Is there a distinctive view from the bicycle?," *Journal of Urban Design*. 2011.
[27] A. Hull and C. O'Holleran, "Bicycle infrastructure: can good design encourage cycling?," *Urban, Plan. Transp. Res.*, vol. 2, no. 1, pp. 369–406, 2014.
[28] "Free A-priori Sample Size Calculator for Structural Equation Models - Free Statistics Calculators." .
[29] N. Urbach and F. Ahlemann, "Structural Equation Modeling in Information Systems Research Using Partial Least Squares Structural Equation Modeling in Information Systems Research Using Partial Least Squares," *J. Inf. Technol. Theory Appl.*, vol. 11, no. 2, pp. 5–40, 2010.
[30] J. F. J. Hair, G. T. M. Hult, C. M. Ringle, and M. Sarstedt, *A Primer on Partial Least Squares Structural Equation Modeling (PLS-SEM) Second Edition*, Second Edi. London: SAGE Publications, Inc, 2017.
[31] P. B. Lowry and J. Gaskin, "Partial least squares (PLS) structural equation modeling (SEM) for building and testing behavioral causal theory: When to choose it and how to use it," *IEEE Trans. Prof. Commun.*, vol. 57, no. 2, pp. 123–146, 2014.
[32] J. Henseler, C. M. Ringle, and M. Sarstedt, "A new criterion for assessing discriminant validity in variance-based structural equation modeling," *J. Acad. Mark. Sci.*, vol. 43, no. 1, pp. 115–135, 2014.
[33] R. B. Darlington and A. F. Hayes, *Regression Analysis and Linear Models*. 2017.
[34] Hu L and Bentler PM, "Fit indices in covariance structure modeling: Sensitivity to underparameterized model misspecification.," *Psychol. Methods*, vol. 3, no. 4, pp. 424–453, 1998.
[35] G. M. Sullivan and R. Feinn, "Using Effect Size—or Why the *P* Value Is Not Enough," *J. Grad.*

Med. Educ., 2012.
[36] E. D. Hahn and S. H. Ang, "From the editors: New directions in the reporting of statistical results in the Journal of World Business," *J. World Bus.*, vol. 52, no. 2, pp. 125–126, 2017.
[37] J. Cohen, *Statistical Power Analysis for the Behavioral Sciences*, 2nd, ed ed. New York, NY.: Lawrence Erlbaum Associates, 1988.
[38] T. Lin, D. Wang, and X. Guan, "The built environment, travel attitude, and travel behavior: Residential self-selection or residential determination?," *J. Transp. Geogr.*, vol. 65, no. October, pp. 111–122, 2017.
[39] H. Verhoeven *et al.*, "Which physical and social environmental factors are most important for adolescents' cycling for transport? An experimental study using manipulated photographs," *Int. J. Behav. Nutr. Phys. Act.*, vol. 14, no. 1, pp. 1–13, 2017.

APPENDIX

Appendix A

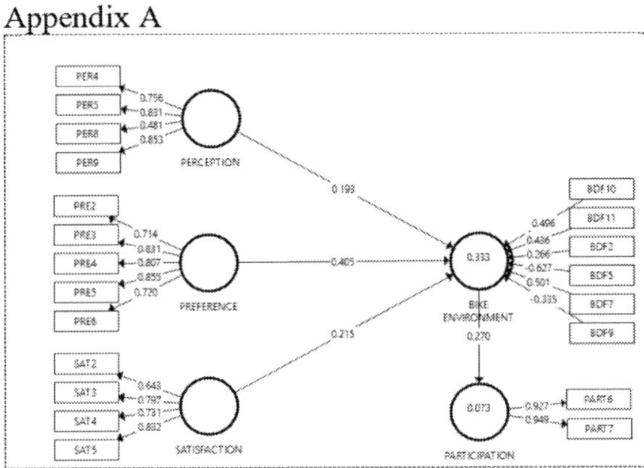

Figure 2. The Measurement Model Result (Reliability And Validity Test).

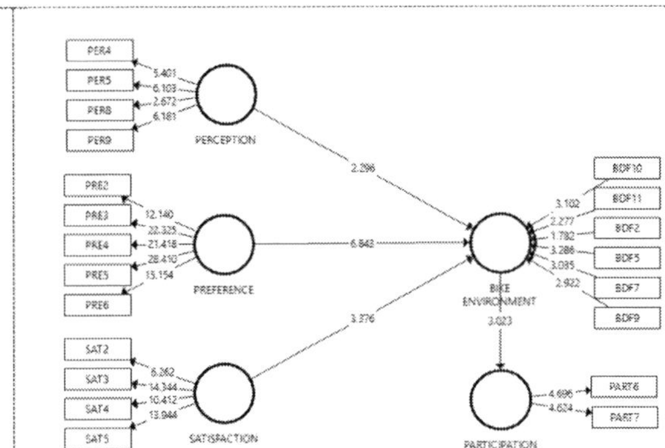

Figure 3. The Bootstrapping Structural Model Result (Hypothesis Testing)

Constructs	Code	Indicators
Perceived perception	**PER4**	There are convenient cycling infrastructure provided in my residential neighbourhood
	PER5	The street network system is continued with other infrastructure including bikeways.
	PER6	The nearest public facilities like a bus stop, school and commercial area are difficult to reach by bicycle
	PER8	The Malaysia climate does not prevent me from doing cycling activity within my residential neighbourhood
	PER9	The natural setting in the residential neighbourhood is pleasant for cycling activity
Perceived preference	**PRE1**	I prefer a variety of slopes and gradient of bikeways
	PRE2	I prefer smooth pavement types and evenness condition
	PRE3	I prefer a dedicated cycle path on each side of the road
	PRE4	I prefer security aspects such as speed limit, speed bump and signage
	PRE5	I prefer direct bikeway system, proximate and easy to access.
	PRE6	I prefer scenic views such as natural elements and greenery
Perceived satisfaction	**SAT1**	I am comfortable cycling within unshaded bikeway during daytime in my residential neighbourhood.
	SAT2	I am pleased the bikeway in my residential neighbourhood that links with the key destination such as parks, school and shops
	SAT3	I enjoy the well-maintained and frees from surface defects such as potholes, root damage and broken up the surface of bikeway
	SAT4	I am satisfied with the bikeway design in term of sufficient width, the suitable material used and excellent facilities provided
	SAT5	I am satisfied with the overall bikeway system that includes personal safety, convenience and enjoyment
	BDF2	The bikeway in the residential neighbourhood needs to provide street lightings to

IOP Conf. Series: Earth and Environmental Science **385** (2019) 012053 doi:10.1088/1755-1315/385/1/012053

Bike built environment		cater to the user at night.
	BDF5	The bikeway within the residential neighbourhood needs to be a well-connected and continuous network.
	BDF7	The bikeway in the residential neighbourhood should provide shaded trees and colourful shrubs along the bikeway.
	BDF9	The bikeway in the residential neighbourhood could not be incorporated into open spaces, green space and parks
	BDF10	The minimum width and surface required for one-way of dedicated bikeway are 2 - 2.5 meter and asphalt concrete.
	BDF11	The sustainable material selection, easy to maintain and durable is the primary consideration for bikeway design
Participation	**PART6**	I will spend about .... when I do cycling activity
	PART7	Travel distance that I prefer for cycling is between
	PART8	I will do cycling for about .... times

IOP Conf. Series: Earth and Environmental Science **385** (2019) 012054 doi:10.1088/1755-1315/385/1/012054

Exploratory Research as the way forward towards a green procurement practices for the construction industry; Research Methodology

A A M Bohari[1], Z A Bidin[1], S L A Rais[1] and M M Saferi[1]
[1]Faculty of Architecture, Planning and Surveying, Universiti Teknologi MARA, Cawangan Sarawak, Kampus Samarahan, Kota Samarahan, Sarawak, Malaysia

Email: asmahalia@sarawak.uitm.edu.my

Abstract. Green construction is a way forward for the construction industry in Malaysia, towards an environmentally responsible industry. One of the important tools to accelerate green construction is the construction procurement. Thus, the concept of a green-oriented procurement has been introduced to integrate green practices into project procurement delivery. However, green procurement is the least discussed and the concept is still very new in the Malaysian construction industry. Thus, this paper describes a proposed strategy of exploring the potential of green procurement for the construction industry. This is based on an on-going research project whose primary objective is to pave a way forward for the key stakeholders who are involved in green building projects.

1. Introduction on the key concepts

Throughout its life cycle, the building industry is related to global environmental issues such as GHG emissions, air pollution and improper waste management. The adverse impact of the construction sector on the natural environment has triggered a growing awareness to embark on towards the greener path. The green construction concept has been introduced as an environmentally friendly approach as it uses sustainable principles in its building techniques and procedures. Underlined in the concept of sustainable development in the Brundtland Report [1], the construction procurement has been pointed as an effective tool to manage adverse environmental issues in construction.

The term procurement refers to product purchasing activities, however, the definition of the term procurement of construction differs slightly from those of products. Construction procurement is explained as "a process of acquiring goods and services for realising a constructed facility according to predefined requirements" [2]. Thus, construction procurement is important because of its influence in the overall project delivery [3]. The term "green procurement" or GP is introduced in the construction industry to fulfil the environmental objective of an organisation [4]. Green procurement builds on the current procurement practices, but it extends through the adoption of sustainability principle.

GP involves the minimisation of the environmental impact throughout the lifecycle of a building from inception until the demolition of the building. Various literatures have highlighted the wider benefits of GP, which include minimising environmental impact, meeting sustainability goals, increasing financial efficiency; reducing the risk of non-compliance with legislation, enhancing organisational reputation and encouraging innovation. GP concerns are with the performance of the construction delivery, the way the project is being delivered as well as the materials used in construction. GP requires active participation and support from the project stakeholders and thus, the awareness among stakeholders on GP is very crucial. Project stakeholders are urged to identify the key green practices that are important for their construction procurement.

Adetunji, Price & Fleming (2008) [5] have revealed the fact that green procurement practices have not managed to capture the interest and attention of construction sector practitioners. Lack of familiarity, knowledge and interest among stakeholders in green procurement pertaining to construction could be the flaws that could lead to low support of the sustainability agenda [6]. It has also been suggested that there must be an established standardised procedures and guidelines for GP to be a framework for the practitioners which currently lacking knowledge and familiarity (refer figure 2).

Thus, this research is conducted with the aim to provide an exploratory study on the potential of GP for building projects. Exploration encompasses three main objectives: the first is to identify the current state of readiness of the stakeholders, the challenges and the success factors, the second is to determine the extent of GP practices in green certified and non-green certified construction projects. and the third objective is to determine the important practices of GP for construction projects.

This paper discusses the progress to date of an ongoing research project that is aimed at exploring the potential of green procurement for the construction industry, and specifically it aims to discuss the research methodology undertaken to ensure the reliability of the research. This involves the selection of suitable research philosophy, approach and strategy. The research methodology of this research will be the pathway towards accomplishing the objectives that are underlined in this research.

2. Methodology

2.1 Understanding research methodology choices for this research

There are two well referred research methodology frameworks available in the built environment; they are the nested method and the research onion [7]. Keraminiyage, (2013) [7] further explains [7] a comprehensive range of possible approaches through a modified model that is based on Saunder's Research Onion and Kagioglou's Nested Approach (Refer Figure 1).

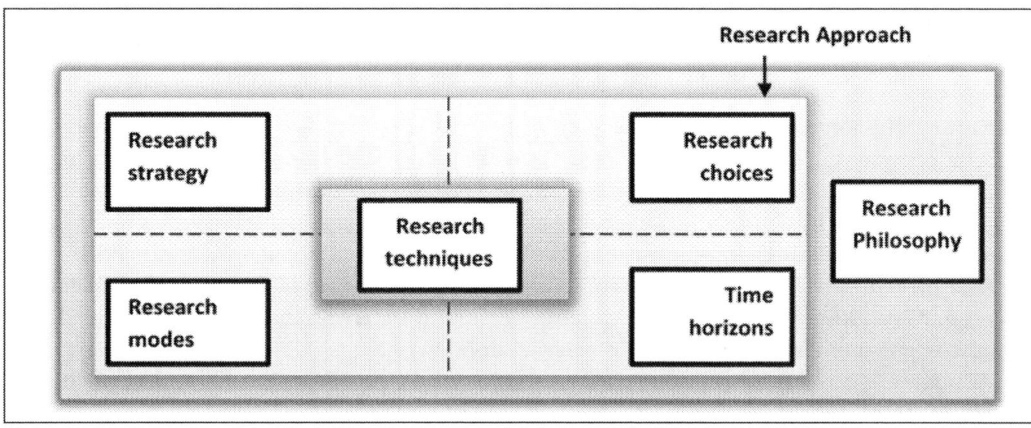

Figure 1. The modified model, Source: [7]

After a thorough review on the choices of research methodology that are available, Table 1 shows the proposed research approach of this research as suggested by [7] in order to achieve the research objectives.

Table 1. Research approach

Research Philosophy	⇨	**Pragmatism**
Research strategy	⇨	Survey
Research choices	⇨	Mixed-method
Time Horizons	⇨	Cross sectional
Technique	⇨	FGD and Questionnaire

2.2 Research philosophy

[8] has mentioned a few views of research paradigms that have been used in the field of modern social science research that can be applied when undertaking research. The research philosophy acts as a guide

to the general approach or strategy that is used to conduct research. Some research uses the postpositivist view (inductive, developed theory and qualitative), some use the positivist (deductive, quantitative data collection and interpretation), and some researchers combine both systems (mixed method), which is called a pragmatic worldview [8], [9] Pragmatism highlights the research problem to be at the central position where the researcher applies all available methods to coin the problem and the possible way out [8]. This research is carried out to deliver a possible solution to a practical problem, namely, to improve the current procurement system for a green project within the Malaysian construction industry. As pragmatism places "the research problem" at the central position, this research would be considered as a pragmatic research (refer figure 2). The pragmatic researcher will use multiple methods of data collection to meet the research objective such as the quantitative and qualitative data collection that can be integrated with the data at different stages [8].

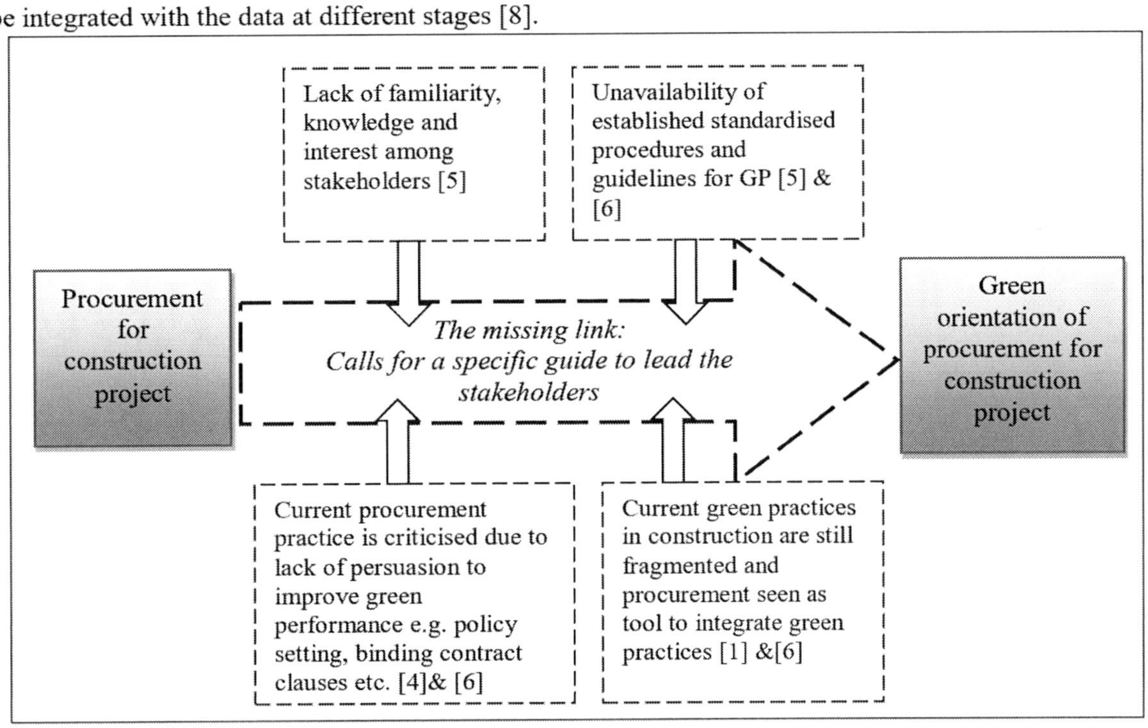

Figure 2 Addressing the research problem

2.3 Research strategy and choices

Considering the various research strategy and choices that are proposed in the literature, the research for this paper is decided based on its' research objectives and aim [8]. This research uses surveys as a data collection technique whereby the approach includes questionnaires, interviews, content analysis, and observations [9]. The surveys adopt quantitative and qualitative approaches and focus on questions such as what, who, where, how much or how many [8], [9]. A mixed method research refers to the mixture of both qualitative and quantitative research methods. This method provides better solutions to the drawbacks of the qualitative and quantitative approaches by the strengths of the other complement rather than contending one with the other. Based on Creswell [8], one of the ways in connecting the data is by connecting the data. The literature review will be the basis to form research direction and the way forward. The qualitative approach will be carried out and then it will be followed by the quantitative phase. Thus, this research involves questionnaires and focus group discussion.

IOP Conf. Series: Earth and Environmental Science **385** (2019) 012054 doi:10.1088/1755-1315/385/1/012054

2.4 Research technique and procedure
To accomplish the research objectives and aim, this research will be conducted in three phases and undertakes the following activities (refer Appendix 1). Two main techniques of data collection will be adopted based on specific procedures: expert discussion through FGD and the questionnaire.

2.4.1 Focus group discussion (FGD)
A focus group discussion (FGD) approach will be conducted to collect qualitative data for this research. Accordingly, preliminary interviews are necessary to help the researcher to confirm, refine and rephrase the research problem and the factors that are derived from the literature. Creswell [8] proposed that "...qualitative research is best suited to address a research problem in which the researcher does not know what the variables are and needs to explore the topic more widely". To gain local context in-depth understanding, the qualitative investigation will be conducted with experienced practitioners and subject matter expert who are policy makers and researchers [10]. The implementation diverges between one context to another to suit local needs.

The key selection criteria for FGD participants are their credential in green construction in Malaysia. The researcher has gathered information from various sources, such as available public information, internet searches, and personal contact with the project's key person. Findings at this stage will reveal nothing conclusive but it may help in defining the problem and can be used to improve the questionnaires being used in the main data collection. The FGD discussion will be recorded and transcribed by the researchers. The written transcripts will be proofread to confirm the validity and reliability. If the discussion uses a local language, in this case the Malay language, the discussion needs to be back-translated to ensure the accuracy of the translation [11]. Data will be analysed using the available qualitative software in the market.

2.4.2 Questionnaire
Term questionnaire refers to ".... the documents that include a series of open and closed questions, to which the respondent is invited to provide answers" [12]. Questionnaires offer an objective means of collecting information about people's knowledge, beliefs, attitudes and behaviour [12]. There are three kinds of research: profiling and descriptive research, predictive and analytical research and lastly, developing and testing measurement scales [12]. This research aims to generate a profile of the samples to examine the key factors of green-oriented procurement for building projects by posing some questions, around what they do and what they think. Questions have been formulated based on findings that are derived from the previous stages, the literature review and preliminary interview. In this research, a questionnaire will be developed through three stages- the pre-drafting process, piloting stage and finalising stage. To ensure that the questions will be clearly understood and that there is no ambiguity among them, a pre-test must be conducted (refer figure 3). The survey is a cross-sectional survey since it has taken place at a single point in time.

The questionnaire will be self-administered and distributed to the respondents randomly to their office address. Before sending the questionnaire, phone calls are made to identify the actual potential respondents and to seek their agreement. As the world is approaching the 4th Industrial Revolution, the online survey is the most used survey method. The online surveys offer a wider access to potential respondents in various locations and saves resources in term of cost and time. Thus, this research will adopt an online survey to facilitate the data collection process. The sample will be drawn based on a purposive sampling method where the respondents are chosen based on the purpose of the survey. Creswell [8] has stated that "purposive sampling does not produce a sample that is representative of a larger population, but it can be exactly what is needed for certain cases - study of an organisation, community, or some other clearly defined and relatively limited group". Data will be analysed using statistical analysis software such as the IBM SPSS Statistics.

IOP Conf. Series: Earth and Environmental Science **385** (2019) 012054 doi:10.1088/1755-1315/385/1/012054

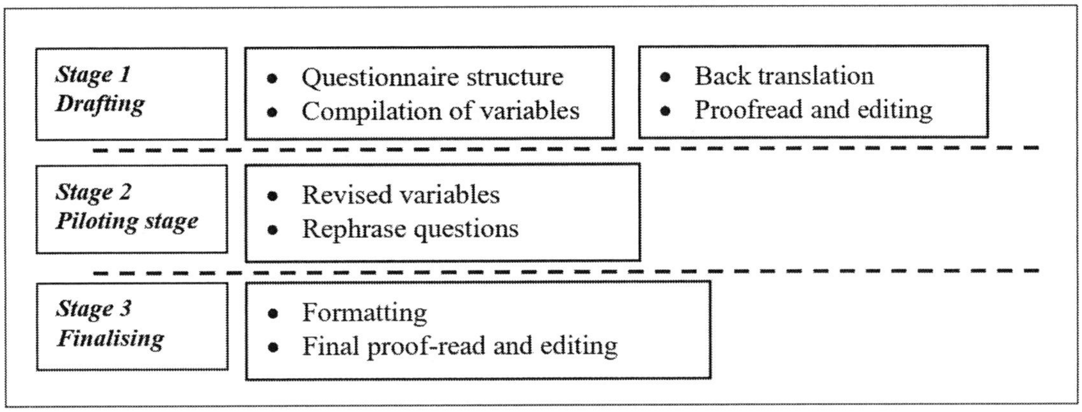

Figure 3. Questionnaire development process

2.5 Research process

The attached Figure 4 as appendix 1 of this paper shows the proposed three stages of the research process that will be undertaken by this research in order to meet the research aim. Stage 1 is where the information that has been gathered is used to carry out and to establish the theoretical and conceptual framework of the research. The FGD will be conducted to gather first-hand information based on local context from experts and experienced practitioners. Stage 2 will be conducted based on the outcome form stage 1 and the main data collection will be carried out. At this stage, the survey using questionnaire will be employed to reach a larger number of potential respondents. Stage 3 is where the outcome from stage 2 is being analysed and discussed in the analysis and recommendation will be made.

3. Preliminary findings from literature review

This section discussed briefly finding from Stage 1 (information gathering) of the research. The outcome from this stage will be further validated through the methodology proposed under sub-section 2 (research methodology) of this paper.

There are ten (10) possible challenges identified through the literature. The possible challenges are

- Lack of knowledge
- Lack of practical tools and information
- Lack of training for procurement officers
- Lack of awareness on the green procurement concept
- Lack of enforcement by government to implement green procurement
- Insufficient policies and regulation promoting green procurement
- Insufficient research and development on green procurement
- Lack of incentive for companies to implement green procurement
- High cost of green product and services
- Limited supply for green product

Five (5) strategies has been identified based on literature analysis. The possible strategies are

- The availability of policy and guideline on green procurement
- Enhancing awareness among all project stakeholder
- Carrot and stick approach
- Client and top management commitment
- Enhancing capacity building among construction players

In order to enhance the implementation of environmental management in the procurement process, there are four main categories of practices identified.

Table 2. Green procurement practices

Improved process for the green-oriented procurement	
1. Green policies and guidelines	• Determine common goals in term of green performance • Establish the scope and interpretation of green policies that encourage green adoption
2. Green products and process	• Selecting materials based on project's green indicators (incorporate into contract) • Promote waste management
3. Environmental evaluation	• Develop evaluation plan; • Integrating monitoring and reporting throughout
4. Stakeholder values, development capacity and competence	• Project briefing, training, Implement and inculcate environmental values, knowledge sharing

4. Conclusion

Green procurement is an important driver for green projects. Past research has highlighted the need for green procurement to accelerate the production of greener construction and the output. Based on an established approach in research methodology, the way forward for this research has been determined in order to meet the research objectives of this research. The research philosophy will lead to the selection of research strategy and research technique. There will be further investigation in the next phases of this research, which is the conceptual validation through FGD and main data collection using questionnaires. This paper reports on the on-going research project pertaining to the exploration of green procurement potential in the Malaysian construction industry and still at the early stage of the research. Thus, it is hoped that in the future this research, which implements a correct scientific investigative methodology, will be able to generate reliable research outcomes, and thus spearhead the GP practices for construction projects and enhance more research pertaining to GP, under different conditions and contexts.

References

[1] WSSD. *Plan of Implementation of the World Summit on Sustainable Development Contents.* Johannesburg Plan Implement. Johannesburg, South Africa, 2002:1-62.

[2] Ruparathna R and Hewage K. Review of contemporary construction procurement practices. *Journal of Management in Engineering.* 2013 31(3):04014038.

[3] Love PE, Skitmore M and Earl G. Selecting a suitable procurement method for a building project. *Construction Management & Economics.* 1998 16(2):221-33.

[4] Zhang X, Wu Y and Shen L. Embedding "green" in project-based organizations: the way ahead in the construction industry?. *Journal of Cleaner Production.* 2015 Nov 16;107:420-7.

[5] Adetunji, I., Price, A. D. F. and Fleming, P. Achieving sustainability in the construction supply chain. *Proceedings of the Institution of Civil Engineers - Engineering Sustainability*, 161(3), 161–172.
[6] Varnäs A, Balfors B and Faith-Ell C. Environmental consideration in procurement of construction contracts: current practice, problems and opportunities in green procurement in the Swedish construction industry. *Journal of Cleaner Production*. 2009 Sep 1;17(13):1214-22.
[7] Keraminiyage K. Research Methodology: what the nested model & the research onion have to say, *PGR Training Seminar*", School of the Built Environment, University of Salford (Online), 2013. Retrieved 1 May 2015 from http://blackboard.salford.ac.uk/bbcswebdav/pid-740704-dt-contentrid-1209404_1/xid-1209404_1
[8] Creswell JW, Creswell JD. Research design: Qualitative, quantitative, and mixed methods approaches. *Sage publications*; 2017 Nov 27.
[9] Creswell JW, Clark VL. Designing and conducting mixed methods research. *Sage publications*; 2017 Aug 31.
[10] Sparrevik M, Wangen HF, Fet AM and De Boer L. Green public procurement–A case study of an innovative building project in Norway. *Journal of Cleaner Production*. 2018 Jul 1;188:879-87.
[11] Khalaila R. Translation of questionnaires into Arabic in cross-cultural research: techniques and equivalence issues. *Journal of Transcultural Nursing*. 2013 Oct;24(4):363-70.
[12] Rowley J. Designing and using research questionnaires. *Management Research Review*. 2014 Mar 11;37(3):308-30.

IOP Conf. Series: Earth and Environmental Science **385** (2019) 012054 doi:10.1088/1755-1315/385/1/012054

Appendix 1

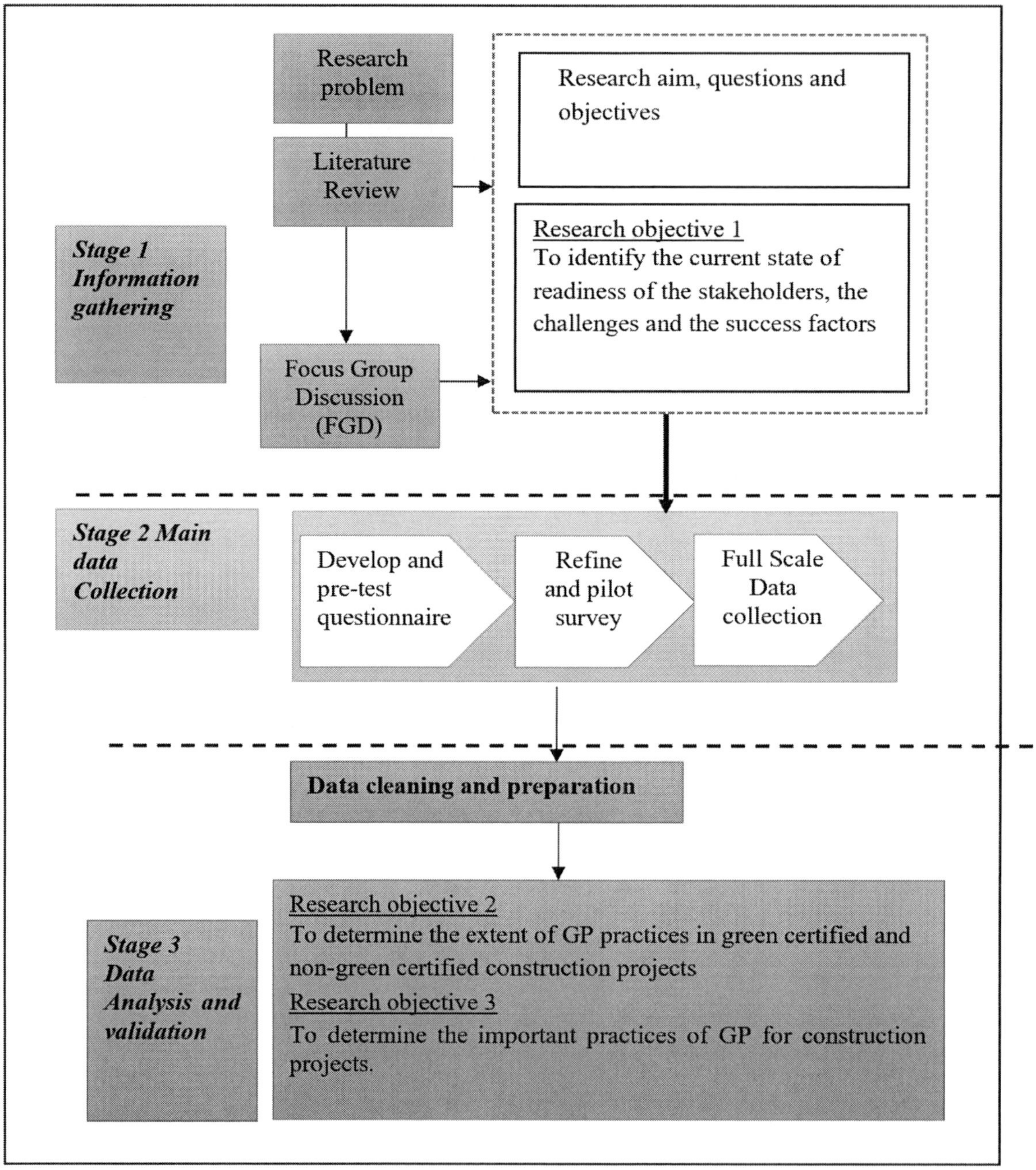

Figure 4 Research process

IOP Conf. Series: Earth and Environmental Science **385** (2019) 012055 doi:10.1088/1755-1315/385/1/012055

Themes and Factors of Construction Safety Management for System Dynamic Model Interactions: A Systematic Review

Nur Azalini Jasni[1], Rumaizah Mohd Nordin[1], Zulhabri Ismail[1] and Nurul Ayuni Abdul Aziz[1].
[1]Centre of Studies of Post Graduate, Faculty of Architecture, Planning and Surveying, Universiti Teknologi MARA (UiTM), Shah Alam, 40450 Selangor, Malaysia

Email: drrumaizah@uitm.edu.my

Abstract. The construction industry has always been regarded as a dangerous industry to work in, and the Malaysian construction industry is not exempted. Most safety and health studies conducted in this industry only view the overall factors contributing to the accidents or focus only on the main factor such as human behaviour. However, accidents actually happen based on the interaction between all the factors of the accidents at a construction site, and thus this study is conducted to identify the factors contributing to the accidents happening at construction sites and to classify them into their themes by using a systematic review method. This method is used to gather the empirical evidence regarding construction safety management by assessing academic journals mostly searched using keywords 'construction safety management', 'themes of accidents', 'factors of accidents' and 'System Dynamic' on academic journal databases known as 'Google Scholar', 'Science Direct' and 'Academia'. The findings of this study identified 18 factors that contribute to the accidents and are classified into four (4) main themes which are human factors, management factors, technical factors and also environmental factors. The themes and factors identified in this study would contribute to develop a construction safety management system using System Dynamic model that will show the complex interactions between all the factors according to their themes.

1. Introduction

The construction industry in Malaysia noted a high number of total accidents with 95 cases reported by June in 2018 alone, and 52 of the cases that ended in casualties [1]. As a complex industry involving many personnel for each project where [2] stated to be inclusive of clients or employers of the project, professionals, contractors, sub-contractors, producers and suppliers, providers and installers, and many other parties or organisations which are relevant in the industry, it is bound to cause several problems, and the main concentration of the industry measures the success rate of a project according to cost, time and quality has led to low regards in safety [3].

Therefore, the government of Malaysia introduced several acts such as Occupational Safety and Health Acts (OSHA), 1994 that implements regulations on safety as well as health aspects [4] and Construction Industry Development Board (CIDB) which was developed to focus on the construction industry quality and health in order to achieve the global competitiveness level [5]. However, to this day, hardly any system developed has been able to handle and reduce the number of accidents reported from all the industries in this country [6].

With that information, many researchers conducted studies regarding the safety in the construction industry to come out with the factors that contribute to the accidents that happen, but many end up focusing on the behaviour of the workers as it is the main factor causing most of the accidents with the highest percentage [7]. As there are not many researches done on the connection of all the factors causing the accidents [8], it is important for this research to be conducted to identify the relations of the factors in order to introduce the best corrective actions to reduce the number of accidents at construction sites.

Hence, System Dynamic (SD) model has been identified as one of the best model to study and understand for a complex system by showing the interrelation between all of its variables [9], is chosen to be developed. To properly come out with the variables for the model, a comprehensive review of previous researches done on the factors of accidents and classification of factors into their themes need to be systematically done using the systematic review methodology.

2. Background

The focus on cost, time and quality as the main driving keys to determine the success rate of a project in the construction industry, especially in Malaysia, has been the reason that the safety and health aspects have been neglected and seen as a taboo subject in the industry [3], the root cause of the accidents that happened need to be identified in order for the effective remedial measures can be applied to prevent such accidents from happening in the future.

Supporting the statement, several theories created by some of the theorists and researchers known as Accidents Causation Model (ACM) explained that every accident happened due to a cause. Heinrich (1959) comes out with the first phase of ACM with the "Domino Theory" which stated that there are five (5) sequential factors that affect any management, creating a domino effect starting with the ancestry or physical environment, followed by fault of a person (carefulness), unsafe act or mechanical or physical condition, which will then cause accidents and lastly, will result in injury [10]. Updating the theory, Reason (2000) comes out with the "Swiss Cheese" model [11], and the latest update on ACM which is the Functional Resonance Accident Model or FRAM by Hollnagel in 2005 explained the several components causing accidents followed by the information have to be identified to prevent accidents in the future [12].

Looking at the construction industry scene, the factors of the accidents studied have identified many factors causing accidents at construction worksites which were categorised by [13] into two groups which are immediate factors and contributing factors with a total of 11 factors extracted. Meanwhile, [14] categorised the factors into unsafe equipment, job site condition, unique nature of the industry, unsafe method, human element, and also management. As there are many factors that can be categorised into different themes, a systematic review is deemed necessary to be conducted. Hence, identifying the factors and themes to further study the interaction between each and every one of the factors is the main focus of this study.

3. Methodology

To properly come out with enough variables to create a System Dynamic (SD) model, systematic review has been adopted in this study where it is a methodological approach to identify, select and appraise literatures agreed to a certain level of quality which are relevant to this study's questions [15].

To start with the review, questions from the objectives of this study are generated, where two questions are identified as follows:

1. What are the factors that cause accidents at a construction site?
2. What are the themes that categorise all the identified factors causing accidents at a construction site?

From the questions, the keywords for the review are identified, which are 'safety management', 'themes of accidents', 'factors of accidents' and also 'System Dynamic' to search for academic journals. Following the keywords generated, the second step includes searching for data via vast databases which are 'Google Scholar', 'Science Direct' and 'Academia' where academic journals that fall under the keywords identified are selected.

The outcome of searching for journals in the databases resulted in a total of 49 journals that consist of researches that mention the factors of accidents at construction sites. From the 49 journals, seven (7) journals are selected which discussed the different factors that contributed to the accidents. The data from the journals selected were then extracted and synthesised as discussed in the Finding. All the steps mentioned will lead to creating the variables to be used to create a System Dynamic model in future research. The steps adapted to create the review for this research is put together into Figure 1 as shown below:

IOP Conf. Series: Earth and Environmental Science **385** (2019) 012055 doi:10.1088/1755-1315/385/1/012055

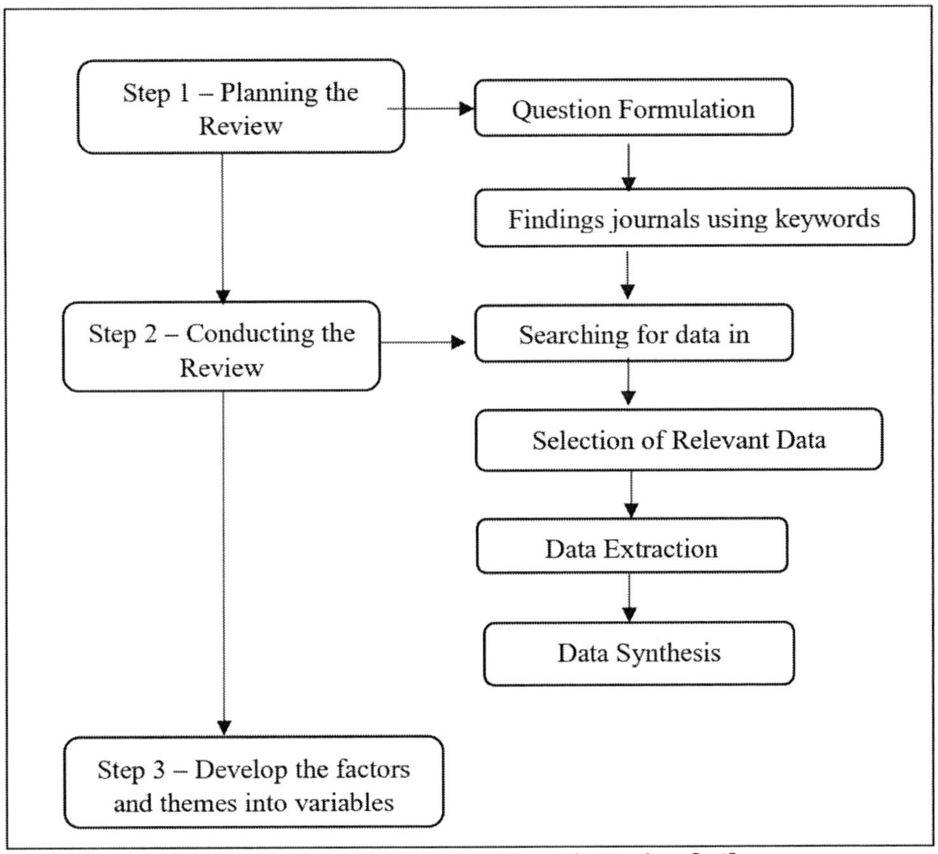

Figure 1. Steps to conduct Systematic Review [16]

4. Finding

Based on the systematic review conducted, seven (7) journals are thoroughly analysed and 18 factors of accidents are identified which are lack of training, unsafe method of handling materials/equipment/operations, fatigue by workers due to overwork, lack of compliance to safety regulation, unsafe behaviour, poor safety awareness among workers/managers, lack of experienced project managers/skilled workers, lack of concern for safety by management, lack of personal protective equipment, defective/poor/unsafe equipment, unsafe condition of workplace, insufficient/lack of housekeeping program, poor site management, poor quality control system, poor illumination, excessive noise, pressure between contractors, procurement & contract problem, nature of the industry and lastly, unforeseen circumstances.

The factors collected from the journals were put into tables and sorted according to the frequencies of each factors mentioned are counted and noted in Table 1 below:

Table 1. Frequencies of the factors mentioned.

Factors	Frequencies
Unsafe behaviour	6
Defective /poor/unsafe equipment	5
Unsafe condition of workplace	5
Lack of personal protective equipment	4
Unsafe method of handling materials/equipment/operations	4
Lack of concern for safety by management	4
Lack of compliance to safety regulation	4
Lack of training	3
Poor site management	3
Poor safety awareness among workers/managers	2
Insufficient/lack/housekeeping program	2
Excessive noise	2
Procurement & contract problem	2
Nature of the industry	2
Lack of experienced project managers/skilled workers	1
Fatigue by workers due to overwork	1
Poor illumination	1
Unforeseen circumstances	1

The factors identified and sorted in Table 1 are then grouped according to their themes, which according to [17], there are four (4) identified themes which are human factors, management factors, technical factors and also environmental factors.

Human Factors
Defined as when human actions or behaviours related to the physical, experience, attitude and also the behaviour of the human themselves that contribute to the accidents [17].

Management Factors
Under this theme, the factors that include management policy, resource management, the management culture and also the safety aspects of the management handled on the site contributing to the accidents are grouped together [17].

Technical Factors
Described as the technical factor or also known as worksite factor involving the jobsite, poor site management, equipment and materials condition and also the tasks during the construction causing the accidents [17].

Environmental Factors
Known as the distant causes that relate to the politics and legislations issues on the construction sites that regards the economic perspective of when the services, materials and also the labour supplies are what causing the accidents [17].

In Figure 1 below, the factors identified from the journals are properly grouped into their themes according to the definition and description by [17]. The figure is known as the Subsystem Model that will be used as variables to develop a System Dynamic (SD) model where the interactions between the factors will be related in a loop.

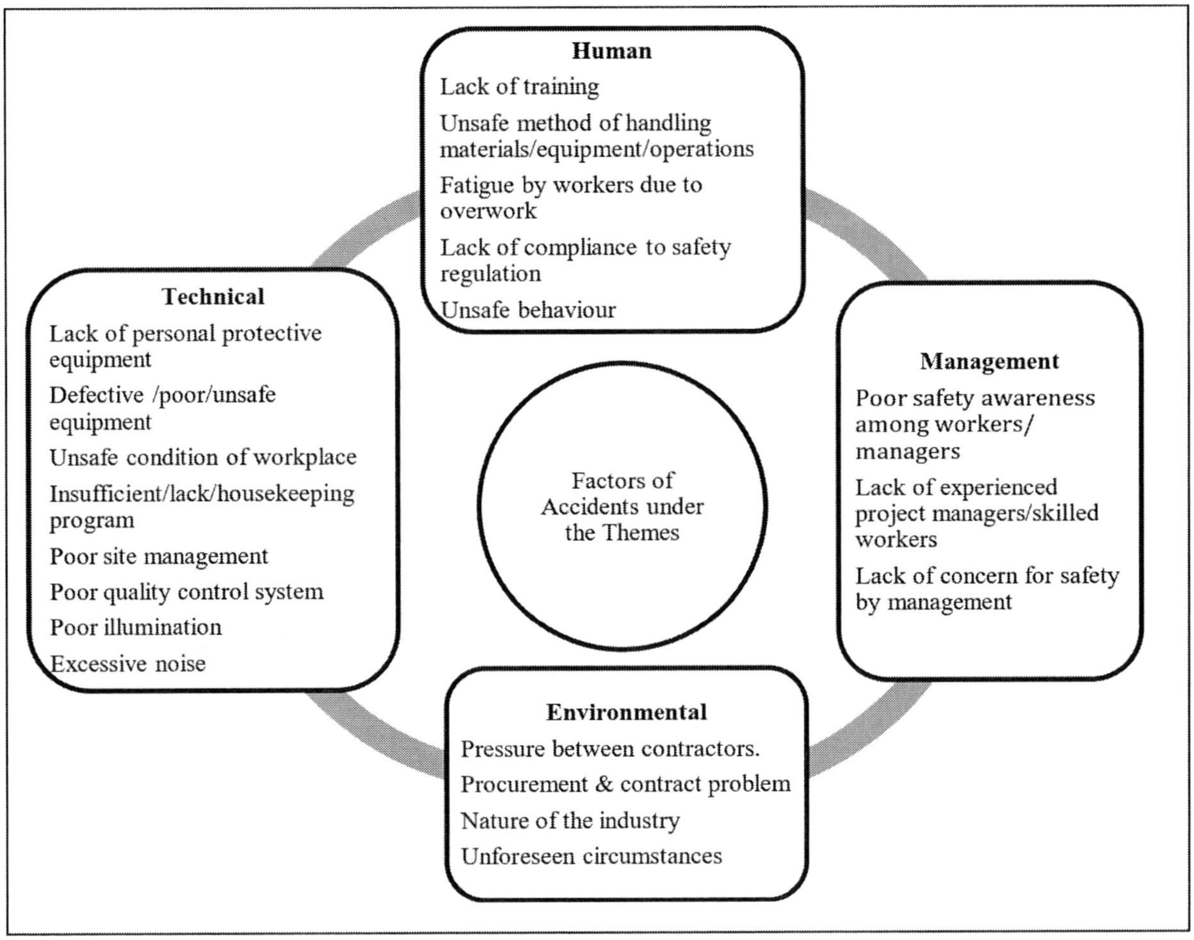

Figure 2. The Subsystem Model with the Themes for the Factors of Accidents

5. Conclusion

Safety management is very important to effectively reduce the number of accidents that happen at a workplace, especially in a complex industry such as the construction industry. By identifying the factors and themes of the accidents that happen, the interaction can be developed in the future as variables in the System Dynamic model.

Based on this study, by using the systematic review method, the findings discovered 18 factors of accidents from the analysis of seven (7) academic journals which are grouped under four (4) themes, known as human factors, management factors, technical factors and lastly environmental factors. The method used in this paper by analysing academic journals on the factors that contribute to the accidents happening at construction sites has enabled the identification of the common factors and grouped accordingly to their themes. The findings from this paper should be able to assist further investigation by providing the ground knowledge on the factors and themes of accidents happening at construction sites.

However, empirical investigation need to be conducted in order to verify the interactions between the factors and themes using the System Dynamic model. The future empirical investigation will assist to understand the interactions between the factors and the themes in order to properly plan the corrective actions to prevent accidents from happening, especially at a dangerous workplace such as a construction site.

Reference

[1] Accidents Statistics by Sector until June 2018 (Investigated) [Internet]. 2018. Available from: http://www.dosh.gov.my/index.php/en/occupational-accident-statistics/by-sector/489-statistics/occupational-accidents-statistics.
[2] Williams O S, Hamid R A and Misnan M S 2017 Analysis of Fatal Building Construction Accidents: Cases and Causes. *Analysis*. 4(8):8030-40.
[3] Williams O S, Hamid R A and Misnan M S 2018 Accident Causal Factors on the Building Construction Sites: A Review. *International Journal of Built Environment and Sustainability*. 5(1).
[4] Rampal K G and Nizam J M 2006 Developing regulations for occupational exposures to health hazards in Malaysia. *Regulatory Toxicology and Pharmacology*. 46(2):131-5.
[5] Construction Industry Development Board (CIDB). Functions Malaysia2018. Available from: http://www.cidb.gov.my/index.php/en/corporate-info/functions.
[6] Asilian-Mahabadi H, Khosravi Y, Hassanzadeh-Rangi N, Hajizadeh E and Behzadan A H 2018 Factors affecting unsafe behavior in construction projects: development and validation of a new questionnaire. *International Journal of Occupational Safety and Ergonomics*.1-8.
[7] Choudhry R M 2014 Behavior-based safety on construction sites: A case study. *Accident Analysis & Prevention*. 70:14-23.
[8] Mohamed S and Chinda T 2011 System dynamics modelling of construction safety culture. *Engineering, Construction and Architectural Management*. 18(3):266-81.
[9] Shin M, Lee H-S, Park M, Moon M and Han S 2014 A system dynamics approach for modeling construction workers' safety attitudes and behaviors. *Accident Analysis & Prevention*. 68:95-105.
[10] Hosseinian S S and Torghabeh Z J 2012 Major theories of construction accident causation models: A literature review. *International Journal of Advances in Engineering & Technology*. 4(2):53.
[11] Reason J, Hollnagel E and Paries J 2006 Revisiting the Swiss cheese model of accidents. *Journal of Clinical Engineering*. 27(4):110-5.
[12] Hollnagel E. FRAM: the functional resonance analysis method: modelling complex socio technical systems: CRC Press; 2017.
[13] Fang D, Xie F, Huang X and Li H 2004 Factor analysis-based studies on construction workplace safety management in China. *International Journal of Project Management*. 22(1):43-9.
[14] Hamid A R A, Majid M Z A and Singh B 2008 Causes of accidents at construction sites. *Malaysian journal of civil engineering*. 20(2).
[15] Khan K S, Kunz R, Kleijnen J and Antes G 2003 Five steps to conducting a systematic review. *Journal of the royal society of medicine*. 96(3):118-21.
[16] Pullin A S and Stewart G B 2006 Guidelines for systematic review in conservation and environmental management. *Conservation biology*. 20(6):1647-56.
[17] Jaafar M H, Arifin K, Aiyub K, Razman M R, Ishak M I S and Samsurijan M S 2018 Occupational safety and health management in the construction industry: a review. *International Journal of Occupational Safety and Ergonomics*. 24(4):493-506.

IOP Conf. Series: Earth and Environmental Science **385** (2019) 012056 doi:10.1088/1755-1315/385/1/012056

Application of Ozone Plasma Technology for Treating Peat Water into Drinking Water

F Ali[1,2,3], D L Lestari[2], and M D Putri[2]
[1]Department of Civil Engineering, Faculty of Engineering, Universitas Indonesia, Depok, 16424, Indonesia
[2]Indonesia Water Institute, Tanjung Barat Indah, South Jakarta, 12530, Indonesia
[3]Ministry of Public Works and Housing, Jl. Pattimura No. 20, Kebayoran Baru, South Jakarta, 12110, Indonesia

Email: firdausali@ymail.com

Abstract. The need of clean water will continue to increase along with the increase of population. However, not all regions in Indonesia have a potential for raw water sources that meet standards to be treated into clean/drinking water, so an alternative water treatment needs to be developed. Peat water is one of water sources that is often found in lowland swamps or in tidal conditions which is much located in Sumatera and Kalimantan. Peat water requires an appropriate technology to be able to treat water into clean water. Peat water treatment using ozone plasma technology is one of technological breakthroughs that can treat peat water into clean water effectively and economically. Peat water used for this study is taken from Sintang, West Kalimantan. Tested Parameters of water quality will be compared with the Regulation of Minister of Health No. 416 Year of 1990. In this study, four parameters in raw water are carried out to be tested after treating with variations in length of contact time of 10, 12, and 15 minutes. Water quality parameters to be tested are turbidity, color, coliform, and permanganate with final quality of treating for each parameter are 5.8 NTU, 1.1 Pt-Co, below 2 MPN/100 mL, and 11.8 mg/L. There is significant decrease in a value of four parameters, which three parameters meeting the maximum limit. Permanganate as parameter that exceed the maximum limit, have high probability of meeting this limit if contact time of treating peat water with ozone plasma is extended.

1. Introduction

Peat water in Indonesia is one of relatively abundant water resources. Results of the study of Geological Resource Center of Ministry of Energy and Mineral Resources report that until 2006 peat land resources in Indonesia covered a total of 26 million ha spread across Kalimantan (50%) and Sumatera (40%), while the rest were spread across Papua and other islands. Peat water has great potential to be treated into clean water, even with the advanced technology can be treated into drinking water. Peat water contains dissolved organic compounds that cause water to become brown and acidic, so it needs special treating before it can be consumed. Advanced technology is needed so that the purpose to utilize the peat water is optimal. Clean water treatment using disinfectants, filtering with membranes, ultra violet, and activated carbon does not guarantee that the water produced has a quality that is in accordance with the quality standards. This is evidenced by the results of some researches that microorganisms still exist in drinking water, especially in water produced in refill drinking water depots [1]. Currently, water treatment technology is actively being developed by oxidizing and decomposing organic pollutants in

IOP Conf. Series: Earth and Environmental Science **385** (2019) 012056 doi:10.1088/1755-1315/385/1/012056

water quickly, effectively and efficiently by using ozone (O_3). Ozone is known as one of the strongest oxidizers that can be produced in two ways: Electrical Discharge, Plasma Corona Discharge which is known as Ozone Plasma. Electrical Discharge is the formation of O_3 oxidizers through the O_2 ionization process using alternating electric current high voltage in the air [2]. Whereas, Plasma Corona Discharge is the formation of O_3 through the O_2 ionization process by using relatively high voltage electricity using low currents that are emitted transversely between plasma cells so that the production costs are relatively lower or cheaper to be applied in water treatment.

Ozone plasma in water can produce various kinds of active ion species that have the potential to decompose the content of organic compounds in water. In addition, plasma can also produce UV light and shock waves that have the potential to significantly decompose organic compounds in water [1]. Drinking water treatment with an ozone plasma system can replace the use of chlorine and activated carbon which has the function of killing microorganisms, and also at the same time can decompose the content of organic compounds. One development and application of ozone plasma technology for water treatment is to use ozone (O_3). In purifying water, ozone and hydroxide radicals formed in the ozone plasma system function as cleaning agents, strong oxidizers, disinfection sterilizers, deodorizers, self-purification, and even color decomposers. By looking at the ozone plasma working system and the characteristics of ozone, this study conducted an in-depth analysis to find out the benefits of using ozone plasma to treat peat water into clean/drinking water, especially in peat areas. This regarding that the use of ozone plasma technology has not implemented in the treatment of peat water, whereas in Indonesia there are many areas experiences clean water crisis, especially in areas that have raw water in the form of peat water.

2. Methodology

2.1. Research Method

This study uses independent variables and dependent variables. The independent variables are peat water sample originating from Sintang, West Kalimantan, and the length of time of contact (t_c) between water in ozone plasma unit. The dependent variables are the value of coliform, turbidity, color, and permanganate. The equipment used in this study is water purification unit with ozone plasma technology. This technology has been widely marketed and is commonly applied for treating clean water. With a series of experiments in the various of t_c, it will be known the effectiveness of the technology in treating the raw water (peat water) into drinking water.

2.2. Contact Mechanism and Ozone Plasma Reaction

Ozone plasma contact mechanism in treating raw water (peat water) is based on the chemical-physical operation process which leads to the formation of OH^- when the plasma unit is in contact with water (H_2O). Plasma units with a relatively high supply of electricity in water will produce ion particle charged with ultrasound energy that are resistant to UV and even shock waves, thus encouraging the formation of OH^- ions in solution [3]. In addition, the chemistry of plasma driven in the gas phase produces reactive oxygen and hydrogen peroxide which are not only radical, but also ozone, which will be needed to break down covalent gas so that it can produce plasma electrons. During the radical gas phase, there will be a diffusion process and changes in Henry's constant. Plasma also interacts with water which directly produces reactive ion species at the interface layer.

Interaction between ozone plasma and liquid will be able to provide an addition of oxidizers which will affect the oxidation level which is relatively much better than conventional methods, including: (1) consumables are not needed, because ozone plasma can be produced from ordinary air or the liquid itself, (2) the level of decomposition of organic compounds is relatively more effective, (3) because consumables are not needed, the costs and infrastructure normally used for consumables can be removed or substituted for the costs of overcoming other obstacles that may be encountered in water treatment, (4) ozone plasma application is inherently modular and can also be used as a final stage in water treatment system that is no different from conventional UV stages for disinfection systems. The use of ozone plasma also allows for the elimination of micro contaminants.

In this ozone plasma system, peat water will be exposed directly with the concentration of O_3 produced through a plasma non-contact tube. By using this system, peat water is treated effectively and efficiently to produce clean water that is suitable for consumption. The mechanism for purifying water with an ozone plasma system begins by injecting oxygen into an area that has a high voltage. O_3 will be produced when passing through the corona plasma unit. When O_3 in the water content passes through a charged tube, radical hydroxide, an oxidizing compound, such as hydrogen peroxide is also produced in the presence of internal high stresses. With this mechanism there will be more production of effective hydroxide radicals. The oxidation ability possessed by O_3 and hydroxide radicals (OH^-) will be able to destroy germs and decompose materials that cannot be degraded by conventional means. The ozone and hydroxide radicals that have been used will then break down into H_2O and O_2 forms which will then be released as treated water which has high dissolved oxygen and clarity that meet clean water standard.

In this study, an experimental treatment of peat water is conducted by varying the t_c, which is 10 minutes, 12 minutes, and 15 minutes. Time variations are made to determine the estimated effective time required by ozone plasma technology to reduce the parameters of coliform, turbidity, color, and permanganate which are the dominant parameters in peat water in general.

3. Results

In this study, tests are carried out on four water quality parameters, which are coliform, turbidity, color, and permanganate. Coliforms in water need to be checked because even though they are not dangerous to humans, coliforms are a good indicator of the presence of pathogenic microorganisms in water. Pathogenic microorganism is a variety of bacteria, parasites, and viruses that have the potential to endanger human health [4]. The presence of coliforms in water samples will be greatly considered because it can indicate the effectiveness of the disinfection process. Chlorine as a disinfectant in the treating of peat water cannot be used because in the chlorine disinfecting process, *trihalomethane* (THM) dangerous compounds will be formed which can be carcinogenic which are triggered by the higher solubility of metals in water because the pH of water is lower [5]. For peat swampy water, it is normal that pH and DO are generally low [6].

The treated water from ozone plasma treatment will be compared with the quality standard stated in the Regulation of The Minister of Health No. 416 Year of 1990 about Water Quality Requirements and Supervision. In testing the parameter of coliform, the raw water has a value of 13 MPN/100 mL, where the allowable limit is 10 MPN/100 mL. After 10 minutes of treating with ozone plasma technology, the coliform level is below 2 MPN/100 mL. By looking at the effectiveness of the ozone plasma technology on water purification, it can be seen that the disinfection process in the ozone plasma works very well.

Turbidity in water can be defined as a decrease in the transparency of a solution because of the presence of suspended or dissolved materials that can cause light to be dispersed, reflected and not transmitted straight. The higher the level of turbidity of water, the more light will be dispersed [7]. Turbidity in water is also caused by colloidal particles, and suspended particles, such as clay, silt soils, fine grains which are divided into organic and inorganic materials, plankton and microscopic organisms. Iron bacteria can also be a source of turbidity. The turbidity also acts as an important indicator for organic pollution, the run-off of suspended material and heavy rain fall in the area [8]. Turbidity removal can be done by giving a coagulant, which is then processed through coagulation and flocculation. In this study, the process is not carried out by given the magnitude of the ability of hydroxide radicals to decompose non-degraded material. This can be seen from the results of experiments that showed a decrease in turbidity level from the previous 50.15 NTU in raw water to 34.6 NTU in water treated with t_c 10 minutes, 12.6 NTU with t_c 12 minutes, and 5.84 NTU with t_c 15 minutes. When compared with the quality standard (25 NTU), it can be seen that purification of water with ozone plasma technology can meet quality standards while being able to cut the use of chemicals as a coagulant which commonly used in the water treatment process.

Natural colors in water occur because of the presence of complex organic molecules that are contained from humus substances such as leaves. The color in the water can also increase concentrated if there is the presence of suspended substances. Color on water usually affects aesthetics compared to health problems [4]. The water in peat swamp is almost black in color. The main reason is the presence of an organic material from peat decomposition [9]. Peat water has a very high color intensity, namely brownish red color, which is caused by a high content of organic matter dissolved mainly in the form of humus acid and its derivatives. Humus acid, derived from the decomposition of organic matter, such as leaves, trees, or wood with various levels of decomposition. In some cases, the color will be higher due to the presence of iron metal which is bound by organic acids dissolved in the water. Technique of color removal can be done in various ways, including: Coagulation-flocculation-sedimentation-filtration, oxidation-reduction, activated carbon, or adsorption-absorption. The technique of color removal basically requires additional media or chemicals that will have a significant effect on the cost of water treatment. In this research, no additional media or chemicals are used at all, only by maximizing the processes that occur in ozone plasma system, where O_3 and OH^- dissolved in water will function to remove color. The color content of the raw water of peat water shows value of 7.8 Pt-Co with maximum limit according to the Regulation of The Minister of Health No. 416 Year of 1990 is 50 Pt-Co. With tc 10minute, treating the water with ozone plasma unit achieves 3.5 Pt-Co and if t_c is increased to 15 minutes, the color level of treated water reaches 1.1 Pt-Co. This shows that treating peat water using ozone plasma has been proven to reduce the color level in the water.

Permanganate number is used to measure the high content of organic matter in water. Permanganate value is the amount of milligram of potassium permanganate needed to oxidize organic matter in 1000 mL of water in boiling conditions [10]. Excessive organic matter in drinking water is not permitted because in addition to causing unwanted colors, smells and tastes, it may also be toxic both directly and after compounding with other existing substances [11]. Permanganate number is closely related to the value of COD (Chemical Oxygen Demand), BOD (Biochemical Oxygen Demand), and DO (Dissolved Oxygen). The relationship between BOD and COD values with permanganate numbers are the three parameters which together indicate the number of organic (and non-organic) compounds dissolved in the waters. Thus, the higher the permanganate number, the higher the COD value and the BOD value.

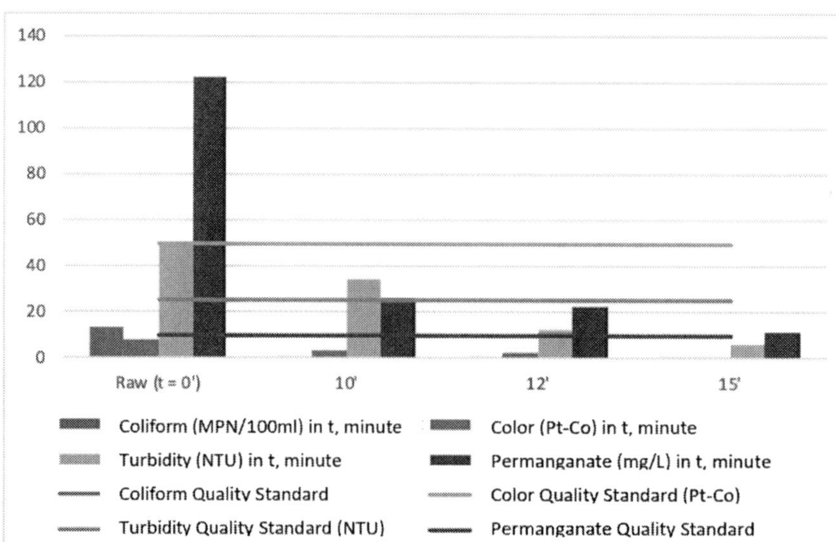

Figure 1. Results of Peat Water Treatment with Ozone Plasma Technology

The more complex water treatment system is needed if the BOD and COD content is high. The permanganate number in the peat water is 122.3 mg/L with allowable limit is 10 mg/L. In this study, it is found that after 10 minutes of treating, the permanganate number can be reduced to 26.3 mg/L. By using t_c 12 minutes, the permanganate number can be reduced to 22.8 mg/L. By extending t_c 3 minutes which is for total of 15 minutes of treating, the final value of permanganate number drops by almost half, reaching 11.8 mg/L. By looking at the results of this experiment, the treating experiments with t_c 15 minutes are able to significantly reduce the level of permanganate in peat water close to the quality standard. It can be said that the permanganate method in acidic condition has a better recovery [12]. This is probably due to the fact that most of the organic matters were oxidized more completely in acidic than in alkaline conditions [13]. To obtain better permanganate numbers, treating by the ozone plasma technology is needed with a t_c of more than 15 minutes or an estimated treating time of no more than 20 minutes. The results of this experimental treatment using ozone plasma technology are showed in Figure 1.

4. Discussion

Peat water which is widely found on the islands of Borneo and Sumatra is a type of water that is commonly found in locations that often occur in tidal conditions, swampy areas, and lowlands. In general, peat water has acidic and brown properties and based on the Government Regulation No. 82 Year of 2001 concerning management of water quality and water pollution control, defines peat water as unsuitable for consumption so it must be treated in advance so that it is suitable for consumption [14], while swampy areas generally only have surface water from peat water. One appropriate solution is to use water treatment plant that can treat the peat water into consumable water. But conventional technology is not appropriate to be used to treat peat water. This is because to treat peat water into consumable water requires high technology and expensive costs so that its affordability by the community will be very low. Water treatment technology using ozone plasma is proven to be able to treat peat water into consumable water.

After treating the peat water by ozone plasma technology with t_c 15 minutes, three parameters tested (coliform, color, and turbidity) are proven to be able to be treated and to meet the clean water quality standards based on the Regulation of The Minister of Health No. 416 Year of 1990. However, there is one parameter that has not met the maximum limit of clean water, which is permanganate number. Even though at t_c 15 minutes the permanganate number is still above the maximum limit, there is a significant decrease and it is highly probable that permanganate number can be below the maximum limit if it is treated with a longer contact time.

5. Conclusion

Based on the results of treating peat water using ozone plasma technology, it is proven that the technology able to treat peat water very effectively compared to the conventional refining technology. At the t_c 15 minutes treating peat water with ozone plasma, 3 of 4 test parameters (coliform, color, and turbidity) can be treated to be below the maximum limit. While another test parameter can be reduced significantly so the value is only slightly exceeding the maximum limit, and there is a high probability that the value can be treated to below the maximum limit by increasing the contact time of peat water in the ozone plasma unit. The use of ozone plasma to treat peat water will be able to reduce the use of chemicals to treat water into clean water so that it will have a significant impact on the cost of treating peat water. Thus, the use of this ozone plasma technology to treat peat water will be an effective cost to be developed and applied in many places on the islands of Kalimantan and Sumatra in Indonesia. This regarding that in the areas that have raw water in the form of peat water, the use of ozone plasma technology can be one of appropriate alternative to treat the raw water into clean/drinking water.

Acknowledgment
This research can be carried out because of the support of tool facilities, laboratories, and staff from Teknikon Surabaya. Thank you to Dwi Lintang Lestari and Marsya Dyasthi Putri from Indonesia Water Institute-Jakarta who have helped with the research and writing of this paper.

References

[1] Hazmi, A., Desmiarti, R., and Waldi, E.P. (2011). Penghilangan Mikroorganisme dalam Air Minum dengan Pulsa Tegangan Tinggi (*Removal of Microorganisms in Drinking Water with High-Voltage Pulses*). Electrical Engineering, Faculty of Engineering, Andalas University.
[2] Yuan, D., Wang, Z., He, Y., Xie, S., Lin, F., Zhu, Y., and Cen, K. (2018). Ozone Production with Dielectric Barrier Discharge from Air: The Influence of Pulse Polarity. *The Journal of the International Ozone Association*, 40 (6): 494-502.
[3] Foster, J.E. (2017). Plasma-based water purification: Challenges and prospects for the future. *Physics of Plasmas*, 24 (5), 055501.
[4] Environmental Protection Agency. (2001). *Parameters of Water Quality: Interpretation and Standards*. Ireland: Author.
[5] Richardson, S., Plewa, M., Wagner, E., Schoeny, R., and Demarini, D. (2007). Occurence, genotoxicity, and carcinogenicity of regulated and emerging disinfection by-products in drinking water: a review and roadmap for research. *Mutation Research/Reviews in Mutation Research*, 636 (1-3), 178-242.
[6] Gasim, M.B., Ismail, B.S., Toriman, E., Mir, S.I., and Chek, T.C. (2007). A Physico-Chemical Assessment of the Bebar River, Pahang, Malaysia. *Global Journal of Environmental Research*, 1 (1): 07-11.
[7] Ziegler, A.C. (2002). Issues Related to Use of Turbidity Measurements as a Surrogate for Suspended Sediment. *Turbidity and Other Sediment Surrogates Workshop, April 30 – May 2, 2002, Reno, NV.*
[8] Yisa, J., and Jimoh, T. (2010). Analytical studied on water quality index of River Landzu. *American Journal of Applied Sciences*, 7 (4): 453-458.
[9] Rosli, N., Gandaseca, S., Ismail, J., and Jailan, M.I. (2010). Comparative Study of Water Quality at Different Peat Swamp Forest of Batang Igan, Sibu Serawak. *American Journal of Environmental Sciences*, 6 (5): 416-421.
[10] Badan Standarisasi Nasional. (2004). *SNI 06-6989.22-2004 Air dan air limbah – Bagian 22: Cara uji nilai permanganat secara titrimetri* (Indonesian National Standard 06-6989.22-2004 Water and wastewater – Part 22: How to test the permanganate value by titrimetry). Serpong: Author.
[11] Soesanto, S.S. (1996). Senyawa Organik Dalam Air Minum (*Organic Compounds in Drinking Water*). *Media Penelitian dan Pengembangan Kesehatan* (Media of Health Research and Development), Vol. VI No. 01.
[12] Goh, C.P., and Lim, P.E. (2006). Potassium permanganate as oxidants in the COD test for saline water samples. *ASEAN Journal on Science and Technology for Development*, 25 (2): 383-393.
[13] Fujimori, K., Ma, W., Kawakami, T.M., and Shibutani, T. (2001). Chemiluminescence Method with Potassium Permanganate for the Determination of Organic Pollutants in Seawater. *Analytical Sciences*, Vol. 17.
[14] Juhra, F., and Notodarmojo, S. (2016). Degradasi Zat Warna pada Air Gambut Menggunakan Metode Fotokatalik ZnO (*Degradation of Dyes on Peat Water Using ZnO Photocatalytic Method*). *Jurnal Teknik Lingkungan* (Journal of Environmental Engineering), 2 (2): 1-10.
[15] Environmental Protection Agency. (2009). National Primary Drinking Water Regulations. *EPA 816-F-09-004*.

Safety Manager Competencies In Managing Construction Projects In Malaysia

H Adnan [1], F N M Yussof [1], F W Jaafar [1], Z Z A Rashid[2], Z Z Abidin [1] and N A Bakhary[3]

[1]Faculty of Architecture, Planning and Surveying, Universiti Teknologi MARA, 40450 Shah Alam, Malaysia
[2]School of Housing, Building and Planning, Universiti Sains Malaysia, Penang.
[3]Faculty of Civil Engineering, Universiti Teknologi MARA, 40450 Shah Alam, Selangor, Malaysia

Email: mimad856@gmail.com

Abstract. This paper aims to identify the core competency required by safety manager in managing construction safety on construction projects and their level of effectiveness. Total of 30 questionnaires received form 100 questionnaires distributed to qualified safety managers work within Klang Valley registered with the Department of Occupational Safety and Health (DOSH) and National Institute of Occupational Safety and Health (NIOSH), Malaysia with various post of safety officer, manager and consultant. It was found that safety managers consume of several important skills and knowledge on general management functions, practices, and procedures regarding safety and occupational health concepts, principles, practices, methods, and techniques based on the working experience on the construction field. The skills and knowledge on general management functions, practices and procedures can be accomplished by presenting personal concerns for the health and safety of employees, conducting the job-training program, participating in the management of safety committees, considering safety in job design, and reviewing the level of work. It has been recommended that safety managers must invest time and energy themselves for own development of skill, knowledge, and other competencies for more efficient professional safety manager.

1. Introduction

Construction is a highly hazardous industry, which consists of a wide range of activities involving construction, alteration, and repair. However, based on National Safety Council [1] construction contributes for approximately 12% of disabling injuries, which is nearly 25% of work-related fatalities. As compared to any other industry, there are substantially more injuries and lost workdays due to injuries or illnesses in construction as the activities involve may expose themselves to serious hazards; such as falling from rooftops, unguarded machinery, being struck by heavy construction equipment and others.

2. Background

Generally, the main duties of safety manager are to comply safety precaution, strategies and policy through site inspection and execute good safety practices to motivate workers for keeping safety measures at the workplace. Safety during a construction project is also fully affected by decisions made during the planning and design stage which depends largely upon education, alertness, and cooperation during the construction process. The need for unique skills and competencies to effectively manage safety arrangements is significantly a great issue, especially in construction industry environments that are described as highly complicated, dangerous and uncertain. These arrangements are also described as multi-sectoral, under-organized, and multi-organizational systems that exclusively require unique management skills and competencies. Thus, it can be proved that the availability of specialized management expertise and skills is the criteria for the successful implementation of safety management to effectively manage the construction projects by the safety manager.

2.1 Responsibilities of a Safety Manager

Construction safety management deals with actions of managers at all levels and enables them to create an organizational goal or setting, where workers will be trained and motivated to perform safe and productive construction work. The system should include the responsibilities and accountabilities, the outline procedures to eliminate potential hazards before they become the factors that contribute to unfortunate accidents [2]. In other words, safety management is a concept that brings extensive approach, allow participation and contribution for all management and workforce level. by taking all the consideration on the actual site risks.

Effective safety management has three main purposes, which is to ensure a safe environment, making the job safe, and making workers safety conscious [3]. The concept of safety management is based on a well-planned system, where the main pillars of this system are the roles and responsibilities of the consultants or participants involved within the specific project. As construction site is the main source of fatalities and accidents in the construction industry, the safety management able to eliminate the hazard risks that usually occur on the construction site [3].

2.2 Core Competencies of Safety Manager

Safety managers play important roles in conducting and managing the safety of the construction site as it is the priority of every personnel in order to achieve the key successful factors in any project. Safety manager performs a significant role in the workplace in regard to safety and health at a workplace, where they are the driving component of the operational aspects of safety management systems. It is vitally important for safety managers to have a clear understanding of their duties and responsibilities within the organization. The three important aspects of safety that are important by every safety manager are the rules and regulation of the workplace, a socio-humanitarian aspect that consider the human lives involved, and lastly is the accidents aspects, which give a direct and indirect impact on costs. Competency is the critical factor in organizational survival that comprises of a set of knowledge, skills, interests, personality,traits, experiences and job-related capabilities, which make the person successful in work [4]. Furthermore, a study on managerial competencies bring out different classifications and approaches which are knowledge-based, skill-based and value-based competencies that should be acquired by safety managers [5]. A connection mechanism between competencies, capabilities, resources, assets, and core competencies are interrelated. For instance, [7]. defined the core competence as a capability, which is central to a firm's value generating activities instead of only ownership of a resource. Core competence somehow differentiates an organization from its competitors [7]. There are six (6) core competencies of the safety manager in the construction industry, which are Occupational, Health and Safety Legislation, Safety Technology, Construction Technology, Risk management, Safety Management and Change Management. Other Competencies required includes Business, Accounting, and Marketing, Management Competencies, Personal Credibility, Information Technology (Web-Based, Building Information Modelling (BIM) and IT- Based Site Monitoring.

3. Aim of the Study and Problem Statement

The safety managers must give more attention to the important elements identified in the study to help them in enhancing performance on the construction sites and reduce the frequency of accidents occurrence. However, there is also a gap in construction industry itself, where there is a lot of issue on improper and ineffective safety management ethics in the construction industry. If the competencies required for performing the safety duty have not been clearly recognized, it will be difficult to decide the required training and experience. It indicates that safety should be embedded or integrated into a management concept within each company level and every part of a cross-organizational project. Therefore, the study aimed to identify and analyse the competency required by safety manager in handling construction safety on sit. In order to achieve the aims, three objectives have been identified, which are to identify the scope of safety manager job duties with emphasis on the safety management criteria on site, to identify and analyses the basic and core competency, knowledge, abilities, and skills required by construction safety manager and to recommend the other competencies required to improve the level of competency of safety manager in construction industry.

4. Methodology

The information and data gathered from the survey were arranged and prepared using the frequency and percentage method and average index method in relation to the targets, objectives, scope, and extent of the study. Two statistical methods were applied, namely descriptive statistic and inferential statistic. The formula being used is:

$$\text{Percentage } (\%) = (n/N) \times 100\% \quad (1)$$

Where, n = Number of respondents
N = Total number of respondents received

100 copies of questionnaires were distributed to the targeted respondents working in the construction industry within Klang Valley. Only thirty percent (30%) of respondents response with complete information. The respondents were among safety consultants, safety officer, safety managers from those who are qualified safety officer from the Department of Occupational Safety and Health (DOSH) and National Institute of Occupational Safety and Health (NIOSH) within Klang Valley, Malaysia.

Questions based on Likert scales were analysed by using average index method, and the outcomes have been presented in the form of an average score or mean. For the Likert scales questions, respondents were needed to give answers in the form of scale 1 to 5. A scale of five (5) categories has been used for the average index method to show priority. The weighting scale for the analysis was based on the thirty (30) respondents. Likert scale was based on 5 points, which was 1 for strongly disagree, 2 for not agree, 3 for fairly agree, 4 for agree and 5 for strongly agreed. The questionnaire also required the demographic profiles of the respondents.

5. Finding

5.1 Participant's Profile

There were 7 questions asked from the respondents that consisted of gender, age, present designation, professional qualification, highest academic qualification, working experience, and number of employees in the company. The questions provided were used for the identification purpose of respondents. It has further reflected the credibility of the respondent by providing input in the questionnaire. The participant's profile was analysed by using frequency and percentage method.

Table 1. Participant Profile

	Frequency (F)	**Percentage (%)**	**Rank**
Gender			
Male	26	87	
Female	4	13	
Age			
26 - 35	13	43	1
36 - 45	8	27	2
Above 45	6	20	3
Below 25	3	10	4
Present Post			
Safety manager	11	37	1
Safety officer	7	23	2
Safety consultant	6	20	3
Others	6	20	4

Professional qualification			
Registered with DOSH	17	57	1
Registered with NIOSH	5	17	2
Others	4	13	3
None of above	4	13	4
Highest Academic Qualification			
Degree	11	37	1
Masters	8	27	2
Diploma	6	20	3
Certificate	3	10	4
PhD	1	3	5
Others	1	3	5
Years of working experience			
Over 10 years	15	50	1
6 - 10 years	8	27	2
3 - 5 years	4	13	3
Below 2 years	3	10	4
No. of employees			
Below 100	10	33	1
100 - 200	11	37	2
Above 200	9	30	3

5.2 Present Job Duties

The present job duties were analysed to identify the job duties competencies on safety and health practices that should be conducted by every competent safety manager.

Table 2. Rank of the Job Duties Competencies Required by Safety Manager

Job Duties	**Mean/ Average Score**	**Overall Rank**
Monitor occupational health and coordinate, review and analyse hazard and risk assessment and environmental impact assessment	4.43	1
Plan, lead and monitor a team of safety committee	4.40	2
Monitor all health and safety matters via coordination with contractors and/or authorities	4.37	3
Conducting safety training to meet the identified training needs	4.33	4
Implementing safety and health policy to meet the statutory, contractual and social obligations	4.30	5
Reviewing procedures for emergency preparedness	4.27	6
Acting as consultant and technical adviser to the safety committee	4.23	7
Issue non-compliance checklist and submit statistics report to the project team	4.20	8
Conduct health and safety committee meeting and follow up with action status	4.17	9
Organising team for safety programmes and activities	4.13	10
Provide urgent response to operational issues from the investigation of non-compliance OHS incidents	4.13	10
Conducting accident and incident investigation and making recommendations for preventing their recurrence	4.10	11
Outline all safe operational measures in order to identify all the relevant hazards	4.07	12
Implementing or reviewing process control programmes	4.07	12
Conduct site inspections and audits in order to identify areas of improvement	4.03	13
Participation in pre-site planning	4.00	14
Provide sufficient safety equipment to all site staff or workers	3.67	15

Table 2 shows that the rank and average score of job duties competencies required by safety manager in implementing safety in construction projects. In order to be a competent safety manager, they must be able to carry out the safety job duties and practices as well. Most of the respondents rated very high mean or average score (Rank 1) on monitor occupational health and coordinate, review and analyses hazard and risk assessment along with environmental impact assessment. It had the highest mean score of 4.43 and highest percentage of sixty percent (60%) on the strongly agree weighting scale as compared to others, which were under the category of strongly agree by most respondents.

Most of the respondents also rated high mean or average score on the plan, lead and monitor a team of the safety committee, where the overall rank is (Rank 2) with a mean score of 4.40. Monitoring of all health and safety matters via coordination with contractors and/or authorities secured overall Rank 3 with a mean score of 4.37, conducting safety training to meet the identified training needs. The overall rank was Rank 4 with a mean score of 4.33. Implementing safety and health policy to meet the statutory, contractual and social obligations was observed with the average score of 4.30 and ranked 5. All these were observed under percentages between forty-three percent (43%) to fifty-three percent (53%). Meanwhile, the results of the analysis showed that the respondents were agreed on reviewing procedures for emergency preparedness with the average score of 4.27 with the overall Rank 6. The other respondents stated that they were also agreed on acting as a consultant and technical adviser to the safety committee, where the overall rank was 7 with a mean score of 4.23. The issue on non-compliance checklist, conduct of health and safety committee meeting, where the overall rank was (Rank 8) with a mean score of 4.20. Conducting health and safety committee meeting and follow up with action status, where the overall rank was 9 with a mean score of 4.17.

It was found that the organizing team for safety programs and activities provided an urgent response to operational issues from the investigation of non-compliance OHS incidents. It received the overall Rank of 10 with a mean score of 4.13 for both job duties. These were all ranked under the same category, which was in the favour by majority of the respondents with the percentage between forty-three percent (43%) to fifty-three percent (53%) on the strongly agree on weighting scale.

It was shown that the respondents were moderately or fairly agreed on conducting accident and incident investigation; and making recommendations for preventing their recurrence, where the overall Rank 11 with a mean score of 4.10. Moreover, outlining all the safe operational measures to identify the relevant hazards and implementing or reviewing process control programmes secured rank 12 with a mean score of 4.07 for both job duties. Site inspections and audits to identify areas of improvement were at rank 13 with a mean score of 4.03 and participation in pre-site planning with the mean or average score of 4.00 (Rank 14). These were all under the category of fairly agree with the percentage between forty percent (40%) to forty-seven percent (47%) on the strongly agree on weighting scale. It was observed that some of the respondents ranked the lowest mean of 3.67 on providing sufficient safety equipment to all site staff or workers with only thirty percent (30%) of percentage for the strongly agree weighting scale.

5.3 Competencies Required by Safety Manager

Table 3 shows that the rank and average score of basic skills competencies required by safety managers in implementing safety in construction projects.

Table 3. Rank of the Basic Skills Competencies Required by Safety Manager.

Basic Skills	**Mean/ Average Score**	**Overall Rank**
Interpersonal communication skill	4.57	1
Problem-solving skill	4.57	1
Analytical skill	4.53	2
Contingency management skill	4.50	3
Information skill	4.47	4
Critical thinking	4.40	5

Conceptual skill	4.37	6
Task management skill	4.33	7
Technical skill	4.30	8
Supervisory skill	4.27	9

Most of the respondents rated very high mean or average score (Rank 1) on interpersonal communication and problem-solving skills, where it had the highest mean score of 4.57 and highest percentage of sixty percent (60%) on the strongly agree weighting scale. It was observed under the category of strongly agree by most respondents. It was followed by the analytical skill, which was also rated with high mean score of 4.53 with Rank 2 on overall rank with a higher percentage of fifty-seven percent (57%) on the strongly agree on weighting scale. Besides that, the results of the analysis from the questionnaire also showed that the respondents were agreed on information and contingency management skills; whereby, contingency management skill obtained the mean or average score of 4.50 with Rank 3. Information skill possess average score of 4.47 with Rank 4, having percentage of fifty-seven percent (57%) for contingency management skill and fifty-three percent (53%) for information skill.

Critical thinking was ranked as fairly agree with a mean of 4.40 with Rank 5 and for overall rank with the percentage of fifty percent (50%) on the strongly agree on weighting scale. Respondents were not agreed on conceptual and task management skills, which were not really required for being a competent safety manager. These were observed with a mean or average score of 4.37, which ranked as Rank 6 and 4.33, which ranked 7 with the percentage of forty-three percent (43%) for conceptual skill and fifty-three percent (53%) for task management skill on the strongly agree weighting scale. However, technical and supervisory skills were ranked as strongly disagree, which were also not required by the safety manager. Moreover, technical skill was ranked 8 and supervisory skill as Rank 9.

5.4 Skills and Knowledge That Should Be Possessed

Table 4 shows that the results obtained from the survey, which has been ranked and finalized according to respondent's preferences on skills and knowledge.

Table 4. Rank of the Skills and Knowledge that Should Be Possessed by Safety Manager

Skills and Knowledge That Should Be Possessed	Mean/ Average Score	Overall Rank
General Management Functions, Practices and Procedures To understand organizational objectives, safety and occupational health principles, regulations, standards and work processes	4.67	1
Safety and Occupational Health Concepts, Principles, Practices, Methods and Techniques Capability to recommend substantive program changes or alternative new courses of managerial action	4.67	1
General Safety Knowledge and Skill To advise or formally instruct supervisors and employees on safety techniques	4.63	2
Knowledge of The Body of Laws, Regulations, Policies and Procedures To sufficiently interpret and explain the reasons and purposes for applying measures and procedures	4.60	3
Organizational Structure To acquire data and information for preparation of accurate reports	4.57	4

It shows that respondents strongly agreed with two highest rated (Rank 1) that safety manager must possess skills and knowledge on general management functions. Also, the practices and procedures on safety and occupational health concepts, principles, practices, methods, and techniques with a mean or average score of 4.67 for both skills and knowledge. From the result, general management functions, practices, and procedures obtained seventy-three percent (73%) of percentage; while safety and occupational health concepts, principles, practices, methods and techniques obtained sixty-seven percent (67%) of percentage for the strongly agree weighting scale. The knowledge on general management functions, practices and procedures are basically focused on understanding organizational objectives, safety and occupational health principles, regulations, standards and work processes. Meanwhile, safety and occupational health concepts, principles, practices, methods and techniques focus on the capability to recommend substantive program changes or alternative new courses of managerial action.

Respondent also agreed on knowledge of general safety and skill, which was to advise or formally instruct supervisors and employees on safety techniques. It was ranked 2nd, having the percentage of sixty-seven percent (67%) for the strongly agree weighting scale. This score was observed in the important range with the mean score of 4.63. But, the knowledge of the body of laws, regulations, policies, and procedures was ranked as unimportant to be a competent safety manager. It was under the category of not agree with only 4.60 mean score, which was ranked 3 with the percentage of sixty-seven percent (67%) for the strongly agree weighting scale for the overall rank. It mainly focused to sufficiently interpret and explain the reasons and purposes for applying measures and procedures. However, knowledge and skill on organizational structure fall under the category of strongly disagree, which was ranked 4 with the lowest mean or average score of 4.57 and lowest percentage of sixty percent (60%) on the strongly agree on weighting scale.

5.5 Core Competencies

Table 5 shows that the result obtained from the survey, which has been ranked and finalized according to respondent's preferences on core competencies.

Table 5. Rank of the Core Competencies Required by Safety Manager

Core Competencies	Mean/ Average Score	Overall Rank
Safety Management Develop safety plans (i.e. safety programme, budgeting, scheduling, assessment, training, measure)	4.73	1
Occupational, Health and Safety Legislation Ensure health and safety are compliance with OHS construction requirements	4.57	2
Risk Management Identification, analysis and evaluation stages in the risk management process	4.57	2
Safety Technology Provides instruction on how to implement and administer a company's safety program	4.33	3
Change Management Develop and implement OHS programmatic changes	4.27	4
Construction Technology Principles of construction including demolition works and erection techniques of plant and equipment	4.17	5

Table 5 shows that the rank and average score of the core competencies that are required by safety manager in implementing safety in high rise construction building. The highest rate of core competency was safety management by the respondents, who strongly agreed that safety managers must possess core competencies in safety management as the key element to be competent with the highest result of mean or average score of 4.73 which is ranked as (Rank 1). This safety management includes developing safety plans such as safety programme, budgeting, scheduling, assessment, training, and measure.

The second highest rated core competency was on occupational, health and safety legislation and risk management with the same mean or average score of 4.57. It was ranked 2 with the percentage of sixty-three percent (63%) on the strongly agree weighting scale for both core competencies. The occupational, health and safety legislation include ensuring health and safety with OHS construction requirements, meanwhile, the risk management includes identification, analysis and evaluation stages in the risk management process. These two elements ranked under the category of agree based on the average index obtained, which also play an important criterion to achieve the competency in managing construction safety by the safety manager.

Besides that, safety technology that focuses on providing instruction on how to implement and administer a company's safety program fall under the category of not agree by the respondents, where competency in this field area is not really required or important in the research with only 4.33 mean score, which was ordered as Rank 3 with only forty-three percent (43%) of percentage on the strongly agree on weighting scale. However, change management and construction technology were ranked under the category of strongly disagree, whereby change management was ranked 4 and construction technology was ranked 5 with the percentage of forty-three percent (43%) and forty percent (40%) on the strongly agree weighting scale. This was also not required at all by the safety manager with the lowest mean or average score of 4.27 and 4.17. This change management includes develop and implement OHS programmatic changes; whereas, the construction technology includes principles of construction including demolition works and erection techniques of plant and equipment.

Table 6. Rank of the Other Competencies Required by Safety Manager

Other Competencies	Mean/ Average Score	Overall Rank
Personal Credibility E.g. professional integrity, meeting commitments, accurate works, self-confidence and providing alternative insights on business issue	4.40	1
Management Competencies E.g. adaptability, building relationships, conflict management, coaching and organizational awareness	4.23	2
Business Knowledge Knowledge of business practices other than safety and health practices e.g. operation, financial, human resource, quality and environmental management practices	4.10	3
Information Technology Competence in new technologies e.g. Web-Based, Building Information Modelling (BIM) and IT-Based Monitoring	4.07	4

Table 6 shows that the respondents strongly agreed (Rank 1) with the highest mean or average score of 4.40 and with the percentage of fifty percent (50%) on the strongly agree weighting scale that safety manager must possess other competencies on personal credibility as the additional competency to be an efficient safety manager. This personal credibility includes professional integrity, meeting commitments, accurate works, self-confidence and providing alternative insights on the business issue.

Besides that, respondents also moderately ranked on management competencies, which were under the category of fairly agree. This element also plays an important role to achieve the competency in managing construction safety by the safety manager with a moderate mean score of 4.23, which was ranked as Rank 2. This management competency mainly focuses on adaptability, building relationships, conflict management, coaching and organizational awareness with thirty-three percent (33%) of percentage on the strongly agree on weighting scale. However, business knowledge and information technology were ranked under the category of strongly disagree, which was not required at all by the safety manager with the lowest result of mean or an average score of 4.10 which was ranked as Rank 3 and 4.07 which was ranked as Rank 4. The business knowledge generally focuses on knowledge of business practices other than safety and health practices such as operation, financial, human resource, quality and environmental management practices. Whereas, it is mainly focused on competency in new technologies for the information technology; such as Web-Based, Building Information Modelling (BIM) and IT-Based Monitoring. Respondents strongly believed that business knowledge and

IOP Conf. Series: Earth and Environmental Science **385** (2019) 012057 doi:10.1088/1755-1315/385/1/012057

information technology were not required or vitally important to possess these competencies in the study.

6. Discussion

The results suggested that the safety managers must be able to carry out the safety job duties and practices and have competencies within the area of their job duties as well to be a competent manager. The results can be compared with the study by [8] which stated that safety manager must be competent and able to monitor occupational health and coordinate, review, and analyses hazard and risk assessment along with environmental impact assessment.

The three tasks were identified as the most important in reducing injury, which include challenging unsafe behaviours and attitudes; monitoring subcontractor activities; and carrying out hazard identification, risk assessments, and control. It was supported by the study [9] which stated that a safety management system, which comprises elements of policy, incentives, training, communication, control, and planning, has been found significantly related to safety, competitiveness, and economic- financial performance. Hence, safety manager must have the ability to recognize and proactively manage hazards and to provide greater understanding of the reasons for conducting certain OHS activities.

6.1 Basic Skills

Respondents strongly agreed that safety managers must possess basic skills in interpersonal communication, problem-solving and analytical skills. The results showed that interpersonal communication and problem-solving skills were the highest skills ranked in the analysis.

Interpersonal Communication Skill

The purpose of this competency is to improve working relationships on site and to reduce conflict. By doing this, it will be easier to collaborate and consult regarding OHS issues. It is because daily communication on site was regarded as critically significant to support OHS and provide the most effective way to communicate about OHS with the workforce. Through communication, any safety issue, problem or matter that arise can be easily communicated through effective communication between safety manager and other workers. Furthermore, according to [10]. safety communications stressed the message of staying focused and analysing the safety standards; further, it takes care of details and raise awareness for every job. Thus, according to all the supported details, it has been assessed that interpersonal communication skill is one of the most important types of basic skill competency that is required by safety manager.

6.1.1 Problem-Solving Skill

Problem-solving skill is the knowledge of how to strategize and implement strategies to handle an immediate crisis; while, at the same time preparing for the possible occurrence [11]. From the result obtained, it is important for safety manager to have this basic skill to immediately solve the problem related to safety as this will have an impact on the performance and competency of safety manager. Thus, as a result with all the supported details, it has been observed that problem-solving skill is one of the most important types of basic skill competency that is required by safety manager.\

6.1.2 Analytical Skill

Analytical skill is a proactive approach that seeks to prevent or mitigate the occurrence of accidents, injuries, and incidence. It has been strengthened by [12]. who stated that the diagnostic and analytical skills enable the manager to design the most appropriate response in a given situation. Meanwhile, according to the Government of Alberta [12], there is a need to incorporate a well thought-out Emergency Response Plan (ERP) into the overall Safety system because there is lack of assurance against certain natural disasters like earthquake and explosions which can spell grave consequences on operational safety. Thus, as a result with all the supported details, it has been demonstrated that analytical skill is one of the most important types of basic skill competency that is required by safety manager.

IOP Conf. Series: Earth and Environmental Science **385** (2019) 012057 doi:10.1088/1755-1315/385/1/012057

6.2 Skills and Knowledge

According to the results, it is important for every safety manager to have knowledge and skill on general management functions, practices, and procedures, which are mainly to understand organizational objectives, safety and occupational health principles, regulations, standards, and work processes.

6.2.1 General Management Functions, Practices and Procedures

Based on [13], stressed that good management practices have a positive impact on overall project safety. This was further strengthened by [14], who stated that the management of health and safety should be an integral part of the management of the work. It has also been reported that whoever is responsible for coordinating the activities of others on site must ensure that health and safety are effectively managed. Thus, as a result with all the supported details, it showed that knowledge and skill on general management functions, practices, and procedures are the most important types of skill and knowledge that should be possessed by safety manager.

6.2.2 Safety and Occupational Health Concepts, Principles, Practices, Methods and Techniques

Safety manager was required to possess skill and knowledge on safety and occupational health concepts, principles, practices, methods and techniques. This is important to create awareness among [15] stated that effective safety knowledge among construction professionals can reduce accidents that directly or indirectly reduce project cost.

6.3 Core Competencies

6.3.1 Safety Management

Safety management includes developing safety plans such as safety programme, budgeting, scheduling, assessment, training, and measure. This implies that an effective safety management system requires the implementation of core structures and processes by key players. Based on the result obtained, [16] stated that an effective safety system management requires management commitment to OHS, assignment of responsibilities, OHS procedures, OHS communication mechanisms, hazard identification, prevention, and control, accident investigation, OHS training, documentation and evaluation of program effectiveness. This is then strengthened by [17] that safety manager should implement and actively manage general EHS management system, including EHS programs that go beyond regulatory compliance. Moreover, they need to develop or refine mission, vision, directives, procedures or policies, and work instructions to ensure regulatory compliance, best practices, and voluntary commitments.

6.4 Other Competencies

The personal credibility includes professional integrity, meeting commitments, accurate works, self-confidence and providing alternative insights on the business issue.

6.4.1 Personal Credibility

Personal credibility is related to the personal behaviour or attitudes of a person in relation to management abilities. This has been supported by [18] who stated that safety awareness has far more to do with personal credibility and attitudes than anything else. Thus, as a result with all the supported details, it has been observed that personal credibility is one of the most important types of another competency that is required by safety manager in this study.

For the basic skills competencies, most of the respondents strongly agreed on interpersonal communication, problem-solving and analytical skills. Effective interpersonal communication is very important as the employee's safety related performance increases when the safety manager explains all operating procedures. Also, when there is organizational commitment to continually improve work processes and to mitigate or reduce risks to a reasonable level among affected employees. The safety managers should communicate the end-user all the time regarding the safety issue on site and must be responsible to find ways of communicating with the clients, staffs and other personnel involved. Besides that, safety manager must also have the basic skill competency in problem-solving skill so that safety problem or accidents arising can be solved immediately with proper safety action plan.

For the core competencies, the respondents strongly agreed on safety management, considering it as one of the core elements for being a competent safety manager. It is important to have the core competency in safety management as this covers the aspects of the organization's safety management system that includes safety officers, safety committee, safety equipment, and policies. This is because implementing successfully safety management practices are associated with lower accident rates on construction site.

7. Conclusion

It is important to realize that the companies or organizations should have competent safety manager in carrying out their job duties, who is responsible to develop and provide active and effective leadership as a safety manager. As a result, safety performance can be enhanced through good working environment and surrounded by high awareness on safety among workers and competence safety personnel on any type of construction project. Safety managers should be able to conduct other safety related job duties; such as proactively identify the potential conflict situations and workplace issues. The manager must provide an open and participative environment for employees to raise workplace issues in accordance with OHS requirements. They should also be expert to monitor, assess, and review the impact of safety training in order to improve the training based on the feedback received. Safety managers are recommended to actively participate in the execution of safety plans. This development is vital in improving their roles and to achieve the standard of competencies in the construction industry. It is also recommended to safety managers to strengthen the basic or core competencies that had been possessed; while, they should also develop other basic or core competency; such as social-interpersonal skills. Besides that, they also need to have the advisory skill to provide safety advices to other workers and personnel regarding safety matter and safety awareness on the site.

References

[1] National Safety Council. (2006).
[2] K. Dorji, and B.H. Hadikusumo (2006), *Safety management practices in the Bhutanese construction industry*, Journal of Construction in Developing Countries, 11 (2006), pp. 53-75.
[3] N. Khalil, H.N. Husin, A.H. Nawawi, and A.E. Hashim,(2008) *Post occupancy evaluation for performance evaluation of building facilities in higher education buildings*. In *Proceedings of the 9th International Conference SENVAR and 2nd ISESEE*, pp. 1-3.
[4] K. Hafeez, N. Malak, and H. Abdelmeguid, (2006), *A framework for TQM to achieve business excellence*, Total Quality Management and Business Excellence, 17, pp. 1213-1229.
[5] R. Olson, A. Grosshuesch, S. Schmidt, M. Gray, and B. Wipfli, (2009) *Observational learning and workplace safety: The effects of viewing the collective behavior of multiple social models on the use of personal protective equipment*, Journal of Safety Research, 40, pp. 383-387.
[6] A. Mooney, *Core competence, distinctive competence, and competitive advantage: What is the difference?*, Journal of education for business, 83 (2007), pp. 110-115.
[7] S. Pilbeam, and M. Corbridge(2010), *People resourcing and talent planning: HRM in practice*. Prentice Hall.
[8] H.C Biggs, V.L. Sheahan, and D.P Dingsdag (2006),, *Improving industry safety culture: The tasks in which safety critical positions holders must be competent*. In *Proceedings CIB99 International Conference on Global Unity for Safety and Health in Construction*, pp. 181-187.
[9] B. Fernandez-Muniz, J.M. Montes-Peon, and C.J. Vazquez-Ordas (2009), *Relation between occupational safety management and firm performance*. Safety science, 47 pp. 980-991.
[10] J.H. Miller (2009), CEOs Who "Get It" Interview. Safety and Health Magazine,
[11] D. Zohar(2002), *Modifying supervisory practices to improve subunit safety: a leadership-based intervention model*. J Appl Psychol, 87 pp. 156.

[12] R.W. Griffin, Fundamentals of managing organizations (2012), *Warsaw: Scientific Publishers PWN*, (2002).Government of Alberta. Building an Effective Health and Safety Management System,
[13] J. M. Wilson, and E.E. Koehn, *Safety management: problems encountered and recommended solutions,* Journal of construction engineering and management, 126 (2000), pp. 77-79.
[14] E. Sawacha, S. Naoum, and D. Fong (1999), *Factors affecting safety performance on construction sites*. International journal of project management, 17 pp. 309-315.
[15] S. Kumar, and V.K. Bansal (2013),*Construction safety knowledge for practitioners in the construction industry*, Journal of Frontiers in Construction Engineering, 2 pp. 34-42.
[16] C. Needleman (2000)., OSHA at the crossroads: Conflicting frameworks for regulating OHS in the United States. *Systematic Occupational Health and Safety Management: Perspectives on an International Development, Pergamon, Oxford*,
[17] T. Arthur (2014),, *Integrating Occupational Health and Safety Systems into a Project Management System*, International Journal of Research in Management and Business Studies, 2 pp. 35-38
[18] A. Suraji, A.R. Duff, and S.J. Peckitt (2001),, *Development of causal model of construction accident causation*. Journal of construction engineering and management, 127 pp. 337-344.

IOP Conf. Series: Earth and Environmental Science **385** (2019) 012058 doi:10.1088/1755-1315/385/1/012058

Application of Biomimicry Architecture Concept on Lapau Panjang Buildings as A Tourism Icon

Zulherman[1], Al Busyra Fuadi[1], Irmaidi Indra[2], I Nengah Tela[1], and Zaitul[3]
[1]Lecturer, Faculty of Civil Engineering and Planning, Bung Hatta University, Padang, Indonesia
[2]Master Student, Faculty of Civil Engineering and Planning, Bung Hatta University, Padang, Indonesia
[3] Lecturer, Faculty of Economics, Bung Hatta University, Padang, Indonesia

Email: zul230174@gmail.com

Abstract. *Lapau-Panjang* / Side-walk-cafe is a supporting building of Padang City tourist attraction facility, where Danau Cimpago area becomes one of tourism objects in Pantai Padang. In this paper the author uses Architecture Biomimicry approach that is the study of Architecture through natural alignment that will be implemented to the built environment, with methodology of literature review / research and 3D modeling. In its application, it will create a form of imitation, where natural forms having measurable geometries are processed or tested to produce certain patterns to be transformed in the building envelope of the study objects, so that it can produce a new charming theme, and can become an icon for Cimpago Lake Tourism Area.

1. Introduction

Biomimicry is an approach to create form exploration innovation to look for solutions to answer challenges faced by humans by imitating any existing pattern and one tested by the nature. Biomimicry aims to reach new products, process and ways in order humans can live and adapt in the earth in long term period. Learning the nature to obtain ideas for solving of these new transdisciplinary problems can obtain new attention from the field of biomimicry [1], and this is supported by a statement, if the nature has willingness than it can be genius in its acts [2].

Biomimicry has been existed for a long time, when Wright and his brother created the first plane, when Da Vinci designed his flying machine, all of which are inspired from the nature. This proves that the nature is more advanced than us and it is tested in reality.

Figure 1. (a). Armadillo Concert Building; (b). Baha'i Building (Wikipedia, 2014)

1. Literature Review

There are approaches emphasizing on this object designing, among others are biological, typology and thematic approaches as well as observation on field conditions. [3] noted that biomimicry design is a design type that can make building skin energy efficient. In addition, [4] argue that biomimicry is furniture design which means that it design the interior from natures. Indeed, [5] state that bimimicry is a problem solving methodology in interior design. [6] give a statement that biomimicry design can enhance sustainable design ecologically. In brief, we can argue that biomimicry design is futureistic design which connected to natures.

2.1. Definition of Biomimicry Architecture

Natural shapes based on simple logic is geometry shapes that can be seen in our surrounding or nature, it can also be found some geometric shapes such as a shape of bee nest as a set of geometric shapes namely hexagon grouping in each side to form a unique unity. But we do not realize to some other things having many geometric shapes that can be explained using simple logics.

Biomimicry is more than to reproduce natural items or system. It does not only design a thing that is considered to be "green" or sustainable. This is the first close examination to an organism or ecosystem, then in its application, it pays attention to basic design principles in the natural solution [7]. The nature has tested-time pattern and solution in our surrounding and bimomimicry is a study and application of natural solution to design challenges. "Nature is imaginative to needs and can solve our many problems until now," [8], founder and president of Biomimicry Institute. Our surrounding world is natural intelligence life encyclopedia giving its beauty – including geometry – sometime it is difficult to reach' by human logics [9].

2.2. Biomimicry as Architecture Criteria

Between biomimicry and biometric, there are new sciences observing materials in the nature and then it aims to result solutions for humans by imitating this design or by taking their inspirations. Biomimicry concept is a new work area selecting natural principles and material design and process, biomimicry can be defined as an 'innovation to be inspired by the nature [10].

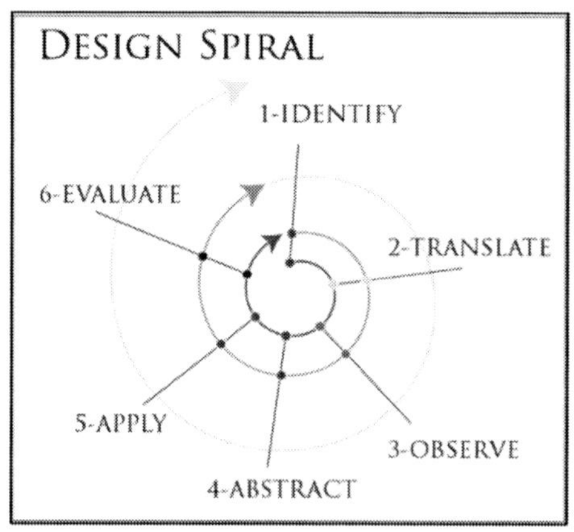

Figure 2. Source: Benyus, Janine, 2014.

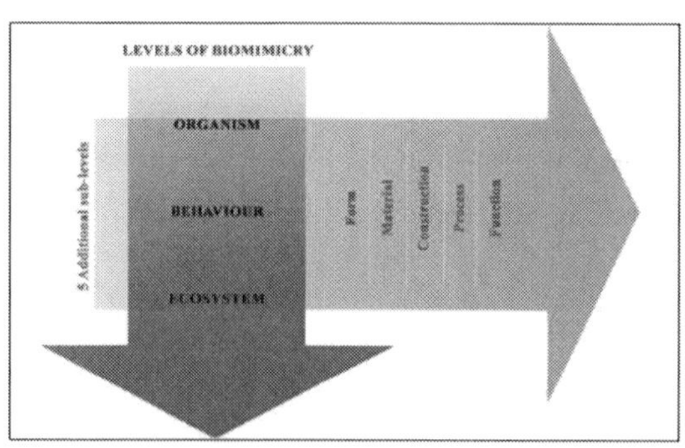

Figure 3. Source: Zari M, 2007; Salma Asraf, 2011.

2.3. Biomimicry level and approach

There are many term definitions related to biometric, as a study field. Biometric is a terminology with a number of synonyms or links to some following other terms.

i. Biomimicry. It was firstly used group of Janine Benyus from United Stated discussing the use of ideas from the nature. Biomimicry is commonly used for progress in technology field.
ii. Bioinspiration. It reveals the most common designs inspired by nature, including the level of abstraction, which is also purely a morphological interpretation.
iii. Biomorphology. The study of construction and organization of living things and organ, tissue and cell components. Morphological structure refers to functional inner design in technology and functional anatomy in biology.

Figure 4. Framework to understand various forms of biomimicry, Source Mahmoud 2012

2.4. Principles of Biomimetic Architecture According to Lebedew

The relationship between architecture, nature and humans that are interdependent with each other like the anatomical structure of the human body, with organs that complement one another, is a model in the philosophy of biomimetic architecture. In this thinking the application of biomimetic architecture is required not only to imitate the forms of living things or from natural forms but must be able to present an environmentally friendly or with the surrounding conditions, in other words buildings that can adapt to living things, comprehensive, architecture with all components and scale, can be seen as something with corresponding characteristics to living thing characteristics. In his thesis, Lebedew proposed a viewpoint system seeing at living system and its criteria (biological criteria) that can be linked or translated into a system of architecture [11].

2. Research method

The object of this research is *Lapau Panjang* Building. *Lapau Panjang* building is located at Muaro, Padang city. This building currently has a business function operared by small medium entreprise (SME). The design of Lapau Panjang building is not attractive and creative. We use the secondary data, in term of detail drawing, function, space, and pattern of building. Besides, we also use the primary data regarding to sociaty perception about future design of Lapau Panjang Building to supplement the primary data. design phases consists of six steps called "design spiral": (i) identify, (ii) translate, (iii) observe, (iv) abstract, (v) apply and (vi) evaluate. In this study, we follow this step to gain the biomimicry design.

3. Result and Discussion

3.1 Scope of Study Review [Identity]

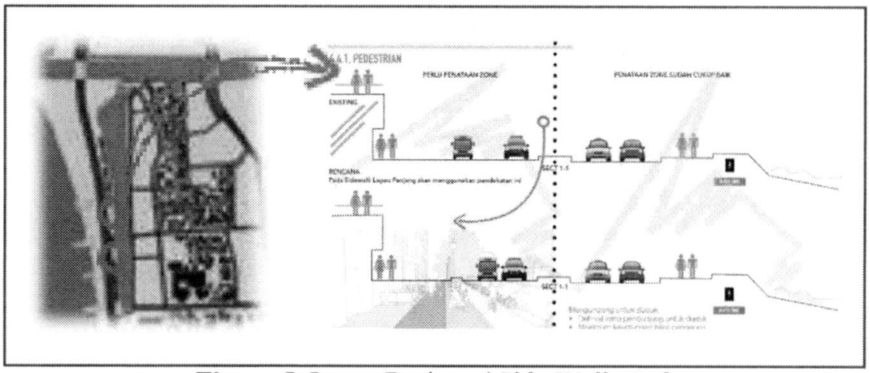

Figure 5. Lapau Panjang / Side-Walk-Cafe

3.2 Theme Transformation Process [Translate]

In this object imitation process, it uses biomimicry architecture theme. It is simply an architecture design concept to be transformed to life or nature criteria into architectural criteria realization or place.

When an architect or designer states that a design is influenced by the nature, he is probably tells about performance: it has organic shapes. Nature is the best teacher in this case, but imitation or being inspired by natural shapes, texture and color cannot be called as biomimetic. Quoting Dr Julian Vincent 'biomimetic must have some biology.' It means that a design must be seen in its some ways to be informed by natural science, not only performance to be the real biomimicry. But, the key to understand the role in architecture so that it can manage to explore its root to natural principle design is a good design sample! Biomimicry is a philosophy approach that can cause new ideas and innovative solutions with many profit potencies, such as from functional or sustainable perspective [12].

The approach through biomimicry design in this study object is through two stages, namely the first stage as comprehensive idea development consisting of 3 aspects, namely understanding on design object, understanding design theme and understanding existing location and sites. The second stage is the application using image-present-test method from Jhon Zeisel. In this stage process, an architect transforms the concept based on the data obtained from comprehensive ideas namely in the first stage. This transformation is initiated by imaging stage (idea concept thinking), followed by presenting stage (presenting concept into Figure or model) and ended by testing (repeated testing). This concept is based on certain testing criteria or through consultation process). Then, it reaches the testing stage, the transformation process is not directly ended, but it is through continuous repeated process such as a spiral. Imitating shape and borrowing shape will later be united in in a summary media or abstract [13]

3.3 Adaption of Building Envelope [Evolute]

Building is a spatial structure established to protect people and their objects from outer environment in case of certain climate conditions such as wine, rain and excessive radiation. Building evolves from primitive structure providing catchment place only for sophisticated structure responding environmental context in which there are many features and elements from the needs for improving life convenience and quality [19]

Building an envelope consisting of basic elements such as window, wall, roof and floor, represents dimension between outer environment and space in the rooms, which significant energy saving can be reached when there is effectively responsive solution designing to specific climate factors [14]. Increasing changing environmental conditions create new challenges to build envelope to accommodate [15]. Resident activities and environmental factors such as air movement, humidity, temperature, sun radiation, air quality, sound influence on building convenience [16].

Regarding building envelope functions as a divider or shield such as for a solution to muffle quite high thermal, design threshold solution utilizes environmental changes and creates media to improve space in a more efficient manner. Solutions such as vernacular buildings that reflect the environment by utilizing wind, radiation, and temperature, increase the energy increase of buildings [17], but this solution is not always air and water resistant. In this case, implementing adaptive solutions that reflect the environmental context can improve the performance of building envelopes, increase the comfort of residents, and potentially reduce energy requirements.

3.4 Implementation & Application

When biomimetic principles are applied during the design process, solutions will mimic the properties of nature. The Life Principle is a filter in which designers test solutions to natural attributes and are the most important step in the "biologization" or programming phase [18]. In applying biomimicry, taking a level or degree is an important step to any study exploration level, there are 3 levels or degrees of biomimicry namely organisms, behaviors, and ecosystems, and there are 5 sub-levels; namely shape, material, construction, process and function. Level selection depends on the complexity of the problem to be solved.

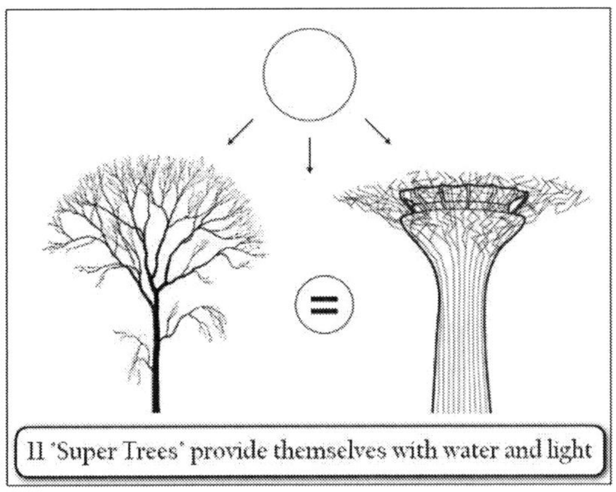

Figure 6. Source Wikipedia 2015

Biomimicry is considered a method to produce innovative design concepts when integrated during the initial design stage, especially the concept design stage. During this stage, possible design concepts are generated and evaluated to choose one design for the detailed one. It is known that the initial design stage is the most contributing to the final product quality. According to the product development paradox, the initial design stage is responsible for around 70% of the final product quality and the final cost, while there is still rare knowledge available to support the initial design phase. The main potential of biomimicry to drive innovative solutions is therefore, when applied during the initial design stage when there is determination of potential solution scope. Then, biomimicry primarily functions as a concept generator of possible design solutions. The design process with biomimicry applications at the initial design stage plays a very important role.

3.5 Model 3D [Modelling]

Modeling function is also referred to as functional analysis or functional decomposition, namely "a description of a product or process in terms of its basic functions needed to achieve its overall function or purpose". During this process, system overall functions are described by smaller sub-functions to change the system's main functions into alternative sub-functions that can be easily handled by the designer. This modeling process explains system architecture, structure and behavior. The generation of functional models during the design process provides many advantages to designers such as explicit responses to customer needs, a comprehensive understanding of design issues, enhanced creativity and innovation.

At this stage, the imitation process is only on one biological object applied in the design, in order to obtain maximum results. To achieve a form with capability of transforming as a certain value or icon in a building requires a long process. To create a level of comfort for visitors and their interests, the chosen biomimicry approach produces an integrated form of building based on the suitability obtained at the imitation state with the surrounding environment, resulting in a complete and real form of the building.

4. Conclusion

After going through a long process in implementing or applying biomimicry in Lapau Panjang, especially on the building envelope, it will bring changes to the building appearance, thus it can attract visitors and increase the activity in the area so that it becomes an icon for the region itself.

The biomimic project is a process of very complex activities in terms of complexity, in which themes are used, transformed or by transferring elements of natural criteria to architectural criteria. Given the condition of the study object, there are many typical masses and with similarity of spatial patterns providing little challenge. On the other hand, this can still be developed further to obtain a better final result, for that matter, it has willing to accept criticism, suggestions for information purposes. It is the authors' hope to be able to continue the biomimicry exposure as a methodology of problem solving so that it will help us find sustainable and effective solutions to the interior, lighting, thermal comfort, energy efficiency, material durability, and productivity.

References

[1] Reich, Y., & Shai, O., 2012. The interdisciplinary engineering knowledge genome. Research in Engineering Design, 23(3), pp. 251-264.
[2] Ralph Waldo Emerson, article 2015
[3] G. A. N. Radwan and N. Osama, "An approach, for energy effecient building skin design," Procedia Environ. Sci., vol. 34, pp. 178–189, 2016.
[4] F. Tavsan and E. Sonmez, "Biomimicry in Furniture Design," Procedia - Soc. Behav. Sci., vol. 197, pp. 2285–2292, 2015
[5] R. M. A. El-zeiny, "Biomimicry as a Problem Solving Methodology in Interior Architecture," Procedia - Soc. Behav. Sci., vol. 50, pp. 502–512, 2012
[6] A. Gamage and R. Hyde, "A model based on Biomimicry to enhance ecologically sustainable design," Archit. Sci. Rev., vol. 55, no. 3, pp. 224–235, 2012
[7] Rasha Mahmoud Ali El-Zeiny, Biomimicry as a Problem Solving Methodology in Interior Architecture, Procedia - Social and Behavioral Sciences 50 (2012)
[8] Benyus, Janine, 2014. A Biomimicry Primer. <http://biomimicry.net/b38 les A_Biomimicry _ Primer_ Janine_Benyus.pdf> inc. 14.3.2015
[9] Benyus, Janine. 2009. Biomimicry: Inovation Inspired by Nature. Harper Collins
[10] Moheb Sabry Aziz & Amr Y. El sherif, 27 November 2015 ,Biomimicry as an approach for bio-inspired structure with the aid of computation, Alexandria University, Faculty of Engineering, Architectural Engineering Department, Egypt
[11] Alviro & Octavianus, 2015. Implementasi Konsep Arsitektur Biomimetik Pada Desain Gelanggang Olahraga Di Minahasa Selatan, ejournal UNSRAT .ac.id
[12] Yael Helfman Cohen · Yoram Reich, Biomimetic Design Method for Innovation and Sustainability
[13] Zeisel, John. 1981. "Inquiry By Design: Tools For Environment-Behavioral Research". Cambridge University Press. Cambridge
[14] Sadineni, S.B.; Madala, S.; Boehm, R.F. Passive building energy savings: A review of building envelope components. *Renew. Sustain. Energy Rev.* 2011, *15*, 3617–363
[15] Zhai, Z.J.; Previtali, J.M. Ancient vernacular architecture: Characteristics categorization and energy performance evaluation. *Energy Build.* 2010, *42*, 357–365
[16] Badarnah, L, Form Follows Environment: Biomimetic Approaches to Building Envelope Design for
Environmental Adaptation, Published: 12 May 2017.
[17] Zhai, Z.J.; Previtali, J.M. Ancient vernacular architecture: Characteristics categorization and energy performance evaluation. Energy Build. 2010, 42, 357–365
[18] Alizadehbirjandi, E., Tavakoli-Dastjerdi, F., St Leger, J., Davis, S.H., Rothstein, J.P.,

Kavehpour, H.P. Ice Formation Delay on Penguin Feathers. Bulletin of the American , Physical Society, 60, 2015.
[19] K. J. Rossin , Biomimicry: nature's design process versus the designer's process , International University of Art and
Design, USA

Effect of reflective building façade on pedestrian visual comfort

N M Ishak[1], W N Hien[2], H S Jenatabadi[3], N A Mustafa[4], E M A Zawawi[1]
[1] Faculty of Architecture, Planning and Surveying, Universiti Teknologi MARA, Shah Alam, Selangor, Malaysia
[2] Department of Building, National University of Singapore, Singapore
[3] Department of Science and Technology Studies, Faculty of Science, University Malaya.
[4] Faculty of Architecture, Planning and Surveying, Universiti Teknologi MARA, Sarawak

Email: norish@uitm.edu.my

Abstract. Building materials have become increasingly diverse and flexible regarding their shapes, colours and texture. These include the use of highly reflective materials for facade (e.g. glass or metal claddings) which becomes a popular concept for high-rise buildings. However, these reflective materials contribute to an increasing number of feedbacks on discomfort associated with reflected glare at the pedestrian's level and the surrounding environment. This study aims to evaluate the impact of reflected glare from the reflective building facade on pedestrians' visual comfort in Kuala Lumpur city area. The relationship between pedestrians and reflected glare issues is determined by applying Structural Equation Modelling (SEM) between male and female. Total of 500 questionnaires is distributed at four different locations. The six-modelled fitting test was tested and within an acceptable range. For both male and female models shows that age significantly affects sensitiveness toward reflected glare.

1. Introduction

The needs to understand the characteristics and properties of building materials is very crucial as it has a direct and indirect impact on the urban environment. The incidence of light on a building surface has cause it to be reflected, absorbed and emitted simultaneously. The exterior parts of the building either vertical or horizontal are the most exposed part by solar radiation. The surface roughness particularly will determine whether the reflected component is in a specular or diffusive manner. Once the source of light is reflected, it has implied thermal and visual performance of a building as well as towards the thermal comfort of the pedestrian [1][2][3][4]. The amount of light being absorbed (albedo) in another hand has significantly affected the surface temperature of a building thus increases the energy consumption and urban heat island problem [5][6][7][8][9].

Over the past decade, metal and glass have become preferred façade material for high rise building in many cities around the world. Aside from the physical properties of these reflective materials, the floor-to-ceiling glass façade has become a popular concept in modern architecture. The argument is that it provides abundant daylight and visual connection to the outdoors as well as the aesthetic point of view. However, there is a high environmental cost to all that glimmer; increased energy consumption due to solar heat gain and glare problem for the indoor and outdoor environment. Also, with higher glazing and reflective fraction, indoor and outdoor glare may possess a risk to the surrounding environment [10][11][12][13][14][15].

IOP Conf. Series: Earth and Environmental Science **385** (2019) 012059 doi:10.1088/1755-1315/385/1/012059

2. Outdoor Glare in Urban Setting

The high number of solar reflective building throughout dense urban developments not only can change the microclimate of the surrounding area but also can create the outdoor glare problem. Theoretically, the high reflective facade is good to keep the intense sunlight into a building, but the outdoor dweller such as pedestrian and motorist will become the victims. The reflected light or glare from vertical surfaces is most apparent on the shaded side of tall buildings, where the sunlight is reflected from the un-shaded light-coloured wall surfaces[16]–[18]. Similarly, concave and convex facades are likely to be more of a dilemma, as it can reflect high angle sunlight to neighbouring buildings and ground surfaces. It is evident that large areas of reflecting material are more likely to cause glare problems than small areas. [19],[20]. The astound case study that showcases this architectural failure is Walt Disney Hall Concert (WDHC) building [21]. The shining façade made of stainless steel has caused the pedestrians, drivers and residential nearby a significant glare and reflected heat problem. Later, even remedial work has been done on the stainless-steel facade; the reflected glare still possesses a risk to the surrounding environment.

The indoor glare has received much attention from the research scholar compared to the outdoor glare. There is still no validated methodology to determine the impact of the outdoor glare in the urban setting particularly for the outdoor dwellers. Several indices can be followed to determine the degree of the glare related problem in building for example Daylight Glare Probability (DGP) or Daylight Glare Index (DGI). Therefore any in-situ measurement that exceeds the permissible limits will be analysed easily in the model. Unlike the outdoor glare, it is yet to be explored on the impact towards the surrounding environment. The study on the human subject has been carried out to prove the existence of outdoor glare problem, but the framework is yet to be improved [22]. A similar study has been done to examine the effect of building façade reflectivity on outdoor visual comfort in the tropical climate [23]. The solar reflectivity studies using simulation tool such as Computational Fluid Dynamic (CFD) has been introduced to study the impact of expected glare from the source building [24]. Moreover, this tool claimed that it accurately predicts the path of reflected light, intensity, and associated temperature increment. Nevertheless, it may require a long process and need a powerful computer that runs it. The other lacking information in this tool is the impact of outdoor dwellers' visual comfort.

Figure 1. Outdoor glare from reflective building facade.

Building materials have become increasingly diverse and flexible regarding their shapes, colours, and texture. The use of highly reflective materials for facade (e.g. glass or metal claddings) has becomes a popular concept for high-rise buildings in Kuala Lumpur (Figure 1). However, these reflective materials contribute to an increasing number of comments associated with reflected outdoor glare problem. In the context of Malaysia's building regulation on daylight reflectance of materials used on building exterior, it is still yet to be implemented. Perhaps, there is no such guideline on building façade reflectance, especially during the design or construction process. For example, Singapore has regulated that the daylight reflectance of materials used on building exterior, from compliance to be less than 20%. It is also mentioned that the external surface of a building must be designed and constructed in a means such that any reflection of sunlight off building external surface does not result in loss of amenity to occupants of other buildings in the vicinity of that building. Thus, this study aims to evaluate the impact of glare from the reflective building facade on pedestrians' visual comfort in Kuala Lumpur city area.

3. Methodology

Structural Equation Modelling (SEM) can hypothesise any kinds of relations and interactions among research variables in a single causal framework. The presented conceptual framework of the study is shown below (Figure 2).

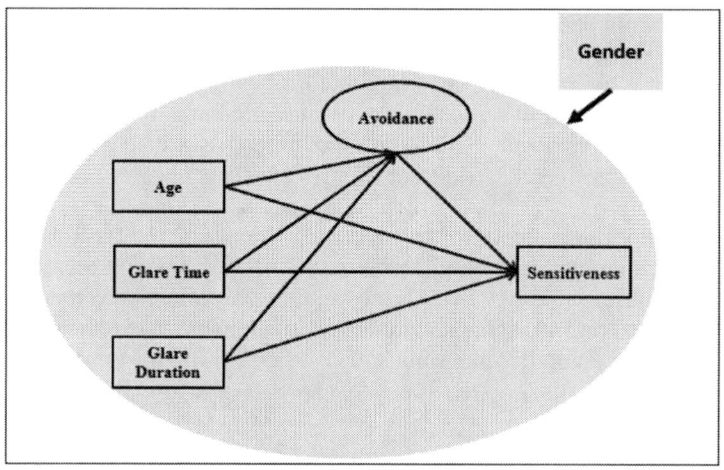

Figure 2. Research Framework

The squares (or rectangles;) and circles (or ellipses;) shown observed and unobserved (latent) variables respectively. In the above Figure 2, age, glare time, and glare duration act as independent variables, avoidance is a mediator, and sensitiveness is the dependent variable. As we can see in Figure 2, age, glare time, gender as a moderator and glare duration have a direct and indirect impact on sensitiveness.

The stratified sampling method has been chosen for the survey. The population surveyed was divided into three different locations (areas). These are respectively Jalan Ampang, Jalan Binjai, Jalan Tun Razak and Persiaran KLCC located in the city of Kuala Lumpur. For each zone, about 125 questionnaires are distributed. Therefore, the sample size is equal to 500 (342 questionnaires are prepared from male and 158 questionnaires for female). Gathering data are done performed between August 2018 until end of November 2018.

4. Data Analysis

The six-model fit test was used to develop the research framework of this study. This framework is based on the influence of independent variables on the dependent variable (sensitiveness) and can be observed in Figure 3. The consistency of predicted and observed data in SEM are lies in all six indices namely, Goodness of Fit Index (GFI), Relative Fit Index (RFI), Incremental Fit Index (IFI), Tucker-Lewis Index (TLI),

Comparative Fit Index (CFI), and Normed Fit Index (NFI). From the study, for all the six indices, it shows that the predicted and observed data are within the acceptable range of above 0.9 hence the research framework is accepted. Figure 4 shows the structural model between male and female. There are seven relations in the research model. As we can see in Figure 4, the structure of the model for the male is different than the female model.

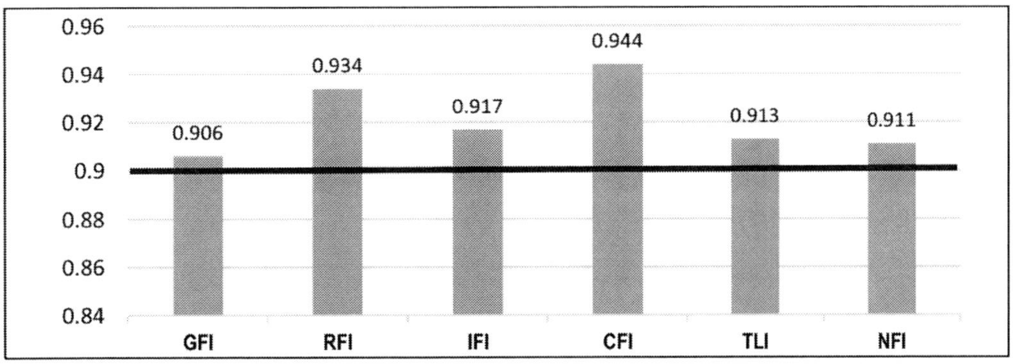

Figure 3. Model fit analysis

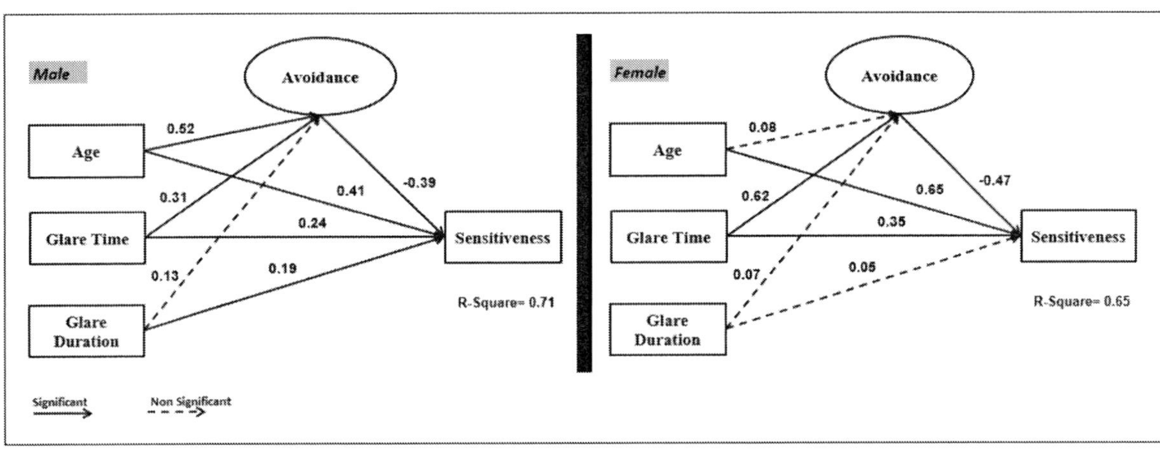

Figure 4. The output of the structural model for male and female.

5. Discussion

This study aims to examine a multi-factorial model for the relationship between pedestrians and outdoor glare issues by applying SEM between male and female. The research framework contains four measurable variables (age, glare time, glare duration, and sensitiveness) and one latent variable (avoidance). The sensitiveness is the main dependent variable; the age of the road users, glare time, and glare duration status are the main independent variables, whereas avoidance is considered the mediator between the dependent and independent variables. Gender is acting as a moderator, which means two different models will be presented in this paper. Moreover, the introduced framework is also designed with improvements from previous modelling studies, using a combination of different relations among research variables. Based on the SEM output which is presented in Figure 4, the R-square of the male model (0.71) is higher than the female model (0.65) which indicates the sensitiveness variation is depending on age, glare time, glare duration, and avoidance. Three independent variables have defined in the research model which are age, glare time, and glare duration. Age in the male model has a significant impact on avoidance (0.52) and sensitiveness (0.41). However, in the female model, age has significant effects on sensitiveness (0.65). Glare time for both models has significant implications on avoidance and sensitiveness. Those impacts, glare time on both avoidance and sensitiveness, for the female model is higher than the male model. In the male model, the impact of glare duration on avoidance is not significant (0.13) but significant towards sensitiveness (0.19). Hence, in the female model glare duration is not a significant effect on both avoidance (0.07) and sensitiveness (0.05). The last relation is about the impact of avoidance on sensitiveness. For both models, this impact is negative and significant, and the value of correlation for a male model (-0.39) is bigger than the female model (-0.47). Another result from SEM analysis is that avoidance, for both models, is a mediator between glare time and sensitiveness. However, mediating of avoidance between age and sensitiveness, glare duration and sensitiveness for both models are rejected.

6. Conclusion

An awareness of the impact of the material's selection between project stakeholders should not be done in isolation. Being in a tropical country, where a large amount of solar insolation occurs, reflected glare from the building façade will have a significant impact on the surrounding environment. Furthermore, heat will also be reflected and cause the ambient temperature to be risen and indirectly promotes the issue of urban heat islands. This study presents the application of SEM in modelling the impact of the reflected outdoor glare from reflective building facade by pedestrians in the city of Kuala Lumpur. It is detrimental to understand the behaviour and properties of building materials before they can be deployed.

Acknowledgement

This research is funded by the Institute of Research Management and Innovation (IRMI), Universiti Teknologi MARA, Shah Alam. (600-IRMI 5/3/LESTARI 020/2018).

References

[1] Tsangrassoulis A, Santamouris M, Asimakopoulos D. Theoretical and experimental analysis of daylight performance for various shading systems. Energy Build [Internet]. 1996 [cited 2013 Nov 16];24:223–30. Available from: http://www.sciencedirect.com/science/article/pii/S0378778896009814

[2] Nikolopoulou M, Baker N, Steemers K. Thermal comfort in outdoor urban spaces: Understanding the Human parameter. Sol Energy. 2001;70(3):227–35.

[3] Santamouris M, Gaitani N, Spanou a., Saliari M, Giannopoulou K, Vasilakopoulou K, et al. Using cool paving materials to improve microclimate of urban areas – Design realization and results of the flisvos project. Build Environ [Internet]. 2012 Jul [cited 2014 Sep 1];53:128–36. Available from: http://linkinghub.elsevier.com/retrieve/pii/S0360132312000376

[4] Sellors J, Nijnatten P a Van, Sanotra P. Measuring the daylight, solar energy and thermal radiation properties of coated glazing. 7th Int Conf Coatings Glas Plast. 2008;1–3.

[5] Taha H. Urban climates and heat islands: albedo, evapotranspiration, and anthropogenic heat. Energy Build. 1997;25(2):99–103.

[6] Akbari H, Konopacki S. Energy effects of heat-island reduction strategies in Toronto, Canada, Energy. 2004;29(2):191–210.

[7] Kolokotsa D, Santamouris M, Zerefos SC. Green and cool roofs' urban heat island mitigatio potential in European climates for office buildings under free floating conditions. Sol Energy [Internet]. 2013;95:118–30.

[8] Santamouris M. Cooling the cities – A review of reflective and green roof mitigation technologies to fight heat island and improve comfort in urban environments. Sol Energy [Internet]. 2014;103:682–703. Available from: http://linkinghub.elsevier.com/retrieve/pii/S0038092X12002447
[9] Synnefa a., Santamouris M, Livada I. A study of the thermal performance of reflective coatings for the urban environment. Sol Energy [Internet]. 2006 Aug [cited 2014 Aug 14];80(8):968–81. Available from: http://linkinghub.elsevier.com/retrieve/pii/S0038092X05002975
[10] Suk JY, Schiler M, Kensek K. Is Exterior Glare Problematic?: Investigation on Visual Discomfort Caused by Reflected Sunlight on Specular Building Facades Online Sustainable Toolshed for Architecture Students View project. 2016 [cited 2018 May 23]; Available from: https://www.researchgate.net/publication/306108182
[11] Suk JY, Schiler M, Kensek K. Development of new daylight glare analysis methodology using absolute glare factor and relative glare factor. Energy Build [Internet]. 2013 Sep [cited 2014 Aug 14];64:113–22. Available from: http://linkinghub.elsevier.com/retrieve/pii/S0378778813002569
[12] Suk JY, Schiler M, Kensek K. Investigation of existing discomfort glare indices using humaN subject study data. Build Environ [Internet]. 2017 [cited 2018 May 22];113:121–30. Available from: https://ac-els-cdn-com.libproxy1.nus.edu.sg/S0360132316303584/1-s2.0-S0360132316303584-main.pdf?_tid=fd1244b1-470c-4682-bb6a-0dfb983851a7&acdnat=1526976722_0eae2335fd80528ad6e4424bc3637d47
[13] Danks R, Good J. Urban Scale Simulations of Solar Reflections in the Built Environment : Methodology & Validation. Urban Scale Simulations Sol Reflections Built Environ Methodol Valid [Internet]. 2016;(May):19–26. Available from: https://www.researchgate.net/publication/303312340_Urban_Scale_Simulations_of_Solar_Reflections_in_the_Built_Environment_Methodology_Validation
[14] Danks R, Good J, Columbia B. Urban Solar Reflection Identification , Simulation , Analysis and Mitigation : Learning From Case Studies Reflections From a Concave. eSim 2016 Build Perform Simul. 2016;309–18.
[15] Danks R, Good J, Sinclair R. Assessing reflected sunlight from building facades: A literature review and proposed criteria. Build Environ [Internet]. 2016;103:193–202.
[16] Hirning MB, Isoardi GL, Coyne S, Garcia Hansen VR, Cowling I. Post occupancy evaluations relating to discomfort glare: A study of green buildings in Brisbane. Build Environ [Internet]. 2013 Jan [cited 2014 Jul 21];59:349–57. Available from: http://linkinghub.elsevier.com/retrieve/pii/S0360132312002429
[17] Gil Vl-R. Evaluation of solar glare from reflective facades : A general method. Light Res Technol. 2015;1–18.
[18] Hirning MB, Isoardi GL, Cowling I. Discomfort glare in open plan green buildings. Energy Build [Internet]. 2014 Feb [cited 2014 Jul 31];70:427–40. Available from: http://linkinghub.elsevier.com/retrieve/pii/S0378778813007603
[19] Schiler M, Valmont E. Microclimatic Impact: Glare around the Walt Disney Concert Hall. Proc Sol World Congr … [Internet]. 2005 [cited 2014 Nov 11]; Available from: http://www.researchgate.net/publication/228838745_Microclimatic_Impact_Glare_around_the alt_Disney_Concert_Hall/file/5046352330b7f80b36.pdf
[20] Schiler M. Examples of Glare Remediation Techniques Four buildings. In: PLEA2009 - 26th Conference on Passive and Low Energy Architecture, Quebec City, Canada, 22-24 June 2009, 2009.
[21] Schiler M. Examples of Glare Remediation Techniques Four buildings. Plea2009. 2009;(June).

[22] Suk JY, Schiler M, Kensek K. Reflectivity and specularity of building envelopes: how materiality in architecture affects human visual comfort Reflectivity and specularity of building envelopes: how materiality in architecture affects human visual comfort. Archit Sci Rev [Internet]. 2017 [cited 2017 Sep 4];60(4):256–65. Available from: http://www.tandfonline.com/action/journalInformation?journalCode=tasr20
[23] Ishak NM, Hien WN, Jenatabadi HS, Ignatius M, Yaman R. The effect of building facade reflectivity on urban dwellers in tropics. In: IOP Conference Series: Earth and Environmental Science. 2018.
[24] Montes-Amoros V. When buildings attack their neighbors: Strategies for protecting against "death rays." CTBUH J. 2015;

IOP Conf. Series: Earth and Environmental Science **385** (2019) 012060 doi:10.1088/1755-1315/385/1/012060

Challenge of Renewable Energy Transition towards Krabi's Sustainable Energy City

Chariya Senpong[1] and Dawan Wiwattanadate[2]
[1]Environment, Development and Sustainability, Interdisciplinary Program Graduate School, Chulalongkorn University, Bangkok, Thailand
[2]Faculty of Engineering, Chulalongkorn University, Bangkok, Thailand

Email: dawan.w@chula.ac.th & mookrawee@hotmail.com

Abstract This paper is a review article collecting relevant data and information for developing a transition model towards a sustainable energy city: a case study of Krabi Province, Thailand. An expanding tourism sector in Krabi is increasing energy demand, which continues to grow year after year. Meanwhile, the energy supply of the province relies on the centralization of the southern and central regional energy system to the national grid transmission system. According to the 2018 National Power Development Plan, Krabi is projected to shift to regional grid decentralization; coupled with increasing its power generation from renewable energy with affordability and community participation. Therefore, Krabi Vision 2020, which is a provincial master plan, would lead the direction of Krabi's plan towards energy transition. Increasing renewable energy investment and disruptive technology to secure the potential of solar, biomass, biogas, wind and mini-hydro is paving the way to shift Krabi to a sustainable energy future. The implementation of a sustainable energy transition across the province, would bring benefits for the local economy, environmental conservation, and energy security. To achieve this energy transition, Krabi requires coordination of smart policy, local and government participation, and sound planning. This review study suggests that the transition towards a sustainable energy future depends on identifying the political barriers, and the adoption of effective management practices.

1. Introduction

Growth of population and urban centers is challenging government and policy makers to create smart cities and shifting to sustainable development goals [1]. Simultaneously, urbanization and regional development can generate unexpected environmental pollution and ecological damage. A shortage of resources and a decrease in biodiversity is challenging regional, national and global organizations to reduce their impact through sustainable development. Sustainability and environmental security in regional development is an important aspect when planning for a harmonious city [2]. The coordination and integration across nations is an essential requirement for energy sustainability strategies [3]

The demand of regional and urban centers is increasing, putting pressure on the supply chain and straining available resources, which have the potential to affect the quality of life. It is the goal of a regional renewable energy transition to reduce a region's impact on the environment and increase people's awareness of their impact on the current climate change crisis. National energy policy will increase a share of renewable energy on the grid, which depends on the potential, regulation enforcement and political will. A sustainable and renewable energy future depends on decentralization of energy generation and the diversification of energy sources.

In the global effort for climate change mitigation renewable energy is one part of the solution of reducing greenhouse gas emissions and climate change impacts [4]. Growing renewable energy sources does not only bring environmental benefits but also positively impacts the economy and energy sectors. Energy markets and policies driving the energy transition is emerging in nations, which targets to achieve a grid powered by renewable energy. The challenge is the transition from fossil fuel to renewable energy generation as a plan for a country and a master plan of regional planning for energy and its management. Even though the global price of renewable energy is continuing to decrease, energy policy is not adequately taking advantage of the high potential of renewable energy to contribute to the existing energy supply. Ambitious policy is needed to increase the share of renewable energy in the energy portfolio.

A global trend of renewable energy has directly affected ASEAN's energy policy. ASEAN Power Grid as a master plan of Asian countries has been one of the intergovernmental targets of efforts and activities to get a closer reach to the goal of reducing regional energy intensity of at least 8% by 2015 (based on 2005 levels), and the collective target of 15% of total installed power capacity from renewable energy sources by 2015. The ministers have also agreed to consider a higher level of commitment in terms of energy intensity reduction and installation of renewable energy beyond 2015 in reference to other international and regional commitments [5]. On the other hand, the ASEAN power grid has not been addressing a concrete policy to achieve that target, and it seems that the supply of electricity for the national grid is still putting the priority on fossil fuel generation.

As Thailand is a key player in the Southeast Asia region, harvesting the potential of renewable energy is dependent on balancing against fossil fuel energy generation. The aim of this research is to investigate the phenomenon of sustainable energy transition at a provincial level that relates to national energy planning. The beginning identifies key aspects of the direction of national energy policy, provincial energy implementation and the transition capacity related to policy makers, institutions and public acceptance.

2. Energy Transition

Transition thinking is to frame the evolution of energy changing in term of traditional to technological development from the dependable fossil fuel to de-carbonisation initiatives [6]. Energy transition can be conceptualized as a socio-technical term being a sort of crossroads between institutional development and policy paradigm [7]. The multi-level perspective is disrupted in terms of structure and demographics of the economy, environmental and social factors, and depends on various factors of legislation, technology and innovation etc.

The unexpected phenomena require understanding the adaptive transition that relies on knowledge, skill and capacity experiencing across initiative. The interconnection of energy transition involves socio-technical, techno-economic, and political actions [8] emerging to deal with complicated situations requiring effective management. This transition process is challenging society which tends to be stubborn when facing major structural changes.[9].

The pattern of energy transition is significant in regards to the potential of renewable energy to be adopted for energy demand and consumption. As various countries move forward technologically, infrastructure and institution practices [10] with relevance to stakeholder perspectives in various states including adopters, non-adopters and potential adopters. Renewable energy market prices reduction, higher rate of return investment, feed-in-tariff agreement policies etc. are all potential factors impacting a renewable energy transition [11][12].

Reduction in negative environmental impacts is a main goal of climate change mitigation and sustainable development. Energy transition becomes a main dimension of city management to deploy it at scale of renewable energy utilization. The potential of renewable energy depends on the interconnection of landscape, management, and investment to secure. City planning priority coordinates with energy transition [13] that illustrates the strategy and indicator to create it long-term. The pathway of energy transition relates to social-spatial in context of distance [14], which refers to cluster and location of land use management.

The transition towards sustainable energy city covers the integrated dimension of geographical, deployed policy and adopters. The disruption takes place in residential, business and governance models

IOP Conf. Series: Earth and Environmental Science **385** (2019) 012060 doi:10.1088/1755-1315/385/1/012060

also approach the justice procedures. Transition management improves a level of public participation and increases the engagement groups [15] to create the public acceptance of change. The emergence of energy transition is from the bottom as a social movement under informal, pluralistic and is politically driven [16].

3. Research Approach

This research demonstrates the investigation of energy demand and supply, coupled with the potential of provincial renewable energy, national energy supply plans and its energy planning to further initiative a mechanism of sustainable energy transition in Krabi province. The aspect of island context energy transition as small utilities, low population density which increase the need to invest more in a decentralised system, energy storage [17] and transition towards renewable energy [18]. Adopting a transition to city sustainability is connected to institutions and stakeholders taking an accountability of, and the pathway to utilize the potentials of energy resources for electricity generation balancing the demand for growth. Political wills for renewable energy is essential to shift the potential compatibility with democratic patterns [19]. Renewable energy transition which takes place in Krabi province to enhance economic, energy and environmental factors as a sustainability framework is based on the important key points of energy transition and mechanism to apply the initiative sustainable energy transition. Reviewing relevant issues of sustainable energy city development in the aspects of energy policy and implementation plan to achieve compliance of national and regional power development plan such as industrial, tourism countries' development. Then, identify a map of provincial and national energy infrastructure systems including the exiting and expanded grid connection and investment; it is necessary to conduct an integrated plan of electricity generation, distribution and fuel sources management.

An initiative mechanism of the sustainable energy transition would cover the interdisciplinary triangle of sustainable development and the energy transition governance as shown in figure below and analysis of the sustainable policy, participatory and planning of energy security. That directly relate to the environmental impact concern with drivers, pressure, state of natural capital and response framework to local and national policy-decisions [20]. The frequency of using indicators like economic, environmental and institutional, and environmental aspects as a tool to approach sustainable creation with the indices for the public, policymakers and scientists to integrate a plan and practice [21][22][23].

Ultimately, to process and deliberate a transition towards sustainable energy future is that engaging the kinds of people experiments from diverse perspective to participate with and understanding its energy transition values [24]. Participatory planning defines organisation and individuals in combination with transition dynamics pathways [25].

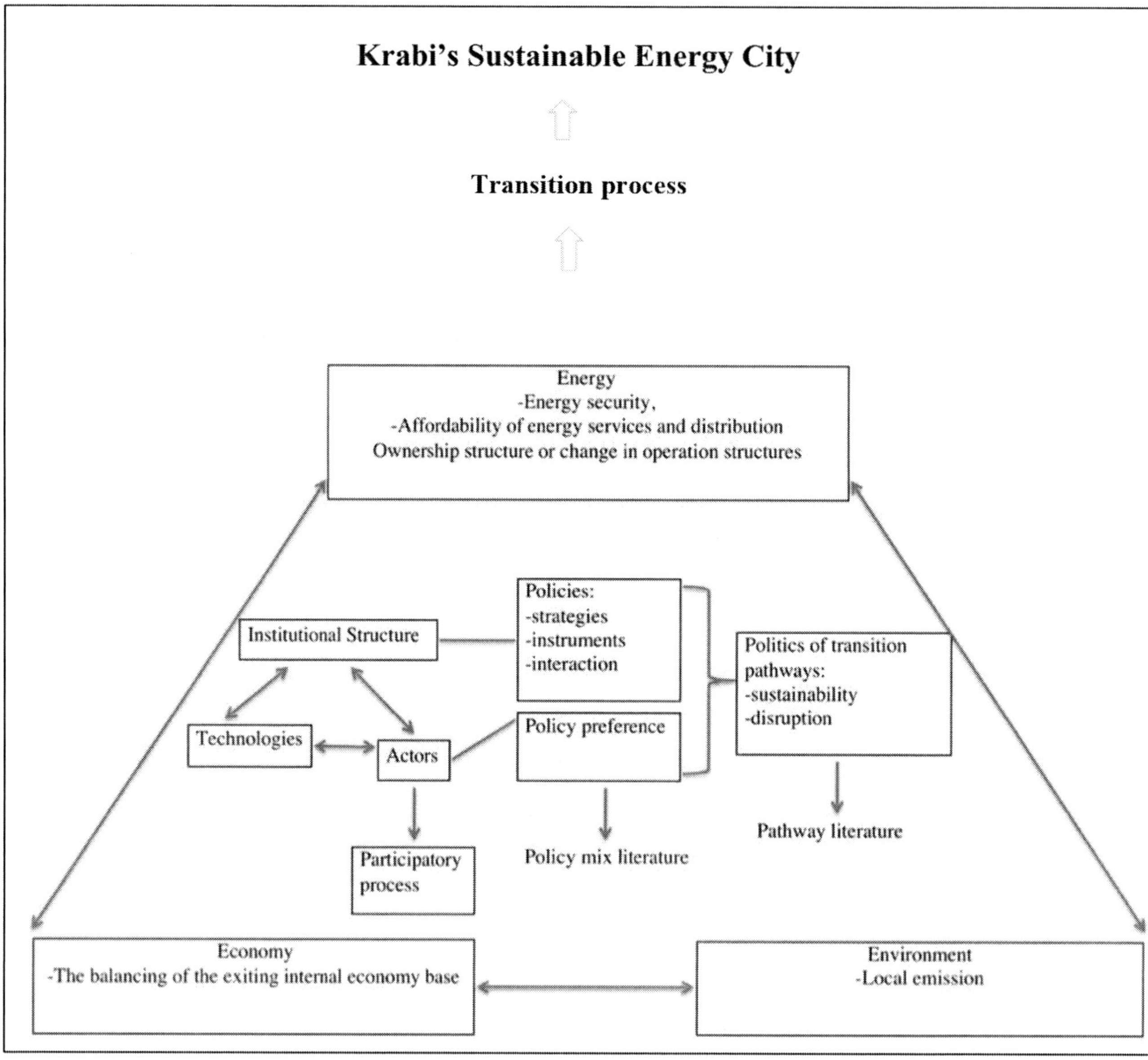

Figure 1: Transition towards Krabi's Sustainable Energy City

Note: Applied from Policies, Actors and Sustainability Transition Pathways: A study of the EU's energy policy mix Marie, L., et al., 2018; and the Interdisciplinary Triangle of Sustainable Energy Development.

Energy sustainability covers main aspects of energy security consisting of availability, accessibility, affordability, and acceptability; however, the process of transition management refers to energy justice and just transition [26][27]. The balances of the internal economy, which relies on tourism, fishery, and agriculture and the external economy concerning environmentally friendly dimension that is targeted to reduce emissions from the point sources are very challenging. The transition process is linked to institutional policies representing key organisations and structural reforms for fundamentally impacting energy creativity [28] and participation to drive the transition direction and initiate the pattern of sustainable energy transition of Krabi province.

Research Methods and Data Collection:

1. Collecting secondary data of cases study of renewable energy transition.
2. Study a system of energy exiting and infrastructure planning and investment.
3. Mapping renewable energy potential of Krabi province.
4. Investigating the procedure of renewable energy transition systems in Krabi province.
5. Study the institution and mechanism of the sustainable renewable energy transition and institution of Krabi province.
6. In-depth interviews with local government agencies, provincial energy authorities and energy agencies, tourism businesses, agriculture, industry, households, and financial sectors etc
7. In-depth interviews with energy policy makers including local, provincial, and national levels.
8. Analysis of the pathway of transition towards Krabi's Sustainable Energy City.

3.1 Thailand Renewable Energy Transition

Thailand's Power Development Plan 2015 indicates the pathway of a country energy planning for 20 years targets; mainly dependent on fossil fuel supply as natural gas base load is about 70 percent. To continually in 2018, Ministry of Energy and energy agencies have proposed the revised power development plan 2018 upon the reasons of the derails of Independent Power Producer (IPP) investments and behaviors of energy consumption changing to more renewable energy increasing, self-generating and peak load at the night time. Referring to PDP report [29] indicates to decentralize the management of fuel sources and power plant in each region, its energy security, distributed energy generating and promote private sector for competitive market.

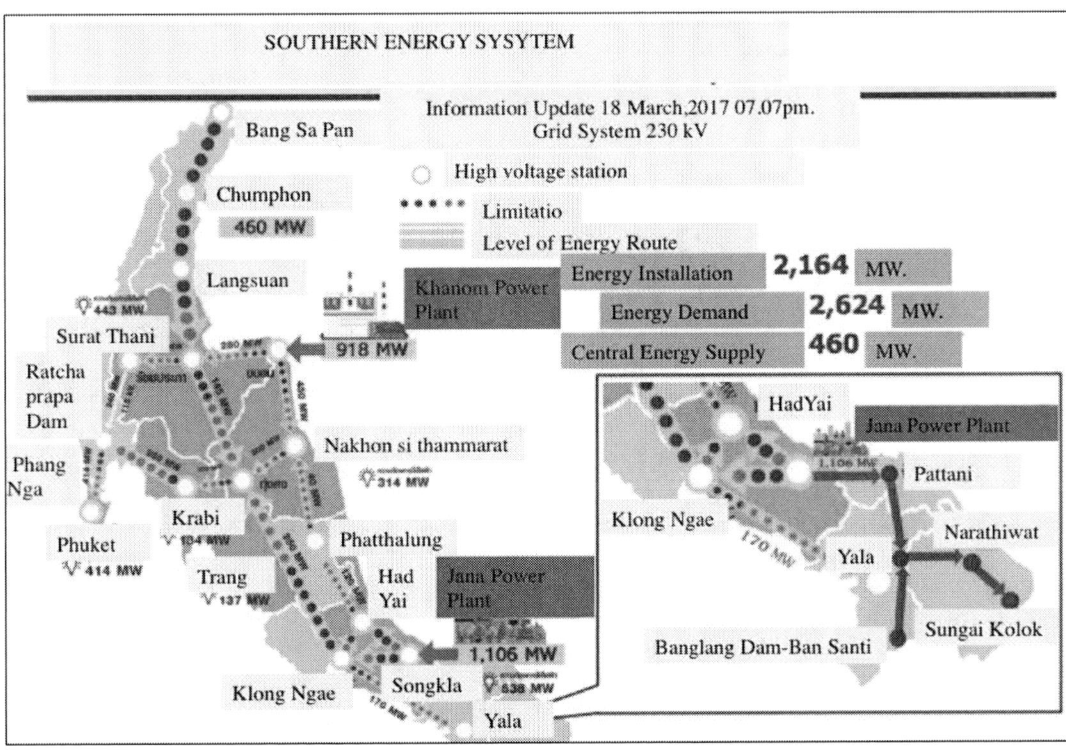

Figure 2: The new version of Thailand's Power Development Plan, EPPO 2018

IOP Conf. Series: Earth and Environmental Science **385** (2019) 012060 doi:10.1088/1755-1315/385/1/012060

The new version of Thailand's Power Development Plan targets renewable energy to supply the grid for the next 20 years with 20,766 megawatts expected to come online from 2019 to 2026. An opening quota of 100 megawatts solar rooftop transmission and later intensive growth after 2027 is planned. The Southern Energy System covers electricity installation of 2,164 megawatts while the regions demand is 2,624 megawatts. Additional supply from central energy transmission leads energy policy makers and agencies promoting new power plants in the southern area to secure the increasing demand in the long term, with public debate on kinds of energy sources and benefits.

The potential of renewable energy in Thailand as a top renewable energy investment destination in Southeast Asia is great, but government policy hinders its development. Relevant policy has been endorsed to help streamline investments and reduce unnecessary regulation. Since 1994, renewable energy installation in country has expanded its capacity reaching 1,000 megawatts in 2006, and increasing to 2,000 megawatts in 2010, in 2016 its generating over 5,000 megawatts. The Renewable Energy Act offers the possibility of easing the transition to renewable energy and its implementation in the country [30]. It has the potential to help encourage the province to research and develop a renewable energy investment plan as a small and independent power producer, highlighting the potentials of producing solar energy from the rooftops of urban homes and rural area, visible institution buildings and communities in the country. It would be empowering a capacity of energy transition on energy revolution, increasing renewable energy and energy efficiency implementation.

There are many benefits that the Renewable Energy Act can offer if it is approved. One of these is showing what people can do in their homes and buildings, and how can they achieve energy independence through solar energy, with the intention of breaking the myths surrounding solar energy e.g. myths on cost-competitiveness, reliability, practicality, etc.

These features will create the perception that producing your own electricity from renewable energy is the path to energy security and energy independence. Since 2013, Thailand has been raising the subject of Renewable Energy Act with the public and later pushing on policy work to the Department of Alternative Energy Development and Efficiency, [31]] finally ending at the Ministry of Energy drafting and National Legislative Assembly. On March 18, 2015 the Committees of the National Legislative Assembly has officially announced the platform of Renewable Energy Law Drafting Public Hearing for renewable energy alignments including renewable energy experts, renewable energy policy working groups, renewable energy investors, civil society and people to raise their finals comments.

In 2014, the National Reform Council [32] passed the resolution on the "Solar Rooftop" expansion scheme - expecting that in the first 5 years (2015-2019) there will be about 100,000 solar rooftop modules (not exceed 10 kilowatt) for households with total installed capacity of 500 MW and in the next 20 years about 1 million solar rooftops with installed capacity of 5,000 MW. If the Thai government is aggressively promoting the investment of solar energy industry [33] including solar PV module production, Thailand is expected to be a winner on this area of renewable energy in the next 10 years. The next step as policy expectation, for Thailand NRC is to push for the renewable energy law to be passed by the parliament to ensure a proper, just, fair, and sustainable supporting mechanism for clean renewable energy development. In terms of renewable energy policy is that Thailand's Alternative Energy Development Plan (AEDP 2012-2022) has promoted production and consumption of renewable energy by setting a challenge target of increasing alternative energy share up to 25% by the year 2022. The national energy planning indicated that the increasing electricity demand would be secured with additional fossil fuel power plants installation as a base load, even fulfilled implementation of the national energy efficiency strategy.

In a country, the barriers of renewable energy installation and its supply on grid are continuing arguments on policy makers and the public. Not only lacking a sufficient support of renewable energy law and regulations but also there are a variety of technology transfers, grid transition systems, technical developments and value of the investment.

To lead urban and regional renewable energy implementation required investigating of the potential of renewable energy supply and integrated on grid into the remaining energy infrastructure and further invests to grid expansion. The potential of renewable energy in country is increasing in a proportion of total power generation, which is different in each regional geographical area, including: solar, biomass, and wind energy. However, the dependable capacity to secure its supply is low compared to an advance renewable energy investment in developed countries and fluctuate flowing supply in some conditions. The implementation of regional renewable energy is just a part of the potential of renewable energy; a planning on energy, environment, and economic aspects are the keys for achievement.

In 2016, Asian countries joint agreement under ASEAN Community to power international trade competition composes of ASEAN Security Community (ASC), ASEAN Economic Community (AEC) and ASEAN Socio-Cultural Community (ASCC). The AEC identified the direction of regional energy development under the umbrella of Head of ASEAN Power Utilities/Authorities (HAPUA), including ASEAN Power Grid with sharing energy resources, effective management and energy security [34]. For long-term implementation, the Electricity Authority of Thailand (EGAT) and Provincial Electricity Authority (PEA) must integrate the transmission system to further supply electricity to neighboring countries with expected increasing demand.

3.2 Krabi's Energy Overview

Krabi is one of 77 provinces in Thailand located at the southern part of Andaman Sea, it has high potential for renewable energy development as the first province of country within the next 10 years. The renewable energy capacity on grid is up to 40 percent, which sourced from biomass and biogas to feed some sub-districts of the province to reach the regional renewable energy utility. However, the potential of other energy sources is going to be researched to reduce the risk of dependent energy supply.

In terms of renewable energy transition in Thailand, it has become the first case development in a country under 'Krabi Vision 2020' [35] and 'Krabi Goes Green', provincial planning has addressed the relevant sustainability of development, and energy planning, increasing renewable energy as a part of that strategy. However, the provincial plan requires the factors and indicators of city growth, economic development, energy security, environmental impact reduction and the climate change risk in relation to the island geography. Krabi has a population of nearly 450,000 people in 8 districts. It is a target of the energy demand supply chain of provincial electricity generation.

Krabi electricity grid management is enforced by Provincial Electricity Authority of Thailand (PEA) to balance the distribution of demand and supply in the province with 10 stations covering the region as shown on the map in figure 2. The provincial electricity transmission is distributed as a monopoly of the Electricity Generating Authority of Thailand (EGAT) and peak load supply from internal renewable energy generating by PEA (Provincial Electricity Authority). Krabi peak load demand of the 3 years is continually increasing between 2015 and 2017, from 134.1 to 144 megawatt as shown in figure 4.

The energy supply for Krabi province is a significant consideration as tourism contributed to visitors of over 6.5 million peoples in 2017 and is expected to grow yearly. As tourism revenue dramatically increases up to 105 million baht in the same period, that shows the strong connection pillars of tourism, agriculture and fishery to secure local economic, energy and environment share.

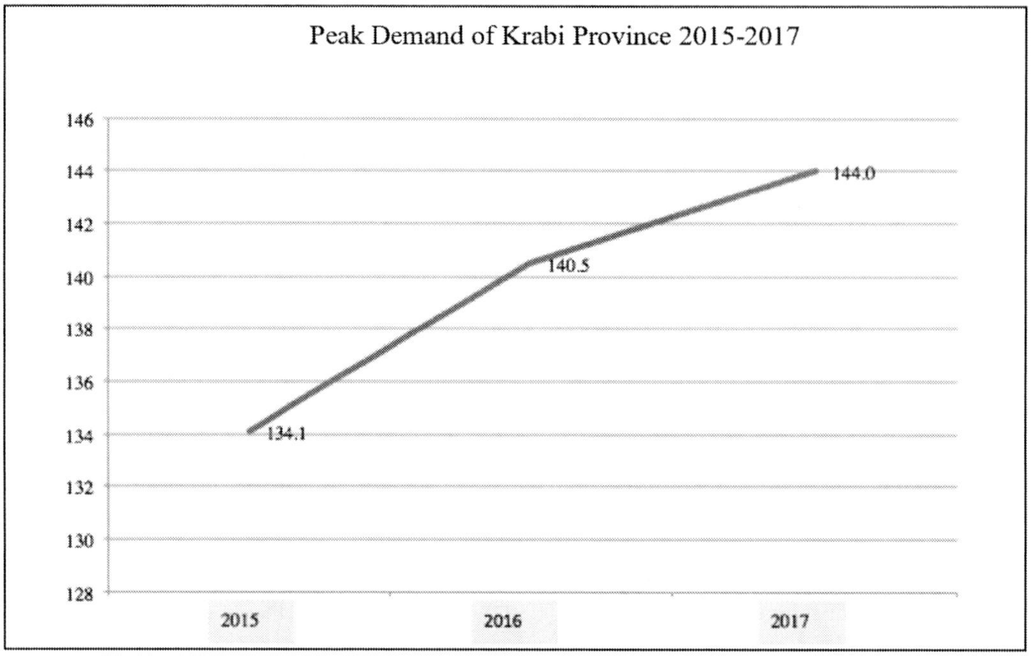

Figure 3: Provincial Electricity Authority, Thailand, 2018

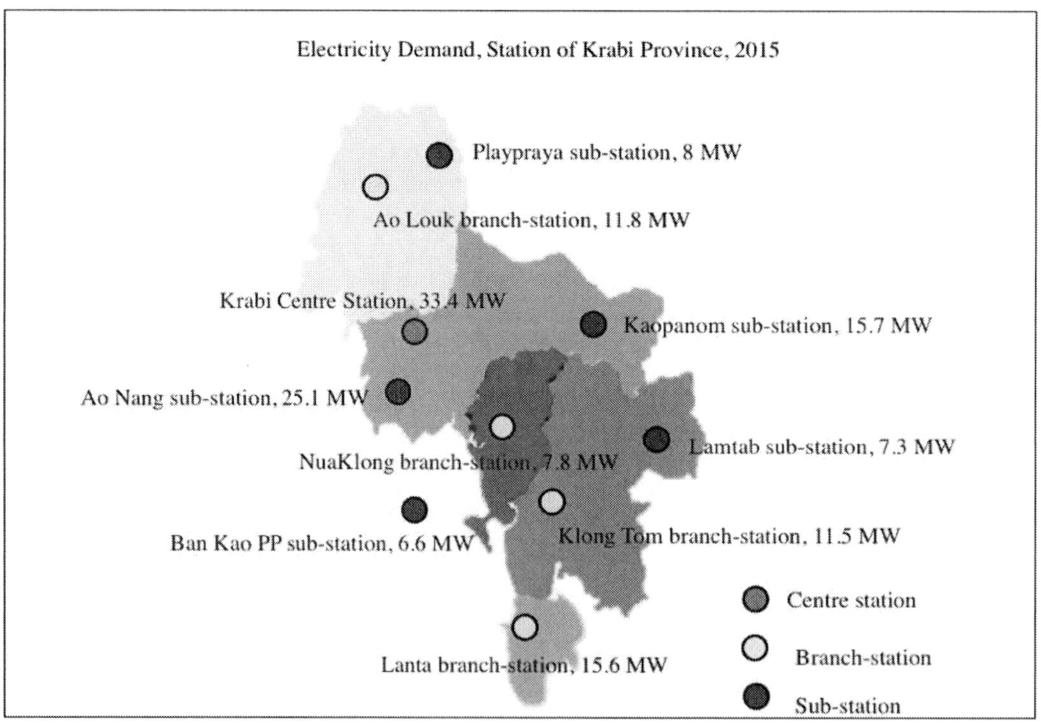

Figure 4: Provincial Electricity Authority of Thailand, 2015

According to PEA, Krabi province is a tourist city with a high number of hotels, resorts and shops where its peak electricity consumption occurs in December as Krabi's peak tourist season. This is different from the overall peak of the country, which lies in the months of April - May, with late evenings as the peak demand hours. The peak hourly consumption behavior of Krabi province is obviously different from non-tourism provinces. Krabi's electricity consumption lies in the 90 -145 megawatt ranges, rising sharply from 7 to 10 am, gradually increasing during the daytime. The highest electricity consumption is observed in the evening. It peaks at 7pm. with 145.64 megawatt of electricity consumed, while the lowest electricity consumption is about 90 megawatt at night from 12 pm. - 6 am.

The average hourly electric consumption comparison during weekends and weekdays of Krabi Province in 2017 shows the 24-hour electricity consumption looks the same throughout the week. Electricity consumption increases rapidly from 90 megawatt to 125 megawatt during 08.30 am. – 10.30 am. Then, a slight increase from 125 to 130 megawatt occurs during the daytime, which increases at 06.00pm. and peaks at 08.30 pm. The peak usage of weekends is higher than that of the weekday's peak at around 3.5 megawatt. The maximum monthly demand for electrical power in Krabi Province in 2017 shows that there are big differences that correlate with Krabi's peak tourism seasons. The lowest electricity consumption occurs in June, with 132.63 megawatt and the lowest average electricity consumption occurs in the range of 132 to 137 megawatt from May to September. However, the electricity demand begins to rise again in September, reaching its peak in December with 145.64 megawatt of electricity and gradually decreases until the end of April when the tourist season ends.

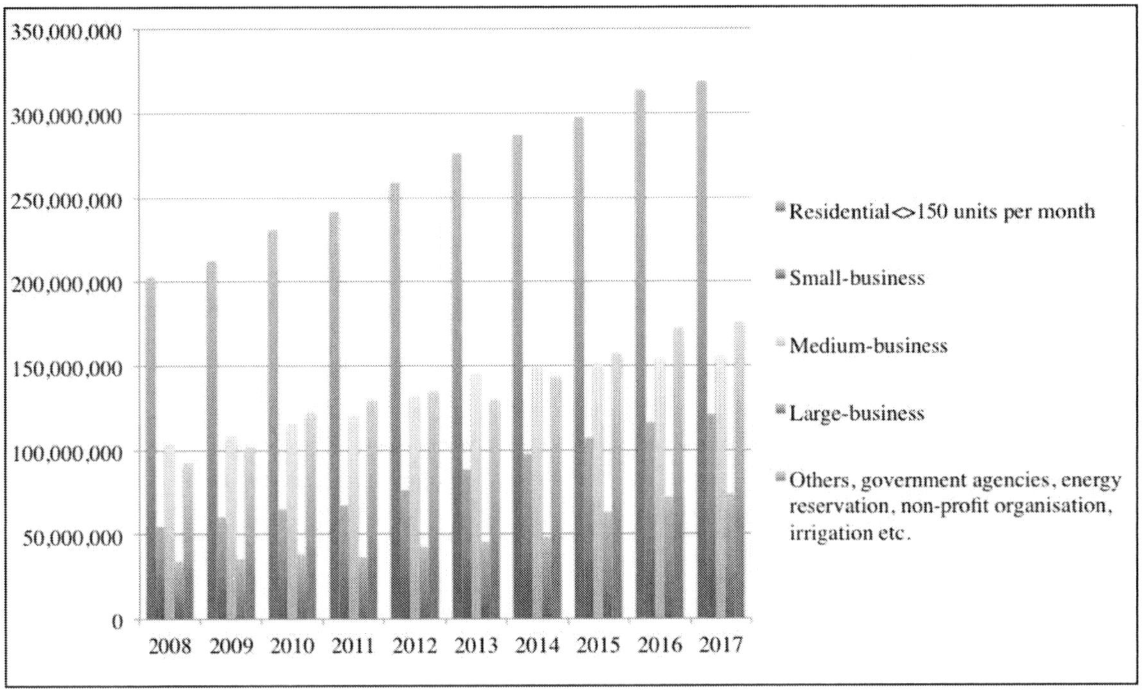

Figure 5: The Statistic of Krabi Electrical Demand-Supply (Unit) 2008-2017

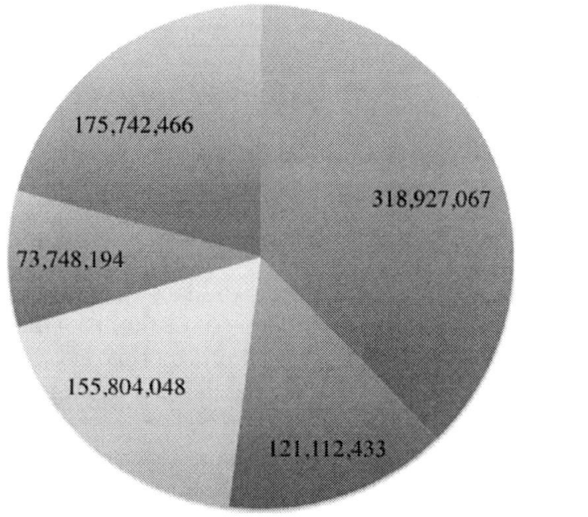

Figure 6 Krabi Province Electricity Demand and Supply (Unit) in 2017, Provincial Electricity Authority (PEA), 2018

The energy demand of Krabi province fluctuates increasing in residential, business and other sectors. Residential energy consumption of around 150 units is the major demand of province and the small, middle business including other sectors and large business respectively. In 2014, Krabi province has officially announced a provincial master plan of 'Krabi Vision 2020: Krabi Goes Green and Green Tourism Plan', and that declaration of Krabi Vision 2020 is the first vision of provincial uptake in the country. In terms of energy systems of the province, it is under the PEA distribution system and also under the Electricity Generating Authority of Thailand (EGAT), which has full authority to manage the national grid system of the country. Renewable energy planning of Krabi is significantly diverse with key tourism destination islands, Ramsar Site Conservation on Krabi Estuary Ramsar site covering main city of Krabi, the biggest seagrass conservation of Thailand and some Environmental Protection Area on land and under the sea which requires a theory of wetland management as human-dominated landscapes threaten or expanded city in complex of provincial development.

3.3 Krabi's Renewable Energy Transition

Since 2004, Krabi province has begun generating electricity from palm wastes. According to the status of renewable energy supply indicates that the most significant sources comes from biomass, biogas and solar energy [36]. Biomass investors are integrating the sharing of internal and external stakeholders, working closely with palm planting business as a majority of provincial agriculture production. Biomass and biogas have the installation capacity approximately 53 megawatts that generate from internal palm and relevant bio-energy fuel sources. And the latest market share is solar farm producing about 18 megawatts however; the limitation of land use becomes a factor of its development. Solar rooftop is expected to be further invested depending on renewable energy policy and the potential of electricity generation. In 2018, solar energy has been projected and official commercial operation date to supply the grid with the following increasing in the context of sub-district cooperative and governmental financial support. Wind energy and mini-hydro energy is expected to be studied as a possible area for investment as well. The installation capacity of wind energy plans to further discuss small onshore wind with speed less than 6 metres per a second under the island geography. In 2010, Prince of Songkla University had a study of the installation capacity of mini-hydro approximately 100-1,5000 kW across

IOP Conf. Series: Earth and Environmental Science **385** (2019) 012060 doi:10.1088/1755-1315/385/1/012060

the province [37] and those additional renewable energy plans to be developed for further energy supply on grid.

Table 1: Krabi's Renewable Energy Status, Provincial Electricity Authority, Thailand, 2019

Company	**Plant**	**Fuel**	**Installation Capacity MW**	**Actual Grid Supply MW**
Wanna Chaideker	Photovoltaic	Solar	0.007	0.007
Thai Green Co.,Ltd.	Photovoltaic	Solar	5	5
Kaset Klongya Cooperative	Photovoltaic	Solar	4.16	4.16
Smart Solar Power Co.,Ltd.	Photovoltaic	Solar	4.64	4.64
Mar Solar Co.,Ltd.	Photovoltaic	Solar	5	5
Solar PV Total			18.81(26.1%)	18.81(31.1%)
Krabi Waste to Energy Co.,Ltd.	Gas engine	Biogas	1.616	1.5
Clean Power Associate Co.,Ltd.	Gas engine	Biogas	1.021	0.99
Sarab Biogas Energy Co.,Ltd.	Gas engine	Biogas	2	2
Thai Integrate Palm Oil Co.Ltd.	Gas engine	Biogas	0.95	0.95
Thai-Indo Palm Oil Factory Co.,Ltd.	Gas engine	Biogas	3.2	1
Namhong Power Co.,Ltd.	Gas engine	Biogas	3.189	3
Modern Green Power Co.,Ltd.	Gas engine	Biogas	3.189	3
Univanit Palm Oil Co.,Ltd.	Gas engine	Biogas	2.856	2.856
Sri Chareoun Palm Oil Co.,Ltd.	Gas engine	Biogas	3.189	3.093
Asean Palm Oil Co.,Ltd.	Gas engine	Biogas	1.65	1
Univanit Palm Oil Co.,Ltd.	Gas engine	Biogas	5.516	2.856
Multi-industry Palm Oil Co.,Ltd.	Gas engine	Biogas	2	1.9
Biogas Total			30.38(42.2%)	24.14(39.9%)
Multi-industry Palm Oil Co.,Ltd.	Steam turbine	Biomass	4.05	1

Sarab Energy Co.,Ltd.	Steam turbine	Biomass	9.5	8.5
Thai Sri Tong Co.,Ltd.	Steam turbine	Biomass	9.24	8
Biomass Total			22.79(31.7%)	17.5(28.9%)
Grand Total			71.98(100%)	60.45(100%)

4. Discussions

Thailand power development is centralization system, majority with fossil fuel or natural gas combined cycle being the main drivers. Meanwhile, domestic natural gas supply from the Gulf of Thailand is becoming insufficient which necessitates the need to import LNG (Liquefied Natural Gas) instead. Upon policy reviews, the proportion of imported electricity is also increasing; with the current trend electricity supply in Thailand will become more import dependent. In terms of renewable energy investment under the new version of Power Development Plan, solar rooftop seems to be significantly increasing. The policy to promote more solar rooftops is challenging and would be an opportunity for a transition towards "Krabi Goes Green" or "Krabi's Sustainable Energy City".

In general, a country plans to diversify sources of electricity generation both from fossil fuels and renewable energy in order to fulfill sustainable energy supplies while maintaining security, economic (electricity cost), environmental and social acceptance. However, which sources of electricity generation are selected depend on policy and potential supply of each province. Moreover, renewable energy purchasing is a challenge in regard to its energy security on the grid. Krabi Vision 2020 collaborates with decentralisation systems of renewable energy, renewable energy policy, disruptive technology and the potential of renewable energy capacity, which present challenges and opportunities for transition towards Krabi's Sustainable Energy City. The growth of Krabi province as a famous tourism destination with an abundance of biodiversity is pushing the local administration and policy-makers to create a secure energy supply, properly manage the environment, manage economic distribution, and mitigate climate change impacts. In total with those perspectives, renewable energy management and smart planning is necessary to reach those indicators of green energy development and green city implementation. The pathway of renewable energy transition to increase the renewable energy on grid has also required some relevant key factors to be considered as following.

4.1 Infrastructure

The potential of energy sources is based on various areas across sub-district territory. Cost of fuel transportation is challenging for energy management in the long term. The limitation of power plant investment is directly connected with the location of fuel sources. Transition towards energy self-reliant systems requires management of infrastructure to supply energy resources for electricity generation. City planning is a key part of this process that can help resolve conflicts of energy investment, and land grabbing as the transition system expands. The potential of Krabi's energy transition is from local power plant development and decentralization of renewable energy investment, differs at each location. Wind energy is expected to utilize across the coastline, in the areas of island or high potential of high winds. In terms of solar investment, the location is more suitable for solar rooftop than solar farm due to land restriction; hence, communities are the targets of solar household producers to feed electricity on the grid. Biomass and biogas are the fundamental energy investment in the province, as Krabi is one of the top palm planting areas in the southern region, which provides electricity generation supply to provincial consumers. The settlement of biomass and biogas plants nearby fuel source supplies assists with their own scale management. And the latest electricity supply plans to invest from municipal solid waste, covering garbage consumption in province. The project is under the responsibility of local governor and business investment.

4.2 Ownership

Energy decentralization means the sharing of electricity management as power dynamics involving local communities, investors and policymakers to make a decision on that transition. The benefits of energy transition would transfer directly from the majority of investors and government to communities' stakeholder. Various models of energy development as sustainable transition are useful to key beneficiaries, compared to the centralization system. The ownership begins from the level of public participation including national policy, provincial policy and investment patterns. Resource owners, power plant holders, public participation, power of relevant policies, implementation and monitoring are the dimension of the ownership word.

4.3 Land use

The local economy is important to take into account when the planning the energy future. The pattern of local development under the umbrella of Vision Krabi 2020 identifies the key provincial income dependent on tourism, agriculture and fishery. The interconnection of the economy and energy transition will dictate the direction of provincial development planning. The balance of limited resources, conflicts and how to be sustainable is under consideration with short and long term transition planning. Land grabbing has some lesson learn from other cities and would impact natural resources, as well as social and economic aspects. The transition from dependent energy city to further self-sufficiency needs to take into account the positive and negative impacts when changing land to produce electricity. Renewable energy transition pathways should eliminate biodiversity and ecosystem exploitation to achieve a green economy and assure fair development [38] .

4.4. Job creation

Renewable energy creates the potential of job and economic growth [39].Renewable energy job creation is relevant to sustainable development goals, directly on goal 8 decent work and economic growth. Referring to the IRENA report [40], Thailand employed some 102,600 people, with most of the jobs in feedstock supply. In 2017, the data from the Office of Agricultural Economics of Thailand (OAE) indicated that a total of 5,826,000 farmers' households planted four key crops including rice, rubber, palm oil and sugar cane that yielded raw materials for biomass energy with a combined value of 688,580 million baht. In 2019, according to the Energy Policy and Planning Office, it is shown that biomass energy production would add up to 41,813 million baht in value, and has the potential to increase the added value of agricultural products at the rate of 6.1% which is equivalent to adding an income of 7,176 baht per household for 5.8 million households.

5. Conclusion

Krabi province is being significantly transformed as the first province of Thailand to exploit local energy resources supplying energy generating systems. Renewable Energy Transition towards Krabi's Sustainable Energy City can be a model to benefit a the provincial economy, distribute its benefits fairly and secure local industries of tourism, fishery and agriculture. As present the renewable energy supply is over 60 percent and continual research and development to reach more capacities including the expansion the vision of energy seller across provincial borders.

This paper illustrates the phenomenon of a renewable energy transition in Krabi, its relation to the provincial planning pattern, provincial energy system, energy investment and further sustainable energy transition. The geography of a city as island and land parts requires the complex transition of land use, infrastructure and ownership. And the sustainability of policy, planning and public participation adapts to be practiced in province when the citizen of a city understand and accept its energy transition.

Acknowledgements
This study is partly funded by Doctorate Scholarship Program " The 100th Anniversary Chulalongkorn University Fund for Doctoral Scholarship", in Environment, Development and Sustainability international program, Graduate School of Chulalongkorn University. All remaining errors are the authors.

References
[1] Nurtaj V, Elchin S, Cihan B, Ceyhun M, 2017. *Transition to renewable energy and sustainable energy development in Azerbaijan*. Renewable and Sustainable Energy Reviews 80 pp 1153-1161
[2] Feng L, Xusheng L, Dan H, Rusong W, Wenrui Y, Dong L, Dan Z, 2009. *Measurement indicators and an evaluation approach for assessing urban sustainable development: A case study for china's Jining City*. Landscape and Urban Planning 90 pp 134-142
[3] Osamah A, Nathalie S, Ayman A, 2018. *Integrating sustainable energy strategy with the second development plan of Kuwait*. Renewable and Sustainable Energy Reviews 82 pp 3430-3440
[4] The Intergovernmental Panel on Climate Change IPCC, 2011. *Summary for policymakers in: IPCC special report on renewable*. Chapter 5 pp 459-468
[5] The ASEAN Secretariat, 2011. *Joint Ministerial Statement of the 29th ASEAN Ministers on EnergyMeeting (AMEM)*, Jerudong, Brunei Darussalam
[6] Michael G, 2018. *Historicizing transitions: The value of historical theory to energy transition research*. Energy Research & Social Science 38 pp 193-198
[7] Philip A, 2016. *Applying institutional theory to the low-carbon energy transition*. Energy Research & Social Science 13 pp 2016-225
[8] Edgar R, Martí R, Alevgul H S, 2017. *A societal metabolism approach to job creation and renewable energy transitions in Catalonia*. Energy Policy 108 pp 551-564
[9] Aleh C, Vadim V Jessica J, Elina B, Benjamin S, 2018. *Integrating techno-economic, socio-technical and political perspectives on national energy transitions: a meta-theoretical framework*. Energy Research & Social Science 37 pp 175-190
[10] Fiona R M, 2018. *Renewable energy and transition-periphery dynamics in Scotland*. Environmental Innovation and Societal Transitions pp 1-9
[11] Jorge B, Rolando F B, Carlo A B, Nora N, 2018. *The renewable energy policy Paradox*.Renewable and Sustainable Energy Reviews 82 pp 1-5
[12] Linda P, Marco R, Andrea T, Silvano C, 2018. *An Agent-based Stock-flow Consistent Model of the Sustainable Transition in the Energy Sector*. Ecological Economics 145 pp 274-300
[13] Xiao-Chen Y, Yan-Jun L, Bing W, Qiao-Hong L, Qing W, 2018. *China's energy transition strategy at the city level: The role of renewable energy*. Journal of Cleaner Production 205 pp 243-246
[14] Ping H and Vanesa C B, 2018. *Interdependence between urban processes and energy Transitions: The dimensions of urban energy transitions (DUET) framework*. Environmental Innovation and Societal Transitions 28 pp 35-45
[15] George G and Megan A F, 2018. *Just transition management: Balancing just outcomes with just processes in Australian renewable energy transitions*. Applied Energy 225 pp 110-123
[16] Jens M and Laurence L D, 2019. *Reimagining energy futures: Contributions from community sustainable energy transitions in Thailand and the Philippines*. Energy Research & Social Science 49 pp 91-102
[17] Conor H and Jeff P, 2018. *Geographies of renewable energy transition in the Caribbean: Reshaping the island energy metabolism*. Energy Research & Social Science 36 pp 165-174

[18] Dirk O and Sven S, 2018. *Spatial transition analysis: Spatially explicit and evidence-based targets for sustainable energy transition at the local and regional scale*. Landscape and Urban Planning 169 pp 1-11
[19] Matthew J B and Jennie C S, 2018. *Political power and renewable energy futures: A critical review*. Energy Research & Social Science 35 pp 78-93
[20] Nancy O, 2006. *Environmental sustainability for urban areas: The role of natural capital indicators*. pp 184-194
[21] Georges A T, Juste R, Jean-Francois L, Paul L, 2010. *Measuring the sustainability of cities: An analysis of the use of local indicators*. Ecological Indicators 10 pp 407-417
[22] Kenneth B, 2012. *Special Section: Economics of urban sustainability: City management and urban environmental indicators*. pp 217-232
[23] Ricardo S M, Cristiano H de O F, Francisco G M F, 2018. *Sustainable and renewable energy supply chain: A system dynamics overview*. Renewable and Sustainable Energy Reviews 82 pp 247-259
[24] Meg H, 2006. *Urban indicators and the integrative ideals of cities*. Urban indicators and the integrative ideals of cities pp 170-182
[25] Enayat A M and Shirin M, 2018. A *participatory exploratory modelling approach for long-term planning in energy transitions*. Energy Research & Social Science 35 pp 205-216
[26] Epaminondas B, 2018. *Sustainable energy development: How can the tension between energy security and energy transition be measured and managed in South Africa?* Journal of Cleaner Production pp 738-753
[27] Jan-Justus A and Charlotte B, Julia T, 2018. *Overcoming energy injustice? Bulgaria's renewable energy transition in times of crisis*. Energy Research & Social Science 42 pp 44-52
[28] Jeffrey B K and Hironobu U, 2018. An institutional analysis of the Japanese energy transition. Environmental Innovation and Societal Transitions 29 pp 126-143
[29] Energy Policy and Planning Office of Ministry of Energy, 2018. *The new version of Thailand Power Development Plan*.
[30] Department of Alternative Energy Development and Efficiency (DEDE), Ministry of Energy, 2013. *The First Draft of Thailand Renewable Energy Law.*
[31] Department of Alternative Energy Development and Efficiency (DEDE), Ministry of Energy, 2012. *Thailand Alternative Energy Development Plan 2012-2022.*
[32] The National Reform Council, 2014. *Thailand Solar Rooftop Expansion Scheme.*
[33] Amornrat L, 2011. *Thailand PV Status Report by Solar Club Thailand, Alternative Energy Development Plan 2012-2022*. Department of Alternative Energy Development and Efficiency (DEDE), Ministry of Energy. pp 21-23
[34] Energy Policy and Planning Office of Ministry of Energy, 2016. *ASEAN Economic Community Policy*. available here http://www.eppo.go.th/images/Power/pdf/AEC_plan.pdf
[35] The Provincial of Krabi, 2014. *Vision Krabi 2020: Krabi Goes Green report*. available here http://www.krabi.go.th/krabi2015/m_vision2020.php
[36] Provincial Electricity Authority, Thailand, 2019. *VSPP-SPP-IPP Status of Krabi Province.*
[37] Payom R, 2010. *Southern renewable energy integration paper*.Water Resource Management and Technology Research Team, Department of Civil Engineering, Faculty of Engineering,Prince of Songkla University.
[38] Alexandros G, Christopher N.H. D, Miguel E, Abubakari A, Tabitha A O, 2017. *Renewable energy and biodiversity: Implications for transitioning to a Green Economy*. Renewable and Sustainable Energy Reviews 70 pp 161-184

[39] Tatyana B and Frederic R, 2018. *Job creation and economic impact of renewable energy in the Netherlands*. Renewable Energy 119 pp 528-538
[40] Renewable Energy and Jobs Annual Review, 2018. The International Renewable Energy Agency (IRENA)

IOP Conf. Series: Earth and Environmental Science **385** (2019) 012061 doi:10.1088/1755-1315/385/1/012061

Barriers of Life Cycle Costing on Construction Consultant Practice in Malaysia

Zulkhairy Affandy Mohd Zaki[1], Abdul 'Izz Mohamad Kamil[1], Mohamad Tajudin Saidin[1], Noor Akmal Adillah Ismail[1] and Nurul Afida Isnaini Janipha[1]

[1]Faculty of Architrecture, Planning & Surveying, Universiti Teknologi Mara Selangor, 40450, Shah Alam, Malaysia

Email: zulkhairy86@uitm.edu.my

Abstract. Life Cycle Costing (LCC) is deliberately known as a vital tool of construction process to optimise the whole life cost from the cradle to grave. LCC also considered as an important factor for a project to meet the equilibrium point of its strategic objectives. In order to implement the LCC, the procedures must undergo certain framework that enables comparative cost assessments to be calculated over specific duration and take account to relevant economic factors of initial, maintenance, operational, salvage value, discount factors, refurbishment and disposal costs. The aim of this paper is to appraise the benefits and barriers of Life Cycle Costing (LCC) practised in the construction industry of Malaysia among the consultant team. The objectives of this paper to explore the benefits and barriers of LCC faced by consultant team in construction projects. To comply with research objectives, a questionnaire survey and interview conducted among the consultant team in Malaysia. 57 participants responded for the survey conducted in this research. From the results, it appears that the implementation of LCC among the professional design team is still limited. Most of the consultant teams had a good knowledge on the processes of analysing LCC, the parameters to be included but still at minimal phases of implementing overall LCC procedures. The lack of reliable data of LCC founded as the main barriers in the Malaysian construction industry. Most of the respondents agreed that the benefit of LCC is the best way to consider the impact of all costs rather than only initial capital costs.

1. Introduction

Life Cycle Cost (LCC) is one of the tool to assure sustainable development to be practiced. LCC is best defined as model that estimates the initial, capital, financing, construction, operation, maintenance, renovation, refurbishment, demolition, and salvage costs over the life span of a building or facility from cradle to grave [6].The analysis of LCC can be used to distinguish the available alternatives that can offer best value for money [5].

2. Literature Review

Construction industry is associated with risks; certainty and uncertainty of cost, time and quality. The construction stakeholders demand for value for money of the product whilst prioritising the sustainability features for the next generations [11]. LCC is not only emphasised on initial capital cost but other costs relatively incurred right after the possession of land [7].

LCC is the commonest economic evaluation tool to assess whole life cost of an asset or parts at the same time achieved the performance criteria. LCC able to visualise the future costs and demonstrate the transparency, accountability, proper planning, cost benefits analyses, and contractual requirement purpose [8]

2.1. Benefits of LCC

The benefits that can be gained from adopting LCC will depend on the objective of the project, LCC exercise and circumstances [10]. The benefits of LCC are:

a. Accountability and transparency of future operational & maintenance costs (including annual sinking fund).
b. Enhanced awareness among the stakeholders of whole life costs.
c. Able to optimise the future cost at early stage (design stages).
d. Delivering value for money in the project.
e. Promoting sustainable development by prioritising performance trade off against cost.
f. Evaluation of best options for whole life of assets or parts.

2.2. Barriers of LCC

The benefits of LCC are obstructed by the list of barriers that exist either externally or internally of the construction projects. The barriers are [12]:

a. Reliability of Capital (CAPEX) and Operational Expenditure (OPEX) data.
b. Lack of guidelines and standard method of LCC.
c. Intangible factors (such as political influence, environment factors and others).
d. The future market conditions – inflation, taxation, interest, future costs.
e. Gaps of knowledge among the owner/investor and consultant team.
f. LCC exercise is 0.2-0.5% expensive.
g. Failure of the design team to visualise the whole life cost.
h. Risk and uncertainty of LCC outcomes.

3. Research Methodology

The method used in obtaining the data are by printed and electronic materials such as journals, books to expedite the literature review. In addition, questionnaire survey distributed to targeted sample of population (consultant team) to aid the empirical study.

3.1. Research Population and Sampling

The professional design practice are the targeted population of this research which includes various background. The population of 200 respondents were identified through companies' email registered with Board of Quantity Surveyor Malaysia, Malaysia Institute of Architects, Malaysia Institute of Planners, and Board of Engineers Malaysia. The distribution to consultants is based on their participation and duty in preparing LCC. The most significant roles in LCC calculation are Quantity Surveyor, Facilities Manager/Building Surveyor and Real Estate Valuer. The ratio of distribution for mentioned consultant team are higher compare to architects, town planner, and others.

3.2 Questionnaire

The survey designed parallel with the objectives of the paper. The first part evaluates the respondents' professional practice, size and year of establishment of the organisation. A choice of multiple answers had been pre-determined to assist the respondents. The second part comprise of questions with five likert-scale answer from strongly agree to strongly disagree about the benefits and barriers of implementing LCC in Malaysian construction industry.

3.3. Method of Data Collection

This paper is focusing on the benefits and barriers of LCC among the consultant team in Malaysia. There are two method of distributing the questionnaire which encompass of face to face survey and online survey. Online survey are done by uploading the questionnaire forms via Google forms.

3.4. Data Analysis

The data collected were analysed by using Statistical Package for Social Science (SPSS). The set of data were analysed in statistical approached which deliver a proper data handling.

4. Findings

4.1. Benefits of Life Cycle Costing

In this section, the data obtained indicates the awareness among the consultant team on the benefit of LCC. The aim of this section in line with the objective of this paper. According to the survey, the mean of top three LCC benefits indicated the same mean (4.42 - agree). However, the ranking was determine by the standard deviation. Most of the respondents agreed that the benefit of LCC is it considered all impacts of all costs rather than only initial capital cost. In the nutshell, it indicates that majority of the consultant team have a decent fundamental knowledge of LCC. Subsequently, the respondents agreed that LCC able to assist in the effective management of assets. Thirdly, they agreed that LCC able to assist in producing earlier actions to generate revenue or lowering the overall cost of an asset.

Generally, all of the LCC's benefits set in the survey achieved standard deviations less than one, which express significant agreement among the respondents. In addition, all of the benefits of LCC obtained the mean more than the scale of four, which represent agree. It proof that most of the respondents in this survey are aware the benefit adopting LCC in their project.

Table 1. Benefits of Life Cycle Costing

Benefits of LCC	Mean	Std. Deviation	Rank
Consider Impact of All Costs	4.42	0.653	1
Effective Management	4.42	0.731	2
Generate Revenue or Lower Costs	4.42	0.801	3
Investment Options Effectively Evaluated	4.37	0.723	4
Accurate and Realistic Assessment	4.35	0.744	5
Framework for Incremental Costs	4.26	0.720	6

4.2. Barriers of Life Cycle Costing Implementation in Malaysian Construction Industry.

Majority of the respondents strongly agree that the utmost barrier is lack of reliable data. Based on the survey conducted, the problem faced by construction player is short of data especially Operational Expenditure (OPEX). However, public and private sectors can obtain the data of Capital Expenditure (CAPEX) from historical database. The data from manufacturer for building components or parts such as Original Equipment Manufacturer (OEM) is the most difficult to obtain. The main barrier of LCC implementation is lack of existing reliable data (mean 4.56), it is strongly supported by research [6]. Next, lack of knowledge on LCC by the client ranked second (mean 4.23), followed by lack of experience by the design consultant team (mean 4.19), lack of interest from the client (mean 4.00) and lack of standard method (mean 3.56). It is contradict with the agreement on lack of standard method, where Public Work Department has published a Guideline of LCC in 2012. It is aim to enhance the awareness among construction player to exercise LCC in their project. The guideline is a good reference to adopt LCC practice. Government also has put initiative by setting a regulation for any public project, which involved with value management approach and worth valued more than RM50 million are mandatorily prerequisite to adopt LCC.

Table 2. Barriers of Life Cycle Costing

Barriers of LCC	Mean	Std. Deviation	Rank
Lack of Reliable Data	4.56	0.780	1
Lack Of Knowledge by the Client	4.23	0.802	2
Lack of Experience by the Professional Team	4.19	0.766	3
Lack of Interest From Client	4.00	0.824	4
Lack of Standard Method	3.56	0.926	5
Intangible Factor	3.12	1.226	6
Lack Of Encouragement By Government	3.05	1.342	7
Risk And Uncertainty of LCC Outcomes	2.96	1.281	8
Unstable Market Condition	2.82	1.241	9

5. Conclusion

It can be presumed that execution of LCC particularly in Malaysia Construction industry as new approach and not widely adopted. Taking into general thought, it is discovered that the practice of LCC among the construction professionals is still limited. The loopholes of implementing LCC are gaps of knowledge among the consultant team [2]. Barriers that exist such as reliability of data, and lack of experience of the consultant team hindered the LCC to be widely adopted in construction project.

6. Recommendations

Government should play the role and put fundamental interest to adopt LCC procedures not only focusing Private Finance Initiative (PFI) but also on others building for government utilisation [1]. It includes for new project or existing building because operational and maintenance no longer play a minor role in life span of a buildings. Therefore, government influential is important to develop LCC assessment. As mentioned in earlier chapter, small improvement or deprivation within the operational phase will give a huge economic impact.

Despite from that, government agencies such Construction Industry Development Board also need to sustenance the effort by setting certain condition, codes, regulation and obligation to the private sectors developers for adopting LCC. Introduction of specific codes for energy usage, which related to LCC should be formulated to encourage the development of LCC in the industry.

Furthermore, government should encourage the parties involved in construction about the awareness of adopting knowledge management in their business activities. It is important for all the stakeholders to access to specific database system that provide reliable data on LCC. The existence data for LCC are not standardised, it is differ from other practice such as quantity surveyor firms in Malaysia able to access the schedule of rates provided by Public Work Department at Rates Online (RATOL) database system.

Private sector clients are categorised as the sponsoring organisation, initiator who has direct intention and responsible to development of a project. Clients have certain factors for embracing long terms economic models into various stages in construction project. Educating the clients on the procedures of LCC is important. Some of the client refused to adopt LCC approach due to incurring cost of managerial aspects. However, the long terms return and savings could be achieved by optimising all associated post-handover cost. Seminar and conferences are some of the recommended mode of educating to encourage the LCC adoption among the investors.

Private sector clients also can assist government towards developing LCC capital and operational expenditures database. The initiatives of developing LCC database in European countries, such as

Sweden are undertake by the non-government organisation. The industry cannot totally rely on the government's efforts.

Professional design team should have a strong fundamental knowledge on LCC procedures (4). Professional design team are responsible to consult the clients and give their professional judgement to optimise the overall cost of the projects. Professional bodies such as Institute of Surveyor Malaysia, Malaysian Institute of Architects, Boards of Engineers Malaysia and other bodies related to construction industry are recommended to encourage their members on the importance of LCC implementation. Besides that, educational efforts should be conducted to enhance the understanding and knowledge on LCC.

Acknowledgement
At the outset, I would like to appreciate with sincerest and deepest gratitude to Allah. I also would like to offer my gratitude to my colleagues for their assistances and guidance. Millions thanks to the respondents that participated in survey. Last, but not least, special thanks are dedicated to my family members and friends for their supports.

References

[1] Abdul Rashid, K 2009 Understanding Private Finance Initiative (PFI), *Selangor Scholarmind Publishing*. 45-91
[2] Arja M, Sauce G, and Souyri B 2009 *External Uncertainty Factors and LCC: A Case Study, Building Research & Information*, **3(37),** 325-334
[4] Azlan Shah A, Syahrul Nizam K, Raha S, and Peng Y.C 2010 *Factors Affecting Housing Maintenance Cost in Malaysia. Journals of Facilities Management*, **Vol 8, No.4,**285-298
[5] Boussabaine A and Kirkham R 2006 *Whole Life-Cycle Costing: Risk and Risk Responses*. John Wiley & Sons.
[6] Davis Langdon Management Consulting 2006 *Literature Review of Life Cycle Costing (LCC)and Life Cycle Assessment (LCA)*
[7] Flanagan R and Jewell C 2005 *Whole Life Appraisal*, Oxford: Blackwell Publishing Ltd.
[8] Langdon D 2010 *Development of A Promotional Campaign for Life Cycle Costing inConstruction Final Report*. United Kingdom: Davis Langdon
[9] Oduyemi O, Okoroh M and Dean A 2014 *Barriers To Life Cycle Costing Usage*. 384-389
[10] Rum N A and Akasah Z A 2012 *Implementing Life Cycle Costing In Malaysian Construction Industry: A Review*, **Volume 6, No. 9 (Serial No. 58),** Journal of Civil Engineering and Architecture, 1202–1209
[11] Swaffield L M and McDonald A M 2008 *The Contractor's Use Of Life Cycle Costing On PFI Projects Engineering. Construction and Architectural Management*, **15,2**, 132-148
[12] Wan Hassan W N H, Zakaria N and Ismail M A 2014 *The Challenges of Life Cycle Costing Application of Intelligent Building in Malaysia Construction Industry*. Journal Design + Built. **Vol 7**

IOP Conf. Series: Earth and Environmental Science **385** (2019) 012062 doi:10.1088/1755-1315/385/1/012062

Consensus Views on the Success Elements for Extension of Time Claim (EoT) in the Malaysian Construction Industry: A Modified Delphi Study

Norazian Mohamad Yusuwan[1], Hamimah Adnan[1], Zul Zakiyuddin Ahmad Rashid[1], Wan Norizan Wan Ismail[1], Noor Aisyah Asyikin Mahat[2]

[1]Faculty of Architecture, Planning and Surveying, Universiti Teknologi MARA Shah Alam, Selangor, Malaysia

[2]School of Housing, Building and Planning, Universiti Sains Malaysia

Email: azianyusuwan@gmail.com

Abstract. Producing a good contractual claim is never an easy task. A good claim not only requires fact and evidences but also demands other important elements that complement each other. This paper reports on a Delphi study focusing on identifying the success elements for Extension of Time (EoT) claim specifically in the Malaysian construction industry. Two rounds of a modified Delphi surveys were conducted and consensus was obtained from twelve experts. The Delphi study has successfully elicit experts' consensus on the elements that perceived to be the most important element in producing a quality EoT claim that may lead to a speedy, successful and amicable settlement of such claim.

1. The Extension of Time (EoT) Claim

Assembling a claim is never an easy task. It should be logically constructed, well-organised and factually convincing to prove that the claimant is innocent while the other party is at fault [1]. The entitlement to extension of time is not simply a matter of producing a list of the events that gave rise to the delays in the construction projects; rather, the contractor must prove how the listed events caused delays and how they impacted the critical path, which then modified the overall duration of the project [2]. The EoT claim can succeed if the contractor can establish causation, liability and damage to each risk event [3]. On a similar note, Abdul-Malak, El-Saadi [4] suggested that to enhance the chances of success in claims, contractors must closely follow the steps stipulated in the contract, provide a breakdown of the alleged additional time, and present sufficient documents. Chappell [5], on the other hand, emphasized that the employer is only bound to meet claims that are based on some express or implied provision of the contract or rule of law, and it is for the contractor to prove the claim. He went on to add that to ensure that his claim will be entertained, the contractor must be able to demonstrate that he has followed the administrative machinery provided in the contract, whereby failure to do so will usually negate his claim.

It is almost impossible to have a construction projects that are free from any contractual claim even if it is successfully executed. The claim document is a medium that provides an opportunity for the claimants to defend and fight for their rights. It is therefore should be readily communicated, understood and remembered [6]. Most importantly, the theme should be the strongest argument in support of the claimant's theory of recovery, duly supported by relevant facts and evidence. In order for the claim to be success or at least to reduce the likelihood of failure and rejection of such a claim, Hewitt [7] laid out the elements that underlie a successful claim, using the abbreviation CEES; C for Cause; E for Effect; E for Entitlement; and S for Substantiation.

It is utmost important that any claim made should have a cause; the event which has given rise to the claim in which one party to the contract has suffered damages for which it should be compensated by the other party [1][7]. Another vital part of the EoT claim is the demonstration that the delay events had an effect on the progress of the entire project. However, demonstrating the impact of delays on the schedule of a project is perhaps the most challenging task to be performed, especially when there is more than one delaying event [7]-[11]. In term of entitlement, it is a golden rule that the time of completion can only be extended if it is permitted by the contract. Therefore, it is vital for the contractor to check for the entitlement to an extension of time as it may vary depending on the type of contract used. Eligibility under the contract is pointless without compliance. To preserve the contractor's right to claim, Hewitt [7] emphasised that any conditions precedent to entitlement to an extension of time must be fully complied with, or in the case of non-compliance, a justification has to be made as to why the conditions should not affect the contractor's entitlement.

In substantiating the EoT claim, it is inherently dangerous to allow facts or assertions to speak for themselves, as the risk is that they will be interpreted in a way that is not intended [12]. Hence, the cause, the effect and the entitlement to an EoT claim not only have to be established and demonstrated, but most importantly, it has to be substantiated. Hewitt [7] referred to substantiation as "proving to a reasonable level that all statements made, points relied on, calculations submitted, and the like are correct". In order to prove or substantiate the claim, it should contain evidence that shall be true, comprehensive and be of the force of law, and that which the lawyer would use in his trial [7][13].

2.0 Research Method: The Delphi Technique

Since its first application over a half of a century ago, the Delphi technique has gained trust amongst researchers as one of the most powerful techniques in eliciting expert opinions. It has been widely utilised in a diversity of research areas and disciplines such as in information system and technology [14][15], tourism [16], health care, environmental management (Danladi, 2015), business and management as well as engineering and construction [17][18]. Conventionally, this technique has been designed as a group communication process aimed at achieving consensus on a specific issues from a selected panel of experts [19][20]. It involves a systematic technique for obtaining expert consensus on a topic being studied, where the experts are required to participate in a multiple iterations or round of surveys [21]. Feedback received will be analysed, and a new questionnaire will be develop to be distribute to the panel of experts in the next round [22]. Following each round, an anonymous summary of the experts' inputs from the previous round will be provided to the experts. The experts will have the opportunity to compare it with their previous response, whereby they are given the chance to maintain or to revise their response in light of the overall results. If common survey tries to identify "what is", the Delphi technique aims to address "what could" or "should be" [19]. In addition , if survey technique is highly dependence on a sample size, it is not so in the case of Delphi technique [15].

Contrary to other method of data collection, the Delphi technique offers several unique features among which are anonymity, iteration with controlled feedback and statistical group response [22]-[24]. The Delphi survey is managed in a totally anonymous way where the panel of experts remain unknown to one another [22][25][26]. As it maintains the anonymity of the participants, it allows participants to freely express their views without any pressure or influence by the others [27][28]. This is the unique characteristic that the Delphi method has in comparison with other method of data collection involving panel of experts such as focus group. The issue of autonomy where one could have a power to driven others in focus group is eliminated in a Delphi method due to its anonymous procedure [29]. Another distinct characteristic of a Delphi technique is that it is iterative in nature. It involves multiple iterations or "round" of inquiry where in each round, participant is provided with feedback that involves new information and expresses the group collective opinion [19][22]. The iterative nature of this method allows participants to rethink and modify their response with the help of the feedback receive from the rest of the experts [25][30]. The questions in a Delphi survey are formulated in such a way that a quantitative and statistical analysis of the answer can be performed [30]. Views from the panel of experts will be statistically analyse and may utilise several approach to report a group response such as the median complemented with minima, maxima, quartiles, and/or the inter-quartile range; or the mean accompanied with standard deviation; or frequency distributions [22].

Classically, the Delphi technique begins with an open-ended questionnaire intends to solicit as many opinions as possible from the expert based on their knowledge, experience and expertise on subject or content area [22][28][31]. As the first round aims to identify issues to be addressed in a subsequent round, open-ended type of questions is judged to be the most appropriate as it helps to increase the richness of the data [32]. Over the years since its first introduction, the Delphi technique has gone through a variety of modifications. The techniques has been modified not only to suit with the nature and objectives of the research but also to reach to a specific aims such as to shorten the process and to guarantee the involvement of the participants throughout the rounds [27][33]. Modified Delphi technique normally does not rely on the expert panel to provide answer in the first round. Rather, the researcher will use other means to collect the answers and presents them to the panel of expert to begin the Delphi process [28]. As Avella [28] further explained, among of the approaches use to collect the initial answer include an extensive review of the relevant literature, a series of interviews with individuals within or outside the study panel, and presents the results of a survey to an external group to the Delphi panel.

2.1 Designing the Delphi Study

Taking into account the context of the study and all the constraints, a modified Delphi technique is suited for this research. The choice was made based on aim of this research; identifying the success elements for EoT claim specifically in the Malaysian construction industry. The two Delphi rounds were conducted to accomplish this aim. To complete the Delphi survey seven stages were involved, started with the selection of experts, followed by the questionnaire development process, first round survey, analysis of first round responses, second round survey, analysis of second round responses and ends with the final report on the Delphi survey.

In order to obtain the most reliable and valuable opinions, the panel of experts for the Delphi survey were carefully selected. The panel of experts consisted of industry professionals identified via semi-structured interviews conducted with the practitioners, i.e. the professional architects and Grade G7 contractors. The interviewees were asked to nominate and recommend other experienced figures within the industry who, in their opinion, have sufficient experience and expertise in dealing with EoT claims. Following the identification of the experts for the Delphi survey through nomination by the other practitioners, the candidates was then being shortlisted based on pre-defined criteria. Among of the criteria devised for a Delphi participant are; practitioners having extensive working experience in the construction industry in Malaysia; practitioners having current, recent or direct involvement in dealing with EoT claims, particularly in private funding projects; and practitioners having sound knowledge and experience with regard to EoT claims. In addition to the expertise required, practitioners should have the willingness and ability to participate during the scheduled Delphi process. Initially, 40 experts were identified and invited via email to participate in the Delphi survey. Of 40 experts contacted, only 17 agreed to participate.

2.2 Delphi Round One- *Rank & Weight*

As this was a modified Delphi survey, the Round 1 questionnaire contained questions with pre-selected items for the panellists to indicate their response. This round was for the panellists to assess the importance of each element in determining the success of EoT claims obtained from the views expressed by the interviewees during the semi-structured interviews and from an extensive literature review, which was later confirmed by the professionals during the pilot test of the questionnaire. Following extensive reviews of previous researches and the semi-structured interviews conducted with industry practitioners, a list of 23 items were listed as elements that contribute to a successful EoT claim. Hewitt [7] simplified the essential elements for a successful claim by using the acronym of CEES: C = cause; E = effect; E = entitlement; and S = substantiation. In this research, Hewitt's guideline was adapted and modified, with one new category being included. The categories were cause, effect, entitlement, substantiation and a new category, presentation. The cause-element consisted of two (2) items, while the effect-element consisted of three (3) items. On the other hand, the entitlement-element, substantiation-element and presentation-element consisted of four (4), one (1) and thirteen (13) items, respectively.

Of the 17 experts who initially agreed to participate, only fifteen 15 responded to the first round. Two (2) experts dropped out due to heavy workload commitments during the survey. Table 1 illustrates the personal profiles of the expert panel in the first round of the Delphi survey. The 15 experts represented a wide distribution of professional figures in the construction industry, with five (5) from the client's organization, six (6) from the consultant's organization, while four (4) were professionals from the contractor's organization. The composition of this group of experts was expected to provide balanced views for the Delphi survey.

Table 1. Personal Profiles of the Delphi Participants

Expert	Current Position	Level of Education	Years of Experience in Construction Industry	Years of Experience Dealing with EoT Claims
1	Project manager	Master	16-20 years	5-10 years
2	Director	Master	More than 25 years	11-15 years
3	Contract Manager	Master	21-25 years	11-15 years
4	Contract Manager	Degree	More than 25 years	21-25 years
5	Contract Manager	Degree	21-25 years	11-15 years
6	Contract Manager	Degree	21-25 years	11-15 years
7	Quantity Surveyor	Degree	16-20 years	11-15 years
8	Project manager	Degree	16-20 years	11-15 years
9	Contract Manager	Degree	More than 25 years	21-25 years
10	Quantity Surveyor	Master	11-15 years	5-10 years
11	Project manager	Degree	21-25 years	11-15 years
12	Contract Manager	Master	21-25 years	11-15 years
13	Project manager	Degree	More than 25 years	21-25 years
14	Quantity Surveyor	Degree	11-15 years	11-15 years
15	Planning Engineer	Master	5-10 years	5-10 years

2.2 Delphi Round Two- *Re-Evaluating the Ratings*

Upon completion of the first round of the survey, the scores of each item, as rated by the panellists, were then recorded and analysed with the aid the SPSS software version 20.0, in which the frequency, mean and standard deviation of each item were computed and the results were provided to the panellists in the second round of the survey. The items were ranked in descending order based on the mean scores from 5 - 1 (extremely important to not important at all), which indicate the importance of the elements in determining the success of the EoT claim as perceived by the panel of experts. If there is a tie between the mean score, the standard deviation of the items will be used to break the tie. The tied items with the lowest standard deviation scores will be ranked first. As the standard deviation indicates the dispersion of variables around the mean values, a smaller standard deviation indicates more agreement on a particular item [34][35].

Since this was the final round of the Delphi survey, the consensus measurement for this research was established. Although there are a wide variety of ways to measure consensus in previous studies, Derus [36] proposed two alternatives on how consensus can best be measured: (a) if the number of samples (N) $\geq$ 35; standard deviation <1.00 (see Seibert [37]), (b) if N<35; consensus is considered to have been achieved on items that received $\geq$80% of panel agreement on the highest two or three points (see [38-

40]. For this research, an agreement criterion of 80% was established as the minimum level considered for consensus. Those items that attracted below 80% agreement in the two highest scales of importance, i.e. 4 = important, 5 = extremely important, were considered as not important and hence did not require any further consideration. Those items that met the criterion for consensus were concluded to be important as the elements that determine the success of EoT claims. Although there was flexibility on how to establish the percentage of agreement, 80% was selected for this research in order to have a high level of consensus and to add credibility to the results [38].

Finally, after the two rounds of a Delphi survey, out of twenty-three (23) elements, twenty (20) elements received the consensus of the experts as the elements that will influence the success of EoT claims. Two rounds of the modified Delphi survey were considered sufficient for this research due to the following reasons:

a. Standard Deviation: The results from Round 2 showed that the standard deviation for those items that met the experts' consensus were very close indeed to those of the first round (Table 2). As the standard deviation describes the dispersion of variables around the mean value, a closer standard deviation indicates a high degree of concentration around the mean value and hence, shows a high level of agreement between the experts.
b. No additional comments had been added by the experts in Round 1 and Round 2.

Overall, it was felt that a third round of the Delphi survey would not add to the indicators provided by the first two rounds. Thus, the Delphi survey was terminated at Round 2.

Table 2. Round 1 and Round 2 Results for Elements for a Successful EoT Claim

	Item	***1st Iteration***			***2nd Iteration***		
		M	***SD***	***Rank***	***M***	***SD***	***Rank***
	Cause-Elements						
1.	It must be an event or a circumstance that has given rise to the claim (i.e. late possession of site, force majeure, exceptionally inclement weather, etc.).	4.47	0.833	2	4.92	0.289	1
2.	Liability for the event. The responsibility for an event must rest either on the employer, his agents (consultants), or it is a neutral event, which is beyond the control of any parties, i.e. force majeure.	4.80	0.414	1	4.92	0.289	1
	Effect-Elements						
1.	Establish link between cause and effect (Delay analysis should demonstrate the effect of the delay event on the contractor's work programme).	4.60	0.507	1	4.92	0.289	1
2.	The narrative should contain details of the work affected, estimated affected duration, status of the work at the time the event occurs, and the consequences of the events should be well described to demonstrate the cause and effect.	4.33	0.617	3	4.42	0.515	2
3.	In the event of concurrent delay, each delaying event should be analysed individually and collectively.	4.47	0.516	2	4.25	0.622	3
	Entitlement-Elements						
1.	Entitlement in the contract; identify the contract provision under which the claim is being made.	4.67	0.487	1	4.92	0.289	1
2.	The contractor must be able to demonstrate that he has used his best endeavours to mitigate/minimise the delay.	4.07	0.961	3	4.92	0.289	1
3.	Any condition precedents to entitlement should be examined in the claim and the contractor must be able to demonstrate that he has complied with such provisions (e.g. notice requirement).	4.40	0.632	2	4.67	0.492	3

	Item	1st Iteration			2nd Iteration		
		M	*SD*	*Rank*	*M*	*SD*	*Rank*
	Substantiation-Element						
1.	Relevant documents of evidence to support the claim such as correspondence, minutes of meetings, site diaries, weather reports, photographs, etc. should be tagged along for the contract administrator's reference and verification.	4.73	0.457	1	4.92	0.289	1
	Presentation-Elements						
1.	The document of claim must be well and professionally presented to make the assessor's job as easy and as pleasant as possible.	4.53	0.639	1	4.67	0.492	1
2.	Make the submission user-friendly such as by providing a clear arrangement and proper labelling to enable the assessor to locate any references, especially when the narrative contains references to other documents of substantiation.	4.40	0.736	3	4.58	0.669	2
3.	Write the narrative on the assumption that the assessor has no previous knowledge of the project.	4.20	0.861	4	4.42	0.515	3
4.	Lead the assessor to a logical conclusion. The document of claim must be properly arranged in such a way that it has a beginning, middle, and end.	4.40	0.632	2	4.42	0.515	3
5.	Make the submission a standalone document by including everything which the assessor will need to refer to.	4.00	0.925	5	4.33	0.492	5
6.	Use the narrative to explain other documents attached as substantiation or in support of the narrative.	3.93	0.798	6	4.17	0.718	6
7.	Any wording, tittles and the like included in the supporting documents must be consistent with the narrative.	3.93	0.798	6	4.17	0.718	6
8.	Avoid the use of acronyms and abbreviations.	3.47	0.915	13	4.00	0.603	8
9.	Avoid the use of legalese and unnecessarily complicated language.	3.80	0.774	9	4.00	0.603	8
10.	Identify quotations correctly and consistently.	3.80	0.861	10	4.00	0.853	10
11.	Present the reference material and documents used as substantiation in a separate volume to the narrative.	3.60	1.183	12	4.00	0.853	10

3. Conclusion

Findings from the Delphi survey with regard to the elements that contribute to the success of EoT claims revealed that the elements of success were closely related to the sufficiency and completeness of the documents to substantiate claims and also the competency of handling claims. Hence, it can be inferred that records and documentation along with the competency of handling claims demand a great deal of attention if a speedy, successful and amicable settlement of EoT claims is to be attained. The chances of success of the EoT claim submitted can be improved if the professionals involved are able to fully grasp what constitutes a good claim. Undeniably, experience is a good teacher, but experience can not only be gained through one's own experience but also by learning from the experiences of others. A willingness to learn is of the utmost importance as knowledge is expanding by the minute. Knowledge can be enhanced through various ways such as by attending seminars, workshops, conferences, and so on, and these are definitely good ways of building competency. Other than that, in order to be competent in managing claims, it is essential for everyone involved to acquire adequate knowledge not only on technical construction but also to acquire a clear understanding of contractual terms and their implications, as these will influence their behaviour towards contract compliances that might influence the way they handle claims and the chances of success.

Acknowledgement
The authors would like to acknowledge the support of funding from the Ministry of Higher Education of Malaysia under the Fundamental Research Grant Scheme (FRGS) for this research. Special thanks also to the reviewers for their invaluable comments on this paper. Not forgetting that special gratitude is given to those industry practitioners who responded and contributed their valuable input in completing the Delphi survey.

References

[1] Kululanga, G., et al., *Construction contractors' claim process framework.* Journal of Construction Engineering and Management, 2001. **127**(4): p. 309-314.
[2] Chong, H.-Y. and Y.-W. Leong, *Legal approach on assessment of contractors entittlement to extension of time.* African Journal of Business Management, 2012. **6**(14): p. 4815-4823.
[3] Alnaas, K.A.A., A.H.H. Khalil, and G.E. Nassar, *Guideline for preparing comprehensive extension of time (EoT) claim.* HBRC Journal, 2014. **10**(3): p. 308-316.
[4] Abdul-Malak, M.A.U., M.M. El-Saadi, and M.G. Abou-Zeid, *Process model for administrating construction claims.* Journal of Management in Engineering, 2002. **18**(2): p. 84-94.
[5] Chappell, D., *Building contract claims.* 5th. ed. 2011: John Wiley & Sons. 523.
[6] Ansley, R.B., T.J. Kelleher, and A.D. Lehman, *Common sense construction law: A practical guide for the construction professional.* 2nd. ed, ed. R.B. Ansley, T.J. Kelleher, and A.D. Lehman. 2001, USA: John Wiley & Sons. 454.
[7] Hewitt, A., *Construction claims & responses: Effective writing & presentation.* 2011: Wiley-Blackwell. 188.
[8] Arditi, D. and T. Pattanakitchamroon, *Selecting a delay analysis method in resolving construction claims.* International Journal of Project Management, 2006. **24**(2): p. 144-155.
[9] Entrusty, *Is the contractor still entittled to extension of time when there is concurrent delay?*, in *Master Builders*. 2006: Malaysia. p. 101-103.
[10] Ndekugri, I., N. Braimah, and R. Gameson, *Delay analysis within construction contracting organisations.* Journal of Construction Engineering and Management, 2008. **134**(9): p. 692-700.
[11] Scott, S., R.A. Harris, and D. Greenwood, *Assessing the new United Kingdom protocol for dealing with delay and disruption.* Journal of Professional Issues in Engineering Education and Practice, 2004. **130**(1): p. 50-59.
[12] Carnell, N.J., *Causation and delay in construction disputes*. 2nd. ed. 2005: Blackwell Publishing. 302.
[13] Chen, C.B. and D.C. Wang, *Study on the construction process control claims management.* Applied Mechanics and Materials, 2012. **120**: p. 385-388.
[14] El-Gazzar, R., E. Hustad, and D.H. Olsen, *Understanding cloud computingadoption issues: A Delphi study approach.* The Journal of System and Software, 2016. **118**: p. 64-84.
[15] Skinner, R., et al., *The Delphi Method Research Strategy in Studies of Information Systems.* Communications of the Association for Information Systems, 2015. **37**.
[16] Konu, H., *Developing nature-based tourism products with customers by utilising the Delphi method.* Tourism Management Perspectives, 2015. **14**: p. 42-54.
[17] Tymvios, N. and J.A. Gambatese, *Direction for Generating Interest for Design for Construction Worker Safety-A Delphi Study.* Journal of Construction Engineering and Management, 2016. **142**(8).
[18] Perrenoud, A.J., *Delphi Approach to Identifying Best Practices for Succession Planning within Construction Firms.* International Journal of Construction Education and Research, 2018.
[19] Hsu, C.-C. and B.A. Sanford, *Delpi Technique*, in *Encyclopedia of Research Design*, N.J. Salkind, Editor. 2012, SAGE Publications.
[20] Chan, A.P.C., et al., *Application of Delphi method in selection of procurement systems for construction projects.* Construction Management and Economics, 2001. **19**(7): p. 699-718.
[21] Hsu, C.C. and B.A. Sandford, *The Delphi technique: Making sense of consensus.* Practical Assessment, Research & Evaluation, 2007. **12**(10): p. 1-8.

[22] Sourani, A. and M. Sohail, *The Delphi Method: Review and Use in Construction Management Research.* INternational Journal of Construction Education and Research, 2015. **11**(1): p. 54-76.
[23] Dalkey, N.C., *The Delphi method: An experimental study of group opinion*. 1969: California.
[24] Awad, M., et al., *Consensus views on the optimum training curriculum for advanced minimally invasive surgery: A delphi study.* International Journal of Surgery, 2018. **53**: p. 137-142.
[25] Manoliadis, O., I. Tsolas, and A. Nakao, *Sustainable construction and drivers of change in Greece: a Delphi study.* Construction Management & Economics, 2006. **24**: p. 113-120.
[26] Diamond, I.R., et al., *Defining consensus: A systematic review recommends methodologic criteria for reporting of Delphi studies.* Journal of Clinical Epidemology, 2014. **67**: p. 401-409.
[27] Skulmoski, G.J., F.T. Hartman, and J. Krahn, *The Delphi method for graduate research.* Journal of Information Technology Education, 2007. **6**(1): p. 1-21.
[28] Avella, J.R., *Delphi Panels: Research Design, Procedures. Advantages and Challenges.* International Journal of Doctoral Studies, 2016. **11**: p. 305-321.
[29] Sahari, S., et al., *A review of Delphi technique in developing human capital disclosure index.* Academy of Accounting and Financial Studie Journal, 2018. **22**(4): p. 1-9.
[30] Landeta, J. and J. Barrutia, *Peopleconsultation to construct the future: A Delphi application.* International Journal of Forecasting, 2011. **27**: p. 134-151.
[31] Custer, R.L., J.A. Scarcella, and B.R. Stewart, *The modified Delphi technique-A rotational modification.* Journal of Vocational and Technical Education, 1999. **15**(2): p. 50-58.
[32] Powell, C., *The Delphi technique: Myths and realities.* Journal of advanced nursing, 2003. **41**(4): p. 7.
[33] Biggs, S.E., et al., *Safety leaders' perceptions of safety culture in large Australasian construction organisation.* Safety Science, 2013. **52**: p. 3-12.
[34] DeLany, B.W., *Entry-level job skills needed by wildlife management professionals*, in *The School of Human Resource Education and Workforce Development* 2004, Louisiana State University and Agricultural and Mechanical College.
[35] Hurme, F.E., *Competencies for rural nursing practice* in *The School of Human Resource Education and Workforce Development*. 2007 Louisiana State University and Agricultural and Mechanical College.
[36] Derus, M.M., *Penunjuk kompetensi yang dianggap kritikal untuk pengurus projek dalam Jabatan Kerja Raya*, in *Quantity Surveying Department*. 2012, Universiti Sains Malaysia: Pulau Pinang, Malaysia.
[37] Seibert, M.J., *The identification of strategic management counseling competencies essential for the small bussiness and technology development center: A modified delphi strategy*, in *Adult and Community College Education Training and Development Program*. 2004, North Carolina State University.
[38] Thomas, C.K., *The role of the public school superintendent in local economic development*, in *Educational Leadership and Policy Studies*. 2002, Virginia Polytechnic Institute and State University: Virginia.
[39] R.Mirra, D., *The Role of the School Superintendent as a Technology Leader: A Delphi Study*, in *Educational Leadership and Policy Studies*. 2004, Virginia Polytechnic Institute and State University: Virginia.
[40] Davidson, J.W., *The necessary components of a staff development program to prepare teachers to teach secondary online classes: A Delphi study* in *Educational Leadership and Policy Studies*. 2005, Virginia Polytechnic Institute and State University: Virginia

IOP Conf. Series: Earth and Environmental Science **385** (2019) 012063 doi:10.1088/1755-1315/385/1/012063

Benefits of Construction Automation and Robotics in the South African Construction Industry

Ayodeji Oke[1], Opeoluwa Akinradewo[1*], Clinton Aigbavboa[1],Olusola Akinradewo[2]
[1] SARChI in Sustainable Construction Management and Leadership in the Built Environment, University of Johannesburg, South Africa
[2] Department of Quantity Surveying, Federal University of Technology, Akure, Nigeria.

Email: opeakinradewo@gmail.com , caigbavboa@uj.ac.za

Abstract. Construction is known for playing important role to the country's economy and construction research is being conducted across the whole world on improving construction industry from the traditional method of construction to a more suitable advanced construction approach. Automation in construction industry are tools for ensuring production in a business without compromising quality of the construction products, which may result from tiredness of workers. The study aimed at assessing the benefits of construction automation and robotics in the South African Construction Industry. Related literatures were reviewed and the study adopted quantitative analysis in which a well-structured questionnaire was developed to retrieve information from respondents within Gauteng province of South Africa. Respondents for the study were Construction Managers, Quantity Surveyors, Architects and Project Managers. Data retrieved was analysed using Mean Item Score and Standard Deviation. The study revealed and concluded that robotics and construction automation results in increasing quality of the construction project by increasing accuracy of components of the construction project, promote design specifications, reduction in the delivery time of a construction project and also make room for achieving standards.

1. Introduction

Construction is known for playing important role to the country's economy and construction research is being conducted across the whole world on improving construction industry from the traditional method of construction to a more suitable advanced construction approach. Construction automation has been implemented in other parts of the world such Malaysia, Japan, and America due to the positive effects that this advanced technology brings to the society in improving construction. These new methods of construction technology had been found to resolve problems associated with conventional method of construction referred to as manual method construction, such as construction waste, lower productivity, and construction accident to the workers that are caused by falling from heights and negligence by the management [1]. This approach of construction automation is very advantageous in construction management because it simplifies the work of consultants or professionals involved in the construction industry by assisting in planning, designing and assuming the life span of the project to be delivered to the client. This approach can also be suitable to countries with high labour cost, with the use of construction automation which can help to eliminate demand that may lead to strikes because they may affect the delivery of the construction project [2].

Automation in construction industry are tools for ensuring production in a business without compromising quality of the construction products, which may result from tiredness of workers. It becomes easy to estimate quantity of materials or products that can be manufactured daily. High production in construction industry means the construction product may be delivered on time. Construction products manufactured under automation process can be produced at higher speeds [2]. Construction automation and robotics is usually applied in the manufacturing of components such as concrete beams, columns and bricks. Technology involved in the production of construction components are lasers (for ensuring quality and for checking dimensions) concrete mixer, cutting and drilling automated equipment etc. The reasons for using prefabricated components is because they are produced under high supervision and controlled environment to maintain temperatures that may affect the quality of the products. Final work is then carried onsite where minimal amount of work is required [3].

2. Literature Review

The motivation for construction automation and robotics came as a result of successes recorded in civil engineering via bridge, tunnel, earthworks and road construction with robotic applications [4]. Outputs of the construction industry has grown very fast in developed countries due to the shift from construction using traditional procedures to Industrialised Building System (IBS) in which prefabrication is carried out in workshops before assemblage is done on the construction site [5]. The construction industry remains conservative when compared to the manufacturing industry in adoption of automation and robotics. This as reported by [6] was as a result of new automatic products not being complementary to the old ones which therefore, doesn't allow their implementation in the construction industry unlike the manufacturing industry. Robotics and automation can be applied to the construction industry in the following ways.

2.1. Construction Monitoring

Construction requires supervision by the construction manager at all times to ensure quality construction delivery to the client. Automation can be applied to improve safety on site and improved quality. Construction automation used for supervision can monitor during the times when managers and safety officer are not available on site or attending meetings to record information about situation on site. Managers can view information at any time because it can be stored and analyzed at a later stage [7].

2.2. Planning and Estimating

Planning and estimation can be complicated which requires an expert project manager. Application software can be applied during this phase of construction. Information technology brings applications that are able to plan and make accurate estimates of cost. With this software, the project manager has less chance of making wrong estimate because breakdown of activities can be automatically produced from the design which makes estimation simpler since each activity can be focused on [8, 7].

2.3. Construction Manufacturing

Construction involves the use of pre-fabricated components, which are manufactured in controlled environment where 100% work is executed using machinery and computer programs to automate the robotics. Automation in the pre-fabrication phase involves cutting, welding, batching, lifting and pre-fabrication of formwork. Automatic cutting grinders and welding robots are required to improve working condition and ensuring high level of safety because less work will be completed by the workers. Cranes are required to carry section of component which requires more labour to lift because they are heavy. Robotics are required for lifting and loading components for storage or transportation so that they can be delivered to site [3].

2.4. Designing

Over the years, construction involves manual designing using 2-dimesional plan. Advanced software can be implemented in the construction industry especially for the architect and engineer to ease designing. This software allows the designer to have a view of the design in 3-dimension and are quicker means of designing while the designer can analyse all the sides of the structure in three-dimensional view. More sophisticated software is used to design and allow the 4-dimesion view of structures. These applications can work as a simulator which assists designers in checking if components do not overlap one another [8].

2.5. *Quality Control*

Construction is concerned with delivering quality products. Automation is used to improve quality of the construction product using highly advanced inspection sensors and lasers. This technology is used to verify work inspected by worker and are mostly used in the pre-fabrication. Testing can be performed on components to monitor the strength of the product in construction using hydraulics, this hydraulics can be connected to a monitor to display information about the components which allows the operator view information on a screen. Lasers and sensors can be linked with computer system to accept and reject if the product fails the test hence automation is applied at higher levels to avoid the failure of structure and unnecessary cost [6, 8].

3. Research Methodology

Due to the nature of this research work, quantitative research method was adopted. A well-structured questionnaire was designed based on literatures reviewed. First section of the questionnaire retrieved the demographic information of the respondents while the second section was designed to assess the opinion of the respondents on the benefits of construction automation and robotics in the South African Construction Industry. Respondents chosen for this research work are construction industry professionals within Gauteng province of South African construction Industry. These respondents include Construction Managers, Quantity Surveyors, Architects and Project Managers. Purposive sampling technique was adopted for the sampling of professionals in order to ensure that they have adequate knowledge of construction automation and robotics. Mean Item Score (MIS) and Standard Deviation (SD) were employed in analysing retrieved data in order to rank the benefits of construction automation and robotics on the South African construction industry.

4. Findings and Discussions

Findings relating to respondents' working experience as shown in figure 1 revealed that 43% have between 1-5 years of working experience, 32% were between 6-10 years of experience, 20% were 11-15 years of experience, 5% were 16-20 years of working experience and 0% have 20 year and above working experience in the construction industry.

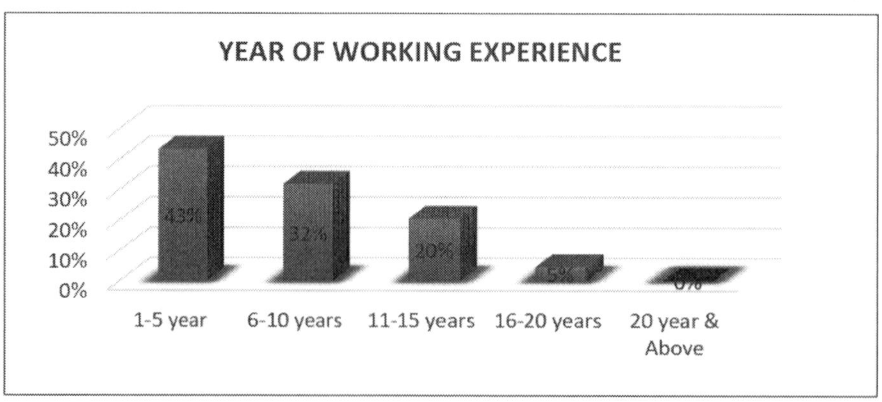

Figure 1. Respondents' years of working experience

Professional qualification of the respondents as shown in figure 2 indicated that 14% were Architect, 30% were quantity surveyor, 20% were construction engineers, 20% were project managers and 16% were construction manager.

Figure 2. Professional qualification of respondents

The findings on the benefits of construction Automation and Robotics on the South African construction industry in table 1 indicated that Increases accuracy of components was rank first with a mean score item of 4.11 and a standard deviation of 0.920, Promotes design specifications was rank second with a mean score item of 4.07 and standard deviation (SD)= 0.873, Increases quality of construction products was rank third with a means score item of 3.98 and standard deviation (SD)=1.045, Reduces the duration of project delivery was rank fourth with a mean score item of 3.89 and standard deviation (SD)= 1.061, Better finished product performance was rank fifth with a mean score item of 3.84 and standard deviation (SD)= 0.914, Standard achieved was rank sixth with a mean score item of 3.84 and standard deviation (SD)=1.010, Cost effectiveness was rank seventh with a mean score item of 3.80 and standard deviation (SD)= 0.978, Eliminates material wastage was rank eighth with a mean score item of 3.77 and standard (SD)= 0.937, Saves Material was rank ninth with a mean score item of 3.77 and standard deviation (SD)= 0.912, Reduces construction accidents was rank tenth with a mean score item of 3.77 and standard deviation (SD)= 1.008, Improves working condition was rank eleventh with a mean score item of 3.77 and standard deviation (SD)= 0.803, Reduces labour cost was rank twelfth with a mean score item of 3.73 and standard deviation (SD)= 0.973, Eliminates injuries was rank thirteenth with mean score item of 3.59 and standard deviation (SD)= 0.996.

Table 1. Benefits of Construction Automation and Robotics

Benefits	MIS	SD	Rank
Increases accuracy of components	4.11	0.920	1
Promotes design specifications	4.07	0.873	2
Increases quality of construction products	3.98	1.045	3
Reduces the duration of project delivery	3.89	1.061	4
Better finished product performance	3.84	0.914	5
Standard achieved	3.84	1.010	6
Cost effectiveness	3.80	0.978	7
Eliminates material wastage	3.77	0.937	8
Saves Material	3.77	0.912	9
Reduces construction accidents	3.77	1.008	10
Improves working condition	3.77	0.803	11
Reduces labour cost	3.73	0.973	12
Eliminates injuries	3.59	0.996	13

Using a 5-point likert scale, the identified benefits were ranked based on their MIS and SD values. All the benefits have a MIS value above 3.00 which is the average score, this shows that respondents are of the opinion that construction automation and robotics has high benefits on the South African construction industry. Most of the benefits has SD value below 1.00 which shows that there is little deviation in the opinion of the respondents concerning each of the benefits ranked. The results were in supports of the research work carried out by [2] whose findings were that Automation helps in labour cost saving, increase productivity, Reduced accidents, cost effective, material saving, and improve working condition, increased accuracy. It is also in agreement with [9] who revealed that automation improve design of the construction product and accurate. The findings were also in tandem with [6] and [8] who asserted that construction automation and robotics helps in quality and reliability, productivity improvement, enhancement of working conditions, safety, savings in labour costs, life cycle cost savings, standardization of components, and simplification of the workforce.

5. Conclusion and Recommendation

The study has succeeded in indicating that robotics and construction automation results in increasing quality of the construction project by increasing accuracy of components of the construction project, promote design specifications, reduction in the delivery time of a construction project and also make room for achieving standards. This study therefore recommends that construction stakeholders and participants should be ready to take financial risks in adopting the use of construction automation and robotics in order for the construction industry to be beneficial of the advantages. For further studies, research can be carried out on the implementation of construction automation and robotics in the South African construction industry.

References

[1] Kim T, Lee UK, Yoo WS, An SH, Cho H, Lee Y, Doh N. Benefit/cost analysis of a robot-based construction automation system. InControl Automation and Systems (ICCAS), 2010 International Conference on 2010 Oct 27 (pp. 616-621). IEEE.

[2] Bock T. Construction automation and robotics. InRobotics and Automation in Construction 2008. InTech.

[3] Pan W, Gibb AG, Dainty AR. Perspectives of UK housebuilders on the use of offsite modern methods of construction. Construction management and Economics. 2007;25(2):183-94.

[4] Vähä P, Heikkilä T, Kilpeläinen P, Järviluoma M, Heikkilä R. Survey on automation of the building construction and building products industry. VTT; 2013.

[5] Oke A, Aigbavboa C, Mabena S. Effects of automation on construction industry performance. In Advances in Engineering Research, Second International Conference on Mechanics, Materials and Structural Engineering (ICMMSE 2017), 2017;102:370-74.

[6] Balaguer C, Abderrahim M. Trends in robotics and automation in construction. InRobotics and Automation in Construction 2008. InTech.

[7] Struková Z, Líška M. Application of automation and robotics in construction work execution. AD ALTA: Journal of Interdisciplinary Research. 2012;2:121-5.

[8] Castro-Lacouture D. Construction automation. InSpringer handbook of automation 2009 (pp. 1063-1078). Springer, Berlin, Heidelberg.

[9] Neelamkavil J. Automation in the prefab and modular construction industry. In26th symposium on construction robotics ISARC 2009;229-306.

IOP Conf. Series: Earth and Environmental Science **385** (2019) 012064 doi:10.1088/1755-1315/385/1/012064

Place-Making a Child-Friendly Urban Environment Through Artworks

T.M.I. Affandi [1,2], **I. Said**[1], **H. Rasidi**[1] **& Z. Za'bar**[2]
[1] Faculty of Built Environment & Surveying, Universiti Teknologi Malaysia, UTM Skudai, 81310 Johor, Malaysia.
[2] Faculty of Architecture, Planning and Surveying, Universiti Teknologi MARA, 40450 Shah Alam, Selangor, Malaysia

Email: irfanworks@gmail.com

Abstract. More recently, concern on children's well-being, health, mobility and upbringing has been increasingly being discussed. The child of today is the city-maker of tomorrow. Rapid urbanization impacts children the most as they have less access to the natural environment in the city. Moreover, in most developing nations including Malaysia, children opinions are not taken into consideration in the development of urban public spaces such as parks, playfields, streets and plazas. Living in an urban environment is not primarily about people relationship but rather the relationship between people, places and spaces. Realizing this, as the Malaysian urban population rapidly increases, the government and local councils have started to rethink and redesign urban spaces to sustain the basic needs and welfare of its citizen. Since neighborhood streets, public spaces, alleys and loose-fit spaces are essential spaces for children's exploration and play, place-making is one of the favorite projects that many have chosen to use to create an urban environment that is child-friendly. According to Project for Public Spaces (2011), place-making is an integrative approach to better plan, design and manage the urban environment. It focuses on the cooperation and contribution of local communities to promote their safety and well-being. This paper attends towards children's artworks impact to create a child-friendly environment in an urban environment. From a review of 50 articles in journals of various discipline which includes Environmental Psychology, Art Education, Urban Sociology, Children Cognitive Functioning and Children Geographies, it is understood that children's participation in place-making an urban environment are influenced to three attributes: (1)artworks as a mode of place-making for a positive emotional sense-of belonging, (2) active participatory of children in creating artworks could cultivate community bond and social skills, and (3) public sphere are spaces for children to exhibit their sense-of-attachment. The review construes that place-making through art can contribute to create a child-friendly environment that is inclusive, healthy, nurturing and sustainable.

1. Introduction

The city-maker of tomorrow is in the living child of today [1]. As Malaysia's urban population grows beyond 71%, it is essential for the people and government to ensure that their living environment and urban spaces are designed and able to sustain the basic needs and well-being of children. In many developing countries including Malaysia, as a result of rapid urbanization, children have limited access to their own natural environment [2]. In addition, adults disregard children's voices as they design the cities public space such as playgrounds, parks, streets and alleys [3]. Adults have a profound idea of what children want and need as they believe that since they were once children. However, children of today are different from before. Children should be given the opportunity to participate in meaningful activities that shape their environment. It is their right [4]. Their views should be taken into consideration in design and planning of their urban environment that will subsequently help the growth and development [5] as well as their psychological well-being [6].

Living with one another in an urban environment is about negotiating between people which includes adults and children, material places and spaces [7]. Giving children the recognition of their capacity to be genuine participants in the planning, development and execution process should be the level of relationships between adults and children [8]. This action illustrates the importance of creating a good environment for the children [9] Accordingly, the Child-Friendly Environment program developed by UNICEF urge communities, all stakeholders and governments to act and value children's right to play [10].

Essentially, a child-friendly environment aims to assure that children are given the right to have a voice about their environment [3] Further research by Horelli [11] established a theoretical framework that describe a more meaningful model of environmental child-friendliness. This framework can be a guide to drafting the municipal and regional plan with regards to children rights.

UNICEF article 31 avows that children has the right to play, leisure and participate in artistic and locale cultural activities. With this child-friendly environment initiative, play is principally very important to children [12], which is both a right and need [13]. Through discovery, as a child's right to play [14] children will develop the mental processes as a way of learning. Play stimulate and advocates children the opportunity to be creative [15]. Play affords physical contact of the environment and social interaction [16]. Through play, children learn more of their urban environment and harness their gross and fine motor skills as well as their abilities through exploring, discovering, doing, succeeding and failing [17]. Therefore, is art considered part of play? Through participatory, can art-making be engaging and responsive towards establishing a child-friendly urban environment?

2. Method

A computerized search through various online databases from SAGE, Scopus, JSTOR and Science Direct were meticulously done on eight disciplines. These disciplines include Environmental Psychology, Environmental and Behavior, Children Geographies, Early Childhood Education. A through search using the following keywords as a guide to the search; Play, art and play, child independent mobility, space, child-friendly environment and public art.

Theoretical paper, reviews, empirical article and chapter in books were also drawn. The literature chosen illustrated the breadth of knowledge about the studies on child and the child-friendly environment. An emphasis of the literature research was carried out on studies and issues pertaining child-friendly environment. A greater importance was placed on literatures concerning research of children's art and play in the built environment. Studies on children's learning experience, social development and physical activity were aspects that was taken into consideration in this review. This review hoped to include the challenges and potentialities of creating a child-friendly urban environment through art-making

3. Results and Discussion

Table 1. Domains and Frequency of Child-Friendly Environment Attributes

Domains	Frequency
Child-Friendly Environment	11
Play	14
Space	10
Child Independent Mobility	25
Art and Play	9
Public Art	21

The above Table 1. indicates a plethora of literatures pertaining child-independent mobility and public art. The body of knowledge leads to the benefits of play and is further elaborated to play in its place and space. It elaborates the importance of space and meanings towards children. These elements create an engaging environment that is child-friendly. However, there is scarce literature of active play in art although its relationship is parallel towards the attributes of creating a child-friendly environment.

3.1 Space

Important areas for children exploration and play in an urban environment includes neighbourhood alleys, streets, public spaces and loose-fit spaces [18] These everyday landscapes allow children to expend their minds, exhaust their energy and relate to their buddies whilst developing their social, cognitive and physical skills and well-being development [19][20]. Being with their buddies in their everyday environment [21] secures their perception of safety [22]. These relationships between buddies, space and environment affords children to develop their capability as social agents and being independent [23].

Designed and designated public space for children that includes parks, playground and school are the only places where children are welcomed and are free. In other places, children are restricted and unwelcomed [24]. It is important that these outdoor spaces have the possibility to afford physical interaction and exploration through congregating when children exercise control of their environment. Loose-fit spaces, public urban spaces and urban neighborhoods afford freedom of play and discovery

3.2 Child Independent Mobility

Of late, children's play has been always dictated and design by adults. This designed play activities deprives children of being self-reliant, and obtain the necessary skills for them to explore and understand their environment [25]. Children should be given the ability to nurture their personal identity and explore on their own. Together with others, they will build a sense of community [26] without supervision of adults [27] Limiting their independence deprives the opportunity to improvise their social life, develop community-based friendships, acquire necessary social skills and street literacy [28] and most importantly their freedom of play [29]. Thus, as a consequence, children are excluded from the common ground for people and adults alike to carry out cultural, social and meaningful activities [30]. Playgrounds as designed play areas can be safe and responsive for children. However, it physically limits the potential for imaginative and free play. However, it can provide a platform for art. Therefore, to expand the definition of play-space, can non-designated spaces for children such as street, parks, playgrounds, squares and other loose-fit spaces be included?

3.3 Play Through Art

Art as well as play is profoundly important to a child's development and education [31]. Art is an avenue for children to be creative and express the emotion, cultures and as a medium of socializing [32]. Henceforth, art is a medium for communicating as well as play. Art is a pervasive force that assist children in shaping their attitudes, values, beliefs and behaviours [33] Children participating in art-making has shown positive social results in schools [34] and the community [35]. Visual art helps illustrate to the audience issues that are currently relevant [36]. Arts is considered part of being child-friendly as it is as natural as play [37]. Conjoining play and art within the context of everyday activities suggest an acknowledgement of new ideas of making public art as being functional. It is then both aesthetically pleasing as well as being functional [31].

3.4 Artworks

Artwork that have artistic value and of a creative expression outside of museum walls is considered as public art [32]. The works include sculpture, mural, bass imprint situated in designated children play-space as well as other public space can stimulate the public. These reactions can be of issues, space and behavior [38][39][40]. Public art and sculptures therefore, have become a representation of identity to communities [41][42]. Place-specific that collaborate-and-create public artworks [32] can be divided into two groups: "listen-and-lead" and "confer-and-defer." Listen-and-lead is when artist are given inputs from communities whereas confer-and-defer on the other hand encompasses artworks designed directly by non-artists. This term can be used to identify those who do not make art professionally such as children. This is an opportune for artist to act as facilitators cooperating between artist and non-artist to create an on-going performance expressions. The beauty of such continuous collaborative engagement will allow these artworks to grow as long as the community takes care of it [36][43].

4. Conclusions

This review paper affirms that art is essentially play. It is creative process that is fun for children. Continuous participatory in art-making by children in their urban environment can potentially enhance children skills, foster community ties and embed a sense of attachment as well as being educational. Therefore, great emphasis should be given to create a child-friendly urban environment through children artworks. The result of this collaboration will definitely help towards children's growth and the creation of an urban environment that is child-friendly

Acknowledgment

The authors would like to thank Universiti Teknologi Malaysia for providing a platform in conducting this study and UiTM for supporting this research.

References

[1] Chawla, L. and Malone, K. (2003) Children in the city: Home, Neighborhood and Community, pp.118-141, Routledge Falmer

[2] Malone, K. (2002) Street Life: Youth, Culture and Competing uses of Public Space. Journal of Environment and Urbanization, 14 (2), pp. 157-168

[3] Riggio, E. (2002), Child Friendly Cities: Good governance in the best interest of the child, Environment and Urbanization, 14(2), pp. 45-58

[4] Van Vliet, W. and Karsten, L. (2015) Child-Friendly Cities in a Globalizing World: Different Approaches and Typology of Children's Role's, Child, Youth and Environ., vol.25, no.2, pp.1-15

[5] Fjurtoft, I. and Sageie (2000) The Natural Environment as a Playground for Children: Landscape Description and Analyses of a Natural Playscape, Landscape and Urban Planning, 48(1), pp. 83-97, Elsevier

[6] Kaplan, S. (1987) Aesthetic, Affect, and Cognition. Environ. Preference from an Evolutionary Perspective. Environ. and Behav., 19(1), pp.3-32, SAGE

[7] Christensen, P and O'Brian, M. (2003) Children in the City: Home, Neighbourhood and Community, Routledge Falmer

[8] UNICEF (1997) Children's Rights and Habitat: Working Towards Child-Friendly Cities, New York

[9] Clark, A. (2004) The Mosaic Approach and Research with Young Children, in: Lewis, Vicky; Kellett, mary; Robinson, Chris; Fraser, Sandy and Ding, Sharon eds. The Reality of Research with Children and Young People, London, UK, SAGE, pp.142-161

[10] Malone, K. (2006) Building a Child Friendly Community, Children's Research Workshops Pilot Study Report, City of Bendigo: Bendigo
[11] Horelli, L. (2007) Constructing a Theoretical Framework for Environmental Child-Friendliness, Child. Youth and Environ., 17(4), pp. 267-292, University of Colorado
[12] Piaget, J. (2007) The child's conception of the world; A 20th century classic of child psychology, Lanham, MD Rowman and Littlefield
[13] Nor Fadzila, A. and Ismail, S. (2012) The Trends and Influential Factors of Children's use of Outdoor Environments; A Review. Procedia – Social and Behavioral Sciences, 38, pp. 204-212, Elsevier
[14] Almon, J. (2003) The Vital Role of Play in Early Childhood Education, Washington Post
[15] Malone, K and Tranter, P.J. (2003) Children's Environmental Learning and the use, design and management of school grounds, Child., Youth and Environ., 13(2), pp.87-137, University of Colorado
[16] Olds, A. R, (1989) Psychological and Physiological Harmony in Child Care Center Design. Children's Environmental Quarterly. 6(4), pp. 8-16 University of Colorado
[17] Opie, I. A, and Opie, P. (1969) Children's Game, New York: Oxford Unversity Press, Oxford
[18] Witten K., Carroll P. (2016) Children's Neighborhoods: Places of Play or Spaces of Fear?. In: Skelton T., Nairn K., Kraftl P. (eds) Space, Place, and Environment. Geographies of Children and Young People, vol 3. Springer, Singapore
[19] Freeman, C and Tranter, P. (2011) Children's Urban Environments: Changing worlds, Earthscan.
[20] Donna Koller & Meredith Farley (2019) Examining elements of children's place attachment, Children's Geographies
[21] Chawla L. (1992) Childhood Place Attachments. In: Altman I., Low S.M. (eds) Place Attachment. Human Behavior and Environment (Advances in Theory and Research), vol 12. Springer, Boston, MA
[22] Geneviève Castonguay, & Sylvie Jutras. (2010). Children's Use of the Outdoor Environment in a Low-Income Montreal Neighborhood. *Children, Youth and Environments, 20*(1), 200-230
[23] Sutton-Smith, B. (1990). School playground as festival. *Children's Environments Quarterly, 7*(2), 3-7.
[24] Percy-Smith, B., & Malone, K. (2001). Making children's participation in neighbourhood settings relevant to the everyday lives of young people. *PLA Notes, 42*, 18–22. London: IIED.
[25] Haider, J. (2007).Inclusive Design: planning public urban spaces for children, Municiple Eng. 160, pp. 83-88
[26] Prezza, M. Alparone, F. Cristallo, C. Luigi, S. (2005) Parental Perception of Social Risk and of Positive potentiality of Outdoor Autonomy for Children, J. Environ. Psychol. 25, pp.437-453
[27] Wyver, S., Tranter, P., Naughton, G., Little, H., Sandseter, E. B. H., & Bundy, A. (2010). Ten Ways to Restrict Children's Freedom to Play: The Problem of Surplus Safety. *Contemporary Issues in Early Childhood, 11*(3), 263–277
[28] Cahill, C. (2000) Street Literacy: Urban teenagers' strategies for negotiating their neighborhood. J. Youth Studies 3(3) pp. 251-277
[29] Wen, L., Kite, J. (2009) Merom, D., Rissel, C., Time spent playing outdoors after school and its relationship with independent mobility. Int. J. Behav. Nutr. Phsychol. Activity, 2009
[30] Anlı Ataöv, & Jawaid Haider. (2006). From Participation to Empowerment: Critical Reflections on a Participatory Action Research Project with Street Children in Turkey. *Children, Youth and Environments, 16*(2), 127-152.
[31] Ryan, E. D., Public art, children, and playgrounds: Design of an earthwork for hidden valley elementary school playground, Children's Environmental Quarterly, Vol. 7, no. 3, pp. 23-31, 1990
[32] Boughton, D. and Rachel, M.(1999) Beyond Multicultural Art Education: International Perspectives, Munster, NY

[33] Chalmers, F. G. (1996) Celebrating Pluralism: Art Education and Cultural Diversity, Paul-Getty Trust
[34] Deasy, Richard J. (2002) Critical Links: Learning In The Arts and Student Academic and Social Development, Washington, D. C.; Council of Chief State School Officers
[35] Catterall, J.S (2002) The Arts and The Transfer of Learning, Washington D.C.; Arts Education Partnership (AEP)
[36] Russell, R. (2004), A Beginner's Guide to Public Art, Art Education: Arts and Humanities vol. 57, no. 4 pp.19
[37] Henderson, M., & Lasley, E. (2014). Creating inclusive classroom through the arts. *Dimensions of Early Childhood*, *42*(3), 11-17
[38] Chang, T. C. (2008). Art and Soul: Powerful and Powerless Art in Singapore. *Environment and Planning A: Economy and Space*, *40*(8), 1921–1943.
[39] Hein, H. (1996) What is Public Art? Time, place and meaning, The journal of aesthetics and art criticism, vol.5, no. 1, 1996
[40] Bach, P. B.(2001) New landmarks: Public art, community, and the meaning of place, Grayson Publishing, 2001
[41] Beunders, H.(2007) The end of arrogance, The advent of persusion: Public art in a multicultural society, social analysis, The international j. of soc. and cult. practice, vol.51, no.1, pp. 42-57
[42] Pinder, D. (2011) Errant paths: The poetics and politics of walking, Environment and Planning
[43] Venda Louise Pollock & Ronan Paddison (2014) On place-making, participation and public art: the Gorbals, Glasgow, Journal of Urbanism: International Research on Placemaking and Urban Sustainability

IOP Conf. Series: Earth and Environmental Science **385** (2019) 012065 doi:10.1088/1755-1315/385/1/012065

The impacts of FOREX fluctuations on international construction projects

Mohd Amizan Bin Mohamed[1], Ayuniza Binti Ahmad[2] and Daud Bin Mohamad[2]
[1] Faculty of Civil Engineering, UiTM Shah Alam, 40450 Shah Alam, Selangor, Malaysia
[2] College of Engineering, Universiti Tenaga Nasional, 43000 Kajang, Selangor, Malaysia

Email: amizan8124@salam.uitm.edu.my

Abstract. The international construction project is a highly valuable market project and many of the contractors would like to be participate. This is because they are misled with the exchange rate (money) and instantly attracted with the profit they assumed will be gained after the project is completed. In fact, without noticing that they will be affected with the FOREX's fluctuations which might complicate the project further. In addition, many new vendors nowadays will try their luck in construction business without making a good judgement on the risks. So, the objective of this study is to identify the impacts of FOREX fluctuations on those perspectives at international level. This study also uses quantitative method through questionnaire survey that were distributed randomly, and the result obtained were from top management and non-top management in construction hierarchy with different level of experiences. The collected data were analysed by using two methods which are Statistical Package for Social Science (SPSS) and Microsoft Excel. The SPSS software is used to process the regression analysis for the impacts of FOREX data of the questionnaire. The Relative Importance Index then is applied to run the data for capability and mitigation actions in determining the ranking of priorities. Overall, the impacts of FOREX fluctuation on international construction projects have been identified with regards to three mechanisms which are business management, technical and financial. These outputs are useful as a guideline for international construction project companies when they are venturing into this challenging market especially in terms of managing/ expediting quality, completion period, and optimising cost of the projects.

1. Introduction

Currency FOREX fluctuations is a natural outcome of the fluctuation of exchange rate system that is the norm for most major economies. The exchange rate of one currency versus the other is influenced by numerous fundamentals and technical factors that are related indirectly without our notice. These include relative supply and demand of the two currencies, economic performances, outlook for inflations, interest rate differentials, capital flows, technical support and resistance levels. As these factors are generally in a state of perpetual flux, currency values fluctuate from one moment to the next. But although a currency's level is largely supposed to be determined by the underlying economy, the tables are often turned, as huge movements in a currency can dictate the economy's fortunes and indirectly will affect the industry involved and give huge impact to company that is related most to international business.

Implementing construction projects in a foreign country is a high-risk business activity [1]. International construction projects normally involve a high level of risks because of the differences in construction practices, working conditions, culture and political, legal, and economic conditions between domestic and overseas markets [2]. Foreign exchange rate (FOREX) risk is considered as

IOP Conf. Series: Earth and Environmental Science **385** (2019) 012065 doi:10.1088/1755-1315/385/1/012065

significant challenge in the international construction business, as business is strongly affected by these fluctuations [3][4].

The objective of this research is to identify the impacts of FOREX fluctuations to international construction project based on business management, technical and financial perspectives.

2. Research methodology

The quantitative method is carried out to determine the impacts of FOREX fluctuations on international business performance. Quantitative methods aim to classify features, count them, and create statistical models to test hypotheses and explain observations of the data gained. According to Chang [5], basic quantitative descriptions can provide a valuable baseline background to help identify issues for a more in-depth, qualitative or quantitative study for an event. Hence, this methodology is suitable because the results and outcomes congruently the objectives of the study.

In this study, a flowchart containing problem statement and objective, literature review, questionnaire, pilot study, data collection, analysis and discussion, conclusion and summary and presentation is used to illustrate the steps of the research methodology. In this flowchart, it can determine where/when to start the research, identify the problems, how to collect data and analyse the data, make the right recommendations and summing up to the conclusions.

2.1. Methodology flowchart

The overall flow, the earliest phase until the final phase, of the methodology process applied in this study is illustrated as below.

The sample size for this study is 300 respondents in construction firms with projects at international level. The main requirement in selection of respondents is that they must be those in the top management group. They could be the contractors, supervisors, general managers, deputy general managers, managers, assistant managers and those involved in international construction projects related to the engineering fields especially at the top level. The reason for this action is because the objectives of this study can only be answered by those impacted or dealing with the impacts of FOREX fluctuations in the real field at international level. Only their experiences, knowledge, and past actions are valid to be utilised in answering the research questions / questionnaire successfully. This then would enable a solid framework to be established regarding to this issue. Presence of the surrounding stakeholders, i.e. top management,project management, project champion, software vendor, and consultants were identified as the root causes driving performance [6, 7]. In addition, the selected companies are those located in Selangor and Kuala Lumpur vicinity since most of the companies related to international projects are located here. Allegedly, the respondents must have had experiences in executing construction projects at the international level. If the respondents' prior knowledge and exposure of FOREX issues is non or low, the data cannot be relied upon them.

Referring to the table, by Krejcie and Morgan [8] and the Educational and Psychological Measurement, 30, 607-610, we can identify and determine sample size for this study based on the population size. Hence, using these guidance, with the population size of 300, the estimated sample size of 169 respondents is indeed sufficed.

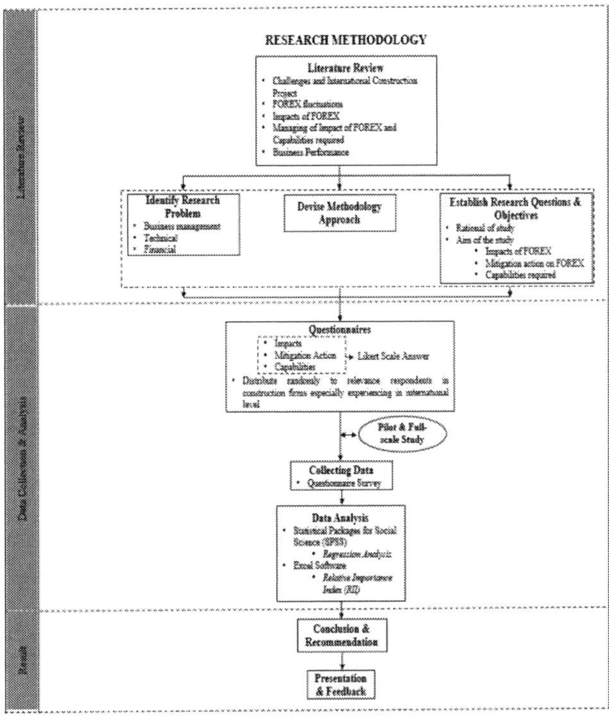

Figure 1. Flow chart.

2.2. Questionnaire design

The questionnaire is an effective way to obtain a data from respondents [9]. According to Saunders et al., questionnaires tend to be used for explanatory research where cause-effect relationships can be examined, or for descriptive studies which involve attitudes and opinions to enable the researcher to investigate the variability in different phenomena [10]. This indicates that questionnaire is a reliable way in getting information. The strategy in questionnaire distribution is to distribute randomly to respondents in construction firm especially those with experiences in construction project at international level possibly at the top management level. About 300 questionnaires survey were distributed to the respondents of the selected areas which are Selangor and Kuala Lumpur because of the arising development of construction industry at those areas. The required data are related to the project whether completed or on-going based on CIDB data. The respondents of the survey questionnaires include contractors, supervisors, general managers, deputy general managers, managers, assistant managers and who are involved in international construction project related to the engineering fields especially in top management. Stated by Akkermans and van Helden [6] and Boonstra [7] presence of the surrounding stakeholders, i.e. top managements, project managements, project champions, software vendors and consultants were identified as the root cause driving performance. They are considered to be the most suitable respondents for the research to explain the decision-making structure and strategic posture of their organisations [11].

IOP Conf. Series: Earth and Environmental Science **385** (2019) 012065 doi:10.1088/1755-1315/385/1/012065

Table 1. Sample size determination.

Sample size determination for most practical situations

Source: Krejcie, R. & Morgan D. (1970). Determining Sample Size for Research Activities, Educational and Psychological Measurement, 30, 607-610.

Where:

N = Population Size
S = Sample Size

Table for Determining Sample Size from a Given Population

N	*S*	*N*	*S*	*N*	*S*
10	10	220	140	1200	291
15	14	230	144	1300	297
20	19	240	148	1400	302
25	24	250	152	1500	306
30	28	260	155	1600	310
35	32	270	159	1700	313
40	36	280	162	1800	317
45	40	290	165	1900	320
50	44	300	169	2000	322
55	48	320	175	2200	327
60	52	340	181	2400	331
65	56	360	186	2600	335
70	59	380	191	2800	338
75	63	400	196	3000	341
80	66	420	201	3500	346
85	70	440	205	4000	351
90	73	460	210	4500	354
95	76	480	214	5000	357
100	80	500	217	6000	361
110	86	550	226	7000	364
120	92	600	234	8000	367
130	97	650	242	9000	368
140	103	700	248	10000	370
150	108	750	254	15000	375
160	113	800	260	20000	377
170	118	850	265	30000	379
180	123	900	269	40000	380
190	127	950	274	50000	381
200	132	1000	278	75000	382
210	136	1100	285	100000	384

N is population size.
S is sample size.

Next, the respondents chose to answer answer by tick or circle they feel the correct one. The questions must be related and reflect effect of the FOREX fluctuation in international construction projects. Then, actions that can mitigate the impacts by considering the possible capabilities and by reviewing relevant literature. This also had been admitted by Govindarajan [12] and Nandakumar et al. [13], they stated that the development of the questionnaire for the quantitative survey started with the review of relevant literature on competitive strategies, decision-making styles and organisational performance to identify the variables, and this was refined by researchers in construction and the built environment to evaluate the content validity. This questionnaire will be using the Likert Scale. Likert Scale falls within the ordinal level of measurement that is, the response categories have a rank order, but the intervals between values cannot be presumed equal, though, as Blaikie points out [14]. This method helps in describing the quantitative value to qualitative value efficiently and to make it amenable to statistical analysis. Generally, the method being used is to ease the researcher to analyse the data efficiently in the last phase of survey. The Likert Scale answers are rank as below:

a) 1 for STRONGLY DISAGREE
b) 2 for DISAGREE
c) 3 for QUITE AGREE
d) 4 for AGREE
e) 5 for STRONGLY AGREE

2.3. Pilot study

Pilot study can be translated as a small preliminary study that being conducted to examine and evaluate the gaps in the questionnaire so some adjustments on the design can be done. This is very essential because every error or misunderstanding on the question that being used in real survey on selected population will cause invalidity in data and the objective will be unachievable. The pilot study also being conducted in advance to obtain expected result from the respondent in full-scale study. The expecting outcome then being used to forecast the future outcome from the population size. It allows the researcher to have initial hypotheses about the study.

For this research, the pilot study was conducted to 110 relevance respondents in construction firms on selected areas especially those who have experiences in international level at top management. The pilot study was conducted to a large number of respondents because the researcher was having a difficulty to get cooperations with the related companies. Hence, unfortunately the result of this pilot study was getting just 3 answered questionnaires from 2 different companies with each of the respondent has different positions in their company and all the respondents come from top management background.

2.4. Method of data analysis

This research is conducted to respondents who are involved in civil engineering field, to testing the questionnaires quality and to identify the feedback from the questionnaire either achieve the goals or not at the pilot study level. If the feedback gives a positive respond. So, after doing the pilot study, the questionnaires are distributed to selected population to get actual result for this study. The data gained will be analysing to get the exact output related to the study. So, data analysis is important part in achieving the objectives and in another meaning content analysis is 'a technique for gathering data, it involves codifying qualitative and quantitative information into pre-defined categories in order to derive patterns in the presentation and reporting of information' [15]. So, for this study there are few effective methods applied to analyse the data in effective and efficient way.

2.4.1. SPSS (Statistical Packages for Social Science. To assist in planning and carry out strategies that will improve the outcome of this study, the Statistical Packages for Social Science (SPSS) software was applied to predict and analyse the collective data. SPSS enables qualitative data to be organised, closely monitored, easily coded, retrieved through direct searching and interrogated to build propositions and theories, due to the systematic recording of the information and data [10]. This software is a comprehensive system for analysing data and aiding the data interpretation process more easily [16]. Using this software, the data transferred automatically from spread sheet into graphic in form as to present the data in a more attractive manner. This methodology is useful in investigating the combination of conditions and pathways that lead to a performance (outcome), and this is particularly relevant in the project-based construction field because such projects frequently involve complex relationships among the variables of interest [17]. The data gathered from the analysing process will be known as raw data and this will be used as our reference from here on.

2.4.2. Regression analysis. Regression is a well known method being used in quantitative research and often being applied to analyse and modelling numbers of variable with relationship. In easy way the regression analysis is to examine the relationship nature of both variables in an effective way. In this analysis the SPSS software was used to ease the process of analysing the variables in a systematic way. It allows multiple models in a single regression command to be specified. In this study the regression method was applied in questionnaire in Part C which explains on impacts of FOREX fluctuations to the international construction project as dependent variable based on three perspectives which is business management, technical and financial that will be independent variable. Thus, the regression analysis was applied to justify which of those rtanked highly in terms of correlation to dependent variables and to test those relationship.

Regression model, basically, specifies the relation of dependent variable (Y) to a function combination of independent variables (X) and unknown parameters (β)

$$Y \approx f(X, \beta)$$

Regression equation can be used to predict the values of 'y', if the value of 'x' is given, and both 'y' and 'x' are the two sets of measures of a sample size of 'n'. The formulae for regression equation would be

$$y^* = a + bx$$

Where,

$$b = \frac{n\sum xy - (\sum x)(\sum y)}{n(\sum x^2) - (\sum x)^2}$$

$$a = \frac{\sum y - b\sum x}{n}$$

Figure 2. Regression equations.

2.4.3. Microsoft Excel. The raw data from the questionnaire in Part D and Part E are keyed-in to Microsoft Excel by using Relative Importance Index (RII) equation. The frequency of each question been determined from Likert scale rating, the formula of Relative Importance Index (RII) then keyed-in to the Microsoft Excel, the value of Relative Importance Index (RII) can be determined and it will be easy to rank the variables on Part D and Part E required that need to be enhanced by the company themselves. This ranking system actually helps to determine the elements in each variable that need to be emphasized strongly in FOREX issues. Last, the data are presented in the form of charts by Excel assist to illustrate the output more clearly.

2.4.4. Relative Importance Index (RII). The Relative Importance Index is the method in arranging the variables according to their relative prior to ranking. Once the ranking has been identified, it leads to knowing the relationship of the variables to the research. To know and rate the numerous causes of project becomes a failure; the RII is widely being adopted metric in AEC risk management in the literature [18-20]. According to those scholars, have been mentioned in their study that RII being adopted to knowing how relevance the impacts of numerous risks to project metrics such as cost, schedule, and performance. For instance, the RII method being was used by Gunduz et al. in their research on determining the essential relative causes of delay in construction project in Turkey [21]. Thus, it shows that the RII method is relevant to be used in this research in determining the significant mitigation actions and capabilities.

$$\textbf{Relative Importance Index} = \frac{\sum w}{AN} = \frac{5n_5 + 4n_4 + 3n_3 + 2n_2 + 1n_1}{5N}$$

Where w is the weighting given to each factor by the respondent, ranging from 1 to 5. For example, n_1 = number of respondents for Little Important, n_2 = number of respondents for Some Important, n_3 = number of respondents for Quite Important, n_4 = number of respondents for Important, n_5 = number of respondents for Very Important). A is the highest weight (i.e. 5 in the study) and N is the total number of respondents. The relative importance index ranges from 0 to 1 (Tam and Le, 2006).

Figure 3. Relative Importance Index equation.

3. Result and discussion

In Figure 4, 11.2% of the respondents do not agree that companies will be affected with the impacts of FOREX meanwhile 88.8% respondents have contradicted answers with the minority opinions. In overall, majority of those companies agreed that their companies are affected by FOREX fluctuations when conducting their projects at international level. This finding is aligns with a statement by Dikmen et al., Han et al., Kim and Hur, and Ling and Hoi in their previous research that is foreign exchange rate (FOREX) risk is considered as significant challenge in the international construction business, as business is strongly affected by these fluctuations [3-4, 22-23]. Thus, it is clearly that FOREX plays a vital role in international market. So, a well preparation is crucial for any company that tends to venture the international market.

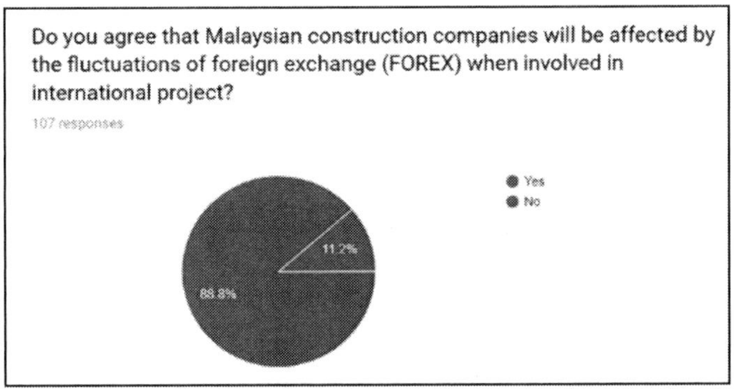

Figure 4. Impacts of FOREX toward Malaysian company.

3.1. *FOREX fluctuations at company level*

The question focused on the effect of FOREX fluctuations at the company level. It purposely asked to identify whether the FOREX issues can influence the company abilities and performance or not. For instance, foreign exchange exposure and economic exposure may be visualized as the overall impact of foreign exchange rate fluctuations on stockholders' wealth or i.e. the market value of the common stock [24, 25]. In this study, 107 respondents conferred different opinions depending on their wisdom and knowledge during executing the project at international level. The result is interpreted in the bar chart below:

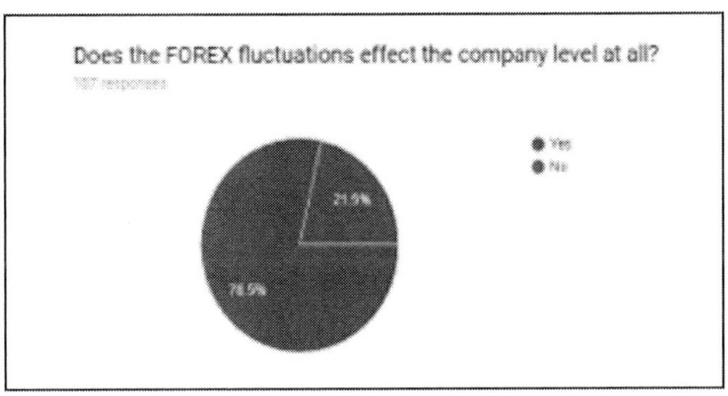

Figure 5. FOREX effect at company level.

Most of the respondents give positive feedbacks on the relationship between FOREX fluctuations and it impacts at the company level. While, the rest do not agreed with the justification. This is because each company has their own strengths and abilities in overcoming the problems occurred. More stable one company is, more easy for the company to cater to the problems. Hence, contradiction of answer in this question because their performance at company level is not affected by the FOREX fluctuations.

But majority of the respondents agreed with the statement that FOREX fluctuations effect their company level. So, it clearly shows that FOREX fluctuations inevitably give impacts at the company level and coincides0 with the statement stated by Menguturk and Karadagli [24, 25].

3.2. FOREX fluctuations at project level

The last question on the company background is related to the company performance at the project level while conducting the project abroad and when it is the FOREX fluctuations occurred. According to Shane et al., Ling and Hoang, and Zhi in their observations state that hierarchy for determination of risk in overseas project can be categorise to four stages which are country, industry, firm, and project stage [26-28]. From the data gathered in the full-scale study of this research, result similarly as previous question. This can be seen in the figure below.

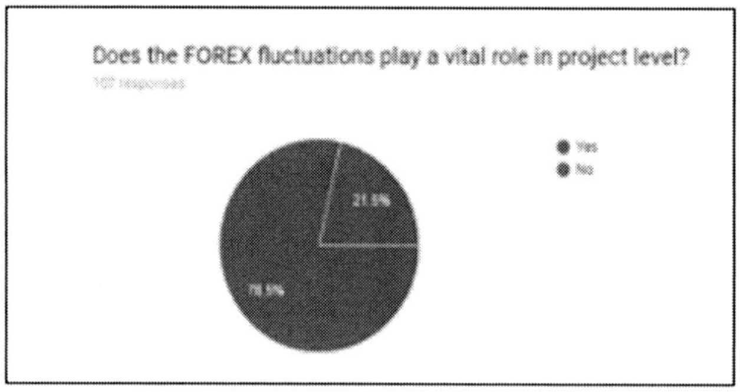

Figure 6. FOREX effect at project level.

Based on Figure 6, it can be intepreted that 74 out of 107 respondents agreed that their company are affected by the FOREX fluctuations at the project level. And this usually will involve high level of risk on the project if related to FOREX conflict. This higher risk causes larger project costs and losses in time compared with domestic projects [29]. Thus, proved that FOREX fluctuations will give bad influence on project level. But another opinion from the rest of the respondents in this research contradict with the figure shown and a justification from Xiang et al. in their previous research. This is because if the management can wisely tackle the problem at the project level properly, it will not affect any progress of the project. This is the ideology being applied by those respondents. Thus, it can be concluded that majority of the respondents gives positive feedback on the justification that FOREX fluctuations affect company at the project level.

3.3. FOREX fluctuation toward international construction projects

Based on the data obtained, it can be determined that the dependency of three perspectives which are business management, technical and financial toward the impact of FOREX fluctuations to the international construction project by using regression analysis in SPSS software. Meanwhile for section D and section E, the data can be tabulated in detailed arrangement based on the frequency of answer given by the respondents in the questionnaire by applying Relative Importance Index in Microsoft excel. The purpose of this action is to see clearly the parameters that the companies need to focus on in order to sustain in the international construction business. Besides, the data are ranked and analysed directly from the real industry's knowledge, wisdom and experiences during their past task in international level. Thus, the data is valid to be consider as lesson and catalyst for other players in construction industry to make improvement.

3.3.1. Impacts of FOREX fluctuations towards international construction projects based on business management, technical and financial perspectives (Top management level). Table 2 illustrates the output of this study after running the SPSS software under linear regression analysis method. The data obtained in the table are the result from 44 respondents that identified categorize in top management level about

IOP Conf. Series: Earth and Environmental Science **385** (2019) 012065 doi:10.1088/1755-1315/385/1/012065

the impacts of FOREX fluctuation that they figured will be the main factors that influence the international construction project in the terms of business management, technical and financial perspectives. This method being undergo because it is crucial in order to examine whether each set of independent variables is a separate for each subpart of the structural model [30]. In addition, by applying this method it allows opportunity to examine the dependency of independent variables toward dependent variable.

Table 2. Linear regression analysis (Top management).

Model Summary

Model	R	R Square	Adjusted R Square	Std. Error of the Estimate
1	.641[a]	.411	.400	.52207

a. Predictors: (Constant), Ave_Technical

ANOVA[a]

Model		Sum of Squares	df	Mean Square	F	Sig.
1	Regression	9.884	1	9.884	36.266	.000[b]
	Residual	14.173	52	.273		
	Total	24.057	53			

a. Dependent Variable: Mean_Impact

b. Predictors: (Constant), Ave_Technical

Coefficients[a]

Model		Unstandardized Coefficients		Standardized Coefficients	t	Sig.
		B	Std. Error	Beta		
1	(Constant)	1.553	.351		4.426	.000
	Ave_Technical	.537	.089	.641	6.022	.000

a. Dependent Variable: Mean_Impact

Excluded Variables[a]

Model		Beta In	t	Sig.	Partial Correlation	Collinearity Statistics
						Tolerance
1	Ave_Management	-.079[b]	-.433	.667	-.061	.348
	Ave_Financial	.168[b]	.847	.401	.118	.291

a. Dependent Variable: Mean_Impact

b. Predictors in the Model: (Constant), Ave_Technical

Based on the research made by Hair et al. from his previous study he found out that the analogous to the assessment of formative measurement models must satisfy tolerance levels below 0.20 or lower with the value of VIF above 5.0 [30]. The summary table indicates that the values of R that represents the simple correlations shows a quite high degree of correlation at 0.641. Meanwhile the values of R2 indicates 41.1% of total variation in the dependant variable which is the impacts of FOREX fluctuation can be be explained by the independent variables which is technical capability. This shows that the technical capability can influence the impacts of FOREX fluctuation and the relationship between those dependant and independent variable during the execution of international construction projects.

The table with ANOVA tabulated data shows the dependant variable which is impacts of FOREX fluctuation is predicted well by the regression model. The regression model statistically predicts the significant outcome variable which the independent variable will has dependency relationship with dependant variable. It proves that the data fits to be applied through the come-out of significant column 'Sig." with value of P is 0.000 which is less than 0.05. As mentioned by Tukur (2008) in his research where to quantify the strength of evidence against null hypothesis $P < 0.05$ (5% significance) is as a standard level for concluding that there is evidence against the hypothesis tested and the regression model valid to use. In addition, the third table, shows the technical capability contributes significantly to the regression model by looking at its significant column value. This coefficient table also provides information to forecast the in fluency of impacts of FOREX fluctuation through technical capability. It can be interpreted through regression equation as:

$$Impacts\ of\ FOREX\ fluctuation = 1.553 + 0.537\ (technical\ capability) \quad (1)$$

Finally, what can be concluded from the last table on the impacts of FOREX fluctuation as a dependent variable is a weak relationship with business management and financial capability as independent variables. This is caused by the expelled relay shown between those variables and somehow, it has strong dependency relationship with technical capability. It can be proved through result at Table 2 that shown strong agreed that the dependency of impacts of FOREX toward technical capability with synchronize P value 0.000 that less than 0.05 and regression equation established through that correlation.

Thus, from this research we can identifiy that the enhancement of technical capability among construction industry player will help them in countering the impacts of FOREX fluctuation. Dikmen et al. [31] mentioned the different sources of competitive advantage and identified technical capability as the top priority for construction firm. This capability is referred to as the ability of a company to perform a variety of direct activities in construction by defining the type, size, and complexity for completing projects [32, 33]. But the slightest degree of chances needs to be considered and be an enhance in order to fill the small gap that maybe become a vast leakage in future if we neglected. This is because it is being supported by previous scholar that mentioned to used resources in effective way and specialist in performing the project within time, cost and with good quality, the management capability is necessarily [34]. The primary factor to choose the management ability is by arranging all the resources through a proper plan which is planning, organizing, leading and controlling to achieving the purpose of study [35, 36]. As mentioned by Ozorhon et al. [37] performance was also measured using multidimensional measures at the financial and operational levels. This finding is parallel with previous studies that revealed Vietnamese and Chinese contractors lagging behind their competitors due to weak financial capacity [38]. Thus, its shown that those two capabilities need to be take concern too even though have low dependency relationship in mitigating impacts of FOREX fluctuations.

4. Conclusion

Based on the data analysed and findings, it can be concluded that the objectives and aims of this study have been achieved successfully. The impacts of FOREX fluctuations on international construction project have been identified and the three perspectives are business management, technical and financial. The correlation is based on their dependency and this is tested through regression analysis. In this research, those three perspectives are independent variables that can stand alone without any support meanwhile for the impacts of FOREX fluctuation that is classified as dependent variable it contradict because of its reliability on independent variables. Through SPSS and RII test, the result shows that technical mechanism has strong relationship with the impact of FOREX and yet weak dependency on business management and financial. Therefore, by knowing and identified these impacts, holistic overview on mitigation actions needed to be considered, to tackle the issue of impacts of FOREX fluctuations based on 3 mechanism highlighted.

Acknowledgments
The authors acknowledge Universiti Tenaga Nasional for funding this research under the UNIIG (J510050787). Also, authors appreciate the efforts of Syahirah Intan Mohd Sheffie for proofreading and editing this paper.

References

[1] Zhang X 2011 Social risks for international players in the construction market: A China study *Habitat Int.* 35 514
[2] Bu-Qammaz A S, Dikmen I and Birgonul M T 2009 Risk assessment of international construction projects using the analytic network process *Canadian J. Civil Eng.* **36** 1170
[3] Han S H, Kim D Y, Jang H S and Choi S 2010 Strategies for contractors to sustain growth in the global construction market *Habitat Int.* **34** 1
[4] Kim H and Hur S 2009 Effect of foreign exchange management on firm performance using genetic algorithm and VaR. *Expert Syst. Appl.* **36** 8134
[5] Chang S E 2010 Urban disaster recovery: A measurement framework and its application to the 1995 Kobe earthquake. *Disasters* **34** 303
[6] Akkermans H and van Helden K 2002 Vicious and virtuous cycles in ERP implementation: A case study of interrelations between critical success factors. *Eur. J. Inform. Syst.* **11** 35.
[7] Boonstra A 2011 Aligning systems, structures and people: Managing stakeholders in enterprise information systems projects *Managing Adaptability, Intervention, and People in Enterprise Information Systems* (Pennsylvania: IGI Global) pp 157-177
[8] Krejcie R V and Morgan D W 1970 Determining sample size for research activities *Educ. Psychol. Meas.* **30** 607
[9] Othman A A E 2015 An international index for customer satisfaction in the construction industry. *Int. J. Construct. Manag.* **15** 33
[10] Saunders M, Lewis P and Thornhill A 2009 *Research Methods for Business Students (5th ed.)* (Harlow UK: Pearson Education)
[11] Pertusa-Ortega E M, Molina-Azorín J F and Claver-Cortés E 2010 Competitive strategy, structure and firm performance: A comparison of the resource-based view and the contingency approach *Manag. Decis.* **48** 1282.
[12] Govindarajan V 1989 Implementing competitive strategies at the business unit level: Implications of matching managers to strategies *Strategic Manag. J.* **10** 251
[13] Nandakumar M K, Ghobadian A and O'Regan N 2010 Business-level strategy and performance: The moderating effects of environment and structure *Manag. Decis.* **48** 907
[14] Blaikie N 2003 Analyzing Quantitative Data: From Description to Explanation (London: Sage)
[15] Guthrie J, Petty R, Yongvanich K and Ricceri F 2004 Using content analysis as a research method to inquire into intellectual capital reporting *J Intellect. Cap.* **5** 282
[16] Allen P J and Bennett K 2010 *PASW Statistics by SPSS: A Practical Guide: Version 18.0* (Boston: Cengage Learning)
[17] Jordan E, Gross M E, Javernick-Will A N and Garvin M J 2011 Use and misuse of qualitative comparative analysis *Construct. Manag. Econ.* **29** 1159
[18] El-Sayegh S M 2008 Risk assessment and allocation in the UAE construction industry *Int. J. Proj. Manag.* **26** 431
[19] Sambasivan M and Soon Y W 2007 Causes and effects of delays in Malaysian construction industry *Int. J. Proj. Manag.* **25** 517
[20] Aibinu A A and Odeyinka H A 2006 Construction delays and their causative factors in Nigeria *J. Construct. Eng. Manag.* **132** 667
[21] Gündüz M, Nielsen Y and Özdemir M 2012 Quantification of delay factors using the relative importance index method for construction projects in Turkey *J. Manag. Eng.* **29** 133
[22] Dikmen I, Birgonul M T and Han S 2007 Using fuzzy risk assessment to rate cost overrun risk in

international construction projects *Int. J. Proj. Manag.* **25** 494
[23] Ling F Y Y and Hoi L 2006 Risks faced by Singapore firms when undertaking construction projects in India *Int. J. Proj. Manag.* **24** 261
[24] Mengütürk M 1994 *International Finance* (Istanbul: Literatür)
[25] Karadagli E C 2015 Economic exposure of emerging market firms. *Act. Prob. Econ.* **166** 67
[26] Shane M S, Stevens M C, Harenski C L and Kiehl K A 2009 Double dissociation between perspective-taking and empathic-concern as predictors of hemodynamic response to another's mistakes *Soc. Cog. Affect. Neurosci.* **4** 111
[27] Ling F Y Y and Hoang V T P 2009 Political, economic, and legal risks faced in international projects: Case study of Vietnam *J. Prof. Iss. Eng. Edu. Pr.* **136** 156
[28] Zhi H 1995 Risk management for overseas construction projects *Int. J. Proj. Manag.* **13** 231
[29] Xiang P, Zhou J, Zhou X and Ye K 2012 Construction project risk management based on the view of asymmetric information *J. Construct. Eng. Manag.* **138** 1303
[30] Hair Jr J, Sarstedt M, Hopkins L and Kuppelwieser V 2014 Partial least squares structural equation modeling (PLS-SEM) An emerging tool in business research *Eur. Bus. Rev.* **26** 106
[31] Dikmen I, Birgonul M T and Kiziltas S 2005 Prediction of organizational effectiveness in construction companies *J. Construct. Eng. Manag.* **131** 252
[32] Warszawski A 1996 Strategic planning in construction companies *J. Construct. Eng. Manag.* **122** 133
[33] Wethyavivorn P 2009 Strategic assets driving organizational capabilities of Thai construction firms *J. Construct. Eng. Manag.* **135** 1222
[34] Ajibade A O 2006 *The Impact of Contractors Management Capability on Construction Project Performance in Nigeria* (Unpublished B. Tech. dissertation, Federal University of Technology, Akure)
[35] Abiola R O 2000 Management implications of trends in the construction costs in Nigeria from 1989-1999 *The Quant. Surveyor* **30** 35
[36] Jolly B, Isa F, Othman S N, Afiq M and Ahmdon S 2016 The influence of management capability, marketing capability and competitive advantage on Malaysian construction project performance *Int. Rev. Manag. and Mark.* **6** 142
[37] Ozorhon B, Arditi D, Dikmen I and Birgonul M T 2010 Performance of international joint ventures in construction *J. Manag. Eng.* **26** 209
[38] Zhao Z Y, Shen L Y and Zuo J 2009 Performance and strategy of Chinese contractors in the international market *J. Construct. Eng. Manag.* **135** 108

IOP Conf. Series: Earth and Environmental Science **385** (2019) 012066 doi:10.1088/1755-1315/385/1/012066

Developing a Research Methodology for Life Cycle Costing Framework for Application in Green Projects

Mysarah Maisham [1], Hamimah Adnan[1], Noor Akmal Adillah Ismail[1] & Noor Aisyah Asyikin Mahat [1]
[1]*Faculty of Architecture, Planning and Surveying, Universiti Teknologi MARA, Shah Alam, Malaysia*

Email: mysarah6375@uitm.edu.my

Abstract. Life cycle costing (LCC) is a method of accessing the total cost of ownership and has been used around the world for the last few decades. Even though many benefits have been associated with LCC, one of them being an effective tool for decision making, its implementation in construction industry is still limited, particularly in green projects. Among the main factors suggested to have hindered application of LCC are lack of knowledge and understanding on LCC, as well as absence of a standard guideline. This paper presents a research methodology adopted for a research aimed to develop a life cycle costing framework for its' application in green projects. The objectives are to investigate the current practice for LCC application in Green projects, and to determine the relationship between LCC principles with factors influencing effective decision making for LCC best practice in Green project. The methodology of the research involves qualitative and quantitative method, and will be validated by an expert panel interview. The paper provides an overview on the research area, particularly to highlight the issue on the application of LCC and its importance to assist decision makers.

1. Introduction

Life Cycle Costing (LCC) has been commonly allude to in sustainable-related studies. Within the context of built environment, life cycle costing is a method used to assess the anticipated economic performance of a building throughout its life cycle which includes: design and construction, operation and maintenance, in addition to disposal [6]. LCC applications in green building have been increasing since sustainable building became prominent at the beginning of 21st century. Through the assessment process of LCC, project stakeholders would be informed of the financial and non-financial gains of environmental and social sustainability initiatives [34]. LCC, has also been associated with green procurement, a procuring process that seek to source goods, services or work with a reduced environmental impact, or in other words, environmentally friendly [13]. In Malaysia's own Government Green Procurement (GGP), LCC has been hailed as an important instrument to support its implementation, and can be considered as one of the awarding criteria for decision making [21]. This paper is part of an on-going research to develop a life cycle costing framework for its' application in green projects. The aim of this paper is to provide an overview on the area of study and to describe the methodology intended to be used by the author to accomplish the research aim.

2. Problem Statement

Quite a number of researches has been done on LCC applications in green building across the world, varying in terms of area of interest. However, there appeared to be a general lack of acceptance of LCC method in the construction industry practices [18] and [33]. Furthermore, [14] believed that not much attention is being given to the real implementation of LCC by the public sector. Therefore, there is a need to investigate to what extend has the Malaysian construction industry applied LCC, particularly in Green projects, including the problems and issues in such application.

Various studies claimed LCC can give great benefits to a development through its application either in conventional projects or even in green projects. Moreover, researchers have specified the important role LCC plays in decision making [27] and [7]. It has also been described as a tool for the comparison and optimization of building solutions [8]. So how does LCC contributes to effective decision making, particularly in Green projects? Understanding on the relationship between LCC and successful decision making is necessary for formulating proper strategies to implement LCC in Green projects. This study will help stakeholders and project team to understand better the advantages gained from the application of LCC in Green projects.

One of the barriers in implementing LCC in construction industry is the lack of knowledge of clients on LCC [24]. The study further listed lack of training on standardized LCC approaches and guidelines is also another problem that limits the use of LCC in the industry. Lack of understanding of the LCC techniques and the absence of a standardized guideline as among the key barriers to wider implementation of LCC [30]. Hence, a guideline or framework is needed to provide clear implementation of LCC in the Malaysian construction industry, particularly for green projects. The findings of this study will not only contribute to filling the gap in knowledge concerning LCC application in Green projects, but also provide valuable reference for stakeholders, project team, and others. Therefore, the aim of this research is to develop a procedural framework for the construction industry practitioners to improve the application of LCC towards effective decision making for Green Project in Malaysia.

3. Literature Review

3.1. The Concept of LCC

The history of LCC can be traced back in the mid-1960s, where the US Department of Defence applied in their procurement technique [2]. Over the years, the application of LCC has expanded to many areas such as product design, buildings, transportation, and technologies [12] and [28]. The term 'life cycle costing' is used interchangeably with other terms such as 'life cycle cost analysis', 'whole life cost', and 'total cost assessment' [35] and [26]. As useful as any, LCC can be defined as "a process to determine the sum of all expenses associated with a product or project, including acquisition, installation, operation, maintenance, refurbishment, discarding, and disposal costs" [37]. A building's economic performance throughout its life cycle, starting with its initial planning and design to construction, operation and maintenance, refurbishment and finally, its demolition, can be evaluated by using LCC [39]. [4] stated LCC consists simple tree of acquisition cost and sustaining cost of the acquisition during its life. On the other hand, the LCC model developed by the Society of Automobile Engineers (SAE), breaks down the costs into five (5) elements which are acquisition cost, operating cost, scheduled maintenance cost, unscheduled maintenance cost, and conversion cost or decommission costs [31].

3.2. LCC Application in Construction Industry

It has been suggested that if properly carried out, LCC will deliver benefits such as transparency of future cost of operations, improve ability to plan future expenditure, increase awareness of total cost improve ability to manipulate and optimise future costs at the design stage, higher chance to achieve and obtain better value for money (VFM) in project, provide competitive alternatives evaluation and better performance trade-offs against cost. However, there still appeared to be a general lack of acceptance of LCC method in the construction industry practices [18] and [33]. The availability and reliability of data such as lifespan, future costs of operation, maintenance, previous consumptions, fixed costs, and investments, is also one of the main challenges for the mainstreaming of the use of LCC [19]. While an investigation by [29] revealed that the key factors limiting wider application of LCC in the UK construction industry were lack of understanding on LCC techniques, absence of standardized guideline and lack of reliable data input for LCC calculations. Nonetheless, this does not mean that the

IOP Conf. Series: Earth and Environmental Science **385** (2019) 012066 doi:10.1088/1755-1315/385/1/012066

industry is not actively promoting LCC as a decision tool for the evaluation of environmental sustainability. Many initiatives have been taken by governments as well as practitioners around the world in the effort of encouraging the application of LCC [9].

3.3. LCC in Green projects

Over the years, studies exploring on environmental and economic analysis of products, technologies and systems has been increasing [38] and [20]. Increment in the research of LCC application in Green projects is also apparent, with variation in the area of interest. This is encouraging news as LCC has long been recognized as one of the foundations for achieving sustainability [16]. Quite a number of researches has been done on LCC applications in green building across the world. For example, [10] had envisaged more LCC applications to profile the performance of green project by assessing the costs and benefits in economic terms. While [32] applied whole-life costing to evaluate different environmentally friendly technological options that relate to construction and facilities management.

There were also studies carried out that attempted to combine the use of LCC with other methods such as Life Cycle Assessment (LCA) [17] and [27] and sensitivity analysis [40] and [23], in order to investigate the implications of critical input variables on LCC for making risk-informed investment decisions concerning the adoption of certain green components. Nonetheless, not many researches have been undertaken to examine LCC implementation in the green scene of Malaysian construction industry. The most current and relevant study was by [25], who attempted to identify the most eco-efficiency scenario in concrete waste management by using LCA modelling together with LCC. Another study carried out a review on the state of cost data inputs of LCC for rigid pavement maintenance and rehabilitation [1]. The research concluded that LCC practice in the relevant area was still at an infancy state due to the difficulties in obtaining quality cost data for a comprehensive and reliable LCC analysis.

[15] emphasized that making environmental decisions is complex and decision makers need practical tools to help them learn about the choices available before deciding on the best sustainable option. [5] highlighted the need to influence critical decisions related to construction and operating costs can be done by applying LCC in the early stages of design process. Apart from that, it was suggested that LCC combined with LCA makes up a valuable tool for decision-makers [7]. [27] stated that a combined-integrated analysis can provide decision makers with a balance set of information to consider both the environment and economy, which is necessary for any green development. He further added that LCC provides vital economic information which help decision-makers fully understand the full costs and benefits involved in the short to long term.

4. Research Methodology

This paper is part of an on-going research to develop a life cycle costing framework for its' application in green projects. Accordingly, the methodology framework proposed for this study will be a mixed-method approach divided into the following four (4) stages:

4.1. Literature Review

The first stage encompasses gathering of relevant literatures from existing publication to be obtained from databases pertaining to scientific and technical research that provides research papers in the area of built environment. The preliminary study will help to identify specific issues and problems in relation to the topic and becomes the basis for extensive literature review and fine tuning for the study. The literature review involves comprehensive understanding on the current state of LCC applied in green projects. The analysis will involve making summary, classification, and comparison of prior research studies.

4.2. Questionnaire Survey

The type of sequential explanatory design of mixed method [11] will be utilised for the research. Hence, a quantitative study will be used to obtain data on the current practice in the use of LCC in green projects including the problems and issues in its application. The data will be gathered through distribution of a set of closed-ended questionnaires to professionals who have been involved in green projects. Respondents from about 300 projects that have been rated green by recognised Malaysia's green rating tool will be identified and gathered. The sample size will depend on the number of projects that have received green certification but will be based on the principle formulated by [22]. Analysis of the data will use SPSS software due to author's familiarity with the software. From the responses, conclusions will be made as to whether or not LCC is frequently applied in green projects, and in cases where LCC was applied, to what extent of its application and whether or not such application had contributed to effective decision making throughout the projects' execution.

4.3. Interview

As mentioned earlier, the research will adopt sequential exploratory method. Thus the next stage is qualitative method of in-depth interview with the purpose of validating data obtained from the quantitative method. The semi-structured interview is meant to get in-depth information in relation to LCC application, the problems and issues, and to understand how LCC had contributed to effective decision making throughout the green projects' execution. From the questionnaire survey, respondents who have been involved in green projects that have applied LCC in any phase of its life cycle, and possess practicing experience of 10 years and more, will be selected as the respondents for the interview session. Generally, at least 5-16 participants will be selected in order allow for a timely intensive analysis for each case to be conducted [36]. The data interpreted by means of content analysis method will then be analysed using Atlas.ti software due to the author's familiarity with this tool.

4.4. Expert Panel

Subsequently, data gathered from both quantitative and qualitative method will be translated into an appropriate framework and will be forwarded to expert panel interview in order to validate the final findings of the framework. Expert panel interviews will complete the triangulation approach to be adopted in this research, which makes the data more reliable and robust. At least 5 experts will be identified for this purpose [3]. Apart from reviewing the findings, the experts will also evaluate the significance of the issues identified from the previous data collection. Once the significance of the findings is validated by the experts, the issues can be properly evaluated based on the expert's evaluation. From here, the framework will be finalized. This framework can act as a reference for the industry and stakeholders and help to improve their understanding on the best approach in implementing LCC in Green projects.

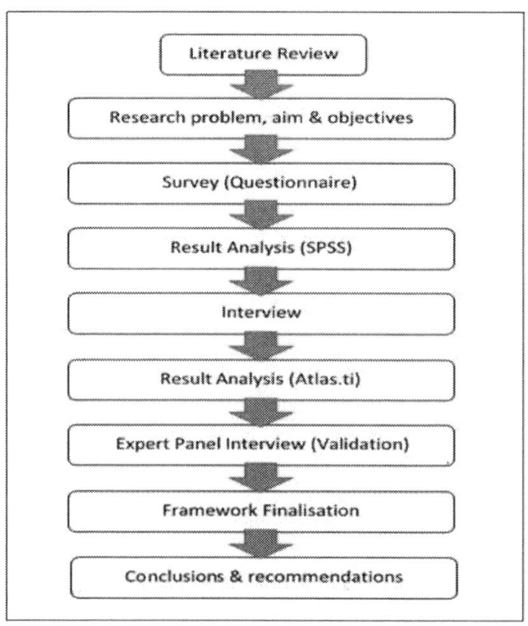

Figure 1: Research methodology framework

5. Conclusion

As a conclusion, it could be summarized that the application of LCC in practice is still limited, especially in green projects, although research trends are increasing in this area. Both public and private sectors are still wedged to the traditional approach of project development, looking towards short term initial cost, rather than the best long-term value. Besides, there is still much work to be done in justifying the need for a conscious shift of focus towards implementing LCC in green projects. This should start by focusing on the clients, who as decision makers play a key role in advocating LCC. Hence, it is crucial for the clients to understand how LCC contributes to making decision effectively. Apart from that, the research aims to develop a life cycle costing framework for its' application in green projects. By adopting the approach discussed in the research methodology section, it is hope that such framework can be established for the construction industry practitioners to improve the application of Life Cycle Costing towards effective decision making for Green Project in Malaysia.

References

[1] Ahmad Jasmi S Z, Ayob M F, Abdul Rashid K, and Mohd Rahim F A 2018 A Review on the State of Cost Data Inputs of Life Cycle Cost (LCC) for Rigid Pavement Maintenance and Rehabilitation in Malaysia *Journal of Design and Built Environment, Special Issue* **1** 2018
[2] Atkinson D 1996 Life cycle costing analysis and its application in the design of South Australian buildings *Master of Applied Science thesis.* Adelaide: University of South Australia
[3] Babbie E R 2011 *Introduction to social research* (5th ed.) (Belmont, Calif: Wadsworth)
[4] Barringer H P and Weber D P 1996 Life cycle costing tutorials *Fifth International Conference on Process Plant Reliability* Gulf Publishing Company (Houston, Texas)
[5] Bogenstatter U 2000 Prediction and optimization of life-cycle costs in early design *Building Research & Information* **28(5/6)** p 376-86
[6] Boussabaine H and Kirkham R J 2004 *Whole life-cycle costing: Risk and risk responses*. Oxford: ZBlackwell
[7] Bruce-Hyrkäs T, Pasanen P, and Castro R 2018. Overview of Whole Building Life-Cycle Assessment for Green Building Certification and Ecodesign through Industry Surveys and Interviews *25th CIRP Life Cycle Engineering (LCE) Conference, 30 April - 2 May 2018, Copenhagen, Denmark, Procedia CIRP* **69** p 178-83
[8] Bull J W 2014 *Life Cycle Costing for Construction* New York: Routledge
[9] Chiurugwi T, Udeaja C, Hogg K and Nel W 2010 Exploration of drivers and barriers to life cycle costing (LCC) in construction projects: professional quantity surveyors' assessment *Ed Tizani, W. Proceedings of the International Conference Nottingham UK* p 1-9
[10] Cole R J and Sterner E 2000 Reconciling theory and practice of life cycle costing *BuildingResearch and Information* **28(5/6)** p 368-75
[11] Creswell J W 2012 *Qualitative inquiry & research design: Choosing among five approaches* (4th ed.) Thousand Oaks, CA: Sage
[12] Eriksson O, Bisaillon M, Haraldsson M, and Sundberg J, 2016 Enhancement of biogas production from food waste and sewage sludge - environmental and economic life cycle performance *Journal of Environment Management* **175** p 33-9
[13] European Commission 2016 *Life-cycle costing* Available at http://ec.europa.eu/environment/gpp/lcc.htm (last accessed: 3 January 2019)
[14] Giacomo M R D, Testa F, Iraldo F and Formentini M 2018 Does green public procurement lead to life cycle costing (lcc) adoption? *Journal of Purchasing and Supply Management* p 1478
[15] Gluch P and Baumann H 2004 The life cycle costing (LCC) approach: a conceptual discussion of its usefulness for environmental decision-making *Building and Environment* **39(5)** p 571-80
[16] Gundes S 2016 The Use of Life Cycle Techniques in the Assessment of Sustainability *Procedia – Social and Behavioural Science* **216** p 916-22
[17] Heijungs R, Settanni E, and Guinée J 2013 Toward a computational structure for life cycle sustainability analysis: unifying LCA and LCC *International Journal on Life Cycle Assessment 2013* **18(9)** p 1722–33

[18] Higham A, Fortune C and James H 2015 Life cycle costing: evaluating its use in UK practice *Structural Survey* **33(1)** p 73-87
[19] Hochschorner E and Noring M 2011 Practitioners' use of life cycle costing with environmental costs - a Swedish study *The International Journal of Life Cycle Assessment* **16** p 897
[20] Islam H, Jollands M, and Setunge S 2015 Life cycle assessment and life cycle cost implication of residential buildings - a review. *Renewable and Sustainable Energy Reviews* **42** p129-40
[21] KeTTHA 2018 *Garis panduan perolehan hijau kerajaan 2.0.* Putrajaya: Ministry of Energy, Green Technology and Water, Malaysia
[22] Krejcie R V and Morgan D 1970 Determining sample size for research activities. *Educational and Psychological Measurement **30*** p 607-10
[23] Langston C 2013 The role of coordinate-based decision making in the evaluation of sustainable built environments *Construction Management and Economics* **31(1)** p 62-77
[24] Lim B T, Zhang W, and Oo B L 2018 Sustainable procurement in australia: quantity surveyors' perception on life cycle costing *International Journal of Integrated Engineering* **10** p 1-6.
[25] Mah C M, Fujiwara T, and Ho C S 2018 Life cycle assessment and life cycle costing toward eco-efficiency concrete waste management in Malaysia *Journal of Cleaner Production* **172** p 3415-27
[26] Mateus R, Neiva S, Bragança L, Mendonça P, and Macieira M, 2013 Sustainability assessment of an innovative lightweight building technology for partition walls - comparison with conventional technologies *Building Environment* **67** p 147-59
[27] Miah J H, Koh S C L and Stone D 2017 A hybridised framework combining integrated methods for environmental Life Cycle Assessment and Life Cycle Costing *Journal of Cleaner Production* **168** p 846-66
[28] Ng B J H, Zhou J, Giannis A, Chang V W C, and Wang J-Y 2014. Environmental life cycle assessment of different domestic wastewater streams: policy effectiveness in a tropical urban environment *Journal of Environment Management* **140** p 60-8
[29] Oduyemi O, Okoroh M and Dean A 2014 Barriers to life cycle costing usage. *Ed Raiden, A.B. and Aboagye-Nimo 30th Annual ARCOM Conference Proceedings* Portsmouth UK
[30] Olubodun F, Kangwa J, Oladapo A, and Thompson J 2010 An appraisal of the level of application of life cycle costing within the construction industry in the UK. Structural Survey **28(4)** p 254-65
[31] Patil, R. B. (2013). Reliability Analysis and Life Cycle Cost Optimization of Band Saw Cutting Machine *Master of Engineering in Mechanical Engineering-Design thesis* Kolhapur: Shivaji University.
[32] Pellegrini-Masini G, Bowles G, Peacock A D, Ahadzi M and Banfill P 2010 Whole life costing of domestic energy demand reduction technologies: Householder perspectives *Construction Management and Economics* **28(3)** p 217-29
[33] Sesana M M, and Salvalai G 2013 Overview on life cycle methodologies and economic feasibility for nZEBs *Building and Environment* **67** p 211-16
[34] Sherif Y S and Kolarik W J 1981 Life cycle costing: concept and practice *Omega* **9(3)** p 287-96.
[35] Shin Y, Cho K, 2015 BIM application to select appropriate design alternative with consideration of LCA and LCCA *Mathematical Problems in Engineering* p 1-14
[36] Smith J A, Flowers P and Larkin M 2009 *Interpretative phenomenological analysis: theory, method and research* Sage: London.
[37] Standards Australia/Standards New Zealand 1999 *Life cycle costing - an application guide, AS/NZS 4536:1999; Standards Australia, Sydney and Standards New Zealand* Wellington p 4
[38] Strazza C, Borghi A D, Costamagna P, Gallo M, Brignole E, and Girdinio P 2015 Life cycle assessment and life cycle costing of a SOFC system for distributed power generation *Energy Conservation and Management* **100** p 64-77
[39] Volkov A, Vitaliy C and Dmitriy K 2014 Life cycle of a building *Advanced Materials Research* **1065-1069** p 2577-80
[40] Wong I L, Perera S, and Eames P C 2010 Goal directed life cycle costing as a method to evaluate the economic feasibility of office buildings with conventional and TI facades *Construction Management and Economics* **28(7)** p 715-35

Productivity Improvement Strategies in Green Construction Project: Formulation of a Theoretical Framework

Noor Aisyah Asyikin Mahat[1], Hamimah Adnan[1], Norazian Mohamad Yusuwan[1] & Mysarah Maisham[1]
[1] Faculty of Architecture, Planning and Surveying, Universiti Teknologi MARA, Shah Alam, Malaysia

Email: nooraisyahasyikin@yahoo.com

Abstract. Productivity is regarded as one of critical importance aspects of construction projects success and persistently deliberated in research topics in the construction industry. Previous researchers have determined different factors that influence construction productivity in conventional construction project, but there is limited research on productivity particularly in green construction projects. To response to the issue, this paper aims to explore a comprehensive literature review to examine the existing knowledge of construction productivity, particularly in green construction projects. Content analysis is used to identify related matters concerning issues and key factors affecting construction productivity, henceforth, a dynamic strategies of productivity improvement are established. Findings from literature are summarized and a theoretical framework of construction productivity improvement strategies is constructed and proposed. The framework highlights the identified factors affecting productivity in green construction projects, as well as determines the improvement strategies to enhance the construction productivity. The framework will be the instrument for the future development and will assist the construction practitioners to plan efficient approaches for achieving productivity growth in green construction projects. An improved understanding on the key factors of the productivity performance would help policy makers and industry to prioritize their policies and actions, whilst keeping of the eminence in Malaysian construction productivity

1. Introduction

The construction industry is an industry of major strategic importance. Its level of productivity has a significant effect on national economic growth and contribute to Gross Domestic Product (GDP)'s growth as an indicator of sustainable economic and development in the country. Under the Construction Industry Transformation Programme (CITP), the Government of Malaysia has placed productivity high on the agenda. According to Construction Industry Development Board [3], (under strategic thrust number three) productivity is the primary engine of growth towards Malaysia's high-income target. As a vital sector to the nation's advancement, the construction industry will lead with high productivity levels through efficient adoption of new technologies and modern practices coupled with high-skilled, highly paid workforce. Productivity was given high priority in the programme because of: (a) a low IBS adoption rate; (b) a largely unskilled workforce; and (c) high proportion of subscale SMEs and Bumiputra firms [3].

Understanding the importance of productivity, the topics has received a great attention of research in construction and engineering management literature, due to the significant influence on the wide economy [1][2][18]. To enhance the construction productivity, a lot of research on strategies to improve construction productivity has been done [12][20][22]. While such studies had improved productivity measurement methods and quantified the impact of influencing factors, there is still a persistent need to explore the strategies for enhancing construction productivity at the industry level [18][25]. A recent study found that construction productivity could be enhanced or hindered by multiple factors within the business environment [12]. The key issues, such as technologies, procurement and workforce, were reflected in the specific political, economic, social and industrial concerns of an economy ought to be the parameter as the strategies in enhancing construction productivity [20].

In the light of the previous research, the present literature of construction productivity are lacks evidence on constraints and strategies concerning productivity improvement particularly in Green Construction Projects (GCP). Such kind of research is important because the construction of green projects differs from that of traditional in terms of the design, materials, and processes [21]. To response to the issues, this paper aims to explore a comprehensive literature review to examine the existing knowledge of construction productivity. Findings from the literature will establish a foundation for dynamic strategies of productivity improvement, particularly in GCP. An improved understanding of the key factors of productivity in GCP would help policy makers and industry to prioritize their policy and action, whilst keeping of the growth of construction productivity performance.

2.0 Literature Review

'Productivity' has been the subject of much debate in the construction industry in many countries for many decades. To formulate and implement measures to improve productivity performance, it is necessary to identify and understand the factors which have an influence on it, as well as the relationships among them [23]. There are many factors that can affect the productivity from the problem of workforce motivation, issues of supervisory, issues of management and project management till weather problems, methods, occupational safety and health, and so forth. In this case, there has been numerous studies about factors that affect productivity in conventional construction projects, however, there is limited research upon productivity in GCP.

Green Building Projects in Malaysia Construction Industry

The construction industry of Malaysia is aware of the importance of sustainable construction to create a high quality living environment for all. As reported in Green Building Index (GBI), as of December 2018, there is 467 numbers of GBI certified Green Building projects in Malaysia. As the target goal in Green Technology Master Plan (2017 – 2030), the numbers will be increased by 550 number of Green Building by 2020, inclusive of green buildings certified by various agencies and organisations such as MyCREST, Green Building Index, Green RE etc. According to [15], green development in Malaysia is still new and researching how these components would contribute towards green development achievement is extremely significant. Henceforth, understanding these variables and uniting all the key components towards green development is critical. Because green buildings use more technologies and material with less environmental impact, they are more complicated than traditional buildings. Moreover, many requirements to achieve a green certificate and shareholders' unfamiliarity with the requirements and technologies tend to lead to cost overruns, project delays, and productivity losses [4] and [16].

Productivity of Green Building Projects

Green building projects generally have more complex designs as compared with traditional building projects [26]. In order to achieve sustainability, green building projects generally require the use of special specifications, materials, construction methods, and building practices [27] and [28]. Moreover, the productivity of the design and construction of green building projects is currently lower than that of traditional projects because practitioners still need time to learn and become proficient in these technologies [11]. Furthermore, unfamiliarity with green technologies and technical difficulties during the construction process can not only affect the project schedule but can also lead to cost increases through rework [29] and [30]. Previous research done by [11] identified the critical factors affecting the productivity of green building construction projects by assessing the likelihood, impact, and criticality of the factors with comparisons against traditional projects. Twenty six (26) factors were identified from a comprehensive literature review and interviews with industry experts. The results indicated that workers' experience, technology, design changes, workers' skill level, and planning and sequencing of work were the top five most critical factors affecting the productivity of green building construction projects.

3.0 Research Methodology

This paper is part of an on-going research to develop a procedural framework for construction productivity improvement in GCP. To accomplish the research, key strategies in productivity improvement particularly in GCP will be investigated. The findings are very important and will be useful as a starting point for future study, especially in formulating a clear guide for the development of upcoming procedural framework. For the purpose of this paper, qualitative analysis technique was used to analyse and interpret the literature. According to [31], content analysis technique is one of the various qualitative techniques that can be used to analyse literatures, whether representing qualitative, quantitative or mixed research that contains numerous sources of qualitative data such as theoretical framework, author's conclusion and interpretation - thereby, justifying within-case qualitative analyses. The process of content analysis involves simultaneous coding of raw data and the construction of categories that capture relevant characteristics of the document's contents [32]. To conduct a literature search, the first step was to identify well known articles relating to productivity in the construction industry. During this process, factors contribute to productivity in GCP as well as key strategies in productivity improvement will be determined. The crucial step in conducting content analysis for this research is to develop a list of strategies to improve productivity, whereby, the references found are mainly from the traditional construction project. The identified attributes were then selected and removed where unneeded, irrelevant and redundant, those that do not contribute to the accuracy of GCP.

4.0 Framework Development Process:

This paper introduced a framework for productivity improvement strategies in GCP. The proposed framework development process is as presented in Figure 1. The framework consists the lists of factors affecting GCP and the strategies to enhance the project productivity throughout construction process. The framework will consider the productivity improvement strategies in Malaysian context, particularly in GCP through the involvement of the local building project stakeholders' expertise, views and preferences in the data collection.

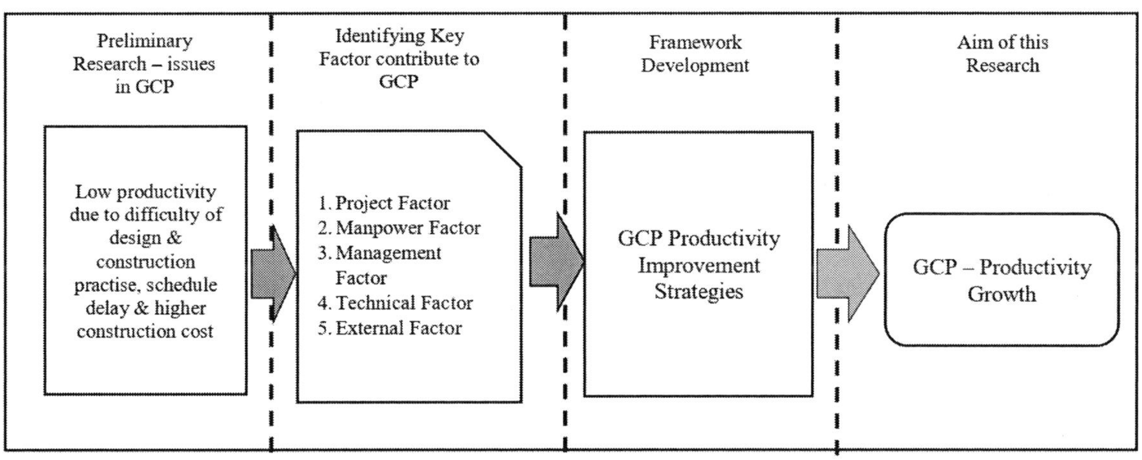

Figure 1. Framework Development Process

The literature findings offered a list of factors contribute to project productivity, and the improvement strategies that can be adopted in GCP. Findings from [11] will be the elementary source of this research to determine the factors affecting productivity in GCP in Malaysia. The research finding by [11] indicates that there are twenty-six (26) factors affecting green construction productivity in Singapore, and the factors were grouped into five major categories – project, manpower, management, technical and external factor. To identify the crucial and critical factors affecting GCP in Malaysia, a questionnaire survey will be distributed among professional who have been involved in GCP. Prior to the distribution of the questionnaire, a pilot study (pre-survey) will be carried out with the experienced researchers and industry specialists to review the appropriateness, structure, readability and validity of the survey questions. Their feedback will be incorporated to revise the survey questions.

Productivity Improvement Strategies

Main contribution of this on-going research is to identify the key strategies of productivity improvement in GCP, therefore, the findings from previous literature on the construction productivity improvement strategies were identified. In this study, author only highlight the findings of productivity improvement strategies of three recent research studies, from year 2016 to 2019. The outcome from the research done by [12] [20] and [23] shows the productivity improvement strategies are limited to conventional construction projects only. However, the outcomes will be addressed and practice in the formulation of a theoretical framework.

A study from [12] investigated the complex interdependence of the factors in driving or hindering construction productivity at the industry, project and activity levels in a systemic manner. The study conceptualised and validated a systemic framework for examining construction industry productivity, and developed three causal loop diagrams (CLDs) for illustrating the dynamic structures that underpin the complex systems of the drivers and constraints. Whilst, [20] adopted PESTEL (political, economic, social, technological, environmental, legal) framework for analysing the macro business environment external to a firm. The research revealed the commonalities of the constraints and strategies shared by the contemporary construction and built environments concerning productivity enhancement through a multi-case lens scrutinizing the construction industries of Singapore, Hong Kong, and the UK. The framework recognizes the potential to explain productivity within the dynamic construction business environment by studying the complexity of the strategies, thereby making a significant contribution to the construction productivity literature. Nevertheless, a study done by [23] recommended ways to help the industry improve productivity and enable the government to devise better policies and programmes. The study suggested that in order to enhance the productivity of Singapore construction firms and practitioners, there is a need to create strategies to have firms compete through performance improvement and have productivity targets and attainments in line with all sectors of the economy.

Critical analysis from the findings of [12] [20] and [23] are summarized and classified into six groups. Proposed recommendation for productivity improvement in GCP will be categorized according to the nature of each factor. The process of grouping of the strategies into its own category was quite challenging as many issues do not fall neatly under one sector. Although the proposed strategies were put under certain categories, it is accepted that other interpretations are possible, as there are conflicts in categorization between published works. Anticipated categories are 1) Government Regulation & Policies, 2) Work Force Development, 3) Project management and administration, 4) Innovative construction technology & technique adoption, 5) Procurement and 6) Effective Site Management. Table 1 signify the detailed of the finding from the [12] [20] and [23] and the proposed categories of strategies aspect.

The key Strategies for Productivity Improvement (as shown in Table 1) will be used as a basis for preparing a discussion guideline in the interview meeting with the experience practitioners in GCP. Result from questionnaire survey and interview meeting then will be triangulated and analysed. Validation of the end-result findings will be reviewed by Expert Panels. Apart from reviewing the findings, the experts will also evaluate the significance of the issues identified from the previous data collection. Once the significance of the findings is validated by the experts, the issues can be properly evaluated based on the expert's evaluation. The research output is in a form of a procedural framework for construction productivity enhancement strategies. This framework can act as a reference for the policy makers and stakeholders to improved understanding of the key drivers of the construction industry's productivity performance, whilst keeping of the eminence in Malaysian construction productivity improvement.

Table 1. Key Strategies for Productivity Improvement

Strategies Aspect	Key Strategy for Productivity Improvement / LR Supporter
1. Government Regulation & Policies	• Policy formation & regulatory requirements [12] • Development of stable and transparent government policies [20]
2. Work Force Development	• Development of sustainable construction workforce [20] • Human resource management [23]
3.Project management and administration	• Planning and design, project management and administration [12] • Project management [23] • Management at company level [23]
4.Innovative construction technology & technique adoption	• Application of new construction technologies and Design standardization [20] • Technology adoption [23]
5.Procurement	• Collaborative procurement [23]
6.Effective Site Management	• Site construction management [12] • Constructability and offsite management [20]

5.0 Conclusion

This paper is part of an on-going research to develop a construction productivity enhancement strategies framework for GCP. The data presented here is primarily based on preliminary review of available relevant literature materials. From the above literature it is concluded that there are many methods of increasing productivity in construction industry. However, the previous research is limited to the conventional construction projects only. It has been noted that increasing the productivity by such above methods for conventional construction projects have reduced cost and time. In the vein of this scenario, this research is intended to identify the key strategies that can sustenance to improve the productivity of GCP. The paper has identified the significant drivers concerning construction productivity in driving or hindering the productivity improvement. The results of this study can also help the practitioners make specific adjustments to traditional project management processes and practices achieving a more productive delivery of green building projects. It is also recommended to establish productivity benchmarks and metrics for different types of green projects which will help practitioners to measure and improve their productivities and competitiveness in green construction.

References

[1] Barbosa, F., Woetzel, J., Mischke, J., Ribeirinho, M. J., Sridhar, M., Parsons, M., Bertram, N.,and Brown, S. (2017). "Reinventing Construction Through A Productivity Revolution." McKinsey Global Institute, (https://goo.gl/1Nqqf8) (Feb. 06, 2017).

[2] Choi, K., and Lee, H. (2016). "Deconstructing The Construction Industry: A Spatiotemporal Clustering Approach To Profitability Modeling." *Journal of Construction Engineering and Management*, 142(10), 04016051.

[3] Construction Industry Development Board (CIDB) (2016),

[4] Construction Industry Institute (CII) (2008). Sustainable design and construction in industrial construction. Austin, TX: Construction Industry Institute (CII).

[5] Dixon, T., McNamara, P., & Newell, G. (2008). The strategic significance of environmental

[6] Dodge Data & Analytics. (2016). World green building trends: Developing markets accelerate global green growth. Smart Market report. Design and Construction Intelligence.

[7] Eastman, C. M., and Sacks, R. (2008). "Relative productivity in the AEC industries in the United States for onsite and off-site activities." *Journal of Construction Engineering and Management*, 134(7), 517-526.

[8] Goodrum, P. M., Zhai, D., and Yasin, M. F. (2009). "Relationship between changes in material technology and construction productivity." *Journal of Construction Engineering and Management*, 135(4), 278-287.

[9] Green, B. (2016). "Productivity in Construction: Creating a Framework for the Industry to Thrive." Chartered Institute of Building, Bracknell, UK,https://goo.gl/aHV3LS, (March 6, Access 2016)

[10] Ho, P. H. K. (2016). "Labour and skill shortages in Hong Kong's construction industry." *Engineering, Construction and Architectural Management*, 23(4), 533-550.

[11] Hwang, B., Zhu, L., and Ming, J. T. T. (2017). "Factors affecting productivity in green building construction projects: The Case Of Singapore." *Journal of Management in Engineering*, 33(3), 04016052

[12] Javed, A. A., Pan, W., Chen, L., and Zhan, W. (2018). "A systemic exploration of drivers for and constraints on construction productivity enhancement." *Built Environment Project and Asset Management*, https://doi.org/10.1108/BEPAM-10-2017-0099.

[13] Kadir, A. M. R., Lee, W. P., Jaafar, M. S., Sapuan, S. M., and Ali, A. A. A. (2005). "Factors affecting construction labour productivity for Malaysian residential projects." *Structural Survey*, 23(1), 42-54.

[14] KeTTHA (2017). Green Technology Master Plan Malaysia 2017-2030 (GTMP), Ministry of Energy, Green Technology and Water (KeTTHA), Putrajaya, Malaysia.

[15] M. I Haloul, S. S Mukhair, 'Study on Effects of Promoting Phases Towards the Implementation of Green Technology in Selangor; Malaysia', International Journal of Advances in Mechanical and Civil Engineering, ISSN: 2394-2827, Volume-3, Issue-4, Aug.-2016

[16] Nalewaik, A., & Venters, V. (2010). Cost benefits of building green. IEEE Engineering Management Review, 38(2), 77–87

[17] O'Connor, J. T., Torres, N., and Woo, J. (2016). "Sustainability actions during the construction phase." *Journal of Construction Engineering and Management*, 142(7), 04016016.

[18] Pan, W. (2018). "Editorial: Rethinking construction productivity theory and practice." *Built Environment and Project Asset Management*, 8(3), 234-238.

[19] Pan, W., and Ning, Y. (2014). "Dialectics of sustainable building: Evidence from empirical studies 1987– 2013." *Building and Environment*, 82(December), 666-674.

[20] Pan, W., Chen, L., and Zhan, W. (2019). "PESTEL analysis of construction productivity enhancement strategies: A case study of three economies." Journal of Management in Engineering, 35(1), 05018013

[21] Shahin Mokhlesian & Magnus Holmén (2012) Business model changes and green construction processes, Construction Management and Economics, 30:9, 761-775, DOI: 10.1080/01446193.2012.6944

[22] Shinde, V. J., & Hedaoo, M. N. (2017). Productivity Improvement in Construction Industry. International Journal of Engineering Technology Science and Research, Vol.4 Issue 12. sustainability by Australian-listed property trusts. Journal of Property Investment & Finance, 26(6), 522-540.

[23] The Singapore Contractors Association Ltd (SCAL) and Singapore Chinese Chamber of Commerce and Industry (2016) , Construction productivity in Singapore: Effective measurement to facilitate improvement.

[24] Australian Construction Industry Forum (2016) Boosting Construction Productivity Policy Paper. Canberra. http://www.acif.com.au/documents/item/786, p. 2.

[25] Yi, W., and Chan, A. P. C. (2014). "Critical Review of Labor Productivity Research in Construction Journals." Journal of Management in Engineering, 30(2), 214-225.

[26] Johnson, S. D. (2000). The economic case for "high performance buildings." Corporate Environmental Strategy, 7(4), 350–361.
[27] Lam, P. T. I., Chan, E. H. W., Poon, C. S., Chau, C. K., & Chun, K. P. (2010). Factors affecting the implementation of green specifications in construction. Journal of Environmental Management, 91 (3), 654–661
[28] Robichaud, L. B., & Anantatmula, V. S. (2011). Greening project management practices for sustainable construction. Journal of Management in Engineering, 27 (1), 48–57.
[29] Hwang, B. G., Thomas, S. R., Haas, C. T., & Caldas, C. H. (2009). Measuring the impact of rework on construction cost performance. Journal of Construction Engineering and Management, 135(3), 187–198.
[30] Hwang, B. G., Zhao, X., & Tan, L. L. G. (2015). Green building projects: Schedule performance, influential factors and solutions. Engineering, Construction and Architectural Management, 22(3), 327–346.
[31] Onwuegbuzie, A. J., Leech, N. L., & Collins, K. M. T. (2012). Qualitative analysis techniques for the review of the literature. The Qualitative Report, 17(56), 1-28.
[32] Merriam, S. (1998). Qualitative research and case study applications in education. California: Jossey-Bass Publisher

IOP Conf. Series: Earth and Environmental Science **385** (2019) 012068 doi:10.1088/1755-1315/385/1/012068

Building Organisational Resilience for the Construction Industy: Strategic Resilience Indicators

Zulkfli Sapeciay[1], Suzanne Wilkinson[2], Seosamh B. Costello[2] and Hamimah Adnan[3]

[1]Faculty of Architecture Planning & Surveying, Universiti Teknologi MARA (UiTM), Malaysia

[2]Department of Civil and Environmental Engineering, The University of Auckland, New Zealand

[3]Faculty of Architecture Planning & Surveying, Universiti Teknologi MARA (UiTM), Malaysia

Email: zul367@salam.uitm.edu.my

Abstract: New Zealand has been affected by a number of natural disasters in recent years. One such example is the 2010 Christchurch earthquake that led to significant impacts to businesses, buildings and critical infrastructure. Construction organisations play an important role in reconstruction programs, and communities rely on their services, particularly post-disaster. Improving the resilience of construction organisations to natural disasters not only minimises the negative consequences to their organisations post-disaster and enhances their organisational performance during business as usual, but also helps to improve community resilience. However, very few studies have developed the concept of organizational resilience and none specifically in the construction sector. This paper identifies strategic resilience indicators for construction organizations. Triangulation analysis of the five studies, in-depth interviews and questionnaire surveys revealed leading indicators of organizational resilience which can help to develop a resilience framework for construction organizations. The findings suggest that the construction industry needs a framework that fosters resilience which itself can potentially enhance organisational capacity to recover quickly after a crisis or disaster.

1. Introduction

The 2011 Christchurch Earthquake, with a 6.2 magnitude, witnessed significant impacts on buildings, lifelines and infrastructure. This resulted in at least 181 fatalities, damage to 100,000 stone and brick buildings, and about 10,000 buildings needing to be demolished and re-built. The New Zealand government established a single body, the Canterbury Earthquake Recovery Authority (CERA), to manage the re-building. Post-disaster, construction organisations are often called upon to manage and carry-out the re-building and reconstruction works, from the demolition of damaged buildings to planning the next phase of rebuilding. Several studies [1][2][3] highlight that in order to ensure that the disaster recovery and reconstruction programs are successfully implemented, it is necessary for construction organisations to be resilient. Enhancing the resilience of construction organisations will help the construction industry to be more resilient. They are also able to respond and recover faster from disasters and continue to contribute their services to the community pre- and post-disaster. However, there is an absence of information on how the resilience practice within an organisation can help enhance the resilience of the construction sector. One possible reason for the lack of information is that the industry consists of various organisations, such as construction clients, planners, architects, contractors, suppliers, consultants and sub-contractors, all of which are unique within their organisational environment [4].

From a construction practitioners' perspective, being resilient might well decide the survival of the affected organisation during a crisis or disaster, and it is imperative that organisations develop strategies based on key indicators, which make their organisation vulnerable to foreseen and unforeseen events.

To begin better understanding of what a resilience organisation may mean to construction organisations, this paper will present a review of existing organisational resilience and its indicators. It also provides an overview; from construction practitioners in New Zealand. Hence, the objective of this paper is to identify set of resilience indicators for construction organizations that will be beneficial from organisational resilience practice. Key indicators of resilience organization are then presented, and five main indicators will be discussed. Finally, an overview of resilience organisation for the construction industry will be given.

2. Organisational Resilience Strategy in New Zealand

There are two distinct strategies identified in managing organisational resilience in New Zealand; Resilient Organisations (ResOrgs) and the New Zealand Ministry of Civil Defence and Emergency Management (MCDEM) strategy. ResOrgs; a group of researchers entrusted by the government to conduct public-good research programmes, has introduced the resilience benchmarking tools for improving organisational resilience. The MCDEM also presented their MCDEM capability assessment tool for evaluating and monitoring organisations' compliance with the CDEM Act and to further to enhance organisational capability towards major crises. Despite both these tools being developed to help organisations in New Zealand to improve their organisational resilience, they are different in specific purpose and approach. On the other hand, the ResOrgs tool looks at the resilience of organisations from a broader business and commercial perspective, the MCDEM tool is more technically focused on CDEM, risk, emergency and business continuity management. The MCDEM tool is based around two National CDEM strategy objectives; 1. Enhance the ability of organisations to prepare for and manage emergencies, and 2. Ensure agencies/organisations are able to function to the fullest possible extent during and after emergencies. The tool uses 12 key performance indicators and its scope is associated with an organisation's risk management plan, emergency response management plan, and business continuity management.

Meanwhile, in 2013 ResOrgs introduced the Resilience Benchmarking tool developed based on the research of [5] and [6]. ResOrgs use 13 indicators of organisational resilience, to help measure the resilience of organisations and they are divided into three interdependent attributes; leadership and culture, networks, and change ready [7]. This tool works by evaluating their resilience strengths and weakness for each indicator, and by benchmarking how they compare to other organisations of a similar size and sector. Other resilience tools reviewed include a benchmarking survey developed for the New Zealand Transport Agency (NZTA) for the purpose of assessing the compliance of Road Controlling Authorities with CDEM 2002 Acts [8], and a framework to measure the resilience of the New Zealand transport system developed for the NZTA in 2014. However, these tools are similar in scope and purpose to the MCDEM capability assessment tool and both the survey and framework have been developed for resilient infrastructure, therefore they were not considered further in this study.

3. An Overview of Key Indicators for Resilience of Organisations

The literature on resilience indicators is vast, contains many methodological approaches and highlights the need for a process of development and validation of indicators that requires a number of specific steps [9]-[11]. Cutter [12] further suggest the stages involve the development of a theoretical framework to provide the basis for variable selection, data weighting and aggregation, and validation. Tanguay [13] extracted relevant indicators from previous studies, interpreting the classification and categorization of indicators based on their frequency. However, most studies on resilience mainly focused on community and societal resilience, city resilience and infrastructure network resilience but they lack information on organisational resilience [2][14]. In 2007, ResOrgs initiated a project aimed at developing a tool for benchmarking the resilience of organizations. ResOrgs utilizes the organisational resilience model; Relative Overall Resilience (ROR) developed by McManus [5] as a conceptual basis for their research and the outcome from Stephenson's work [6] in identifying indicators for resilient organisations. After considering the work of McManus [15][16][5], Stephenson [6] and Lee [17], ResOrgs identified 13 key indicators of organizational resilience which have been used in the Resilience Benchmark Tools and are being implemented by ResOrgs for benchmarking the resilience of organizations in New Zealand. These 13 indicators are divided into 3 groups; leadership and culture, networks, and change ready. They help to measure the resilience of organizations, to monitor processes over time, and to compare resilience strengths and weaknesses against other organizations [7]. In 2013, ResOrgs introduced the 'Resilience Benchmark Tool' for improving organisations' understanding of resilience and its impact on organizational performance. It provides organizations with a tool for evaluating their resilience strengths and weaknesses for each of these indicators, and for benchmarking how they compare to other organizations of a similar size or in a similar sector. Meanwhile, the CDEM Capability Assessment tool is based around two national CDEM strategy objectives and 12 key performance indicators, and the scope is associated with an organisation's risk management plan, emergency response management plan, and business continuity management for MCDE in 2014.

The key organisational resilience indicators introduced by [17] of ResOrgs will be used as a stating point in this study. Their applicability in the construction sector will be determined, they will be modified as appropriate, and they will then be used to assess the resilience of construction organisations in New Zealand. The decision to start with the ResOrgs tool and resilience indicators in this study was based on the experts and practitioners' opinion collected from a preliminary study. In addition, the Resorgs tool is considered the benchmark tool used by leading engineering and lifeline organisations in assessing their organisational resilience. It was decided to emphasis on the ResOrgs tool and its indicators as a baseline from which to develop a resilient organisations framework for the construction sector. This decision was influenced by the results of the preliminary study where construction practitioners and experts acknowledged that ResOrgs indicators were applicable and measurable for the construction sector. It is intended that these indicators can be used in setting a benchmark for measuring the resilience of construction organizations.

4. Methodology

The research methodology in this study consists of a triangulation analysis of literature review, questionnaire survey and in-depth interview to identify the most important indicators related to resilience of organisations in the construction industry. First, a comprehensive literature review was conducted to identify the most frequent resilience indicators mentioned in five studies related to resilient organisation. This study applied the methodology used by [13] and [18] in their research on selecting resilience indicators for cities. Initially, extended literature review on the five studies revealed 72 indicators related to organisational resilience. These indicators were compiled. The indicator recurrence or the indicators that overlap, the description of each indicator was reviewed. For example, this study considered the risk management planning and the planning strategies as the same indicators because they are both sides of the same issue. After classification of indicators, 27 indicators were selected to observe the frequency of use of these indicators in measuring the resilience of organisations by using Nvivo. The purpose is to compare the most frequent indicators used in previous studies related to organisational resilience with the results from questionnaire surveys and interviews undertaken in this research study.

Second, apart from extended review on the five studies, this study collected respondents' views on whether the current resilience assessment tools and key resilience indicators were applicable to construction organizations. An electronic questionnaire survey was undertaken to elicit their views. Based on the feedback from construction practitioners and resilience experts in New Zealand, the researcher decided initially to use the 13 resilience indicators developed by Resilient Organisations (ResOrgs) in the questionnaire survey and interviews. The samples in this survey were building and civil engineering contractors in Auckland, Christchurch and Wellington and in total 118 respondents were selected, 50 answered which is a response rate of 42% and according to [19] and [20], the response percentage is acceptable. The questionnaires was composed of both closed-ended and open-ended questions and were divided into two sections; Section A consisted of general questions to identify the profile of respondents and organisations, and to get an overall view of their organisational resilience or disaster management practice pre- and post-disaster. Section B aimed to find out the respondents' views and opinions on key resilience indicators. Since an extensive literature review on the organisational resilience indicators has been conducted by several studies [5][6][7][15][16] [21], their findings form the basis for this questionnaire. The respondents were asked to rank the most important 13 indicators developed by ResOrgs in managing organisational resilience for the construction sector. A five-point Likert scale was used to capture the importance of the resilience indicators. Table 1 shows that from the completed replies, approximately two-thirds were from sub-contractors, while the others come from main contractors. Since more than 95 per cent of construction organisations in New Zealand are small to medium enterprises (SMEs) or sub-contractors which are usually affected by disasters and crises [22], the views from main contractors and sub-contractors should serve to establish an overview on the current practice of disaster management within the sector.

Table 1. Characteristics of respondents from the questionnaire survey

	Respondent characteristics	No. of Respondents (total 50)	(%)
Designation	Director	25	50.0
	General Manager	13	26.0
	Project Manager	7	14.0
	Contract Manager	5	10.0
Years of experience	0-5	0	0.0
	6-10	1	2.0
	11-15	1	2.0
	16-20	8	16.0
	21 and above	40	80.0
Sub-sector	Building and structure	21	42.0
	Civil engineering	29	58.0
Type of organisation	Main contractor	14	28.0
	Sub-contractor	36	72.0

Third, in addition to the literature review and questionnaire survey, semi-structured interviews were conducted with 23 construction practitioners from construction client and contractor organizations. The primary objective of the interviews was to find out the experts' views and opinions on the existing resilience assessment tools. They were mainly project senior executives of the organisations, usually directors, senior managers and senior contract managers. Each interview lasted approximately 30 minutes and covered a number of topics related to organizational resilience in the construction sector. During the course of the interviews, the construction practitioners were questioned about their views on current practice of organisational resilience, existing resilience assessment tools and key resilience indicators. The interviews were also used as the basis for appraising the resilience indicators for the construction industry and their applicability in developing an assessment tool for construction organizations. The interview transcripts are subjected to qualitative analysis using Nvivo software. This software identified key themes in the literature review, questionnaire surveys and in-depth interview which helped to identify the most important key resilience indicators regarding construction organizations. Table 2 shows the characteristics of respondents from the interview sessions.

Table 2. Characteristics of respondents from the semi-structured interviews

	Respondent characteristics	No. of Respondents (total 23)	(%)
Designation	Director	5	21.7
	General Manager	4	17.4
	Project Manager	10	43.5
	Contract Manager	4	17.4
Years of experience	21 and above	23	100.
Type of	Client	4	17.4
	Main contractor	8	34.8
	Sub-contractor/SMEs	11	47.8

Comparison of the results from the literature review, questionnaire surveys and in-depth interviews were carried out using the triangulation method. Black [23] stated that triangulation can improve reliability by eliminating judgmental bias. Similarly, in the context of construction management research, triangulation as an appropriate research approach to understand phenomena that influence organizations in the construction industry. The findings proved useful in the identification of the most significant indicators. The most significant resilience indicators will be validated in a follow-up survey involving 23 representatives from the same construction practitioners in the interview seasons. Finally, this paper will present the 5 most important resilience indicators for construction organisations.

5. Results and Discussion

5.1 Strategic key resilience indicators

Comparison of the resilience studies reveals that there are different approaches to describing indicators or creating dimensions for resilience indicators. Most resilience indicators describe tools of prevention, characteristics of the situation and critical infrastructure regardless of dimension of resilience [13], [18]. Most resilience studies aim to integrate specific objectives and priorities in disaster management perspectives. Due to limited studies in organisational resilience, this study focused on only 27 indicators from five studies related to resilience of organisations. One of the objectives of this study is to observe the frequency of the use of these indicators. Table 3 presents the 11 indicators that were used 3 times or more in the five studies. These most frequent indicators were identified as significant indicators in developing assessment tools to measure the resilience of organisations. These can be used in developing an assessment tool for enhancing the resilience of construction organizations.

Table 3. Indicators used three times or more in the five studies

Frequency	Indicator	Dimension
5	Planning strategies	Change ready
	Capability & capacity of internal resources	Network
4	Roles & responsibilities	Leadership & culture
	Organisations connectivity	Network
	Silo mentality / Breaking silo	Network
	Leadership	Leadership & culture

	Emergency management plan, Business continuity management	Change ready
3	Staff engagement and involvement	Leadership & culture
	Devolved & responsive decision making	Leadership & culture
	Participation in exercise	Change ready
	Information & leverage knowledge	Network

Apart from literature review on the five studies, a survey was conducted by using the questionnaire survey in exploring whether the construction practitioners were familiar with organizational resilience, 84 per cent of the respondents state that they were aware of the terms. Contrary from the five studies, this survey decided to use 13 indicators developed by ResOrgs New Zealand based on the feedbacks from preliminary survey. Table 6 presents the respondents' views on the degree of importance of the 13 indicators developed by ResOrgs New Zealand, using a five point Likert scale. The purpose is to determine the significance of these indicators to be used on the proposed resilience strategy for the construction sector.

Table 4. Mean rating of key indicators of construction organizational resilience (questionnaire survey)

Indicators	Importance (Mean Rating)
Leadership	4.84
Staff engagement	4.50
Decision making	4.47
Situation awareness	4.34
Planning strategies	4.31
Internal resources	4.22
Proactive posture	4.06
Effective partnership	3.81
Innovation and creativity	3.81
Unity of purpose	3.78
Stress testing plan	3.69
Breaking silo	3.59
Leveraging knowledge	3.47

Results from the survey noted that 'leadership' was ranked the highest mean rating of the organizational resilience indicators with a 4.84 mean rating. This was followed by 'staff engagement', 'decision making', 'situation awareness' and 'planning strategies' receiving 4.50, 4.47, 4.34 and 4.31, respectively. These five highest rated indicators can be regarded as the most important indicators in determining the resilience score in construction organizations. Usually, a boundary will be defined for marking the indicators as significant or non-significant. The researcher decided to focus on the top five significant indicators to develop a resilience framework for construction organisations based on the feedback from construction practitioners and experts on how far a construction SMEs is capable of measuring their resilience. In addition to the literature review and questionnaire surveys, in-depth interviews were conducted in this study. In the interview sessions, 100 percent of respondents were construction practitioners experienced in crises or emergencies, especially after the 2011 Christchurch earthquake, and were very familiar with disaster resilience or disaster management. This provides confidence that the opinions and suggestions from respondents are of value and appropriate to be analyzed in the context of organizational resilience towards disasters. Apart from the respondent's views on the applicability of resilience benchmarking tools developed by ResOrgs and MCDEM for the construction industry, a secondary objective of the interview was to get the practitioners to rank the

IOP Conf. Series: Earth and Environmental Science **385** (2019) 012068 doi:10.1088/1755-1315/385/1/012068

indicators developed by ResOrgs. Table 5 shows that majority of the interviewers state that 'decision making' (mean = 4.83) is considered by most of the interviewees to be the most important indicator in managing construction organisations during crisis. Roles of 'Leadership' (mean = 4.70) occupied the second position in the list while 'staff engagement' (mean = 4.61), 'internal resources' (mean = 4.48) and 'planning strategies' (mean = 4.43) take the third, fourth and fifth position.

Table 5: Mean rating of key indicators of construction organizational resilience (in-depth interviews)

Indicators	Importance (Mean Rating)
Decision making	4.83
Leadership	4.70
Staff engagement	4.61
Internal resources	4.48
Planning strategies	4.43
Effective partnership	4.39
Situation awareness	4.35
Leveraging knowledge	4.17
Proactive posture	4.13
Innovation and creativity	4.04
Unity of purpose	4.04
Breaking silo	3.70
Stress testing plan	3.65

As expected, the results from the in-depth interview sessions had a similar pattern to the findings from questionnaire surveys, especially on the most important resilience indicators. This may be due to the fact that the majority of respondents from interviews and surveys were experienced in reconstruction works after the 2011 Christchurch earthquake, and are aware with the important of disaster resilience for the construction industry. From the triangulation analysis of the data from the literature review (five studies), in-depth interviews and questionnaire surveys, 16 potential resilience indicators for construction organisations have been identified. These 16 potential indicators were re-circulated to the same 23 construction practitioners in the second round interview. The order of the list of indicators for each interviewer was randomized as part of the process in controlling the bias in this study. The respondents were then asked to re-assesses their rating on the most significant indicators actually for developing resilience framework for construction organisation. Table 6 summarizes the result of the second round interviews, which showed the percentage on the important of each resilience indicators for construction organizations.

Table 6: Percentage of respondents indicating the important of resilience indicators for the construction industry

Indicators	Percentage (%)	Ranking
Leadership	91.3	1
Planning strategies	78.3	2
Internal resources	60.9	3
Decision making	56.5	4
Staff engagement	52.2	5
Situation awareness	47.8	6
Leveraging knowledge	43.5	7
Roles & responsibilities	39.1	8=
Breaking silo	39.1	8=
Organisation connectivity	34.8	9=

Unity of purpose	34.8	9=
Participation in exercise	30.4	10=
Effective partnership	30.4	10=
Proactive posture	26.1	11
Innovation and creativity	21.7	12
Stress testing plan	17.4	13

Table 6 revealed that, in the event of a natural disaster, leadership was identified as the most important indicator for construction organisations. The interviewees, who consist of construction experts, stated that emphasis should be placed on the role of committed management and leaders to keep their organisations functioning throughout the post-disaster phases. Furthermore, communities rely on services provided by construction organisations in major restoration and reconstruction activities post-disaster [24]. In the first round of interview sessions, the respondents also were also asked whether the 13 resilience indicators developed by ResOrgs were related to organisational resilience within the construction industry. 32 per cent of the respondents agreed, while 24 per cent viewed the indicators as possibly related, and 44 per cent, the rest of respondents, were uncertain. The possible reason for the survey result can be the perception that resilience in the construction industry only focuses on physical properties [11] or hard resilience elements which consist of 'robustness and redundancy' [25]. Through the interviews, construction experts supported this view and suggested that the existing 13 indicators only cover the elements that measure how organisations manage or respond to disasters. Construction practitioners and experts have suggested a greater emphasis on the technical element when dealing with measuring construction organisational resilience towards natural disasters, supported by the existing organisational element. However, based on the five studies of organisational resilience, questionnaire surveys and in-depth interviews, this study presents key resilience indicators which help construction organisations to enhance their resilience. The majority of experts agreed that the top five indicators identified as most important for assessing resilience of the construction industry were, in rank order; 1) leadership, 2) planning strategies, 3) internal resources, 4) decision making, and 5) staff engagement. These indicators will be discussed briefly to demonstrate their significant for construction organisations and the industry as a whole.

6. Conclusion

This study provides an investigation into resilience of organisations within the construction industry. Based on the survey and interviews, it is clear that this is not currently achieved especially for construction SME's. Consequently, there is a need for resilience management strategies and measurement tools to assess the resilience of organisations in the construction sector. In response, this study used a triangulation analysis method to identify the most important indicators which can help create a guideline or framework to enhance the resilience of construction organisations. A propose framework will developed and then validated in a follow-up interview involving a representative number of the same construction practitioners as part of further research. This study also noted that the indicators introduced by ResOrgs and MCDEM were difficult to use by construction practitioners which need a framework and measurement tool particularly for construction organisation in assessing their organisational resilience. The results from this study identifies that the leading and dominant indicators can help enhance the resilience of construction organisations. While the decision makers and the industry are concentrating on finding a suitable framework for post-disaster recovery and reconstruction, critical indicators of resilience within organisations should be understood by leaders and construction practitioners. These key indicators can be used to develop, evaluate, and execute plans or strategies for pre- and post-disaster for a successful disaster recovery.

Acknowledgement

This paper detail study funded by the Ministry of Higher Education Malaysia and Universiti Teknologi MARA Malaysia. Their support is gratefully acknowledged.

IOP Conf. Series: Earth and Environmental Science **385** (2019) 012068 doi:10.1088/1755-1315/385/1/012068

References

[1] A. Chang-Richards, S. Wilkinson, R. Seville, and D. Brunsdon, "Resourcing of the Canterbury rebuild: Case studies of construction organisations," *Resilience Organisations Research Report 2013/01*, New Zealand, Resilient Organisations. 2013.

[2] Z. Sapeciay, S. Wilkinson, S.B. Costello, and M.N. Sajoudi, "Potential key indicators of organisational resilience for the construction industry," in *The 38th Australasian Universities Building Education Association Conference,* 2013 Auckland, New Zealand.

[3] S. Wilkinson, A.Y. Chang-Richards, Z. Sapeciay, and R. Organizations, "Investment in the construction sector to reduce disaster risk management," 2014. [Online]. Available: http://www.preventionweb.net/english/hyogo/gar/2015/en/bgdocs/inputs/Wilkinson%20et%20al., %202014.%20Investment%20in%20the%20construction%20sector%20to%20reduce%20disaster %20risk%20management%20.pdf. [Accessed Nov. 2014].

[4] Z. Sapeciay, S. Wilkinson, and S.B. Costello, "Building Organisational Resilience for the Construction Industry: New Zealand Practitioners' Perspective," in *International Journal of Disaster Resilience in the Build Environment,* Vol. 8, Issue 1, pp. 98-108. 2017.

[5] S.T. McManus, "*Organisational Resilience in New Zealand,*" Ph.D. dissertation, University of Canterbury. 2008.

[6] A. Stephenson, "Benchmarking the resilience of organisations," Ph.D. Dissertation, University of Canterbury. 2010.

[7] Resilient Organisations, "Resilient Organisations resilience benchmark tool," 2012. [Online]. Available: http://www.resorgs.org.nz/Current-Research/the-resilience-benchmark.html. [Accessed Feb. 2016].

[8] A. Dantas, and S. Giovinazzi, "Benchmarking the readiness of road controlling authorities to meet their obligations under the Civil Defence and Emergency Management (CDEM) Act 2002," New Zealand Transport Agency Research Report 409, New Zealand, Resilient Organisations. 2010.

[9] M. Freudenberg, "Composite indicators of country performance: A critical assessment," *OECD Science, Technology and Industry Working Papers*, no. 2003/16, OECD Publishing, France. 2003.

[10] M. Nardo, M. Saiana, A. Saltelli, dan S. Tarantola, *Handbook on constructing composite indicators: Methodology and user guide.* OECD Publishing, France. 2008.

[11] S.L. Cutter, C.G. Burton, and C.T. Emrich, "Disaster resilience indicators for benchmarking baseline condition", *Journal of Homeland Security and Emergency Management,* vol. 7. iss. 1. article 51. 2010.

[12] S.L. Cutter, L. Barnes, M. Berry, C. Burton, E. Evans, E. Tate, and J. Webb, "A place-based model for understanding community resilience to natural disasters," in Global Environmental Changes, vol. 18. Pp. 598-606. 2008.

[13] G.A. Tanguay, J. Rajaonson, J. Lefebvre, and P. Lanoie, "Measuring the sustainability of cities: An analysis of the use of local indicators," 2019. [Online]. Available: http://www.sciencedirect.com.ezproxy.auckland.ac.nz/science/article/pii/S1470160X09001277#, [Accessed Jan. 2016].

[14] L. Irajifar, T. Alizadeh, and N. Sipe, "Disaster resiliency measurement frameworks: State of the Art": *Proceedings of the 19th CIB World Building Conference*, May 5-9, 2013, Brisbane, Australia. 2013.

[15] S. McManus, E. Seville, D. Brunsdon, and J. Vargo, "Resilience management: A framework for assessing and improving the resilience of organisations," Resilience Organisation Research Report

2007/01. New Zealand, Resilient Organisations. 2007.

[16] S. McManus, E. Seville, J. Vargo and D. Brunsdon. "Facilitated Process for Improving Organizational Resilience," *Natural Hazard Review,* vol. 9, pp. 81-90. 2008.

[17] A.V. Lee, J. Vargo, and E. Seville, "Developing a Tool to Measure and Compare Organisations' Resilience," *Natural Hazard Review,* Vol. 14, pp. 29-41. 2013.

[18] J.M. Normandin, Therrien, G.A. Tanguay, "City strength times of tribulance: Strategic resilience indicators," 2009. [Online]. Available: http://archives.enap.ca/bibliotheques/2012/09/030314256.pdf. [Accessed Jan. 2015].

[19] J. Yu, and H. Cooper, "A quantitative review of research design effects on response rates to questionnaires," in *Journal of Marketing Research,* vol. 20, no.1, pp. 36-44. 1983.

[20] M.K. Malhotra, and V. Grover, "An assessment of survey research in POM: from construct to theory," *Journal of Operation Management*', vol. 16, pp. 407-425. 1998.

[21] A. Stephenson, R. Seville, J. Vargo and D. Roger "Benchmark resilience: A Study of the resilience organisations in Auckland region." Resilience Organisation Research Report 2010/03b. New Zealand, Resilient Organisations. 2010.

[22] T. Hatton, E. Seville, and J. Vargo, "Improving the resilience of SMEs: Policy and practice in New Zealand," Resilient Organisations Research Report 2012/012 October 2012. New Zealand, Resilient Organisations. 2012.

[23] T.R. Black, *Evaluating social science research: An introduction.* Sage Publications, London. 1993.

[24] Z. Sapeciay, S. Wilkinson, and S.B. Costello, "Understanding the resilience of construction organizations," in *International Conference on Building Resilience 2013,* Sri Lanka.

[25] X. Miao, D. Banister, (2012). "Coping with natural disaster through resilience," 2012. [Online]. Available: http://www.tsu.ox.ac.uk/pubs/1059-miao-banister.pdf. [Accessed Aug. 2016].

Ceramic Dust Based Binder for Sustainable Stabilisation of Low Flood-Plain Soil

Mohamad Nidzam Rahmat[1], Sharina Ibrahim[1] and Norsalisma Ismail[1]
[1]Faculty of Architecture, Planning and Surveying, Universiti Teknologi MARA, 40450 Shah Alam. Selangor. Malaysia.

Email: mnidzam676@uitm.edu.my

Abstract. The main objective of this study is to investigate the potential use of ceramic dust, by-products from manufacturing industry as a binder in the stabilisation of Laterite Soil. Laboratory investigation involving fabrication of soil cylinders of laterite soil stabilised with 100% CD and combination of CD with traditional binder Lime and PC at 50:50 ratio at 10%, 20% and 30% stabiliser dosage. The compacted cylinders were cured for 7, 28, 60 and 180 days prior for testing. The results from this study illustrated that all key parameters of compressive strength, durability and linear expansion were within the acceptable engineering standard of soil stabilisation. When using blended binders, it further enhances the engineering properties of stabilised material. From the environmental and sustainability, this cementitious technology anticipated that this new innovative stabiliser can be used in low flood-plain soil to mitigate the effect of flooding.

1. Introduction

The most serious problems in the field of foundation engineering are the changes of water content of some clay soils. These changes may cause undesirable movement of structures particularly for road pavements and building foundations. Since decades, soil engineers have been reported the problems correlated with low-load bearing soils. Thus, in order to produce an improved soil material which has the desired engineering properties, it is essential for the soil to gone thorough alteration of their engineering properties. Either by doing the alteration by mechanical or chemical, this process is called as soil stabilisation. The purpose of soil stabilisation not only limited to enhance the load-bearing of the soil capacity but also improve shear strength, permeability, enhance soil resistance to the weathering process and traffic usage to meet specific engineering projects requirement.

The treatment of these low load-bearing capacity soils is not always easy or economical especially dealing with soft subgrade soil is one of most major problems moreover, with the presence of water due to flood disaster will tremedously affect the nature and stabilty of this soils that significantly contribute to the changeable behavior of its condition [1]. In order to cater this issue, there is a need for soils alteration to improve the engineering properties. The conventional methods such as removing the soil and change for remediation is costly and unsustainable. The best way is to use locally available materials through sustainable cementitious technology with are more sustainable [2]-[5].

Although the conventional soil stabilisation using lime and/or cement is well established, there is also a need of alternative cementitous technologies which are more sustainable, environmentally friendly economical and relatively cheap or free. At present, searching for the best stabiliser to overcome problems occur by the soft soils are becoming the main concern. Instead of using chemical product, utilising the industrial waste by-products may offer more economical alternatives for a wide range application of soil stabilisation [1]. There are various industrial by-products that being utilised for stabilisation of soils such Copper slag [6]-[8], Tile Waste [9], Steel Slag [10], Saw Dust Ash [11], Foundry Sand Waste [8] are some of the prominent waste materials which has been successfully utilised for soil stabilisation. This paper aims to investigate on the innovative soil stabilisation approach utilising industrial wastes Ceramic Dust as based for binders for the application on the low-flood plain.

2. Materials and Methodology

Laterite Clay was used as the target material for the investigation. It was taken from a construction site in Shah Alam, Selangor. The soil is known to have high content of secondary oxides of iron or aluminum or both. Additionally, it is nearly of bases and primary silicates, but it may contain large amount of quartz and kaolinite. The mineralogical studies by Arinze (2015) showed that LC contains Iron Oxide minerals such as goethite ($HFeO_2$); lepidocrocite ($FeO[OH]$) and hematite (Fe_2O_2).

Table 1. Some Common Properties of Laterite Soil (LS)

	Moisture Content (%)	Liquid Limit (%)	Plastic Limit (%)	Plasticity Index (%)	Max. Dry Density (Mg/m3)	Compressive Strength (kN/m2)
Laterite Soil	6.41	41.4	23.5	17.5	1.66	220-623

2.1 Stabilisers

Ceramic Dust (CD) was used as main stabiliser supplied by Guocera Tile Industries Sdn. Bhd. in the form of sludge. It was then dried at room temperature and crushed into powder form. Portland Cement (PC) were manufactured by YTL Cement Sdn. Bhd., Malaysia and Lime was supplied by MCB Industry Sdn. Bhd. Table 2. Shows some oxide composition of material used in this investigation

Table 2. Oxide composition of Target material and stabilisers

Oxides Composition (%)	LS	CD	Lime	PC
SiO_2	26.51	45.26	0.9	9.67
Al_2O_3	33.26	18.32	0.15	4.70
Fe_2O_3	21.56	13.7	25.97	25.29
CaO	0.3	8.36	69.09	49.38
MgO	1.13	5.59	0.46	4.21
$CaCO_3$	-	56.77	2.20	-

2.2 Mix Compositions

LS were stabilised using 100% CD as binder and the combination of CD witl traditional stabiliser Lime and PC (CD:Lime and CD:PC) 50:50 ratio at 10%, 20% and 30% stabiliser dosage. The mix design composition and mix code are presented in Table 3.

Table 3. Mix Design Composition

Target Material	Mix Code	Stabiliser	Ratio (%)	Dosage (%)

IOP Conf. Series: Earth and Environmental Science **385** (2019) 012069 doi:10.1088/1755-1315/385/1/012069

	LCD	CD	100	10, 20, 30
Laterite Soil	LCD_L	CD:Lime	50:50	10, 20, 30
(LS)	LCD_{PC}	CD:PC	50:50	10, 20, 30

2.3 Methodology

Soil cylinder of 50mm diameter and 100mm in height were fabricated from dried LS mixed with the different stabilisers as illustrated in Table 4. All cylinder sample were fabricate based on formula; T + sT + W(T+sT) = 380g. Where T = Target Soil, s = Stabiliser, W = Water. The blended dry materials comprising of LS, CD with Lime or PC were mixed thoroughly prior to adding pre-determined moisture contents based on a proctor compaction test as accordance to BS 1377 to establish the Optimum Moisture Content (OMC) and Maximum Dry Density (MDD) of the material. After mixing these materials were placed in a steel mould fitted with a collar then subjected to a static compression using a hydraulic compacter. The cylinder was then extruded using hydraulic jack. The cylindrical specimens were wrapped with cling film and placed in plastic containers for 7, 28, 60 and 180 curing period. Three cylindrical specimens per mix composition were subjected to Unconfined Compressive Strength (UCS) test as compliance with BS 1377-7:1990. One cylinder sample was set-up in a Perspex container for linear expansion test. For durability Index test, two cylindrical specimens were weighted before completely soaked in water for 4-days to investigate the effect of flash flooding. This was followed by another UCS test and the strength values compared with baseline values obtain from UCS test conducted on the un-soaked specimen equivalent as a percentage for durability index (DI) using formula:

$$DI = \frac{UCS\ value\ for\ soaked\ specimen\ (S)}{UCS\ value\ for\ unsoaked\ specimen\ (US)}\ X\ 100 \quad (1)$$

UCS test were performed using CBR Test machine with capacity of 30kN with dynamic mode 200 points per second (Figure 1)

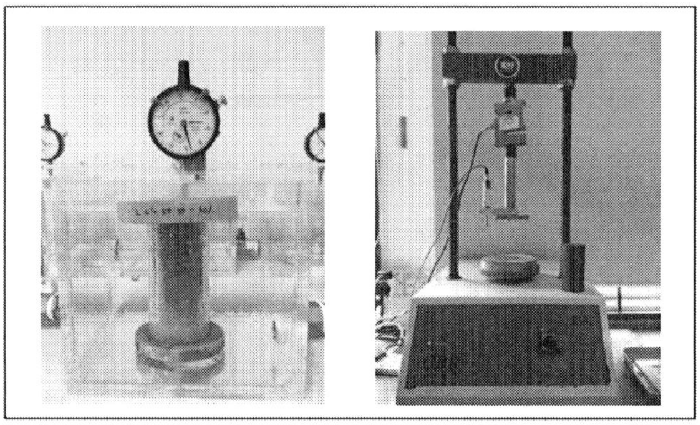

Figure 1. Apparatus for Linear Expansion and UCS Test

3. Results and Discussion

3.1. Control System LS+100% CD

Figure 2 (a), (b) and (c) shows the UCS, Durability Index (DI) and Linear Expansion (Lexp) test of LS+100% CD only as control system. It is clearly shown that stabilised LCD did not perform in both strength development, durability index and imbibe sufficient of water through linear expansion. Increasing the stabiliser dosage from 10% to 30% marginally increase the UCS with prolong curing period to 180 days. Durability Index is also recorded below 10% and Linear expansion were about 2.2% all stabiliser dosage used. This is hypotheses as because there is insufficient lime to produce significant amounts of cementing reaction products allows for modification, with little or no stabilisation during the curing process.

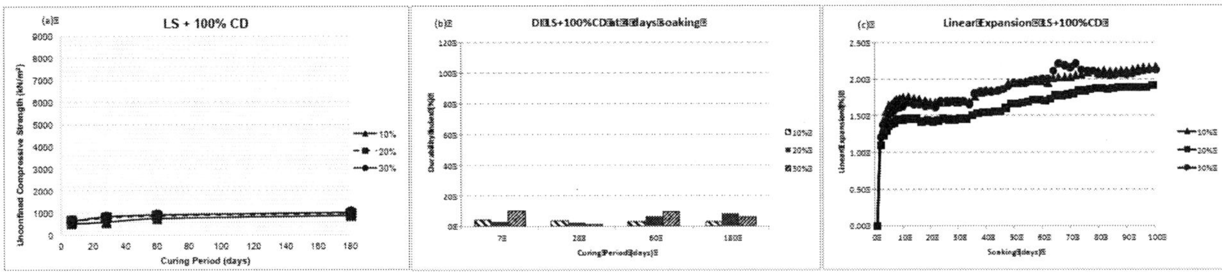

Figure 2. UCS, DI and Lexp test for control system LS+100% CD

3.2. *Unconfined Compressive Strength (UCS)*

Figure 3(a) and (b) illustrate the UCS for LS stabilised with CD combined with Lime (LCD_L) and CD combined with PC (LCD_P) at (50:50) ratio respectively. Both systems using identical stabiliser dosage of 10%, 20% and 30%. Generally for both systems, strength developed tremendously as stabiliser dosage and curing period increased. Early strength of 7 days curing for LCD_L was lower at about 1000 kN/m^2 for all stabiliser dosage compared to LCD_P, which recorded about 2236 kN/m^2 for 30% stabiliser dosage. The strength developed dramatically along all curing period especially at 180 days of curing. At prolonged curing period of 60 and 180 days LCD_L system shown better strength values compared to LCD_P system. Lime in the system enhances the formation of strength contributing silicate (C-S-H) and aluminate hydrates (C-A-H) from the hydration of clay-lime system.

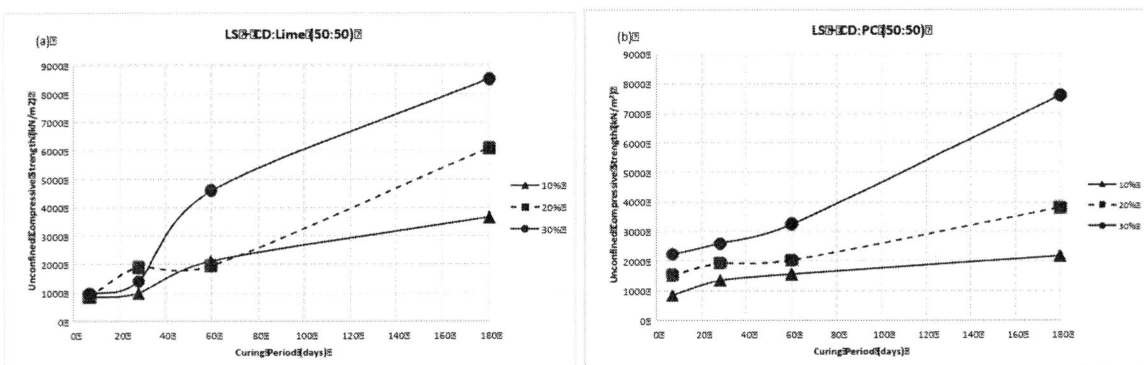

Figure 3. UCS vs Curing Period for (a) LCD_L and (b) LCD_P system.

3.3. *Durability Index (DI)*

Figure 4(a) and (b) shows the Durability Index of LCD_L and LCD_P systems respectively. For durability Index test, specimens were cured for 7, 28, 60 and 180 days prior to fully soaked in the water for 4 days. This is to simulate of flash flood in the low flood-plain situation. Generally the DI for both systems is identical to UCS test results which increasing the amount of stabiliser, increased the DI for both system and increased curing period marginally increased the DI. To compare two systems, again LCD_L system recorded higher DI, means that stabilisation of LS with lime is more durable in case of flooding but LCD_P is more consistent. Incorporation of 30% stabiliser dosage gives higher DI in both systems.

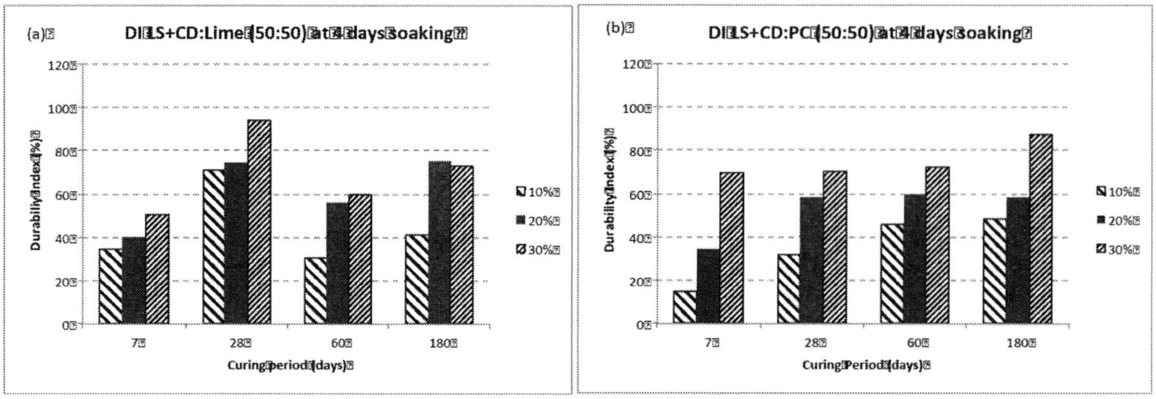

Figure 4. Durability Index for (a) LCD_L and (b) LCD_P system.

3.4. Linear Expansion (Lexp)

Figure 5(a) and (b) illustrate the linear expansion of LCD_L and LCD_P systems. Linear expansion was monitored for 100 days. Readings were recorded every day for the first 10 days and then weekly readings were taken. For both systems, high rate of absorption occurs during the first 10 days of soaking. LCD_L system recorded higher linear expansion, which is about 2.2% compared to LCD_P system that only absorb about 1.2% of water.

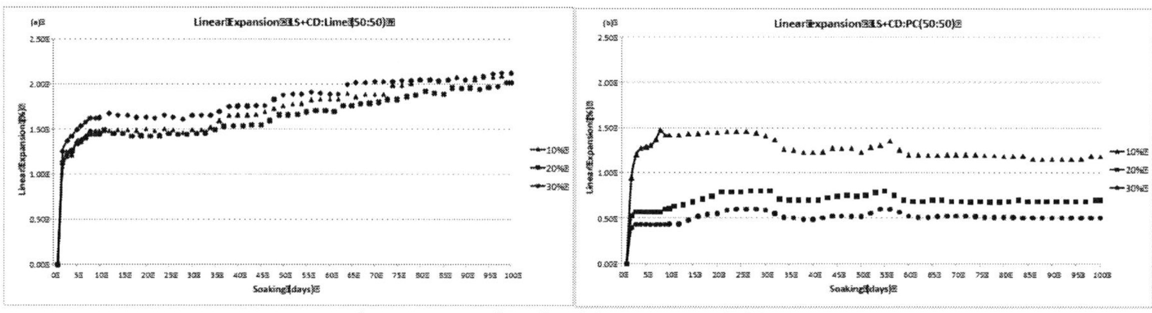

Figure 5. Linear expansion for (a) LCD_L and (b) LCD_P system

4.0 Conclusion

The introduction of CD as blended stabilisers CD:lime, CD:PC to LS clearly increases the compressive strength of the clay soils. Increased amounts of stabiliser dosage from 10% to 30% generally resulted in increased unconfined compressive strength and durability index with increasing curing period for stabilised LS especially in the prolonged curing of 180 days. Over a 100-days period, stabilied LCD_P system exhibited less expansion compared to those specimens LCD_L system and specimen stabilised using 100% CD. This study has shown that Ceramic dust based binders can be produced using a combination of CD, blended with Lime or PC with appropriate ratios. It also indicated that there is an increasing number of potential non-traditional sustainable soil stabilising agents for the better strength and durability achievement of low-load bearing capacity soil. Laboratory investigation of cylindrical soil specimens of LCD_L and LCD_P system recorded better engineering performance compare to stabilised LS with CD alone.

Acknowledgments

The authors would like to thank Universiti Teknologi MARA (UiTM) Shah Alam, Malaysia for sponsoring this research 600-RMI/MyRA 5/3/ LESTARI (khas) 239/2017 and the Faculty of Architecture, Planning and Surveying for the technical assistance and use of laboratory testing facilities. Thank you also goes to TTDI Alam Impian, Guocera Tile Industries Sdn. Bhd. and

YTL Slag Cement Sdn. Bhd. for encouragement and unlimited supply of Laterite Clay, Ceramic Dust and Lime respectively.

References

[1] Fauzi, A., Djauhari, Z., & Fauzi, U. J. (2014). Soil Engineering Properties Improvement by Utilization of Cut Waste Plastic and Crushed Waste Glass as Additive.

[2] Lim, S. M., Lim, A. J. M. S., Ismail, B., & Wijeyesekera, D. C. (2012). Critical review of innovative soil road stabilization techniques.

[3] M.A. Saleem, S.M. Kazmi, S. Abbas, (2017), Clay bricks prepared with sugarcane bagasse and rice husk ash – a sustainable solution, MATEC Wev of Conference, EDP Science, 2017 pp 03001

[4] S.M. Kazmi, S. Abbas, M.I. Nehdi, M.A. Saleem, M.J. Munir, (2017), Feasibility of using waste glass sludge in production of eco-friendly bricks, J. Material Civil Eng 29 (8) (2017) pp 04017056.

[5] G.H.M.J.S. De Silva, B.V.A. Perera (2018), Effect of waste rice husk ash (RHA) on structural, thermal and acoustic properties of fired clay bricks. J. Material Civil Eng 18 (2018) pp 252-259.

[6] E. Ravi, R. Udhayasakthi, T. Senthil Vadivel, (2016) "Enhancing the Clay Soil Characteristics using Copper slag Stabiliation",Journal of Advances in Chemistry, Vol.12, Issue No. 26, December 2016.

[7] Mohammed A. Qureshi, Hevin M. Mistry, Vikas D. Patel, (2015), "Improvement in Soil Properties of Expansive Soil by Using Copper Slag", International journal of Advance Research in Engineering in Science and Technology, Vol.2, Issue No.07, July 2015

[8] P. Rajendra Kumar, P. Suresh Praveen Kumar, G. Maheswari, (2017), "Laboratory Study of Black Cotton Soil Blended with Copper Slag and Fly Ash", International Journal of Innovative Research in Science, Issue No.2, February 2017

[9] K.P. Summaya, Mohammed Refeequedheen. K, Sameer V.T, Firoz, Khais .P.T, Jithin. K, (2016), "Stabilization of expansive soil Treated with Tile Waste", International Journal of Civil Engineering, Vol. 3, Issue No. 03, March 2016.

[10] A. Mohan chand, V Ramesh babu, B. Ramesh babu, K. Niveditha, (2017), "Behaviour of Black Cotton Soil with Addition of Copper Slag and Steel Slag", International Research Journal of Engineering and Technology, Issue No. 01, Jan-2017

[11] Wajid Ali Butt, Karan Gupta, J.N Jha, (2016), "Strength behaviour of Clayey soil Stabilized with Saw Dust Ash", International Journal of Geo-Engineering, Issue No. 21 October – 2016

The mediating effects of cost estimates reliability on BIM adoption: SEM model analysis

Noor Akmal Adillah Ismail[1], Mysarah Maisham[1], Zulkhairy Affandy Mohd Zaki[1], Raja Rafidah Raja Muhammad Rooshdi[1] and Shaza Rina Sahamir[1]
[1]Faculty of Architecture, Planning and Surveying, Universiti Teknologi MARA Shah Alam, Selangor, Malaysia

Email: noorakmaladillah@gmail.com

Abstract. This paper presents the Structural Equation Modelling (SEM) analysis to evaluate the mediating effects of cost estimates reliability towards the Building Information Modelling (BIM) technology adoption. Using results from a survey by questionnaire, SEM analysis employing SPSS-AMOS application is to assess the relationships amongst the studied constructs (Improved Information, Perceived Benefits, BIM Adoption and Cost Estimates Reliability) by testing three hypotheses. The SEM model was established with its validation through the assessment on its uni-dimensionality, validity, reliability, and fitness index, prior to the hypotheses testing. There are two effects possibly involved in the constructs' relationships, namely direct and indirect effects, in which were examined using Sobel Mediation Test. The results showed that from all mediating relationships developed amongst the constructs, only Cost Estimates Reliability construct demonstrates its role as partial mediator in the direct relationship between Improved Information and BIM adoption constructs. Whilst the other two hypotheses failed the test, indicating no mediation is possible in between the related constructs relationships. The analysis confirmed that the BIM capabilities could provide better mechanism in understanding project information to establish more reliable cost estimates. It could then improve the estimators' skills and knowledge throughout the whole cost estimating process, hence potentially motivating them towards adopting the BIM technology.

1. Introduction

Nowadays, the use of BIM technology in the construction industry has been widely employed, claimed help improve productivity in construction practices. There are several studies acknowledged many benefits of adopting BIM especially in improving planning and design including cost and time saving, quality and performance improvement, and collaboration and communication enhancement. Particularly, in preparing construction cost estimates, the application of BIM tools also assists the Quantity Surveyors' tasks in establishing more reliable project costs.

The innovation of BIM would possibly upgrade the traditional method of cost estimating in encountering constant errors, inaccuracies, omissions and ethical flaws [1]. With BIM 3D models visualizing the construction sequences by project stages, it allows the Quantity Surveyors to manipulate more accurate information to subsequently produce more reliable cost estimates [2]–[5]. The provision of digital database furnished by BIM platform restores the limitation of conventional 2D drawings usage, hence allowing all disciplines in project team to more effectively integrate information for cost estimating purposes [6][7]. Also, data coordination that could be accomplished by BIM tools authorizing multiple views of building models, could help to evaluate components clashes and discrepancies in related building structure [7][8].

On the other hand, the BIM technology diffusion is designated by the perceived benefits of the professionals in the construction industry through the utilization of BIM [9]. Therefore, it could be presumed that the benefits obtained from BIM implementation in assisting the cost estimates preparation by the Quantity Surveyors might lead them to adopt the technology in their practice. Accordingly, by demonstrating SEM analysis, this paper aims to examine the mediating effects of cost estimates reliability on BIM adoption amongst the Malaysian Quantity Surveyors as cost estimators.

2. Research methods

For this paper, data were gathered through a survey using questionnaire. The pre-tested questionnaire was distributed amongst the Quantity Surveyors registered with the Royal Institution of Surveyors Malaysia (RISM). From the total of 1140 registered members on the list, 294 were chosen as samples from the overall population [10], by employing random sampling technique [11]. A satisfactory response rate was remarked at 68.5% [12][13] where 202 Quantity Surveyors responded and their feedback were usable for further analysis. Likert scale measurement was used for the questionnaire, requiring the respondents to rate their opinions towards the effects of BIM on cost estimates reliability, consequently affects the adoption of the technology amongst them. Four main constructs were assessed namely *BIM perceived benefits*, *BIM improved information*, *cost estimates reliability* and *BIM adoption* with three hypotheses incorporated: (H1) Perceived Benefits mediates the relationship between Improved Information and BIM Adoption; (H2) Cost Estimate Reliability mediates the relationship between Improved Information and BIM Adoption; (H3) Cost Estimate Reliability mediates the relationship between Improved Information and Perceived Benefits. The hypotheses were established upon few technology adoption theories from the literature namely Roger's Innovation Diffusion Theory (IDT)[14], Theory of Reasoned Action (TRA)[15], Theory of Planned Behaviour (TPB)[16], Technology Acceptance Model (TAM)[17], and Unified Theory of Acceptance and Use of Technology (UTAUT)[18]. The SEM method engaging SPSS-AMOS was used to evaluate the studied constructs and their mediating effects relationships. Prior to testing those mediating effects, the final SEM model was developed in previous study [19] as shown in Figure 1.

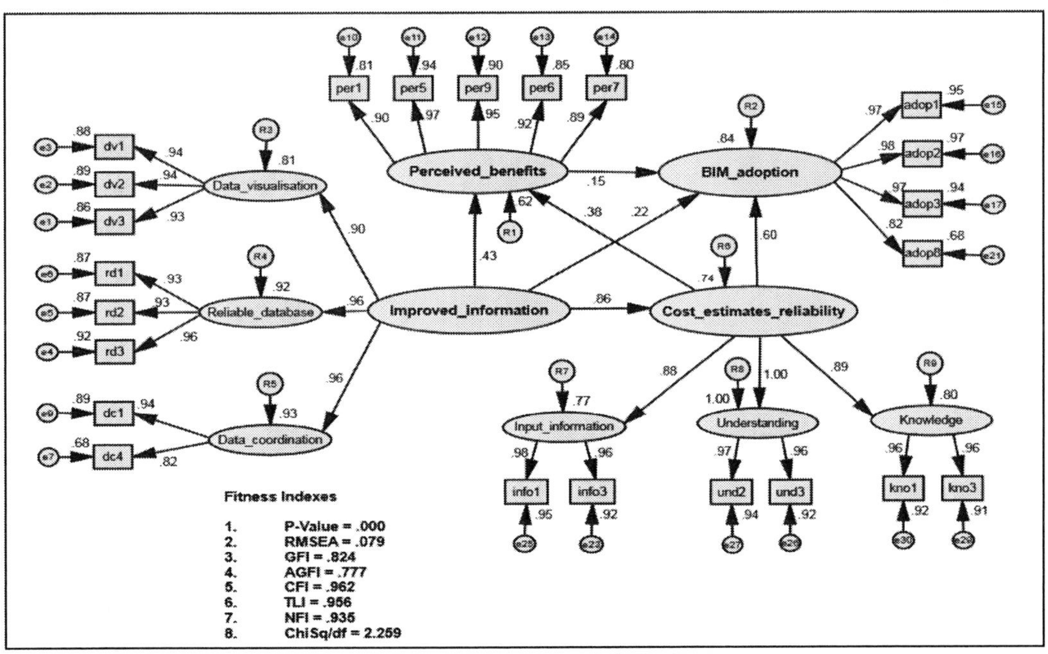

Figure 1. SEM model [19]

IOP Conf. Series: Earth and Environmental Science **385** (2019) 012070 doi:10.1088/1755-1315/385/1/012070

The mediating effects between the constructs were analysed to confirm the hypotheses H1, H2 and H3 within the model relationships. In analysing a mediator amongst the constructs involved, there were two effects incorporated in the relationships; direct effect and indirect effect. Direct effect exists directly between the independent construct and dependent construct, while, the indirect effect comes from the independent construct to the dependent construct but goes indirectly through the mediating construct. The mediating effects were tested using the Sobel Mediation Test. According to this test, to confirm whether mediating effects exist in the tested relationship, the value of total indirect effects must be higher that the direct effect. This is due to when the mediator enters the relationship and has an effect indirectly on the relationship; the direct effect will be reduced. Some of the direct effects have shifted to the mediator. The mediating effect acts as 'partial mediation' if the value of direct effect is still significant, even though it is reduced. Instead, if the value of direct effect is no longer significant, it is called 'complete mediation'. However, no mediation occurs if the direct effect value is larger than the total of indirect effects when calculated.

3. Results and discussion

Figure 1 was simplified as Figure 2 for the mediating effects evaluation purpose. Figure 2 illustrates the direct effects and indirect effects developed from the constructs' relationships, with their respective regression weights derived from the established SEM model [14]. The standardised regression weights and their significance for each path can be referred to in Table 1.

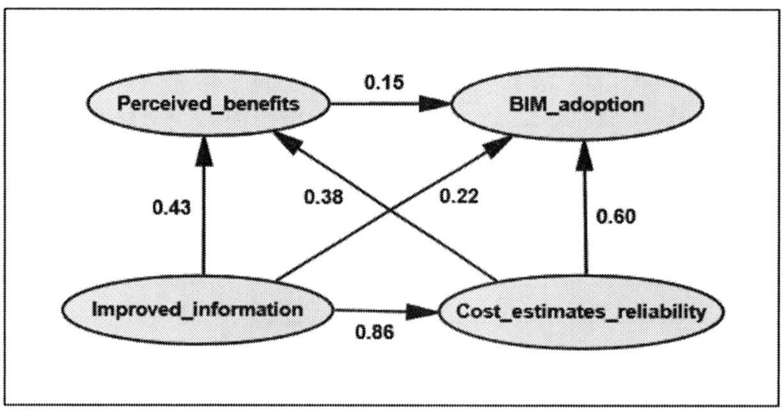

Figure 2. Direct and indirect effects in the constructs' relationships

Table 1. The Regression Path Coefficient between the constructs and its significance

Construct	Path	Construct	Estimate	S.E	C.R.	P-value	Result
Cost Estimate Reliability	<---	Improved Information	**0.862**	0.067	12.582	***	Significant
Perceived Benefits	<---	Improved Information	**0.434**	0.117	4.041	***	Significant
Perceived Benefits	<---	Cost Estimate Reliability	**0.382**	0.117	3.627	***	Significant
BIM Adoption	<---	Perceived Benefits	**0.149**	0.054	2.824	0.005	Significant
BIM Adoption	<---	Cost Estimate Reliability	**0.601**	0.088	7.685	***	Significant
BIM Adoption	<---	Improved Information	**0.219**	0.084	2.873	0.004	Significant

**** indicate high significance at < 0.001*

Table 1 shows the paths and its coefficient (in bold) which indicate the effects of every construct on its respective measuring constructs. For example, the path coefficient of Improved Information to Cost Estimate Reliability is 0.862. This value indicates that for every unit increase in Improved Information, its effects would contribute 0.862 units of increase in Cost Estimate Reliability. The effects of Improved Information to Cost Estimate Reliability are significant with P-value of less than 0.05.

There are two mediators analysed in the model; Perceived Benefits construct and Cost Estimates Reliability construct. The Perceived Benefits construct potentially gives mediating effect in the direct relationship between Improved Information as an independent construct and BIM Adoption as a dependent construct (H1). The Cost Estimate Reliability construct may influence two direct relationships between Improved Information and BIM Adoption constructs (H2), and between Improved Information and Perceived Benefits constructs (H3).

The first mediator to be tested is the Perceived Benefits construct as a mediator towards the direct relationship of Improved Information and BIM Adoption constructs (H1). The main hypothesis of this relationship is that Perceived Benefits mediates the relationship between Improved Information and BIM Adoption. Based on values and relationships in Table 1, Figure 3 describes the procedure for testing the mediating effect of Perceived Benefits towards the relationship. As presented in Figure 3, the mediation test for the Improved Information-Perceived Benefits-BIM Adoption shows that no mediation occurs in the relationship because the total value of indirect effect (0.06) is lower than the value of direct effect (0.22).

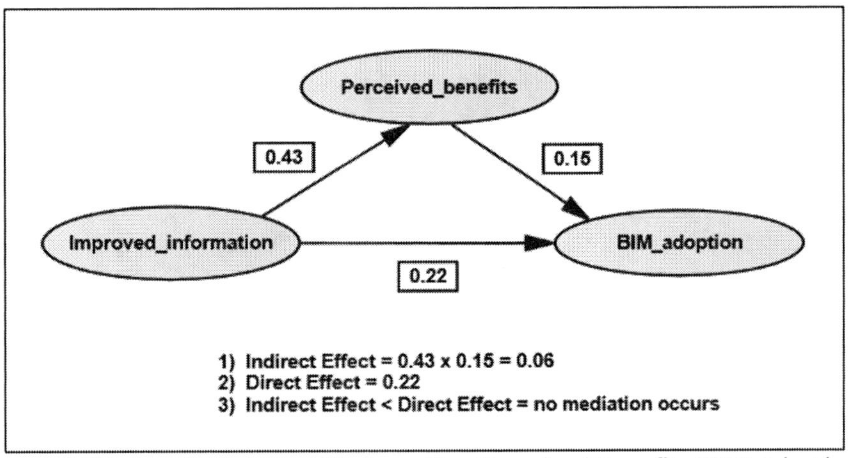

Figure 3. The mediation test for Improved Information-Perceived Benefits-BIM Adoption relationship

The second mediator to be tested in the model relationship is the Cost Estimate Reliability construct. This construct potentially mediated two direct relationships; the Improved Information-Cost Estimates Reliability-BIM Adoption relationship and the Improved Information-Cost Estimates Reliability-Perceived Benefits relationship. Firstly, the Cost Estimates Reliability construct is tested as a mediator of the Improved Information-Cost Estimates Reliability-BIM Adoption relationship (H2). The standardised regression weights and their probability values are obtained from Table 1 to indicate the significance for the paths involved in the relationship. Figure 4 represents the mediation test for the relationship involving the direct and indirect effects in the paths. The mediation test for the Improved Information-Cost Estimate Reliability-BIM Adoption confirms that mediation occurs in the relationship. The total value of indirect effect (0.51) is higher than the value of direct effect (0.22). The type of mediation involved in this relationship is partial mediation, since the effect in the relationship (Improved Information to BIM Adoption) is still significant (refer Table 1) after the mediator (Cost Estimates Reliability) is included in the model.

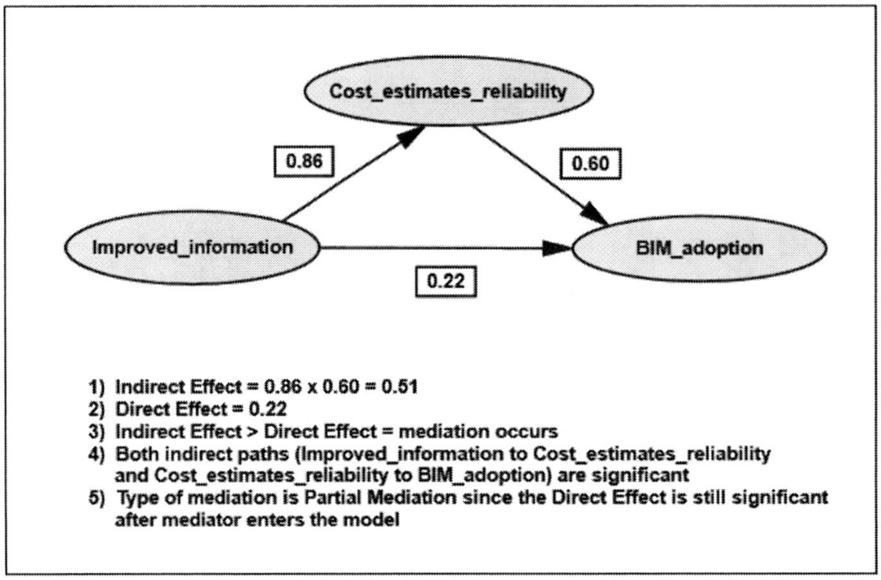

Figure 4. The mediation test for Improved Information-Cost Estimates Reliability-BIM Adoption relationship

Next, the mediating effect of the Cost Estimate Reliability construct towards the direct relationship of Improved Information and Perceived Benefits (H3) was tested. The hypothesis tested was that Cost Estimate Reliability mediates the relationship between Improved Information and Perceived Benefits. The standardised regression weights and their probability values indicating the paths' significance are previously shown in Table 1. Based on the values and relationships in Table 1, Figure 5 below describes the procedure for testing the mediating effect of Cost Estimate Reliability towards the relationship. As presented in Figure 5, the mediation test for the Improved Information-Cost Estimate Reliability-Perceived Benefits shows that no mediation occurs in the relationship. The total value of indirect effect (0.33) is lower than the value of direct effect (0.43).

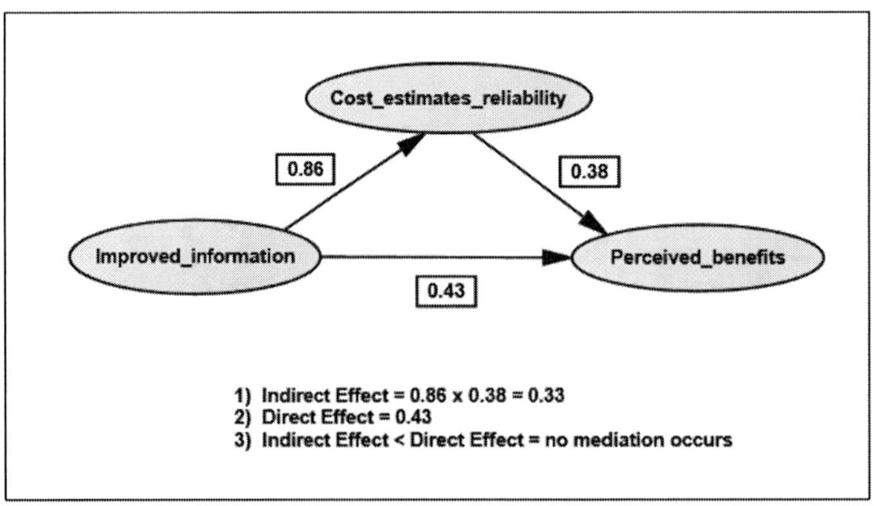

Figure 5. The mediation test for Improved Information-Cost Estimate Reliability-Perceived Benefits relationship

The mediation test conducted on potential mediators in the model, which are Perceived Benefits and Cost Estimate Reliability, is verified. From the mediating relationships developed in the model, only Cost Estimate Reliability construct has a role as partial mediator in the direct relationship between Improved Information and BIM Adoption constructs. The mediation test confirms the hypotheses H1, H2 and H3 as described earlier. The result of the overall mediation tests is summarised as in Table 2 below.

Table 2. The hypotheses result for mediation tests

Hypotheses Statements	Types of Mediation	Results on Hypotheses
H1: Perceived Benefits mediates the relationship between Improved Information and BIM Adoption	No mediation	Not supported
H2: Cost Estimate Reliability mediates the relationship between Improved Information and BIM Adoption	Partial mediation	Supported
H3: Cost Estimate Reliability mediates the relationship between Improved Information and Perceived Benefits	No mediation	Not supported

The results in overall have demonstrated significant relationships in between cost estimates reliability and BIM adoption. It has been manifested through the mediating effects in the model that the capabilities of BIM tools contributing towards improved project information to generate more reliable project costs, could possibly become the main reason why the Quantity Surveyors as estimators mechanize BIM operation in their practices.

4. Conclusion

This paper employed SEM using SPSS-AMOS application to examine the mediating effects within the relationships amongst the constructs in SEM model, namely Improved Information, Perceived Benefits, Cost Estimate Reliability and BIM Adoption. The mediating effects of the constructs were assessed through three relationships (H1, H2 and H3). Prior to testing hypotheses for the mediating effects, a SEM model was established in the previous study prescribing the possible relationships involved in the model. In conclusion, Structural Equation Modelling (SEM) approach was used to establish not only general causal effects relationships, but also to determine statistically the mediating effects amongst the constructs relationships within the model. The findings highlighted that Cost Estimates Reliability construct possesses a role as partial mediator in between the direct relationship of Improved Information and BIM Adoption constructs. Evidently, the results collaterally indicate that BIM application providing better project information, could lead towards the development of more reliable cost estimates. Subsequently, it could become the motivating factor for the Quantity Surveyors as estimators to adopt the technology in their practice, foreseeing that the BIM tools would facilitate them to perform cost estimating tasks more effectively. It has been theorised that the more the given perceived relative advantage, the better the adoption rate will be. Along the same lines as perceived benefit, TAM termed perceived usefulness to describe how the individual's beliefs that technology could help them perform their job better may lead to the rejection or acceptance of the technology [17]. Similarly, in UTAUT, they combine the constructs of relative advantages of IDT and perceived usefulness in TAM to form the measure of performance expectancy in evaluating technology adoption [18]. They define the performance expectancy as individuals' beliefs that using the technology at some point will assist them to increase job performance. Meanwhile, beliefs function as behavioural determinants to adopting a technology [17] and TPB reiterates that it is to be determined by a person's intentions and actions. Accordingly, in TRA [15] added that to understand what factors contribute to people's behaviour whether to adopt or not to adopt a technology, there is a need to assess their beliefs, the grounds that affect their attitudes towards behaviour. It is significant that the findings of this paper could afford some data in developing related research frameworks for further research, specifically in integrating BIM innovation with cost estimating practice within the construction industry.

References

[1] O. A. Olatunji and W. D. Sher, "A comparative analysis of 2D computer-aided Estimating (CAE) and BIM estimating procedures," in *Handbook of Research on Building Information Modeling and Construction Informatics: Concepts and Technologies*, IGI Global, 2010, pp. 170–189.

[2] N. Thurairajah and D. Goucher, "Advantages and challenges of using BIM: A cost consultant's perspective," in *49th ASC Annual International Conference Proceedings, California Polytechnic State University (Cal Poly)*, 2013.

[3] A. Z. Sheth and S. M. Malsane, "Building Information Modelling, a tool for green built environment," in *All India Seminar on Innovation in Green Building Technology, Green Build 2014*, 2014.

[4] R. Eadie, H. Odeyinka, M. Browne, C. Mckeown, and M. Yohanis, "An analysis of the drivers for adopting Building Information Modelling," *J. Inf. Technol. Constr.*, vol. 18, pp. 338–352, 2013.

[5] C. Furneaux and R. Kivvits, "BIM – Implications for Government," CRC for Construction Innovation, Brisbane, 2008.

[6] InfoComm International, "Building Information Modeling (BIM) Guide," 2011.

[7] R. P. Kumanayake and R. M. P. S. Bandara, "Building Information Modelling (BIM); How it improves building performance," in *International Symposium on Ensuring National Security Through Reconciliation & Sustainable Development*, 2012, pp. 357–365.

[8] D. A. Campbell, "Building Information Modeling: The web 3D application for AEC," in *Web3D 2007*, 2007, pp. 173–177.

[9] W. Lu, Y. Peng, Q. Shen, and H. Li, "A generic model for measuring benefits of BIM as a learning tool in construction tasks," *J. Constr. Eng. Manag.*, no. 139, pp. 195–203, 2012.

[10] R. V Krejcie and D. Morgan, "Determining sample size for research activities," *Educ. Psychol. Meas.*, vol. 30, pp. 607–610, 1970.

[11] J. W. Creswell, *Education research: Planning, conducting and evaluating quantitative and qualitative research*, 4th ed. Boston: Pearson, 2012.

[12] Y. Baruch, "Response rate in academic studies - A comparative analysis," *Hum. Relations*, vol. 52, no. 4, pp. 421–438, 1999.

[13] T. P. Johnson and J. S. Wislar, "Response rates and nonresponse errors in surveys," *JAMA J. Am. Med. Assoc.*, vol. 307, no. 17, pp. 1805–1806, 2012.

[14] E. M. Rogers, *Diffusion of innovations*, 4th ed. New York: Free Press, 1995.

[15] I. Ajzen, "Martin Fishbein's legacy: The Reasoned Action Approach," *Ann. Am. Acad. Pol. Soc. Sci.*, vol. 640, no. 1, pp. 11–27, 2012.

[16] I. Ajzen, "The Theory of Planned Behavior," *Organ. Behav. Hum. Decis. Process*, pp. 179–211, 1991.

[17] F. D. Davis, "Perceived usefulness, perceived ease of use, and user acceptance of information technology," *MIS Q.*, vol. 13, no. 3, pp. 319–340, 1989.

[18] V. Venkatesh, M. G. Morris, G. B. Davis, and F. D. Davis, "User acceptance of information technology: Toward a unified view," *MIS Q.*, vol. 27, no. 3, pp. 425–478, 2003.

[19] N. A. A. Ismail, N. H. Idris, H. Ramli, R. R. R. Muhammad Rooshdi, and S. R. Sahamir, "The relationship between cost estimates reliability and BIM adoption: SEM analysis," *IOP Conf. Ser. Earth Environ. Sci.*, vol. 117, no. 1, 2018.

IOP Conf. Series: Earth and Environmental Science **385** (2019) 012071 doi:10.1088/1755-1315/385/1/012071

The Readiness Of The Stakeholders In The Implementation Of Low Carbon Cities Framework (Lccf) In An Urban Area: Methodology Of Research

Siti Kartina Juhari[1], Dasimah Binti Omar[1], Oliver Ling Hoon Leh[1], Siti Mazwin Kamarudin[1], Marlyana Azzyati Marzukhi[1].

[1]Faculty of Architecture, Planning and Surveying, Universiti Teknologi MARA, Shah Alam, Selangor, Malaysia.

Email: kartinauitm@gmail.com

Abstract. In the awareness of global climate challenges, Malaysian local authorities are inspired to apply Low Carbon City Framework (LCCF) to mitigate the carbon emission at city, project and regional scale. Effective implementation measure is believed will enable the private sector to emulate the LCCF implementation at a given scale. However, within the boundary of intense industrialization 4.0, it is observed that the LCCF implementation measures in local authorities are moving slowly. LCCF Checklist is readied, however, its effectiveness is unknown. The aim of this paper is to seek the readiness of local authorities and related stakeholders in implementing the LCCF. The methodology of this research is a mixed method namely qualitative and quantitative approaches. The survey methods are by interview, questionnaire and observation. Town planners, who will be interviewed, are the subject matter expert in managing the planning permission submission for the development control of their areas. Descriptive statistical analysis will be used to show the willingness of the stakeholders namely the developers and planning consultants in implementing the LCCF. Cross-tabulation and correlations tests will be used to improve the understanding of the willingness of the stakeholders. The contribution of this research will gauge the readiness of the major construction industry stakeholder in applying the LCCF at a major local authority in Malaysia. It is also may reflect a large portion perception of industry on the LCCF implementation and its readiness at local authorities' level. This research is anticipated to be a very useful tool through the LCCF Checklist to assist local authority in achieves sustainable development and environment.

1. Introduction

The current global complex environmental problems have resulted in environmental degradation which has caused, among other things, global warming. Environmental problems have been recognized and acknowledged at the United Nations Conference held in Rio de Janeiro in 1992, which composed in Agenda 21. However the global system of environmental management which focuses on lifestyle and industrialization is moving slowly in the right direction [5]. Among the approach is the Low Carbon City Framework (LCCF). Human activities in cities contribute 80% of greenhouse gas (GHG) emissions which are mainly responsible for the current climate change threat [12]. In this sense, LCCF is one of the most important tools applied in local authorities as guiding blueprint for low carbon cities amongst local authority's community. Low Carbon City relates to carbon minimization in all sectors by developing a society that emits GHG only in an amount that can be absorbed by nature and achieving a lifestyle that realizes the richer quality of life. Health and interaction with nature as well as promoting nature-friendly technologies such as the utilization of biomass is expected in near future by manifesting

IOP Conf. Series: Earth and Environmental Science **385** (2019) 012071 doi:10.1088/1755-1315/385/1/012071

in LCCF [4]. Despite that, it is acknowledged that challenges exist which recorded in local authority lab report of LCCF implementation [15].There were also recurrences of reports in planning permission submission stage in this matter [15]. The aim of this paper is to develop the research methodology to study the readiness of the stakeholders in the implementation of LCCF in an urban area with the focus on the implementation by the local authority in processing the planning permission.

2. Low Carbon Cities (LCC)

In general, Low Carbon Cities (LCC) is defined as a city that comprises of societies that consume sustainable green technology, green practices and emit relatively low carbon or GHG as compared with present-day practice to avoid the adverse impacts on climate change [8]. According to the Chinese Research Academy of Environmental Sciences, a low carbon city leads to low carbon economics and society along with a sustainable form of development. There are two aspects in a low carbon city conception, namely; (a) low carbon economies which increase energy, water efficiency and reduce carbon emission based on efficiency in use of resources and green technology; and (b) low carbon consumption which reduce carbon emission from all aspects of city living which include recycling, protecting the natural environment, maintaining green areas in the city and increasing carbon sink [7]. These aspects were translated into LCCF in Malaysia [13].

2.1 The Necessity to Develop Low Carbon City (LCC)

Cities, in general, generate very huge carbon dioxide emission and are responsible for consuming two-thirds of the world's energy and generating over 70% of its greenhouse gas emissions. About 2.6 billion tCO_2e is generated by the world's 50 cities annually [14]. Currently, half of the world's population lives in cities and this is expected to reach 70% by 2050. Also, cities account for over 67% of the energy-related global greenhouse gases, expected to rise to 74% by 2030 [14]. The key to creating low carbon cities in the field of urban planning is to freely come up with ideas to create a complementary and mutually supportive situation in which people's lives are improved and they have a better environment in which to live [9]. Low Carbon Cities had been implemented globally and defined slightly different names as a reference to a province, city, municipality, or community that pursues a systematic process to achieve GHG emission reductions [2]. There were a few case studies in Asia and Europe. Issues were addressed, the approach was taken contextualized locally. The rapid urbanization causes rapid environmental degradation demanded a new development pattern with less energy consumption. The new concept of Low Carbon City namely the eco-city was among the answer. Implementing the concept require the change in the development mode from only focusing on economic growth to an equal focus on the social and environmental aspects in the planning process had taken place in Shenzhen, China [11]. Thailand's "Low Carbon City" initiative aimed to help achieve reductions in GHG emissions and catalyse this shift to a low carbon society. Another example was the UK is Oxford City Council. In 2008 City Council created a Carbon Management Plan aimed to reduce the City Council's own emissions by 25% by 2011 with its measures [10]

2.1 Low Carbon Cities Framework (LCCF) in Malaysian Local Authority

The LCCF is only applied to participating local authority in Malaysia (UPEN, 2016). This is cascading from the Malaysian government is cognizant of the effects of global warming and is committed to combating this global phenomenon. The nation's commitment was announced to the global community. In order to reduce carbon footprint in Malaysia, the Prime Minister pledged commitment at the 15th United Nations Framework Convention on Climate Change (COP15) 2009 in Copenhagen, Denmark. Malaysia has committed to reducing its carbon dioxide emission intensity based on GDP by 40 per cent by 2020, as compared to 2005 levels; conditional upon transfer of technology and finance from developed nations [7].Ministry of Energy, Green Technology and Water (2015) proposed strategic planning for green technology as one of the machinery for economic growth which includes the LCCF. Low Carbon Cities Framework & Assessment System (2011) highlights the LCCF is a performance-

IOP Conf. Series: Earth and Environmental Science **385** (2019) 012071 doi:10.1088/1755-1315/385/1/012071

based system that captures the actual environmental impact of development in terms of total carbon emission. The framework provides quantitative carbon equivalent as a result of human activities in cities so that there may be awareness towards low carbon emission levels can be reduced. The framework serves as a guide to users on a pertinent element that contributes to the reduction of carbon emission. It comprises for four (4) key elements, 13 performance criteria and 35 sub-criteria which provides information helping the stakeholders to identify areas in which they could target an overall carbon reduction. The construction industry stakeholders had started to explore the existing framework for carbon reduction and established the policy by respective industry stakeholders [3]. The initiative for LCCF had started in 2011 simplified the LCCF is a system developed by the Ministry of Energy, Green Technology and Water (KeTTHA). The purpose of this system is to assist industry stakeholders such as developers, local councils, town planners, non-governmental organizations (NGO's) and the public to lower the levels of carbon emission in cities towards achieving sustainable urban developments. It serves as a guide that will propel stakeholders for cities, townships and neighbourhoods to re-assess their priorities in the planning and developing of new projects, as well as strategies that can be taken by existing cities, townships and neighbourhoods in reducing their carbon emission levels. Besides serving as a comprehensive guide, the LCCF also has an inbuilt carbon calculator with carbon equivalents that would help stakeholders assess their current baseline levels of the cities, townships and neighbourhood and target their intended levels. Green Technology applications for Low Carbon Cities (GTALCC) is a project to facilitate the implementation of low carbon initiatives and to showcase a clear and integrated approach to low carbon development in Malaysia [14]. Among the evidence is the emergence of green rating tools such as Green Building Index (GBI), Malaysia Carbon Reduction and Sustainability Tool (MyCREST) and Green Real Estate (GreenRE) from various stakeholders of the local construction industry. These are the rating systems adapting to the Malaysian context which upkeeps the LCCF system at larger scale simultaneously. In this sense, LCCF is hoping to holistically strategize the implementation of environmental sustainability in the Malaysian construction industry interlinking green rating tools and green procurement implementation.

2.3 LCCF Assessment System in Malaysian Local Authority

The assessment system of LCCF encompasses carbon calculator guiding user determine their current baseline of carbon emission [7]. Various approaches are recommended in the framework to achieve a certain reduction level to the user. The calculator will be used again to reassess the carbon emission levels to see if the user has obtained a good reduction level. The built-in calculator equipped with carbon emission factors enables the user to calculate based on their requirements. The LCCF is only applying to the local authority who participate which mean not all local authority is obliged. This is among the challenges identified in making LCCF implementation effective. The assessment is carried out by the planning department of the respective local authority specified in the department control unit. It is made by reviewing the planning permission documents requirements submitted by the submitting person of newly developing projects at One Stop Centre (OSC) counter. All pertinent matter was carried forward to the Planning Permission meeting together with primary stakeholders of the proposed projects.

2.4 LCCF Checklist and Planning Permission

The implementation LCCF Checklist in a local authority have a round table session. Interview with the local authorities to get the idea and discussed about issues, concerns and constrains in incorporating LCCF into the process and procedures of development application [6]. The assessment takes place in the development planning permission stage as part of the need statement for the future project undertaking somehow may need revision. In addition, there is a dearth of literature critically discussing the LCCF checklist globally much less the Malaysian LCCF checklist. Vagueness in understanding and implementing LCCF in local authority much less the construction industry is identified in promoting LCCF. In general "planning permission" is a procedure in which the applicant must submit an application to the local planning authority for approval before starting any development on the land or building as provided under section 21A of the [Development Proposal (LCP)] and section 21B [Layout

Plan] (Act 172, 1974). Development control is the process of regulating the use of land and buildings by the local planning authority. Definition of development in Section 2 of the Town and Country Planning Act 1976 (Act 172) is defined as The carrying out of building, engineering, mining, industrial or other operations in, on, over or under land or the making of any material in the use of any buildings or other land or part of or to subdivide or amalgamate the land; and; "development" should be defined accordingly' [1]. There are five(5) major local authorities in Malaysia including Putrajaya, IRDA, MPSJ, MPK and MBSA had embraced the idea. These local authorities strike a similarity which is locating at the heart of their municipalities. LCCF only take place in Selangor, Malacca and Johor but not Malaysia as a whole. Rapid urbanisation in these areas and access of information are believed as the driver of these municipalities to join the LCCF.

3. LCCF Checklist

There is no literature discussing LCCF strategy formulation and implementation in other local authority which rapidly urbanising such as Negeri Sembilan and Pulau Pinang or in the Malaysia East states. There is also the recurrence of the report by Submitting Person (SP) on the misunderstanding of LCCF checklist criteria in the planning permission stage. Internally some planning department officers within local authority argue the checklist is not effective. External stakeholder namely planning consultant and developer representative issued the same problems. LCCF Checklist as a tool for planning permission in development control. It depends on the LCCF Criteria. The approach is City Based (mitigating all the criteria as stated within the LCCF) and One System (mitigating one criteria or not all the criterias in the LCCF). The four elements of LCCF are further categorised into 13 performance criteria and 35 sub-criteria, each of which provides specific intents towards carbon reduction targets. The criteria as listed in Table 1 below:

Table 1. Element, Performance Criteria and Sub Criteria for LCCF

4 Element	Indicator	13 Performance Criteria	35 Sub Criteria
Urban Environment (UE)	• Site Selection • Urban Form • Urban Greenery & Environmental Quality	3 Performance Criteria	14 Sub Criteria
Urban Transportation (UT)	• Reduction Use of Private Motorised Transport on Urban Road Network • Increase in Public Transport • Mode Shift from Private to Public Transport and Non-Motorised Transport • Use of Low Carbon Transport • Improvement to Level of Service of Road Links and Junctions • Utilisation of Transit-Oriented-Development (TOD) Approach	6 Performance Criteria	11 Sub Criteria
Urban Infrastructure (UI)	• Infrastructure Provision • Waste • Energy • Water	4 Performance Criteria	10 Sub Criteria
Building (B)	• Sustainable Energy Management Low Carbon Buildings	2 Performance Criteria	5 Sub Criteria

Source: KeTTHA (2011)

4. Research Methodology

The selected case study area is the Subang Jaya Municipal Council. It is consisting of five (5) zones, namely Subang Jaya, Kinrara, Puchong, Seri Kembangan and Putra Permai. However, this research is only focusing on Subang Jaya, Sunway, Putra Heights, USJ and Puchong. The study will apply an

IOP Conf. Series: Earth and Environmental Science **385** (2019) 012071 doi:10.1088/1755-1315/385/1/012071

exploratory research design (non-experimental research design) to achieve the research aim and objectives. Research methods that are applied in this research are qualitative and quantitative approaches. A quantitative data in this study will be collected in the first stage by using the structured questionnaire to identify the emphasized data on local authority and stakeholders, namely developer, consultant, and technical agencies which related with One Stop Centre (OSC) meeting. Respondents for the questionnaire survey which be selected by using stratified sampling. Qualitative data collection is for the interview with town planning officers in local authority and other stakeholders, i.e. developers and consultants. The quantitative data will be analysed with the system using Statistical Packaging for Social Science (SPSS version 16). The descriptive analysis will be used because it can explore the stakeholders' attention and willingness to implement the LCCF Checklist. This research is going to explore 4 elements of LCCF Checklist, i.e. the stakeholders' willingness and perception, effort, knowledge and pro-active attitude in implementing LCCF or Low Carbon Cities. Cross-tabulation and correlation tests will be used to further analyse the factors and relationship of the variables. This research is focused to study the criteria and readiness of LCCF implementation by stakeholder and to investigate the achievement of LCCF of the aspects of Economy, Social and Environment.

Figure 1. Conceptual Framework: Relationship Between Variables

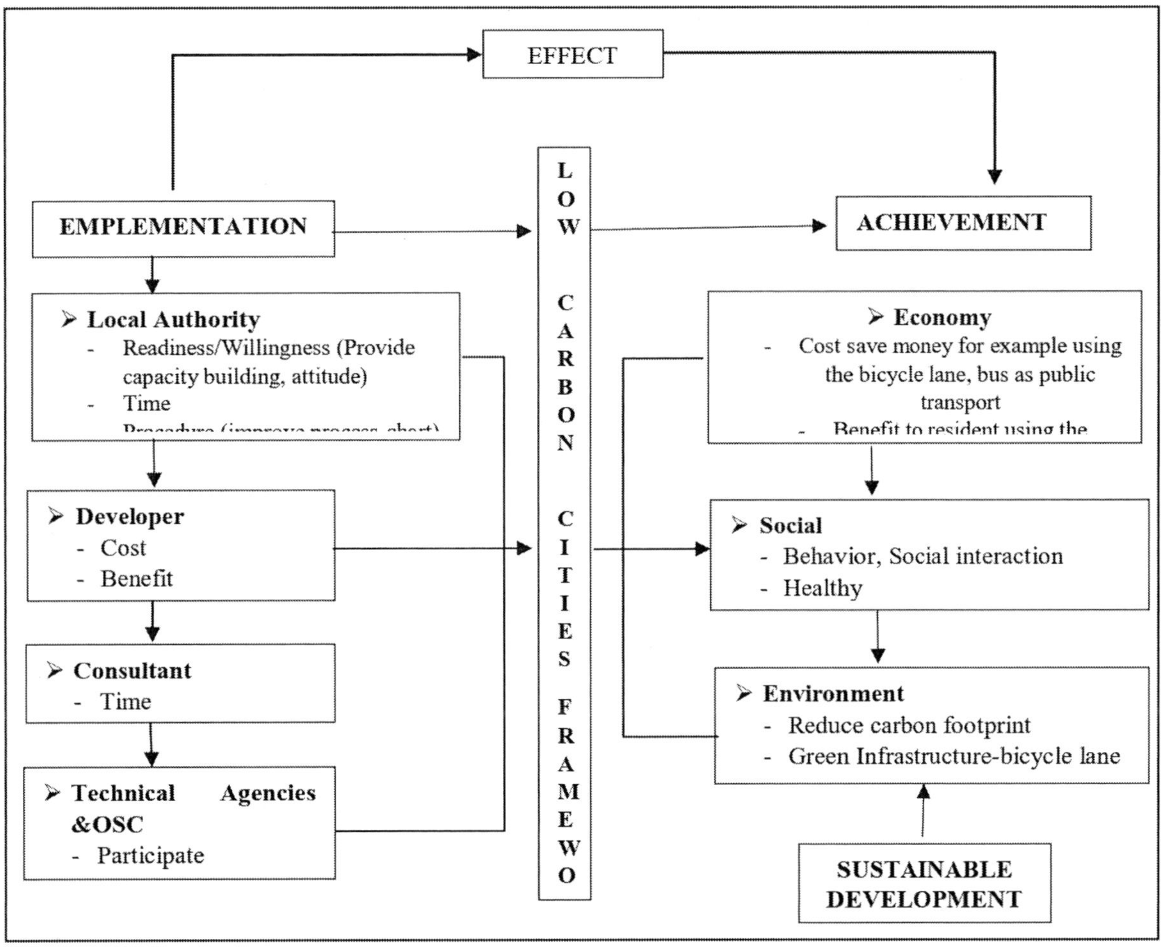

Figure 2. Research flow

Aim ⇨ Objectives ⇨ Analysis ⇨ Finding ⇨ Potential to develop tool for LCCF Checklist

- Quantitative
- Qualitative

5. Conclusion

In conclusion, this article had reviewed the existing phenomena for the implementation of LCCF in development control. It is compelled that challenges and issues exist in multifaceted of policy implementation of LCCF in Malaysia. Therefore, it is in need of extensive research in this particular matter. Local authority plays a major role of manifesting LCCF in newly develop the project but few challenges must be the highlight. First, LCCF criteria must be studied and detailed to ease the understanding and implementation of local authority and stakeholder of the construction industry. The recurring reports by the practising planner in the participating local authorities and stakeholders of the industry must be heard and look into. This is common in environmental solution by bringing every major stakeholder on board which local authority is the moderator.

Acknowledgements

The authors would like to thank Universiti Teknologi MARA (UiTM) for the support and partly funding the study and publication through the LESTARI research grant [(600-IRMI/MYRA 5/3/LESTARI (K) (237/2017)]. The authors are also thankful to all the departments, organisations, and individual who had contributed to this study.

References

[1] Act 172. (2001). *Town and Country Planning Act 1976 [Act 172]* (Malaysia) p 9

[2] Asia Leds Partnerships. (2013). *Case Studies on Low Emission Development, Thailand's LowCarbon City Initiatives, Thailand (Electronic Materials)*, Asia-LEDS-Partnership-Case-Study-Thailand-Low-Carbon-City-Initiative-March-2013) p 1

[3] Dato' Sri Peter Chin Fah Kui (2011). *LCCF book* (Minister of Energy, Green Technology and Water, Malaysia)

[4] Ho Chin Siong. (2011). *Low Carbon cities – the way forward for real estate development at Conference on Mixed use development, Crown Plaza Mutiara, Kuala Lumpur, October 17-18, 2011 (Electronic Materials)* (International Journal of Real Estate Studies, Volume 8, Number 1) p 1

[5] Ho Chin Siong, Abdul-Azeez and Isiaka Adeyemi, (2013) *International Journal of Real Estate Studies,* (Universiti Teknologi Malaysia, Faculty of Built Environment, UTM Johor, Malaysia) (International Journal of Real Estate Studies, Volume 8, Number 1) p 1

[6] Malaysian Institute Of Planners (MIP) (2015), *Application Of LCCF Checklist And Calculator In Achieving Low Carbon Property Development* (Grand Dorsett Hotel, Subang Jaya, Malaysia)

[7] Ministry of Energy, (2011) Green Technology and Water, Malaysia (KeTTHA), *Low Carbon Cities Framework Book* (Pusat Pentadbiran Kerajaan Persekutuan, Putrajaya, Malaysia)

[8] Ministry Of Energy, Green Technology and Water (KeTTHA) (2017). *Low Carbon Cities Framework* (Pusat Pentadbiran Kerajaan Persekutuan, Putrajaya, Malaysia)

[9] Onishi, T. and Kobayashi, H. (2011). *Low Carbon Cities - The Future of Urban Planning. Master's Program in Sustainable Regeneration Series* (Department of Urban Engineering, Graduate School of Engineering, the University of Tokyo)

[10] Paul Wedgwood. (2015) *A Strategic Framework For Low Carbon Cities* (General Manager Scotland)

IOP Conf. Series: Earth and Environmental Science **385** (2019) 012071 doi:10.1088/1755-1315/385/1/012071

[11] Ruben Cales. (2014). *Shenzhen Low Carbon City: A transformation of Concept and Planning Process* (University of Amsterdam, Graduate School of Social Science, Department of Human Geography, Planning & International Development Studies)

[12] Satterthwaite, d. (2008). *Cities contribution to global warming: notes on the allocation of greenhouse gas emmisions* (Environmentand urbanization,. USA)

[13] Siti Kartina Juhari, (2018), *Implementation Of Low Carbon Cities Framework (LCCF) In Local Authority: The Malaysia's Case,* (International Conference On Education, Islamic Studies And Social Sciences Research (ICEISR 2018), Padang, Sumatera Barat, Indonesia)

[14] Steve Anthony Lojuntin, (2018), *Green Technology Application For The Development Of Low Carbon Cities (GTALCC) Project*, (SEDA, Malaysia)

[15] The World Bank (2010) *Cities and Climate Change; An urgent agenda* (Volume 10 December 2010 Washington.)

[16] UPEN, (2016) Selangor State *Government* State *Economic* Planning Unit, *Workshop On LCCF Application in Planning Permission In Selangor* (Selangor,Malaysia)

IOP Conf. Series: Earth and Environmental Science **385** (2019) 012072 doi:10.1088/1755-1315/385/1/012072

Fundamental Valuation of Construction Stocks: A Content Analysis from Property Developers

M S A Nordin[1], N Ab Rahim[1] and H Adnan[1]

[1]Faculty of Architecture, Planning and Surveying, Universiti Teknologi MARA, 40450 Shah Alam, Selangor, Malaysia

Email: saifulalizan@uitm.edu.my

Abstract. The expansion of the Malaysian Construction Companies requires a holistic understanding on the intrinsic values as well as the nature of growth and value features of stocks in order to effectively invest in this sector. The paper aims to assess the intrinsic values of the Malaysia construction companies using fundamental methods of valuation. Content Analysis has been employed to the 10 property developers that has been listed under the top 10 Malaysian property developers in year 2018 identified as Company A to J. Based on the data derived from the companies' annual report, Bursa Malaysia and stock-related websites, intrinsic value of the construction stock is assessed using three fundamental methods of valuation comprising Price Earning Multiple, Relative Price Earning as well as the Dividend Growth Model. The results suggested that there is a huge inconsistencies of intrinsic values across the first two methods of valuation where most of the stocks are in the state of undervalued while valuation derived from the Dividend Growth Model signifies the overvaluation state of all stocks' opening prices. This research is expected to provide a preliminary evidence on the intrinsic values of Malaysian Construction Companies as similar analysis and indicators can be used in reviewing the characteristics as well as fundamentally valuing the construction stock in both small and larger investment portfolios.

1. Introduction

Share trading become a new phenomenon among corporate entities in Malaysia after public listed companies on the Bursa Malaysia (formerly known as Kuala Lumpur Stock Exchange) began to issue stock options to its executives as a way of compensation in 1980s [1]. As a result, discussion on intrinsic value in acquiring mispriced asset became a major debate among economist, academicians and investors. This is because investors have started to realize the benefits of investing in the stock market, especially when its return on investment obtained is said to be higher than other asset classes.

Meanwhile, the construction industry is one of the sectors that has become the main attraction of investors due to its encouraging investment performance. In Malaysia, construction companies carry out their business activities through the collection of land banks in potential areas to be developed with various construction projects according to the needs and the demand at the respective locality. The construction projects involved focus not only on high-end and medium-cost development projects, but also on low cost development as well as public infrastructure. For example, in order to increase the number of affordable housing in Malaysia, the Malaysian government under the Ministry of Housing and Local Government as well as the Ministry of Economic Affairs have urged property developers to assist the government in providing shelters to the B40 group i.e. the low-income earners.

In order to encourage publics to invest in their shares, the company's performance needs to be scrutinized. The variation and volatility in the construction sectors are largely driven by its underlying performance of the previous, existing and proposed development. In addition, the construction-related research and development (R&D) in the context of property market research, consumers' preference and construction products' quality plays an important role. According to Garcia and Oliveira [2], company that emphasising the R&D will have a rapid growth of development and earnings than other companies in the market. The establishment of construction stocks allow public to indirectly participate in the national construction industry even with the small initial cost of capital.

Numerous studies pertaining to the construction and other stock sectors are related to the effects of the stock prices in respond to the macro and micro-economic factors, risk and returns structures, stocks management as well as portfolio diversification for example by [3 [4] [5] [6]. Other studies in identifying the effects of taxation on Malaysian stock market index volatility, [7] found that the volatility of the Malaysian stock market index is persistent during the GST announcement and highly persistent after the implementation. The study however generalize all sectors in the stock market. Other example from Indonesia, [8] found that interest rate, inflation rate, exchange rate, and GDP growth rate, as composite variables, have a significant influence on real estate and property companies' stock price.

This study however present the application of fundamental valuation particularly in valuing the construction stocks. It served as an alternative to the technical analysis of the stock by focusing on the real value of the company that simultaneously minimizing the speculative elements in the analysis.

2. Literature Review

2.1. Efficient Market Hypothesis in Stock Market

As stock market is a leading indicator, the most common issues in stock market is market efficiency in terms of operational, allocative and informational. The Efficient Market Hypothesis (EMH) suggests that the stock are accurately priced by the market so that the return is justified by the amount of risk that the security has. In other words, investors would not be able to obtain an abnormal return as the publicly information reflects the true value of stock. There are three levels of market efficiency starting from lowest which is weak-form, semi-strong form and strong-form. According to [9] the weak-form EMH states that the market is efficient in such a way that price, volume information, and stock price movements cannot be predicted using historic information. This weak-form EMH asserts that the stock returns are time invariant, that is, there is no identifiable short-term-based pattern [9]. Therefore under this levels, the higher the ability of people to assess the stock information, the greater the profit they can earn.

However market is not always efficient. The achievements of Warren Buffets in sustaining the returns as well as strategies in acquiring a mispriced asset based on an intrinsic value thus denying the principles of EMH. Besides that, the lack of long time series data and infrequent trading make the valuation of stock market quite challenging especially when involving an Initial Public Offerings (IPO). For example, [10] stated that the crisis in 1988 did not allow for the development of the stock options in Malaysia as much as in Japan. This may be due the slow responsiveness to the stock information. In comparing the Malaysian stock with Japan, [10] also reveals that the issues related to stock options in Malaysia are not that big as in Japan, because the amount is not large and material. However In the case of semi-strong efficiency, both fundamental analysis and technical analysis enables investors to produce excess returns as the stock prices response quickly to the newly available data in the market [2].

2.2. Value vs Growth Investing

There is no denying that stock market is the most preferred investment and discussion on value vs. growth investing has caught the attention of many investors in determining the appropriate selection of equity that match with the investment style and investors' requirements. [11] defined growth stock as a stock that possess a higher earnings per share (EPS) above the rate of other common stocks and the growth is expected to continue in future. Growth investors are looking at the return from the capital appreciation thus concentrating on the companies that have long-term earnings growth prospects [2]. [2]

further added that investors are willing to pay more to buy growth stocks, which causes them to reflect higher price ratios in order to reflect the market expectations.

Value stock is identified as a stock that having a lower price as compared to its fundamentals which is EPS, book value per share, cash flow per share, net tangible asset per share as well as dividend yield [11]. Value investors will normally dealing with the stock at a discount relative to industry peers, low price ratios and high dividend yields as well as a firm balance sheet operating in a secured investment environments [2]. The application of fundamental methods of stock valuation will enables investors to identify the characteristics of the stocks as the information used in the fundamental valuation models are the main indicators in determining the growth and value stock.

2.3. Intrinsic Values Via Fundamental Valuation

Unlike direct property valuation which heterogeneous, stock valuation is quite straight forward as all information are readily available in the market. Stock valuers, analyst or even investors can assess relevant information via Bursa Malaysia to see the trading prices (opening price and closing price) as well as company's website or annual report to see any updates on organizational or business structure that may affect the stock performance in the market. The application of intrinsic values as an indicators to the stock prices would benefit investors in choosing the right stock to be invested in.

Intrinsic value of stocks is determined by a careful fundamental analysis. Fundamental analysis is a method of assessing the intrinsic value of a company's stock by looking at the historical accounting and financial data. This information can be obtain freely from published annual report, official announcements and company's website. However certain information such as investment and financial plan of a company is privately available where it only can be obtain through a private enquiries on a company.

According to [12], accounting information, especially the annual reports, and direct contact with company management, represent the most important and most useful sources of information to financial analysts, although there is a clear shift in the relative importance of these sources over time. Survey-based studies on financial analysts indicate that analysts consistently emphasise the long-term view over the short-term in stock evaluation and selection [12]. In investigating the relationship between accounting information and stock returns of selected Indian stocks [13] found that all individual accounting signals have a positive correlation with future stock returns. However the functionality of fundamental analysis might be differ in different market place. In contrast, other study by Iqbal et al. [14] however found that the fundamental analysis cannot predict stock returns in Pakistani listed companies. This is because the market information in Pakistan is not efficient as it cannot reflects the true value of stock.

By assuming the market information is efficient, many models have been constantly develop and improved over time in determining the intrinsic value of stock. However the authors only discuss on three main approaches that will be used in analyzing the selected share namely P/E Multiples, Relative P/E and Dividend Growth Model. The application of these models is briefly explains in the methodology section.

3. Methodology

3.1. Research Approach

Content Analysis has been employed to the 10 property developers identified as Company A to J where all companies has been listed in the Top 10 Property Developers' Awards (TPDA) 2018 by The Edge and have been randomly sorted in this study. TPDA is awarded after both quantitative and qualitative evaluation by industry peers with all scores and rankings are audited by Deloitte Malaysia [15]. Based on the data derived from the companies' annual report, Bursa Malaysia and stock-related websites, intrinsic value of the construction stock is assessed using three fundamental methods of valuation comprising Price Earning (P/E) Multiple, Relative P/E as well as the Dividend Growth Model. The date of valuation is effective as at March 1, 2019 where data on opening prices are concurrent with the period of the study while other content inputs including P/E, EPS and Dividend per Share is taken at the recent date as announced by the companies for the financial year of 2018. The process of content analysis is summarized in Figure 1.

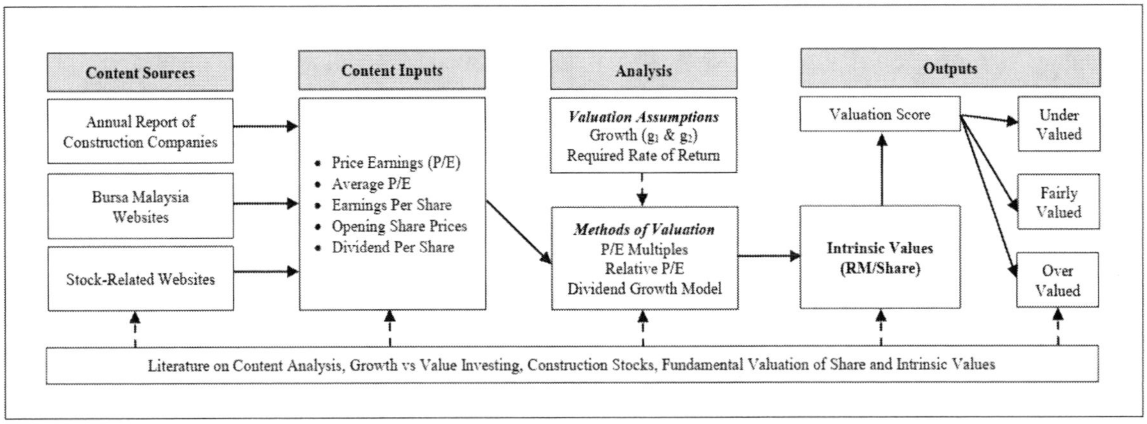

Figure 1. Process of Content Analysis in the Study

The P/E is a good indicators to the financial strength of the firm. By using the simple earnings capitalization model is employed to investigate the association between price and earnings across profit and loss firms, Papadaki and Siougle [16] signifies a negative P/E relation for the company that report losses and a positive P/E relation for the company firms that report profits. Even though the stocks' intrinsic values can be established via P/E-related analysis, there is a need for other indicators particularly on dividend side. Even though the ratios derived from the P/E form the essence of fundamental analysis, investor sentiment may also contribute to the movements in P/E ratios [17]. Jitmaneero [17] also empirically proved that the P/E ratio is not only driven by fundamental factors but also investor sentiment to a greater extent. Therefore dividend can be an alternative or complimentary to the fundamental valuation as there is a positive relation is found between dividend yield and stock price changes, and a negative relation between dividend payout ratio and stock price changes [18].

3.2. Methods of Valuation

3.2.1. Price Earnings Multiples. Price Earnings (P/E) Multiples is a popular technique where the company's expected EPS are multiplied by the average P/E ratio for the industry to estimate the company's share value. The methodological strength of P/E Multiples is, it is considered superior to the use of book and liquidation values since it considers expected earnings. According to Wang and Ahammad [12], analyst compare the market P/E and share price in the same industry to come out with an estimated reasonable price interval before recommendation on investment being made. If the intrinsic value of share is higher than the share price, it is implies the shares are undervalued. As mention earlier the expected earning is a main determinant under this method. The expected changes in EPS, expected return on equity, and prospects for the relevant industry are also considered the most important variables over the long-term besides giving a greatest weight over the short term [12].

3.2.2. Relative Price Earnings. Relative P/E is a measure of stock performance in relation to the average performance of the market. The similar concept of average P/E industry is adopted from P/E Multiples method by assuming that the selected companies represent the selected industry. Indication under this approach suggested that 1.00 is the fair value of the relative P/E to the stock market. Therefore if the relative P/E to market is exceed 1.00, the stock is said to be overvalued and otherwise if it less than 1.00.

IOP Conf. Series: Earth and Environmental Science **385** (2019) 012072 doi:10.1088/1755-1315/385/1/012072

3.2.3. Dividend Growth Valuation Model. Valuation of common stocks by using a dividend valuation model is based on the present value (PV) of the expected streams of the future dividend. There are three types of dividend valuation models namely Zero Dividend Growth Model, Constant Dividend Growth Model and Variable Dividend Growth Model. The computation of stock value under Zero Dividend Growth Model is characterised by its constant dividend payment for a perpetual time period. The formula is also simple as PV for the perpetuity can be calculated by dividing the dividend payment with the discount rate. Meanwhile under the Constant Dividend Growth Model, it is assume that the dividend is increase at a constant rate and the model can be expressed as $P_o = D_1/(k-g)$, where P_0 is the current price of share, D_1 is the dividend per share for 1-year in advance, k is the required rate of return as well as g i.e. the growth rate. However as the stock market is imposed to the financial volatility, the distributed dividend cannot be perpetually constant. Hence the Two-Stage Dividend Growth Model under the variable dividend model is the most appropriate model in determining the stock's value as it take into account the changes in expected growth rates. The model assumes that the stock prices will initially grow at a rate g_1 for T years and subsequently $g_2 < k$ during the second stage or perpetual period. The same model is dictated as follows;

$$P_0 = \frac{D_0(1+g_1)}{k-g_1}\left[1-\left(\frac{1+g_1}{1+k}\right)^T\right]+\left(\frac{1+g_1}{1+k}\right)^T\frac{D_0(1+g_2)}{k-g_2} \quad (1)$$

where;

P_0 = Current Price of Share
D_0 = Current Dividend per Share Received
k = Required Rate of Return
g_1 = First Stage Growth Rate
g_2 = Second Stage/Perpetual Growth Rate

4. Results and Discussion

Table 1 depicts the summary of intrinsic values derived from the three methods of valuation. Opening prices of the stock is based on the quoted opening prices listed at Bursa Malaysia as at March 1, 2019. Intrinsic values is determine by analysing all relevant information using respective valuation models. Percentage *+%* and *-%* at each cells represent the variance of valuation with the opening share prices traded in the market. Finally the valuation scores column referring to the state of the opening price as compared to intrinsic values with the maximum score of 3.

Based on the data analysed, three companies were reported at issuing a penny stock where the opening prices is less than RM1.00 per unit namely Company B, C and F. In terms of fundamental valuation, there is a huge inconsistencies of intrinsic values across the first two methods of valuation where most of the stocks are in the state of undervalued. Company A and F recorded a fairly valued score based on the first and second method where the variance ranging from -0.06% to -1.67%. Company C and H however recorded an overvaluation figure where values derived from all methods are lower than the stock opening price at RM0.86 per unit and RM1.82 per unit respectively. No single procedure is conclusively precise and therefore, combined valuation model is more informative by providing better and more accurate estimations of stock values [19] which is in this study, stock values vary from the points of view of P/E and distributed dividends.

The percentage changes ranging from -58.94% to -94% in the Dividend Growth Model suggested that all companies are issuing relatively low dividends which resulting to the lower intrinsic values. These were supported by Yoo [20] where each simple multiple valuation is likely reflects the common information and also encompasses an incremental information that is beneficial for the improvement of the valuation accuracy. In this study, the values derived from the Dividend Growth Model reflects the real performance of the dividend distributions. The results also suggested that all shares analysed are likely to lies within the growth stock where the best strategy to be adopted is growth investing strategy. From fundamental perspectives, undervalued stock are a good stock to buy as market has yet to realise its real value besides its future potential growth and vice-versa for the overvalued stock. Meanwhile, fairly valued stock referring to the stock that bears only a small or no variance at all as compared to the market values represented by the opening share prices.

Table 1. The Intrinsic Values of the Top 10 Malaysian Property Developers

No	Malaysian Property Developers	Opening Stock Prices (RM/unit)	Intrinsic Values (RM/Unit) [%: variance to the opening prices]			Valuation Scores (3/3)		
			Method 1: P/E Multiples	Method 2: Relative P/E	Method 3: Dividend Growth Model	Under Valued	Fairly Valued	Over Valued
1	Company A	2.9400	2.9152 [-0.84%]	2.9381 [-0.06%]	0.5475 [-81.38%]	0/3	2/3	1/3
2	Company B	0.8650	1.7853 [+106.39%]	1.3745 [+58.90%]	0.146 [-83.12%]	2/3	0/3	1/3
3	Company C	0.8600	0.6737 [-21.67%]	0.5825 [-32.27%]	0.2737 [-68.17%]	0/3	0/3	3/3
4	Company D	1.6000	2.1098 [+31.87%]	1.8698 [+16.86%]	0.4562 [-71.49%]	2/3	0/3	1/3
5	Company E	1.0000	1.1988 [+19.88%]	1.1933 [+19.93%]	0.4106 [-58.94%]	2/3	0/3	1/3
6	Company F	0.8200	0.8099 [-1.23%]	0.8063 [-1.67%]	0.0912 [-88.88%]	0/3	2/3	1/3
7	Company G	2.2900	2.8677 [+25.23%]	2.5193 [+10.01%]	0.3540 [-84.54%]	2/3	0/3	1/3
8	Company H	1.8200	0.7962 [-56.25%]	0.5236 [-71.23%]	0.2436 [-86.62%]	0/3	0/3	3/3
9	Company I	2.2800	3.2383 [+42.03%]	2.5851 [+13.38%]	0.1369 [-94.00%]	2/3	0/3	1/3
10	Company J	1.6400	2.0700 [+26.22%]	1.8432 [+12.39%]	0.3194 [-80.52%]	2/3	0/3	1/3

Sources: The Authors' Analysis (2019)

5. Conclusion

This study has presented an estimate intrinsic values of the selected construction stocks. As the study involving the application of valuation model, the intrinsic values derived are highly dependent on the available market data as at the period of the study. Several assumptions have been made in arriving at the dividend growth rate (g_1 and g_2) as well as the required rate of return (k) under the Dividend Growth Model. Therefore it should be noted that the result only provides a preliminary evidence on the intrinsic values of construction stocks in relation to its listed/opening prices and should not be treated as a consultation to specific investment decision. The volatility nature of the stock market also provides a lagging effect to the valuation outcome as it requires an up-to-date valuation based on the latest market data.

References

[1] Ariff, M., & Nassir, A.M. (1998). Stock pricing in Malaysia: corporate financial and investment management. UPM Press.

[2] Garcia, M. T. M., & Oliveira, R. A. A. (2018). Value versus growth in PIIGS stock markets. Journal of Economic Studies, 45(5), 956-978.

[3] Hiang L. K., Ooi, J., & Gong, Y. (2005). Cross-market dynamics in property stock markets: Some international evidence. Journal of Property Investment & Finance, 23(1), 55-75.

[4] Zull Kepili, E. I., & Masron, T. A. (2016). Malaysia property sector: Performance analysis and portfolio diversification benefits within sub-sector. International Journal of Housing Markets and Analysis, 9(4), 468-482.

[5] Kan, Y. Y. (2017). Macroeconomic environment of bull markets in Malaysia. Qualitative Research in Financial Markets, 9(1), 72-96.

[6] Ung, L. J., Brahmana, R. K., & Puah, C. H. (2018). Brokerage fee, ownership expropriation and earnings management of Malaysian property companies. Property Management, 36(4), 461-482.
[7] Haron, R., & Ayojimi, S. M. (2018). The impact of GST implementation on the Malaysian stock market index volatility: An empirical approach. Journal of Asian Business and Economic Studies.
[8] Simbolon, L., & Purwanto. (2018). The Influence of Macroeconomic Factors on Stock Price: The Case of Real Estate and Property Companies. In Global Tensions in Financial Markets, 29 (pp. 19-39).
[9] Hadi, A. R. A., Pyeman, J., & Mahmood, W. M. W. (2011). A Quest for Small-Firm Effect: Evidence from KLSE Second Board. IUP Journal of Financial Economics, 9(3), 28.
[10] Yamashita, K., Hanefah, H. M. M., & Noguchi, A. (2010). Stock options rules in Malaysia and Japan: A comparative analysis. Journal of International Business Research, 9, 99.
[11] Pyeman, J., Hadi, A. R. A., Yahya, M. H., Zakariah, S. & Ahmad, I. (2013). Value versus growth investing. Penerbit UiTM Press
[12] Wang, J., & Ahammad, M. F. (2012). Private information acquisition and stock evaluation by Chinese financial analysts. International Journal of Management, 29(1), 117.
[13] Venkatesh, C. K., & Ganesh, L. (2013). Fundamental analysis and stock returns: Indian evidences. International Journal of Applied Financial Management Perspectives, 2(1), 248.
[14] Iqbal, N., Khattak, S. R., & Khattak, M. A. (2013). Does fundamental analysis predict stock returns? Evidence from non-financial companies listed on KSE. Knowledge Horizons. Economics, 5(4), 182.
[15] The Edge (2018, October 30). The Edge Malaysia Property Excellence Awards. S P Setia is Malaysia's Top Property Developer. Retrieved from http://tepea.my/2018/index.php
[16] Papadaki, A., & Siougle, G. (2007). Value relevance of price, earnings and book values in the Athens Stock Exchange. Managerial Finance, 33(5), 309-320.
[17] Jitmaneeroj, B. (2017). Does investor sentiment affect price-earnings ratios?. Studies in Economics and Finance, 34(2), 183-193.
[18] Hussainey, K., Oscar Mgbame, C., & Chijoke-Mgbame, A. M. (2011). Dividend policy and share price volatility: UK evidence. The Journal of risk finance, 12(1), 57-68.
[19] Tiwari, R., & Singla, H. K. (2015). Do combining value estimates increase valuation accuracy? Evidence from Indian chemical industry. Journal of Accounting in Emerging Economies, 5(2), 170-183.
[20] Yoo, Y. K. (2006). The valuation accuracy of equity valuation using a combination of multiples. Review of Accounting and Finance, 5(2), 108-123.

IOP Conf. Series: Earth and Environmental Science **385** (2019) 012073 doi:10.1088/1755-1315/385/1/012073

Factors lead to overhang in affordable housing: A content analysis in qualitative approach

N Ab Rahim[1], S M Adzhar[1], N Basrah[1], R Ab Majid[1] and S Mustafar[1]

[1]Faculty of Architecture, Planning and Surveying, Universiti Teknologi MARA, Shah Alam, Selangor, Malaysia

Email: norbayaabrahim@gmail.com

Abstract. Housing issue is one of the key challenges and has its social repercussions if not well managed. The numerous affordable housing programs have been introduced and intended to increase the supply of affordable housing. The programs are meant to alleviate the affordable housing issue in the country. Although the lacking units of affordable housing persists, the slow market trends have also led to an increase in overhang units. The number of property overhang partly in affordable housing units were reported increased continuously year over year and does not attract the target market nor able to cater the housing needs of the targeted house buyers. Industry players were of the opinion that the perceived slow demand in low cost housing units are mainly due to difficulty in securing financing, unsuitable location as well as a shift in demand towards better quality projects and other factors. The purpose of this study is to identify the current overhang scenario in affordable housing and factors that lead to the issues. There is a need to have an in-depth study in addressing the issues. Otherwise, the overhang will exert pressure on prices and rental in the market and gives impact in near future.

1. Introduction

Housing issue is one of the major challenges for the government and has its social repercussions if it is not well managed. Part of the initiatives taken by the Federal and State Governments to increase the supply of lacking units including introducing projects such as Perumahan Rakyat 1 Malaysia (PR1MA), Residensi Wilayah (formerly Rumah Wilayah Persekutuan-RUMAWIP), Rumah Selangorku, Program Perumahan Rakyat (PPR), MyHome 2, Rumah Mesra Rakyat Plus (RMRplus). The housing program is meant to alleviate the housing affordability issue as well. Affordable housing is defined as an adequate housing in quality and location, and not so costly that prevents its occupants from meeting other basic living needs [1]. The standard definition for affordability in housing is that households should pay not more than 30% of their income for housing including utilities. The term is referring to the households' capacity in meeting the costs of housing while maintaining the capability to encounter other basic of living costs [2].

The introduction of more affordable housing in Malaysia such as PR1MA and Residensi Wilayah housing programs were initiated with the main objective to provide sufficient units of affordable housing as to cater the demand especially for the middle-income class earners. However, despite of the lacking of affordable housing, the issue of property overhang in Malaysia is worsening where the report reveals that the statistic is increasing year by year. On the other hand, there are also issues about the lack of supply of affordable housing which has becoming the common news in the property industry. According to the National Property Information Centre (NAPIC), there were a total of 29,227 overhang residential units in Malaysia as at the first half of 2018, valued at RM17.24bil [3]. Part of the figure represents the affordable housing statistics. This comes to the scenario where the success rate of affordable housing depends on some factors which affect the take-up rate and the overhang statistics. Hence, the aim of this paper is to identify the current scenario of affordable housing and to determine factors that affecting overhang in affordable housing in the country.

2. Literature Review

2.1 Affordable Housing Landscape in Malaysia

Providing affordable housing especially in big cities remains to be a challenging task since the price of housing units are continually increasing while the supply of units is still inadequate to meet the demand of the middle-income group. The supply of affordable housing in Malaysia is still insufficient especially at the urban area. The undersupply of housing is particularly serious in the affordable housing segmentation [5]. The shortage and lack of supply of affordable housing in Malaysia, especially in Klang Valley is a renowned issue and widely discussed where the housing market in Malaysia has not been able to provide an adequate supply of affordable housing for the masses [5].

In Malaysia, more than 20 national and state level agencies are involved in the provision of affordable housing. One of the program introduced is PR1MA with the aims is to establish, plan, develop, construct and maintain affordable lifestyle housing for middle-income households in urban centres. The government plans to build more affordable housing in the cities and towns in Malaysia. The PRIMA project was launched on 4 July 2011 by the former Prime Minister, Najib Razak. The program is managed by the government owned company known as Perbadanan PR1MA Malaysia under the Ministry of Housing and Local Government.

Perbadanan PR1MA Malaysia was established under the PR1MA Act 2012 with the focus to plan, develop, construct and maintain high-quality housing with lifestyle concepts for middle-income households in key urban centres. PR1MA homes come in various types and sizes within an integrated community sensibly designed to suit different household needs with the price between RM100,000 to RM400,000. Earmarked for development in key strategic urban areas nationwide, PR1MA is open to all Malaysians with a monthly household income between RM2,500 to RM15,000 and these houses are sold about 20%-30% lower than the market prices. The funding of PR1MA projects is from the government. It is learned that PR1MA needs about 12,500 acres of land in accomplishing the mandate to build 500,000 units by 2020 [6].

Residensi Wilayah or formerly known as Rumah Mampu Milik Wilayah Persekutuan (RUMAWIP) is the Malaysian government's initiative to build 80,000 affordable housing units throughout Kuala Lumpur, Putrajaya and Labuan by the year 2020. Up to 80,000 units of affordable houses were targeted to be built in the Federal Territory under the RUMAWIP scheme in 2018, with 50,000 units to be built in Kuala Lumpur. The classification of affordable housing is known as properties that intended to cater for the median population specifically to the middle-income groups and are priced between RM150,000-RM300,000 (up to RM350,000 in prime areas) and with a build-up of at least 800 sq. ft. (excluding the balcony) and with 3-bedrooms. The criteria of affordable housing have to be priced below RM300,000 and any price above RM300,000 is not considered as affordable housing. This criterion is set by the government in 2019. This is different from social housing where the definition of social housing is defined as low cost to medium cost housing which developed for the lower income groups or those classified as in the urban poverty groups. In order to be qualified for this, typical household income need to be below RM3,000 monthly. The social housing usually has 1-2 bedrooms. The price is ranging between RM35,000-RM60,000 with built-up of 650 sq. ft.

The housing affordability in Malaysia is a function of house price and income. The median house price was 4.4 times median annual household income in 2014 which signifies to 'seriously unaffordable' state [7]. The affordability levels are different from each state as dictated by supply and demand. The lack of definition in threshold of maximum affordable price levels in the country has led to the use of varied pricing structure in affordable housing. The house prices are depending highly on location, hence some states in the country have more affordable housing markets than other states. Kuala Lumpur and Penang are both severely unaffordable with median multiple of 5.4x and 5.2x [8].

Based on NAPIC statistic, over a 10-year period, the overhang value has increased to more than 100% from RM4.48bil in 2008 to RM10.08bil as at March, 2017. Although the number of residential launched have dropped compared to 2008, the overhang value has risen dramatically, due to high property prices. The houses priced between RM200,000-RM400,000, are considered in affordable range, make up 28% of total overhang as at 1Q17, rising from 25.4% at 4Q16. From the highlighted data, it indicates that housing market is still trapped in the consolidation mode on a weakening bias [3].

However, despite of the launches of more affordable housings, report shows that overhang in affordable housing is in worrying state. Based on the statistic obtained in PR1MA's website as of 31 March 2018, PR1MA had approved 268,537 housing units for development with various stages of

IOP Conf. Series: Earth and Environmental Science **385** (2019) 012073 doi:10.1088/1755-1315/385/1/012073

construction. However, the take up rate is only 11.5% with the total value of 16,000 units. In 2018, The Housing and Local Government Ministry recorded a property overhang of RM22.5 billion worth in unsold houses nationwide, with 73% of the statistic comprises affordable houses priced at RM250,000 and below [4]. A substantial portion of those statistic is from government affordable housing projects that has failed to gain a buying traction. In 2015, residential property with the price of RM500,000 and above were accounted for less than one third of the housing property overhang statistics. On the flipside, the house with price range RM300,000 or less became half of the overhang numbers. Furthermore, house with price of RM100,000 and below were included for about 15% of the overhang statistic. This shows the statistics of overhang is involving the affordable housing segmentation [11].

The terms of 'property overhang' (including residential or housing) as indicated in Malaysia Property Market Report have been changed for three times since the year of 1999, 2000 and 2003. In 1999, overhang is referred as the total number of housing units that was remained unsold after it was launched for sale on or after 1 January 1997. In year of 2000 to 2002, the term of `property overhang' refers as property that remained unsold for more than 9 months after it was launched for sale on or after 1 January 1997 (completion units, under construction and not built).The term also changed with effective from 2003, whereby 'property overhang' has obtained a new definition which is to only include the completed housing unit with Certificate of Fitness for Occupation (CFO) by local authorities and remained unsold for more than 9 months after it was launched for sale on or after 1 January 1997 [4]. Table 1 illustrates the residential property overhang in Malaysia.

Table 1. Residential Property Overhang Units in Malaysia

Price Range (RM)	2015	2016	2017	1H2018
50,000 or less	414	456	1,023	873
50,000-100,000	1,348	1,299	1,430	1,672
100,001-150,000	603	822	1,056	914
150,001-200,000	907	975	2,069	2,670
200,001-250,000	888	920	1,303	2,320
250,001-300,000	1,213	1,358	1,648	1,935
300,001-400,000	1,006	1,472	4,469	4,983
400,001-500,000	846	1,438	2,585	2,516
500,001-1,000,000	1,967	3,943	5,612	7,525
More than 1,000,000	971	2,109	3,543	3,819
Total	**10,163**	**14,792**	**24,738**	**29,227**

Source: NAPIC (2019)

2.2 Factors Lead to The Overhang of Affordable Housing

The related factors that lead to the overhang units in affordable housing are economic, political and environmental. The search for factors that relates to affordable housing has been discussed widely. It has been argued that one of the main affecting factors is due to its poor location and unattractiveness to aspire homeowners [6]. However, for some affordable housing, such as low-cost units, the lack of demand for low-cost units could be due to a mismatch in location and product type [6]. This reflects the general cautious mode of buyers, dampened by high cost of living, job worries and the lack of income growth due to the softening economic conditions in recent years. The unattractive location factor of the affordable housing projects such as the distance from work places and low transport connectivity has led to this overhang issue.

The overhang and unsold units are reported did not attract the target market nor cater to the housing needs of the targeted house buyers [6]. Imperfection of the housing market which is distorted by external influences such as unethical speculation by housing developers has contributed to the market failure on

IOP Conf. Series: Earth and Environmental Science 385 (2019) 012073 doi:10.1088/1755-1315/385/1/012073

the affordable housing. It is learned that some of the overhang issue were originated from the weakness in the planning approval system by the local authorities whereas the housing applications were being permitted without the consideration of the actual demand. The approval process at the Local Planning Authorities and State Authorities which focus on the technical requirements without taking an equal consideration of the actual demand also contributes to the issue of overhang and oversupply [14].

Some of the industry players pointed the overhang issues was due to Bumiputra quota which varies and not just from state to state, but also between zones in a state [10]. There is also a social stigma in young Malaysian in purchasing a designated affordable housing projects such as PR1MA and Residensi Wilayah that deter a chunk of targeted demographic. Meanwhile, the Association of Valuers, Property Managers, Estate Agents and Property Consultants in the Private Sector Malaysia (PEPS) were in the opinion that the high unsold was due to the factors of the indiscriminate building by developers, lack of market studies and financial feasibility studies, no coordination on planning among local authorities, indiscriminate approvals by the various local authorities, delay in gazetting of local plans, which lead to uncontrolled development, the higher cost, and artificial demand .

The inability to secure end financing is another factor that worsen the issue [9]. The two main issues that led to unsold of affordable housing units are identified as end financing and unreleased Bumiputra units [12]. The financing issue include lower margin of financing offered, buyer's income ineligibility and also adverse credit history.

2.3 Critical Success Factors (CSFs) of Affordable Housing

The Critical Success Factor (CSFs) were proposed by Daniel in 1961 and popularised by Rockart's in 1979 [16]. The CSFs method has been widely implemented and has been used in different fields of researches as to determine on the most critical factors influencing success. The CSFs defined as those characteristics, conditions or variables that, when properly sustained, maintained, or managed, can have a significant impact on the success of a firm competing in particular industry [17]. In affordable housing, there are factors that are identified as the Success Factors. The need of thorough and complete knowledge of the CSFs, will able to produce good result. Without applying the CSFs concept, the organisation's performance will not able to produce the expected results. The identified CSFs in affordable housings are 'time, cost and quality management', 'satisfaction, health and environmental safety', 'user affordability and design consideration' and 'cost of individual units and technology' [13].

Meanwhile, some researchers highlight that there are other seven CSFs for affordable housing which are known as the 'institutional framework for public housing', 'availability of competent personnel', 'effective project management', 'good maintenance management practice', 'appropriate design and good location', 'effective housing finance system' and 'adequate political support' [14]. First time buyers in Malaysia are considering several factors in choosing their first unit. The related factors are the demographic, financial, location, structure, neighbourhood and economic. The identified major factors that influence property buyer's in choosing a living place are location, building size, type of unit, the number of bedrooms, design and layout, amenities, facilities, view, density, developer's reputation, promotional effort, security, and convenience [15].

Among the researchers, there is no exact consensus on what constitutes projects success. This is because every project type may have different criteria in determining success [14]. In measuring a successful housing programs, it is not merely measuring the supply of the affordable housing. Other criteria are to be included to ensure the programs are able to meet the need of the household especially on the accessibility and also the affordability term. In Nigeria, four criteria for measuring public housing product success namely meeting the project purpose, end user's satisfaction, environmental impact and aesthetic appearance of the property [14]. A good affordable housing can be proven by the take up rate by the purchasers and no issues on unsold units of the affordable housing.

The policies and programs are key to the affordable housing success. This can be done through the implementation of proactive preservation and the policies development for affordable housing. By enacting to the policies, government can stem the associated issues of like the placement and accessibility to transit, employment hub and neighbourhood amenities. The housing programs have to be designed to address needs across all income segments and accounts for changes that may occur in residents' circumstances and also the city demographic. Therefore, the policies and programs have to be comprehensive and can be addressed systematically by setting targets, employing cost reduction levers and strengthening local delivery [24].

Vast literatures discussed on the nature, the measurements, and the barriers of the issue. In the supplier's side, more developers should focus on the affordable housing, and try to make these properties available to certain quarters of the community, and to incorporate with a range of community amenities in these developments. In term of the demand and supply side of housing, the interventions have been largely on the demand side while the supply side have been direct provision of low-cost housing or subsidizing the cost of housing. The land costs and its use, the planning policy, the construction cost are the driven factors in supply [1].

The CSFs are the crucial factors of any housing development as it will be able to determine the success rate. It is significantly being used in order to present or to identify a few key factors which the focus should be given by the organisations in order to succeed.

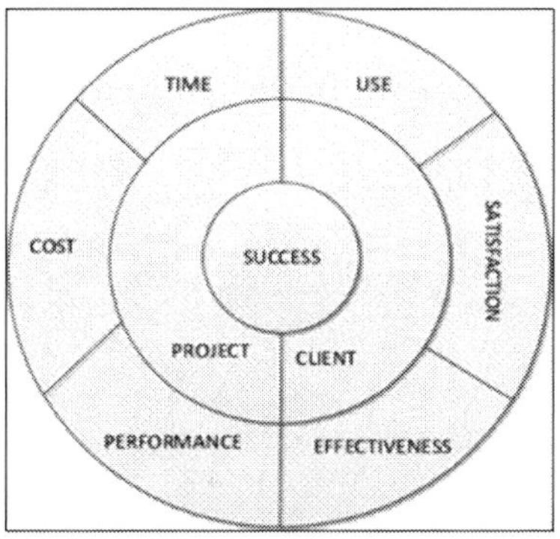

Figure 1. Model of Project Success Criteria
Source: Pinto and Slevin, (1988)

Figure 1 shows a model developed by Pinto and Slevin (1988) in order to assess on the project success. This model comprises of two components specifically on 'Projects and Client'. The first component consists of time, cost and performance aspects. The second component includes three main criteria namely use, satisfaction and effectiveness [14].

The criteria for project success are measured by satisfaction from clients, meet all the projects' objectives and requirements, meet the agreed budget, deliver the key project or products on time, able to add value to the organisation, meet the quality requirements, and have sense of professional satisfaction for the team [22]. The success of a project is measured by product and project quality, timeliness, budget compliance and degree of customer satisfaction. The success also must starts with two other measures known as the measure of effectiveness and measure of performance [23]. Three most important criteria for measuring project management in public housing are client's satisfaction, completing the projects on time and achieve the specified quality standard [14].

The success of the project can be described as a complex and often illusory construct, but nonetheless it is of crucial importance for an effective project implementation [21]. One of the factors in affordable housing is the location and access. The accessibility factor distresses the opportunity and able to ease on the individual at a given location that participates in their activity [18]. One of the house buyer preference in choosing units for own staying is accessibility. The house buyers are not willing to pay more than required for transportation to workplaces, hospital, parking and others [25]. The accessibility factor also include the occasion of diversity of facilities and the amenities. These two items definitely securing the lifestyle and its quality.

In order to meet the increasingly affluent and discerning house buyers, developers instead of just offering dream homes in prime locations, they should provide intangible benefits in the neighbourhood that just as sought after by today's house buyers, such as a sense of security in the landscape compound, a feeling of harmony with one's surroundings, and an infrastructure which supports the lifestyle of house buyers [19].

The efficiency of public amenities will able to reinforce the quality and the wellbeing of society. The convenience access to the service of transportation is one of the vital problems for low income earners. An improper transportation service will be one of the obstacle for the employees to success. Good quality of public transport and job accessibility are the key success for employees to perform well [20]. This could result in a desegregated urban poor area that has less-than-equal access to public services and facilities, including schools and heath care, and challenge sustainable local development in economic inequality. Those employees need to commute for long distances as they are unable to find affordable housing near to their job place. As they have to spend excessive amount of time to travel, they will be less productive and more frustrated compared to others.

The importance of public service accessibility for affordable housing is also driven by spatial equity, a notion associated with the degree to which services or amenities are equally distributed across different economic and political groups. In comparing to economically advantaged households, low-income households are limited to access public facilities which influenced substantially by the allocation policies of affordable housing. In the absence of government intervention, the allocation of dwellings is driven by 'market forces' determined by the interactions of buyers and sellers, each with their own interests. In general, depending on affordability, households can make decisions based on housing size and other characteristics when choosing housing locations. This does not mean that market allocation always leads to equitable outcomes, but it does suggest that public provision and allocation of housing replace household choice with some forms of local government bureaucratic decision based on comprehensive consideration of expenditures and benefits. An affordable housing household attach greater importance to commuting convenience due to the remote residence locations [7].

Other factors are future potential, speculation, financial loan, physical attraction and demography. The CSFs are the few key areas where 'things must go right' for the business to flourish and for the manager's goals to be attained. The CSFs create impacts on the affordable housing. This concept applicable in defining the CSFs of affordable housing.

The CSFs are the key elements to ensure the future success of the affordable housing projects and able to prevent the overhang issues especially in affordable housing market segmentation. There are several factors as listed in influencing on the successes of the affordable housing. In relation to the impact of branding, the earlier research was done by Ashworth and Voofd (1994) which suggested that the geographical marketing mixes are defined as promotional measures, spatial functional measures, organisational measures and also financial measures [26]. This is applicable to the affordable housing. The scope and effectiveness of property marketing is largely determined by the selection and application of the appropriate combination of these measures. The role of branding is increasing important to the property industry. Property developer need to increase the roles and importance of the branding as the branding should no more be used as a marketing gimmick but as a driving force in becoming the direction and focus in any property company's strategic planning.

There are three key factors considered as in realising the success factor of affordable low-cost housing project namely community involvement, affordable loan and integration of economic activities. However, in figuring out the success factor of affordable housing, an upgrading method which is citywide, programmatic, channelled through and managed by government, and combines a community demand orientation with supply side rules of access are the keys to lead the affordable housing to become success. The explicit longer time goal such as infrastructure upgrading strategy are also the success criteria of affordable housing. The hampering sales of affordable housing would include the inconvenient location, lack of accessibility and the unsuitable design and product [8].

An affordable housing process takes into account on the accessibility measures for intending occupants, adequacy and quality, availability measures and affordability measures. There are six categories that have to be satisfied when providers deliver affordable housing, namely appropriateness of the dwelling, housing and social mix, tenure choice, and location, the quality of environmental planning, design and cost. This includes the delivery speed that maybe the most important factors in affordable housing project success. In affordable housing, the value of social capital or known as technology is crucial as it can produce economic benefits and should it is neglected, it is able to create the economic disadvantages. The satisfaction of low-cost residents, from where it was found that it is critical to improve neighbourhood facilities to increase residents' satisfaction [18].

The appropriate location is one of the most important CSFs in developing a successful affordable housing. In developing a successful affordable housing, the projects must be built where residents are able to reach jobs in reasonable commuting times, access to school and vital services and able to connect with their society around them. Hence, the right location is one of the factor to be considered. If the decision is wrong, the affordable housing projects will not be a success regardless of how well the construction, operations and financing are managed. The travel time to workplace is one of the factors to be considered by homeowner. The house buyers prefer the residential units with similar price and size with better connectivity to the city centre. The connectivity and distance factor of affordable housing units are also contributing to the property overhang issue.

Table 2. The Factors Affecting Overhang in Affordable Housing

	Factors lead to overhang in affordable housing	Authors
1	Unattractive location	Ling, 2017
2	Price (Escalating house prices in major cities)	Ling, 2017
3	Mismatch of the product, location and price	Ling, 2017
4	Imbalance of property market	Ling, 2017
5	Imperfection of the housing market distorted by external influences such as unethical speculation by housing developers contributed to market failure	Rameli et.al, 2007
6	End financing and unreleased Bumiputra units	Soam, 2018
7	Time, cost and quality management, satisfaction, health and environmentally friendly, user affordability and design consideration and individual units and technology	Kwofie et. Al, 2012
8	Supply and demand (Mismatch in demand and supply of affordable houses)	KRI, 2016
9	The accessibility	Li, 2018
10	Poor integrated planning and implementation of affordable housing provision	Ali et. al, 1988
11	Social Stigma in purchasing an affordable housing	Kamaruddin, 2018
12	Indiscriminate building by developers, lack of market studies and financial feasibility studies, no coordination on planning among local authorities, indiscriminate approvals by various authorities, delay in gazetting local plans	PEPS, 2018
13	Affordability and design consideration	Mukhija, 2014
14	Community involvement, affordable loan and integration of economic activities	Crisp , 2016
	Technology and costs	
15	Institutional framework of public housing, availability of	Mammela, 2018
16	competent personnel	Mukhtar, 2017
17	Effective project management and good maintenance practise	Mukhtar, 2017

18	Effective housing finance system and adequate political support	Anyanyu, 1991
19	Demographic, structure, neighbourhood and economic	Oliver, 2005
20	Connectivity, distance and commuting time, low transport connectivity	Wu, 2017

3. Methodology
An extensive literature review was conducted as to achieve the aim of the research. This involves the discussions, the thorough study of various published sources such as articles and journals, other research papers and other research materials. This research will be using the content analysis and thematic analysis as the preliminary method. Content analysis is a research method for making replicable and valid inferences from data to their context, with the purpose of providing knowledge, new insights, a representation of facts and a practical guide to action. In order to achieve the objectives as to establish a structural modelling of CSFs in affordable housing, the research interview will be conducted as to explore for the views, beliefs, and experiences of individuals on the specific matters. The data will be collected through the interviews with the relevant authorities and suppliers' side such as PR1MA, Residensi Wilayah programs, Real Estate and Housing Developers 'Association (REHDA) as respondents. The interview method provides a deeper understanding of social phenomena and then would be obtained from purely quantitative methods such as questionnaires.

4. Result and Findings
The purpose of this study is to develop a CSFs structural modelling for the affordable housing projects and involving three (3) phases of research strategy. The theoretical frameworks have formed 20 listed factors and managed to identify the related success factors. The outcome of the research is shown in Table 2. The study reveals the related factors on the successful of affordable housing in general. The focus of the research is mainly in urban areas. The analysis is of the related journals are done accordingly. The main keyword on the journals are overhang, the success factors, affordable housing and urban areas.

5. Conclusion and Discussion
The key roles of the research are about the establishment of a framework for the successful affordable housing. This research aims to design and develop a framework on affordable housing at urban areas in Malaysia. This framework will able to determine factors that leads to the overhang issues in affordable housing with the focus on the affordable housing project that are located at the urban areas. By developing a structural modelling on the CSFs of affordable housing, it is intended to figure out the CSFs in affordable housing as to reduce the issue of overhang in affordable housing. The determined of the CSFs is able to provide an informative guideline to both developers and purchasers. The establishment of the framework is hoped to gives benefit to the public and also to the industry players.

Acknowledgements
The authors would like to acknowledge the Universiti Teknologi MARA (UiTM) for the financial support under the LESTARI grant (600-IRMI 5/3/LESTARI (007/2018)).

References

[1] UN-Habitat (2011). Affordable Land and Housing in Asia. UNON, Publishing Services Section, Nairobi.
[2] Schwartz, M. and Wilson, E. (2008), Who Can Afford to Live in a Home? A Look at Data from the 2006 American Community Survey, US Census Bureau, Washington, DC.
[3] National Property Information Centre (NAPIC) (2014a), Key Data 2014: Property Market Report,available at: www.jpph.gov.my.

[4] National Property Information Centre (NAPIC) (2018), Property Market Status Report. Fourth Quarter/ Q4 Year 2017 [Property Overhang Q4 2017], Ministry of Finance Malaysia.
[5] Cheah, S. L, Almeida, S., Muhamad Shukri, Lim, L. S. (2017) Economics Department November, Imbalances in the Property Market.
[6] Thean L. C. (2017) retrieved from https://www.thestar.com.my/business/business News/2017/07/15/housing-the-nation/.
[7] Boon, H. Y., Xin H. N., (2018) Housing Affordability in Malaysia: Perception,Price Range, Influencing Factors and Policies, International Journal of Housing Markets and Analysis, 11(3), 476-497
[8] Khazanah Research Institute (KRI) (2015), The State of Household, Khazanah Research Institute, Kuala Lumpur,
[9] Ghani, A.S. (2008), Neighbourhood Factors in Private Low-cost Housing in Malaysia, Habitat International, 32, 485-493
[10] Au, F. Y. (2016) retrieved from https://www.edgeprop.my/content/half-overhang-units-are-homes-priced-rm300000-and-below
[11] National Property Information Centre (NAPIC) (2016), Property Market Status Report. Fourth Quarter/ Q4 Year 2015 [Property Overhang Q4 2015], Ministry of Finance Malaysia.
[12] Rehda's Property Industry Survey 1H2018 (2018) retrieved from https://www.nst.com.my/news/nation/2018/10/419988/big-number-affordable-homes-unsold
[13] Adinyira, E., Botchway, E. and Kwofie, T. E. (2012), Determining Critical Project Success Criteria for Public Housing Building Projects (PHBPS) in Ghana
[14] Musa, M., Amirudin, R, Sofield, T. & Mohamad, I., (2017). Critical Success Factors for Public Housing Projects in Developing Countries: A Case Study of Nigeria. Environment, Development and Sustainability. 19(5), 2039.
[15] Chua, A. K. B. (1998). Market Research for Condominium Development. Unpublished B.Sc. project, Universiti Teknologi Malaysia
[16] Rockart, J.F. (1979), "Chief Executives Define Their Own Data Needs", Harvard Business Review, March-April.
[17] Leidecker, J.K, Bruno, A. V. (1984) Identifying and Using Critical Success Factors. Long Range Planning, 17, 23-32.
[18] N.I. Mohamad, N.M. Tawil, I.M. Usman and M.M. Tahir (2014) Identifying Critical Factors of Sale Failure on Commercial Property Types, Shop House Using Multi Attribute Variable Technique
[19] Tan, T. H. (2011). Neighbourhood Preferences of House Buyers: The case of Klang Valley, Malaysia. International Journal of Housing Markets and Analysis International Journal of Housing Markets and Analysis. 4, 58-69.
[20] Housing Affordability: A 21st Century Problem (ahuri final report no. 105). Sydney: Australian Housing and Urban Research Institute Sydney Research Centre
[21] Pinto, J. K. & Slevin, D. P. (1988). "Critical Success Factors in Effective Project mplementation." Project management handbook, D.I. Cleland and W. R. King, eds, Van Nostrand Reinhold, New York, 479-512.
[22] Scullen, J. (2016) https://johnscullen.com/how-to-apply-project-success-criteria/
[23] Alleman, G.B. (2014). Performance-Based Project Management Increasing the Probability of Project Success., New York: AMACOM.
[24] Woetzel, J., Ram, S., Mischke, J., Garemo, N. and Sankhe, S. (2014), "A Blueprint For Addressing The Global Affordable Housing Challenge", Executive Summary, 5-40
[25] Olanrewaju, A. & Tan C. W., (2017) "An Exploration of Determinants of Affordable Housing Choice", International Journal of Housing Markets and Analysis, 10(5), 703-723
[26] Ashworth, G.J. & Voogd, H. (1994). Marketing and Place Promotion. In: Gold, J.R. & Ward, S.V. (eds.) Place Promotion: the Use of Publicity to Sell Towns and Regions. Wiley, Chichester, 39-52.

AUTHOR INDEX

Institute of Physics
Dirac House, Temple Back
Bristol BS1 6BE UK

ISSN: 1755-1307
ISBN 978-1-7138-0253-2

CURRAN ASSOCIATES INC.
proceedings.com